ANNUAL REVIEW
OF GENETICS

EDITORIAL COMMITTEE (1977)

A. CAMPBELL
A. D. KAISER
A. G. KNUDSON, JR.
B. MINTZ
R. D. OWEN
H. L. ROMAN
F. H. RUDDLE
L. M. SANDLER

Responsible for the organization of Volume 11
(Editorial Committee, 1975)

A. CAMPBELL
R. I. DEMARS
A. G. KNUDSON, JR.
R. D. OWEN
H. L. ROMAN
L. M. SANDLER
D. SCHWARTZ
C. YANOFSKY

Production Editor	T. HASKELL
Indexing Coordinator	M. A. GLASS
Subject Indexer	B. BREWER

ANNUAL REVIEW OF GENETICS

HERSCHEL L. ROMAN, *Editor*
The University of Washington, Seattle

ALLAN CAMPBELL, *Associate Editor*
Stanford University, Stanford

LAURENCE M. SANDLER, *Associate Editor*
The University of Washington, Seattle

VOLUME 11

1977

ANNUAL REVIEWS INC.   4139 EL CAMINO WAY   PALO ALTO, CALIFORNIA 94306

ANNUAL REVIEWS INC.
Palo Alto, California, USA

COPYRIGHT © 1977 BY ANNUAL REVIEWS INC., PALO ALTO, CALIFORNIA. ALL RIGHTS RESERVED. No part of this book may be reproduced in any form or by any means without permission in writing from the publisher.

International Standard Book Number: 0-8243-1211-2
Library of Congress Catalog Card Number: 67-29891

Annual Reviews Inc. and the Editors of its publications assume no responsibility for the statements expressed by the contributors to this Review.

REPRINTS

The conspicuous number aligned in the margin with the title of each article in this volume is a key for use in ordering reprints. Available reprints are priced at the uniform rate of $1 each postpaid. The minimum acceptable reprint order is 10 reprints and/or $10.00, prepaid. A quanity discount is available.

PRINTED AND BOUND IN THE UNITED STATES OF AMERICA

CONTENTS

L. C. Dunn and His Contribution to T-Locus Genetics, *Dorothea Bennett*	1
Genetic Polymorphisms in Human Blood, *Eloise R. Giblett*	13
Genetic Control of Sporulation, *Roy H. Doi*	29
Mutations Affecting Fitness in Drosophila Populations, *Michael J. Simmons and James F. Crow*	49
Prospects for Plant Genome Modification by Nonconventional Methods, *Andris Kleinhofs and Ram Behki*	79
DNA Rearrangements in Procaryotes, *Peter Starlinger*	103
Immunogenetics of Cell Surface Antigens of Mouse Leukemia, *Lloyd J. Old and Elisabeth Stockert*	127
Genetics of Structure and Function of Bacterial Flagella, *Tetsuo Iino*	161
Sister Chromatid Exchange, *Sheldon Wolff*	183
The Genetic Structure of RNA Tumor Viruses, *Peter K. Vogt and Sylvia S. F. Hu*	203
Regulation of Gene Expression in Eucaryotes, *Bert W. O'Malley, Howard C. Towle, and Robert J. Schwartz*	239
Interactions Between Host and Viral Genomes in Mouse Leukemia, *Richard Steeves and Frank Lilly*	277
Genetics of Bacterial Ribosomes, *Masayasu Nomura, Edward A. Morgan, and S. Richard Jaskunas*	297
Genetics of Cellular Differentiation: Stable Nuclear Differentiation in Eucaryotic Unicells, *T. M. Sonneborn*	349
Genetic Recombination in Bacteria, *A. Eisenstark*	369
Behavioral Genetics in Bacteria, *John S. Parkinson*	397
Invertebrate Neurogenetics, *Samuel Ward*	415
Illegitimate Recombination in Bacteria and Bacteriophage, *Robert A. Weisberg and Sankar Adhya*	451
INDEXES	
Author Index	475
Subject Index	495
Cumulative Index of Contributing Authors, Volumes 7–11	502
Cumulative Index of Chapter Titles, Volumes 7–11	503

ANNUAL REVIEWS INC. is a nonprofit corporation established to promote the advancement of the sciences. Beginning in 1932 with the *Annual Review of Biochemistry,* the Company has pursued as its principal function the publication of high quality, reasonably priced Annual Review volumes. The volumes are organized by Editors and Editorial Committees who invite qualified authors to contribute critical articles reviewing significant developments within each major discipline.

Annual Reviews Inc. is administered by a Board of Directors whose members serve without compensation.

BOARD OF DIRECTORS
1977

Dr. J. Murray Luck
Founder Emeritus, Annual Reviews Inc.
Department of Chemistry
Stanford University

Dr. Joshua Lederberg
President, Annual Reviews Inc.
Department of Genetics
Stanford University Medical School

Dr. James E. Howell
Vice President, Annual Reviews Inc.
Graduate School of Business
Stanford University

Dr. William O. Baker
President
Bell Telephone Laboratories

Dr. Sidney D. Drell
Deputy Director
Stanford Linear Accelerator Center

Dr. Eugene Garfield
President
Institute for Scientific Information

Dr. William D. McElroy
Chancellor
University of California, San Diego

Dr. William F. Miller
Vice President and Provost
Stanford University

Dr. John Pappenheimer
Department of Physiology
Harvard Medical School

Dr. Colin S. Pittendrigh
Director
Hopkins Marine Station

Dr. Esmond E. Snell
Department of Microbiology
University of Texas at Austin

Dr. Harriet Zuckerman
Department of Sociology
Columbia University

Annual Reviews are published in the following sciences: Anthropology, Astronomy and Astrophysics, Biochemistry, Biophysics and Bioengineering, Earth and Planetary Sciences, Ecology and Systematics, Energy, Entomology, Fluid Mechanics, Genetics, Materials Science, Medicine, Microbiology, Nuclear Science, Pharmacology and Toxicology, Physical Chemistry, Physiology, Phytopathology, Plant Physiology, Psychology, and Sociology. The *Annual Review of Neuroscience* will begin publication in 1978. In addition, two special volumes have been published by Annual Reviews Inc.: *History of Entomology* (1973) and *The Excitement and Fascination of Science* (1965).

L. C. DUNN AND
HIS CONTRIBUTION
TO T-LOCUS GENETICS

❖3112

Dorothea Bennett

L. C. Dunn
1893–1974

L. C. DUNN AND
HIS CONTRIBUTION
TO T-LOCUS GENETICS

❖3112

Dorothea Bennett
Developmental Genetics, Sloan-Kettering Institute for Cancer Research
New York, New York 10021

Leslie Clarence Dunn, who died on March 19, 1974, was a geneticist's geneticist and a biologist's biologist. He had an ingrained awareness of all of the different things that we call genetics and a strong feeling that genetics forged links between all matters of biological interest. This quality made it possible for him to talk on equal terms with almost anyone interested in a biological problem of any kind—this applied both to neighbors bothered by their inability to grow tasty tomatoes in their backyard and to colleagues concerned about genetic bases for racial discrimination. He was fascinated by what we call genetic variation, in whatever form it manifested itself, and I think this made him a true naturalist; he was equally interested in man, mouse, and garden flowers in the sense that they presented interesting biological problems with genetic correlates that were just as real to him although their overall importance might be very different. This catholicism is reflected in the biological material that he worked with during his professional career; it ranged from mice to chickens to man and Drosophila, and even included, after retirement, an excursion into morning glory genetics. To some extent his own broad interests dissatisfied him, because he felt they made him an amateur in everything and a professional in nothing. Clearly, this was not so; what he considered amateurism was partly a reflection of his ability to use genetics as a connecting link among the biological sciences in general.

Dunn started in science because he was interested in natural history, and he pursued it for the same reason. The fact that he got paid for doing it seemed like a lucky accident to him, and the fact that his work brought him eminence as a scientist meant little to him. He certainly never sought importance or position, he never played politics, and he seemed completely free from pretense of any kind. As his student and later his colleague, the most important message I took from him was to follow science first, to do my job as I thought right, and never to worry about whether the result would be "interesting," "topical," "grant-worthy," or a step toward personal advancement. This attitude no doubt reflected the fact that Dunn started in science in a time that was quieter and gentler than now, when it was easier to do science single-handedly without serious worries about either competition or support for elaborate and expensive laboratories; but it also reflected standards of integrity and honesty that were rarely matched then or now. This attitude is fortunately infectious; all of my own first students of course knew Dunn, and he took a personal interest in them as scientific grandchildren. They heard his message and still persist in being almost cantankerously obstinate in doing science for the sake of science alone, avoiding the kind of expediencies that might have tainted them in his eyes. And now they tell their own students, Dunn's great-grandchildren, who never knew him, to follow the same principles. This is one of Dunn's legacies that I think is as important as his contributions to knowledge.

It is the story of these contributions, however, that constitutes the bulk of this memoir. The development of genetics has been so rapid that it comes as a surprise to realize that Dunn's initial entry into biology occurred less than 10 years after the rediscovery of Mendel's principles. Thus, as a student he was exposed to these ideas as something excitingly new that required general confirmation and testing. Probably because of his naturalistic inclination, his bent seemed to lead him always to tackle these new problems at the level of the whole organism and particularly with reference to development and evolution.

When Dunn entered Dartmouth College in 1911, he already considered himself a biologist with good theoretical and practical training. From early childhood he had spent Saturday and Sunday afternoons working with his favorite person, his grandmother, in her garden, and had learned the satisfaction of steady application to simple tasks that gave insight into living things. A high school teacher, Dr. Marie Walcott, had amplified his love for gardens into an interest in taxonomic botany and through that, had given him an awareness of the powerful beauty of science. Dunn's high school biology notebook was apparently such an impressive document that Dartmouth excused him from freshman biology courses, and he began an intensive study of botany. By the time he was a junior, though, Dunn had found Punnett's *Mendelism* and was apparently immediately fascinated with the exactness of the rules governing the behavior of hereditary factors, since until then most of his experience had been with the descriptive aspects of biology. During the winter of the same year, Morgan's *Heredity and Sex* appeared; this captured Dunn's interest entirely, and his path into genetics was fixed. He goaded his major professor, Professor John H. Gerould, and other graduate students into organizing a seminar to discuss Punnett's and Morgan's books and to read some of the papers on genetics that were appearing in the journals. In the Easter recess of that year Dunn went to New York to see Morgan at Columbia, to inquire whether there was any prospect of working in his laboratory either before or after completing his college degree. But Morgan was apparently unimpressed, perhaps even dismayed, by the young botanist's ignorance of systematic zoology and lack of any solid training in zoology. At any rate he held out no good hope, since his space was already overcrowded and his few posts already committed to Columbia graduate students.

Disappointed, Dunn returned to Dartmouth to try to remedy his zoological defects, and to look for a geneticist who would accept him as a graduate student. In time he made an arrangement with W. E. Castle, who, since 1901, had been engaged in testing the ideas promulgated by Mendel in both the new laboratory animal, Drosophila, and mammals.

Dunn's research activities with Castle seem to have been mostly of the search and explore type; this presumably represented attempts to become comfortable with a new way of thinking, as well as of generating and analyzing data. He bred mice and defined some coat-color traits, he classified different types of spotting variants, he studied segregation ratios. But the only real questions they asked proved to be red herrings. Briefly, they tried to test the hypotheses that linkage relations between genes could be altered by selection, and that genes themselves could be changed in a directed way. These were quite reasonable questions for the time since genes were largely being investigated as statistical units in segregation analyses, and little or nothing was known about what they were composed of, how they were associated, or how they controlled phenotypic traits. So it seemed reasonable to Dunn and Castle that selection might well enhance the degree to which certain markers traveled together, and Dunn set out to demonstrate this notion in Drosophila. Unfortunately, Morgan's group was also testing the same ill-fated hypothesis, and they published while Dunn's work was still in progress. He shifted then, but apparently with little real enthusiasm, to a similar study of linked genes in both mice and rats.

This work was disturbed by restlessness aroused by the worsening, in late 1916, of the First World War; at that time Dunn joined the Harvard regiment as a volunteer to receive military instruction from French and British officers. When the United States entered the war, Dunn's university work ceased entirely and he went to officer's training camp and then to France. When he returned in 1919 he completed his dissertation and made arrangements to begin work as a geneticist at the Storrs Agricultural Experiment Station in Storrs, Connecticut.

There, he was the only member of the small station staff who was not also a member of the staff of the Connecticut Agricultural College which then had about 350 students. The director of the Storrs Agricultural Experiment Station was Dr. Edward H. Jenkins, a chemist trained at Yale University who was also the director of the Connecticut Agricultural Experiment Station at New Haven. That experiment station was the first in the United States (founded in 1877) to be devoted only to testing and research. Dr. Jenkins hoped to develop a similar research program at Storrs, and wanted Dunn to devote himself entirely to research in genetics, especially as applied to poultry breeding. The freedom to develop his own program in the new area of genetics, just exactly as he wished, was the motivating factor that took Dunn to Storrs.

At Storrs, Dunn embarked on an extensive program of inbreeding by brother-sister matings, and on a study of the factors influencing "hatchability" of eggs, that is, a study of embryonic mortality. The inbreeding experiments, stimulated by the study of inbreeding in corn at the New Haven Experimental Station, were intended to measure changes in various elements of vigor, in morphological characters, and in egg production, as well as to produce inbred lines of chickens that at that time did not exist. The results, published mostly in technical journals and bulletins, were not of great consequence, since most "inbred" lines died out after two generations or so. A few families survived and were turned over to Henry Wallace where they formed foundation stock for the Hybred Corn Company's introduction of crossbred chickens.

Again, this was a period when Dunn did a good deal of seemingly superficial work that nevertheless gained him a wide knowledge of the practice and the theory of genetics in a familiar species. He studied plumage colors, egg production, egg characteristics, and growth rates; almost every measurable characteristic of a chicken was at one time or another measured and related to specific genes or genetic background. By 1924, Dunn began to home in on the subject matter that was to form the basis for much of his future work, namely lethal factors in embryos. He became especially interested in so-called parrot-beak embryos, which provided the first analyzed cases of chondrodystrophy in birds. In collaboration with Walter Landauer, who provided a thorough study of the anatomy and development of chondrodystrophic embryos, it was demonstrated that virtually identical chondrodystrophic phenotypes could be due to heredity or to purely nongenetic causes. They found the same two sets of factors to operate in causing the condition known as rumplessness in fowls to be a gene mutation in some cases and nongenetic alterations in development in others. This collaboration not only resulted in a lifelong friendship between the two men but, interestingly, established a kind of reciprocal approach to problems of development that each was to follow through the rest of his career. Dunn chose what he thought was the simplest path, namely the analysis of single-gene mutations that produced developmental aberrations that should be easy to analyze. Landauer elected to deal with all the complexities of nongenetic variation, which led eventually to the development of the field of chemical teratology.

Dunn's course diverged briefly in 1927 from his growing interest in developmental genetics. On his first sabbatical leave he worked for a time at the Naples Zoological Station on invertebrate development without finding an interesting problem. Then he spent some months at Edinburgh with Frank Crew where he began observations on selective fertilization in fowl and also returned to the mouse to study an enzyme step (dopa-oxydase) in the determination of spotting patterns. The latter problem proved interesting to him, and he went to Berlin to pursue it further with Richard Goldschmidt at the Kaiser Wilhelm Institute for Biology. No significant advances in research were made during that stay, but Dunn was strongly influenced by the atmosphere of scientific attachment to fundamental problems that he found at the Institute, where Carl Correns (then director), Goldschmidt, Max Hartman, Otto Warburg, and others met in seminars filled with vigorous debate. This atmosphere apparently revived Dunn's memories of graduate school days and evoked both a desire to return to a more academic environment as well as the impetus to make a fresh start on problems of what he was by then calling *developmental genetics*. Goldschmidt was a leader in this new field, and had just published his *Physiologische Theorie der Vererbung* which attempted to interpret the effects of genes on development in terms of relative rates of different physiological processes. The reasoning behind this concept came first from studies on intersexuality in the gypsy moth, where crosses between geographical races often produced intersexes. The physiological mechanism that Goldschmidt proposed was that sexuality was a genetic trait and that each individual begins life with the capacity to develop into either a male or a female. This led him to the conception of a balance mechanism for sex determination, and to proposals for such mechanisms in development generally. Dunn wanted to follow this lead because he thought such an approach might unite genetics and embryology in a mechanistic way.

The opportunity for a new start came quickly. The sabbatical year was curtailed by the sudden death of Dunn's father-in-law, and the family returned to Storrs. For his new endeavors Dunn brought with him chickens from Scotland and Germany, and stocks of mice with suspected mutant genes that he had collected from dealers and fanciers all over England. Hardly had he settled these animals in Storrs when he was approached by Gary Calkins, professor of zoology at Columbia University, to fill a vacancy in that department that arose as a result of the retirement of E. B. Wilson and the departure of T. H. Morgan, C. B. Bridges, and A. H. Sturtevant to found a new laboratory at the California Institute of Technology. This was not the first offer that Dunn had had, of course, but he had declined all others without investigation and without regrets, since he had been eager to continue research in the free and untrammeled atmosphere of his small laboratory in the country. But the Columbia offer came at just the right time; it promised a return to an academic environment, and perhaps a broader scope for the new direction that Dunn wanted to take in research. In any case he accepted what he considered a flattering offer for a full professorship in 1928.

On moving to Columbia, Dunn focused his attention on a genetical and developmental analysis of pattern formation in the mouse. The establishment of pattern is of course the essential feature of development, and Dunn thought it could be analyzed most effectively by studying its genetic control. For this purpose he returned to studies of spotting patterns in mice, with an eye to testing Goldschmidt's central themes that mutations alter the rates of processes in development and that the whole of development implies a merging of many subpatterns, all of which must be physiologically integrated with one another. But since he was returning to Columbia, which had such strong associations with Drosophila even though Morgan was gone, Dunn could not resist expanding his work to Drosophila as well. So he began to collect Drosophila mutations such as *Minute* and *Bar*, which could be interpreted as affecting developmental rates, with the thought of analyzing each of them separately, and then in combination. The Drosophila work was never as close to his heart as was the mouse work, probably because he never found the fruit fly as interesting as a mammal—and what could be examined in Drosophila could generally be done as well in the mouse; if work with the mouse was more difficult and more time consuming it was nevertheless more interesting.

In his sabbatical of 1932–1933, Dunn made one last attempt to combine work on Drosophila and on mice, going to the University of Oslo to do Drosophila experiments with Otto Mohr and to work on mice with Christina Bonnevie. The experiments with Drosophila proved the general point that there was quite a strict proportionality between the retardation induced by various *Minute* mutations and the size of the *Lobe* eye, and therefore were supportive of Goldschmidt's ideas. This set of experiments incidentally led them to detect an apparent back mutation, in which a white-eye mutation reverted to honey, or part way to wild type. This represented a wholly new concept, because until then mutation had been thought to be a one-way process. Dunn worked hard to nail this point down, and derived from it a lasting interest in gene structure and organization as related to mutation, matters that had previously never been of great concern to him because he had felt that gene structure was far less interesting than gene function.

The mouse aspect of Dunn's work in Oslo may have been particularly significant, and foreshadowed events to come. It was his first real excursion into abnormal embryology in the mouse, and it involved a mutation that produced a short tail, the so-called Shaker Short. Bonnevie and Dunn produced only descriptive studies of this mutant, and Dunn was never satisfied that they had got a real explanation or interpretation of the defects. Nevertheless, this work probably set him up for an interest in the T-locus mutations that had just come his way before he went to Oslo; the stocks were left in the care of a young student, Paul Chesley.

The first T mutation was reported in 1927 by a Russian surgeon, Nelly Dobrovolskaia-Zavadskaia. A cancer researcher studying the effects of radiation on mice, she found in the offspring of an irradiated male two categories: short-tailed and normal-tailed. She proved that the short-tailed offspring carried a new dominant mutation, which she named in Greek *Brachyury*. When she bred two short-tailed animals together she was puzzled because she could never get a true-breeding short-tailed line. She made the correct assumption, but had no proof for it, that the *Brachyury* (T) mutation was lethal when homozygous.

Shortly after reporting on T, Dobrovolskaia-Zavadskaia was joined in Paris by another young Russian émigré, N. Kobozieff. The two pursued the study of T by crossing short-tailed animals to two other apparently normal strains of mice, one a French laboratory line, and one descended from a wild mouse that Zavadskaia had trapped while on holiday near the Spanish border. To their surprise a striking abnormality appeared in the offspring of both crosses: taillessness. They inbred the offspring from each cross and were further confounded when, this time, the abnormal phenotype bred true! At this point, Dobrovolskaia-Zavadskaia came to the United States to lecture for some Russian refugee organizations, and appeared at Dunn's laboratory at Columbia looking for a solution to her puzzle. The best solution she could see was "to give up the confusing tails and return to my proper field which is cancer research." Dunn was fascinated with the problem and answered, "If you are going to give them up, just give them to me."

The short-tailed and tailless mice arrived at Columbia in 1931. The first job was to give a thorough test to the idea that T was lethal in homozygotes. Zavadskaia had found some crippled short-tailed animals, partly paralyzed with hemiplegia, and she thought they might be homozygotes. But as soon as Dunn and his student Chesley began to breed the stock, they became convinced that the homozygotes were dying as embryos and that the crippled animals were extreme variants caused by modifier genes or nongenetic developmental fluctuations. The first task then was to find a class of dying embryos that represented the expected Mendelian proportion of 25%. They found them almost immediately by dissecting pregnant females at midgestation. At 10 days of pregnancy some of the embryos had strikingly abnormal traits; instead of having truncated tails they had truncated bodies that ended abruptly just behind the forelimbs. This abnormality was then traced further back in time at half-day intervals, and Dunn and Chesley learned that they could first recognize the abnormal class at 8 days. At that time the embryology of the mouse had not been well described for that stage, so the two had to work out a schedule for normal development, and had to study the normal embryology of the mouse for comparision with their mutant. Eventually they determined that the fundamental defect in T/T embryos depended on a failure of notochord differentiation. Chesley's dissertation describing this work is now recognized as a classic of its kind; it was not only the first thorough study of the effects of a lethal gene in mammals, but also the first clear demonstration of genetic interference with an inductive system. It laid the basis for a whole series of studies of the effects of genes on early development in mammals.

Having confirmed Zavadskaia's suspicion that the failure of $T/+$ mice to breed true was due to lethality of T-homozygotes, Dunn then went on to analyze the genetics of taillessness. First he tested and confirmed Zavadskaia's observation that matings between two tailless mice of the same origin did in fact produce only tailless offspring; he concentrated on the line derived from the Spanish mouse because its wild origin intrigued him. An analysis of litter size in those matings provided the clue that a balanced lethal system was operating; in other words, in crosses of short by short, where he had already shown that one category of embryos died, he had found litter size at birth reduced by one quarter. In litters from two tailless animals, the progeny size was reduced by one half. He surmised that the tailless mice were double heterozygotes for two lethals—the *Brachyury* gene that they had already defined, and a new one, apparently a recessive, which at first was merely inferred.

The main puzzle now took quite a different form, since the two lethals involved appeared to represent alleles at the same locus. There was one precedent for a balanced lethal system, described by H. J. Muller, but in that case the two lethals involved showed occasional crossing-over in heterozygotes and were clearly not allelic. Dunn's problem now was to explain why, although all the data collected indicated that the two genes in question were strict genetic alternatives and did not show crossing-over, they showed complementation and produced a viable heterozygote that carried two lethal genes at the same locus. An unprecedented situation, this made him suspect that genes—at least these genes—might be bigger things than were supposed, and that different alleles controlled different sets of processes in early development. He began to call them *pseudoalleles*, which is a term still used for this and other genetic systems whose genetic structure is not well understood.

In any case the situation quickly became more complicated. Dunn then analyzed both tailless lines that Zavadskaia had sent: Line A derived from the laboratory stock and Line 29 from the Spanish mouse. Each produced only one kind of offspring (tailless) that survived to birth, and in each the homozygous *Brachy* mutation killed one quarter of the embryos, and in each another lethal apparently killed another quarter. But those other lethals proved to be different in the two tailless lines. This was shown when Dunn crossed Line A by Line 29 and found with gleeful surprise that a class of normal-tailed offspring was present at birth. It was quickly shown by breeding tests that these normal-tailed animals carried, as predicted, *no* wild-type alleles at the T locus, but rather two different lethals. Again, this raised the problem of complementation between two lethal alleles, and again Dunn suggested an explanation based on their controlling different sets of embryonic processes. This was possible to test, since if correct, one would expect the two recessives to affect the embryo in different ways, perhaps by killing it at different times or in different ways.

Just at this juncture in 1935 when it was clear that the unraveling of the T locus would require much embryological work as well as genetic analysis, Dunn was fortunate to be joined by an associate, Salome Gluecksohn-Schoenheimer, who had trained with Spemann in the classical German school of amphibian embryology. With her background she was able very efficiently to study the embryonic effects of the two different recessive lethals, and soon showed that, as suspected, the lethals had different effects when homozygous. Both of them interrupted development very early on; in one case the homozygous embryos died almost immediately after implantation, in the other the embryos failed even to implant. The main point for Dunn was that the embryological evidence confirmed the genetic hypothesis, that these were two lethals related by allelism but different in function.

The clear evidence was that the T locus contained sets of multiple alleles, whose complementary interactions suggested to Dunn a complexity on the part of genes or loci or both, and led him to want to know more about what multiple allelic systems represented. Although such systems had been recognized and studied since the early part of the century and had served to dispel the primitive "presence-absence" notion of the relationship of a Mendelian dominant to its recessive counterpart, little was known about them except that, as could be read from Curt Stern's (1930) *Multiple Allelie*, there were no general rules. Dunn strongly suspected that T-locus mutations were not alleles in the normal sense, but that they served as variants within a region that had some structural and functional unity. He set out to dissect that region by collecting more mutations for embryological and recombinational analysis in attempts to get an estimate of the total variation it contained. This point will be returned to later.

While the pure genetics of his system remained a major concern to Dunn from the 1930s on, he was necessarily caught up in another strange functional aspect of the T locus. His very first breeding analyses of tailless mice had turned up a fact that was no less than astonishing. Male tailless mice did not obey Mendel's rules. In matings to wild-type females, males of the genotype Dunn symbolized as T/t usually produced 80% or more of normal tailed $(+/t)$ offspring, instead of the 1 normal:1 short ratio expected. Females, on the other hand, behaved normally. The first question that needed asking was whether this was due to a defect in transmission of T or a superiority on the part of t. T was already known to segregate normally from $T/+$ males, and Dunn soon showed that in $t/+$ males the recessive lethal gene was transmitted in the same high ratios as from T/t males. This was an especially interesting point, since it meant, apparently, that a *lethal* mutation had some clear advantage over its normal allele. In his characteristic way Dunn systematically eliminated any trivial explanations he could think of; he showed that the effect was truly genotypic and not phenotypic and ruled out possibilities of modifier genes, differential viability, and errors in classification. He was also very perspicacious in calling the phenomenon distortion of *transmission* rather than *segregation* because he did not want to limit himself to thinking of meiotic disturbances; since transmission was actually all that he was measuring, he wanted to leave entirely open the question of mechanism. In any case it was clear that he had on his hands a major departure from conventional genetics, for which no explanation could be offered, but nevertheless a departure that would have serious implications for evolutionary theory. It was not realized at the time, nor in fact is it even widely appreciated today, how important an evolutionary force nonrandom transmission of gametes can be. As has been shown by R. C. Lewontin, a student of Th. Dobzhansky and a colleague of Dunn who became interested in this question during his stay at Columbia, even minor transmission distortions represent a potential selective advantage well able to swamp the obvious and well-known selective pressures exerted on deleterious genes.

This implication of this new evolutionary force by no means evaded Dunn, especially since he remembered that one of his original lethal *t* alleles had come from a wild Spanish mouse. It occurred to him right away that the unlikely existence of a lethal gene in a wild population might be due to the protective influence of a high transmission of that allele through males. The next obvious step was to find out whether any other wild mouse populations carried such peculiar alleles. In the laboratory of Howard Schneider at the Rockefeller Institute, Dunn found a population of mice that had been derived some five years earlier from wild ancestors trapped in New York and in Philadelphia. He tested them by matings to *Brachy* heterozygotes and the first few litters proved his point; heterozygosity for *t* was present. Testing of a large number of mice showed that the proportion of heterozygotes was surprisingly high, in the neighborhood of 50%. Dunn derived several tailless lines from inbreeding and made another astonishing observation; the population contained not just one but two different *t* mutations, one lethal and one semiviable. The semiviable produced complete sterility in homozygous males, so in an evolutionary sense it was also lethal. Both alleles had, as was expected, the very high transmission ratios necessary to maintain them in the population.

This intriguing situation led Dunn into mouse population genetics both as practiced in the field and as simulated by mathematical formulae and computers. He managed to collect mice from a good geographic spread of North America, from Japan, and from Europe. In virtually all populations that were adequately sampled there were lethal or semilethal *t* alleles (never viables) with high transmission ratios, and heterozygotes in high frequencies that often approached 50%. There was no question that *Mus* as a genus was polymorphic for *t* mutations and therefore in some not understood way these genes had to be considered part of the normal genome of the mouse. Dunn and his students, T. Prout and D. Bruck, began to try to reconcile the two selective factors that had to be involved, namely embryonic mortality and transmission advantage. Applying Hardy-Weinberg statistics that took both factors into account, they came to a conclusion Dunn found rather absurd. Nor only *should* most wild populations have t mutations but they should also have them in even *higher* proportions than were actually found; in fact, the alleles should come close to fixation, that is, 90–100% heterozygotes. Since Dunn felt that data properly collected from natural populations could not lie, and his populations had only 50% or so heterozygotes at most, he looked for what was lacking in Hardy-Weinberg statistics. The obvious parameter to suspect was population size, which of course in Hardy-Weinberg calculations is infinite. Not much was known about population structure in wild mice at the time, but Dunn's "amateur" naturalism led him to suspect that they probably lived in very small mating units, e.g. one's kitchen or someone else's barn. It proved too difficult to analyze the real situation quickly or accurately, so with the computer-oriented Lewontin, filched temporarily from Dunn's colleague Theodosius Dobzhansky, computer models were built to simulate stochastic mating processes in small populations. They showed that in small populations of no more than 10 individuals there were indeed restrictions on the degree of heterozygosity the population could attain without running too serious a risk of extinguishing itself by fixing the mutant allele. In fact, when a starter population containing t mutations in any frequency was allowed to produce more than a couple of hundred generations it virtually always died out. But, since real wild populations did nevertheless nearly all have t alleles, this suggested that the model was again falling short of reality. The implication was clear nevertheless that mice do live in small populations, and this has since been measured and confirmed. The element Dunn thought most likely missing from the stochastic model was migration from one small deme to the next, most probably in the form of a wandering male carrying with him a t allele with an infectiously high transmission ratio.

To return now to the purely genetic aspects of the T locus, it must be pointed out that in the course of maintaining the two original tailless lines, which were thought to be true-breeding, it became clear that not even this rule was to be followed in its entirety. The balanced lethal system broke down every now and then, and about one in every 500–1000 offspring was not tailless, but completely normal. These "exceptions" as Dunn called them were put through breeding analysis and proved, like complementing Line A/Line 29 double heterozygotes, to carry two different recessive t mutations. But in the first few cases analyzed, one of the recessive genes was different from any so far studied; it was viable when homozygous but nevertheless interacted with T to give a tailless phenotype, and thereby, by definition, was a mutation at the same locus. Since Dunn by then had no evidence from extensive data on the breeding of $T/+ \times T/+$ that T or $+$ mutated to t alleles, his interpretation was that the recessive lethal t mutations in the tailless stock had undergone further mutation to a viable state. Later, it was shown that a comparable process of mutation could give rise also to new lethals. The constant generation of new t mutations from preexisting ones produced a plethora of material to be sorted out and categorized as lethal or viable, and measured as to transmission ratio. As a general rule it appeared that viables had also changed with respect to their effects on transmission ratio; if they affected it at all, and some did not, they gave abnormally low ratios, thus producing yet another puzzle. Further breeding tests were done to make sure that the new class of *t* mutations comprised also genetic alleles of *t* lethals as well as *T*, and again no recombination occurred in any combination tested.

At any rate, Dunn began to suspect that he might be dealing not with a conventional *gene*, but with a *region* of genetic material. He thereupon began a search for other genes—particularly tail mutations—that might map in that same region. He systematically tested for linkage with *T* any mutation that turned up in his lab, any that he could collect from other labs, and any that he could buy from fanciers. The first linked marker (*Kink*) that he found was obtained from "The Sunshine Mousery of Manetee, Florida, The Home of America's Fanciest Mice." The marker, which came as a pleiotropic effect associated with the waltzing behavior that interested the fanciers, provided new insights in two rather different ways.

First, linkage estimates with *Kink* revealed still another genetic peculiarity associated with *t* mutations. Whereas the initial studies done with *T* alone showed the *T*-*Kink* distance to be about six units, experiments done with *T* and *Kink* on one chromosome and a lethal *t* mutation on the other immediately revealed that *t* mutations suppressed recombination over the whole distance between the markers. This of course raised further doubts about the allelism of *T* and *t*; since the genetic definition of alleles is that they show no recombination, it was patently impossible to test for, or argue, allelism in the case of a gene that prohibits recombination. For example, although *Kink* and *t* showed no functional interaction at all, they also behaved as genetic alleles in segregation analysis. The recombination suppression associated with lethal *t* alleles of course raised further questions as to whether they were induced chromosome aberrations rather than mutations. Dunn investigated this point cytologically as well as he could at the time, but found no evidence for either inversion or duplication-deficiency changes, nor have any been found with the increasingly sensitive karyological methods available today. Linkage studies with *Kink* revealed another difference between viable and lethal *t* mutations; viables did not act as crossover suppressors. This provided one reassuring fact; at least the allelism claimed for *T* and viable *t*'s could be assumed to stand, and perhaps by analogy the case for allelism of lethal *t*'s was strengthened.

The second important implication that Dunn saw in the linkage of *T* to *Kink* and then to a second tail mutation, *Fused*, was that the spatial juxtaposition of these loci on the chromosome and their similarity of effects might not be fortuitous. This was quite the opposite view from that generally held, since Drosophila geneticists had by and large not discovered any clusters of different genes with related functions. In fact, the (1945) paper by Dunn and Ernst Caspari, "Neighboring Loci with Similar Effect on Development," may have been the first expression of that idea—that genes are not distributed randomly on chromosomes. Dunn was impressed later by the suggestion by E. B. Lewis that such situations might well arise from tandem duplications followed by mutational divergence; he thought that this was an obvious mechanism for increasing gene number as well as enhancing the sophistication of genetic regulatory systems. It is interesting that the same ideas, and even some data to support them, are now being applied to the same chromosome in the mouse, but in terms of genes determining cell surface antigens, not tails.

Kink was a marker suitable for these first analyses of recombination, but it was difficult to work with over the long haul, since two mutations affecting the adult phenotype of the tail were necessarily not always easy to discriminate. The situation was much improved when Mary Lyon discovered, and sent to Dunn, another marker in the region of T; this mutation, called *tufted*, affected a quite different trait, hair growth, but was very close to the locus of *Kink*. Dunn (as well as of course Mary Lyon who was also interested in the T locus) immediately began to use the marker to analyze T-locus genetics. They found, as would have been expected, that the t alleles that suppressed recombination with *Kink* also suppressed recombination with *tufted*. More important, both found that new viable exceptions generated by lethal t mutations all involved a recombinational event between the loci of T and *tufted*. This suggested very strongly that in fact t mutations comprised regional chromosomal differences, not single-gene mutations, since apparent crossovers could separate not only lethal factors from viable ones but also the element responsible for high transmission.

In the course of his work on the T locus, Dunn and his associates isolated and defined well over 100 recessive t mutations. More than 20 lethals of independent origin were found to fall into six complementation groups, each of which affected early embryonic development at a different and very specific stage. Dunn thought that the T locus represented an important center for controlling the first steps in establishing the patterns of embryogenesis, and that once its genetic complexity was unravelled, it might serve as a model for understanding gene organization and gene regulation in mammals. This promise may be fulfilled; work that was just under way at his death has shown that T locus genes appear to participate in development in a critical way by specifying cell surface components that guide embryonic cells through their first steps in differentiation. More than that is something that would have pleased the evolutionist in him very much; it seems that the role of T-locus genes is not limited to the mouse, but that similar genes, conserved through evolution, operate in the development of all mammals.

I have tried in this brief memoir to give a picture of Dunn as a man and scientist, and to concentrate on what I thought were the important factors that led him first to genetics and then to such an intensive and elegant concentration on one small region of mouse chromosome. In doing this I have relied on my own impressions gained during our long friendship, and on autobiographical material he left behind, particularly his Columbia University "Oral History" which was made available to me by his wife Louise. The memoir is sadly lacking in many respects because it follows only one thread, and gives no impression of his many other interests and accomplishments. I am tempted to do justice to him by discussing these: his books, which served to introduce generations of students and laymen to principles of genetics, his interest in human variation and his clear view of race, and his crusading for political freedom were all important aspects of his life. But, in fact, without a whole volume or maybe two to do it in, the job is just too big to attempt. Anyway, the story outlined here is probably the one that he would have thought most interesting.

GENETIC POLYMORPHISMS IN HUMAN BLOOD

❖3113

Eloise R. Giblett

Puget Sound Blood Center, Seattle, Washington 98104 and University of Washington School of Medicine, Seattle, Washington 98195

CONTENTS

INTRODUCTION	13
THE ALLOANTIGENS	14
Red Cell Allotypes (Blood Group Antigens)	14
White Cell Allotypes	15
Plasma Protein Allotypes	16
Immunoglobulins	16
β Lipoprotein	17
ELECTROPHORETIC VARIANTS	17
Plasma Components	17
Complement polymorphisms	17
Vitamin D-binding protein	19
Cellular Components	19
The polymorphic enzymes in blood cells	19
USEFULNESS OF GENETIC MARKERS IN BIOLOGICAL STUDIES	19
Mutation Mechanisms	19
Human Chromosome Mapping	20
Clonal Origin of Tumors	21
Fate of Bone Marrow Grafts	21
Detection of Cellular Mosaicism	21
Associations with Disease	22
Red cell antigens	22
Histocompatibility antigens	23
Enzyme deficiencies	23
CONCLUSION	23

INTRODUCTION

The term *genetic polymorphism* is applicable whenever a chromosome locus has two or more alleles with frequencies in large populations of more than one percent. Since blood is the most readily available tissue, information about human genetic polymorphism is based largely on studies of blood specimens. Phenotype designations are made either by immunological or biochemical methods. Differences in cellular or molecular antigenic structure, i.e. the alloantigens, are detected with specific antibodies. Differences in molecular charge and size are demonstrated by electrophoretic techniques. Many polymorphisms have been found in this way (1–4), and controversy about their biological and evolutionary significance continues unabated. It is not the purpose of this brief review to focus on that unsettled issue, but rather to bring the list of human genetic polymorphisms up-to-date and to describe some ways they have been used for studying biological problems.

THE ALLOANTIGENS

Red Cell Allotypes (Blood Group Antigens)

The latest edition of Race & Sanger (5) contains a thorough review of the serology and genetics of human red cell allotypes. Their biochemistry has also been reviewed (6, 7). The most interesting recent discoveries in this field are related to the biochemistry of the P blood group system and to the association of certain red cell antigen phenotypes with disease states. The latter subject is discussed in a later section of this review.

The biochemical nature of most blood group antigens is not known, but it is well established that the H, A, B, Lewis, and P system determinants are carbohydrates (8–11). The "immunodominant" sugars responsible for these specificities are placed in a definite order by glycosyl transferases, which are themselves the products of specific alleles.

Glycolipids on the red cell membrane consist of the sphingolipid, ceramide, and attached oligosaccharide chains of varying length and sequence (12, 13). As shown in Table 1, these molecules can be divided into two different series depending on their resemblance to globoside or paragloboside (14–16). Globoside itself has P antigenic specificity, and its precursor, trihexosylceramide, has P^k specificity. On the other hand, the P_1 antigen has a structure related to paragloboside, as do the H, A, and B antigens. This and other information (17) demonstrates that the genes responsible for P and P_1 specificities are almost certainly not allelic, and that the substrates for their products are not the same. The P gene product is apparently a β-GalNAc transferase that uses trihexosylceramide as substrate, while the P_1 gene product is an α-Gal transferase with paragloboside as substrate. Further details are available elsewhere (17, 18).

Carbohydrates are also involved in determining the structure of M and N antigens (19). However, since the precise biochemistry is not resolved, the alleged nonallelism of their respective genes is still open to some question (20, 21). Unfortunately, while a sophisticated genetic model has been proposed to account for the complexities of Rh antigens (22), there is very little biochemical information about this important genetic system. Some evidence suggests that the Rh specificities reside in the red cell protein moiety (23). If this supposition is correct, substitutions of amino acids at more than one site on the polypeptide could explain Rh antigen complexity, analogous to the multiplicity of the HLA allotypes.

Table 1 Two kinds of glycosphingolipids with blood group specificities[a]

Structure	Antigen specificity	Chemical name
Globoside series		
βGal (1→4) Glc-Cer	—	Lactosylceramide
αGal (1→4) βGal (1→4) Glc-Cer	P[k]	Trihexosylceramide
βGalNAc (1→3) αGal (1→4) βGal (1→4) Glc-Cer	P	Globoside
Paragloboside series		
βGal (1→4) βGlcNAc (1→3) βGal (1→4) Glc-Cer $\uparrow$ 2 1 α-L-Fuc	H	
αGalNAc (1→3) βGal (1→4) βGlcNAc (1→3) βGal (1→4) Glc-Cer $\uparrow$ 2 1 α-L-Fuc	A	
αGal (1→3) βGal (1→4) βGlcNAc (1→3) βGal (1→4) Glc-Cer $\uparrow$ 2 1 α-L-Fuc	B	
αGal (1→4) βGal (1→4) βGlcNAc (1→3) βGal (1→4) Glc-Cer	P_1	
βGal (1→4) βGlcNAc (1→3) βGal (1→4) Glc-Cer	—	Paragloboside

[a]Symbols used are Cer, ceramide; Gal, galactose; Fuc, fucose; Glc, glucose; GalNAc, N-acetylgalactosamine; and GlcNAc, N-acetylglucosamine.

White Cell Allotypes

So far, the best-defined antigens on white cells are those concerned with histocompatibility: the HLA antigens. At least four closely linked loci, designated as *A, B, C,* and *D,* determine these antigens. The first three are identified serologically, and the fourth by mixed lymphocyte culture (MLC) testing. The genetics and biology of the entire major histocompatibility complex (MHC) have been recently reviewed (24, 25).

Alloantibodies used for HLA serological testing are obtained from human subjects who have been exposed to white cells on multiple occasions by pregnancy or transfusion. The molecules bearing the HLA determinants are glycoproteins. Since complete sequencing has not yet been done, the precise biochemical basis for the individual determinants is unknown.

The MLC test depends upon the ability of lymphocytes having one *HLA-D* genotype to stimulate lymphocytes with another *HLA-D* genotype, as measured by uptake of tritiated thymidine (26). In recent practice, X-irradiated cells known to be homozygous for at least the *D* locus are used as the stimulating or "typing" cells (27). The amount of tritiated thymidine incorporated into DNA by the cells being typed acts as a measure of their tranformation and thus reflects the difference between the two cell lines. A low or absent response implies sharing of identical antigens.

Current lists of all HLA antigens and their haplotype frequencies are given in a recent international workshop report (28), which also contains details about the significance of these antigens in tissue transplantation. (For some additional information about HLA, see section on chromosome mapping.)

Another way to detect white cell surface antigens is to use antisera raised against human B lymphocytes (29). Such antisera can be found in immunized human subjects (30) or can be prepared by injecting rabbits with papain-digested cells obtained from human splenic tissue (31). The antigenic determinants, which are detected by cytotoxicity testing, are located on B cells but not on T cells, granulocytes, platelets, or red cells. They are independent of the immunoglobulin intrinsic to B-cell membranes, and are probably analogous to the Ia immune response antigens found in mice (32). The human B-cell antigens are thought to be either the serological equivalents of the HLA-D locus products detected by MLC, or else the products of closely linked genes. Research in this area is very active, because of its implications in genetic control of the human immune response (33).

Plasma Protein Allotypes

It is theoretically possible for transfusions containing human plasma to stimulate alloantibodies against any antigenic determinants present on molecules of the donor's plasma proteins but not present in the recipient. However, only the immunoglobulin and low-density lipoprotein molecules have been clearly demonstrated to be alloantigenic, and their allotypes are detected by specific human antibodies. Two other antigenic determinants have been detected by antibodies induced in rabbits by human serum injection: Lp^a on lipoprotein (34, 35) and Xm on α_2 macroglobulin (36) molecules.

IMMUNOGLOBULINS The human immunoglobulin allotypes represent amino acid substitutions at various sites on the constant domains of the heavy and light polypeptide chains (37). The term *Gm* originally meant simply *gamma* but it has come to be applied to the heavy chains of all four IgG subclasses, representing four tandem gene loci. Thus, the G1m determinants occur only on the heavy chain of IgG1 molecules, G2m on IgG2, G3m on IgG3, and G4m on IgG4.

The amino acid sequences associated with some of the Gm determinants are known (38–42). For example, the specificity of G1m(1) is determined by Arg-Asp-Glu-Leu at positions 355 to 358 on the $\gamma 1$ chain (the heavy chain of IgG1). Another determinant is called G1m(non-1) because it can occur as an alternative to G1m (1) on the $\gamma 1$ chain but is always found on the $\gamma 2$ and $\gamma 3$ chains. Its specificity is determined by Arg-Glu-Glu-Met at positions 355 to 358 of $\gamma 1$ and at analogous positions on $\gamma 2$ and $\gamma 3$. Presumably G1m(1) evolved from G1m(non-1) by two separate mutations, since they differ by two amino acid substitutions.

The presence of an arginine residue at position 214 in the constant domain of the Fd fragment of $\gamma 1$ determines G1m(3) specificity, while substitution of lysine at that site determines G1m(17) specificity. The large number of detectable Gm haplotypes and their differences among populations make this genetic system nearly as valuable as the HLA system for differentiating ethnic groups (38).

There are only two known IgA subclasses, called IgA1 and IgA2, and only IgA2 has known allotypes. A2m(1) and A2m(2) are its two determinants (43, 44). An additional determinant which is allotypic on IgA2 but isotypic on IgA1 molecules has also been described (45). In molecules with A2m(1) specificity, the light chains are not covalently bound to the heavy ($\alpha 2$) chains but are linked to each other (46). A selective deficiency of all IgA molecules occurs in about 0.1% of people (47), who are thus able to form antibodies against IgA. In addition, a small number of transfused people form allotype-specific anti-IgA. Both kinds of antibodies have been implicated in an unusual form of transfusion reaction (48), a characteristic apparently not shared by the other immunoglobulin alloantibodies.

Allotypes of the immunoglobulin light chains are limited to those of the κ subtype. Three determinants, originally given the designation InV, are now called Km(1), Km(2), and Km(3). The Km(1) determinant is associated with the presence of a leucine residue at position 191 of the k chain constant region, while Km(3) has valine in that position (49). Km(2) specificity depends on both the leucine at position 191 and an alanine residue at position 153 (50, 51). While light chains of the λ subtype have no known allotypes, they can be divided into isotypes common to all people. Amino acid substitutions at positions 154 and 191 of the constant region are associated with the Oz and Kern antigenic determinants which suggest the existence of at least three structural gene loci for this region of the λ chain (52, 53).

β LIPOPROTEIN β lipoprotein allotypes, called Ag (for antigen), are detected by precipitin tests using the serum of human subjects immunized by pregnancy or transfusion (54, 55). At least ten Ag determinants have been reported, possibly representing five pairs of alternative antigens (56). These antigens could represent five different β lipoprotein molecular classes, analogous to the five immunoglobulin classes or four IgG subclasses.

ELECTROPHORETIC VARIANTS

Plasma Components

Most of the plasma protein electrophoretic variants (listed in Table 2) have been thoroughly described in previous reviews (1, 2, 4, 57–59). Aspects of their evolution have also been discussed (60).

Because the structural gene loci of several of the complement components appear to be closely linked to the *HLA* loci in the major histocompatibility region, they are of especial genetic interest. The subject has been recently reviewed by Alper & Rosen (61) but a brief outline here seems warranted.

COMPLEMENT POLYMORPHISMS At least four of the complement components (C2, C3, C4, and C6) are genetically polymorphic, as is properdin factor B (also known as Bf or earlier, as glycine-rich β-glycoprotein, GBG, or C3 proactivator, C3PA). There is sufficient C3 protein (about 100 mg/100 ml) to be visualized directly after staining electrophoresed serum. However, the other four proteins are present in much lower concentrations, requiring the use of immunofixation techniques, i.e. precipitation with specific antibodies after electrophoresis (62).

In the C3 system, there are two major alleles called $C3^F$ and $C3^S$ to characterize the fast and slow electrophoretic mobility of their respective protein products (63,

Table 2 Plasma protein electrophoretic polymorphisms[a]

Protein	Locus name
Haptoglobin	$Hp\ \alpha$
Transferrin	Tf
Vitamin-D binding protein	Gc (for group-specific component)
Ceruloplasmin	Cp
α_1 antitrypsin	Pi (for protease inhibitor)
α_1 acid glycoprotein	Oro (for orosomucoid)
β_2 glycoprotein I	—
Properdin factor B	Bf
Complement	
Second component	$C2$
Third component	$C3$
Fourth component	$C4$
Sixth component	$C6$
Enzymes	
Pancreatic amylase	AMY_2
Cholinesterase	E_2

[a] For references, see Harris (1), Giblett (3), and, for the complement components, those given in the text.

64). The fast- and slow-moving proteins are inherited as autosomal codominants, so that both are visible on stained starch or agar gels after electrophoresis of heterozygote serum. In all populations tested, the frequency of $C3^S$ is greater than that of its allele, varying in American populations from about 0.77 in Caucasians to 0.94 in Afro-Americans and 0.99 in those of Oriental origin (61). Several other rare alleles have also been reported, including some associated with C3 deficiency. Homozygotes or compound heterozygotes for the latter variants are subject to severe bouts of bacterial infections (65). The C3 locus is apparently *not* linked to the *HLA* loci (66).

Polymorphism of properdin factor B also involves two major alleles, Bf^S and Bf^F, with varying gene frequencies depending on racial origin (67). Close genetic linkage of the *Bf* and *HLA* loci is established (68, 69).

Genetic polymorphisms involving C2, C4, and C6 have only recently been reported. Heterozygote patterns for C2 are found in about 5% of people tested (70). C4, which consists of three different polypeptide chains (71), shows inherited electrophoretic variation, presumably representing allelism at only one structural gene locus (72). The genes determining these variants are apparently allelic to a rare $C4^0$ gene with no detectable product; thus, a single polypeptide chain is presumably involved. C4 in man is analogous to the Ss protein described in mice (73) which is determined by a locus on the mouse major histocompatibility complex (MHC), just as C4 is linked to the MHC in man (72). The C6 polymorphism is best demonstrated by isoelectric focusing of serum with subsequent lysis of sensitized sheep cells at the sites of C6 activity (74). Electrophoretic heterogeneity of homozygote as well as heterozygote patterns complicates phenotyping. The two major alleles in the system have frequencies of about 0.60 and 0.40 in the racial groups tested. Like the loci of *C2, C4,* and *Bf,* the *C6* locus is linked to the *HLA* loci (61).

VITAMIN D-BINDING PROTEIN Genetic variation of a plasma protein called group specific component or Gc was first described nearly 20 years ago (75). However, it was not until 1974 that the function of this protein as the major vitamin D-binding molecule was appreciated. Addition of radioactive vitamin D to serum followed by electrophoresis revealed radioactivity patterns identical with those of the Gc protein (76). The *Gc* structural gene locus is known to be closely linked to that of albumin (77), another protein with a carrier function. The biochemical structure of vitamin D-binding protein is only partially resolved (78) and there is as yet no information about amino acid sequence.

Cellular Components

Hemoglobin, by far the most prevalent protein in blood cells, has been studied more extensively than any other human protein. It is genetically polymorphic in many populations (79). The other polymorphic proteins detected in human blood cells by electrophoresis of cell extracts are enzymes, demonstrated by specific stains for catalytic activity. An excellent recent manual by Harris & Hopkinson (80) describes methods for detecting the enzymes, and gives their tissue distribution, subunit structure, inheritance, geographic distribution, and gene frequencies.

THE POLYMORPHIC ENZYMES IN BLOOD CELLS The top section of Table 3 lists those polymorphic enzymes which can be detected by electrophoresis of red cell hemolysates. Many of these enzymes are also present in leukocytes, but the convenience of preparing hemolysates makes red cells the usual source. Details about most of these enzymes have been extensively reviewed (80–82). Those reported during the past five years are GPT (83), ESD (84), UMPK (85), GPX (86), αGLU (87), αFUC (88), $ACON_S$ (89), HK_3 (90), GLO (91), and CDA (92).

Harris (1) has used the extent of genetic variation observed among human enzymes to calculate the overall degree of allelic variation, concluding that at least 30% of loci have two or three allelic genes with fairly common frequencies.

USEFULNESS OF GENETIC MARKERS IN BIOLOGICAL STUDIES

Mutation Mechanisms

Studies of the aberrant hemoglobins have provided valuable clues about the kinds of genetic mutations that occur in men. As described in recent reviews (93, 94), analyses of globin chain amino acid sequences have shown that most of these hemoglobin variants are due to single base-pair substitutions in the DNA of the globin genes. However, some variants represent single or multiple base-pair deletions or insertions, frameshifts, errors in gene termination, or fusion secondary to unequal crossing-over.

Table 3 Blood cell enzyme electrophoretic polymorphisms[a]

Enzyme	Locus name	EC number
Red cells		
Acid phosphatase 1	ACP_1	3.1.3.2
Adenosine deaminase	ADA	3.5.4.4
Adenylate kinase	AK_1	2.7.4.3
Carbonic anhydrase 2	CA_2	4.2.1.1
Diaphorase (NADPH dependent)	DIA_2	1.6.*.*
Esterase D	ESD	3.1.1.1
Galactose-1-P-uridyl transferase	$GALT$	2.7.7.12
Glucose-6-P dehydrogenase	Gd	1.1.1.49
Glutamic-pyruvic transaminase	GPT	2.6.1.2
Glutathione peroxidase	GPX	1.11.1.9
Glutathione reductase	GSR	1.6.4.2
Glyoxalase I	GLO	4.4.1.5
Peptidase A	$PEPA$	3.4.11.*
Peptidase C	$PEPC$	3.4.11.*
Peptidase D	$PEPD$	3.4.13.9
Phosphoglucomutase 1	PGM_1	2.7.5.1
Phosphoglucomutase 2	PGM_2	2.7.5.1
Phosphogluconate dehydrogenase	PGD	1.1.1.44
Uridine monophosphate kinase	$UMPK$	2.7.4.*
White cells		
Aconitase (soluble)	$ACON_S$	4.2.1.3
Cytidine deaminase	CDA	3.5.4.5
α-L-fucosidase	αFUC	3.2.1.51
α-glucosidase	αGLU	3.2.1.20
Glutamic-oxaloacetic transaminase (mitochondrial)	GOT_M	2.6.1.1
Hexokinase 3	HK_3	2.7.1.1
Malic enzyme (mitochondrial)	ME_M	1.1.1.40
Phosphoglucomutase 3	PGM_3	2.7.5.1

[a] Nomenclature taken from reference 98.

Human Chromosome Mapping

Considerable progress has recently been made in determining the chromosomal assignment and approximate location of human gene loci. Since the review of this subject in 1973 by McKusick & Chase (95), the proceedings of three international workshops on gene mapping have been published (96–98) and a more up-to-date map will be available after a 1977 meeting in Winnipeg.

Several loci representing the genetic polymorphisms have been assigned to specific chromosomes (summarized in 96–98). Chromosome one contains the loci for both Rh and Duffy blood group systems as well as for the polymorphic enzymes 6PGD, UMPK, PGM, Amy-2, Pep C, and αFuc. The *ABO* locus is near the tip of the long arm of chromosome nine, where it is closely linked to the AK_1 locus. The ACP locus is on chromosome number two, PGM_2 on number four, ESD on number 13, and ADA on number 20. Of greatest immunogenetic interest is the major histocompatibility complex located on chromosome 6. The occurrence in this region of genes related to the structure of several complement components was mentioned earlier. Also on chromosome 6, quite closely linked to the HLA-D locus, is the locus for glyoxalase-1 (GLO), and (nearer to the centromere) the PGM_3 locus.

Clonal Origin of Tumors

In accordance with the X-inactivation hypothesis of Lyon and Beutler, each somatic cell in females contains only one genetically active X chromosome, derived from one or the other parent. Thus, when a tumor develops from a single cell in a female who is heterozygous for some X-linked genetic marker (such as the enzyme G-6-PD), that tumor should contain only the maternally derived or the paternally derived gene for that marker. A recent review of the extensive work done in this area (99) shows that such clonal origin is characteristic of most human tumors, as well as related disorders such as chronic myelocytic and lymphocytic leukemia, myelofibrosis, paroxysmal nocturnal hemoglobinuria, and cold agglutinin disease. Burkitt's lymphoma also has a clonal origin. Since the evidence strongly suggests that this tumor is caused by a virus, it appears that either a single cell responds to the viral invasion by undergoing oncogenic change or else many cells respond, but only one escapes immunological surveillance (99).

Fate of Bone Marrow Grafts

Patients with aplastic anemia, leukemia, or severe combined immunodeficiency disease are the major candidates for bone marrow transplantation, especially from their HLA compatible siblings. When the donor is of the opposite sex, karyotyping for XX and XY can determine whether the cells that subsequently populate the peripheral blood are of donor origin. When the donor is of the same sex, any of the blood cell genetic markers that differ in donor and recipient can be used to follow the fate of the grafted tissue (100). When an identical twin is available for such a graft, no differentiating markers are detectable, but the assumption of monozygosity can be confirmed to a high degree of probability by testing both twins and their parents for all markers.

Detection of Cellular Mosaicism

In rare instances, two human egg nuclei are fertilized by two sperm nuclei, but the result, instead of twins, is the development of a single individual with two separate somatic cell lines. Since the original description in 1962 of dispermic chimerism (101), 20 more examples have been found (5). Most of the individuals with this condition have been detected because the father contributed an X chromosome to one cell line and a Y chromosome to the other, with resultant gonadal abnormalities. The blood cell populations are distinguishable by their different genetic markers, such as blood groups, HLA antigens, and cell enzymes. Since the chimerism affects all somatic tissues, eye color differences have also been noted, as well as patches of light and dark skin when one parent is heterozygous for genes determining pigmentation.

Another kind of twin chimerism involves only blood cells. It occurs when vascular anastomoses form between dizygotic twins in utero, resulting in permanent grafting of hemopoietic tissue in one or both children. First diagnosed in bovine twins, this condition has since been detected in 20 human sets (5). Because the chimerism is not present in other tissues, no gonadal changes occur, there is no pigment heterogeneity, and the karyotypes of skin fibroblasts show no evidence of chimerism.

Associations with Disease

RED CELL ANTIGENS The molecules carrying most of the red cell allotypic determinants are glycolipids or glycoproteins intrinsic to the cell membrane. It is likely that these molecules have functional, as well as structural, roles. The best evidence for this assumption lies in the relationship of the Duffy blood group system to infestation by malarial parasites. Miller and his colleagues (102, 103) have shown that red cells lacking Fy^a or Fy^b determinants are very resistant in vitro to invasion by *Plasmodium knowlesi* merozoites, and that Fy(a-b-) individuals bitten by *P. vivax*-infested mosquitoes fail to develop malaria. Since about 70% of Afro-Americans have the Fy(a-b-) phenotype, it is likely that the proportion is even greater in some areas of Africa, where there is probably a resultant high degree of malaria resistance. The underlying mechanism is not yet proven, but the evidence points to the Duffy antigenic determinants as receptors for attachment of the merozoite prior to entry into the red cell.

Red cells lacking the common Duffy antigens have no apparent structural alteration. However, there are two other red cell phenotypes—both very rare—in which absence of the common antigens is associated with variable degrees of membrane abnormalities and increased red cell destruction. One of these phenotypes is called *Rh null* because the red cells do not react with any of the antibodies with Rh specificity (5). The other phenotype is named *McLeod,* for the person in whom it was first found (104). Red cells of this rare type react very weakly with antibodies specific for common antigens in the Kell blood group system. They also are reported to have a distinctive membrane defect (105). There is evidence that Kx, a very common X-linked antigen normally present on both red cells and neutrophiles, is associated with the autosomally determined Kell system antigens (106). Children with the X-linked form of chronic granulomatous disease (CGD) sometimes have the McLeod red cell type but sometimes do not (107). According to Marsh et al (108), the Kx antigen is missing from the neutrophiles in both forms of X-linked CGD, but only those CGD patients with the McLeod type lack red cell Kx. Furthermore, individuals with the McLeod red cell type but normal neutrophile function have the usual amount of neutrophile Kx. While the underlying biochemical genetics is certainly obscure, it is clear that molecules bearing Kell system antigenic determinants have important structural and probably, functional roles. Furthermore, if X-linked CGD is indeed caused by deficiency of an NAD or NADP oxidase (109, 110), that ectoenzyme molecule might itself be the Kx antigen carrier.

HISTOCOMPATIBILITY ANTIGENS A large number of human disease states, particularly those related to immune dysfunction, are reported to have associated increased or decreased frequencies of HLA antigens. The details about these associations have been recently reviewed (111, 112). The strongest HLA and disease association known is that between ankylosing spondylitis (AS) and HLA-B27: about 90% of Caucasians with this disease have the B27 antigen as opposed to about 8% of normal controls. The estimated frequency of AS in the USA population is nearly two million (112), many times greater than was previously estimated, because of undetected cases. Adult rheumatoid arthritis, on the other hand, is associated with an HLA-D specificity called Dw4. Some of the other diseases having significantly increased association with HLA antigens (reviewed in 112) are multiple sclerosis with Dw2, myasthenia gravis with B8, psoriasis vulgaris with Bw37, dermatitis herpetiformis with B8, juvenile diabetes with Dw3, Addison's disease also with Dw3, subacute thyroiditis with Bw35, and idiopathic hemachromatosis with A3.

Space limitations preclude further discussion here of this rapidly expanding area of research. The mechanisms underlying these associations are unknown, and therefore open to much speculation.

ENZYME DEFICIENCIES Inherited enzyme deficiencies are associated with a wide variety of human inborn errors of metabolism (1, 113). The polymorphic enzymes, like other proteins, have rare genetic variants in addition to their more common allele products. In some cases, these variants have very low or absent catalytic activity. Thus, individuals who either are hemizygotes (in the case of X-linked loci) or are homozygotes or compound heterozygotes (in the case of autosomal loci) for such deficient alleles usually develop some clinical manifestation. Deficiencies of enzymes involving glycolysis (such as the X-linked G6PD) have been known for years to cause increased red cell destruction. More recently, deficiency of another polymorphic enzyme, adenosine deaminase, was found to cause severe combined immunodeficiency disease. In that particular instance, the enzyme defect was detected serendipitously while testing the blood of a bone marrow graft candidate for genetic markers (114). It is quite conceivable that similar testing for other polymorphic enzymes will eventually uncover additional deficiencies causing inherited disease states.

CONCLUSION

Much of our understanding about human molecular biology has been derived, directly or indirectly, from studies of polymorphic components of blood cells and plasma. Recent demonstrations of associations between the antigenic determinants on blood cells and the development of disease have focused attention on the functional activities of cell membrane components. Also, in studying the polymorphic enzymes in blood, detection of "silent" alleles has uncovered hitherto unsuspected associations between those enzymes and specific tissue function. Research in these areas is flourishing and should soon produce information fundamental to our comprehension of mechanisms underlying disease susceptibility and pathophysiology.

Literature Cited

1. Harris, H. 1975. *The Principles of Human Biochemical Genetics.* New York: Am. Elsevier. 473 pp. 2nd ed.
2. Giblett, E. R. 1975. Red cell genetic polymorphisms: Their usefulness in some studies of human biology. In *The Red Blood Cell,* ed. D. M. Surgenor, 2:935-57. New York: Academic
3. Giblett, E. R. 1969. *Genetic Markers in Human Blood.* Oxford: Blackwell. 629 pp.
4. McKusick, V. A. 1970. Human genetics. *Ann. Rev. Genet.* 4:1-46
5. Race, R. R., Sanger, R. 1975. *Blood Groups in Man.* Oxford: Blackwell. 659 pp. 6th ed.
6. Watkins, W. M. 1974. Blood group substances: Their nature and genetics. In *The Red Blood Cell,* ed. D. M. Surgenor, 1:294-361. New York: Academic. 2nd ed.
7. Hakomori, S., Kobata, A. 1974. Blood group antigens. In *The Antigens,* ed. M. Sela, 2:80-141. New York: Academic
8. Watkins, W. M. 1966. Blood group substances. *Science* 152:172-81
9. Watkins, W. M., Morgan, W. T. J. 1976. Immunochemical observations on the human blood group P system. *J. Immunogenet.* 3:15-27
10. Marcus, D. M. 1969. The ABO and Lewis blood-group system: immunochemistry, genetics and relation to human disease. *N. Engl. J. Med.* 280:994-1006
11. Naiki, M., Marcus, D. M. 1974. Human erythrocyte P and P^k blood group antigens: Identification as glycosphingolipids. *Biochem. Biophys. Res. Commun.* 60:1105-11
12. Hakomori, S. 1975. Fucolipids and blood group glycolipids in normal and tumor tissue. *Prog. Biochem. Pharmacol.* 10:167-96
13. Marcus, D. M., Schwarting, G. A. 1976. Immunochemical properties of glycolipids and phospholipids. *Adv. Immunol.* 23:203-40
14. Naiki, M., Fong, J., Ledeen, R., Marcus, D. M. 1975. Structure of the human erythrocyte blood group P_1 glycosphingolipid. *Biochemistry* 14:4831-37
15. Naiki, M., Marcus, D. M. 1975. An immunochemical study of the human blood group P_1, P and P^k glycosphingolipid antigens. *Biochemistry* 14:4837-41
16. Marcus, D. M., Naiki, M., Kundu, S. K. 1976. Abnormalities in the glycosphingolipid content of human P^k and p erythrocytes. *Proc. Natl. Acad. Sci. USA* 73:3263-67
17. Marcus, D. M., Naiki, M., Kundu, S. K., Schwarting, G. A. 1977. Immunochemical studies of the human blood group P system. *Proc. 5th Int. Convocation Immunol.* In press
18. Giblett, E. R. 1977. Some perspectives on blood group genetics and immunology. *Proc. 5th Int. Convocation Immunol.* In press
19. Springer, G. F., Desai, P. R. 1975. Human blood group MN and precursor specificities: Structural and biological aspects. *Carbohydr. Res.* 40:183-92
20. Dahr, W., Uhlenbruck, G., Bird, G. W. G. 1975. Influence of free amino and carboxyl groups on the specificity of plant anti-N. *Vox Sang.* 28:389-91
21. Lisowska, E., Duk, M. 1975. Effect of modification of amino groups of human erythrocytes on M, N and N_{VG} blood group specificities. *Vox Sang.* 28:392-97
22. Rosenfield, R. E., Allen, F. H., Rubenstein, P. 1973. Genetic model for the Rh blood group system. *Proc. Natl. Acad. Sci. USA* 70:1303-7
23. Green, F. A. 1972. Erythrocyte membrane lipids and Rh antigen activity. *J. Biol. Chem.* 247:881-87
24. Bach, F. H., van Rood, J. J. 1976. The major histocompatibility complex—genetics and biology. *N. Engl. J. Med.* 295:806-13, 872-78, 927-36
25. Snell, G. D., Dausset, J., Nathenson, S. 1976. *Histocompatibility.* New York: Academic. 401 pp.
26. Dupont, B., Hansen, J. A., Yunis, E. J. 1976. Human mixed-lymphocyte culture reaction: Genetics, specificity and biological implications. *Adv. Immunol.* 23:108-202
27. Mempel, W., Grosse-Wilde, H., Baumann, P., Netzel, B., Steinbauer-Rosenthal, I., Scholz, S., Bertrams, J., Albert, E. D. 1973. Population genetics of the MLC response: Typing for MLC determinants using homozygous and heterozygous reference cells. *Transplant. Proc.* 5:1529-34
28. Kissmeyer-Nielsen, F., ed. 1975. *Histocompatibility Testing 1975.* Rep. VI Int. Histocompatibility Workshop Conf. Copenhagen: Munksgaard. 1035 pp.
29. Wernet, P. 1976. Human Ia-type alloantigens: Methods of detection, aspects of chemistry and biology, markers

30. for disease states. *Transplant. Rev.* 30:270–98
30. Winchester, R. J., Fu, S. M., Wernet, P., Kunkel, H. G., Dupont, B., Jersild, C. 1975. Recognition by pregnancy serums of non-HLA alloantigens selectively expressed on B lymphocytes. *J. Exp. Med.* 141:924–29
31. Billing, R., Rafizadeh, B., Drew, I., Hartman, G., Gale, R., Terasaki, P. 1976. Human B-lymphocyte antigens expressed by lymphocytic and myelocytic leukemia cells. *J. Exp. Med.* 144:167–78
32. Shreffler, D. C., David, C. S. 1975. The H-2 major histocompatibility complex and the I immune response region: Genetic variation, function and organization. *Adv. Immunol.* 20:125–95
33. McDevitt, H. O., Delovitch, T. L., Press, J. L., Murphy, D. B. 1976. Genetic and functional analysis of the Ia antigens: Their possible role in regulating the immune response. *Transplant. Rev.* 30:197–235
34. Berg, K. 1968. The Lp system. *Ser. Haematol.* 1:111–36
35. Sing, C. F., Schultz, J. S., Shreffler, D. C. 1974. The genetics of the Lp system. II. A family study and proposed models of genetic control. *Ann. Hum. Genet.* 38:47–56
36. Berg, K., Bearn, A. G. 1966. An inherited X-linked serum system in man. The Xm system. *J. Exp. Med.* 123:379–97
37. Grubb, R. 1970. *The Genetic Markers of Human Immunoglobulins.* New York: Springer. 152 pp.
38. Steinberg, A. G. 1969. Globulin polymorphisms in man. *Ann. Rev. Genet.* 3:25–52
39. Natvig, J. B., Kunkel, H. G. 1973. Human immunoglobulins: Classes, subclasses, genetic variants and idiotypes. *Adv. Immunol.* 16:1–60
40. Kunkel, H. G., Kindt, T. 1975. Allotypes and idiotypes. In *Immunology and Immunodeficiency,* ed. B. Benacerraf, pp. 55–80. Baltimore: Univ. Park Press
41. Nisonoff, A., Hopper, J. E., Spring, S. B. 1975. *The Antibody Molecule,* pp. 346–406. New York: Academic
42. Mage, R., Lieberman, R., Potter, M., Terry, W. D. 1973. Immunoglobulin allotypes. In *The Antigens,* ed. M. Sela, 1:300–77. New York: Academic
43. Vyas, G. N., Fudenberg, H. H. 1969. Am(1), the first genetic marker of human immunoglobulin A. *Proc. Natl. Acad. Sci. USA* 64:1211–16
44. van Loghem, E., Wang, A. C., Shuster, J. 1973. A new genetic marker of human immunoglobulins determined by an allele at the $\alpha 2$ locus. *Vox Sang.* 24:481–88
45. van Loghem, E., de Lange, G., Kolstinen, J. 1976. The first iso-allotype of human IgA proteins. An antigenic determinant occurring as allotype in IgA2 subclass and as isotype in IgA1 subclass. *Scand. J. Immunol.* 5:161–64
46. Jerry, L. M., Kunkel, H. G., Grey, H. M. 1970. Absence of disulfide bonds linking the heavy and light chains; a property of a genetic variant of $\gamma A2$ globulins. *Proc. Natl. Acad. Sci. USA* 65:557–63
47. Ammann, A. J., Hong, R. 1971. Selective IgA deficiency: Presentation of 30 cases and a review of the literature. *Medicine* 50:223–36
48. Vyas, G. N., Holmdahl, L., Perkins, H. A., Fudenberg, H. H. 1969. Serologic specificity of human anti-IgA and its significance in transfusion. *Blood* 34:573–81
49. Milstein, C., Munro, A. J. 1970. The genetic basis of antibody specificity. *Ann. Rev. Microbiol.* 24:335–58
50. Baglioni, C., Zonta, L. A., Cioli, D., Carbonara, A. 1966. Allelic antigenic factor Inv(a) of the light chains of human immunoglobulins: Chemical basis. *Science* 152:1519–21
51. Milstein, C. P., Steinberg, A. G., McLaughlin, C. L., Solomon, A. 1974. Amino acid sequence change associated with genetic marker Inv(2) of human immunoglobulin. *Nature* 248:160–61
52. Gibson, D., Levanon, M., Smithies, O. 1971. Heterogeneity of normal human immunoglobulin light chains. Nonallelic variation in the constant region of lambda chains. *Biochemistry* 10:3114–22
53. Hess, M., Hilschmann, N., Rivat, L., Rivat, C., Ropartz, C. 1971. Isotypes in human immunoglobulin lambda chains. *Nature New Biol.* 234:58–61
54. Allison, A. C., Blumberg, B. S. 1965. Serum lipoprotein allotypes in man. *Prog. Med. Genet.* 4:176–201
55. Hirschfeld, J. 1972. The Ag system: Present concepts and immunogenetic models. In *Protides of the Biological Fluids,* ed. H. Peeters, 19:157–60. Oxford: Pergamon
56. Bütler, R., Brunner, E., Morganti, G. 1974. Contribution to the inheritance of the Ag groups, a population genetic study. *Vox Sang.* 26:485–96

57. Giblett, E. R. 1974. Haptoglobin. In *Structure and Function of Plasma Proteins,* ed. A. C. Allison, 1:55–72. London: Plenum
58. Giblett, E. R. 1974. Transferrin. In *Physiological Pharmacology,* ed. W. S. Root, N. I. Berlin, 5:555–68. New York: Academic
59. Fagerhol, M. F., Laurell, C.-B. 1970. The Pi system—inherited variants of serum α_1-antitrypsin. *Prog. Med. Genet.* 7:96–111
60. Buettner-Janusch, J. 1970. Evolution of serum protein polymorphisms. *Ann. Rev. Genet.* 4:47–68
61. Alper, C. A., Rosen, F. S. 1976. Genetics of the complement system. *Adv. Hum. Genet.* 7:141–88
62. Alper, C. A., Johnson, A. M. 1969. Immunofixation electrophoresis: A technique for the study of protein polymorphisms. *Vox Sang.* 17:445–52
63. Alper, C. A., Propp, R. P. 1968. Genetic polymorphism of the third component of human complement. *J. Clin. Invest.* 47:2181–91
64. Azen, E. A., Smithies, O., Hiller, O. 1969. High-voltage starch-gel electrophoresis in the study of post-albumin proteins and C'3 (β1c-globulins) polymorphism. *Biochem. Genet.* 3:215–28
65. Alper, C. A., Colten, H. R., Rosen, F. S., Rabson, A. R., MacNab, G. M., Gear, J. S. S. 1972. Homozygous deficiency of C3 in a patient with repeated infections. *Lancet* 2:1179–87
66. Gedde-Dahl, T., Teisberg, P., Thorsby, E. 1974. C3 polymorphism: Genetic linkage relations. *Clin. Genet.* 6:66–72
67. Alper, C. A., Boenisch, T., Watson, L. 1972. Genetic polymorphism in human glycine-rich beta-glycoprotein. *J. Exp. Med.* 135:68–80
68. Allen, F. H. 1974. Linkage of HL-A and GBG. *Vox Sang.* 27:382–84
69. Teisberg, P., Olaisen, B., Gedde-Dahl, T., Thorsby, E. 1975. On the localization of the Gb locus within the MHS region of chromosome no. 6. *Tissue Antigens* 5:257–61
70. Alper, C. A. 1976. Inherited structural polymorphism in human C2. Evidence for genetic linkage between C2 and Bf. *J. Exp. Med.* 144:1111–15
71. Müller-Eberhard, H. J. 1975. The complement system. In *The Plasma Proteins,* ed. F. W. Putnam, 1:393–432. New York: Academic. 2nd ed.
72. Teisberg, P., Åkesson, I., Olaisen, B., Gedde-Dahl, T., Thorsby, E. 1976. Genetic polymorphism of C4 in man and localization of a structural C4 locus to the HLA gene complex of chromosome 6. *Nature* 264:253–54
73. Lachmann, P. J., Grennan, D., Martin, A., Demant, P. 1975. Identification of Ss protein as murine C4. *Nature* 258:242–44
74. Hobart, M. J., Lachmann, P. J., Alper, C. A. 1975. Polymorphism of human C6. See Ref. 55, 22:575–80
75. Hirschfeld, J., Jonsson, B., Rasmuson, M. 1960. Inheritance of a new group-specific system demonstrated in normal human sera by means of an immunoelectrophoretic technic. *Nature* 185:931–33
76. Daiger, S. P., Schanfield, M. S., Cavalli-Sforza, L. L. 1975. Group-specific component (Gc) proteins bind vitamin D and 25-hydroxy-vitamin D. *Proc. Natl. Acad. Sci. USA* 72:2076–80
77. Weitkamp, L. R., Rucknagel, D. L., Gershowitz, H. 1966. Genetic linkage between structural loci for albumin and group specific component (Gc). *Am. J. Hum. Genet.* 18:559–71
78. Bearn, A. G., Bowman, B. H., Kitchin, F. D. 1964. Genetical and biochemical considerations of the serum group-specific component. *Cold Spring Harbor Symp. Quant. Biol.* 29:435–42
79. Lehmann, H., Huntsman, R. G. 1974. *Man's Haemoglobins.* Philadelphia: Lippincott. 478 pp. 2nd ed.
80. Harris, H., Hopkinson, D. A. 1976. *Handbook of Enzyme Electrophoresis in Human Genetics.* New York: Am. Elsevier
81. Hopkinson, D. A., Harris, H. 1971. Recent work on isozymes in man. *Ann. Rev. Genet.* 5:5–32
82. Hopkinson, D. A., Edwards, Y. H., Harris, H. 1976. The distributions of subunit numbers and subunit sizes of enzymes: A study of the products of 100 human gene loci. *Ann. Hum. Genet.* 39:383–411
83. Chen, S.-H., Giblett, E. R., Anderson, J. E., Fossum, B. L. G. 1972. Genetics of glutamic-pyruvic transaminase: Its inheritance, common and rare variants, population distribution and differences in catalytic activity. *Ann. Hum. Genet.* 35:401–9
84. Hopkinson, D. A., Mestriner, M. A., Cortner, J., Harris, H. 1973. Esterase D: A new human polymorphism. *Ann. Hum. Genet.* 37:119–37
85. Giblett, E. R., Anderson, J. E., Chen, S.-H., Teng, Y.-S., Cohen, F. 1974. Uridine monophosphate kinase: A new ge-

netic polymorphism with possible clinical implications. *Am. J. Hum. Genet.* 26:627–35
86. Beutler, E., West, C. 1974. Red cell glutathione peroxidase polymorphisms in Afro-Americans. *Am. J. Hum. Genet.* 26:255–58
87. Swallow, D. M., Corney, G., Harris, H., Hirschhorn, R. 1975. Acid α-glucosidase: A new polymorphism in man demonstrable by "affinity" electrophoresis. *Ann. Hum. Genet.* 38:391–406
88. Turner, B. M., Turner, V. S., Beratis, N. G., Hirschhorn, K. 1975. Polymorphism of human α-fucosidase. *Am. J. Hum. Genet.* 27:651–61
89. Slaughter, C. A., Hopkinson, D. A., Harris, H. 1975. Aconitase polymorphism in man. *Ann. Hum. Genet.* 39:193–202
90. Povey, S., Corney, G., Harris, H. 1975. Genetically determined polymorphism of a form of hexokinase, HK III, found in human leucocytes. *Ann. Hum. Genet.* 38:407–16
91. Kompf, J., Bissbort, S., Gussmann, S., Ritter, H. 1975. Polymorphism of red cell glyoxalase 1 (EC-4.4.1.5) a new genetic marker in man. *Humangenetik* 27:141–43
92. Teng, Y.-S., Anderson, J. E., Giblett, E. R. 1975. Cytidine deaminase: A new genetic polymorphism demonstrated in human granulocytes. *Am. J. Hum. Genet.* 27:492–97
93. Ranney, H. M., Lehmann, H. 1975. The hemoglobinopathies. See Ref 2, 2:874–908
94. Weatherall, D. J., Clegg, J. B. 1976. Molecular genetics of human hemoglobin. *Ann. Rev. Genet.* 10:157–78
95. McKusick, V. A., Chase, G. A. 1973. Human genetics. *Ann. Rev. Genet.* 7:435–73
96. Bergsma, D., ed. 1974. *New Haven Conf. 1973: 1st Int. Workshop Human Gene Mapping. Birth Defects: Orig. Artic. Ser. X:3.* New York: Natl. Found. 216 pp.
97. Bergsma, D., ed. 1975. *Rotterdam Conf. 1974: 2nd Int. Workshop Human Gene Mapping. Birth Defects: Orig. Artic. Ser. XI:3.* New York: Natl. Found. 310 pp.
98. Bergsma, D., Ed. 1976. *Baltimore Conf. 1975: 3rd Int. Workshop Human Gene Mapping. Birth Defects: Orig. Artic. Ser. XII:7.* New York: Natl. Found. 452 pp.
99. Fialkow, P. J. 1976. Clonal origin of human tumors. *Biochim. Biophys. Acta* 458:283–321

100. Storb, R., Thomas, E. D., Weiden, P. L., Buckner, C. D., Clift, R. A., Fefer, A., Fernando, L. P., Giblett, E. R., Goodell, B. W., Johnson, F. L., Lerner, K. G., Neiman, P. E., Sanders, J. E. 1976. Aplastic anemia treated by allogeneic bone marrow transplantation: A report on 49 new cases from Seattle. *Blood* 48:817–53
101. Gartler, S. M., Waxman, S. H., Giblett, E. R. 1962. An XX/XY human hermaphrodite resulting from double fertilization. *Proc. Natl. Acad. Sci. USA* 48:332–35
102. Miller, L. H., Mason, S. J., Dvorak, J. A., McGinniss, M. H., Rothmann, I. K. 1975. Erythrocyte receptors for *Plasmodium knowlesi* malaria: Duffy blood group determinants. *Science* 189:561–63
103. Miller, L. H., Mason, S. J., Clyde, D. F., McGinniss, M. H. 1976. The resistance factor to *Plasmodium vivax* in blacks. *N. Engl. J. Med.* 295:302–4
104. Allen, F. H., Krabbe, M. R., Corcoran, P. A. 1961. A new phenotype (McLeod) in the Kell blood-group system. *Vox Sang.* 6:555–60
105. Wimer, B. M., Marsh, W. L., Taswell, H. F. 1976. Clinical characteristics of the McLeod blood group phenotype. *Blood* 48:959 (Abstr.)
106. Marsh, W. L., Oyen, R., Nichols, M. E. 1976. Kx antigen, the McLeod phenotype and chronic granulomatous disease: Further studies. *Vox Sang.* 31:356–62
107. Giblett, E. R., Klebanoff, S. J., Pincus, S. H., Swanson, J., Park, B. H., McCullough, J. 1971. Kell phenotypes in chronic granulomatous disease: A potential transfusion hazard. *Lancet* 1:1235
108. Marsh, W. L., Oyen, R., Nichols, M. E., Allen, F. H. 1975. Chronic granulomatous disease and the Kell blood groups. *Br. J. Haematol.* 29:247–62
109. Hohn, D. C., Lehrer, R. I. 1975. NADPH oxidase deficiency in X-linked chronic granulomatous disease. *J. Clin. Invest.* 55:707–13
110. Iverson, D., DeChatelet, L. R., Spitznagel, J. K., Wang, P. 1977. Comparison of NADH and NADPH oxidase activities in granules isolated from human polymorphonuclear leukocytes with a fluorometric assay. *J. Clin. Invest.* 59:282–90
111. Svejgaard, A., Platz, P., Ryder, L. P., Staub-Nielsen, L., Thomsen, M. 1975.

HLA and disease associations—a survey. *Transplant Rev.* 22:1–44
112. Sasazuki, T., McDevitt, H. O., Grumet, F. C. 1977. The association between genes in the major histocompatibility complex and disease susceptibility. *Ann. Rev. Med.* 28:425–52
113. Stanbury, J. B., Wyngaarden, J. B., Fredrickson, D. S., ed. 1972. *The Metabolic Basis of Inherited Disease.* New York: McGraw-Hill. 1788 pp. 3rd ed.
114. Giblett, E. R., Anderson, J. E., Cohen, F., Pollara, B., Meuwissen, H. J. 1972. Adenosine deaminase deficiency in two patients with severely impaired cellular immunity. *Lancet* 2:1067–68

GENETIC CONTROL OF SPORULATION

❖3114

Roy H. Doi

Department of Biochemistry and Biophysics, University of California, Davis, California 95616

CONTENTS

INTRODUCTION	29
BACTERIAL SPORULATION SYSTEM	30
GENETICS OF BACTERIAL SPORULATION	32
Early Sporulation Genes	33
Late Sporulation Genes	34
Sporulation Operons	35
TRANSCRIPTION DURING SPORULATION	35
Early Sporulation Functions	36
Messenger RNA Populations During Sporulation	37
RNA Polymerase Mutations Causing Asporogeny	38
Enzymology of RNA Polymerase of Sporulating Cells	40
CONCLUSIONS	43

INTRODUCTION

Bacterial sporulation represents a relatively simple case of cellular differentiation. During spore formation a temporal sequence of events leads to the development of a cellular form distinct from the vegetative cell in morphology, physical and biochemical properties, and physiological function (1–8). Since one of the central goals in developmental genetics is to clarify the relationships of the macromolecular components of the cell during the differentiation process, several approaches are being taken to elucidate the genetic complexity of sporulation and to determine the biochemical mechanisms involved in regulating the temporal expression of the genome.

A particularly favorable aspect of bacterial sporulation is the availability of mutants that are affected only in sporulation. Most of these mutants can grow perfectly well in the vegetative phase and cannot be distinguished in any biochemical way from the wild-type cells during growth. Thus an analysis of "sporulation mutants" can be focused on a particular function that is required only during differentiation. This review focuses briefly on current understanding of the genetics of sporulation and discusses several current approaches which have great potential for providing information about the regulation of transcription and the role of RNA polymerase during sporulation. Most of these studies have been carried out with *Bacillus subtilis* mutants which can be analyzed by transformation and transduction.

BACTERIAL SPORULATION SYSTEM

Bacterial sporulation is initiated at the end of logarithmic growth when certain nutrients in the growth medium are depleted (9). A sequence of morphological events ensues, which is illustrated schematically in Figure 1. The morphological stages are numbered from 0 to VII (10), and the complete development of a mature spore takes about 7–8 hr. The end of the log phase of growth is designated as T_0 and every hour subsequent to T_0 is called T_1, T_2, T_3, etc. There is reasonably close

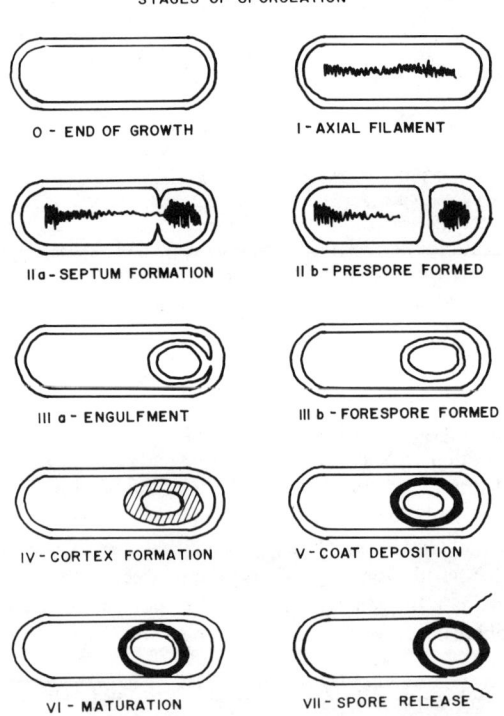

Figure 1 Schematic illustration of sporulation stages in *Bacillus*.

GENETIC CONTROL OF SPORULATION   31

but not exact correlation between the stage number and the hour after the end of the log phase of growth.

The fact that sporulation is initiated at the end of growth has made the analysis of sporulation-specific functions more difficult. A number of genes are released from catabolite repression at the end of the growth phase, and it is difficult to differentiate between these derepressed gene functions and those that are specific for sporulation. Release from catabolite repression is necessary for sporulation, but it is not sufficient, since conditions which cause release from catabolite repression do not necessarily permit sporulation (11). A final cell division occurs at about $T_{1.5}$ followed by the formation of the asymmetric spore septum in each of the final daughter cells (12). The period from T_0 to T_2 covers the final vegetative functions and the transition to the prespore state, and sporulation can be considered to be initiated at about T_1.

Several cytological and biochemical factors have been correlated with successful initiation of sporulation. Since the formation of the DNA axial filament (stage I) is not readily or consistently visualized, recent reviews on the subject have considered all the early sporulation stages up to septum formation (stage II) as stage 0 (13, 14) and this recent convention is also followed in this review. However, a final round of DNA replication and cell division is necessary prior to forespore septum formation (12, 15, 16), and there is a critical period during the final round of DNA replication at about T_1 which controls the initiation of sporulation (17). There are many additional unknown factors involved in the initiation phase, since not only is prespore septum formation asymmetrical, but no cell wall is laid down between the membranes of developing prespore and the mother cell, an event which would occur in ordinary cell division (Figure 1, stage IIb). The absence of the cell wall allows the engulfment of the prespore by the mother cell to form the double-membraned forespore at stage III (Figure 1).

The production of antibiotics and extracellular proteases has also been intimately linked with the initial stages of sporulation (6). Their production differs from the synthesis of a number of other extracellular products during the stationary phase in that the regulation of their synthesis is closely coordinated with the sporulation process. Most antibiotic-less and protease-less mutants are also asporogenous and result from mutations with pleiotropic effects (6).

The production of highly phosphorylated nucleotides has been analyzed at T_0 to T_2. A rapid production of ppGpp (MSI) and pppGpp (MSII) was noted around T_1 followed by an equally rapid decrease in these compounds (18). A rapid drop in the rate of rRNA synthesis was also noted at the time when MSI and MSII were synthesized (19); these observations support the recent contentions that MSI and MSII regulate rRNA synthesis (20, 21). However, there does not appear to be any correlation between the production of MSI and MSII with the initiation of sporulation, since a relaxed mutant that did not produce MSI and MSII was able to sporulate normally (18). However, in *B. subtilis* four other highly phosphorylated nucleotides (HPN) have been detected at about T_0 to T_1. These HPNs include ppApp (HPNI), pppApp (HPNII), ppZpUp (HPNIII; Z is an unidentified sugar moiety), and pppAppp (HPNIV) (22, 23). A membrane fraction has been isolated from sporulating cells which can synthesize HPNI and HPNIV in vitro from ATP.

The synthesis of HPNI and HPNIV is independent of added RNase, Pronase, KCN, or sodium azide. Phosphorylated metabolites of glucose including glucose-1-phosphate, glucose-6-phosphate, and glucose-1,6-diphosphate which are known to repress sporulation in vivo inhibited the synthesis of HPNI and HPNIV by the membrane fraction in vitro (22). It was also shown that HPNIII and HPNIV were synthesized in vivo only under conditions that favored sporulation, e.g. when glucose or phosphate was depleted from the growth medium. The correlation of the appearance of HPNIII and HPNIV with conditions favoring the initiation of sporulation has suggested to Rhaese (22) that these HPNs may trigger the sporulation events. The acceptance of this hypothesis will depend on finding mutants that cannot make these HPNs and simultaneously cannot sporulate. Also the role of these HPNs in regulating gene transcription directly or indirectly will have to be established. It should be mentioned that significant levels of cAMP have not been observed in most *Bacilli* tested (24, 25); however, cGMP has been found (24, 26).

The complexity of the developmental process has been accentuated by the observation that spore coat proteins that cover the forespore at stage V are synthesized as early as stage II (27, 28, 29). Aronson & Fitz-James (30) have calculated the amount of spore coat peptide that would have to be synthesized at the reduced rate of protein synthesis during sporulation to cover the developing forespore; from their calculations it is not unreasonable that the synthesis of spore coat protein occurs from a very early stage in sporulation. The spore coat protein that is produced early in sporulation would have to be protected until stage V from protein turnover, which occurs at a very high rate in sporulating cells (31). This poses several interesting problems in the interpretation of the sequential appearance of sporulation-specific gene products and the correlation between the time of gene transcription and the use of the gene product at any particular morphological stage.

Besides the many enzymological changes that occur during sporulation (2, 32), the pattern of DNA-binding proteins of log phase and sporulating cells has been analyzed by Hoch and his colleagues (33, 34). Both quantitative and qualitative changes in DNA-binding proteins were observed during sporulation. Although the exact nature of these proteins has not been determined because of the small recoverable quantities, it is possible that these proteins could regulate some aspects of gene transcription during growth and sporulation. This aspect is discussed later with respect to the Spo0 mutants.

Thus a number of cytological and biochemical events have been associated with the early sporulation process. The subsequent sections deal with the genetic studies associated with these processes and the current understanding of differential transcription and the transcriptional apparatus in sporulating cells.

GENETICS OF BACTERIAL SPORULATION

The relative ease in selecting asporogenous mutants and the availability of transduction and transformation analyses of *B. subtilis* have facilitated the genetic studies of sporulation. The usual method of selection has relied on the fact that wild-type cells form pigmented colonies on agar plates, whereas asporogenous cells form colonies that appear larger, unpigmented, and translucent. However, there are cells that are asporogeneous or oligosporogenous (cells that sporulate at very low frequency) and that can still form pigments (14). This latter situation could result in the lack of selection of these types of mutants.

Asporogenous mutants (Spo) have been defined by the last morphological stage to which they can develop (10) (see Figure 1). Their phenotypes are designated as Spo0, SpoI, SpoII, etc, indicating morphological development to stages 0, I, II, etc. Since axial filament formation may be the result of a physical reaction of DNA to a change in the chemical environment (35), both stage 0 and stage I phenotypes are currently considered to be in stage 0. Stage II or septum formation is the first stage at which sporulation can be distinguished cytologically. Different mutants within a particular stage such as stage 0 are designated as Spo0a, Spo0b, Spo0c, etc, with a, b, and c indicating different phenotypic properties of mutants blocked at stage 0.

Early Sporulation Genes

The complexity of the early stage of sporulation is demonstrated by the fact that stage 0 mutants have been mapped in nine different loci. (*The term locus as applied to sporulation genes does not imply a single gene, but a cluster of genes*); *the terms "locus" and "operon" will be used interchangeably* (13, 36). The genotypes for these stage 0 mutants have been designated as *spo0A*, *spo0B*, etc, with A, B, etc, designating different genetic loci. These *spo* loci are scattered throughout the *B. subtilis* chromosome, and no evidence has been obtained for any nonchromosomal *spo* locus. The map location of *spo* genes also indicates that there is no correlation between the sequence of spore gene expression and its location from the replication origin. Mutations in the *spo0* loci result in pleiotropic effects such as loss of production of antibiotics and extracellular protease and loss of transformability (37, 38).

The pleiotropic nature of the stage 0 mutations has been analyzed and a complex picture emerges which suggests that membrane functions could be affected by some of these mutations (39–41). In these studies *spo0A* cells were plated on antibiotics that function at the membrane level or on phages and selected for resistance against these factors. Those cells that were found to be resistant to the antibiotics or phages were shown to have recovered some properties of the wild-type sporulating cells, such as protease and antibiotic production and transformability, in a hierarchical fashion (39, 41); however, they were still asporogenous. Some cells selected for resistance to antibiotics simultaneously gained some resistance to phages (40). All these partial revertants still retained their *spo0A* genotype. Some of these partial suppressor mutations resulted in temperature-sensitive suppression and some suppressed nonsense mutations (39); therefore, these partial suppressor genes probably code for proteins, possibly with functions in the membrane. The stage 0 mutants appear to have a lesion(s) which can be compensated for *physiologically* by the products of these suppressor genes. From these studies, Ito (40) concluded that the *spo0A* and *spo0B* loci did not code for the structural genes for the extracellular proteases nor the enzymes involved in antibiotic production; since the cells remained asporogenous even if they produced proteases and antibiotics, the *spo0* loci controlled other factors required for sporulation.

An analysis of the DNA-binding proteins from stage 0 mutants and the wild type has revealed a most interesting pattern (33). In the wild-type cells, dramatic shifts occur in DNA-binding proteins both qualitatively and quantitatively during sporulation. An analysis of log phase mutant cells revealed surprisingly that stage 0 mutations caused an accumulation of DNA-binding proteins even during *growth* of these mutant cells. Suppression of the *spo*A mutation relieved the accumulation of these proteins. That the *spo*0 mutation affected the protein pattern of cells in the log phase suggested that the products of the stage 0 loci were functional not only during sporulation but also during vegetative growth. Since the first morphological evidence of sporulation is the asymmetric formation of septum and this may be only a modification of normal cell division (42), it is possible that some of the *spo*0 genes are indeed involved in normal membrane functions. Mutations in *spo*0 genes can be tolerated perhaps during growth, but not during sporulation which occurs under physiological stress. The pleiotropic nature of the *spo*0 mutations may be explained by their possible role in membrane functions. The other explanations for their pleiotropic effects are that some of the *spo*0 genes may be early regulator functions which affect the expression of a number of developmental genes or that the *spo*0 genes are early developmental structural genes whose products are required for the expression of subsequent developmental genes.

Late Sporulation Genes

During stages II through V (see Figure 1), several morphological and biochemical events occur which can be used to determine the phenotype. Several biochemical markers appear such as the synthesis of alkaline phosphatase, glucose dehydrogenase, and dipicolinic acid, and the resistance to chemicals and heat (43, 44). Cytological markers include forespore formation, cortex formation, and refractility (10).

There are seven stage II, five stage III, seven stage IV, and five stage V loci (13). Mapping data indicate that the loci for late sporulation genes are scattered throughout the chromosome. However, there is a particularly large cluster of late sporulation genes located between the *phe* to *lys* segment of the map (13, 14). Also a number of late sporulation genes are located near the *ura* to *met*C and the *aro*I to *pur*A regions (14). An analysis of the sequence of late genes on the map shows no direct relationship between the replication origin, the gene loci, and the stages of morphological development.

A clustering of genes for specific stages also has been noted, particularly of genes for stages II and III (14). The different mutants that affect a particular stage of development have slightly different phenotypes suggesting that different specific functions are altered in these mutants. *None of the late sporulation gene functions have been identified to date.* Only the gross morphological stage at which development has been arrested can be recognized in these mutants.

There are some events occurring during late sporulation that have been studied biochemically such as the appearance of alkaline phosphatase, glucose dehydrogenase, dipicolinic acid (DPA), spore cortex, and spore coat protein (43). However, the interesting aspect of these studies is that although the normal spore may have up to 15% of its dry weight as DPA (45), mutants have been isolated that form DPA-less spores (46) which are refractile and heat resistant, but which have an abnormal dormancy property, i.e. they tend to germinate very readily. Mutants that form cortex-deficient spores have also been isolated (47). Thus it is interesting that sporulation can continue in the absence of the complete expression of the normal complement of sporulation genes.

Spore coat deposition occurs quite late in sporulation at about stage V. Yet much evidence has accumulated that at least some of the coat proteins are made as early as stage I or II (27–29). Post-translational regulation may play an important role in spore coat development, since the spore coat appears to be made first in a soluble precursor form which then is converted to a more insoluble form (27; L. Munoz and R. H. Doi, unpublished observation). Thus in the case of spore coat protein and other sporulation products (48) morphological development late in sporulation depends directly on an early sporulation-specific gene which continues to be expressed throughout sporulation. Thus morphological development is not necessarily coincident with the sequential expression of sets of genes.

Sporulation Operons

Piggot (36) has defined "operons" for sporulation in the following manner: two Spo mutations are in separate operons (*a*) if they are separated by an auxotrophic marker, (*b*) if they are unlinked by transformation, and (*c*) if they cause blocks at different morphological stages of sporulation. By use of these criteria and the data from various studies, he found a total of nine stage 0 operons, seven stage II operons, five stage III operons, seven stage IV operons, and five stage V operons for a minimal total of 33 operons (13, 36). A further extension of these analyses suggested that the most probable number of sporulation operons was within limits of 33 and 59 (49). Thus the actual number of sporulation genes probably numbers in the hundreds.

The inherent difficulty in most of the genetic studies of sporulation to date has been the lack of identification of the primary products for the *spo* loci. This fact and the pleiotropic nature of all *spo* mutations has hampered the biochemical approach. In order to overcome the problems presented in the random selection of Spo mutants, an approach has been undertaken recently (50, 51) to gain understanding of specific functions required for sporulation. The basic method has been to select for antibiotic-resistant mutants which can grow normally but are affected only during sporulation, i.e. they are blocked at a particular stage of sporulation because of the defect caused by the mutation that rendered them antibiotic resistant. If mutants are selected for resistance to antibiotics with known sites of activity, the lesion caused by the mutation to a particular enzyme or to a structural protein can be confirmed by a relatively few biochemical tests. With this general approach, recent studies have been undertaken to gain a better understanding of the role of DNA-dependent RNA polymerase during sporulation.

TRANSCRIPTION DURING SPORULATION

The essential role of transcription during sporulation has been documented by studies using inhibitors of transcription and temperature-sensitive RNA polymerase mutants. Antibiotics that specifically inhibit RNA polymerase activity are capable of inhibiting sporulation at all stages (52, 53) indicating that continued transcription throughout sporulation is needed and that all or part of the essential transcription machinery is related to that found in vegetative cells. Pulse-labeling and inhibitor studies have indicated that the mRNA made during sporulation has a short half-life and that the synthesis of many sporulation-associated enzymes are also dependent on mRNA with short half-lives (53–56). Furthermore, a temperature shift-up of temperature-sensitive RNA polymerase mutants inhibited the appearance of enzymes normally synthesized by sporulating cells and sporulation itself (52). Therefore, continued transcription is necessary throughout the sporulation process.

Transcription during sporulation has been studied by four general approaches: (a) analysis of early sporulation functions; (b) analysis of messenger RNA populations during sporulation; (c) selection of RNA polymerase mutants that are altered specifically in sporulation functions; and (d) enzymological analysis of the RNA polymerase present in sporulating cells. Each of these approaches alone has limitations, but recently enough data have been obtained to provide a basis for a more thorough genetic and biochemical analysis of initiation and the sequential stages of spore formation.

Early Sporulation Functions

Several in vivo studies utilizing phage infection have revealed that the transcription function of early sporulating cells are altered. Yehle & Doi (57) noted that different responses were obtained after infection of sporulating cells depending on the phage species used. One type of phage as represented by phage $\beta 3$ replicated very poorly if at all in sporulating cells resulting in the phage DNA being incorporated into the developing spore. The other type of phage as represented by phage $\beta 22$, would infect sporulating cells and, although there was a delay in the latent period and the replication cycle proceeded at a slower rate than in vegetative cells, eventually the infected host cell produced a low yield of phage and lysed without having the opportunity to sporulate. An analysis of the messenger RNA being produced during infection of sporulating cells revealed that very little or no $\beta 3$ RNA was being produced whereas a normal complement of $\beta 22$ mRNA was produced. Thus in the case of phage $\beta 3$ it appeared that its DNA could not be transcribed efficiently during sporulation. This and a subsequent analysis of phage Φe (58) led to the interesting hypothesis proposed by Losick & Sonenshein (59, 60) that an alteration of the transcription machinery during sporulation was causing abortive phage infection during the sporulation cycle with the implication that phage promoters were similar to vegetative cell promoters which were transcribed less efficiently during sporulation. Other studies have indicated that some phages are able to develop much later into the sporulation sequence than others and that this was a function of the amount of information carried by the phage genome (61). As later discussion brings out, it is difficult to correlate the phage infection studies precisely with transcriptional machinery changes that occur during sporulation.

The initiation of sporulation appears to be particularly sensitive to netropsin, an inhibitor of RNA synthesis (62). Netropsin, a polypeptide antibiotic that binds to AT-rich regions of DNA, does not inhibit growth at low concentrations but prevents sporulation (63). It appears to be inhibiting sporulation specifically since catabolite-repressed enzymes can be derepressed in its presence and the general pattern of RNA and protein synthesis is not affected drastically in cells which are in the stationary phage (64). It has been proposed that netropsin may be binding to sporulation-specific DNA sites which are particularly rich in A-T bases. The preferential inhibition of sporulation by ethidium bromide also suggests that the DNA compositions of sporulation genes may be different from that for log phase genes (65).

Messenger RNA Populations During Sporulation

Pulse-labeling patterns of sporulating cells indicate that the rate of RNA synthesis at different periods of sporulation changes dramatically (19, 51, 66, 67). At the end of log growth (at T_0) there is a sudden drop in the rate of RNA synthesis by approximately 90%. This may be related to the high concentration of ppGpp and pppGpp which is found at that time (18), since these nucleotides inhibit rRNA synthesis (20, 21). As sporulation proceeds, there are several periods of high and low rates of RNA synthesis (67) which could reflect turning on and off of certain sets of sporulation genes.

The studies of mRNA populations of sporulating cells have been carried out primarily by competition DNA-RNA hybridization studies (68, 69). These studies have shown that sporulation-specific mRNA are transcribed from both heavy and light strands of DNA in a sequential manner from stages I through V (66, 70, 71). By the late stages of sporulation 40% of the hybridizable mRNA is qualitatively of the sporulation-specific type. However, even at the late stages of sporulation, 60% of the mRNA is still qualitatively similar to vegetative phase mRNA. A small percentage of the log phase genes is also turned off during sporulation, and this is reflected in both the DNA-RNA hybridization experiments and in the pattern of protein synthesis during early sporulation (72).

The mRNA isolated from dormant spores also comprises sporulation-specific mRNA in addition to vegetative phase mRNA (73). These data indicate that differential expression of the genome is taking place within the developing forespore as well as the mother cell and that proteins are being made for sporulation or germination. Unique small basic proteins have been observed in *B. megaterium* spores which are degraded during germination (74, 75).

The mRNA studies indicate that sporulation is a much more complex process than growth, since the sequential expression of sporulation genes is superimposed on the expression of log phase genes. The pattern of mRNA present in sporulating cells also indicates that two types of transcriptional apparatus may exist simultaneously: one type would be similar to that found in log phase cells and capable of transcribing log phase genes and the other type(s) would be capable of transcribing sporulation-specific genes.

The limitations inherent in these mRNA studies are the lack of direct evidence that the mRNAs are indeed sporulation-specific, the inability to study a gene-specific mRNA, and the lack of quantitation of mRNA species. These studies have, however, established that the pattern of mRNA synthesis during sporulation is distinct from that found in log phase cells and that genes not expressed during growth are being turned on in a sequential manner during morphogenesis.

RNA Polymerase Mutations Causing Asporogeny

One of the inherent difficulties with many of the Spo mutants which have been studied to date is that these Spo mutants were randomly selected for their asporogenic phenotype. Not one of these mutations has been linked to a direct sporulation-specific function and many or most are pleiotropic mutations (13, 14). In order to identify and correlate specific functions with the sporulation process and to understand the regulation of gene transcription during sporulation, RNA polymerase mutants have been selected that have provided much useful preliminary information.

RNA polymerase mutants have been selected for resistance to rifampicin, an antibiotic which inhibits RNA polymerase activity of $B.$ $subtilis$. Rifampicin inhibits the initiation step of RNA synthesis by binding directly to the enzyme; mutations in the β subunit of the enzyme can prevent this binding (76). The order of the markers in $B.$ $subtilis$ has been determined by three factor crosses as Cys-Rif^R (Stv)^R-Std^R-Str^R (77–79). The β gene is located near the origin of DNA replication and located between cysA14 and str^r. The map locations for α, σ, and β' are not known in $B.$ $subtilis$.

The phenotypes resulting from selection of rifampicin-resistant mutants can be classified into the following groups:

Class I   These Rif^R mutants can grow and sporulate normally at all temperatures. Spores are morphologically and functionally normal (50).

Class II These Rif^R mutants can grow normally and sporulate, but the resulting spores have an altered morphology (50, 80).

Class III These Rif^R mutants can grow normally but produce normal spores with reduced frequency (i.e. they are oligosporogenic) or at reduced rates (59, 78, 81, 82).

Class IV These Rif^R mutants can grow and sporulate normally at permissive temperatures, and can grow but not sporulate at high temperatures, i.e. *they are temperature-sensitive only during sporulation*. The temperature sensitivity is expressed at different morphological stages depending on the mutation (51, 67, 78, 83, 84).

The different phenotypes must arise from mutations at a number of close sites in the β subunit gene and depending on the site, different functions of the enzyme are affected during sporulation.

The Class IV mutants are particularly interesting since they provide the opportunity of exploring the functions of the RNA polymerase that are affected only during initiation and the subsequent stages of sporulation. These mutants grow at the same rate as the wild type during the log phase of growth indicating that the RNA polymerase can function perfectly for vegetative functions. However, depending on the mutation, the mutants are blocked at morphological stages from 0 to IV as observed by electron and light microscopy (67, 83). The pattern of RNA synthesis is dramatically altered in these mutants when compared to the wild type especially after the morphological stage at which the mutant has been blocked (51, 67). However, in all cases a significant amount of RNA synthesis continues even at the blocked stage; this suggests that only part of the transcription mechanism is being inhibited by the mutational event. In fact, in DNA-RNA hybridization studies the RNA obtained from mutants grown at the nonpermissive temperature to a period equivalent to Stage IV was able to compete with part of the Stage IV RNA obtained from wild-type sporulating cells (67); however, protein synthesis was slightly unbalanced as evidenced by the accumulation of large crystal structures in some of these mutant cells even at the permissive temperature (85). These and similar studies on enzyme synthesis (51, 86) indicate that specific and not global transcriptional function is altered in these mutants.

Several reasons can be given to explain the phenotype of these rifampicin-resistant mutants. At the permissive temperature the conformation of the enzyme is suitable for normal transcription of both log phase and sporulation genes. At the nonpermissive temperature the mutant enzymes are still able to recognize all log phase gene promoters, since the growth rates of the mutants are identical with that of the wild type. However, it is during sporulation that the sporulation-specific function of the mutant RNA polymerase is inhibited by the improper conformation imposed by the high temperature. A simple explanation, which is difficult to show experimentally, is that the mutant RNA polymerase may be functioning slightly less efficiently than the wild-type enzyme and that the reduced rate of transcription results in a total but slight metabolic imbalance. This does not disrupt growth but does prevent sporulation which requires greater metabolic balance than growth, since it occurs under stress when the medium is deficient.

The other major explanation for the phenotype of these mutants is based on the assumption that the RNA polymerase has specific functions during sporulation which differ from its function during the growth phase and that it has to interact with sporulation-specific "effectors" (51, 67). The change in conformation caused by the high temperature would prevent its interaction with sporulation-specific "effectors," which could be small molecules, σ-like factors, antiterminator factors, modifying enzymes (e.g. kinases), or other regulatory macromolecules that alter its specificity and allow it to transcribe sporulation-specific genes. A streptovaricin-resistant RNA polymerase mutant which has a mutation close to the rifampicin locus also has a phenotype similar to that exhibited by the Class IV mutants (87); it was of interest that no streptolydigin-resistant RNA polymerase mutants were conditionally temperature sensitive during sporulation. Since streptolydigin only inhibits elongation and not initiation, it is possible that this function is identical in both vegetative and sporulation RNA polymerase molecules. The availability of these Class IV mutants will allow an investigation of the structure and function of sporulation RNA polymerase.

Enzymology of RNA Polymerase of Sporulating Cells

The RNA polymerase core enzyme from *B. subtilis* vegetative cells has the usual four subunits designated $\alpha_2\beta\beta'$ (60, 88, 89). The molecular weights of the subunits are 45,000 for α, 150,000 for β, and 160,000 for β'. The core associates with σ factor (90), which has a molecular weight of 55,000, to form the holoenzyme ($\alpha_2\beta\beta'\sigma$). The vegetative holoenzyme is also associated with a number of other small polypeptides with molecular weights of 21,000 (91), 11,000 (92), and 9,500 (92). Variable amounts of these subunits are found with the holoenzyme suggesting that all holoenzyme molecules may not contain these subunits or that these subunits are easily dissociated from the holoenzyme during purification. Both of these explanations appear plausible and the differences observed between various investigators have to be considered on this basis. The functions particularly of the core-associated small polypeptides are uncertain although the 21,000 dalton species appears to be involved in promoter recognition in SPO1 phage-infected cells (91).

The DNA-RNA hybridization studies and the analysis of the conditionally temperature-sensitive RNA polymerase sporulation mutants suggest strongly that a sequential expression of sporulation genes is occurring. The direct enzymological approach has been taken to determine what the role of vegetative holoenzyme may be during sporulation and whether any modification of RNA polymerase structure may be occurring during sporulation.

In a series of thought-provoking papers the role of σ factor during early sporulation of *B. subtilis* has been analyzed. These studies have been based in part on the observation that phage Φe infection is inhibited during early sporulation (58) and the idea that this abortive infection may be caused by a change in the transcription specificity of the RNA polymerase during early sporulation (59). In fact a major change was noted in the in vitro activity of sporulation RNA polymerase which had greatly reduced activity on Φe DNA when compared to the vegetative phase RNA polymerase (59). The initial hypothesis (60) that a major modification of the β subunit was the cause of this specificity change was subsequently shown not to be the case, since the modification of the β subunit was shown to be the result of proteolytic cleavage during enzyme purification (93). All enzymological studies with sporulating cells have been plagued by the extremely high levels of proteolytic activity, many of unknown specificity. In these studies it was also noted that σ factor had an apparent lower affinity for the core of sporulating RNA polymerase (60, 93) which suggested an explanation for the lower activity of the enzyme on Φe DNA.

In investigations to determine the cause for the apparent lower σ affinity to sporulation core, it was found that the σ factor concentration in sporulating cells was similar to that found in vegetative cells (94), that the σ from sporulating cells was capable of stimulating and binding to vegetative core enzyme (95), that the sporulation σ factor had similar antigenic and chemical properties with the vegetative σ factor (96, 97), and that sporulation σ factor would stimulate sporulation core activity on Φe DNA template (96). All these studies indicated that σ factor of sporulating cells was similar or identical with the σ factor found in vegetative cells. The apparent lowered affinity of σ to the sporulation core could therefore be due either to some modification in the sporulation core that reduced its affinity to the σ factor or to some other "inhibitory" factor that prevented binding of the σ factor to a normal core. It seemed unlikely that the sporulation core had been modified, since under the proper conditions the σ factor was able to stimulate the sporulation core (95). So far no direct biochemical evidence has been obtained for any inhibitory factor for σ activity, although a physiologically unstable σ inhibitor was suggested by chloramphenicol inhibition studies (95). In these studies the antibiotic treatment of sporulating cells restored the ability of the sporulation RNA polymerase to transcribe Φe DNA in vivo and allowed the isolation of σ associated with core. An apparent half-life of 11 min at 37°C was determined for the putative inhibitor activity. It was proposed that this σ inhibitor was rapidly depleted in the absence of protein synthesis. Attempts to isolate this inhibitor have not been successful to date.

Another candidate suggested for the putative inhibitor molecule was a protein that was found associated with the core of sporulation RNA polymerase (98, 99). This protein has a molecular weight of 85,000 and appears initially at about T_1 (100); it could function by binding to the core and preventing the association of core with the σ factor; however, its turnover characteristics and inhibitory properties have not been reported as yet.

Since many of the earlier studies on the σ functions were done under conditions in which the activity of proteolytic enzymes was not completely inhibited (93), it is possible that some of the results could be explained in another manner. For instance a partial proteolysis of the core could decrease its affinity to the σ factor; σ factor could also be highly susceptible to proteolysis when not associated with core. From our experience (101), crude and partially purified extracts of sporulating cells are highly contaminated with proteases; furthermore, the use of the protease inhibitors phenylmethysulfonyl fluoride or diisopropylfluorophosphate with metal chelating agents does not ensure that all proteolytic activities in extracts from sporulating cells are inhibited (102); enzyme that has been partially degraded by proteases still retains enzymatic activity and can still bind to a DNA-cellulose column with a lower affinity than the undegraded enzyme (101). Any model of transcriptional modification based on loss of σ activity would have to account for the ability of phage to replicate until T_6 (61), the high level of active σ present in sporulating cells (96), the significant transcription of vegetative genes during sporulation (66, 71), and the presence of significant amounts of vegetative holoenzyme in sporulating cells (103).

Another major way that RNA polymerase specificity may be altered during sporulation is to modify the structure of the RNA polymerase core by either enzymatic modification of the core subunits (104) or the association of additional subunits to the core (91, 105, 106). These modifications could facilitate the recognition of sporulation gene promoters. In this regard no gross changes in the basic subunits of the RNA polymerase core have been observed at any stages of sporulation (89, 93, 103). However, these analyses have been based primarily on electrophoretic mobility of the subunits through a SDS-polyacrylamide gel which does not allow detection of minor structural modification such as phosphorylation or adenylylation. Therefore the possibility of minor modifications of the core subunits is still open to question.

Some data are available indicating that the core structure of RNA polymerase can be modified by association of the core with polypeptides other than σ factor. At T_3, which is equivalent to about stage II–III when the forespore is formed (Figure 1), two forms of RNA polymerase have been identified (99, 103). One form of RNA polymerase appears to be identical with the holoenzyme found in vegetative cells with composition $\alpha_2\beta\beta'\sigma$ (Enzyme I) and the other new form which is found only in sporulating cells has the composition $\alpha_2\beta\beta'\delta^1$ (Enzyme II). The new δ^1 subunit has a molecular weight of 28,000 and has been designated as δ *factor* or *differentiation factor*. Enzyme II has a higher affinity for the DNA-cellulose (103) and phosphocellulose (99) columns than Enzyme I. Enzyme II activity is affected differently by KCl than Enzyme I (99) and it has a specific activity greater than the σ-containing enzyme on several templates (103). The affinity of δ^1 to the core is greater than σ factor with core, since it is very difficult to remove δ^1 from the core by the usual methods; furthermore, when *B. subtilis* DNA is used as the template, the antibiotic netropsin, which binds to A-T–rich regions of DNA, inhibits Enzyme II much more than Enzyme I (T. Nakayama and R. H. Doi, unpublished observations), suggesting that the two forms of RNA polymerase may have different binding sites on DNA. The σ form comprises about 80–85% of the total RNA polymerase and the δ form only 15–20% based on core subunit content. Another polypeptide with a molecular weight of 85,000 has also been found associated with the δ^1 enzyme under certain conditions of enzyme purification (99). No specific role is known for this polypeptide.

At T_5, preliminary evidence has been found that the cell contains not only vegetative type holoenzyme and the δ^1 enzyme, but another δ-containing enzyme with the composition $\alpha_2\beta\beta'\delta^2$ (Enzyme III) (103). The molecular weight of δ^2 is about 20,000. This form of the enzyme also has a higher affinity for the DNA-cellulose column than the σ enzyme.

The presence of the σ-containing enzyme in sporulating cells appears reasonable since both mRNA (66) and proteins (2, 72) of the vegetative phase type are found in sporulating cells. That is, vegetative genes continue to be expressed actively during sporulation. The role of the new δ-containing enzymes is purely conjectural at this time; however, their appearance depends on the normal sporulation process (103) and they may play a crucial role in the transcription of sporulation-specific genes. The definitive correlation of these new forms that are produced sequentially during sporulation to the sequential appearance of mRNAs and to the blocked stages of the conditionally temperature-sensitive RNA polymerase mutants would be particularly important.

Studies with *B. thuringiensis* have indicated a much more complex change in the RNA polymerase pattern (107). Many of them appear to be due to limited proteolytic cleavages in vivo (108). It is difficult to compare the results from two different species, since variations in conditions during the purification procedures for even the same species result in different RNA polymerase preparations and properties (99, 103). This is one of the most critical aspects that will have to be investigated with extreme care in future studies with RNA polymerase, since it is likely that a number of RNA polymerase-associated polypeptides may be lost by current purification procedures.

Since differential transcription mechanisms in prokaryotic systems include positive and negative controls (109), antitermination (110), attenuation (111), and new (112, 113) or modified (91, 105, 106) RNA polymerase molecules with new promoter specificities, all of these potential mechanisms will have to be considered in any final analysis of transcriptional control during sporulation.

CONCLUSIONS

The analysis of mRNA synthesis, of conditional temperature-sensitive RNA polymerase sporulation mutants, and of the structure and function of RNA polymerase of sporulating cells all point to transcriptional control as a major site of regulation during bacterial morphogenesis. The large number of independent mutational sites that affect the initial stage (stage 0) of sporulation demonstrates the complexity of the transition from growth to morphological development. This particular stage should be investigated in terms of transcriptional control and RNA polymerase activity in greater detail, now that suitable RNA polymerase mutants and enzymological methods are available.

The demonstration of new forms of RNA polymerase in sporulating cells indicates that some of the early sporulating functions are involved in modifying RNA polymerase structure. Two basic reasons can be proposed for the modification of RNA polymerase structure during sporulation which depend on whether promoter sites for sporulation genes are identical with or different from those for vegetative genes. If the sporulation promoters are outside the normal spectrum of vegetative promoters ordinarily recognized by the vegetative RNA polymerase holoenzyme, the RNA polymerase modifications may be changing the promoter recognition specificity of the enzyme. However, if the promoters for sporulation genes are identical with those for vegetative genes, then the modification of structure is not changing its promoter specificity, but altering its interaction with putative regulatory factors which permit the RNA polymerase either to bind to the sporulation gene promoters or to allow it to read through the blocks presented by negative effectors or special termination sites.

Thus a direct test of these two basic models is to isolate DNA fragments which contain vegetative genes or sporulation genes and test them as templates for vegetative and sporulation RNA polymerases. If the vegetative holoenzyme can transcribe RNA from both types of genes in vitro, then the promoters are probably quite similar; regulation of transcription would probably involve other factors. If the vegetative enzyme cannot utilize sporulation templates, it would imply that either sporulation promoters are outside the spectrum of vegetative promoters or that the promoters are similar to vegetative promoters; however, "factors" are required to allow transcription of these sporulation genes. These factors could include metabolic products which interact with DNA-binding proteins similar to cAMP-CAP system (109) or core-modifying factors (91, 92). If the vegetative enzyme can only abortively initiate RNA synthesis from sporulation genes, some mechanism related to release from termination (110) or attenuation (111) could be involved in regulation. In any case, the technical advances to this point indicate that a number of genetic and biochemical tests are now feasible to answer basic questions related to transcription specificity of sporulation genes.

ACKNOWLEDGMENTS

A number of helpful discussions were held with Terrance Leighton, Shirley Halling, and Tatsuo Nakayama. The author has been supported by research grants PCM74–12137–A02 from the National Science Foundation and GM–19673 from the National Institute of General Medical Sciences.

Literature Cited

1. Dawes, I. W., Hansen, J. N. 1972. Morphogenesis in sporulating *Bacilli. CRC Crit. Rev. Microbiol.* 1:479–520
2. Doi, R. H., Sanchez-Anzaldo, F. J. 1976. Complexity of protein and nucleic acid synthesis during sporulation of *Bacilli.* In *Microbiology—1976,* ed. D. Schlessinger, pp. 145–63. Washington DC: Am. Soc. Microbiol.
3. Hanson, R. S., Peterson, J. A., Yousten, A. A. 1970. Unique biochemical events in bacterial sporulation. *Ann. Rev. Microbiol.* 24:53–90
4. Murrell, W. G. 1967. Biochemistry of the bacterial endospore. *Adv. Microbiol. Physiol.* 1:133–251
5. Freese, E. 1972. Sporulation of *Bacilli,* a model of cellular differentiation. *Cur. Top. Dev. Biol.* 7:85–124
6. Schaeffer, P. 1969. Sporulation and the production of antibiotics, exoenzymes, and exotoxins. *Bacteriol. Rev.* 33:48–71
7. Kornberg, A., Spudich, J. A., Nelson, D. L., Deutscher, M. P. 1968. Origin of proteins in sporulation. *Ann. Rev. Biochem.* 37:51–78
8. Kaneko, I., Doi, R. H., Santo, L. Y. 1974. Bacterial sporulation and germination. *Cell* 6:154–76
9. Schaeffer, P., Millet, J., Aubert, J.-P. 1965. Catabolic repression of bacterial sporulation. *Proc. Natl. Acad. Sci. USA* 54:704–11
10. Schaeffer, P., Ionesco, H., Ryter, A., Balassa, G. 1963. La sporulation de *Bacillus subtilis:* Etude genetique et physiologique. *Colloq. Int. CNRS* 124:553–63
11. Laishley, E. J., Bernlohr, R. W. 1966. Catabolite repression of "three sporulation enzymes" during growth of *Bacillus licheniformis. Biochem. Biophys. Res. Commun.* 24:85–90
12. Leighton, T., Khachatourians, G., Brown, N. 1975. The role of semiconservative DNA replication in bacterial cell development. In *ICN-UCLA Symp. Mol. Cell. Biol.* III. *DNA Synthesis and Its Regulation,* ed. M. Goulian, P. Hanawalt, C. F. Fox, pp. 677–87. Menlo Park, Calif.: Benjamin
13. Piggot, P. J., Coote, J. G. 1976. Genetic aspects of bacterial endospore formation. *Bacteriol. Rev.* 40:908–62
14. Hoch, J. A. 1976. Genetics of bacterial sporulation. *Adv. Genet.* 18:69–98
15. Dawes, I. W., Kay, D., Mandelstam, J. 1971. Determining effect of growth medium on the shape and position of daughter chromosomes and on sporulation in *Bacillus subtilis. Nature* 230:567–69
16. Mandelstam, J., Sterlini, J. M., Kay, D. 1971. Sporulation in *Bacillus subtilis.* Effect of medium on the form of chromosome replication and on initiation to sporulation in *Bacillus subtilis. Biochem. J.* 125:635–41
17. Mandelstam, J., Higgs, S. A. 1974. Induction of sporulation during synchronized chromosome replication in *Bacillus subtilis. J. Bacteriol.* 120:38–42
18. Rhaese, H.-J., Dichtelmuller, H., Grade, R. 1975. Studies on the control of development. Accumulation of guanosine tetraphosphate and pentaphosphate in response to inhibition of protein synthesis in *Bacillus subtilis. Eur. J. Biochem.* 56:385–92
19. Hussey, C., Losick, R., Sonenshein, A. L. 1971. Ribosomal RNA synthesis is turned off during sporulation of *Bacillus subtilis. J. Mol. Biol.* 57:59–70

20. Van Ooyen, A. J. J., Gruber, M., Jørgensen, P. 1976. The mechanism of action of ppGpp on rRNA synthesis *in vitro. Cell* 8:123–28
21. Reiness, G., Yang, H. L., Zubay, G., Cashel, M. 1975. Effects of guanosine tetraphosphate on cell-free synthesis of *Escherichia coli* ribosomal RNA and other gene products. *Proc. Natl. Acad. Sci. USA* 72:2881–85
22. Rhaese, H.-J., Groscurth, R. 1976. Control of development—role of regulatory nucleotides synthesized by membranes of *Bacillus subtilis* in initiation of sporulation. *Proc. Natl. Acad. Sci. USA* 73:331–35
23. Rhaese, H.-J., Grade, R., Dichtelmuller, H. 1976. Studies on the control of development. Correlation of initiation of differentiation with synthesis of highly phosphorylated nucleotides in *Bacillus subtilis. Eur. J. Biochem.* 64:205–13
24. Clark, V. L., Bernlohr, R. W. 1972. Catabolite repression and the enzymes regulating cyclic adenosine 3',5'-monophosphate and cyclic guanosine 3',5'-monophosphate levels in *Bacillus licheniformis. Spores V*:167–73
25. Setlow, P. 1973. Inability to detect cyclic AMP in vegetative or sporulating cells or dormant spores of *Bacillus megaterium. Biochem. Biophys. Res. Commun.* 52:365–72
26. Bernlohr, R. W., Haddox, M. K., Goldberg, N. D. 1974. Cyclic guanosine 3':5'-monophosphate in *E. coli* and *B. licheniformis. J. Biol. Chem.* 249:4329–31
27. Aronson, A. I., Fitz-James, P. C. 1968. Biosynthesis of bacterial spore coats. *J. Mol. Biol.* 33:199–212
28. Uchida, A., Kadota, H., Schaeffer, P. 1976. Appearance of spore coat protein in the cell extracts of *Bacillus subtilis* asporogenic mutants. *J. Bacteriol.* 126:1342–43
29. Wood, D. A. 1972. Sporulation in *Bacillus subtilis*. Properties and time of synthesis of alkali-soluble protein of the spore coat. *Biochem. J.* 130:505–14
30. Aronson, A. I., Fitz-James, P. 1976. Structure and morphogenesis of the bacterial spore coat. *Bacteriol. Rev.* 40:360–402
31. Doi, R. H. 1972. Role of proteases in sporulation. *Curr. Top. Cell. Regul.* 6:1–20
32. Freese, E., Fujita, Y. 1976. Control of enzyme synthesis during growth and sporulation. See Ref. 2, pp. 164–84

33. Brehm, S. P., Le Hegarat, F., Hoch, J. A. 1975. Deoxyribonucleic acid–binding proteins in vegetative *Bacillus subtilis:* Alterations caused by stage 0 sporulation mutants. *J. Bacteriol.* 124:977–84
34. Brehm, S. P., Le Hegarat, F., Hoch, J. A. 1974. Developmental modulation of DNA-binding proteins of *Bacillus subtilis* during sporulation stages. *J. Bacteriol.* 120:1443–50
35. Mandelstam, J. 1973. DNA replication and induction of sporulation in *Bacillus subtilis. Colloq. Int. CNRS* 227:115–17
36. Piggot, P. J. 1973. Mapping of asporogenous mutations of *Bacillus subtilis:* A minimum estimate of the number of sporulation operons. *J. Bacteriol.* 114:1241–53
37. Brehm, S. P., Staal, S. P., Hoch, J. A. 1973. Phenotypes of pleiotropic-negative sporulation mutants of *Bacillus subtilis. J. Bacteriol.* 115:1063–70
38. Michel, J. F., Millet, J. 1970. Physiological studies on early-blocked sporulation mutants of *Bacillus subtilis. J. Appl. Bacteriol.* 33:220–27
39. Ito, J., Mildner, G., Spizizen, J. 1971. Early blocked asporogenous mutants of *Bacillus subtilis* 168. I. Isolation and characterization of mutants resistant to antibiotics produced by sporulating *Bacillus subtilis* 168. *Mol. Gen. Genet.* 112:104–9
40. Ito, J. 1973. Pleiotropic nature of bacteriophage tolerant mutants obtained in early-blocked asporogenous mutants of *Bacillus subtilis* 168. *Mol. Gen. Genet.* 124:97–106
41. Guespin-Michel, J. F. 1971. Phenotypic reversion in some early blocked sporulation mutants of *Bacillus subtilis:* Isolation and phenotype identification of partial revertants. *J. Bacteriol.* 108:241–47
42. Hitchins, A. D., Slepecky, R. A. 1969. Bacterial sporulation as a modified procaryotic cell division. *Nature* 223:804–7
43. Warren, S. C. 1968. Sporulation in *Bacillus subtilis*. Biochemical changes. *Biochem. J.* 109:811–18
44. Milhaud, P., Balassa, G. 1973. Biochemical genetics of bacterial sporulation. IV. Sequential development of resistance to chemical and physical agents during sporulation of *B. subtilis. Mol. Gen. Genet.* 125:241–50
45. Powell, J. F. 1953. Isolation of dipicolinic acid (pyridine-2:6-dicarboxylic acid) from spores of *Bacillus megaterium. Biochem. J.* 54:210–11

46. Zytkovicz, T. H., Halvorson, H. O. 1972. Some characteristics of dipicolinic acid-less mutant spores of *Bacillus cereus, Bacillus megaterium,* and *Bacillus subtilis. Spores* V:49–52
47. Pearce, S. M., Fitz-James, P. C. 1971. Sporulation of a cortexless mutant of a variant of *Bacillus cereus. J. Bacteriol.* 105:339–48
48. Young, M. 1976. Use of temperature sensitive mutants to study gene expression during sporulation in *Bacillus subtilis. J. Bacteriol.* 126:928–36
49. Hranueli, D., Piggot, P. J., Mandelstam, J. 1974. Statistical estimate of the total number of operons specific for *Bacillus subtilis* sporulation. *J. Bacteriol.* 119:684–90
50. Doi, R. H., Brown, L. R., Rodgers, G., Hsu, Y.-P. 1970. *Bacillus subtilis* mutant altered in spore morphology and in RNA polymerase activity. *Proc. Natl. Acad. Sci. USA* 66:404–10
51. Leighton, T. J. 1973. An RNA polymerase mutation causing temperature-sensitive sporulation in *Bacillus subtilis. Proc. Natl. Acad. Sci. USA* 70:1179–83
52. Leighton, T. J., Doi, R. H. 1971. The stability of messenger ribonucleic acid during sporulation in *Bacillus subtilis. J. Biol. Chem.* 246:3189–95
53. Leighton, T. 1974. Further studies on the stability of sporulation mRNA in *B. subtilis. J. Biol. Chem.* 249:7808–12
54. Horn, D. T., Aronson, A. I., Golub, E. S. 1973. Development of a quantitative immunological assay for the study of spore coat synthesis and morphogenesis. *J. Bacteriol.* 113:313–21
55. Tipper, D. J., Pratt, I. 1970. Cell wall polymers of *B. sphaericus* 9602. II. Synthesis of the first enzyme unique to cortex synthesis during sporulation. *J. Bacteriol.* 103:305–17
56. Linnett, P. E., Tipper, D. J. 1976. Transcriptional control of peptidoglycan precursor synthesis during sporulation in *Bacillus sphaericus. J. Bacteriol.* 125:565–74
57. Yehle, C. O., Doi, R. H. 1967. Differential expression of bacteriophage genomes in vegetative and sporulating cells of *Bacillus subtilis. Virology* 1:935–47
58. Sonenshein, A. L., Roscoe, D. H. 1969. The course of phage φe infection in sporulating cells of *Bacillus subtilis* strain 3610. *Virology* 39:265–76
59. Losick, R., Sonenshein, A. L. 1969. Change in the template specificity of RNA polymerase during sporulation of *Bacillus subtilis. Nature* 224:35–37
60. Losick, R., Shorenstein, R. G., Sonenshein, A. L. 1970. Structural alteration of RNA polymerase during sporulation. *Nature* 227:910–13
61. Kawamura, F., Ito, J. 1974. Bacteriophage gene expression in sporulating cells of *Bacillus subtilis* 168. *Virology* 62:414–25
62. Wartell, R. M., Larson, J. E., Wells, R. D. 1974. Netropsin. A specific probe for A-T regions of duplex deoxyribonucleic acid. *J. Biol. Chem.* 249:6719–31
63. Keilman, G. R., Tanimoto, B., Doi, R. H. 1975. Selective inhibition of sporulation of *Bacillus subtilis* by netropsin. *Biochem. Biophys. Res. Commun.* 67:414–20
64. Keilman, G. R., Burtis, K., Tanimoto, B., Doi, R. H. 1976. Effect of netropsin on the derepression of enzymes during growth and sporulation of *Bacillus subtilis. J. Bacteriol.* 128:80–85
65. Rogolsky, M., Nakamura, H. T. 1974. Sensitivity of an early step in the sporulation of *Bacillus subtilis* to selective inhibition by ethidium bromide. *J. Bacteriol.* 119:57–61
66. Sumida-Yasumoto, C., Doi, R. H. 1974. Transcription from the complementary deoxyribonucleic acid strands of *Bacillus subtilis* during various stages of sporulation. *J. Bacteriol.* 117:775–82
67. Sumida-Yasumoto, C., Doi, R. H. 1977. Ribonucleic acid polymerase mutants of *Bacillus subtilis* conditionally temperature sensitive at various stages of sporulation. *J. Bacteriol.* 129:433–44
68. Doi, R. H., Igarashi, R. T. 1964. Genetic transcription during morphogenesis. *Proc. Natl. Acad. Sci. USA* 52:755–62
69. Aronson, A. I. 1965. Characterization of messenger RNA in sporulating *Bacillus cereus. J. Mol. Biol.* 11:576–85
70. Yamakawa, T., Doi, R. H. 1971. Preferential transcription of *Bacillus subtilis* light deoxyribonucleic acid strands during sporulation. *J. Bacteriol.* 106:305–10
71. DiCioccio, R. A., Strauss, N. 1973. Patterns of transcription in *Bacillus subtilis* during sporulation. *J. Mol. Biol.* 77:325–36
72. Linn, T., Losick, R. 1976. The program of protein synthesis during sporulation in *Bacillus subtilis. Cell* 8:103–14
73. Jeng, Y.-H., Doi, R. H. 1974. Messenger ribonucleic acid of dormant spores

of *Bacillus subtilis*. *J. Bacteriol.* 119: 514–21
74. Setlow, P. 1975. Identification and localization of major proteins degraded during germination of *Bacillus megaterium* spores. *J. Biol. Chem.* 250: 8159–67
75. Setlow, P. 1975. Purification and properties of some unique low molecular weight basic proteins degraded during germination of *Bacillus megaterium* spores. *J. Biol. Chem.* 250:8168–73
76. Linn, T., Losick, R., Sonenshein, A. L. 1975. Rifampin resistance mutation of *Bacillus subtilis* altering the electrophoretic mobility of the beta subunit of ribonucleic acid polymerase. *J. Bacteriol.* 122:1387–90
77. Haworth, S. R., Brown, L. R. 1973. Genetic analysis of ribonucleic acid polymerase mutants of *Bacillus subtilis*. *J. Bacteriol.* 114:103–13
78. Sonenshein, A. L., Cami, B., Brevet, J., Cote, R. 1974. Isolation and characterization of rifampin-resistant and streptolydigin-resistant mutants of *Bacillus subtilis* with altered sporulation properties. *J. Bacteriol.* 120:253–65
79. Harford, N., Sueoka, N. 1970. Chromosomal location of antibiotic resistance markers in *Bacillus subtilis*. *J. Mol. Biol.* 51:267–86
80. Korch, C. T., Doi, R. H. 1971. Electron microscopy of the altered spore morphology of a ribonucleic acid polymerase mutant of *Bacillus subtilis*. *J. Bacteriol.* 105:1110–18
81. Sonenshein, A. L., Losick, R. 1970. RNA polymerase mutants blocked in sporulation. *Nature* 227:906–9
82. Pun, P. P. T., Murray, C. D., Strauss, N. 1975. Characterization of a rifampin-resistant, conditional asporogenous mutant of *Bacillus subtilis*. *J. Bacteriol.* 123:346–53
83. Santo, L., Leighton, T. J., Doi, R. H. 1973. Ultrastructural studies of sporulation in a conditionally temperature-sensitive ribonucleic acid polymerase mutant of *Bacillus subtilis*. *J. Bacteriol.* 115:703–6
84. Rothstein, D. M., Keeler, C. L., Sonenshein, A. L. 1976. *Bacillus subtilis* RNA polymerase mutants temperature-sensitive for sporulation. In *RNA Polymerase*, ed. R. Losick, M. Chamberlin, pp. 601–16. New York: Cold Spring Harbor Lab.
85. Santo, L. Y., Doi, R. H. 1973. Crystal formation by a RNA polymerase mutant of *Bacillus subtilis*. *J. Bacteriol.* 116:479–82
86. Hoganson, D. A., Irgens, R. L., Doi, R. H., Stahly, D. P. 1975. Bacterial sporulation and regulation of dihydrodipicolinate synthase in ribonucleic acid polymerase mutants of *Bacillus subtilis*. *J. Bacteriol.* 124:1628–29
87. Leighton, T. 1977. New types of RNA polymerase mutations causing temperature-sensitive sporulation in *Bacillus subtilis*. *J. Biol. Chem.* 252:268–72
88. Avila, J., Hermoso, J. M., Vinuela, E., Salas, M. 1971. Purification and properties of DNA-dependent RNA polymerase from *Bacillus subtilis* vegetative cells. *Eur. J. Biochem.* 21:526–35
89. Orrego, C., Kerjan, P., Manca de Nadra, M. C., Szulmajster, J. 1973. RNA polymerase in a thermosensitive sporulation mutant (ts-4) of *Bacillus subtilis*. *J. Bacteriol.* 116:636–47
90. Shorenstein, R. G., Losick, R. 1973. Purification and properties of the sigma subunit of RNA polymerase from vegetative cells of *B. subtilis*. *J. Biol. Chem.* 248:6163–69
91. Pero, J., Nelson, J., Fox, T. D. 1975. Highly asymmetric transcription by RNA polymerase containing phage-SP01-induced polypeptides and a new host protein. *Proc. Natl. Acad. Sci. USA* 72:1589–93
92. Duffy, J. J., Geiduschek, E. P. 1975. RNA polymerase from phage SP01-infected and uninfected *Bacillus subtilis*. *J. Biol. Chem.* 250:4530–41
93. Linn, T. G., Greenleaf, A. L., Shorenstein, R. G., Losick, R. 1973. Loss of the sigma activity of RNA polymerase of *Bacillus subtilis* during sporulation. *Proc. Natl. Acad. Sci. USA* 70:1865–69
94. Tjian, R., Losick, R. 1974. An immunological assay for the sigma subunit of RNA polymerase in extracts of vegetative and sporulating *Bacillus subtilis*. *Proc. Natl. Acad. Sci. USA* 71:2872–76
95. Segall, J., Tjian, R., Pero, J., Losick, R. 1974. Chloramphenicol restores sigma factor activity to sporulating *Bacillus subtilis*. *Proc. Natl. Acad. Sci. USA* 71:4860–63
96. Tjian, R., Stinchcomb, D., Losick, R. 1974. Antibody directed against *Bacillus subtilis* factor purified by sodium dodecylsulfate slab gel electrophoresis. Effect of transcription by RNA polymerase in crude extracts of vegetative and sporulating cells. *J. Biol. Chem.* 250: 8824–28

97. Duie, P., Kaminski, M., Szulmajster, J. 1974. Immunological studies on the sigma subunit of the RNA polymerase from vegetative and sporulating cells of *Bacillus subtilis. FEBS Lett.* 48:214–17
98. Greenleaf, A. L., Linn, T. G., Losick, R. 1973. Isolation of a new RNA polymerase-binding protein from sporulating *Bacillus subtilis. Proc. Natl. Acad. Sci. USA* 70:490–94
99. Linn, T., Greenleaf, A. L., Losick, R. 1975. RNA polymerase from sporulating *Bacillus subtilis*. Purification and properties of a modified form of enzyme containing two sporulation polypeptides. *J. Biol. Chem.* 250:9256–61
100. Greenleaf, A. L., Losick, R. 1973. Appearance of a RNA polymerase-binding protein in asporogenous mutants of *B. subtilis. J. Bacteriol.* 116:290–94
101. Fukuda, R., Keilman, G., McVey, E., Doi, R. H. 1975. RNA polymerase pattern of sporulating *Bacillus subtilis*. *Spores* VI:213–20
102. Nakayama, T., Munoz, L., Doi, R. H. 1977. Procedure to remove protease activities from *Bacillus subtilis* sporulating cells and their crude extracts. *Anal. Biochem.* 78:165–70
103. Fukuda, R., Doi, R. H. 1977. Two polypeptides associated with ribonucleic acid polymerase core of *Bacillus subtilis* during sporulation. *J. Bacteriol.* 129:422–32
104. Mailhammer, R., Yang, H. L., Reiness, G., Zubay, G. 1975. Effects of bacteriophage T4-induced modification of *E. coli* RNA polymerase on expression *in vitro. Proc. Natl. Acad. Sci. USA* 72:4928–32
105. Duffy, J. J., Petrusek, R. L., Geiduschek, E. P. 1975. Conversion of *Bacillus subtilis* RNA polymerase activity *in vitro* by protein induced by phage SP01. *Proc. Natl. Acad. Sci. USA* 72:2366–70
106. Spiegelman, G. B., Whiteley, H. R. 1974. *In vivo* and *in vitro* transcription by RNA polymerase from SP82-infected *B. subtilis. J. Biol. Chem.* 249:1483–89
107. Klier, A. F., Lecadet, M.-M., Dedonder, R. 1973. Sequential modifications of DNA-dependent RNA polymerase during sporogenesis in *Bacillus thuringiensis. Eur. J. Biochem.* 36:317–27
108. Klier, A., Lecadet, M.-M. 1974. Evidence in favour of the modification *in vivo* of RNA polymerase subunits during sporogenesis in *Bacillus thuringiensis. Eur. J. Biochem.* 47:111–19
109. De Crombrugghe, B., Chen, B., Anderson, W., Nissley, P., Gottesman, M., Pastan, I. 1971. Lac DNA, RNA polymerase and cyclic AMP receptor protein, cyclic AMP, lac repressor and inducer are the essential elements for controlled lac transcription. *Nature New Biol.* 231:139–42
110. Roberts, J. W. 1976. Transcription termination and its control in *E. coli*. In *RNA Polymerase*, ed. R. Losick, M. Chamberlin, pp. 247–71. New York: Cold Spring Harbor Lab.
111. Bertrand, K., Korn, L., Lee, F., Platt, T., Squires, C. L., Squires, C., Yanofsky, C. 1975. New features of the regulation of the tryptophan operon. *Science* 189:22–26
112. Chamberlin, M., McGrath, J., Waskell, L. 1970. New RNA polymerase from *E. coli* infected with bacteriophage T7. *Nature* 228:227–31
113. Clark, S., Losick, R., Pero, J. 1974. New RNA polymerase from *Bacillus subtilis* infected with phage PBS2. *Nature* 252:21–24

MUTATIONS AFFECTING FITNESS IN DROSOPHILA POPULATIONS[1]

✦3115

Michael J. Simmons
Department of Genetics and Cell Biology, University of Minnesota, St. Paul, Minnesota 55108

James F. Crow
Department of Genetics, University of Wisconsin, Madison, Wisconsin 53706

CONTENTS

INTRODUCTION	50
THE MEASUREMENT OF VIABILITY	50
NEWLY ARISEN VIABILITY MUTANTS	52
Mutation Rates and Homozygous Effects on Viability	52
Heterozygous Effects on Viability	55
VIABILITY MUTANTS FROM EQUILIBRIUM POPULATIONS	56
Detrimental and Lethal Loads	56
Viability Mutants and Epistasis	58
Heterozygous Effects on Viability	59
VIABILITY MUTANTS AND FITNESS	62
Viability and Fertility As Components of Fitness	62
New Mutants and Fitness	63
Homozygous Effects on Fitness of Mutations from Equilibrium Populations	64
Heterozygous Effects on Fitness of Mutations Affecting Viability	67
SOME UNRESOLVED PROBLEMS	69
The Mutation Load	69
Fitness Variance Maintained by Balanced Selection	70
Very Nearly Neutral Mutations	70
Overdominance in Homozygous Backgrounds, Coupling-Repulsion Effects, and Optimum Heterozygosity	70
Hybrid Dysgenesis	74
CONCLUSIONS	74

[1]Supported in part by the National Institutes of Health (GM–22038).

INTRODUCTION

The aim of this review is limited: to summarize what is known about the impact of mutation on the fitness of Drosophila populations. As the ultimate raw material for evolution, mutation lies at the base of any discussion of evolutionary mechanisms. However, we are concerned here mainly with a by-product of this, the recurrent production of harmful mutations and their elimination by natural selection. This is what has been called *purifying selection* (1). Haldane (2) called it the price that a species pays for the privilege of evolution.

We report only experiments in which viability or some other component of fitness, or total fitness, is measured and experiments on the mutational contribution to this. Knowledge is now sufficient to make a rather accurate assessment of the rate at which mutations occur and of the way they are eliminated from the population. We can show that mutation makes a large contribution to the standing variance of population fitness, but we cannot say how large a contribution is made by other factors, such as balancing selection.

This review does not consider chromosome or isozyme polymorphism, geographical structure, speciation, or ecological factors. Nor does it consider population genetic theory. It is restricted to the input of new mutations and their elimination by natural selection, and to the mutation load thereby created.

THE MEASUREMENT OF VIABILITY

Viability is the chance that a zygote survives to the age of reproduction. Drosophila mutants affecting this trait are usually studied in segregation tests in which flies of a genotype of interest are contrived to hatch in the same culture as marked flies that serve as a viability standard. The frequency with which the interesting genotype emerges is a function of its viability and the Mendelian mechanism. The flies that emerge may be homozygous or heterozygous for viability-affecting mutants; in turn, the mutants may be newly arisen—either spontaneous or induced—or derived from an equilibrium population. Whatever the case, studies involving viability mutants often make use of special chromosomes that contain inversions and dominant markers. The inversions permit mutant-bearing chromosomes to be passed through females without recombinational breakup; of course, there is no recombination in Drosophila males.

One kind of manipulation made possible by the inversions is shown in Figure 1. The figure illustrates a procedure whereby an entire chromosome, in this case the *D. melanogaster* second, is made homozygous. The wild male in generation 1 comes from nature. The female to which it is mated carries two different second chromosomes, each with a dominant marker (Cy = Curly wings and Pm = Plum-colored eyes) and inversions. The Cy chromosome has inversions that effectively suppress crossing-over throughout the second chromosome. Thus, in generation 3 when Curly brothers and sisters are mated with one another, the wild second chromosome for which they are heterozygous (this comes from nature) is transmitted to the offspring as a unit. Two thirds of them will be like their parents ($Cy/+$); the other one third will be homozygous for the entire $+$ chromosome. The Cy/Cy flies die. If the $+$ chromosome carries viability-depressing mutants, then the proportion of non-Curly flies in the cultures should be reduced; fewer of these will survive relative to the $Cy/+$ class. However, since the Cy marker itself has an effect on viability, a deviation from the expected ratio of 2 Curly to 1 wild is not necessarily evidence for the presence of viability mutants on the wild chromosome. To control for the effect of the Cy marker or other viability-affecting properties of the chromosome, males and females descended from different wild-caught males, and therefore carrying different wild chromosomes, are mated in generation 3, so that in generation 4 the non-Curly flies carry random combinations of second chromosomes. These are therefore as heterozygous as a random fly freshly caught in nature. In experiments such as these, the ratio of non-Curly to Curly can be used as an index of the viability of the homozygous and heterozygous flies; often the index is standardized by the mean of the heterozygotes, so that the "normal" viability is 1.

In the homozygosis experiment, some of the cultures produce no non-Curly flies. Such cultures indicate the presence of at least one lethal mutant on the chromosome being tested. It is not unusual, especially in very large cultures, for a few flies of the lethal genotype to survive. Usually these are less than 3% and create a classification problem in small cultures, for they may or may not appear. In practice, a chromosome is classified as lethal if its standardized homozygous viability is 0.1 or less. This criterion of lethality makes for greater reproducibility of results; moreover, the few escapees that hatch in the incompletely lethal cultures are usually very weak, and in nature would have little chance of reproducing. Chromosomes that give more than 10% but less than 50–60% of the viability of the randomly constituted heterozygotes are called *semilethals*; as a class these are rare. Chromosomes with viability greater than 50–60% are called *quasinormals*; though free of drastic lesions, they carry mutants with minor effects on viability. Together with the semilethals, such mutants are referred to as *detrimentals*.

The total impact of the detrimentals on the viability of homozygotes can be quantified by applying the concept of genetic load. Although the word comes from Muller (3), the concept was first applied to human population data by Morton, Crow & Muller (4), and subsequently to Drosophila data by Greenberg & Crow (5). With A, B, and C as the mean viabilities of heterozygotes, nonlethal homozygotes, and all homozygotes, respectively, the detrimental load, D, is calculated as $ln(A/B) = lnA - lnB$; the total load, T, is obtained as $lnA - lnC$, and the lethal load, L, as the difference between these two ($T - D$). For some purposes the detrimental load is partitioned into two components, one measuring the impact of semilethal mutants (D_s), the other that of very mildly deleterious ones (D_m). The algebraic details are given by Temin et al (6). Loads have the property of being additive; they represent the summation of the effects of individual viability mutants. Conceptually, this is equivalent to dispersing the mutants into separate individuals, ascertaining their homozygous effects on viability, and summing them.

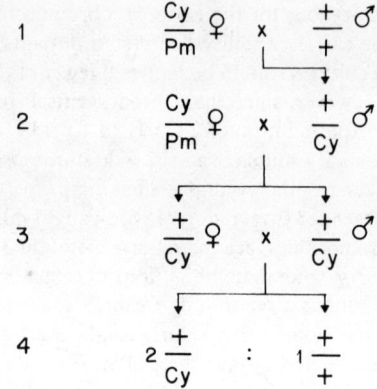

Figure 1 A scheme for making chromosomes homozygous.

NEWLY ARISEN VIABILITY MUTANTS

Mutation Rates and Homozygous Effects on Viability

Lethal mutations on the second chromosome of *D. melanogaster* occur spontaneously at a rate of 0.005–0.006 per generation (7–10). This rate is consistent with that for the X chromosome when the lengths of the chromosomes are taken into account. The number of loci on the second chromosome, assuming that this corresponds to the number of bands on the salivary chromosomes (11), is about 2000. This gives a rate of $2.5-3.0 \times 10^{-6}$/locus/generation. The actual rate may be slightly higher if only those loci capable of mutating to the lethal condition are counted; however, detailed studies of specific chromosomal regions (11) suggest that these are a very small minority. The lethal rate compares with $6-7 \times 10^{-6}$ mutations per locus per generation for visibles (12), and with 4.3×10^{-6} mutations per locus per generation for electrophoretically detectable variants (13, 14). These were determined by patiently counting individual new mutants as they appeared in appropriate experiments.

Unlike lethals, detrimental mutants cannot be counted directly. Their effects are small and variable—all lethals reduce viability by the same amount, 100%—so objective detection and classification are impossible. There is no way of identifying, from viability depression, the number of detrimentals a chromosome carries; it might contain a few with fairly large effects, or a multitude with very small ones. As a result, the procedure for estimating the rate at which detrimentals occur is more complicated than mere mutant counting. It requires an experiment in which mutations are accumulated on sheltered chromosomes over many generations. Though the effects of individual mutants may be too small to be detected, even in a large experiment, the total effect of mutants accumulated over many generations is a quantity that can be measured. Changes in this and related quantities over time can be used to estimate the rate at which the mutants occur.

Three experiments of this sort have been done (9, 15, 16). In each, mutations were allowed to accumulate on many wild second chromsomes of *D. melanogaster* for 25–40 generations. All the chromosomes in an experiment were originally derived from a single chromosome, but they accumulated mutations independently. The chromosomes were maintained in heterozygous condition with *Pm*, and always transmitted through males to avoid recombination. To minimize selection, a single *Pm/+* male was randomly chosen from each chromosome line every generation and mated to *Cy/Pm* females, and the progeny were reared under optimal conditions. Periodically the chromosomes were made homozygous using the procedure shown in Figure 1. This permitted the homozygous viabilities to be determined at successive times while mutations accumulated.

Lethal mutations occurred at the expected rate, about 0.006 per chromosome per generation. In addition, the mean viability of the nonlethal chromosomes declined in a linear fashion. In theory, the rate of decline per generation, M, should equal the average viability effect of a mutant, $\langle s \rangle$, times the whole chromosome mutation rate, Σu, so, $M = \langle s \rangle \Sigma u$. The mean effect, $\langle s \rangle$, is a weighted one, the weights being simply the rates at which mutants with particular effects occur; thus $\langle s \rangle = \Sigma us/\Sigma u$. While the mean viability was found to decrease, the genetic variance for viability among chromosomes increased, also approximately in a linear fashion. In theory, the increase in genetic variance per generation, V, should be roughly equal to $\Sigma us^2 = \langle s^2 \rangle \Sigma u$, where $\langle s^2 \rangle = \Sigma us^2/$ (9, 17). Since $\langle s^2 \rangle = \langle s \rangle^2 + V_s$, where V_s is the variance of the effects of different detrimental mutants, $V = (\langle s \rangle^2 + V_s)\Sigma u = \langle s \rangle^2 (1 + C^2)\Sigma u$; C is the coefficient of variation of the s's. Thus $\Sigma u = M^2 (1 + C^2)/V$, so if C^2 is assumed to be zero, the detrimental mutation rate for the entire chromosome can be estimated. Since zero is the smallest possible value for C^2, the estimate is necessarily a minimum. The results for the three experiments are given in Table 1.

The figures in Table 1 under the heading "All nonlethals" come from the analysis of quasinormal and semilethal chromosomes; those under the heading "Quasinormals only" were obtained by excluding the semilethals. The quasinormal estimates are probably more reliable because semilethal chromosomes unduly inflate the value of V, to which the estimated mutation rate is highly sensitive. The mean of the three determinations is 0.124 mutations per second chromosome per generation. On a per locus basis, this translates into 6.2×10^{-5}, roughly 25 times the lethal mutation rate.

The estimate is a minimal one because the variance in the effects of the detrimental mutants is certainly not zero, whereas the estimating procedure assumes that it is. Mukai et al (9) suggested that the effects of detrimentals might be exponentially distributed; since the variance of an exponentially distributed quantity is equal to its mean squared (thus, $C^2 = 1$), under this hypothesis the mutation rate can be unequivocally estimated and is twice the minimum estimate; the result, using all the nonlethal chromosomes in the experiments of Mukai et al (9), was 0.118 mutations per second chromosome per generation. The semilethal chromosomes were assumed to represent the long tail of the exponential distribution.

Table 1 Mutation rates and homozygous effects of detrimental mutations

All nonlethals				Quasinormals only[a]				Number of generations	Reference
$\langle s\rangle \Sigma u$	Σu	$\langle s\rangle$	$V_s \times 10^{-4}$	$\langle s\rangle \Sigma u$	Σu	$\langle s\rangle$	$V_s \times 10^{-4}$		
—	—	—	—	0.0038	0.141	0.027	1.09	25	Mukai (15)
0.0044	0.059	0.075	1.42	0.0040	0.172	0.023	1.36	40	Mukai et al (9)
0.0021	0.020	0.105	2.75	0.0017	0.058	0.030	2.30	40	Ohnishi (16)

[a] Quasinormals are defined as those chromosomes with viability at least 0.6 normal.

With the observed decline in viability per generation and the associated increase in the genetic variance, the average effect of an individual mutant can also be estimated. The formula is $\langle s\rangle = V/(1 + C^2)M$. This time the estimate is a maximum, since C^2, which is assumed to be zero, appears in the denominator of the estimator. The results are presented in Table 1. Using the figures involving the quasinormals only, the mean value is 0.027; thus, the average detrimental mutant lowers homozygous viability by 3% or less. A third parameter, V_s, can be estimated by noting that $(V/2M)^2 < V_s$; the results are also given in Table 1.

To summarize: the rate of lethal mutations per second chromosome is 0.005–6, or about 3×10^{-6} per locus. The rate of mildly deleterious mutants is at least an order of magnitude higher, the minimum estimate being about 6×10^{-5}. Of course, the rate may be much higher if there is a large class of mutants with viability effects that are neutral, or very nearly so. For the same reason, the estimated viability reduction per homozygous mutant, about 3%, is a maximum.

The above estimation procedures have been extended to the study of induced mutations. Ohnishi (16) monitored the decline in viability and increase in genetic variance for chromosomes treated every generation with low concentrations of EMS; treatment was carried out for as many as 30 generations. Lethals occurred at a rate of 0.13 per chromosome per mM EMS; in contrast to spontaneous mutations, detrimentals were not much more frequent, the rate being 0.17 per chromosome per mM EMS. The maximum average effect of an individual detrimental, and the maximum variance of the effect were estimated to be 0.09 and 0.002, respectively. These results are based on quasinormal chromosomes only. The conclusion to be drawn is that EMS-induced mutants have a different distribution of viability effects than spontaneous ones. For the spontaneous mutations, $\langle s\rangle(1 + C^2) = 0.03$; for the EMS-induced ones, the same quantity equals 0.09. Either the mean effect of an EMS-induced mutant is greater than that of a spontaneous mutant, or C^2 is appropriately larger, or both. Furthermore, the per generation decline in viability for spontaneous detrimentals is about 0.0033 (the average of the values in Table 1). This is approximately equal to the mild detrimental load for new mutants. Since the decline in viability per generation due to lethals is 0.005 (simply their rate of occurrence), the ratio of the detrimental and lethal loads is about 0.7 for spontaneous mutants (0.5 in Ohnishi's study). For EMS-induced mutants, the ratio is much lower, about 0.2–0.4 (16, 18, 19). This suggests that EMS-induced mutations are more drastic in their effects than spontaneous ones; the lethal class is relatively more frequent for EMS than for spontaneous, perhaps because a larger fraction of the mutants are small deletions.

X-ray–induced mutations seem to be even more drastic than EMS-induced, though the data are rather limited. Friedman (20) found that the $D:L$ ratio for radiation-induced mutants on the *D. melanogaster* X chromosome was 0.125, indicating the predominance of lethals. This agrees with the widely held belief that X-ray-induced mutations result primarily from chromosome breaks.

Heterozygous Effects on Viability

Chromosomes carrying newly arisen lethal mutations have been tested for effects on viability in heterozygous condition. The experiments have varied in detail, but for the most part, have shared a common strategy: Lethal heterozygotes were contrived to hatch in cultures with marked flies; the frequencies of the two types were used to establish the viabilities of the lethal heterozygotes, relative to the marked class. For controls, crosses that segregated lethal-free flies and marked ones were performed; the viabilities of the lethal-free flies were obtained from these. Since the same marker was scored in both sets of crosses, the results were directly comparable.

Yoshikawa & Mukai (21) found that the viability of flies heterozygous for spontaneous lethal mutants was lowered by 2–4%, depending on the genetic background in which the lethals were tested; the average decrease was 2.7%. In another experiment, they found that the viability reduction for flies carrying two complementary lethal chromosomes was twice that for those carrying only one lethal. The reduction per lethal mutant was 0.8%. However, this figure may be somewhat underestimated since the experiment was done so that the marked flies segregating with the lethal heterozygotes sometimes carried the lethal mutants under test. This was not the case in the control crosses. Thus, allowing for a viability-reducing effect among the marked flies of the lethal cultures, the actual heterozygous effect could be twice the published estimate. The heterozygous effect of a mutant can be thought of as a function of its homozygous effect (s) and its degree of dominance (h). The desired function is hs. For lethals $s = 1$, so the heterozygous viability reduction observed in these experiments is an estimate of the average degree of dominance, $\langle h \rangle$. Yoshikawa & Mukai (21) estimated the genetic variance of h to be 0.0027.

Ohnishi (22) calculated the degree of dominance for spontaneous and EMS-induced lethals to be 0.036–0.056; however, for the reason mentioned above, these values could be somewhat underestimated. Tobari (23) measured the heterozygous effects of lethal chromosomes extracted from a laboratory population which had been maintained for about a year. At 25° and at 17°, viability was reduced by 2.5%, but at 29° it was increased by 1.5%, indicating that at that temperature the lethals were overdominant.

X-ray–induced lethals have been studied by Stern et al (24), Kitagawa (25), and Simmons (26). In all cases, deleterious heterozygous effects were reported. Kitagawa (25) explored the possibility that such lethals might interact with one another, resulting in a lower heterozygous viability than would be expected on the basis of independent action; he synthesized chromosomes carrying 0–3 X-ray–induced lethals and tested them in combination with one another. Heterozygous viability was reduced by 1.5% per lethal mutant; however, there was evidence for some synergistic interaction. Whatever their nature, and however they came into being, spontaneously or through the aegis of a mutagen, Drosophila lethal mutations on the average are clearly harmful in heterozygous condition.

Tests of newly arisen mildly deleterious mutants have been much more difficult to perform and the results more difficult to interpret. There is general agreement that in crosses of the type, $Cy/+_i \times Cy/+_j$, where $+_i$ and $+_j$ stand for two different chromosomes each carrying different minor viability-depressing mutants, the $+_i/+_j$ type has reduced viability. The amount of dominance estimated by Mukai & Yamazaki (27) was about 0.35. However, this estimate depends on the extent to which this partial dominance is expressed in the Cy heterozygotes.

It would seem to be a simple matter to obtain correlations between the effects of homozygotes and heterozygotes for the same chromosome and thereby estimate the partial dominance, but experiments of different design have yielded inconsistent results, some even indicating overdominance. We return to this question later in the section, "Some Unresolved Problems."

VIABILITY MUTANTS FROM EQUILIBRIUM POPULATIONS

Detrimental and Lethal Loads

When Drosophila chromosomes extracted from natural or long-standing laboratory populations are made homozygous, a depression in viability results. Customarily this depression has been expressed in terms of genetic loads. These reflect the incidence of lethals, severe detrimentals, and mild detrimentals in the populations from which the chromosomes were taken. However, the loads are not simply functions of mutant frequencies; rather they confound frequencies with viability effects. For the lethals there is no problem because their effects are constant—100% viability reduction—so $\langle s \rangle = 1$. Thus, the lethal load is only a function of lethal frequencies—the sum of these over all loci. However, for detrimentals, the frequencies and effects cannot be disentangled. The detrimental load is equal to the product of the two, summed over all loci: $D = \Sigma q_i s_i$, where q_i is the frequency of the ith detrimental mutant in the population. For the mild detrimental load, D_m, the summation is over the mild mutants only.

In their original paper on this subject, Greenberg & Crow (5) summarized detrimental and lethal loads for a variety of Drosophila species. Since the appearance of their paper, additional determinations have been made, and we summarize them in Table 2. In many cases, the entries in the table are the averages of independent load determinations. The loads and load ratios in italics were taken directly from the literature; all others have been calculated by us from raw data extracted from the papers. In doing the calculations, we have transformed viabilities presented as proportions of wild-type flies in test cultures into appropriate ratios; the data so treated are identified with a percentage sign in the column headed "Viability index." This has been done because calculations with proportions underestimate the detrimental load (5). All other data were originally presented in the form of ratios. In constructing the table, we made the criterion of lethality uniform, (less than 10%

Table 2 Homozygous viability loads for populations assumed to be in equilibrium[a]

Chromosome and reference	Viability[b] index	Population	N	D_m	D	L	$D_m:L$	$D:L$
D. melanogaster II								
Band (93); Band & Ives (45)[c]	%	USA	469	0.252	0.425	0.253	0.996	1.690
					(0.280)	*(0.250)*		*(1.118)*
Dawood (94)	%	Egypt	301	0.106	0.241	0.294	0.361	0.820
Greenberg & Crow (5)	ratio	USA and cage	465		*0.180*	*0.287*		*0.627*
Hoenigsberg et al (95)	%	Hungary	961	0.095	0.103	0.088	1.080	1.170
Hoenigsberg et al (95)	%	Colombia	1133	0.180	0.283	0.256	0.703	1.105
Hoenigsberg et al (96)	%	Colombia	2693	0.135	0.379	0.405	0.333	0.936
Kenyon (51)	ratio	USA (cage)	116		0.181	0.147		1.231
Kosuda (30)	ratio	Japan	289	*0.058*	*0.097*	*0.166*	*0.349*	*0.584*
Mukai & Yamaguchi (42)	ratio	USA	691		*0.334*	*0.501*		*0.667*
Temin (41)	ratio	USA and cage	1083		*0.147*	*0.262*		*0.563*
Temin et al (6)	ratio	USA	1855	*0.095*	*0.157*	*0.247*	*0.385*	*0.636*
Tobari (49)	%	cage	159	0.242	0.343	0.245	0.988	1.400
Watanabe (97)[d]	%	Japan	323	0.064	0.163	0.132	0.485	1.235
Watanabe & Ohnishi (46)[c,e]	%	Japan	200		*0.144*	*0.375*		*0.383*
Watanabe et al (98); Watanabe (97)[c,d]	%	Japan	905	0.109	0.266	0.168	0.649	1.583
					(0.128)	*(0.167)*		*(0.767)*
Watanabe et al (98)[c–e]	%	Japan	1377		*0.231*	*0.382*		*0.605*
Unweighted mean (our figures)				0.134	0.236	0.247	0.633	1.017
D. melanogaster III								
Kosuda (30)	ratio	Japan	289	*0.174*	*0.213*	*0.315*	*0.552*	*0.676*
Temin et al (6)	ratio	USA	1855	*0.122*	*0.175*	*0.240*	*0.508*	*0.729*
Watanabe et al (39)	ratio	USA (cage)	502		*0.464*	*0.665*		*0.697*
Unweighted mean				0.148	0.284	0.407	0.530	0.701
D. pseudoobscura II								
Dobzhansky et al (28)	%	USA	113	0.147	0.269	0.206	0.714	1.437
Marinkovic (38)	%	USA	256	0.188	0.319	0.199	0.945	1.603
Wills (52)	ratio	USA	48		0.176	0.209		0.842
								(0.855)
Unweighted mean				0.167	0.246	0.204	0.830	1.298
D. pseudoobscura III								
Dobzhansky et al (28)	%	USA	104	0.302	0.352	0.226	1.336	1.558
D. willistoni II								
Malogolowkin-Cohen et al (29)	%	NE South America	103	0.146	0.380	0.386	0.378	0.984
D. willistoni III								
Malogolowkin-Cohen et al (29)	%	NE South America	95	0.237	0.506	0.184	1.288	2.750
Grand unweighted mean							0.704	1.113

[a] N = number of chromosomes made homozygous; D_m = mild detrimental load; D = total detrimental load; L = lethal load; numbers in italics come directly from the literature; all others were calculated from raw data.
[b] Viability index — either percentage or ratio.
[c] In the authors' calculations, viabilities were not transformed into ratios.
[d] Criterion of lethality was less than 1% heterozygote viability.
[e] Excluded from calculation of mean.

of the viability of the controls) except where indicated. Similarly, the criterion for defining quasinormal chromosomes was standardized (50% of the control viability). The detrimental loads recalculated for the data originally given in proportions may be slightly underestimated, and the corresponding lethal loads slightly overestimated. The reason for this is that in almost every case where the data were given in proportions, only the viabilities of heterozygotes, quasinormal homozygotes, and all homozygotes were presented. The calculation of D requires the viability of the nonlethal homozygotes. To obtain this we divided the viability of all homozygotes by the frequency of nonlethal chromosomes, and assumed that all the lethals were completely penetrant. This is not strictly correct, but the resulting bias is trivial. Notice, however, that the calculation of D_m is not affected by this procedure.

For summary statistics we have used unweighted averages as representative of the loads and their ratios. Usually loads based on relatively few chromosomes were determined more accurately than those based on many because individual viability tests were replicated. In the experiments with larger numbers of chromosomes, replication was not the rule. So, given the variability in procedures and statistical treatments, the unweighted means are probably best.

Since the most extensive data are for the second chromosome of *D. melanogaster*, and since we shall have recourse to these in a later section, we call attention to them now; the average loads are $D_m = 0.134$, $D = 0.236$, and $L = 0.247$, and the average load ratios are $D_m{:}L = 0.633$ and $D{:}L = 1.017$.

Viability Mutants and Epistasis

In the preceding sections we have ignored the possibility of epistasis, which, if pervasive, could reduce the impact of mutation on the population. One way of assessing the significance of this is to study the consequences of different degrees of inbreeding. Dobzhansky, Spassky & Tidwell (28) and Malogolowkin-Cohen et al (29) did this with *D. pseudoobscura* and *D. willistoni*, respectively. Their procedure was to estimate the proportion of zygotes surviving to the adult stage in cultures where the parents were related in different degrees; the degree of relationship was quantified by Wright's coefficient of inbreeding, F. In both experiments the logarithm of the proportion of survivors declined linearly over the range of F studied; however, when the relationship was extrapolated to complete inbreeding ($F = 1$), the predicted decrease in viability was less than the total viability load obtained by chromosome homozygosis. This suggested that at extreme levels of inbreeding, deleterious mutants could interact synergistically with one another. However, the evidence was little more than suggestive because different criteria of survival and different environments were used in the two parts of the experiments.

Another procedure for studying epistasis involves the simultaneous homozygosis of two chromosomes. The intention of the procedure is to see whether flies homozygous for two chromosomes are appreciably less viable than a model of independent action would predict. The method has the advantage that with suitable crosses, double and single homozygotes can be reared in the same culture, thus making conditions for their survival uniform; moreover, the performance of these flies can be measured relative to a common standard. Using this method, Temin et al (6) searched for epistatic interactions between the second and third chromosomes of *D. melanogaster;* the mild detrimental loads for these chromosomes were 0.095 and 0.122, respectively. A load measuring the interaction between them was 0.021, and was not statistically significant. Kosuda (30) repeated the experiment, but on a smaller scale, and found a larger interchromosomal load (0.081), which was significant. An earlier study with *D. pseudoobscura* by Spassky, Dobzhansky & Anderson (31) also revealed a significant interaction component (0.094). Together, the experiments suggest the existence of some interchromosomal epistasis.

Although Temin et al (6) found no significant interactions between chromosomes, they did detect interactions among mutants on the same chromosomes. In their experiments chromosomes were made completely homozygous, but they also contrived chromosomes to be homozygous for half their genes. The detrimental loads observed for the complete homozygotes were more than twice as large as those detected for the half-homozygotes, indicating synergism at the higher level of inbreeding.

The chromosomes analyzed by Temin et al (6) came from an equilibrium population. Mukai (32) focused on the effects of chromosomes that carried newly arisen mutations; he monitored the decline in viability of homozygotes for these chromosomes over a span of 60 generations. During the first 40 or so, viability declined in a linear fashion, but thereafter it fell off more rapidly. Over the entire 60 generations, a quadratic component was significant. This indicated synergistic interaction among newly arisen mutants with respect to viability.

As for epistasis with respect to total fitness, very little is known. Latter & Robertson (33), who studied the "competitive index" of flies inbred to different degrees, found no evidence of it.

These data suggest that interactions are not great. However, all that show a significant effect point in the same direction: the interaction is synergistic. The effect of such synergistic interaction in reducing the mutation load has been discussed by Crow (34). The homozygous mutation load may be reduced, perhaps by as much as 50% but more likely less, by viability interactions.

Heterozygous Effects on Viability

We first summarize work that has been done with lethal chromosomes from equilibrium populations. In every experiment, the procedure has been to compare the viabilities of lethal heterozygotes with those of flies carrying nonlethal chromosomes derived from the same population.

Three recent experiments have indicated that lethal chromosomes are overdominant, at least under certain conditions. Dobzhansky & Spassky (35) reported that lethal second chromosomes of *D. pseudoobscura* from populations in the American Southwest conferred a viability advantage on their heterozygous carriers, providing that the lethals were tested in a native ("coadapted") genetic background. The data were reanalyzed by Anderson (36), who noted that in such a background the viabilities of lethal heterozygotes were improved by 4%. However, in alien backgrounds the lethals lowered viability by 5%. Both results were statistically significant; however, we should note that during the experiments the tested chromosomes were passed through females several times, albeit balanced by inversions. Nevertheless, Chovnick (37) has pointed out the danger of this procedure. Earlier work by Marinković (38) in which a less complicated, but perhaps more reliable genetic scheme was used, showed no heterotic effects. In fact, Marinković reported significant deleterious effects, though he did not control the background of chromosomes other than the second. Watanabe, Yamaguchi & Mukai (39) also found evidence that lethals are overdominant. They studied *D. melanogaster* third chromosomes from North Carolina, and noted that the viabilities of lethal/nonlethal flies were significantly higher than those of nonlethal/nonlethal ones. However, flies carrying two complementary lethal chromosomes had intermediate viabilities. The results do not seem to make sense. In another study with *D. melanogaster,* Band (40) reported the heterozygous superiority of naturally occurring lethal second chromosomes, though only under one set of conditions; under another, she found that the same chromosomes were deleterious.

Most work with lethal second chromosomes of *D. melanogaster* has revealed deleterious heterozygous effects. Kitagawa (25) found that chromosomes bearing various numbers of lethal mutants extracted from a Japanese population lowered heterozygous viability; the reduction was 2.5% per lethal mutant. Temin (41) reported that flies heterozygous for lethals from natural and cage populations were 5% less viable than controls. Mukai & Yamaguchi (42) calculated a viability reduction of 1.2–2.4% for lethal chromosomes from a North Carolina population (the higher figure allows for the effects of the lethals in the class of flies used as a viability standard). When these lethal chromosomes were subdivided into those with inversions and those without, it was found that the former group showed negligible viability effects; however, in the latter, heterozygous viability was reduced by 1.8–3.5%. Earlier, Yoshikawa & Mukai (21) had shown that lethal chromosomes from a natural population lowered viability by 0.7%, or, on a per mutant basis, by 0.57%. Hiraizumi & Crow (43) and Watanabe & Oshima (44) have reported deleterious effects for lethals from equilibrium populations; however, equivocal results have been obtained by other investigators (23, 40, 45, 46). Prout (47) studied lethals from natural populations of *D. willistoni* and found them to be deleterious in heterozygous condition.

The heterozygous effects on viability of individual detrimental mutants from an equilibrium population cannot be measured. However, when the viabilities of random combinations of chromosomes from such populations are ascertained—these are usually called "heterozygous" viabilities—along with those of the corresponding homozygotes, the average degree of dominance of the mutants with respect to viability can be estimated. The estimator is the regression coefficient obtained when the sum of the viabilities of the homozygotes corresponding to each heterozygote is used as the independent variable to predict heterozygous viability. The rationale for this statistical procedure was given by Mukai et al (9): At any one locus, a chromosome either carries a viability mutant or it does not. Suppose the frequency of a mutant (a) at the ith locus is q, and $p = 1-q$. Then in random combinations of chromosomes there will be a Hardy-Weinberg distribution for this locus: $p^2(AA)$, $2pq(Aa)$, and $q^2(aa)$. The deleterious effects on viability of these genotypes will be 0, hs, and s, respectively. Their mean and variance will be $sq(2hp + q)$ and $2pqh^2s^2$. Now consider the two homozygotes corresponding to each of the above genotypes. The sums of their deleterious viability effects will be 0 for $AA + AA$, s for $AA + aa$, and $2s$ for $aa + aa$. The mean and variance of these will be $2qs$ and $2pqs^2$; however, the variance is approximate because terms in q^2 have been ignored, the tacit assumption being that in an equilibrium population q will be small. The covariance between the viability effects of the randomly combined chromosomes and the sum of the effects of the corresponding homozygotes is $2pqhs^2$ (again terms in q^2 have been ignored). The covariance, divided by the variance of the sum of homozygote viabilities, is equal to the regression coefficient. Thus,

$$b = \Sigma 2pqhs^2/\Sigma pqs^2, \qquad 1.$$

where the summation is over all relevant loci; but this is the average dominance of the detrimental mutants in an equilibrium population, $\langle h_e \rangle$, each h being weighted by the variance of the homozygous viability effects. For partially dominant mutants at equilibrium, $q = u/chs$ (c is a parameter that relates selection on viability to selection on total fitness) and $p \doteq 1$. Then, if s/c and $1/h$ are uncorrelated,

$$\langle h_e \rangle = \Sigma u/(\Sigma u/h) = h_n'; \qquad 2.$$

h_n' is the harmonic mean of the degrees of dominance of newly arising mutations, each being weighted by its corresponding mutation rate. In the actual calculations, genetic variances and covariances should be used.

The procedure is valid for random combinations of quasinormal chromosomes; however, Mukai & Yamaguchi (42) extended it to combinations of quasinormals and lethals; the homozygous effects of the lethals were simply ignored. Unfortunately, a great deal of work involving the simultaneous determination of homozygous and heterozygous viabilities has been analyzed incorrectly (38, 45, 48, 49). In these cases, viabilities of constituent homozygotes were averaged instead of summed (not a problem), but the contributions of lethal chromosomes were included in the calculations.

The results of three properly analyzed studies dealing with *D. melanogaster* chromosomes collected from North Carolina populations are presented in Table 3. For the second chromosome, with the exception of the estimate for the inversion-bearing chromosomes, the dominance values are all similar, being about 0.2. Mukai (50) predicted a value of this magnitude from the properties of newly arisen mutants. The results and the prediction make sense because the mutants with the largest heterozygous effects—those with values of h close to 0.5—should be eliminated more quickly from a population than mutants with milder effects. At equilibrium, the less dominant mutants should outnumber the more dominant ones, so the mean dominance should be less than that for new mutants (this is a biological translation of the algebraic fact that the harmonic mean is always less than the arithmetic mean).

Table 3 Estimated degrees of dominance of mildly detrimental mutants from equilibrium populations

Chromosome analyzed	All chromosomes	Inversion chromosomes	Inversion-free chromosomes	Reference
II	0.21 ± 0.11	—	0.21 ± 0.15 0.21 ± 0.06	Mukai et al (9)
II	0.29 ± 0.07	0.49 ± 0.08	0.18 ± 0.08	Mukai & Yamaguchi (42)
III	0.40 ± 0.09	—	0.40 ± 0.10	Watanabe, Yamaguchi & Mukai (39)[a]

[a] These values were computed by using only chromosomes with high viabilities. See reference for details.

At face value the average dominance of mutants on the *D. melanogaster* third chromosome is different from that of the mutants on the second. However, Watanabe, Yamaguchi & Mukai (39) pointed out several reasons it might be overestimated. We should also note the danger of obtaining spuriously high dominance values simply because experiments of this sort tend to be stretched out over long periods of time, and as conditions change in the laboratory, homozygous and heterozygous viabilities might change pari passu; the resulting positive correlation could be significant, but of course it would be genetically irrelevant. Kenyon (51) observed just such a phenomenon. A more perplexing discrepancy concerns the difference between the estimates for inversion-bearing and inversion-free second chromosomes (42). No explanation for this has been offered.

Wills (52) and Strickberger (53) studied homozygous and heterozygous viabilities of *D. pseudoobscura* third chromosomes. They found evidence of partial dominance; in Wills' study, $\langle h_e \rangle = 0.18$–$0.35$, and in Strickberger's, $\langle h_e \rangle = 0.08$–$0.35$.

To summarize: Estimates of the dominance of detrimental mutants from equilibrium populations are obviously quite variable, but most of them are in the range of 0.20–0.35. New mutants, on the other hand, exhibit greater dominance, 0.35–0.50 (see below). The difference between these estimates undoubtedly reflects the work of natural selection.

VIABILITY MUTANTS AND FITNESS

Viability and Fertility As Components of Fitness

Thus far we have discussed the effects of viability mutations on *viability*. However, it is not unreasonable to expect that a mutation which lowers the chance of surviving to the adult stage also impairs a fly's ability to find a mate and to reproduce. We designate mating success and reproduction jointly as *fertility*. The significance of this component of fitness has been documented in several experiments, which we briefly summarize.

Anderson & Watanabe (54) studied the fertilities of different karyotypes of *D. pseudoobscura;* specifically they allowed two karyotypes of one sex to compete for the favors of a single karyotype of the opposite sex. Data from male and female competitions were obtained; in both cases these consisted of the karyotype frequencies of the competitors, and of their offspring. The latter were determined by salivary gland analysis of progeny reared under optimal conditions, so viability differences were minimized. By comparing parental and filial karyotype frequencies it was possible to quantify the intensity of selection. Anderson & Watanabe (54) found striking differences among the selective values of the various karyotypes, and also noted that they were frequency dependent. They concluded that the effect of a karyotype on fertility was at least as pronounced as its effect on viability.

Bundgaard & Christiansen (55) investigated the components of selection in an experiment involving marked fourth chromosomes of *D. melanogaster*. Since there is negligible recombination in the fourth chromosome, the markers, though physiologically nonallelic, behaved formally as alleles. The selection process was divided into three components: zygotic selection (survival from egg to adult), sexual selection (success in finding a mate), and fecundity selection (egg laying). The experimental strategy was to monitor changes in genotype frequencies over the course of one carefully defined generation. From the data, Bundgaard & Christiansen (55) deduced that the most important component of selection—the one causing the largest fraction of the change in marker frequency per generation—was success in finding a mate, in particular, male mating success. Comparison of the composition of the initial and mated adult populations demonstrated this. Fecundity selection and viability selection were relatively trivial components of the whole process.

Earlier work with fourth chromosome markers by Prout (56, 57) had suggested that viability plays a minor role in selection. However, in experiments of a different sort, Polivanov & Anderson (58) obtained evidence that viability and fertility contributed equally to the selective process. In these, the frequency of a dominant marker was monitored in *D. melanogaster* populations maintained on a schedule of discrete generations. From frequency changes measured over different parts of the life cycle, the components of fitness were calculated. Even though the estimates varied over time, the importance of fertility was clearly demonstrated.

Knight & Robertson (59) followed a different tack. They measured the "competitive index" of a stock relative to a tester strain. The procedure was to mix flies of both sexes and strains in predetermined proportions, and allow them to mate. Their progeny were than classified and counted. The tester strain was chosen so that purebreds and crossbreds could be distinguished. Given this, the competitive index of the stock under test was computed as the number of purebreds of that type, divided by the number of purebreds of the tester type; this ratio was duly adjusted for initial parental frequencies. Knight & Robertson (59) observed a startling difference between the competitive indices of three inbred wild stocks and the viabilities of corresponding stocks homozygous for their second chromosomes. No doubt, part of the difference was due to the fact that mutants throughout the genome affected the competitive index, but only those on chromosome II affected the viability. However, the competitive index includes fertility, and part of the discrepancy must be attributed to it.

New Mutants and Fitness

The effects of new mutations on total fitness have been little studied, the only relevant experiments being those of Simmons, Sheldon & Crow (19), Mitchell (60), and Mitchell & Simmons (61). In these, mutations were induced with EMS on suitably marked *D. melanogaster* chromosomes. Their frequencies were monitored in populations kept on a schedule of discrete generations, and the rate at which the frequencies changed was used to estimate the effects of the induced mutations. Each experiment was designed so that the tested chromosomes were preserved from recombinational breakup. Simmons, Sheldon & Crow (19) and Mitchell & Simmons (61) achieved this by preventing them from passing through females. In the experiments of Mitchell (60), the chromosomes did pass through females, but were prevented from crossing-over by the use of an adequate balancer.

Mitchell (60) and Mitchell & Simmons (61) studied EMS-treated X chromosomes. In segregation tests they measured the viability-reducing effects of the chromosomes in homozygous and hemizygous condition; however, the primary purpose of their work was to estimate the reduction in total fitness. Mitchell (60) did this for heterozygous females (to keep the treated chromosomes heterozygous she used an X-Y translocation, the X chromosome of which contained crossover-suppressing inversions), and Mitchell & Simmons (61) did so for hemizygous males (using attached-X chromosomes to maintain treated X's in the male line). The results showed that treated chromosomes reduced male fitness 3–4 times as much as male viability: Moreover, in heterozygous females, the chromosomes had as much effect on fitness as they did in homozygous condition on viability alone.

Simmons, Sheldon & Crow (19) studied EMS-induced mutations on the second chromosome; homozygous effects on viability and heterozygous effects on fitness were ascertained, the former in segregation tests, the latter in population experiments (an X-2 translocation was used to keep the tested second chromosomes permanently heterozygous in males). The results agreed with those of Mitchell (60): The heterozygous effects of the treated chromosomes on fitness were about the same as their homozygous effects on viability.

The experiments with EMS demonstrated that fertility plays a larger role than viability in changing the frequencies of mutation-bearing chromosomes. This could mean that new mutations with effects on viability have much greater effects on fertility, with the consequence that their rate of elimination from a population is essentially determined by the latter. It could also mean that viability and fertility mutations occur independently, but that the fertility mutants have more impact. Unfortunately, because the chromosome, and not the individual mutation, was the unit of study, there is no way of favoring one explanation over the other.

Homozygous Effects on Fitness of Mutations from Equilibrium Populations

In an earlier section we discussed the homozygous effects on viability of chromosomes extracted from equilibrium populations; we now turn to their homozygous effects on fitness as a whole.

The pioneering work was done by Sved & Ayala (62). It dealt with second chromosomes of *D. pseudoobscura* which had been extracted from a population in the American Southwest; subsequently, the same type of experiment was performed with other material, including the second and third chromosomes of *D. melanogaster* (63–65) and the second chromosome of *D. willistoni* (66).

The experimental procedure was simple; using the marked-inversion technique, wild chromosomes were made homozygous. Quasinormal homozygotes were then introduced into population cages along with heterozygotes carrying a marker chromosome and the same wild chromosome as the homozygotes. The marker was dominant and homozygous lethal, and was also associated with crossover-suppressing inversions. This last feature prevented the tested chromosomes from being broken up by recombination during the experiments. The frequencies of the two types of flies—wild homozygotes and marker heterozygotes—were monitored in the populations. Different strains of homozygotes were tested independently, and usually the tests were replicated. In conjunction with these, random combinations of wild chromosomes (so-called heterozygotes) were mixed with marked flies to establish control populations. The fitness of the marked flies was determined relative to the randomly constituted heterozygotes. Together these gave the fitness of the homozygotes relative to the random heterozygotes.

In separate experiments, the viabilities of the homozygotes were measured relative to the marked class. This was done so that the homozygous viability depression could be compared with the corresponding reduction in fitness as a whole.

In these experiments, the marker chromosome was homozygous lethal. Despite this, it coexisted with the wild chromosome in most of the homozygote populations, indicating superiority of the marked flies over their wild competitors. The marker's equilibrium frequency was used to estimate the fitness of the homozygotes, relative to the marked class. In a minority of the homozygote populations, the marker was eliminated; for these the relative fitnesses were estimated by measuring the rate of elimination. Since the marker was eliminated in all the control populations, this procedure also had to be used to estimate the fitness of the marked class, relative to the wild heterozygotes.

The experimental results are given in Table 4. In each experiment the number of tested homozygotes was small; nevertheless, there are reasons to have confidence in the results. First, replicate populations converged to the same equilibrium, even though they were established with very different initial frequencies. Thus, the results were reproducible. Second, estimation of the homozygote fitness, relative to the marked class, depended only on the equilibrium frequency of the marker (when one was attained). The path to equilibrium, no matter how erratic, made no difference. On the other hand, the fitness of the marked flies, relative to the randomly constituted heterozygotes, was estimated with much less certainty, for in that case, the method made use of selection dynamics rather than selection statics; to the extent

Table 4 Homozygous effects on fitness of chromosomes from natural populations

Species and chromosome	Source	Number homozygotes tested	Homozygote viability[a]	Homozygote fitness[a]	Marker fitness[b]	Homozygote fitness[b]	Reference
D. pseudoobscura II	American Southwest	18		0.61	0.60	0.36	Sved & Ayala (62)
D. melanogaster II	NSW, Australia	25	0.73	0.28 (0.34)[c]	0.50	0.14 (0.17)[c]	Sved (63)
D. melanogaster II	California	23	0.78[d]	0.13 (0.24)[c]	0.77	0.10 (0.18)[c]	Tracey & Ayala (65)
D. melanogaster III	California	14	0.82[d]	1.19	0.27	0.32	Tracey & Ayala (65)
D. melanogaster III	NSW, Australia	14	0.75	0.19	0.50	0.10	Sved (64)
D. willistoni II	Brazil	15		2.55	0.13	0.34	Mourao et al (66)

[a] Relative to marked flies.
[b] Relative to random heterozygotes.
[c] In these averages, artifactual negative fitness values were set equal to zero.
[d] Pooled estimates of two sets of tests, and expressed as ratios rather than proportions: corresponding viabilities of random heterozygotes were 1.08 (II) and 1.33 (III).

that fluctuations occurred in the course of the experiments—random or otherwise—we must have less confidence in those estimates. Furthermore, the estimates were based on discrete generation theory, but the populations reproduced continuously. For cases in which homozygous fitnesses were estimated by rate-of-elimination methodology, these reservations also apply.

The product of the homozygous fitness (relative to the marked class) and the fitness of the marked class (relative to the random heterozygotes) is equal to the fitness of the homozygotes, relative to the random heterozygotes. This is the quantity of interest. It ranges from an astonishing low of 0.10 to 0.36. Since the chromosomes tested in these experiments were more or less quasinormal with respect to viability, their homozygous effects were much greater for fitness as a whole than they were for viability alone. The values obtained by Tracey & Ayala (65) for the *D. melanogaster* second chromosome illustrate this. Relative to the marked class, homozygous viability was reduced by 22%, but homozygous fitness fell by 87%. Tracey & Ayala (65) also measured the viabilities of randomly constituted heterozygotes; the homozygotes were 71% as viable as these, but, as Table 4 indicates, only 10% as fit. The fitness reduction was therefore three times as great as the viability reduction.

Unfortunately, because the chromosome, and not the gene, was the experimental unit, we cannot say that the viability mutants carried by the tested chromosomes had pleiotropic effects on other components of fitness. The depression in fitness observed in these experiments could be explained by postulating two independent sets of genes, one influencing viability, the other fertility. Some insight can be gained by correlating the viabilities and fitnesses of the chromosomes studied. Sved (63) did this, and after excluding one nearly lethal chromosome, found the correlation to be 0.27 (not significant). However, considering the sources of error, the fact that the value was positive is encouraging. A reasonable hypothesis is that mutations affecting viability also affect fertility, but by how much we do not know.

The simultaneous study of viability and fertility has not been very rewarding. Temin (41) attempted to correlate the viability and fertility of second chromosome homozygotes of *D. melanogaster*; fertility was operationally defined as the fraction of flies producing offspring, and was studied separately for males and females. Nonsignificant correlations were found. Marinković (38) examined the viability and fecundity (number of eggs laid) of second chromosome homozygotes of *D. pseudoobscura*; again, the correlation was not significant. However, poorly fecund homozygotes were less viable than very fecund ones. Watanabe & Oshima (67) studied the viability of second chromosome homozygotes of *D. melanogaster* in conjunction with a trait they called *productivity*; this was measured as the number of progeny produced by a homozygote in matings with a laboratory strain. Male and female productivities were measured separately. For male productivity, there was a significant positive correlation with viability, but for female productivity, there was none. The work was extended by Watanabe & Ohnishi (46) who observed that homozygotes which were male or female sterile had lower viabilities than fertile homozygotes. Thus, though there is some evidence that viability and fertility are correlated, it is weak. One reason for this may be that mating success has been totally ignored in the experiments. Bundgaard & Christiansen (55) concluded that this component of fertility was all-important, and we suspect that if viability mutants are indeed pleiotropic, their effects might be observed in properly designed studies of mating success.

Heterozygous Effects on Fitness of Mutations Affecting Viability

Direct measurements of the homozygous effects on total fitness of individual viability mutations have not yet been made. However, when indirect methods are used, their effects on fitness in the heterozygous state can be estimated. Fortunately, these are the effects that most interest us, since harmful mutants that are partially dominant will never be frequent enough to become homozygous in a large randomly mating population. Rather, they will have their impact in heterozygous condition.

First we consider lethals. Mukai et al (9) argued that the dominance of lethals found in a natural population was 0.024; this is the dominance for fitness, not just viability. It was calculated by comparing the lethal loads for new and equilibrium mutants on the second chromosome of *D. melanogaster*, the former being 0.005–0.006, and the latter 0.247 [taken from Temin et al (6)] (coincidentally, the value 0.247 is also the average lethal load for the *D. melanogaster* second chromosome shown in Table 2).

A different kind of analysis was presented by Crow & Temin (7); they elaborated earlier work by Dobzhansky & Wright (68) by bringing together data on lethal mutation rates, lethal frequencies in natural populations, and rates of allelism; from these they estimated the average dominance of single lethal mutants to be 0.0154. However, they were careful to note that if there is appreciable local inbreeding in Drosophila populations, the actual value could be lower.

Additional evidence concerning the dominance of lethals was obtained by Murata (69), who studied the frequency distribution of lethals in small populations. He calculated the dominance with respect to fitness to be 0.038; considering the uncertainties associated with his methods, the value is not inconsistent with those of Mukai et al (9) and Crow & Temin (7).

Experiments by Tobari & Murata (70–72) have provided estimates of the dominance for fitness of newly arisen lethals; they first studied the increase in the frequency of lethal second chromosomes in irradiated populations of *D. melanogaster*, and after ceasing the irradiation, monitored the diminution of the induced lethal load. During the accumulation of radiation-induced lethals, the dominance with respect to fitness was estimated to be 0.073. In the generations immediately following cessation of the irradiation, it was 0.033, and in later generations, it was close to 0.02. Even if the estimates are not taken literally, they make good sense: The most dominant lethals were eliminated quickly, leaving a residuum of less deleterious mutants. Given the possible sources of error, the data on the dominance of lethals are remarkably consistent. We conclude that lethals in equilibrium populations reduce fitness by about 2% in heterozygous condition.

In an earlier section we noted that the $D_m : L$ ratio for new mutations on the *D. melanogaster* second chromosome is about 0.7; the corresponding value for an equilibrium population is 0.633 (from Table 2). The approximate equality of these implies that mildly detrimental mutants are eliminated as rapidly as lethals from a population. Because the mild detrimentals have only 2-3% the viability effect of a lethal, they must be more dominant, or more pleiotropic, or both.

To estimate the impact of mildly detrimental viability mutants on the fitness of a population, we present an argument proposed by Crow (17), alternative versions of which can be found in Crow (73) and Mukai et al (9). The mild detrimental load associated with newly arising viability mutations on the *D. melanogaster* second chromosome is 0.0033 (from Table 1). In theory this equals $\Sigma u_i s_i$. The mild detrimental load for mutants on second chromosomes from an equilibrium population is 0.134 (from Table 2), and this equals $\Sigma q_i s_i$, where q_i is the equilibrium frequency of a mutant at the ith locus. An alternate way of writing this is $\Sigma u_i t_i s_i$, where t_i is the length of time (in generations) that a new mutant persists in the population at equilibrium (it should be evident that $q_i = u_i t_i$). The ratio of the equilibrium load to the load for new mutants is $\Sigma u_i t_i s_i / \Sigma u_i s_i = \langle t \rangle$, the mean persistence of a mutant in the population at equilibrium. But if a mutant persists an average of $\langle t \rangle$ generations, then each generation the probability of its elimination is $1/\langle t \rangle$. Substituting the actual load values, we find that the mean persistence of a mutant in a natural population is 40 generations (and the average fitness disadvantage of a mutant in a natural population is $1/40 = 0.025$). This, of course, is the effect of a mutant on fitness in heterozygous condition. Remarkably, it is very nearly the same as the effect on viability in homozygous condition (0.027, from Table 1). Since the degree of dominance for viability for chromosomes from natural populations is probably 0.3 or less (see below), and $(0.3)(0.027) = 0.008$ is only a third as large as 0.025, we conclude that viability mutants also affect fertility, perhaps twice as much as they do viability. This is consistent with direct measurements of whole-chromosome viability and fitness effects. Furthermore, the heterozygous effect on fitness of a detrimental obtained by this method is about the same as the heterozygous fitness effect of a lethal; however, since the mutation rate for detrimentals is at least 25 times greater than that for lethals, the detrimentals have much more impact on the fitness of a population.

The argument is not airtight. One problem is that the equilibrium detrimental load is underestimated, since the reference point from which it is measured is a fly with a random pair of chromosomes, not one free of mutations. However, the load for new mutants is probably also underestimated since the viability standard is a fly heterozygous for the mutants being tested in homozygous condition; if these have appreciable heterozygous effects (and we must suppose they do), then homozygous viability is overestimated. Fortunately, it is the ratio of loads that enters into the argument, and this is probably not as much troubled by biases as the loads themselves.

To summarize: The overall conclusion is that mildly deleterious mutants have a much greater dominance for fitness than do drastic mutants, which are more nearly recessive. In fact, the absolute heterozygous effect on fitness of the two types is, within the precision of the methods of measurement, essentially the same. Thus, the impact on the population of a mild mutant is no different from that of a drastic mutant, since neither is very likely to become homozygous. Since mild mutants are much more frequent, their total impact is much greater.

SOME UNRESOLVED PROBLEMS
The Mutation Load

If we accept at face value the mutation rate estimates of Table 1, mildly detrimental mutations occur with a frequency of about 0.1 per chromosome per generation, which amounts to 0.25 per gamete. According to the Haldane (2) principle, the fitness reduction at equilibrium for partially dominant mutations is twice the total mutation rate and would therefore be about 50%.

Clearly there is no such reduction in viability. Drosophila eggs frequently have a probability of 0.95 of producing larvae that survive to adulthood; so the viability mutation load is of the order of 5% or less, rather than 50%. How can the high mutation rate be reconciled with mutation load theory?

One way is by epistasis. Synergistic interactions of mutants can decrease the mutation load, but, as mentioned above, the extent of the interaction for homozygous viability is such as to reduce it at most by a factor of about 2 (34, 74). We know much less about the interaction of heterozygous effects. Since they are smaller, we would expect the interactions to be less. Of course, Drosophila in nature live under much more rigorous conditions and greater competition for larval survival than in most of the experiments that we have reported. However, attempts to see whether interactions were greater in crowded than uncrowded cultures gave negative results (6). So synergism by itself does not appear to be more than a partial solution to the problem.

A second explanation is that most of the selection acting on natural populations of Drosophila is concerned with components of fitness other than viability. The data reported in this review support the idea that other components, particularly fertility and mate competition, are more important components of fitness differences than pre-adult survival. We have no measures of the variance of total fitness, but it is not unlikely that it is very much greater than that for viability. If so, much of the mutation load may result in reduced fertility rather than viability. Very little is known about epistasis for fertility components; if it is large, the mutation load could be reduced still further.

The seemingly very high rate of mutation of minor mutants is still a puzzle. We should notice that the methodology is not what might be desired—it depends on measures of genetic variance, which are notoriously capricious, and on quite indirect calculations. Yet most of the obvious biases are such as to cause the estimate to be too small rather than too large.

It is not impossible to accommodate this large a mutation rate by assuming that there is considerable epistasis, especially for fertility factors, and that most selection is on the less well understood components of fitness, such as fertility and mate selection, rather than viability. A full and proper cost-accounting of mutation is not yet possible, but it is clear that recurrent mutation exerts a large deleterious effect.

The species pays a high price for the privilege of mutation and the evolutionary possibilities deriving from it.

Fitness Variance Maintained by Balanced Selection

Despite an enormous amount of work and many words written on the subject, the relative contributions of mutationally maintained loci and those maintained by balancing selection to the population variance in fitness are still not known (1, 75). Attempts to partition the genetic variance for fitness into additive and dominance components are fraught with difficulties, and the experiments (76) have been only partially successful.

The existence of a large mutational component to the hidden variability disclosed by chromosome homozygosis is well established. How large, relative to this, the balanced component is remains unsolved. It is, of course, a commonplace that overdominant or frequency-dependent mutants, even if very rare, could make large contributions to the variance of an equilibrium population. We should also note that the depression in fitness observed when chromosomes are made homozygous (62–66) is large enough to accommodate many balanced polymorphisms. Perhaps isozyme variants are maintained in this fashion (65).

Very Nearly Neutral Mutations

The methodology discussed in this review cannot deal with mutants that have no effect on fitness. It has been suggested (and vociferously championed) that isozyme variants are selectively neutral (77). The question is by no means settled, but it does seem clear that mutants that have very little selective differences among them are common in natural populations.

From the standpoint of this review, neutral mutants are largely irrelevant. We are concerned here with the effects of mutation on fitness; if a mutant has zero effect on fitness, there is nothing about it that the methodology of this review can reveal. The other side of the coin is that such mutants, if they exist, are of no concern in a discussion of fitness.

However, the realization that mutants with very small effects on fitness exist and that they are partially dominant has considerable evolutionary significance. The larger the number of factors involved, the more possibilities there are for fine-tuning by selecting among genotypes that differ ever so slightly. Furthermore, the more genes that are involved in a trait, the less Mendelian noise there is; segregation effects largely cancel out. Finally, the more nearly intermediate the heterozygote is between the two homozygotes and the less epistasis, the more responsive the system is to selection; the total genetic variance is little more than the additive component—a system very favorable to efficient response to natural selection.

Overdominance in Homozygous Backgrounds, Coupling-Repulsion Effects, and Optimum Heterozygosity

Wallace (78, 79) argued that in a homozygous background, radiation-induced mutants are overdominant for fitness, at least for viability. Mukai & Yamazaki (27) reported that the viability effects were different when a group of mutants were on a single chromosome than when they were distributed equally among two homologues. Mukai (80) also suggested that heterozygous mutants increase viability until some 8 to 11 have accumulated at which point additional mutants are deleterious. All these are contrary to conventional genetic assumptions, yet there is some support for all of them.

To determine the heterozygous effects of mildly detrimental mutants, two types of experiments have been performed. In one, quasinormal chromosomes have been combined with a chromosome presumed to be free of viability-depressing mutations; since all the mutants were carried by one member of the chromosome pair, the resulting flies were called *coupling heterozygotes*. In the second type of experiment, mutant-bearing chromosomes have been combined randomly with one another, the resulting flies being called *repulsion heterozygotes*. The viabilities of both types of heterozygotes were measured relative to marked flies; in the case of the repulsion test, and sometimes in the case of the coupling test, these flies also carried the mutants whose effects were being studied. It has been customary to assume that such flies were unaffected by the mutants, i.e. that the marker chromosome was completely dominant. This is a dubious assumption; for a discussion see (76). In addition, the viabilities of the homozygotes that corresponded to the heterozygotes were determined relative to the marked flies. Analysis of homozygote and heterozygote viabilities permitted inferences to be made about the degree of dominance of mildly detrimental mutants.

The results of coupling and repulsion experiments differ. Newly arisen detrimental mutants appear to be overdominant in coupling phase, but partially dominant in repulsion. A summary based on studies of spontaneous detrimentals is presented in Table 5. All the studies involved mutants that were accumulated on the second chromosome of *D. melanogaster* in the manner described earlier; the number of generations during which the mutations were accumulated is given in the table. The degree of dominance presented for each experiment is a crude one, the method of calculation being described at the bottom of the table.

The coupling tests of Mukai and co-workers (27, 80–82) showed overdominance (negative values of $\langle h \rangle$), providing the "original" chromosome from which the mutant ones were derived was used as the mutation-free homologue. When this was replaced by an unrelated or distantly related chromosome, the mutants behaved in a partially dominant manner. Of course, the "original" chromosome was not really used in the coupling tests—it could not be preserved during the term of the experiment. Instead, at generation 32 a chromosome with high homozygous viability—practically the same as the original chromosome—was used as the homologue in the coupling tests; presumably it carried very few viability-reducing mutations. At generation 60, and thereafter, a high viability chromosome was extracted from a population started in generation 33 with a few chromosomes of fairly high viability; this chromosome was then used as the test homologue. The nonisogenic chromosomes used in generation 32 were also chosen because of their high homozygous viabilities.

Mukai's experiments (27, 80–82) indicated that the viabilities of heterozygotes were negatively correlated with those of the corresponding homozygotes. A more

Table 5 Estimated degrees of dominance of mildly detrimental mutants of spontaneous origin

Generation	Nature of test[a]	$\langle h \rangle$[b]	Reference
32	Coupling with "original" chromosome	−0.22	Mukai, Chigusa & Yoshikawa (81)
32	Coupling with nonisogenic chromosomes		Mukai, Chigusa & Yoshikawa (82)
	Chromosome from same population	0.09	
	Chromosome from different population	0.13	
32	Repulsion	0.34	Mukai & Yamazaki (27)
52	Repulsion	0.35	
60	Coupling with "original" chromosome	−0.32	
78	Coupling with "original" chromosome	−0.09	Mukai (80)
85	Coupling with "original" chromosome		
	Chromosome III heterozygous	−0.15	
	Chromosome III homozygous	−0.09	
40	Coupling with "original" chromosome	0.49	Ohnishi (22)
40	Repulsion	0.40	

[a] Coupling test: Chromosomes that had accumulated mutations were combined with a chromosome presumed to be free of mutants to constitute *coupling heterozygotes*. Repulsion test: Chromosomes that had accumulated mutations were randomly combined with one another to constitute *repulsion heterozygotes*.

[b] For coupling tests, $h = (1 - H_e) / (1 - H_o)$, where H_e = standardized mean viability of coupling heterozygotes and H_o = standardized mean viability of homozygotes. For repulsion tests, $h = (1 - H'_e) / 2(1 - H_o)$, where H'_e = standardized mean viability of repulsion heterozygotes. Only chromosomes with homozygous viability at least 0.6 normal were studied in these experiments.

careful examination showed that the relationship between heterozygous viability and homozygous viability was not linear, but parabolic; heterozygous viability increased as homozygous viability decreased, but when homozygous viability declined to 30–50% of normal, heterozygous viability also began to decline. Mukai (80) dubbed the phenomenon *optimum heterozygosity*; he hypothesized that the viability of a coupling heterozygote improved with every additional mutant until 8 to 11 had been sustained. Thereafter it declined. It is possible that the chromosomes which exhibited post-saturation decline carried semilethal mutants, the dominance effects of which might be similar to those of lethals. In any event, an attempt to abolish the overdominance of detrimentals on the second chromosome by introducing appreciable heterozygosity in the third was unsuccessful (80). This result essentially destroyed the idea of optimum heterozygosity.

The results of the coupling tests performed by Ohnishi (22) were not in agreement with those of Mukai. In fact, they indicated that detrimental mutants were essentially additive in their effects on viability in coupling condition ($\langle h \rangle = 0.5$). Ohnishi observed that the viability of coupling heterozygotes declined during the first 20 generations of mutation accumulation, but held steady thereafter. Several determinations of the degree of dominance were made during the accumulation process, and the pooled result is given in Table 5 ($\langle h \rangle = 0.49$). There was a technical difference between the experiments of Ohnishi and Mukai. In the latter, the marker class, which served as a viability standard, was free of mutants whose effects were being measured, but in the former it was not. This might account for the discrepant results.

In addition to spontaneous mutants, Ohnishi (22) studied EMS-induced ones; these showed partial dominance in coupling condition ($\langle h \rangle = 0.27$). Mukai (18) also studied the dominance of EMS-induced mutations, and in the main, found partial dominance, the value of $\langle h \rangle$ being similar to that for lethals. Unpublished observations by R. G. Temin indicate that the degree of dominance for EMS-induced detrimentals in coupling condition is about 0.2. This finding is based on experiments of a design different from those of Mukai and Ohnishi.

Repulsion tests for the heterozygous effects of detrimentals have consistently shown partial dominance (Table 5). Moreover, when the homologue in a coupling test was nonisogenic with the mutant-bearing chromosomes—a kind of repulsion test—partial dominance was detected (82). The degree of dominance observed in the repulsion tests was quite high, 0.3–0.5 (22, 27), indicating that in this condition newly arisen detrimentals are much more dominant than lethals. The average heterozygous effect on viability of a single detrimental mutant can be estimated by multiplying the average degree of dominance by the average homozygous viability effect: $(0.35)(0.03) = 0.01$, practically the same as the heterozygous effect of a lethal. The values of $\langle h \rangle$ given in Table 5 for the repulsion tests are crude, but more refined calculations give essentially the same results. For example, Mukai & Yamazaki (27) found that the distribution of viabilities of repulsion heterozygotes was bimodal, one mode representing combinations of two mutant chromosomes, the other representing combinations of a mutant chromosome with a mutant-free one. The latter group, which consisted of de facto "coupling" heterozygotes, had a higher mean viability than the former, comprised of bona fide repulsion heterozygotes. Moreover, when the viabilities of the heterozygotes were correlated with the sums of the viabilities of their constituent homozygotes, the coupling cluster gave a significantly negative value, while the repulsion cluster gave a significantly positive one. Thus, the coupling heterozygotes that were accidentally generated in the repulsion test showed overdominance, confirming the results of deliberate coupling tests. The true repulsion heterozygotes showed partial dominance. For these, the degree of dominance was calculated by regressing the heterozygote viabilities on the sums of the viabilities of the constituent homozygotes. The result was $\langle h \rangle = 0.39$. The data were analyzed in a different way by Mukai (50), giving $\langle h \rangle = 0.43$; he also calculated the genetic variance of h, the result being 0.044.

R. G. Temin (unpublished observations) studied the heterozygous effects of EMS-induced detrimentals in coupling condition with a nonisogenic homologue. The degree of dominance varied from 0.06 to 0.24. Ohnishi (22) studied EMS-induced detrimentals in repulsion configuration and found that $\langle h \rangle = 0.27$. Thus, EMS-induced mutants seem to be less dominant than spontaneous ones; however, the mean homozygous effect on viability of an EMS-induced detrimental is likely to be larger than the corresponding effect of a spontaneous mutant (16). Thus, the lower dominance is compensated for by a higher homozygous effect, so the average heterozygous effect is about the same.

For radiation-induced detrimentals, heterozygous effects cannot be measured reliably (26). Either the mutants occur frequently but have miniscule effects, or their effects are fairly large, but the mutants themselves are rare. Several experiments in which irradiated chromosomes were tested for heterozygous effects, but not for homozygous effects, showed overdominance, but only in certain genetic backgrounds (78, 79, 83, 84). Discrepant results were obtained by other investigators

(85–90). Wallace & Kass (91) have attempted to explain overdominance in a homozygous background with a model of gene regulation.

In view of the equivocal nature of the experimental evidence, these issues—overdominance of new mutants in homozygous background, optimum heterozygosity, and coupling-repulsion effect—remain unresolved. It is worth saying, however, that none of the three would have much influence on the genetics of natural populations, for the conditions under which these hypotheses apply are not found in nature to any appreciable extent.

Hybrid Dysgenesis

Kidwell, Kidwell & Sved (92) have called attention to a group of related phenomena that they term *hybrid dysgenesis*. These include a high mutation rate, male recombination, distorted segregation ratios, chromosomal aberrations, and nondisjunction. All are associated nonreciprocally in F_1 hybrids, in particular in hybrids between laboratory and natural populations.

Chromosomes from natural populations that produce these effects when placed in combination with chromsomes from laboratory populations are being found increasingly. Laboratory marker chromosomes have been found frequently of the type that interact with chromsomes from males caught in natural populations. If these produce the phenomena named above in any frequency, they could be a source of error in many of the kinds of experiments reported here. They would be particularly important when small effects are being studied. Since studies of heterozygous effects depend on measuring very small effects that become significant only when enormous numbers of Drosophila are reared, it is quite possible that the noise introduced by these aberrancies may be sufficient to mask the effect being studied. This is one possible explanation of the confusing inconsistencies reported above.

CONCLUSIONS

The mutation rate of recessive lethals is 0.005–0.006 per second chromosome per generation, or about 2.5–3.0×10^{-6} per locus. The rate for mutations causing mild deleterious effects is at least an order of magnitude higher, a minimum estimate being about 0.10 per chromosome, or 6×10^{-5} per locus. The rate could, of course, be much higher if there is a large class of mutant genes that have zero effect on fitness, or nearly so. The estimated average homozygous effect of mildly deleterious mutants is 0.02–0.03, i.e. a 2 or 3% reduction in viability. For the same reasons that the mutation rate estimate is minimum, this is maximum.

Lethal mutations are not completely recessive; heterozygotes for lethal mutants have a viability decrease of 1–3%. On the other hand, heterozygotes for mildly deleterious genes have a depression of 30–50% of the homozygous effect. Thus, the milder the effect of a mutant, the greater its dominance.

The effect of a mutant on total fitness is considerably greater than its effect on viability alone. In fact, the heterozygous effect of a minor mutant on fitness is about the same as its homozygous effect on viability.

The absolute effect on fitness in heterozygotes is approximately the same for mutants that are mildly deleterious in homozygotes as for mutants that are lethal when homozygous. Neither type of mutant very often becomes homozygous in natural populations; hence, the effects on population fitness of mutants that are mild or drastic when homozygous are essentially the same. Since the milds are so much more numerous, their impact on the population is much greater.

The very high mutation rate of mildly deleterious genes suggests a large mutation load, of the order of 50% fitness reduction. There is clearly no such reduction for viability. The most likely explanation of how the species tolerates such a high mutation rate now appears to be a combination of epistasis, permitting more mutant eliminations for a given number of "genetic deaths," and a large genetic variance in fertility and success in competition for mates. Further work is needed before this is understood.

Literature Cited

1. Lewontin, R. C. 1974. *The Genetic Basis of Evolutionary Change.* New York: Columbia Univ. Press. 346 pp.
2. Haldane, J. B. S. 1937. The effect of variation on fitness. *Am. Nat.* 71:337–49
3. Muller, H. J. 1950. Our load of mutations. *Am. J. Hum. Genet.* 2:111–76
4. Morton, N. E., Crow, J. F., Muller, H. J. 1956. An estimate of the mutation damage in man from data on consanguineous marriages. *Proc. Natl. Acad. Sci. USA* 42:855–63
5. Greenberg, R., Crow, J. F. 1960. A comparison of the effect of lethal and detrimental chromosomes from Drosophila populations. *Genetics* 45:1153–68
6. Temin, R. G., Meyer, H. U., Dawson, P. S., Crow, J. F. 1969. The influence of epistasis on homozygous viability depression in *Drosophila melanogaster. Genetics* 61:497–519
7. Crow, J. F., Temin, R. G. 1964. Evidence for the partial dominance of recessive lethal genes in natural populations of Drosophila. *Am. Nat.* 98:21–33
8. Wallace, B. 1968. Mutation rates for autosomal lethals in *Drosophila melanogaster. Genetics* 60:389–93
9. Mukai, T., Chigusa, S. I., Mettler, L. E., Crow, J. F. 1972. Mutation rate and dominance of genes affecting viability in *Drosophila melanogaster. Genetics* 72:335–55
10. Ohnishi, O. 1977. Spontaneous and ethyl methanesulfonate-induced mutations controlling viability in *Drosophila melanogaster.* I. Recessive lethal mutations. *Genetics.* In press
11. Judd, B. H., Young, M. W. 1974. An examination of the one cistron: one chromomere concept. *Cold Spring Harbor Symp. Quant. Biol.* 38:573–79
12. Schalet, A. 1960. *A study of spontaneous visible mutations in Drosophila melanogaster.* PhD thesis. Indiana U., Bloomington. 76 pp.
13. Mukai, T. 1970. Spontaneous mutation rates of isozyme genes in *Drosophila melanogaster. Drosophila Inf. Serv.* 45:99
14. Tobari, Y. N., Kojima, K. 1972. A study of spontaneous mutation rates at ten loci detectable by starch gel electrophoresis in *Drosophila melanogaster. Genetics* 70:397–403
15. Mukai, T. 1964. The genetic structure of natural populations of *Drosophila melanogaster.* I. Spontaneous mutation rate of polygenes controlling viability. *Genetics* 50:1–19
16. Ohnishi, O. 1977. Spontaneous and ethyl methanesulfonate-induced mutations controlling viability in *Drosophila melanogaster.* II. Homozygous effect of polygenic mutations. *Genetics.* In press
17. Crow, J. F. 1977. Minor viability mutants in Drosophila. *Genetics.* In press
18. Mukai, T. 1970. Viability mutations induced by ethyl methanesulfonate in *Drosophila melanogaster. Genetics* 65:335–48
19. Simmons, M. J., Sheldon, E. W., Crow, J. F. 1977. Heterozygous effects on fitness of EMS-treated chromosomes in *Drosophila melanogaster.* Submitted
20. Friedman, L. D. 1964. X-ray-induced sex-linked lethal and detrimental mutations and their effects on the viability of *Drosophila melanogaster. Genetics* 49:689–99
21. Yoshikawa, I., Mukai, T. 1970. Heterozygous effects on viability of spontaneous lethal genes in *Drosophila*

22. Ohnishi, O. 1977. Spontaneous and ethyl methanesulfonate-induced mutations controlling viability in Drosophila melanogaster. III. Heterozygous effects of polygenic mutations. *Genetics.* In press
23. Tobari, I. 1966. Effects of temperature on the viability of heterozygotes of lethal chromosomes in *Drosophila melanogaster. Genetics* 53:249–59
24. Stern, C., Carson, G., Kinst, M., Novitski, E., Uphoff, D. 1952. The viability of heterozygotes for lethals. *Genetics* 37:413–49
25. Kitagawa, O. 1967. Interactions in fitness between lethal genes in heterozygous condition in *Drosophila melanogaster. Genetics* 57:809–20
26. Simmons, M. J. 1976. Heterozygous effects of irradiated chromosomes on viability in *Drosophila melanogaster. Genetics* 84:353–74
27. Mukai, T., Yamazaki, T. 1968. The genetic structure of natural populations of *Drosophila melanogaster*. V. Coupling-repulsion effect of spontaneous mutant polygenes controlling viability. *Genetics* 59:513–35
28. Dobzhansky, T., Spassky, B., Tidwell, T. 1963. Genetics of natural populations. XXXII. Inbreeding and the mutational and balanced genetic loads in natural populations of *Drosophila pseudoobscura. Genetics* 48:361–73
29. Malogolowkin-Cohen, C., Levene, H., Dobzhansky, N. P., Simmons, A. S. 1964. Inbreeding and the mutational and balanced loads in natural populations of *Drosophila willistoni. Genetics* 50:1299–1311
30. Kosuda, K. 1971. Synergistic interaction between second and third chromosomes on viability of *Drosophila melanogaster. Jpn. J. Genet.* 46:41–52
31. Spassky, B., Dobzhansky, T., Anderson, W. W. 1965. Genetics of natural populations. XXXVI. Epistatic interactions of the components of the genetic load in *Drosophila pseudoobscura. Genetics* 52:653–64
32. Mukai, T. 1969. The genetic structure of natural populations of *Drosophila melanogaster*. VII. Synergistic interaction of spontaneous mutant polygenes controlling viability. *Genetics* 61:749–61
33. Latter, B. D. H., Robertson, A. 1962. The effects of inbreeding and artificial selection on reproductive fitness. *Genet. Res.* 3:110–38
34. Crow, J. F. 1970. In *Mathematical Topics in Population Genetics,* ed. K. Kojima, pp. 128–77. Berlin: Springer. 400 pp.
35. Dobzhansky, T., Spassky, B. 1968. Genetics of natural populations. XL. Heterotic and deleterious effects of recessive lethals in populations of *Drosophila pseudoobscura. Genetics* 59:411–25
36. Anderson, W. W. 1969. Genetics of natural populations. XLI. The selection coefficients of heterozygotes for lethal chromosomes in Drosophila on different genetic backgrounds. *Genetics* 62:827–36
37. Chovnick, A. 1973. Gene conversion and transfer of genetic information within the inverted region of inversion heterozygotes. *Genetics* 75:123–31
38. Marinković, D. 1967. Genetic loads affecting fecundity in natural populations of *Drosophila pseudoobscura. Genetics* 56:61–71
39. Watanabe, T. K., Yamaguchi, O., Mukai, T. 1976. The genetic variability of third chromosomes in a local population of *Drosophila melanogaster. Genetics* 82:63–82
40. Band, H. T. 1963. Genetic structure of populations. II. Viabilities and variances of heterozygotes in constant and fluctuating environments. *Evolution* 17:307–19
41. Temin, R. G. 1966. Homozygous viability and fertility loads in *Drosophila melanogaster. Genetics* 53:27–46
42. Mukai, T., Yamaguchi, O. 1974. The genetic structure of natural populations of *Drosophila melanogaster*. XI. Genetic variability in a local population. *Genetics* 76:339–66
43. Hiraizumi, Y., Crow, J. F. 1960. Heterozygous effects on viability, fertility, rate of development, and longevity of Drosophila chromosomes that are lethal when homozygous. *Genetics* 45:1071–83
44. Watanabe, T. K., Oshima, C. 1970. Persistence of lethal genes in Japanese natural populations of *Drosophila melanogaster. Genetics* 64:93–106
45. Band, H. T., Ives, P. T. 1963. Genetic structure of populations. I. On the nature of the genetic load in the South Amherst population of *Drosophila melanogaster. Evolution* 17:198–215
46. Watanabe, T. K., Ohnishi, S. 1975. Genes affecting productivity in natural populations of *Drosophila melanogaster. Genetics* 80:807–19

47. Prout, T. 1952. Selection against heterozygotes for autosomal lethals in natural populations of *Drosophila willistoni. Proc. Natl. Acad. Sci. USA* 38: 478-81
48. Dobzhansky, T., Krimbas, C., Krimbas, M. G. 1960. Genetics of natural populations. XXIX. Is the genetic load in *Drosophila pseudoobscura* a mutational or a balanced load? *Genetics* 45:741-53
49. Tobari, I. 1966. Effects of temperature on the viabilities of homozygotes and heterozygotes for second chromosomes of *Drosophila melanogaster. Genetics* 54:783-91
50. Mukai, T. 1969. The genetic structure of natural populations of *Drosophila melanogaster.* VIII. Natural selection on the degree of dominance of viability polygenes. *Genetics* 63:467-78
51. Kenyon, A. 1967. Comparison of frequency distributions of viabilities of second with fourth chromosomes from caged *Drosophila melanogaster. Genetics* 55:123-30
52. Wills, C. 1966. The mutational load in two natural populations of *Drosophila pseudoobscura. Genetics* 53:281-94
53. Strickberger, M. W. 1972. Viabilities of third chromosomes of *Drosophila pseudoobscura* differing in relative competitive fitness. *Genetics* 72:679-89
54. Anderson, W. W., Watanabe, T. K. 1974. Selection by fertility in *Drosophila pseudoobscura. Genetics* 77:559-64
55. Bundgaard, J., Christiansen, F. B. 1972. Dynamics of polymorphisms. I. Selection components in an experimental population of *Drosophila melanogaster. Genetics* 71:439-60
56. Prout, T. 1971. The relation between fitness components and population prediction in Drosophila. I. The estimation of fitness components. *Genetics* 68: 127-49
57. Prout, T. 1971. The relation between fitness components and population prediction in Drosophila. II. Population prediction. *Genetics* 68:151-67
58. Polivanov, S., Anderson, W. W. 1969. Selection in experimental populations. II. Components of selection and their fluctuation in two populations of *Drosophila melanogaster. Genetics* 63: 919-32
59. Knight, G. R., Robertson, A. 1957. Fitness as a measurable character in Drosophila. *Genetics* 42:524-530
60. Mitchell, J. A. 1977. Fitness effects of EMS-induced mutations on the X chromosome of *Drosophila melanogaster.* I. Viability effects and heterozygous fitness effects. *Genetics.* In press
61. Mitchell, J. A., Simmons, M. J. 1977. Fitness effects of EMS-induced mutations on the X chromosome of *Drosophila melanogaster.* II. Hemizygous fitness effects. *Genetics.* In press
62. Sved, J. A., Ayala, F. J. 1970. A population cage test for heterosis in *Drosophila pseudoobscura. Genetics* 66:97-113
63. Sved, J. A. 1971. An estimate of heterosis in *Drosophila melanogaster. Genet. Res.* 18:97-105
64. Sved, J. A. 1975. Fitness of third chromosome homozygotes in *Drosophila melanogaster. Genet. Res.* 25:197-200
65. Tracey, M. L., Ayala, F. J. 1974. Genetic load in natural populations: Is it compatible with the hypothesis that many polymorphisms are maintained by natural selection? *Genetics* 77: 569-89
66. Mourao, C. A., Ayala, F. J., Anderson, W. W. 1972. Darwinian fitness and adaptedness in experimental populations of *Drosophila willistoni. Genetica* 43:552-74
67. Watanabe, T. K., Oshima, C. 1973. Fertility genes in natural populations of *Drosophila melanogaster.* II. Correlations between productivity and viability. *Jpn. J. Genet.* 48:337-47
68. Dobzhansky, T., Wright, S. 1941. Genetics of natural populations. V. Relations between mutation rate and accumulation of lethals in populations of *Drosophila pseudoobscura. Genetics* 26: 23-51
69. Murata, M. 1970. Frequency distribution of lethal chromosomes in small populations of *Drosophila melanogaster. Genetics* 64:559-71
70. Tobari, I., Murata, M. 1970. Effects of x-rays on genetic loads in a cage population of *Drosophila melanogaster. Genetics* 65:107-19
71. Murata, M., Tobari, I. 1973. Changes in frequency of lethal second chromosomes in experimental populations of *Drosophila melanogaster* with radiation histories. *Jpn. J. Genet.* 48:349-59
72. Murata, M., Tobari, I. 1976. Changes in frequency and allelism of recessive lethals in experimental populations of *Drosophila melanogaster* with radiation histories. *Jpn. J. Genet.* 51:27-37
73. Crow, J. F. 1968. In *Population Biology and Evolution,* ed. R. Lewontin, pp. 71–

86. Syracuse, NY: Syracuse Univ. Press. 205 pp.
74. Kimura, M., Maruyama, T. 1966. The mutation load with epistatic gene interactions in fitness. *Genetics* 54: 1337–51
75. Dobzhansky, T. 1970. *Genetics of the Evolutionary Process.* New York: Columbia Univ. Press. 505 pp.
76. Mukai, T., Cardellino, R. A., Watanabe, T. K., Crow, J. F. 1974. The genetic variance for viability and its components in a local population of *Drosophila melanogaster. Genetics* 78: 1195–1208
77. Kimura, M., Ohta, T. 1971. Protein polymorphism as a phase of molecular evolution. *Nature* 229:467–69
78. Wallace, B. 1958. The average effect of radiation-induced mutations on viability in *Drosophila melanogaster. Evolution* 12:532–52
79. Wallace, B. 1963. Further data on the overdominance of induced mutations. *Genetics* 48:633–51
80. Mukai, T. 1969. The genetic structure of natural populations of *Drosophila melanogaster.* VI. Further studies on the optimum heterozygosity hypothesis. *Genetics* 61:479–95
81. Mukai, T., Chigusa, S., Yoshikawa, I. 1964. The genetic structure of natural populations of *Drosophila melanogaster.* II. Overdominance of spontaneous mutant polygenes controlling viability in homozygous genetic background. *Genetics* 50:711–15
82. Mukai, T., Chigusa, S., Yoshikawa, I. 1965. The genetic structure of natural populations of *Drosophila melanogaster.* III. Dominance effect of spontaneous mutant polygenes controlling viability in heterozygous genetic backgrounds. *Genetics* 52:493–501
83. Mukai, T., Yoshikawa, I., Sano, K. 1966. The genetic structure of natural populations of *Drosophila melanogaster.* IV. Heterozygous effects of radiation-induced mutations on viability in various genetic backgrounds. *Genetics* 53: 513–27
84. Maruyama, T., Crow, J. F. 1975. Heterozygous effects of x-ray-induced mutations on viability of *Drosophila melanogaster. Mutat. Res.* 27:241–48
85. Pandey, J. 1975. Further studies on heterozygous effects of radiation on viability of *Drosophila melanogaster. Mutat. Res.* 27:249–53

86. Falk, R. 1961. Are induced mutations in Drosophila overdominant? II. Experimental results. *Genetics* 46:737–57
87. Falk, R. 1967. Fitness of heterozygotes for irradiated chromosomes in *Drosophila. Mutat. Res.* 4:805–19
88. Falk, R. 1967. Viability of heterozygotes for induced mutations in *Drosophila melanogaster.* III. Mutations in spermatogonia. *Mutat. Res.* 4:59–72
89. Falk, R., Ben-Zeev, N. 1966. Viability of heterozygotes for induced mutations in *Drosophila melanogaster.* II. Mean effects in irradiated autosomes. *Genetics* 53:65–77
90. Falk, R., Rahat, A., Ben-Zeev, N. 1965. Viability of heterozygotes for induced mutations in *Drosophila melanogaster.* I. Irradiated X-chromosomes. *Mutat. Res.* 2:438–51
91. Wallace, B., Kass, T. L. 1974. On the structure of gene control regions. *Genetics* 77:541–58
92. Kidwell, M. G., Kidwell, J. F., Sved, J. A. 1977. Hybrid dysgenesis in *Drosophila melanogaster:* a syndrome of aberrant traits including mutation, sterility and male recombination. Submitted
93. Band, H. T. 1964. Genetic structure of populations. III. Natural selection and concealed genetic variability in a natural population of *Drosophila melanogaster. Evolution* 18:384–404
94. Dawood, M. M. 1961. The genetic load in the second chromosome of some populations of *Drosophila melanogaster* in Egypt. *Genetics* 46:239–46
95. Hoenigsberg, H. F., Castro, L. E., Granobles, L. A., Idrobo, J. M. 1969. Population genetics in the American tropics. II. The comparative genetics of Drosophila in European and neo-tropical environments. *Genetica* 40:43–60
96. Hoenigsberg, H. F., Granobles, L. A., Castro, L. E. 1969. Population genetics in the American tropics. IV. Temporal changes effected in natural populations of *Drosophila melanogaster* from Colombia. *Genetica* 40:201–15
97. Watanabe, T. K. 1969. Frequency of deleterious chromosomes and allelism between lethal genes in Japanese natural populations of *Drosophila melanogaster. Jpn. J. Genet.* 44:171–87
98. Watanabe, T. K., Watanabe, T., Oshima, C. 1976. Genetic changes in natural populations of *Drosophila melanogaster. Evolution* 30:109–18

# PROSPECTS FOR PLANT GENOME MODIFICATION BY NONCONVENTIONAL METHODS

✣3116

Andris Kleinhofs[1]

Program in Genetics and Department of Agronomy and Soils, Washington State University, Pullman, Washington 99164

Ram Behki

Chemistry and Biology Research Institute, Research Branch, Canada Agriculture, Ottawa, Ontario, K1A 0C6, Canada

CONTENTS

INTRODUCTION	79
UPTAKE OF DNA BY PLANT TISSUES	80
UPTAKE OF DNA BY PLANT CELLS, PROTOPLASTS, AND POLLEN	83
UPTAKE OF PHAGE PARTICLES BY PLANT CELLS AND PROTOPLASTS	87
UPTAKE OF ORGANELLES BY PLANT PROTOPLASTS	89
HERITABLE CHANGE OF PLANT PHENOTYPE	91
SOMATIC CELL FUSION	93
CONCLUDING REMARKS	95

INTRODUCTION

The essence of plant breeding has been and is genetic recombination and selection. In the past 20 years mutagenesis has been added as a tool for the generation of new variability (1–3). Otherwise, plant breeders have had to rely on sexual processes to bring together desired genes, genetic recombination to reshuffle them, and their own powers of observation and ingenuity to select the desirable phenotypes. These methods have been exceedingly successful, as indicated by the steadily increasing crop yields over the past 100 years. Nevertheless, we may be approaching a yield plateau

[1] Information Paper. College of Agriculture Research Center, Washington State University, Pullman.

from which a further increase is more and more difficult to achieve. In addition, ever increasing pressures for insect and disease resistance, enhanced nitrogen utilization, increased photosynthetic efficiency, and improved nutritional and industrial quality are being felt by today's plant breeders. Consequently, there is a growing interest in the application of techniques of genetic engineering to plant breeding problems.

In this review we critically examine the prospects for genetic engineering in plants. Specifically, we evaluate the published information with the goal of identifying sound bases of knowledge upon which future research can be built and the more questionable information that needs to be reexamined. We feel that this type of a review is urgently needed in the field of plant genetic engineering in spite of the many reviews that have already been written (4–8). In addition to the above reviews, readers are also referred to the proceedings of several recent symposia for more information about this topic (9–13).

In this paper we define genetic engineering as any nonconventional method of genetic manipulation dealing with the transfer of genes between plants and from other organisms to plants. These nonconventional techniques can be further divided into molecular and cellular processes. In the molecular process we include methods that involve the use of purified or protected DNA, while in the cellular process we include methods that deal with organelle or intact chromosome transfer and cell fusion.

The successful introduction of DNA into plant cells so as to modify the plant's genetic make-up permanently is a highly complex process. For a systematic evaluation of this process, we divide it into five biochemical steps: (*a*) uptake, (*b*) integration, (*c*) replication, (*d*) expression, and (*e*) transmission.

All of these processes, except integration, are essential for successful genetic modification of plant genomes. Integration of the transferred genetic information into the plant's genome would be desirable since it would stabilize it and facilitate its replication. Clearly, however, it is not an absolute prerequisite since numerous living cells, including plant cells, possess functional and stable nonintegrated DNA molecules, that is, chloroplast and mitochondrial DNA. The various steps in genetic engineering are not always clearly differentiated in the literature. We have organized our review along different lines. Nevertheless, the reader should keep these steps in mind, and we refer to them throughout the article.

UPTAKE OF DNA BY PLANT TISSUES

The uptake of DNA by plants has been demonstrated in numerous systems and by numerous techniques (14–37). Thus, there seems to be a general agreement that exogenous DNA is taken up by germinating embryos or by excised seedlings. The fate of the exogenous DNA following uptake, however, has remained a subject of considerable controversy. Several workers (15–18, 22–31, 34, 35) have reported that exogenously applied DNA is not only taken up by plants but also is integrated into the host genome and replicated with it. Other investigators (20, 21, 32, 33, 36–40) have observed neither integration nor replication of the exogenous DNA and have concluded that the ultimate fate of the absorbed DNA is degradation and use of the degradation products for endogenous DNA synthesis. The evidence for integration and replication of the exogenous DNA is based mainly on isopycnic centrifugation in cesium chloride (CsCl) gradients. Two major types of experiments have been conducted. They can be referred to as the *integration* and the *replication* experiments. In the integration experiment, radioactive labeled DNA of density different from the host DNA is administered to the seedlings. After a period of metabolism, the total seedling DNA or DNA from dissected organs is isolated and analyzed by CsCl density gradient centrifugation. The occurrence of a radioactive DNA band at the donor DNA density is taken as evidence for the uptake of that DNA. The appearance of a radioactive DNA band approximately intermediate in density between the host and donor DNA is taken as evidence for integration of the radioactive donor DNA with the nonradioactive host DNA (17). The intermediate density DNA band is further analyzed by sonication and denaturation. Shearing by sonication shifts the radioactivity of the intermediate density DNA band back to donor DNA density. Denaturation shifts the intermediate density DNA band to higher density as would be expected for the shift from double-stranded to single-stranded DNA. The shift in density upon sonication is a very crucial factor in these experiments. It demonstrates that the integrated radioactive donor DNA fragments can be released from the nonradioactive host DNA and return to the donor DNA density. The denaturation experiment demonstrates that the donor and host DNA fragments cannot be released by denaturation and therefore are not held together as the complementary partners of a double-stranded DNA molecule.

Experiments of this nature have been used to demonstrate the integration of *Micrococcus lysodeikticus* DNA in barley root DNA (17) and the integration and translocation of *Escherichia coli* and *Streptomyces coelicolor* DNA in *Arabidopsis* (22). Similar arguments have been presented for the integration of T4 DNA into *Matthiola incana* seedlings (30, 35), although the new "intermediate" density DNA band in this case was considerably heavier (1.724 g/cm^3) than either the host (1.698 g/cm^3) or the donor (1.694 g/cm^3) DNA.

Recently Gradmann-Rebel & Hemleben (35) have provided evidence based on DNA-DNA hybridization for the presence of T4 DNA sequences in the high density (1.724 g/cm^3) DNA band. Evidence based on the appearance of an intermediate density DNA band has also been presented for the integration of ^{3}H-adenine-BUdR labeled homologous DNA into *Matthiola incana* seedling genome (31).

The "replication" experiment is conducted in a similar manner except that nonradioactive donor DNA of different density from the host DNA is used. After a period of metabolism, ^{3}H-thymidine is applied, and the host DNA is isolated and analyzed by CsCl density gradient centrifugation. The occurrence of a radioactive intermediate density DNA band is taken as evidence of replication of the donor-host DNA complex (15, 18). The denaturation and sonication analysis of this replicating intermediate density DNA fraction indicates a homogenous double-stranded molecule that shifts to higher density on denaturation, but separates into two radioactive peaks of approximately donor and host DNA density on shearing by sonication. These results are explained by an end-to-end linkage of double-stranded donor and host DNA molecules (18). Experiments of this type have been used to demonstrate the replication of bacterial-plant DNA complexes in tomato (15), barley (18), and *Arabidopsis* (22, 23). Variations on these experiments include exposing the cut ends of tomato and eggplant seedlings to bacterial suspensions (25, 27). In these experiments, as in those where isolated DNA was used (15), an intermediate density radioactive DNA band was recovered. Even more surprisingly, self-replicating bacterial DNA was also recovered from the plant cells.

Attempts to reproduce these results by others either exactly with the same hosts and donor DNA or in other plant systems have been unsuccessful (20, 21, 32, 33, 39, 40). Hotta & Stern (21) attempted to reproduce the integration experiments of Ledoux & Huart (17) with barley and the experiments of Anker & Stroun (16) with tomato. They concluded that labeled micrococcal DNA absorbed by young tomato shoots was either degraded and reutilized by the host for its own DNA synthesis or it remained intact but chemically unassociated with the host nuclear DNA. Similar results were obtained with barley. Intermediate density DNA peaks were observed only under conditions of X irradiation or low humidity. These intermediate density DNA molecules did not replicate and were not located in the cell nuclei. Bendich & Filner (20) investigated the uptake and integration of *Pseudomonas aeruginosa* DNA by pea seedlings. They concluded that partially degraded bacterial DNA can be taken up into plants, but is not found in the cell nuclei. Kleinhofs (32) and Kleinhofs et al (33) have attempted to repeat the integration and replication experiments with barley, tomato, and peas with completely negative results. Intermediate density DNA peaks were observed only when axenic conditions broke down, and these peaks were clearly shown to be due to bacterial contamination. These authors criticized the use of buoyant density in CsCl as the sole criterion for characterization of a DNA fraction. Artifacts in CsCl density gradients (41) as well as bacterial contamination, unknown plant DNA viruses, or unknown plant DNA satellites may account for unusual DNA peaks.

Kleinhofs (37) attempted to detect integration of bacterial DNA in barley seedling DNA by sensitive DNA–DNA reassociation techniques. Although apparent uptake was observed, integration of the exogenous DNA in the plant genome was not detected. Kado & Lurquin (40) also were unable to detect replication of *Agrobacterium tumefaciens* DNA in *Phaseolus aureus* seedlings by reassociation techniques. In spite of the sensitivity of these techniques, which is several orders of magnitude greater than that of the CsCl technique (42), the results were completely negative.

Recently, Hanson & Chilton (39) have succeeded in clarifying some of the results with the tomato system. They were able to reproduce the results of Stroun and co-workers (25–27) in that they observed three buoyant density maxima in CsCl gradients of DNA obtained from tomato seedlings exposed to viable *A. tumefaciens* followed by ^{3}H-thymidine labeling. These DNA peaks corresponded to the host plant DNA density, the *A. tumefaciens* DNA density, and an intermediate DNA density. Treatment of the plant with chloramphenicol prior to and during labeling eliminated the *A. tumefaciens* DNA density peak, suggesting that it was due to residual surviving *A. tumefaciens* cells in the tomato seedlings. The intermediate density DNA fraction was further analyzed by sonication and rebanding in CsCl.

These experiments demonstrated that the homogeneous intermediate density DNA does indeed separate into lighter and heavier components upon shearing by sonication or in a French pressure cell. Analysis by DNA filter hybridization and by reassociation kinetics failed to show any detectable homology of this DNA with the bacterial donor DNA. Homology with the host plant DNA was established. Thus, these results, even though similar to those reported earlier by Stroun and co-workers (25–27), clearly emphasize that the conclusions of integration and replication of bacterial DNA with plant DNA based solely on isopycnic centrifugation in CsCl cannot be taken as definitive.

The one report (35) claiming integration of T4 DNA in the plant genome, based on DNA-DNA hybridization, must be viewed with skepticism. The very low counts recovered and the lack of appropriate melting temperature controls suggest that artifacts may account for the presented results.

The failure to obtain confirming evidence for the integration and replication of exogenous DNA in plants prompted us to reexamine the phenomenon of exogenous DNA uptake by plants. As indicated earlier, apparent uptake of exogenous DNA by plants has been demonstrated in all cases where it was looked for. The question of exogenous DNA penetration inside the cell membrane or inside the nuclear membrane has been considered by only a few investigators. As pointed out by Hotta & Stern (21), the removal of donor DNA from intercellular spaces and from vascular tissues is extremely difficult. Hotta & Stern also observed that nuclei isolated from seedlings in the presence of excess calf thymus DNA contained no significant amount of radioactivity at the bacterial DNA density peak while nuclei isolated in the absence of excess calf thymus DNA contained large amounts of radioactive DNA at the donor bacterial DNA density. They concluded that the bacterial DNA might be present in the cytoplasm of the cells and that it became associated with the nuclei only during the isolation procedure. These results did not rule out the possibility that the exogenous DNA may have been adsorbed to the cell membrane. In spite of these very definitive experiments, no other investigators have taken the precaution of isolating plant cell nuclei in the presence of excess calf thymus or other DNA in order to suppress nonspecific binding of the donor DNA to the nuclear membrane. Thus, it seems possible that the apparent uptake of exogenous DNA by plants represents, at least to a certain extent, persistent exogenous DNA bound to plant cell walls or membranes in DNase resistant form.

In summary, it can be concluded that the case for uptake, integration, and replication of exogenously supplied DNA in intact plants is at best tentative. To be convincing, future experiments need to be extended over longer time periods, at least one generation.

UPTAKE OF DNA BY PLANT CELLS, PROTOPLASTS, AND POLLEN

Hotta & Stern (21) reported that lily meiotic cells do not incorporate appreciable amounts of either homologous or heterologous DNA supplied exogenously. Bendich & Filner (20) were able to detect the uptake of tobacco or *P. aeruginosa* DNA by tobacco cells in shake cultures provided that the cells were first washed and treated with pronase to eliminate DNase secreted by the cells. In these experiments 0.5% of the exogenously supplied DNA was found to be associated with the nuclear fraction, thus indicating real uptake. Unfortunately, precautions against possible nonspecific binding of the exogenous DNA to the nuclear membrane (21; see discussion in previous section) were not reported.

Heyn & Schilperoort (38) also working with tobacco cells in culture were concerned about the problem of differentiating between DNA that had penetrated the cell membrane and that which is simply adsorbed to the cell wall. In attempts to resolve this difference, they isolated protoplasts from DNA-treated cells before isolation of the total cell DNA for analysis on CsCl gradients. They found that a considerable amount of the DNA that appeared to be taken up by the cells was merely adsorbed to the cell wall and eliminated with the protoplast isolation procedure. Protoplasts from both DEAE-dextran–pretreated cells and nonpretreated cells were isolated and analyzed. In the case of nonpretreated cells, very little if any high molecular weight *A. tumefaciens* DNA was detected in the protoplasts. In the case where the cells were pretreated with DEAE-dextran, some high molecular weight *A. tumefaciens* DNA was found in the protoplasts. This was calculated to represent only 0.145% of the high molecular weight DNA that was tightly bound to the cell walls. This DNA was assumed to be inside the cell, although no efforts were made to differentiate between cell membrane-bound DNA and that which had actually penetrated the membrane. Nevertheless, it seems clear that most of the DNA that is presumed to be taken up by plant cells is actually bound to the cell wall, trapped between the cell wall and membrane, or bound to the cell membrane. No indication for integration of the exogenous bacterial DNA in the plant DNA was observed.

Kado & Lurquin (44) have also studied the fate of *A. tumefaciens* DNA in tobacco cells. These experiments were somewhat different from the previously discussed experiments in that (*a*) callus cultures rather than suspension cultures were used, (*b*) the cells were grown for four days in the presence of the exogenous DNA compared to the shorter times of incubation used in the studies of Bendich & Filner (20) and Heyn & Schilperoort (38), and (*c*) DNA hybridization techniques were used in addition to CsCl density gradient analysis. No radioactive *A. tumefaciens* DNA could be detected in the tobacco cells after four days of incubation in the presence of exogenous *A. tumefaciens* ^{3}H-DNA by either CsCl density gradient analysis or DNA-DNA filter hybridization. In the experiments where nonradioactive *A. tumefaciens* DNA was used followed by ^{3}H-thymidine labeling, no replication of the exogenous *A. tumefaciens* DNA was detected by DNA-DNA filter hybridization or by DNA sequence enrichment experiments. Although the sensitivity claimed for the DNA filter experiments is probably excessive as a result of the failure to include radioactive tobacco DNA in the calibration curve (32, 45), these experiments nonetheless are very definitive and indicate that if any exogenous bacterial DNA is taken up by the tobacco cells it is not present after four days. No integration or replication of this DNA was detected with these very sensitive techniques.

Lurquin & Hotta (46), in similar experiments, investigated the fate of *E. coli* and *Micrococcus lysodeikticus* DNA in callus cells of *Arabidopsis thaliana*. After seven-day incubation of callus tissue in the presence of the bacterial DNA, total cell DNA was isolated and analyzed by CsCl density gradients and DNA reassociation kinetics. No bacterial DNA sequences were detected in these tests. Thus again, the fate of exogenous DNA incubated with plant cells seems to be degradation and reutilization rather than integration and replication.

Lurquin & Behki (47) in a very detailed study of the fate of exogenous DNA in *Chlamydomonas reinhardi* cells concluded that *E. coli* ^{3}H-DNA supplied to vegetative cultures of wild-type or CW15 (lacking cell wall) cells could bind to the cell wall or cell membrane. The extent of this binding decreased with time and was to a large degree (over 90%) DNase sensitive. Nevertheless, about 0.01% of the bacterial DNA remained irreversibly associated with the cells when they reached stationary phase. The irreversible binding of the donor bacterial DNA to *Chlamydomonas* cells could be increased by treatment of the cultures with polycations such as DEAE-dextran, poly-L-lysine, and poly-L-ornithine. The acid-insoluble radioactivity irreversibly bound to wild-type cells consisted mainly of oligonucleotides with a small proportion present as less depolymerized donor DNA. No radioactivity was found to be associated with the recipient high molecular weight *Chlamydomonas* DNA as would be expected if integration of the radioactive *E. coli* DNA had taken place. It should be noted that *Chlamydomonas* utilizes thymine derivatives (generated by degradation of donor DNA labeled in the thymine moiety) to a very limited extent and almost exclusively for the synthesis of chloroplast DNA (48). This feature allowed the investigators to discriminate between DNA uptake and degradation-reutilization processes.

Protoplasts may offer an advantage over whole cells as exogenous DNA receptors because of the absence of the cell wall (43). No advantage was evident in *Chlamydomonas* CW15 (lacking cell wall) cells for exogenous DNA uptake, but in that case high DNase activity was found in the growth medium (47). It has been reported that soybean protoplasts have relatively low DNase activity (49). Experiments on exogenous DNA uptake by protoplasts have been few. Ohyama, Gamborg & Miller (50) studied the uptake of *E. Coli* ^{14}C-DNA by *Ammi visnaga* protoplasts. They found that a considerable amount (0.6–2.8%) of exogenous *E. coli* DNA was taken up by the protoplasts. Of this amount, about 20% was acid precipitable. These experiments unfortunately relied exclusively on acid precipitability as the criterion for uptake of macromolecular DNA. It is well known that even relatively small oligonucleotides can be precipitated by trichloroacetic acid and thus we have no way of knowing what percentage of these molecules were sufficiently large to be biologically significant. Also still unresolved remains the problem of adsorption versus real uptake of the exogenous DNA. Although DNase treatments after uptake incubation were used in these experiments, the question could have been answered more definitively, at least with respect to uptake in the nucleus, by isolation of the protoplast nuclei in the presence of excess calf thymus DNA (21).

Another study on the uptake of DNA by protoplasts (51, 52) employed double labeled *Petunia hybrida* DNA and *P. hybrida* mesophyll protoplasts as the host cell.

In these experiments nuclei were isolated (but not in the presence of excess heterologous DNA) and shown to contain about 0.04% of the supplied radioactivity, while the cytoplasmic DNA contained about 0.007% of the supplied radioactivity. The isolated DNA was shown to have approximately the same $^3H/^{14}C$ ratio as the donor DNA. Endogenous DNA synthesis presumably did not occur although it was not rigorously excluded.

A thorough study on the factors affecting the uptake of radioactive single-stranded bacteriophage DNA by tobacco mesophyll protoplasts was recently conducted (53). The amount of DNA taken up was based on radioactivity associated with the protoplasts after extensive DNase treatments. Endogenous DNA synthesis was excluded by inhibitor studies. Uptake of DNA was greatly enhanced by the presence of poly-L-ornithine and Zn^{2+} in the medium. Up to 30% of the input DNA was taken up within 1 hr, and 70% of this was recovered in the cytoplasm fraction of protoplast homogenate. The DNA recovered from the cytoplasm was partially degraded, but about 30% retained the size of intact DNA. The problem of DNA adsorption to the cell membrane or uptake in the nucleus was not dealt with.

These experiments on plant protoplasts strongly suggest, although they do not rigorously prove, exogenous DNA uptake. Unfortunately, there are no published reports to date (February 1977) on the long-term fate of exogenous DNA in plant protoplasts. Such experiments are needed to clearly establish the fate of any exogenous DNA that might penetrate plant protoplasts and to assess the potential of this technique for gene modification in plants.

It is appropriate to mention here that there is rigorous experimental evidence for the uptake of another self-replicating nucleic acid molecule by plant protoplasts. We are referring here to the infection of host plant protoplasts by purified plant viral RNA molecules (53–59). These experiments clearly illustrate that plant protoplasts can take up nucleic acid macromolecules and that these molecules are capable of replication within the plant protoplasts. It must be remembered, of course, that these RNA molecules have evolved to replicate in the host plant cells. Nevertheless, these types of experiments provide some hope that it may be possible to introduce nucleic acid molecules in plant protoplasts in such a manner that they are maintained in their functional form. Unfortunately, these observations have not yet been extended to the DNA plant viruses.

The use of pollen as an exogenous DNA vector has been proposed by Hess et al (60). If exogenous DNA or information-carrying phage could be introduced in pollen grains prior to the use of this pollen in fertilization, it seems possible that the exogenous genetic information could be incorporated in the zygote during fertilization. Some preliminary work along these lines has been accomplished by Hess (61), but it seems too early to critically evaluate the potential of this approach.

In conclusion, it appears that the uptake of exogenous DNA by protoplasts offers perhaps the best chance for success in introducing exogenous DNA in plant cells. The long-term fate, intracellular compartmentalization, and function of DNA taken up by plant protoplasts, remains to be investigated. There is no evidence for the exogenous DNA integration or replication in plant cells.

UPTAKE OF PHAGE PARTICLES BY PLANT CELLS AND PROTOPLASTS

The preceding discussion has dealt primarily with physical methods of detecting exogenous DNA uptake, integration, and replication. Such techniques, no matter how sensitive, cannot detect the transfer of one or few genes. For this purpose the biological expression of a gene unique to the exogenous genetic material and not found in the host cell can provide a most potent and sensitive detection method, particularly if coupled with a rigorous selection procedure. The next section of this review deals with experiments where biological expression of a unique exogenous DNA is used as a tool to detect gene transfer.

In most of the experiments to be discussed in this section, transducing phages were used as the donor material and plant cells in culture as the recipient. The use of phage clearly has some advantages over the use of total bacterial DNA, including increased concentration of the relevant genes, protein coat protection of the DNA, and the susceptibility of plant cells to infection by plant viruses.

Doy and co-workers (62–64) have reported on experiments where haploid cell lines of *Lycopersicon esculentum* and *Arabidopsis thaliana* were exposed to $\phi 80$ or λ phage carrying *gal*, *lac*, or *sup*F genes. Cells inoculated with $\lambda \text{p}gal^+$ were able to survive and grow slowly on 2% galactose medium while controls inoculated with $\lambda \text{p}gal^-$, $\phi 80$, $\phi 80\text{p}lac^+$, or no phage died. Similar results were observed with callus cultures inoculated with $\phi 80\text{p}lac^+$ and grown in lactose medium. Some of the callus cultures survived for three or four subcultures on lactose medium. It should be noted, however, that a few control calluses also survived and grew slowly on lactose medium. The expression of gene *z*, coding for β-galactosidase, was confirmed in the $\phi 80\text{p}lac^+$-inoculated cultures by β-galactosidase activity and by the use of antiserum specific for *E. coli* β-galactosidase. Doy, Gresshoff & Rolfe (63) showed that the β-galactosidase activity in callus cultures inoculated with a mixture of $\phi 80\text{p}lac^+/\lambda gal^+$ (PFU $10^9/10^9$) started 40 days after inoculation, reached a very high level by the fiftieth day, but declined to nearly control levels by the seventieth day after inoculation.

Very interesting results were obtained by inoculating callus cultures growing on optimal glucose medium with $\phi 80 sup\text{F}^+$ carrying an amber suppressor. Soon after inoculation the treated cells stopped growing, indicating that the $sup\text{F}^+$ amber suppressor was somehow not compatible with plant cell growth.

Doy and co-workers coined the term *transgenosis* to define the process of transfer and all processes, such as transcription and translation, needed for the phenotypic change of the plant cells by the transducing phage. Integration of foreign genes into the plant genome is not implied and neither is the continuous stability of the transferred information. More recently, Doy (65) has suggested that the term *transgenosis* may prove useful as a general description of asexual transfer, expression, and inheritance of genetic information between donor and recipient organisms widely separated by evolution.

Carlson (66) reported the expression of two T3-specific enzymes in barley protoplasts inoculated with T3 phage. Carlson observed the appearance of both T3 RNA polymerase and S-adenosylmethionine–cleaving enzyme activities within a short time, i.e. 12 hr, after inoculation. In the case of T3 RNA polymerase, the peak activity was reached 24 hr after inoculation and declined to background levels by 96 hr. The S-adenosylmethionine–cleaving enzyme activity reached a broad maximum at about 72 hr and then started to decline slowly. At the end of the experiment at 120 hr, considerable enzyme activity still remained.

Johnson, Grierson & Smith (67) reported that inoculation of cultured sycamore cells with λplac^+5 bacteriophage resulted in the ability of the cells to grow on lactose. In subsequent publications describing more detailed experiments (68, 69), the observed growth of λplac^+5-inoculated cells was demonstrated to be due primarily to an increase in cell volume. In experiments where cell number was used as a measure of growth, it was shown that control cells do not divide while those inoculated with λplac^+5 undergo a few rounds of cell division. This response is variable, but generally occurs between four and eight weeks after placing the cells on lactose. Cell division is not sustained, however, and the cell count may actually decline in subsequent weeks. Subculturing in fresh lactose medium did not stimulate division of λplac^+5–treated cells.

Further experiments demonstrated that there was a transient increase in β-galactosidase activity in cells transferred to lactose. The maximum enzyme activity was reached about eight days after the transfer to lactose medium and the greatest activity occurred in λplac^+5–treated cells. The galactosidase activity in extracts from λplac^+5–treated cells did not show specific inhibition by β-phenylthiogalactoside under conditions where 50% of the bacterial enzyme was inhibited. Furthermore, no precipitin reaction was observed when extracts from λplac^+5–treated cells were reacted with bacterial β-galactosidase antisera in the Ouchterlony plate immunodiffusion or immunoelectrophoresis procedure.

Based on the above discussion, the critical question still remains whether there is any definitive evidence that foreign DNA, heterologous or homologous, can be expressed within plant cells. The available evidence is clearly still very slight. Carlson's work (66) was described only in the barest outline and there has been no follow-up published. The experiments of Grierson and co-workers (67–69) do not provide conclusive evidence that the *E. coli* β-galactosidase is synthesized in the sycamore cells. The evidence of Doy and co-workers (62, 63) is based on the use of antibody heat-protection assay, but they do not provide any evidence for a positive precipitin reaction between the antibody and the extract from treated plant cells. More recently, Doy (65) has concluded that the long-term phenotype of growth on lactose is not evidence of inheritance and expression of *E. coli* gene z. The β-galactosidase levels in these long-term cultures have remained low and the enzyme is not characteristically *E. coli* in the immunological test. It was suggested that the plant cells in long-term culture adapted to growth on lactose by mutational or epigenetic changes. Such lines could be selected from nontreated plant cell cultures.

In summary, the available evidence suggests that bacterial genes can be transferred by transducing phage to plant cells. There they appear to be briefly expressed. There is certainly no evidence for long-term survival of such foreign genes in plant cells. The observation that foreign genes can be expressed in plant cells, even if only briefly, is in itself very significant, but it would be desirable to obtain more rigorous evidence for this phenomenon and to elucidate factors for prolonged expression and its eventual stabilization as a part of the host cell genome.

UPTAKE OF ORGANELLES BY PLANT PROTOPLASTS

The recent advances in protoplast technology have enabled transplantation experiments to be performed with plant organelles. Carlson (66, 70) reported on the transplantation of normal green chloroplasts to protoplasts isolated from leaf mesophyll cells of a maternally inherited variegating albino mutant of *Nicotiana tabacum* described by Burk & Grosso (71). Plants were regenerated from the albino protoplasts containing green chloroplasts. Whole green plants were presumably recovered.

This work has been criticized by Potrykus (72, 73) who studied the uptake of green chloroplasts by albino petunia protoplasts. He found that uptake of green chloroplasts by albino protoplasts led to cells with mixed plastids containing from 1–20 green chloroplasts and 50–100 white plastids. Although plants were not regenerated in these experiments, Potrykus concluded that such cells should give rise to variegated plants and not completely green plants. He further suggested that the results obtained by Carlson may be due to the existence of periclinal chimeras of the type green/white/white in phenotypically white leaves.

Recently, Kung et al (74) investigated the genetic makeup of a hybrid plant derived after incorporation of *N. suaveolens* chloroplasts into protoplasts from the white tissue of a maternally inherited variegating albino mutant of *N. tabacum* cv Xanthi nc (71). In this case, the presence of chloroplast and nuclear genes can be determined by analyzing the polypeptide composition of Fraction 1 protein of these plants. Fraction 1 protein (ribulose diphosphate carboxylase) consists of two different subunits referred to as the large and small subunits. The large subunit is coded by chloroplast DNA (75) and the small subunit is coded by nuclear DNA (76). In the uptake experiment reported by Kung et al (74), green calluses were observed to arise at an approximate rate of 2.0×10^{-4} while the reversion rate was determined to be about 10^{-8}. Unfortunately, attempts to regenerate plants from the green calluses were unsuccessful except for a single individual. This individual was a variegated plant of abnormal morphology. The plant was also sterile and consisted of cells with abnormal chromosome numbers ranging from 61–66. The somatic chromosome numbers of *N. suaveolens* and *N. tabacum* are 32 and 48 respectively.

Analysis of the polypeptide composition of the Fraction 1 protein from the regenerated variegated plant showed that it contained large subunits of both *N. suaveolens* and *N. tabacum*. This clearly indicates that chloroplasts from *N. suaveolens* were taken up by the host protoplasts. Examination of the small subunits also showed the presence of both *N. suaveolens* and *N. tabacum* Fraction 1 protein genes. This indicates that a nucleus as well as chloroplasts must have been taken up by the *N. tabacum* protoplast that gave rise to this variegated plant.

Bonnett & Eriksson (77) studied the transfer of algal (*Vaucheria dichotoma*) chloroplasts to carrot (*Daucus carota*) protoplasts. Chloroplast incorporation efficiency of 16% was obtained by the use of polyethylene glycol. The intracellular localization of the algal chloroplasts in carrot protoplasts was demonstrated (78). Conditions favoring protoplast aggregation promoted chloroplast uptake. The chloroplasts inside the protoplasts were not surrounded by a limiting membrane of carrot origin. Some possessed intact outer chloroplast membranes (78). Chloroplasts from *Petunia hybrida* behave somewhat differently in *Parthenocissus tricuspidata* protoplasts (79). The envelope of intact chloroplasts readily fused with the protoplast plasmalemma and established continuity between chloroplast stroma and protoplast cytoplasm. Broken chloroplasts became localized in membrane-bounded vesicles in the cytoplasm of the higher plant protoplasts (79).

Potrykus & Hoffmann (80) studied the uptake of isolated *P. hybrida* nuclei by protoplasts of *P. hybrida*, *Nicotiana glauca*, and *Zea mays*. About 0.5% of the protoplasts contained 1–5 nuclei detected by fluorescence staining with ethidium bromide. The location of the nuclei inside the protoplasts was demonstrated by following the position of the nucleus in the microscope while rolling the protoplast under the coverslide. In some cases the protoplasts burst under slight pressure and the incorporated nucleus was extruded together with the original organelles. The experiments, however, were not carried out under sterile conditions and no plants were regenerated.

Another approach to the transfer of genetic functions from microorganisms to plants has been through the uptake of intact microorganisms by plant protoplasts. Davey & Cocking (81) reported the uptake of *Rhizobium leguminosarum* by pea protoplasts. More recently, Davey & Power (82) have demonstrated that protoplasts isolated from *Parthenocissus tricuspidata* cells can take up yeast cells, yeast protoplasts, and blue-green algal cells. Electron microscopy showed that the microorganisms became localized in membrane-bounded vesicles in the cytoplasm of the higher plant protoplasts. Burgoon & Bottino (83) demonstrated the uptake of the blue-green algae *Gloeocapsa* into maize and tobacco protoplasts. To date, long-term maintenance of microbial cells in eucaryotic cells has been demonstrated only with *Azotobacter vinelandii* in the mycorrhizal fungus *Rhizopogon* sp. (84).

The experiments discussed in this section provide some hope for the genetic manipulation of plant genomes by the protoplast and organelle transfer techniques. There seem to be two major technical obstacles to immediate or near-future application of this technique to practical problems. First, the successful regeneration of fertile plants from protoplasts has been achieved in only seven genera: *Nicotiana* (85), *Daucus* (86), *Petunia* (87, 88), *Asparagus* (89), *Brassica* (90), *Datura* (91), and *Ranunculus* (92). These successes suggest that it is only a matter of refinement of techniques before other genera join the list; nevertheless, to date this has not been accomplished in the major crop plants. Second, it still remains to be demonstrated that chloroplasts or nuclei can be successfully transferred and plants recovered between species or genera that cannot be crossed sexually (see also the section on cell fusion). The chloroplast transfer is much more important than nuclear transfer, since under normal sexual crossing in higher plants only the female parent contributes chloroplast genes. Thus, there is no potential for chloroplast gene recombination in sexual crosses, and chloroplast transfer could open new horizons in plant breeding that are presently not available. Clearly though, substantial new research is needed to determine the potential and extent of chloroplast gene recombination.

HERITABLE CHANGE OF PLANT PHENOTYPE

Transformation of plants with purified DNA has probably been attempted in numerous laboratories, but negative results are not commonly published. Coe & Sarkar (93) reported their negative attempts with maize. Ledoux and co-workers (23, 29, 94–96) have published on extensive transformation experiments with thiamine auxotrophs in *Arabidopsis* using heterologous bacterial DNA. Mutations known in *Arabidopsis* at three loci are concerned with the synthesis of thiamine (97). Homozygous recessive seeds are produced on mutant seedlings supplemented with thiamine either in sterile culture or in soil. The seeds are sterilized and allowed to germinate in a pool of bacterial DNA (usually ~ 1 mg/ml) for 3–4 days. The seedlings are then washed and transferred to sterile medium without thiamine. Control seedlings show chlorophyll deficiency in the cotyledons and new leaves and normally die in about three weeks. It is claimed that the corrected plants grow slowly with a more or less normal green color and that about one third of these produce fertile siliques. Using these techniques, Ledoux and co-workers (29, 94–96) have reported correcting several mutants at the three loci, using *E. coli*, *Agrobacterium tumefaciens*, *Bacillus subtilis*, *Micrococcus lysodeikticus*, *Streptomyces coelicolor*, and calf thymus DNAs. Negative results have been obtained with phage T7 and 2C DNA. A more important negative control was obtained with *thi*A$^-$ (thiazole-requiring) *E. coli* strain P678 DNA.

The corrected plants have been studied in subsequent generations and in crosses of several types. Upon selfing, the corrected plants produced only homozygous corrected progeny. Upon crossing with wild-type plants, the corrected phenotype appeared to be recessive in that all F_1 progeny were normal green. In the F_2 generation the seedlings segregated in unusual ratios for normal, leaky, and lethal phenotypes. Upon crossing with the original mutant, the corrected phenotype behaved as if it were dominant in that all the F_1 seedlings were phenotypically normal green. Again the F_2 seedlings segregated for normal, leaky, and lethal phenotypes. Reciprocal crosses indicated that the "correction" was nuclear rather than cytoplasmic since it could be transferred either through the male or female gametophyte. The segregation ratios have not yielded any definitive information about the nature of the corrected mutants (94, 95).

The experiments described here on the transformation of *Arabidopsis* thiamine auxotrophs appear to be very thorough and have produced very interesting data. The possibility that some nonspecific effect may be involved is suggested by the high frequency (12%) of corrected py^- plants obtained with calf thymus DNA (29). It is not clear whether calf thymus DNA has the DNA sequences to correct the py^- locus, but even if it does the frequency of correction seems very high compared to DNAs that would be expected to have a much higher frequency of the relevant genes (i.e. bacterial DNA). The other problem is that to our knowledge these results have not been reproduced, even though several laboratories, as well as one of us (AK), have tried.

Recently, Rédei et al (98) have carried out classical genetic analysis of several of the "corrected" mutants from Ledoux's laboratory. They reported that all the evidence (allelism, linkage, other chromosome markers, nutritional requirement, temperature sensitivity, dormancy) indicates that the variegated "corrected" mutants are very similar to, most likely identical with, Langridge's mutant 1018-6. Thus they were unable to find any evidence for DNA-mediated correction and concluded that the simplest interpretation of the facts is that the "corrected mutants" are mechanical contaminants either of Langridge's mutant 1018-6 or other mutants or the wild type.

Hess (99–103) has reported on transformation experiments with white-flowered *Petunia hybrida* using DNA from a red-flowering line as the donor. Very high frequencies, up to 27%, of transformed plants were reported. Surprisingly, treatment of the white-flowered line 34d10 with DNA isolated from its own leaves resulted in 5% and 9% of the plants producing flowers with slight anthocyanin coloration. Treatment of the DNA from the red-flowered line with DNase reduced, but did not destroy, the transforming activity. The occurrence of the faint red coloration in the white homozygous line as a result of treatment with its own DNA was ascribed to the effect of environmental factors. Genetic analysis of some of the presumably transformed plants indicated that they were mostly homozygous for the anthocyanin-inducing alleles (101). In subsequent experiments (102), five homozygous and three heterozygous transformed plants were obtained from about 13,000 plants treated with DNA from the red-flowering line. These experiments have been criticized by Bianchi & Walet-Foederer (104) on the basis that environmental changes can account for occurrence of pigmented flowers and that somatic mutations occur in the embryonic cell layers with a frequency in the same range as that claimed to occur in DNA-treated plants (102). In addition, the presence of three embryonic cell layers in the eight-day-old seedlings treated with DNA by Hess precludes the occurrence of homozygous and uniformly colored transformants; that is, mosaic flowers should occur if transformation of one or a few cells takes place in one of the embryonic cell layers.

Transformation in *Petunia* does not need to be done with intact plants in the future. The system of haploid protoplast isolation and regeneration to intact plants is now available (87, 88). It is hoped that future transformation work with *Petunia* will be clearer and easier to interpret.

Transformation of the *waxy* endosperm character in barley to the wild-type allele by treatment of developing barley embryos with DNA from wild-type plants has been reported (105). Results obtained in these experiments indicate that only "slightly deproteinized highly polymeric DNA from barley endosperm material" was successful in bringing about this transformation. Actually only one treated plant, No. 506/19, produced wild-type pollen grains in a high frequency. No controls with similarly prepared DNA from *wx* plants were reported. In a follow-up paper, Soyfer et al (106) analyzed the progeny of some of the treated plants. Particular attention was devoted to plant No. 506/19, which showed over 99.5% wild-type pollen in the first generation. Of 28 plants produced by plant No. 506/19, 23 showed all or nearly all wild-type pollen and 5 showed all or nearly all *waxy* pollen. The plants producing wild-type pollen were two-rowed type (as the DNA donor plants) and plants producing *waxy* pollen grains were six-rowed type (as the recipient plant). The correlation between *waxy* and six-row types, surprising, since the genes are located on different chromosomes, was not explained.

Transformation of genes for fruiting habit (*Fa* vs *fa*) and fruiting direction (*Up* vs *up*) was reported in red peppers, *Capsicum annuum* L. (107). When seeds of Tochigisantaka (T) cultivar (*upup, fafa, YY, C_1C_1*) were treated with DNA from Kiiro (K) cultivar (*UpUp, FaFa, yy, c_1c_1*) a *Fa*- variant was obtained in the selfed generation after DNA treatment. This variant maintained the *upup* (erect fruiting habit) trait, thus excluding seed contamination as a possible source of error. Another variant was obtained in which both (*up* and *fa*) genes were changed to that of the DNA donor (*Up* and *Fa*). This change, however, was obtained in grafting experiments, and the authors have previously reported (108) that such changes can be brought about by ordinary grafting of the T cultivar on the K cultivar stock, thus making the interpretation still more difficult. The authors concluded that the hereditary changes obtained by grafting could be due to DNA transfer from the stock to the scion. No explanation was offered as to why these dominant traits did not express themselves until the second generation after DNA treatment.

In summarizing this section we conclude that the present science of transformation in plants is still vague and poorly defined. More and better-controlled experiments are needed to make the results convincing and reproducible in other laboratories, utilizing gene markers that can be specifically identified at the protein or RNA level as well as other markers to permit identification of seed mixtures and outcrossing. An important control is the utilization of identical DNA preparations, but deleted in the genetic material for which the transformation is attempted. Such controls need to be performed under conditions identical with the transformation experiments and in similar numbers in order to demonstrate that the observed changes are due to specific genetic information and not a nonspecific DNA phenomenon. Nonspecific mutation and physiological effects of DNA are known (109).

SOMATIC CELL FUSION

The previously discussed systems all involved the transfer of information carrying molecules or organelles to plant cells. Genetic engineering in plants could also be brought about by the fusion of two somatic cells carrying different genetic information. Such a process would be closely analogous to the sexual cell fusion process, but would hopefully circumvent the sexual incompatibility barriers. It is not feasible to completely review here the extensive literature on protoplast isolation, culture, and fusion. The readers are referred to previous reviews (59, 66, 100–115) and recent symposia proceedings (116). We concern ourselves primarily with the application of protoplast techniques to the prospects for generation of new plant types.

It is now relatively easy to obtain large quantities of protoplasts from organs of many plant species and from cultured cells (112, 113, 117). The limited success of somatic cell hybridization for generating new plants is primarily due to the lack of suitable selection mechanisms (genetic markers) for the isolation of heterokaryons or hybrids from the protoplast fusion mixture and to our inability to regenerate plants from protoplasts of many plant species. Protoplasts from an increasing number of plant species, including cereals, are being cultured to assess their morphogenetic potential. Plating efficiency of protoplasts isolated from most species except *Nicotiana tabacum,* however, remains low (110, 112).

Initially, protoplasts were fused in the presence of sodium nitrate (110, 118). Improved efficiency of tobacco protoplast fusion at high calcium concentration (0.05 M) and at high pH (10.5) was then reported (119). Kao & Michayluk (120) and Eriksson et al (112) found polyethylene glycol (PEG) to be an efficient inducer of protoplast fusion. This agglutinating agent has proven so effective for fusion work that it has been used universally and successfully in many systems, for example, in interkingdom cell fusion (121–123), in the uptake of organelles or microorganisms by plant protoplasts (see section on uptake of organelles by plant protoplasts), and in the fusion of bacterial protoplasts (124). The greater efficiency of protoplast fusion significantly increases the possibility of nuclear fusion.

The regeneration of hybrid plants after protoplast fusion has been achieved with *Nicotiana glauca* and *N. langsdorffii* (118, 125) tobacco mutants (126, 127) and *Petunia* species (128). In each case the parent plants were sexually compatible and the somatic hybrids obtained could be compared with those produced by sexual crosses. The promise of obtaining a hybrid plant of sexually incompatible species by protoplast fusion has so far remained elusive.

Carlson, Smith & Dearing (118) succeeded in fusing *N. glauca* and *N. langsdorffii* protoplasts to obtain the interspecific hybrid. Selection of fusion product was based on differential growth of the hybrid and parent cells under the cultural conditions used. Progeny from this fusion product were recently analyzed for the presence of nuclear and cytoplasmic parental genes by isoelectric focusing of the Fraction 1 protein components (74). The small subunit polypeptides (indicative of nuclear genes) of both parental species were found. The *N. glauca* large subunit polypeptides (indicative of chloroplast genes) were also found, but there was no indication of the presence of the large subunit polypeptides of *N. langsdorffii*. Either the *N. langsdorffii* chloroplast genes coding for the large subunit of Fraction 1 protein were not present or they were not expressed. Unfortunately, the progeny of only one parasexual hybrid was examined; thus it was not possible to conclude whether this is a general phenomenon. Further studies are needed to evaluate cytoplasmic characteristics of somatic hybrids.

Smith, Kao & Combatti (125) have recently repeated the *N. glauca-N. langsdorffii* somatic hybridization experiment and carried out detailed analysis of 23 regenerated hybrid plants. They found that they contained a high and variable chromosome number. Instead of the earlier reported amphiploid number of 42 (118), the chromosome number in the hybrids ranged from 56 to 64. They attributed this upward shift as arising primarily from triple fusion, followed by chromosome losses during callus growth and subsequent plant development. The occurrence of aberrant chromosome number (129) and karyotype instability, particularly in plants regenerated from callus cultures of genetically tumor-prone *Nicotiana* species, is known (130).

Melchers & Labib (126) and Melchers, Keller & Labib (127) produced somatic hybrids by fusion of protoplasts from two different chlorophyll-deficient, light-sensitive haploid mutants of *N. tabacum*. The selection of the hybrid tissue was based on genetic complementation so that the fusion products produced normal green calli that were easily distinguishable from the mutant calli. The somatic hybrids were also obtained by normal sexual crosses for comparison and were found to be identical with those produced by protoplast fusion. Since chlorophyll-deficient mutants are widespread among higher plants, a selection system based on their use might prove of significant value in somatic hybridization studies.

Power et al (128) recently produced a hybrid plant from protoplast fusion of *Petunia hybrida* with *Petunia parodii*. They employed a selection system based on the naturally existing differential sensitivity of heterospecific protoplasts to specific antimetabolites.

Products of intergeneric and interfamilial protoplasts fusion have been obtained and the heterokaryocytes observed to undergo a few cell divisions (131–133). Hybrid calli formation has not been usually obtained. The failure to obtain hybrid calli has been attributed primarily to the absence of a selection system. Other reasons, not yet understood in molecular terms, may reside in the incompatibility of heterospecific protoplasts upon fusion. This may be a phenotypic as well as a genotypic phenomenon.

Power et al (134) unsuccessfully tested the recovery of a somatic hybrid from the fused protoplasts of *Petunia hybrida* and *Parthenocissus tricuspidata*. They attributed the observed elimination of the petunia complement in *Petunia-Parthenocissus* fusion product to the asynchrony of mitotic cycle times and phases. Two recent encouraging reports (135, 136) describe the production of interfamilial hybrid callus from soybean-*Nicotiana* fusion. These experiments employed mechanical isolation of the fusion product. The hybrids were initially unstable, losing the greater proportion of the *Nicotiana* chromosomes. The remaining segments of *Nicotiana* chromosomes appeared to become synchronous with the soybean cell cycle. This experiment, if reproducible, would demonstrate at least the principle that sexual incompatibility of distantly related species can be bypassed by the somatic hybridization technique. The morphogenetic potential of the reported hybrid, however, remains to be determined.

The fusion of somatic cells for the purpose of genetic modification or generation of new crop plants is an appealing technique, but it still remains to be developed and tested. It is clear that fusion of cells from species that can be sexually crossed is possible. This will probably soon be demonstrated for other genera in addition to *Nicotiana* and *Petunia*. This technique is of great interest for developmental and gene regulation studies as well as for organelle genome recombination. The application of the techniques to intergeneric fusion will require development of additional technology and an understanding of the basic plant development and genetic processes.

CONCLUDING REMARKS

It is interesting to speculate about the future of genetic engineering in plants. The early work claiming exogenous DNA uptake, integration, and replication in plants is most probably wrong. Nevertheless, it provided a stimulus for interest and research in this field. Some of the present-day excessive speculations about accomplishments in the near future probably also will not be borne out. Will they serve as stimulants for future accomplishments or will they serve to deafen other scientists and laymen to the potential of these genetic engineering techniques? Clearly we do not foresee the application of any of the discussed techniques to practical plant improvements in the near future (0–5 years). Some approaches may yield valuable input in the intermediate future (5–15 years). For example, recently it has been discovered that *Agrobacterium*-mediated tumor induction in plants proceeds via the transfer of a small piece of *Agrobacterium* plasmid to the plant genome (137, 138). This naturally occurring phenomenon may well be exploited to transfer genes from other organisms to plants. Similarly, it seems feasible that the natural plant DNA viruses may well be exploited as carriers of genetic information from unrelated organisms to the viral hosts. The engineering of a nonvirulent carrier virus seems to be possible; the stabilization and expression of the viral plus donor DNA hybrids in the plant cell may be more difficult (139).

Another area where we anticipate significant contributions in the intermediate future is in the alteration of the plant cytoplasmic genomes. These small, in relation to the nuclear genome, pieces of genetic information have been cut off from the benefits of modulated gene recombination in plant breeding by the evolution of the specialized male gametes (sperm) in plants. Nevertheless, they control numerous functions important to plant productivity. The accomplishments in protoplast fusion and uptake of particles by protoplasts will enable us to accomplish recombination and therefore presumably improve the function of the cytoplasmic genomes. Much remains to be learned, however, about the genetics and biochemistry of the cytoplasmic genomes, and the technology of crop plant protoplasts needs to be developed before we can apply these techniques to solving practical plant improvement problems.

Our review is critical in nature because from experience we know of the many pitfalls involved in this research. It is hoped, however, that this will not discourage serious investigators from pursuing knowledge in genetic engineering of plants, but rather help to switch the ratio of "noise" to "signal" in a meaningful way. In a science that up to the present has been overburdened with an excessive "noise" level, this is badly needed.

Literature Cited

1. Sigurbjörnsson, B. 1971. Induced mutations in plants. *Sci. Am.* 224:86–95
2. Gustafsson, Å. 1975. Mutations in plant breeding—A glance back and a look forward. In *Biomedical, Chemical, and Physical Perspectives, Proc. 5th Int. Cong. Radiat. Res.*, ed. O. F. Nygaard, H. I. Adler, W. K. Sinclair, pp. 81–95. New York: Academic. 1381 pp.
3. Nilan, R. A., Kleinhofs, A., Konzak, C. F. 1977. The role of induced mutation in supplementing natural variability. In *Proc. Conf. Genet. Basis Epidemics Agric.*, ed. P. R. Day, pp. 367–84. New York: NY Acad. Sci. 400 pp.
4. Ledoux, L. 1965. Uptake of DNA by living cells. *Prog. Nucleic Acid Res. Mol. Biol.* 4:231–67
5. Hess, D. 1972. Transformationen an höheren Organismen. *Naturwissenschaften* 59:348–55
6. Heyn, R. F., Rörsch, A., Schilperoort, R. A. 1974. Prospects in genetic engineering of plants. *Q. Rev. Biophys.* 7:35–73
7. Johnson, C. B., Grierson, D. 1974. The uptake and expression of DNA by plants. *Curr. Adv. Plant Sci.* 9:1–12
8. Bottino, P. J. 1975. The potential of genetic manipulation in plant cell cultures for plant breeding. *Radiat. Bot.* 15:1–16
9. Ledoux, L. G. H., ed. 1971. *Informative Molecules in Biological Systems.* Amsterdam: North-Holland. 466 pp.

10. Ledoux, L., ed. 1972. *Uptake of Informative Molecules by Living Cells.* Amsterdam: North-Holland. 416 pp.
11. Ledoux, L., ed. 1975. *Genetic Manipulations with Plant Material.* New York: Plenum. 601 pp.
12. Markham, R., Davies, D. R., Hopwood, D. A., Horne, R. W., eds. 1975. *Modification of the Information Content of Plant Cells.* Amsterdam: North-Holland. 350 pp.
13. Rubenstein, I., Phillips, R. L., Green, E., Desnick, R. J., eds. 1977. *Molecular Genetic Modification of Eucaryotes.* New York: Academic. In press
14. Stroun, M., Anker, P., Charles, P., Ledoux, L. 1966. Fate of bacterial deoxyribonucleic acid in *Lycopersicon esculentum. Nature* 212:397–98
15. Stroun, M., Anker, P., Ledoux, L. 1967. DNA replication in *Solanum lycopersicum esc.* after absorption of bacterial DNA. *Curr. Mod. Biol.* 1:231–34
16. Anker, P., Stroun, M. 1968. Bacterial nature of radioactive DNA found in tomato plants incubated in the presence of bacterial DNA-^{3}H. *Nature* 219:932–33
17. Ledoux, L., Huart, R. 1968. Integration and replication of DNA of *M. lysodeikticus* in DNA of germinating barley. *Nature* 218:1256–59
18. Ledoux, L., Huart, R. 1969. Fate of exogenous bacterial deoxyribonucleic acids in barley seedlings. *J. Mol. Biol.* 43:243–62
19. Anker, P., Stroun, M., Gahan, P., Rossier, A., Greppin, H. 1971. Natural release of bacterial nucleic acids into plant cells and crown gall induction. See Ref. 9, pp. 193–200
20. Bendich, A. J., Filner, P. 1971. Uptake of exogenous DNA by pea seedlings and tobacco cells. *Mutat. Res.* 13:199–214
21. Hotta, Y., Stern, H. 1971. Uptake and distribution of heterologous DNA in living cells. See Ref. 9, pp. 176–86
22. Ledoux, L., Huart, R., Jacobs, M. 1971. Fate of exogenous DNA in *Arabidopsis thaliana.* Translocation and integration. *Eur. J. Biochem.* 23:96–108
23. Ledoux, L., Huart, R., Jacobs, M. 1971. Fate of exogenous DNA in *Arabidopsis thaliana.* II. Evidence for replication and preliminary results at the biological level. See Ref. 9, pp. 159–75
24. Stroun, M., Anker, P., Gahan, P., Sheikh, K. 1971. Bacterial infection due to natural release of nucleic acids from bacteria into plant cells. See Ref. 9, pp. 187–92
25. Stroun, M., Anker, P., Gahan, P., Rossier, A., Greppin, H. 1971. *Agrobacterium tumefaciens* ribonucleic acid synthesis in tomato cells and crown gall induction. *J. Bacteriol.* 106:634–39
26. Stroun, M., Anker, P., Cattaneo, A., Rossier, A. 1971. Effect of the extent of DNA transcription of plant cells and bacteria on the transcription in plant cells of DNA released from bacteria. *FEBS Lett.* 13:161–64
27. Stroun, M., Anker, P. 1971. Bacterial nucleic acid synthesis in plants following bacterial contact. *Mol. Gen. Genet.* 113:92–98
28. Ledoux, L., Huart, R. 1972. Fate of exogenous DNA in plants. See Ref. 10, pp. 254–76
29. Ledoux, L., Huart, R., Jacobs, M. 1972. Fate and biological effects of exogenous DNA in *Arabidopsis thaliana.* In *Way Ahead in Plant Breeding,* ed. F. G. H. Lupton, G. Jenkins, R. Johnson, pp. 165–84. Dorking: Adlard
30. Rebel, W., Hemleben, V., Seyffert, W. 1973. Fate of T_4 phage DNA in seedlings of *Matthiola incana. Z. Naturforsch.* 28:473–74
31. Hemleben, V., Ermisch, N., Kimmich, D., Leber, B., Peter, G. 1975. Studies on the fate of homologous DNA applied to seedlings of *Matthiola incana. Eur. J. Biochem.* 56:403–11
32. Kleinhofs, A. 1975. DNA-hybridization studies of the fate of bacterial DNA in plants. See Ref. 11, pp. 461–77
33. Kleinhofs, A., Eden, F. C., Chilton, M. D., Bendich, A. J. 1975. On the question of the integration of exogenous bacterial DNA into plant DNA. *Proc. Natl. Acad. Sci. USA* 72:2748–52
34. Ledoux, L. 1975. Fate of exogenous DNA in plants. See Ref. 11, pp. 479–98
35. Gradmann-Rebel, W., Hemleben, V. 1976. Incorporation of T4 phage DNA into a specific DNA fraction from the higher plant *Matthiola incana. Z. Naturforsch.* 31:558–64
36. Kleinhofs, A., Ulrich, T. H. 1976. Uptake and fate of exogenous DNA in plants. In *Barley Genetics III, Proc. 3rd Int. Barley Genet. Symp.,* ed. H. Gaul, pp. 11–19. Munich: Verlag Karl Thiemig. 849 pp.
37. Kleinhofs, A. 1977. Physical and biological studies of DNA uptake by plants. See Ref. 13. In press

38. Heyn, R. F., Schilperoort, R. A. 1973. The use of protoplasts to follow the fate of *Agrobacterium tumefaciens* DNA on incubation with tobacco cells. *Colloq. Int. CNRS* 212:385–95
39. Hanson, R. S., Chilton, M. D. 1975. On the question of integration of *Agrobacterium tumefaciens* deoxyribonucleic acid by tomato plants. *J. Bacteriol.* 124:1220–26
40. Kado, C. I., Lurquin, P. F. 1975. Studies on *Agrobacterium tumefaciens*. IV. Nonreplication of the bacterial DNA in mung bean (*Phaseolus aureus*). *Biochem. Biophys. Res. Commun.* 64:175–83
41. Ledoux, L., Charles, P. 1972. On the use of the preparative CsCl gradients. See Ref. 10, pp. 29–46
42. Lurquin, P., Mergeay, M., Van Der Parren, J. 1972. Mathematical appendix. Banding of depolymerized DNA's in CsCl density gradients studied by computer-aided simulations. See Ref. 10, pp. 45–50
43. Cocking, E. C. 1973. Plant cell modifications: Problems and perspectives. *Colloq. Int. CNRS* 212:327–41
44. Kado, C. I., Lurquin, P. F. 1976. Studies on *Agrobacterium tumefaciens*. V. Fate of exogenously added bacterial DNA in *Nicotiana tabacum*. *Physiol. Plant Pathol.* 8:73–82
45. Farrand, S. K., Eden, F. C., Chilton, M. D. 1975. Attempts to detect *Agrobacterium tumefaciens* and bacteriophage PS8 DNA in crown gall tumors by DNA·DNA-filter hybridization. *Biochim. Biophys. Acta* 390:264–75
46. Lurquin, P. F., Hotta, Y. 1975. Reutilization of bacterial DNA by *Arabidopsis thaliana* cells in tissue culture. *Plant Sci. Lett.* 5:103–12
47. Lurquin, P. F., Behki, R. M. 1975. Uptake of bacterial DNA by *Chlamydomonas reinhardi*. *Mutat. Res.* 29:35–51
48. Swinton, D. C., Hanawalt, P. C. 1972. In vivo specific labeling of *Chlamydomonas* chloroplast DNA. *J. Cell Biol.* 54:592–97
49. Holl, F. B. 1973. Cellular environment and the transfer of genetic information. *Colloq. Int. CNRS* 212:509–16
50. Ohyama, K., Gamborg, O. L., Miller, R. A. 1972. Uptake of exogenous DNA by plant protoplasts. *Can. J. Bot.* 50:2077–80
51. Hoffmann, F., Hess, D. 1973. Die Aufnahme radioaktiv markierter DNS in isolierte protoplasten von *Petunia hybrida*. *Z. Pflanzenphysiol.* 69:81–83
52. Hoffmann, F. 1973. Die Aufnahme doppelt-markierter DNS in isolierte Protoplasten von *Petunia hybrida*. *Z. Pflanzenphysiol.* 69:249–61
53. Suzuki, M., Takebe, I. 1976. Uptake of single-stranded bacteriophage DNA by isolated tobacco protoplasts. *Z. Pflanzenphysiol.* 78:421–33
54. Aoki, S., Takebe, I. 1969. Infection of tobacco mesophyll protoplasts by tobacco mosaic virus ribonucleic acid. *Virology* 39:439–48
55. Bancroft, J. B., Motoyoshi, F., Watts, J. W., Dawson, J. R. O. 1975. Cowpea chlorotic mottle and brome mosaic viruses in tobacco protoplasts. See Ref. 12, pp. 133–60
56. Sarkar, S., Upadhya, M. D., Melchers, G. 1974. A highly efficient method of inoculation of tobacco mesophyll protoplasts with ribonucleic acid of tobacco mosaic virus. *Mol. Gen. Genet.* 135:1–9
57. Takebe, I., Aoki, S., Sakai, F. 1975. Replication and expression of tobacco mosaic virus genome in isolated tobacco leaf protoplasts. See Ref. 12, pp. 101–17
58. Watts, J. W., Cooper, D., King, J. M. 1975. Plant protoplasts in transformation studies: Some practical considerations. See Ref. 12, pp. 119–31
59. Takebe, I. 1975. The use of protoplasts in plant virology. *Ann. Rev. Phytopathol.* 13:105–25
60. Hess, D., Gresshoff, P. M., Fielitz, U., Gleiss, D. 1974. Uptake of protein and bacteriophage into swelling and germinating pollen of *Petunia hybrida*. *Z. Pflanzenphysiol.* 74:371–76
61. Hess, D. 1975. Uptake of DNA and bacteriophage into pollen and genetic manipulation. See Ref. 11, pp. 519–37
62. Doy, C. H., Gresshoff, P. M., Rolfe, B. G. 1973. Biological and molecular evidence for the transgenosis of genes from bacteria to plant cells. *Proc. Natl. Acad. Sci. USA* 70:723–26
63. Doy, C. H., Gresshoff, P. M., Rolfe, B. G. 1973. Time-course of phenotypic expression of *Escherichia coli* gene Z following transgenosis in haploid *Lycopersicon esculentum* cells. *Nature New Biol.* 244:90–91
64. Gresshoff, P. M. 1975. Theoretical and comparative aspects of bacteriophage transfer and expression in eukaryotic cells in culture. See Ref. 11, pp. 539–49
65. Doy, C. H. 1975. The transfer and expression (transgenosis) of foreign genes in plant cells, reality and potential. In *The Eukaryote Chromosome*, ed. W. J. Peacock, R. D. Brock, pp. 447–57. Can-

berra: Aust. Natl. Univ. Press. 500 pp.
66. Carlson, P. S. 1973. The use of protoplasts for genetic research. *Proc. Natl. Acad. Sci. USA* 70:598–602
67. Johnson, C. B., Grierson, D., Smith, H. 1973. Expression of λplac5 DNA in cultured cells of a higher plant. *Nature New Biol.* 244:105–6
68. Grierson, D., McKee, R. A., Attridge, T. H., Smith, H. 1975. Studies on the uptake and expression of foreign genetic material by higher plant cells. See Ref. 12, pp. 91–99
69. Smith, H., McKee, R. A., Attridge, T. H., Grierson, D. 1975. Studies on the use of transducing bacteriophages as vectors for the transfer of foreign genes to higher plants. See Ref. 11, pp. 551–63
70. Carlson, P. S. 1973. Towards a parasexual cycle in higher plants. *Colloq. Int. CNRS* 212:497–505
71. Burk, L. G., Grosso, J. J. 1963. Plasmagenes in variegated tobacco. *J. Hered.* 54:23–25
72. Potrykus, I. 1973. Transplantation of chloroplasts into protoplasts of *Petunia*. *Z. Pflanzenphysiol.* 70:364–66
73. Potrykus, I. 1975. Uptake of cell organelles into isolated protoplasts. See Ref. 12, pp. 169–79
74. Kung, S. D., Gray, J. C., Wildman, S. G., Carlson, P. S. 1975. Polypeptide composition of Fraction 1 protein from parasexual hybrid plants in the genus *Nicotiana*. *Science* 187:353–55
75. Chan, P. H., Wildman, S. G. 1972. Chloroplast DNA codes for the primary structure of the large subunit of Fraction I protein. *Biochim. Biophys. Acta* 277:677–80
76. Kawashima, N., Wildman, S. G. 1972. Studies on Fraction I protein. IV. Mode of inheritance of primary structure in relation to whether chloroplast or nuclear DNA contains the code for a chloroplast protein. *Biochim. Biophys. Acta* 262:42–49
77. Bonnett, H. T., Eriksson, T. 1974. Transfer of algal chloroplasts into protoplasts of higher plants. *Planta* 120:71–79
78. Bonnett, H. T. 1976. On the mechanism of the uptake of *Vaucheria* chloroplasts by carrot protoplasts treated with polyethylene glycol. *Planta* 131:229–33
79. Davey, M. R., Frearson, E. M., Power, J. B. 1976. Polyethylene glycol-induced transplantation of chloroplasts into protoplasts: An ultrastructural assessment. *Plant Sci. Lett.* 7:7–16
80. Potrykus, I., Hoffmann, F. 1973. Transplantation of nuclei into protoplasts of higher plants. *Z. Pflanzenphysiol.* 69:287–89
81. Davey, M. R., Cocking, E. C. 1972. Uptake of bacteria by isolated higher plant protoplasts. *Nature* 239:455–56
82. Davey, M. R., Power, J. B. 1975. Polyethylene glycol-induced uptake of microorganisms into higher plant protoplasts: An ultrastructural study. *Plant Sci. Lett.* 5:269–74
83. Burgoon, A. C., Bottino, P. J. 1976. Uptake of the nitrogen fixing blue-green algae *Gloeocapsa* into protoplasts of tobacco and maize. *J. Hered.* 67:223–26
84. Giles, K. L., Whitehead, H. 1976. Uptake and continued metabolic activity of *Azotobacter* within fungal protoplasts. *Science* 193:1125–26
85. Takebe, I., Labib, G., Melchers, G. 1971. Regeneration of whole plants from isolated mesophyll protoplasts of tobacco. *Naturwissenschaften* 58:318–20
86. Grambow, H. J., Kao, K. N., Miller, R. A., Gamborg, O. L. 1972. Cell division and plant development from protoplasts of carrot cell suspension cultures. *Planta* 103:348–55
87. Durand, J., Potrykus, I., Donn, G. 1973. Plantes issues de protoplastes de Pétunia. *Z. Pflanzenphysiol.* 69:26–34
88. Power, J. B., Frearson, E. M., George, D., Evans, P. K., Berry, S. F., Hayward, C., Cocking, E. C. 1976. The isolation, culture and regeneration of leaf protoplasts in the genus Petunia. *Plant Sci. Lett.* 7:51–55
89. Bui-Dang-Ha, D., Mackenzie, I. A. 1973. The division of protoplasts from *Asparagus officinalis* L. and their growth and differentiation. *Protoplasma* 78:215–21
90. Kartha, K. K., Michayluk, M. R., Kao, K. N., Gamborg, O. L., Constabel, F. 1974. Callus formation and plant regeneration from mesophyll protoplasts of rape plants (*Brassica napus* L. cv. Zephyr). *Plant Sci. Lett.* 3:265–71
91. Schieder, O. 1975. Regeneration von haploiden und diploiden *Datura innoxia* Mill. mesophyll-protoplasten zu pflanzen. *Z. Pflanzenphysiol.* 76:462–66
92. Dorion, N., Chupeau, Y., Bourgin, J. P. 1975. Isolation, culture and regeneration into plants of *Ranunculus sceleratus* L. leaf protoplasts. *Plant Sci. Lett.* 5:325–31
93. Coe, E. H. Jr., Sarkar, K. R. 1966. Preparation of nucleic acids and a ge-

netic transformation attempt in maize. *Crop Sci.* 6:432–35
94. Ledoux, L., Huart, R., Jacobs, M. 1974. DNA-mediated genetic correction of thiamineless *Arabidopsis thaliana*. *Nature* 249:17–21
95. Ledoux, L., Huart, R., Mergeay, M., Charles, P., Jacobs, M. 1975. DNA mediated genetic correction of thiamineless *Arabidopsis thaliana*. See Ref. 12, pp. 67–89
96. Ledoux, L., Huart, R., Mergeay, M., Charles, P., Jacobs, M. 1975. DNA mediated genetic correction of thiamineless *Arabidopsis thaliana*. See Ref. 11, pp. 499–517
97. Rédei, G. P. 1975. *Arabidopsis* as a genetic tool. *Ann. Rev. Genet.* 9:111–27
98. Rédei, G. P., Acedo, G., Weingarten, H., Kier, L. D. 1977. Has DNA corrected genetically thiamineless mutants of Arabidopsis? In *Cell Genetics in Higher Plants, Proc. Int. Training Course, Szeged, Hungary,* ed. D. Dudits, G. L. Farkas, P. Maliga, pp. 91-94. Budapest: Akad. Kiadó
99. Hess, D. 1969. Versuche zur Transformation an höheren Pflanzen: Induktion und konstante Weitergabe der Anthocyansynthese bie *Petunia hybrida. Z. Pflanzenphysiol.* 60:348–58
100. Hess, D. 1969. Versuche zur Transformation an höheren Pflanzen: Wiederholung der Anthocyan-Induktion bei *Petunia* und erste Charakterisierung des transformierenden Prinzips. *Z. Pflanzenphysiol.* 61:286–98
101. Hess, D. 1970. Versuche zur Transformation an höheren Pflanzen: Genetische Charakterisierung einiger mutmaßlich transformierter Pflanzen. *Z. Pflanzenphysiol.* 63:31–43
102. Hess, D. 1972. Versuche zur Transformation an höheren Pflanzen: Nachweis von Heterozygoten in Versuchen zur Transplantation von Genen für Anthocyansynthese bei *Petunia hybrida*. *Z. Pflanzenphysiol.* 66:155–66
103. Hess, D. 1973. Transformationsversuche an höheren Pflanzen: Untersuchungen zur Realisation des Exosomen-Modells der Transformation bie *Petunia hybrida. Z. Pflanzenphysiol.* 68:432–40
104. Bianchi, F., Walet-Foederer, H. G. 1974. An investigation into the anatomy of the shoot apex of Petunia hybrida in connection with the results of transformation experiments. *Acta Bot. Neerl.* 23:1–6
105. Turbin, N. V., Soyfer, V. N., Kartel, N. A., Chekalin, N. M., Dorohov, Y. L., Titov, Y. B., Cieminis, K. K. 1975. Genetic modification of the *waxy* character in barley under the action of exogenous DNA of the wild variety. *Mutat. Res.* 27:59–68
106. Soyfer, V. N., Kartel, N. A., Chekalin, N. M., Titov, Y. B., Cieminis, K. K., Turbin, N. V. 1976. Genetic modification of the waxy character in barley after an injection of wild-type exogenous DNA. Analysis of the second seed generation. *Mutat. Res.* 36:303–10
107. Nawa, S., Yamada, M., Ohta, Y. 1975. Hereditary changes in *Capsicum annuum* L. III. Induced by DNA treatment. *Jpn. J. Genet.* 50:341–44
108. Ohta, Y., Chuong, P. V. 1975. Hereditary changes in *Capsicum annuum* L. I. Induced by ordinary grafting. *Euphytica* 24:355-68
109. Gahan, P. B., Anker, P., Stroun, M., Jacob, K. 1969. DNA-induced chromosome damage in *Vicia faba. Caryologia* 22:307–10
110. Cocking, E. C. 1972. Plant cell protoplasts—isolation and development. *Ann. Rev. Plant Physiol.* 23:29–50
111. Cocking, E. C. 1975. Protoplasts as genetic systems. See Ref. 11, pp. 311–27
112. Eriksson, T., Bonnett, H., Glimelius, K., Wallin, A. 1974. Technical advances in protoplast isolation, culture and fusion. In *Tissue Culture and Plant Science,* ed. H. E. Street, pp. 213–31. London: Academic. 502 pp.
113. Gamborg, O. L., Constabel, F., Fowke, L., Kao, K. N., Ohyama, K., Kartha, K., Pelcher, L. 1974. Protoplast and cell culture methods in somatic hybridization in higher plants. *Can. J. Genet. Cytol.* 16:737–50
114. Gamborg, O. L., Constabel, F., Kao, K. N., Ohyama, K. 1975. Plant protoplasts in genetic modifications and production of intergeneric hybrids. See Ref. 12, pp. 181–96
115. Hess, D., Potrykus, I., Donn, G., Durand, J., Hoffmann, F. 1973. Transformation experiments in higher plants: Prerequisites for the use of isolated protoplasts. (isolation from mesophyll and callus cultures, uptake of proteins and DNA, and regeneration of whole plants). *Colloq. Int. CNRS* 212:343–51
116. Street, H. E., ed. 1974. *Tissue Culture and Plant Science.* London: Academic. 502 pp.
117. Vasil, I. K. 1976. Progress, problems

and prospects of plant protoplast research. *Adv. Agron.* 28:119–60
118. Carlson, P. S., Smith, H. H., Dearing, R. D. 1972. Parasexual interspecific plant hybridization. *Proc. Natl. Acad. Sci. USA* 69:2292–94
119. Keller, W. A., Melchers, G. 1973. Effect of high pH and calcium on tobacco leaf protoplast fusion. *Z. Naturforsch.* 28:737–41
120. Kao, K. N., Michayluk, M. R. 1974. A method for high-frequency intergeneric fusion of plant protoplasts. *Planta* 115:355–67
121. Ahkong, Q. F., Howell, J. I., Lucy, J. A., Safwat, F., Davey, M. R., Cocking, E. C. 1975. Fusion of hen erythrocytes with yeast protoplasts induced by polyethylene glycol. *Nature* 255:66–67
122. Dudits, D., Rasko, I., Hadlaczky, G., Lima-de-Faria, A. 1976. Fusion of human cells with carrot protoplasts induced by polyethylene glycol. *Hereditas* 82:121–24
123. Jones, C. W., Mastrangelo, I. A., Smith, H. H., Liu, H. Z., Meck, R. A. 1976. Interkingdom fusion between human (HeLa) cells and tobacco hybrid (GGLL) protoplasts. *Science* 193:401–3
124. Schaeffer, P., Cami, B., Hotchkiss, R. D. 1976. Fusion of bacterial protoplasts. *Proc. Natl. Acad. Sci. USA* 73:2151–55
125. Smith, H. H., Kao, K. N., Combatti, N. C. 1976. Interspecific hybridization by protoplast fusion in *Nicotiana*. Confirmation and extension. *J. Hered.* 67:123–28
126. Melchers, G., Labib, G. 1974. Somatic hybridization of plants by fusion of protoplasts. I. Selection of light resistant hybrid of "haploid" light sensitive varieties of tobacco. *Mol. Gen. Genet.* 135:277–94
127. Melchers, G., Keller, W., Labib, G. 1975. Somatic hybridisation. See Ref. 12, pp. 161–68
128. Power, J. B., Frearson, E. M., Hayward, C., George, D., Evans, P. K., Berry, S. F., Cocking, E. C. 1976. Somatic hybridisation of *Petunia hybrida* and *P. parodii*. *Nature* 263:500–2

129. Guo, C. 1972. Effects of chemical and physical factors on the chromosome number in *Nicotiana* anther callus cultures. *In Vitro* 7:381–86
130. Sacristán, M. D., Melchers, G. 1969. The caryological analysis of plants regenerated from tumorous and other callus cultures of tobacco. *Mol. Gen. Genet.* 105:317–33
131. Constabel, F., Dudits, D., Gamborg, O. L., Kao, K. N. 1975. Nuclear fusion in intergeneric heterokaryons. A note. *Can. J. Bot.* 53:2093–95
132. Kao, K. N., Constabel, F., Michayluk, M. R., Gamborg, O. L. 1974. Plant protoplast fusion and growth of intergeneric hybrid cells. *Planta* 120:215–27
133. Reinert, J., Gosch, G. 1976. Continuous division of heterokaryons from *Daucus carota* and *Petunia hybrida* protoplasts. *Naturwissenschaften* 63:535–36
134. Power, J. B., Frearson, E. M., Hayward, C., Cocking, E. C. 1975. Some consequences of the fusion and selective culture of *Petunia* and *Parthenocissus* protoplasts. *Plant Sci. Lett.* 5:197–207
135. Kao, K. N. 1977. Chromosomal behavior in somatic hybrids in soybean—*N. glauca*. *Mol. Gen. Genet.* 150:225–30
136. Wetter, L. R. 1977. Isoenzyme patterns in soybean-*Nicotiana* somatic hybrids. *Mol. Gen. Genet.* 150:231–35
137. Nester, E. W., Chilton, M. D., Drummond, M., Merlo, D., Montoya, A., Sciaky, D., Gordon, M. P. 1977. Search for bacterial DNA in crown gall tumors. In *Impact of Recombinant Molecules on Science and Society, 10th Ann. Miles Symp.*, ed. R. F. Beers, Jr., E. G. Bassett, pp. 179–88. New York: Raven
138. Sciaky, D. 1977. *Plasmids of Agrobacterium tumefaciens and their role in crown gall tumorigenesis.* PhD thesis. Washington State Univ., Pullman. 130 pp.
139. Szeto, W. W., Hamer, D. H., Carlson, P. S., Thomas, C. A. Jr. 1977. Cloning of cauliflower mosaic virus (CLMV) DNA in *Escherichia coli*. *Science* 196:210–12

DNA REARRANGEMENTS IN PROCARYOTES

♦3117

Peter Starlinger
Institut für Genetik der Universität zu Köln, Weyertal 121, D-5000 Köln 41, Federal Republic of Germany

CONTENTS

INTRODUCTION	103
DNA INSERTIONS AND THE CHROMOSOMAL ABERRATIONS ASSOCIATED WITH THEM	104
Temperate Bacteriophages	104
Bacteriophage λ and φ80	104
Bacteriophage P2	105
Bacteriophage Mu	105
Plasmids	106
The F plasmid	106
R plasmids	106
Col plasmids	107
P1	107
Insertions of Nonreplicating DNA Sequences	108
IS elements	108
Transposons	109
SPONTANEOUS CHROMOSOMAL ABERRATIONS	110
Duplications	110
Deletions	112
Other Chromosomal Aberrations	113
PHYSIOLOGICAL REACTIONS LEADING TO DNA REARRANGEMENTS	115
DNA REARRANGEMENTS THAT HAVE ACCUMULATED DURING EVOLUTION	115
DNA Rearrangements of Small Chromosomal Segments	115
Are Procaryotic Chromosomes Formed by Repeated Duplications of an Ancestral Chromosome?	116
CONCLUDING REMARKS	116

INTRODUCTION

Chromosomal aberrations have been studied in eucaryotes for a long time. In procaryotes, during the early days of bacterial genetics, attention was focused on the study of point mutations, which were characterized at the molecular level. It has become clear, however, that other mutations, including deletions, duplications, inversions, and transpositions, also occur. An additional class of aberrations is the fusion of two (bacterial, plasmid, or phage) chromosomes without any loss of genetic material during this event (1).

These rearrangements are most probably products of enzymatic reactions. Very little is known about the specificity of these enzymes and their mechanisms of action. They are active even in strains that are deficient in general, homologous recombination. In many cases, the endpoints of chromosome rearrangements are not randomly distributed in the chromosome regions investigated. The term *illegitimate recombination* was coined for these events (2). Often, aberrations occur in the vicinity of preformed genetic elements that are able to insert at different sites of DNA molecules. These elements are temperate bacteriophages, plasmids, and a new class of transposable genetic elements called IS elements and transposons.

The interactions of these elements with each other and with the bacterial chromosome and the chromosomal rearrangements occurring in their vicinity are discussed in the first section of this article.

In the second section, spontaneous aberrations are discussed. Though it is assumed that these arise by a variety of mechanisms, the question whether some of them may be associated with the presence of an as yet unidentified transposable DNA element is examined.

It will become clear during the following discussion that chromosomal aberrations are not rare, when compared with point mutations. The question arises, therefore, why chromosomal rearrangements do not lead to a quick scrambling of linkage relationships during evolutionary time spans. These questions are discussed in the concluding section of this article.

DNA INSERTIONS AND THE CHROMOSOMAL ABERRATIONS ASSOCIATED WITH THEM

Temperate Bacteriophages

BACTERIOPHAGE λ AND $\phi 80$ The literature until 1971 has been thoroughly reviewed (2). λ is usually integrated at a unique major attachment site. In addition, a large, but limited number of minor integration sites are present on the *Escherichia coli* chromosomes (3, 4). Integration and excision occur via a phage-coded enzyme system. The major attachment site is asymmetric and differs between bacterial and phage DNA. Extended homology is not used for the integration, but a small homologous region seems to be present between the asymmetric parts of the integration site (5, 6).

Inexact excision of phage DNA occurs. Constraints of head-filling require that the λ chromosomes formed by this inexact excision are usually defective, if they are detected in mature phages. Linkage of genes to the attachment site of λ (or $\phi 80$) allows the formation of transducing phages for these genes (7).

Inexact excision of λ from secondary attachment sites leads to the formation of transducing phage carrying genes linked to these sites (3).

Still other transducing phages are created, if bacterial genes are transposed into the vicinity of the major attachment site of λ with the help of F'-particles (see below) (8). The enzymes responsible for the formation of transducing bacteriophage and the genes coding for these enzymes are not known. Neither bacterial *rec* mutations nor the phage mutations *red, int,* or *xis* abolish the formation of transducing phages (2).

While one endpoint of the bacterial DNA carried by the transducing phage is always at the attachment site, the other endpoint is variable in a population of phage isolated independently. This distribution is nonrandom in λdgal phages (9). Bacteria surviving induction of a prophage and the imprecise excision of a transducing phage carry deletions (10–13). Since no constraints of phage head-filling are operative, the deletions are often larger than the amount of bacterial DNA integrated into phage chromosomes. Some of these deletions have one endpoint within the *gal* operon, and the distribution of these endpoints is nonrandom and corresponds to that of the preferred endpoints of the bacterial DNA segments carried by the transducing phages (12). Endpoints of deletions that have one terminus within prophage DNA retained in the bacterial chromosome upon induction of excision-deficient phages are also nonrandom (13). However, random endpoints have been found in the *ara* operon, when λ prophage was excised from an unusual location near *ara* (11).

BACTERIOPHAGE P2 Deletions are formed adjacent to prophage P2 (14). The formation of deletions, called *eduction,* requires a functional *P2 int* gene (15).

BACTERIOPHAGE MU The literature has been reviewed recently (16–18). Bacteriophage Mu is integrated randomly into recipient DNA (19). Its integration causes a mutation, if it occurs in a structural gene (20). The mutations caused by the integration of Mu are polar, if the mutated gene is part of an operon (19, 21). In about 15% of the cases, a deletion is caused by the integration of Mu (22, 23). Precise excision of an integrated Mu is very rare in the wild type (20, 21) but occurs readily with *X* mutants of the phage (24, 25).

Bacteriophage Mu can mediate the integration of circular DNA into the bacterial chromosome or into plasmids. Also, DNA segments from the chromosomes can be transposed. In both cases, the transposed DNA segment is flanked at both termini by a direct repeat of the prophage Mu (26, 27). This method can be used to transpose any segment of bacterial DNA to a broad host range plasmid and thus to mobilize this DNA for interspecies crosses (28–30). In addition, DNA adjacent to a prophage Mu can also be inverted. In this case, the inverted segment of bacterial DNA seems to be flanked by two copies of prophage Mu DNA which are inverted relative to each other. An ingenious and simple mechanism has been proposed to explain all of these aberrations. It is assumed that transposition or integration of Mu requires an enzyme system capable of recognizing both the right and the left terminus of Mu DNA. If, by mistake, this enzyme system recognizes one end of one prophage and the other end not of the same prophage copy but of its sister copy in or behind a replication fork, and if these two ends are treated as usual in a transposition or integration event, all the aberrations described above will occur (28, 29).

Plasmids

THE F PLASMID The properties of F have been reviewed elsewhere (31–35). The integration of F into the *E. coli* chromosome is nonrandom. At a variety of sites, integration has been observed repeatedly. At these sites, integration occurs via recombination between short, homologous DNA sequences carried on both the *E. coli* chromosome and the F-DNA (34, 36–45). Two of these sequences have been identified as IS elements IS*2* and IS*3* (see below) (46, 47). This recombination is (at least partially) independent of the bacterial *rec* system (34, 38, 48–51). Since in most instances, F integration does not cause the mutation of a gene, it has not been shown directly that exact excision is possible. Indirect evidence indicates that exact excision is easily detected when F had integrated via recombination between γ-δ sequences carried on both the F plasmid and the chromosome, but not if integration occurred via IS2 or IS3 (34, 38, 52).

Inexact excision has been observed frequently. It gives rise to F' particles. Some of these carry bacterial material located on both sides of the integrated F plasmid. Another class of F' particles has one endpoint within the bacterial DNA and the other one within the F-DNA. This latter endpoint is usually at the terminus of one of the IS sequences involved in integration. This process is termed *half site–specific recombination* (34). The enzymes responsible for this excision and the subsequent ring formation are not known.

As a result of *incompatibility*, two (genetically marked) F particles cannot coexist within the same bacterial cell. Selection for bacterial markers carried by two different F' particles often leads to the isolation of fused F' particles. These are not simple fusions, but usually carry a large deletion of F material. The endpoints of the deleted material are often endpoints of the IS elements carried by the F' particles. Thus, the fusion event may be associated with site-specific recombination or half site-specific recombination (53–56).

The fusion of different F' particles offers possibilities for creating novel arrangements of bacterial genes. Integration of a bacteriophage into these fused F' particles allows the isolation of novel types of transducing bacteriophages (53–56). In one particular isolate of a $\phi80gal$, the *gal* operon is inverted relative to its wild-type orientation (56). Integration of F'8*gal* into the chromosome by recombination between the plasmid-borne and the chromosomal *gal* operon leads to a complicated series of events that result in the formation of enlarged F'-factors and an inversion of one of the *gal* operons flanking the integrated plasmid (57–59).

R PLASMIDS Detailed reviews of the biology of R plasmids have appeared elsewhere (60–62). Many R plasmids are able to integrate into the bacterial chromosome. This can be shown by chromosome mobilization, by integrative suppression of *dnaA* mutations, or by both (63–72).

In most cases, the mechanism of integration is not known. Integration of *R100-1* gives rise to stable Hfr strains, many of which have their transfer origin in the *lac pro* region (68), where many Hfr strains have their origin (34, 44, 45). Since *R100* carries both IS2 and IS3, the mechanism of integration may be similar to the formation of Hfr strains by the F plasmid.

R1-drd 19 seems to have a unique integration site at 27 min of the *E. coli* K12 map (63, 64). The site seems also to be present on the *Salmonella typhimurium* map, but because of an inversion of this chromosomal region relative to the *E. coli* K12 map (74), it has a transfer direction opposite to that found in *E. coli* K12 (70). The Hfr strains formed are unstable, as can be seen by the fact that in most matings the whole R factor is transferred, thus resembling the behavior of F'-strains rather than of Hfr strains. An as yet unidentified integration sequence may be responsible. This hypothesis is supported by the observation that plasmid mutants unable to transfer the *trp* region of *E. coli* have a lower molecular weight (70).

Some R plasmids have a broad host range. RP4, originally isolated in a *Pseudomonas* strain, is easily transferred by conjugation into most gram-negative species (75, 76). This plasmid is able to integrate into the *E. coli* chromosome (77). Another very similar *Pseudomonas* plasmid, R68, mobilizes the chromosome of *Pseudomonas aeruginosa,* strain PAT, at high frequency, but it is 10^4 times less effective with strain PAO. Variants of it, however, are able to integrate into this strain also (78). These variants differ from the original strain by an increase in molecular weight of 1.5 M daltons. This increase may be due to the addition of an IS element (79).

Another fusion event involving R factors is the reversible dissociation and fusion of the DNA segment known as the resistance transfer factor (RTF) and the segment carrying most resistance determinants (r-det). This phenomenon is observed with several fi⁺ R-plasmids in *Proteus mirabilis* (80–84). The IS element IS*1* located at the two junctions between RTF and r-det may be involved in dissociation and fusion (46, 85).

In *E. coli*, the isolated r-det is not seen as a dissociation product of R100, but an independent r-det plasmid is seen in cells, in which R100 or RTF is integrated into the chromosome (73). A similar dissociation upon integrative suppression has been observed with an F-like hemolysin plasmid (67). A direct repeat of a 0.4 Kb sequence has been observed at the two termini of a tetracycline-resistance gene in *Streptococcus faecalis.* This gene is capable of amplification (86, 87).

COL PLASMIDS Col plasmids are plasmids coding for colicins. The extensive literature on these plasmids has been recently reviewed (88). Col V, belonging to the F-like plasmids, is able to form Hfr strains and does so preferentially near gene *xyl* on the *E. coli* linkage map, transferring the chromosomes in clockwise manner (89). No Hfr strains could be detected in clones of cells harboring Col Ib (65), though this plasmid was able to suppress *dnaA* mutations. Since integrative suppression by other I-like R-plasmids has been reported to yield Hfr strains (72), the ability to form Hfr strains may be due to the presence of an integration sequence on the latter I-like R-plasmid.

Col plasmids have a remarkable ability to recombine with R-plasmids and F-plasmids, and also to take up chromosomal genes (90).

P1 P1 is a temperate bacteriophage, which, however, does not integrate into the bacterial chromosome. Its chromosome replicates as a plasmid in the lysogenic state. Variants of P1 have been observed, which have incorporated certain bacterial genes (91–93). Since P1 does not integrate into the bacterial chromosome, this phenomenon is not understood. It may be related to the transposition of transposons (see below).

On the whole, the ability of plasmids to integrate into chromosomes and to mobilize these seems to depend on properties independent of the plasmid phenotypes like incompatibility and sex pili. This ability may be linked to the presence of IS elements, of which some are usually found on all members of a certain plasmid class, while others are observed on a single representative of a class only. The distribution of known IS elements on F-like plasmids studied by heteroduplex methods follows this pattern (37, 46).

Plasmids integrated into a bacterial chromosome sometimes undergo structural changes. A complete inversion of the integrated F is postulated to explain the change from one Hfr to another with the same transfer origin but the opposite orientation of transfer (94). More complicated rearrangements including total and partial inversions, deletions, and duplications must be invoked to explain the properties of Hfr strains derived from a parental Hfr strain created by the insertion of an F-like Col factor in an unusual location (95).

Insertions of Nonreplicating DNA Sequences

Insertion of small, nonreplicating, transposable DNA sequences into bacterial chromosomes, or into the DNA of temperate bacteriophages or plasmids has been observed repeatedly. The original site of the inserted material may either be the same chromosome, or another chromosome replicating in the same cell. Some of these DNA sequences are devoid of a recognizable phenotype, except for effects at the site of integration. These are called *IS (inserted sequence) elements.* Other insertions carry gene(s), most often coding for the resistance to an antibiotic. These are called *transposons*.

IS ELEMENTS IS elements have been reviewed in detail elsewhere (96–98). Presently, 5 IS elements have been characterized in *E. coli*. Their size varies between 800 and ~1500 nucleotide pairs (99–104). They occur in several copies each on the *E. coli* chromosome and some of them also occur on its plasmids and on temperate bacteriophage (46, 47, 85, 105–107), and on the chromosome of *Salmonella typhimurium* and *Citrobacter freundii* (108; H. Saedler, personal communication). Their transposition into a gene abolishes its function (109–112), leading to a mutation. Most of these mutations revert to wild type, indicating the exact integration and excision of the IS element. Reversion cannot be enhanced by mutagens. Both integration and excision are independent of *recA* (111, 113). The presence of an inserted segment of DNA can be detected by length measurements on the DNA (114), buoyant density measurements on transducing bacteriophages (21, 99, 113, 115), DNA-DNA (105) or RNA-DNA hybridization studies (101, 116), and by inspection of heteroduplex molecules in the electron microscope (46, 47, 85, 105–107).

In addition to exact excision, secondary chromosomal aberrations are frequently detected in the vicinity of IS elements. Deletions (with nonrandom endpoints) are detected adjacent to IS*1* (117, 118), IS*3* (119). They do not remove the IS element.

One endpoint is (within the limits of resolution of genetic methods) at the end of the IS element. The other endpoint may be as far as several hundred to several thousand nucleotide pairs away. In a strain carrying a deletion adjacent to IS*1*, secondary and even tertiary deletions can be detected (118). Deletion formation adjacent to IS*1* is strongly temperature-dependent (117). This distinguishes deletion formation from exact excision, which shows no temperature dependence. An *E. coli* mutant has been isolated, in which the deletion formation adjacent to IS*1* is strongly decreased (120). In this mutant, exact excision, leading to the wild type of the mutated gene, is not impaired.

Duplications adjacent to an IS*2* integrated in the *galOP* region have been detected by their effect of relieving the polarity of IS*2* on *gal* operon expression. The *gal* operon is apparently fused to a promoter outside the *gal* operon, and thus synthesis of *gal* enzymes becomes constitutive in the duplication mutants (121–123; M. Fiandt, A. Ahmed, and W. Szybalski, personal communication).

Integration of IS elements shows a certain site specificity, which is intermediate between the pronounced specificity of λ and the complete lack of specificity of Mu. The degree of specificity varies for different IS elements. Only one integration site has been observed so far for IS*4* in gene *galT* of *E. coli* (4, 10, 100, 104). This integration site is also a secondary integration site of bacteriophage λ (4). Little is known about the integration specificity of IS*3*. IS*1* and IS*2* seem to be less specific than IS*4* and have been found in several locations. IS*1* has been isolated repeatedly in the operator-promoter region of the *gal* operon and IS*4* in *galT* at apparently the same site, but in both possible orientations (99, 100, 104, 124). The site in *galOP* is also an integration site for IS*2*. In both cases, the site is defined at the level of resolution of both genetic crosses and heteroduplex molecules in the electron microscope. At the level of nucleotides, the IS elements in this cluster are inserted a few nucleotides apart from each other [Chadwell and Starlinger, quoted in (125)].

TRANSPOSONS Transposons are unique DNA structures consisting of one or several genes coding for the resistance to an antibiotic that can be transposed from one chromosome to another in the same cell via illegitimate recombination (85, 126–153). Transposons are to be designated by the letters *Tn,* followed by the isolation number and (in parentheses) the abbreviation for the antibiotic, to which resistance is conferred, e.g. *Tn1(Ap)* (154). In some, but not all instances, transposons are bordered by a repeated DNA sequence occurring at both of its termini (85, 133, 136–138, 147–149, 152). In some cases, these sequences are inverted relative to each other; in one case, *Tn9(Cm)*, the sequence is repeated directly (149). This sequence has been identified as IS*1*. The inverted repeat bordering the tetracycline transposon *Tn10* has been recognized as IS*3* (85). Integration of transposons into a gene causes a mutation of this gene (137, 143). Most of the mutations revert to wild type, indicating that the insertion has occurred without substitution. More often, however, an inexact excision occurs. This can lead to deletions adjacent to the transposon (119, 129, 143, 145, 151).

In addition to the deletions described, duplications and inversions have been observed in the vicinity of the tetracycline transposon *Tn10* (145).

Different transposons show varying degrees of site specificity of integration (135–137, 141, 143, 145).

Internal deletions of *Tn2* created in vitro by the action of nucleases have lost the ability to transpose and may have lost a gene(s) involved in this process (147, 148). In the case of the ampicillin transposon *Tn1,* one copy of the transposon present on plasmid DNA prevents the integration of a second *Tn1* transposon in position *cis* (150).

Most of the aberrations formed in the vicinity of transposons or IS sequences can be explained by the same mechanism which was proposed for chromosomal rearrangements associated with bacteriophage Mu. This is not possible in cases where a duplication of bacterial DNA is formed in which the duplicated segments of bacterial DNA are not separated by the transposon or IS elements. This is so in the case of the *gal* duplications adjacent to IS2 in the leader sequence of the *gal* operon. In this case, the bacterial DNA has most probably been separated from the IS element, as can be deduced from the phenotype (122). These duplications may arise by different mechanisms, as is discussed in the section on spontaneous duplications.

SPONTANEOUS CHROMOSOMAL ABERRATIONS

In the following section, spontaneous chromosomal aberrations are described. Some of these may be associated with transposable DNA elements, either present in the vicinity and not yet identified, or inserted there prior to the formation of these aberrations. I mention circumstantial evidence of this kind, where it exists.

It would be surprising, however, if there were no other mechanisms available for the formation of such aberrations as duplications, deletions, and inversions which, unlike transpositions, can be formed in one elementary act formally resembling a crossing-over without homologous pairing (transposition requires two such elementary acts). It has been suggested that a mispairing involving short regions of homology, stabilized by protein interaction, might serve as the site either of a deletion or of an inversion, or (if two sister chromosomes in a replication fork are involved) of a duplication in one and a deletion in the other sister chromosome (155, 156).

In most cases, it is not clear whether any DNA sequence can serve as the breakage point in the formation of aberrations, or whether the enzymes involved have a certain sequence specificity. Nonrandomness of endpoints of the aberrations in a certain region of the chromosome may be an indication of sequence specificity of the enzymes involved.

Duplications

Duplications have been observed in a variety of loci in *E. coli, Salmonella typhimurium* and their bacteriophages, and also in *Bacillus subtilis* and *Diplococcus pneumoniae.* In some cases, the duplications could be selected by the need for an increased gene dosage (157–162). In other cases, duplications were detected, because they are able to maintain a heterogenotic state, e.g. two genes for transfer RNA, of which one copy codes for a suppressor tRNA, while the other one codes for an essential tRNA (163, 164). Duplications have also been found, in which a polar effect was relieved by separating a certain gene from a stop signal and attaching it to another promoter (155, 165). A further method to detect duplications is a transduction cross between complementing pairs of mutants (166). Their distinction from true recombinants is possible by the instability of the duplications. Unstable transformants have been used to detect duplications in *Pneumococcus* (167–169) and in *Bacillus subtilis* (156, 170–175).

Duplications have been regularly observed in transduction experiments with phage P1 in *E. coli,* in which one end of the segment to be transduced does not find a pairing partner in the recipient as a result of a deletion. In these cases, an illegitimate fusion event (e.g. circularization of the transducing fragment), followed by recombinational insertion can be used to explain the ensuing structures. Replication during the integration event must be postulated if recipient alleles are found duplicated in the recombinants (93, 176, 177). Duplications have also been investigated in phages T4 and λ. In T4, duplications have been selected as sets of complementing *rII* mutations that cannot recombine because of overlapping deletions in that region (178–181). The unstable duplications in T4 can be stabilized by introducing a deletion extending into a duplicated region from either side. If the deletion removes indispensable genetic information, loss of the duplication via recombination is selected against (182, 183). In bacteriophage λ, the introduction of duplications has been selected for in strains carrying deletions that leave the DNA too short to be effectively packaged into the phage head (184–188).

The duplications described so far have been shown or presumed to be tandem duplications. Duplications are called tandem, if the duplicated segments are not separated by unique DNA, even if the genes studied are separated by an (also duplicated) DNA segment. Where heteroduplex analysis is possible, the tandem duplications are detected by inspection of DNA molecules prepared from one strand carrying the duplication and one strand devoid of it. In this situation, the insertion loop is not observed in a fixed position. Different molecules carry it in different positions which are found in a region, whose length is the same as that of the single-stranded loop. This is explained by the assumption that the loop can be situated at any site in the duplicated region (185). In the absence of heteroduplex analysis, the tandem character of a duplication is inferred from its instability interpreted as recombinational loss of a duplicated segment, which can be observed if one of the segments is marked by a recessive allele (155, 156, 160–164, 169). In nontandem duplications, recombinational loss of one of the duplicated segments would eliminate the unique DNA separating these segments, giving rise to a (potentially lethal) deletion.

The transduction of a tandem duplication into a nonduplicated recipient is possible, even if the duplicated region is larger than the DNA molecule carried by the transducing phage. It is assumed that transfer of a segment carrying the junction between the duplicated segments, followed by its out of register pairing with two sister chromosomes and a double crossover, gives rise to the duplication in the recipient (156, 163, 175).

A duplication of gene *argH* in *E. coli,* in which the duplicated segment is translocated to a nearby location, has been described (155). Several duplications of the distal genes of the *trp* operon of *E. coli* may also be of this type (189). A duplication of phage genes in ϕ80*dsu3* is the result of a complicated aberration, in which part of the bacterial DNA is translocated within the phage chromosome (190).

The duplications described so far have been selected as spontaneous mutations or as consequence of recombination between a transposition strain and its parental strain. Duplications can also be induced by mutagens causing misrepair (191–192).

The formation of *arg* duplications in *E. coli* and of duplications in λ is independent of *recA* (and in the latter case of phage genes *red, int,* or *xis*) (155, 184–188), while the formation of duplications in the *trp* region of *S. typhimurium* was found to depend on a functional allele of *recA* (165). It is interesting that the nontandem duplication of an *argH* reverts in *recA*, though at a reduced frequency. This might indicate a transposon structure with direct repeats at the termini, allowing loss of this structure both by homologous and by site-specific recombination (155).

The frequency of duplications is usually reported to be around 10^{-4} to 10^{-5} (159–163). A lower value of 10^{-6} to 10^{-7} is reported for the *arg* region of *E. coli* (155). In the *his* region of *S. typhimurium,* duplications are found at a frequency of 10^{-4} to 10^{-3} (166). In *B. subtilis,* 0.5% of transformants of the *ilvA4* mutant to *ilv*$^+$ are duplications (156, 170). A frequency of 10^{-5} is reported in λ, and this is increased to 10^{-3} in a λ-ϕ80 hybrid, probably a result of the presence of a ϕ80 gene mapping on the right arm of the chromosome (187). Duplications are rare (5×10^{-8}) in T4, but in this case, a deletion has to be selected simultaneously, if maturation is to occur (180).

The size of the duplicated region varies from around the size of a gene (155) to one third of a chromosome (161, 171). In several regions of bacterial or bacteriophage chromosomes, nonrandomness of duplication endpoints has been observed (155, 161, 166, 181). In λ, however, the endpoints of duplications are found to be distributed randomly (188). In this latter case, however, duplications were selected in any position of the chromosome rather than in a specific genetic region, and thus the number of duplications in a given region of the chromosome may have been too small to detect a nonrandom distribution. A certain type of unstable mutation of gene *argF*, stabilized in a *recA* background and therefore presumed to be tandem duplications, is isolated in one strain of *E. coli,* but not in another. This might indicate the participation of specific DNA sequences (IS elements?) in the formation of the chromosomal rearrangement (162).

Deletions

Deletions have been widely studied in bacteria and their phages. The literature until 1970 is reviewed in (2). More recently, spontaneous deletions have been studied in *E. coli* for ribosomal protein genes (193), for gene *zwf* coding for glucose-6-phosphate dehydrogenase (194), and also in phage λ (195) and in *S. typhimurium* for the chlorate D region (196).

The frequency of deletions was to 10^{-6} to 10^{-9} per cell plated in the cases reviewed in (2). A frequency of 10^{-4} per cell plated has been reported for the *zwf* region (194). 10^{-8} to 10^{-7} deletions per cell plated were found for the ribosomal protein genes in most strains investigated, but in one particular bacterial strain, 2.5×10^{-5} deletions per cell plated were detected (193). In most cases, no data are available about the size of deletions and the randomness of endpoints. The frequency of deletions for a certain gene is thus generally lower than the frequency of duplications. This is not necessarily an argument for different mechanisms of formation of these two aberrations, because a large fraction of deletions may not be detectable as a result of the loss of indispensable gene functions. Therefore, models are not ruled out, according to which deletions and duplications are produced in one event in a replication fork (155).

Deletions are induced by various mutagens in *E. coli* (197–202), and its bacteriophage T4 (203) and in *B. subtilis* (204).

By use of mutants involved in the repair of damage to DNA, the mechanism of spontaneous or induced deletion formation has only partially been elucidated. The formation of spontaneous deletions is stimulated by mutations in gene *polA* (205) and has not been found to be affected by mutations in genes *recA, B, C, rep, uvrA, B, C, ras,* or *endo I* (2). However, the stimulation of deletion formation after UV irradiation is more pronounced in *uvrA* or *B* mutants, while *recA* did not affect the UV-stimulated deletion formation. X rays stimulated deletion formation equally well in the wild-type and *uvrA* or *B* or *recA* mutants. The pattern of deletions formed (short vs long), however, is different in spontaneous and induced deletions (200–202).

Unfortunately, similar data are not available for the formation of duplications. Pronounced similarities or dissimilarities of the pattern of formation of both aberrations could provide an argument for or against their formation in one event.

The finding of strain differences in the formation of deletions in the same gene region (199, 206, 207) may be due to the presence of a deletogenic IS element in one strain, which is missing in the other. The *trp* deletions reported in (206, 207) are formed in a temperature-sensitive reaction, similar to deletion formation adjacent to IS*1* (117).

The observation of frequent secondary deletions adjacent to about one half of the rare primary deletions in the *gal* region of *E. coli* (118) is compatible with the assumption that IS*1* insertion is associated with the formation of primary deletions. Alternatively, the new junction of DNA sequences in the rare primary deletion may have similarities to the junction of IS*1* with adjacent DNA.

Other Chromosomal Aberrations

Spontaneous transpositions of a genetic segment within the bacterial chromosome have been described. Translocation of *argH* and possibly also of *trpCBA* have been mentioned above (155, 189). The translocation of part of the *trp* operon has also been observed in *B. subtilis* in mutant *trpE26* (172).

When the transposition strain was crossed with its parental strain, unstable duplications arose. These were interpreted to be the result of genetic recombination between the identical segments carried in different positions in the two chromosomes.

It is conceivable that the very large duplications observed in *S. typhimurium* also arise by recombination between homologous regions carried in different parts of the chromosome (161).

In the *his* operon of *S. typhimurium*, a mutant, in which the control region and part of the first gene are deleted, gives rise to secondary mutants, in which the distal genes of the operon are expressed. These have been interpreted to be transpositions to a plasmid, in which the genes are linked to a plasmid promoter. While these revertants are found easily, no transposition to a chromosomal site has been found (208) [see (141) for analogy].

In *E. coli*, transpositions of large segments of the chromosome to new positions have been seen after treatment with X rays or nitrogen mustard (209, 210). In one case, the transposition strain was found to be unstable and to create mutants frequently, when crossed with the parental strain (209). It is possible that secondary chromosomal aberrations (deletions and unstable duplications) are formed, when the transposed segment of the donor DNA recombines with the same segment in its original position in the recipient, as was analyzed in more detail in mutant *trpE26* in *B. subtilis* (173–175).

A translocation of bacterial genes carried by a transducing $\phi 80$ within the phage chromosome has been described (190).

A spontaneous inversion of a large segment of the *B. subtilis* chromosome has been observed in mutant *trpE26*. This inversion is adjacent to the transposition of a chromosome segment from its original position near the *aroB* gene to a new location near the gene *thr*. The transposition is stable unless a large duplication is created which includes the transposed segment and the adjacent segment between the new and the old position of the transposed segment. In these duplication strains, the transposed segment is (rarely) lost and concomitantly the inversion is restored to its original orientation (173–175).

The authors explain the results by assuming that homologous sequences are present at the ends of the segments involved in the aberrations mentioned above. If one considers the orientation of such regions (IS elements, prophage?), it is possible to explain the correlation between transposition and inversion. If the transposed sequence is a transposon, it may be flanked by inverted repeats. Inversion of an adjacent segment may occur between a sequence inside the transposon and an inverted repeat of it at the distal end of the segment to be inverted. In this case, the inverted sequences flanking the transposon will be converted to direct repeats and their excision would be possible only after the reversal of the adjacent inversion.

Inverted repeats, many of them of the length of known IS elements, are detected frequently in *E. coli* DNA by heteroduplex techniques (211, 212).

An inversion of a segment of bacterial DNA adjacent to an integrated $\phi 80$ prophage of a length of approximately 24 kb must have occurred either preceding the formation of $\phi 80$ *psu3* or at the time of formation of transducing phage, because the *su3* gene is inverted relative to its order in the normal bacterial chromosome (213).

It is not known whether the few instances of translocations and inversions reported are due to a rare occurrence of these, or to an insufficient technique of detection. The latter possibility must be considered seriously, because inversions might either recombine well (and then be mistaken as point mutations) or badly (in which case they would score as deletions). The heterozygosity of transposition-duplications might go undetected, because these would not be as unstable as tandem duplications, which are lost by homologous recombination. If, however, transposition is indeed rare as compared to tandem duplications, this may be an indication that some mechanisms capable of creating duplications and deletions may not be operative in the creation of transpositions and inversions.

PHYSIOLOGICAL REACTIONS LEADING TO DNA REARRANGEMENTS

While the events described in the preceding sections are rare, and can be considered to be mutations, chromosomal rearrangements are sometimes occurring at a much higher frequency and are possibly reflecting events occurring normally during the life cycle of bacteria or bacteriophage. Excision of prophage upon induction is one of these examples.

Bacteriophage Mu carries a DNA sequence, called *G-loop,* which inverts very frequently, when the bacteriophages are induced from the prophage state. Inversion seems to occur by a *rec*-independent recombination event at a small inverted DNA sequence repeat flanking the G loop. The length of this inverted repeat is approximately 50 nucleotide pairs. The same DNA sequence occurs in the DNA of bacteriophage P1, where it is flanked by a longer inverted repeat of approximately 600 nucleotide pairs length. The literature has been reviewed recently (18).

DNA REARRANGEMENTS THAT HAVE ACCUMULATED DURING EVOLUTION

DNA Rearrangements of Small Chromosomal Segments

Bacteria carry several copies of the transcription unit for the ribosomal RNA, which is transcribed in the direction, 16S, 23S, and 5S RNA. In *E. coli,* the number of rRNA transcription units is estimated to be between 5 and 10. These transcription units are not tandem duplications, but some of them are clustered within limited regions of the chromosome. The literature is reviewed in (215). Though the sequence differences of rRNA molecules seem to be confined to relatively few positions, larger differences are found in spacer regions. In two instances, a gene coding for a particular transfer RNA has been found in the spacer between the gene for 16S and 23S RNA. One of these tRNAs accepts Glu, the other Ile (216, 217). It is not yet known, whether these *tRNA* genes are confined to specific tRNA transcription units, or whether they can be moved around by recombination.

The gene coding for elongation factor *EF-Tu* in *E. coli* is present twice on the *E. coli* chromosome, one copy mapping near *rif* and the other near *str* (218).

While some speculations have been brought forward concerning the functional role of the multiplication of genes coding for products that are needed in large amounts during protein biosynthesis, no such explanation offers itself for another case: Genes *argF* and *argI* code for enzymes catalyzing the same reaction and are duplications, as has been shown by heteroduplex mapping of the DNA (219).

A pair of closely linked (tandem?) genes for tyrosine tRNA is present on the *E. coli* chromosome (220).

While duplications can be detected within a single bacterial strain, a deletion or inversion that has occurred long ago during evolution can only be detected by comparison of the genetic map of related bacteria. The best-known example is provided by the linkage maps of *E. coli* and *S. typhimurium*, which show great similarities, but one large inversion in the *trp-cysB* region and the presence of some genes in one species that are missing in the other. The literature on these species differences has been recently reviewed (221). In the latter cases, it is not known whether the gene missing in one species has been lost by a deletion from a common ancestor possessing the gene, or whether it has been added to the other species by transposition. The finding of two copies of IS3 in inverted order on both sides of the *lac* operon in *E. coli* (34) might indicate that the *lac* operon was inserted into the chromosome of an ancestor by a transposition event similar to the insertion of Tn10 (Tc).

Are Procaryotic Chromosomes Formed by Repeated Duplications of an Ancestral Chromosome?

The distribution of genome sizes of different procaryotes has suggested the possibility that more complex chromosomes have evolved by the duplication of a small ancestral chromosome of the size found in *Mycoplasma* and that the duplicated regions have evolved since into separate genes (222). Stronger indications for a mechanism of this kind have been brought forward both for *Streptomyces coelicolor* (223) and *E. coli* (224) by an analysis, which seems to indicate that similar functions have a greater than random chance to map either 180° apart on the circular map of the former or 90° or 180° apart on the circular map of the latter organism.

CONCLUDING REMARKS

It is clear from the evidence available that chromosomal rearrangements are not rare, when compared with point mutations. Why then are chromosomes as stable as is observed in nature? The chromosomes of *E. coli* and *S. typhimurium* have undergone base changes in many of their nucleotides, each single change occurring at frequencies about 10^{-9}/generation, and about 10^5–10^6 such independent events are necessary to create the diversity between these species. The number of events necessary to scramble the bacterial chromosome must be much smaller, and still the overall arrangement of these chromosomes is conserved remarkably well. Apparently strong evolutionary forces must be at work. Can these be understood?

The least difficulties arise with deletions. The loss of genetic information will, in most cases, be of disadvantage to the bacteria. Also, tandem duplications will not be assembled at increasing frequency, since general recombination tends to remove them. The difficulties arise with transpositions and inversions. These have been observed repeatedly in the laboratory even if they have not been selected for. Two extreme alternatives can be discussed for the conservation of chromosomal arrangements:

1. Each gene has been transposed frequently and has finally been trapped in a favorable position. This position may be determined by the linkage to neighboring genes, which is conserved against genetic recombination and thus allows alleles to stay together, the gene products of which are better suited to perform a given physiological role than the combination of the products of different alleles.

2. Transpositions and inversions are not random events, creating new junctions between any two nucleotides with equal probability. Rather, they occur often at preformed sequences, especially at inverted repeats, as has been observed in the case of transposons. In this case, the scrambling of genes would preferentially occur in fixed modules of genetic material. The number of possible arrangements may be much smaller, and it is more easily conceivable that one of this limited number of combinations conveys a certain advantage to the cells. Moreover, it can be hypothesized that the creation of one transposition or inversion may trigger subsequent similar events, as in the case of the *trpE26* mutation of *B. subtilis* (173–175). A possible mechanism for such a triggering has been discussed, and the ensuing cascade of aberrations may be more disadvantageous to the cell than a single event.

In addition, transposition will also create instabilities, if the transposition strain enters into recombinational exchanges with the parental strain.

While the mechanisms have been demonstrated in certain bacterial strains, their generality remains to be shown. In any case, it will be a challenging task for geneticists to find the functional role of both the mechanisms for the creation of aberrations and the mechanisms counteracting the scrambling of the bacterial chromosome.

The techniques for the investigation of new gene combinations are now at hand. DNA can be joined in new combinations both in vitro and in vivo, and it has already been suggested that such recombinant DNA molecules be introduced into cells, where they will pair with different regions of the chromosome and thus create unstable duplications in defined regions of the chromosome (175). Extension of this method to the use of molecules that carry inversions of DNA fragments relative to their position on the chromosome will lead to the creation of inversions, and the failure to detect such inversions in some regions of the chromosome may be a first clue to the existence of regions, where a disturbance of the chromosome map is of disadvantage to the cells.

It should be mentioned in conclusion that transposable genetic elements have also been described in higher organisms, though they have not yet been investigated by biochemical means. The best known of these are the controlling elements in maize, described in the pioneering studies by B. McClintock. Many more have been characterized by now (97, 98, 225). Directed transpositions seem to play a role in the differentiation of antibody-forming cells (226).

It is conceivable that a DNA rearrangement can be produced at a certain time in development. This could be passed on clonally and thus would be well suited to preserve changes in gene expression. Coupled rearrangements, like overlapping inversions, may even link events that are separated in time by many cell generations, like the "presetting" effects described in maize. It may well be that the main functional roles of DNA rearrangements are yet to be discovered and it is hoped that the knowledge gained by the study of procaryotes will be helpful in this respect.

AKNOWLEDGMENT

I would like to thank Drs. M. Achtman, L. Anagnostopoulos, J. Besemer, R. Ehring, S. Falkow, N. Glansdorff, N. Kleckner, G. Mosig, H. Riley, and H. Saedler for communicating unpublished results and for critically reading the manuscript.

Literature Cited

1. Reanney, D. 1976. Extrachromosomal elements as possible agents of adaptation and development. *Bacteriol. Rev.* 40:552–90
2. Franklin, N. 1971. Illegitimate recombination. In *The Bacteriophage Lambda,* ed. A. D. Hershey, pp. 175–94. Cold Spring Harbor, NY: Cold Spring Harbor Lab.
3. Shimada, K., Weisberg, R. A., Gottesman, M. E. 1972. Prophage *lambda* at unusual chromosomal locations. I. Location of the secondary attachment sites and the properties of the lysogens. *J. Mol. Biol.* 63:483–503
4. Shimada, K., Weisberg, R. A., Gottesman, M. E. 1973. Prophage λ at unusual chromosomal locations. II. Mutations induced by bacteriophage λ in *Escherichia coli* K-12. *J. Mol. Biol.* 80:297–314
5. Shulman, M., Gottesman, M. 1973. Attachment site mutants of bacteriophage λ. *J. Mol. Biol.* 81:461–82
6. Shulman, M., Mizuuchi, K., Gottesman, M. M. 1976. New att mutants of phage λ. *Virology* 72:13–22
7. Campbell, A. 1962. Episomes. *Adv. Genet.* 11:101–45
8. Beckwith, J. R., Signer, E. R., Epstein, W. 1966. Transposition of the *lac* region of *E. coli. Cold Spring Harbor Symp. Quant. Biol.* 31:393–401
9. Pfeifer, D., Rosypal, S. 1969. Deletion mapping of galactose mutants with transducing λ phages. *Mol. Gen. Genet.* 104:116–23
10. Shapiro, J. A., Adhya, S. L. 1969. The galactose operon of *E. coli K-12.* II. A deletion analysis of operon structure and polarity. *Genetics* 62:249–64
11. Schleif, R. 1972. Fine-structure deletion map of the *E. coli* L-arabinose operon. *Proc. Natl. Acad. Sci. USA* 69:3479–84
12. Pfeifer, D., Hirsch, H. J., Bergmann, D., Hamlaoui, M. 1974. Non-random distribution of deletion endpoints in the *gal* operon of *E. coli. Mol. Gen. Genet.* 132:203–13
13. Marchelli, C., Ghelardini, P., Nasi, S. 1976. Deletions induced by heat treatment of *E. coli K12* lysogenic for λ prophage. *Genetics* 82:161–68
14. Sunshine, M. G., Kelly, B. 1971. Extent of host deletions associated with bacteriophage P2-mediated eduction. *J. Bacteriol.* 108:695–704
15. Sunshine, M. G. 1972. Dependence of eduction on *P2 int* product. *Virology* 47:61–67
16. Howe, M. M., Bade, E. G. 1975. Molecular biology of bacteriophage *mu. Science* 190:624–32
17. Couturier, M. 1976. The integration and excision of bacteriophage *mu-1. Cell* 7:155–63
18. Bukhari, A. I. 1976. Bacteriophage mu as a transposition element. *Ann. Rev. Genet.* 10:389–412
19. Bukhari, A. I., Zipser, D. 1972. Random insertion of *mu-1* DNA within a single gene. *Nature New Biol.* 236:240–43
20. Taylor, A. L. 1963. Bacteriophage-induced mutation in *Escherichia coli. Proc. Natl. Acad. Sci. USA* 50:1043–51
21. Jordan, E., Saedler, H., Starlinger, P. 1968. 0°- and strong-polar mutations in the *gal* operon are insertions. *Mol. Gen. Genet.* 102:353–63
22. Howe, M. M., Zipser, D. 1974. Host deletions caused by the integration of

bacteriophage *mu-1. Am. Soc. Microbiol. Abstr.* No. 208, p. 235
23. Cabezón, T., Faelen, M., de Wilde, M., Bollen, A., Thomas, R. 1975. Expression of ribosomal protein genes in *Escherichia coli. Mol. Gen. Genet.* 137:125–29
24. Bukhari, A. I. 1975. Reversal of mutator phage *mu* integration. *J. Mol. Biol.* 96:87–99
25. Bukhari, A. I., Taylor, A. L. 1975. Influence of insertions on packaging of host sequences covalently linked to bacteriophage *mu* DNA. *Proc. Natl. Acad. Sci. USA* 72:4399–4403
26. Faelen, M., Toussaint, A., Couturier, M. 1971. *Mu-1* promoted integration of a λ-*gal* phage in the chromosome of *E. coli. Mol. Gen. Genet.* 113:367–70
27. Van de Putte, P., Gruijthuijsen, M. 1972. Chromosome-mobilization and integration of *F*-factors in the chromosome of *recA* strains of *E. coli* under the influence of bacteriophage *mu-1. Mol. Gen. Genet.* 118:173–83
28. Faelen, M., Toussaint, A. 1976. Bacteriophage *mu-1:* A tool to transpose and to localize bacterial genes. *J. Mol. Biol.* 104:525–39
29. Toussaint, A., Faelen, M., Bukhari, A. 1977. *Mu*-mediated transposition as part of the *mu*-life cycle. In *DNA Insertions, Plasmids and Episomes,* ed. A. Bukhari, J. Shapiro, S. Adhya. Cold Spring Harbor, NY: Cold Spring Harbor Lab. In press
30. Faelen, M., Toussaint, A., van Montagu, M., van den Elsacker, S., Engler, G., Schell, J. 1977. In vivo genetic engineering: The *mu*-mediated transposition of chromosomal DNA segments onto transmissible plasmids. See Ref. 29
31. Scaife, J. 1967. Episomes. *Ann. Rev. Microbiol.* 21:601–38
32. Willetts, N. 1972. The genetics of transmissible plasmids. *Ann. Rev. Genet.* 6:257–68
33. Achtman, M. 1973. Genetics of the *F* sex factor in enterobacteriaceae. *Curr. Top. Microbiol. Immunol.* 60:79–121
34. Davidson, N., Deonier, R. C., Hu, S., Ohtsubo, E. 1975. The DNA sequence organization of *F* and *F*-primes and the sequences involved in *Hfr* formation. In *Microbiology—1974,* ed. D. Schlessinger, pp. 56–65. Washington DC: Am. Soc. Microbiol.
35. Achtman, M., Skurray, R. 1977. Redefinition of the mating phenomenon. In *Microbial Interaction.* London: Chapman & Hall. In press
36. Sharp, P. A., Hsu, M. T., Ohtsubo, E., Davidson, N. 1972. Electron microscope heteroduplex studies of sequence relations among plasmids of *Escherichia coli.* I. Structure of *F*-prime factors. *J. Mol. Biol.* 71:471–97
37. Sharp, P. A., Cohen, S. N., Davidson, N. 1973. Electron microscope heteroduplex studies of sequence relations among plasmids of *Escherichia coli.* II. Structure of drug resistance (*R*) factors and *F* factors. *J. Mol. Biol.* 75:235–55
38. Ohtsubo, E., Deonier, R. C., Lee, H. J., Davidson, N. 1974. Electron microscope heteroduplex studies of sequence relations among plasmids of *Escherichia coli.* IV. The *F* sequences in *F14. J. Mol. Biol.* 89:565–84
39. Lee, H. J., Ohtsubo, E., Deonier, R. C., Davidson, N. 1974. Electron microscope heteroduplex studies of sequence relations among plasmids of *Escherichia coli.* V. ilv^+ deletion mutants *F14. J. Mol. Biol.* 89:585–97
40. Ohtsubo, E., Lee, H. J., Deonier, R. C., Davidson, N. 1974. Electron microscope heteroduplex studies of sequence relations among plasmids of *Escherichia coli.* VI. Mapping of *F14* sequences homologous to $\phi 80dmetBJF$ and $\phi 80dargECBH$ bacteriophages. *J. Mol. Biol.* 89:599–618
41. Deonier, R. C., Ohtsubo, E., Lee, H. J., Davidson, N. 1974. Electron microscope heteroduplex studies of sequence relations among plasmids of *Escherichia coli.* VII. Mapping the ribosomal RNA genes of plasmid *F14. J. Mol. Biol.* 89:619–29
42. Ohtsubo, E., Soll, L., Deonier, R. C., Lee, H. J., Davidson, N. 1974. Electron microscope heteroduplex studies of sequence relations among plasmids of *Escherichia coli.* VIII. The structure of bacteriophage $\phi 80d3ilv^+su^+7$ including the mapping of the ribosomal RNA genes. *J. Mol. Biol.* 89:631–46
43. Anthony, W. M., Deonier, R. C., Lee, H. J., Hu, S., Ohtsubo, E., Davidson, N. 1974. Electron microscope heteroduplex studies of sequence relations among plasmids of *Escherichia coli.* IX. Note on the deletion mutant of *F,FΔ(33–43). J. Mol. Biol.* 89:647–50
44. Deonier, R. C., Davidson, N. 1976. The sequence organization of the integrated *F* plasmid in two *Hfr* strains of *Escherichia coli. J. Mol. Biol.* 107:207–22
45. Deonier, R. C., Oh, G. R., Hu, M. 1977. Further mapping of IS*2* and IS*3* in the *lac-purE* region of the *Escherichia coli*

K-12 genome: Structure of the *F*-prime ORF203. *J. Bacteriol.* 129:1129–40
46. Hu, S., Ohtsubo, E., Davidson, N., Saedler, H. 1975. Electron microscope heteroduplex studies of sequence relations among plasmids: Identification and mapping of the insertion sequences IS*1* and IS*2* in *F* and *R* plasmids. *J. Bacteriol.* 122:764–75
47. Hu, S., Ptashne, K., Cohen, S. N., Davidson, N. 1975. The $\alpha\beta$ sequence of *F* is IS*3*. *J. Bacteriol.* 123:687–92
48. Clowes, R. L., Moody, E. E. M. 1966. Chromosomal transfer from recombination-deficient strains of *E. coli K12*. *Genetics* 53:717–26
49. Willetts, N. S., Broda, P. H. 1969. The *Escherichia coli* sex factor. In *Bacterial Episomes and Plasmids, CIBA Found. Symp.*, pp. 32–48
50. Broda, P. H., Meacock, P. 1971. Isolation and characterization of *Hfr* strains from a recombination-deficient strain of *Escherichia coli*. *Mol. Gen. Genet.* 113:166–73
51. Broker, T. R. 1977. Recombination models for the inverted DNA sequences of the gamma-delta segment of *E. coli* and the G-segment of phages *mu* and *P1*. See Ref. 29
52. Broda, P., Meacock, P., Achtman, M. 1972. Early transfer of genes determining transfer functions by some *Hfr* strains in *Escherichia coli K12*. *Mol. Gen. Genet.* 116:336–47
53. Press, R., Glansdorff, N., Miner, P., de Vries, J., Kadner, R., Maas, W. K. 1971. Isolation of transducing particles of $\phi 80$ bacteriophage that carry different regions of the *Escherichia coli* genome. *Proc. Natl. Acad. Sci. USA* 68:795–98
54. Palchaudhuri, S. R., Mazaitis, A. J., Maas, W. K., Kleinschmidt, A. K. 1972. Characterization by electron microscopy of fused *F*-prime factors in *Escherichia coli*. *Proc. Natl. Acad. Sci. USA* 69:1873–76
55. Palchaudhuri, S. R., Maas, W. K. 1976. Fusion of two *F*-prime factors in *Escherichia coli* studied by electron microscope heteroduplex analysis. *Mol. Gen. Genet.* 146:215–31
56. Besemer, J., Kubai, D. F. 1976. Isolation and characterization of $\phi 80dgal$ transducing phages that carry *gal* operator promoter insertion mutations. *Mol. Gen. Genet.* 148:79–92
57. Adelberg, E. A., Bergquist, P. L. 1972. The stabilization of integration by genetic inversion: A general hypothesis. *Proc. Natl. Acad. Sci. USA* 69:2061–65

58. Jamieson, A. F., Bergquist, P. L. 1977. Plasmid replication and *Hfr* formation in strains of *Escherichia coli* carrying *seg* mutations. *Mol. Gen Genet.* 150: 171–82
59. Bergquist, P. L., Jamieson, A. F. 1977. Genetic inversion in the formation of an *Hfr* strain from a temperature-sensitive *F'gal* strain. *J. Bacteriol.* 129: 282–90
60. Watanabe, T. 1969. Transferable drug resistance. The nature of the problem. See Ref. 49, pp. 81–101
61. Meynell, G. G. 1972. *Bacterial Plasmids.* London & Basingstoke: Macmillan
62. Falkow, S. 1975. *Infectious Multiple Drug Resistance.* London: Pion
63. Pearce, L. E., Meynell, E. 1968. Specific chromosomal affinity of a resistance factor. *J. Gen. Microbiol.* 50:159–72
64. Cooke, M., Meynell, E. 1969. Chromosomal transfer mediated by derepressed *R* factors in *F' Escherichia coli K12*. *Genet. Res.* 14:79–87
65. Moody, E. E. M., Hayes, W. 1972. Chromosome transfer by autonomous transmissible plasmids: The role of the bacterial recombination (*rec*) system. *J. Bacteriol.* 111:80–85
66. Moody, E. E. M., Runge, R. 1972. The integration of autonomous transmissible plasmids into the chromosome of *Escherichia coli K12*. *Genet. Res.* 19:181–86
67. Goebel, W. 1974. Integrative suppression of temperature-sensitive mutants with a lesion in the initiation of DNA replication. Replication of autonomous plasmids in the suppressed state. *Eur. J. Biochem.* 43:125–30
68. Nishimura, A., Nishimura, Y., Caro, L. 1973. Isolation of Hfr strains from R^+ and *Col V2$^+$* strains of *Escherichia coli* and derivation of an *R'lac* factor by transduction. *J. Bacteriol.* 116: 1107–12
69. Sotomura, M., Yoshikawa, M. 1975. Reinitiation of chromosome replication in the presence of chloramphenicol under an integratively suppressed state by R6K. *J. Bacteriol.* 122:623–28
70. Hedèn, L. O., Meynell, E. 1976. Comparative study of *R1*-specific chromosomal transfer in *Escherichia coli K-12* and *Salmonella typhimurium LT2*. *J. Bacteriol.* 127:51–58
71. Hedèn, L. O., Rutberg, L. 1976. *R* factor-mediated polarized chromosomal transfer in *Escherichia coli C*. *J. Bacteriol.* 127:46–50

72. Datta, N., Barth, P. 1976. *Hfr*-formation by *I* pilus-determining plasmids in *Escherichia coli* K-12. *J. Bacteriol.* 125:811–17
73. Chandler, M., Silver, L., Frey, J., Caro, L. 1977. Suppression of an *E. coli dnaA* mutation by the integrated *R* factor *R100.1:* II. Generation of small plasmids following integration. *J. Bacteriol.* In press
74. Sanderson, K. E., Hall, C. A. 1970. *F*-prime factors of *Salmonella typhimurium* and an inversion between *S. typhimurium* and *Escherichia coli. Genetics* 64:215–28
75. Datta, N., Hedges, R. W., Shaw, E. J., Sykes, R. B., Richmond, M. 1971. Properties of an *R* factor from *Pseudomonas aeruginosa. J. Bacteriol.* 108:1244–49
76. Olsen, R. H., Shipley, P. 1973. Host range and properties of an *R* factor from *Pseudomonas aeruginosa R* factor *1922. J. Bacteriol.* 113:772–80
77. Unger, L., Sokatch, J. R., Martin, R. R. Cited in 78
78. Haas, D., Holloway, B. W. 1976. *R* factor variants with enhanced sex factor activity in *Pseudomonas aeruginosa. Mol. Gen. Genet.* 144:243–51
79. Jacob, A. E., Cresswell, J. M., Hedges, R. W. 1977. Molecular characterization of the *P* group plasmid *R68* and variants with enhanced chromosome mobilizing ability. *Fed. Eur. Microbiol. Soc. Lett.* In press
80. Nisioka, T., Mitani, M., Clowes, R. 1969. Composite circular forms of *R* factor deoxyribonucleic acid molecules. *J. Bacteriol.* 97:376–85
81. Cohen, S. N., Miller, C. A. 1970. Nonchromosomal antibiotic resistance in bacteria. III. Isolation of the discrete transfer unit of the *R* factor *R1. Proc. Natl. Acad. Sci. USA* 67:510–16
82. Silver, R. P., Falkow, S. 1970. Studies on resistance transfer factor deoxyribonucleic acid in *Escherichia coli. J. Bacteriol.* 104:340–44
83. Rownd, R., Mickel, S. 1971. Dissociation and reassociation of *RTF* and *r*-determinants of the *R* factor *NR 1* in *Proteus mirabilis. Nature New Biol.* 234:40–43
84. Rownd, R., Perlman, D., Goto, N. 1975. Structure and replication of *R* factor DNA in *Proteus mirabilis. Microbiology—1974,* ed. D. Schlessinger, pp. 76–94. Washington DC: Am. Soc. Microbiol.
85. Ptashne, K., Cohen, S. N. 1975. Occurrence of insertion sequence (IS) regions on plasmid deoxyribonucleic acid as direct and inverted nucleotide sequence duplications. *J. Bacteriol.* 122:776–81
86. Yagi, Y., Clewell, D. B. 1976. Plasmid-determined tetracycline resistance in *Streptococcus faecalis:* Tandemly repeated resistance determinants in amplified forms of *pAMα1* DNA. *J. Mol. Biol.* 102:583–606
87. Yagi, Y., Clewell, D. B. 1977. Identification and characterization of a small sequence located at two sites on the amplifiable tetracycline resistance plasmid *pAMα1* in *Streptococcus faecalis. J. Bacteriol.* 129:400–6
88. Hardy, K. G. 1975. Colicinogeny and related phenomena. *Bacteriol. Rev.* 39:464–515
89. Kahn, P. L. 1968. Isolation of high-frequency recombining strains from *Escherichia coli* containing the *V* colicinogenic factor. *J. Bacteriol.* 96:205–14
90. Fredericq, P. 1969. The recombination of colicinogenic factors with other episomes and plasmids. See Ref. 49, pp. 163–78
91. Luria, S. E., Adams, J. N., Ting, R. C. 1960. Transduction of lactose-utilizing ability among strains of *E. coli* and *S. dysenteriae* and the properties of the transducing phage particles. *Virology* 12:348–90
92. Stodolsky, M. 1973. Bacteriophage *P1* derivatives with bacterial genes: A heterozygote enrichment method for the selection of *P1dpro* lysogens. *Virology* 53:471–75
93. Rae, M. E., Stodolsky, M. 1974. Chromosome breakage, fusion and reconstruction during P1 dl transduction. *Virology* 58:32–54
94. Berg, C. M., Curtiss, R. III 1967. Transposition derivatives of an *Hfr* strain of *Escherichia coli K-12. Genetics* 56:503–25
95. Kahn, P. L. 1969. Evolution of a site of specific genetic homology on the chromosome of *Escherichia coli. J. Bacteriol.* 100:269–75
96. Starlinger, P., Saedler, H. 1972. Insertion mutations in microorganisms. *Biochimie* 54:177–85
97. Starlinger, P., Saedler, H. 1976. IS-elements in microorganisms. *Curr. Top. Microbiol. Immunol.* 75:111–52
98. Bukhari, A., Shapiro, J., Adhya, S., eds. 1977. *DNA Insertions, Plasmids and Episomes.* New York: Cold Spring Harbor Lab.
99. Hirsch, H. J., Saedler, H., Starlinger, P. 1972. Insertion mutations in the control region of the galactose operon of *E. coli.*

II. Physical characterization of the mutations. *Mol. Gen. Genet.* 115:266–76
100. Fiandt, M., Szybalski, W., Malamy, M. H. 1972. Polar mutations in *lac, gal* and phage λ consist of a few DNA sequences inserted with either orientation. *Mol. Gen. Genet.* 119:223–31
101. Hirsch, H. J., Starlinger, P., Brachet, P. 1972. Two kinds of insertions in bacterial genes. *Mol. Gen. Genet.* 119:191–206
102. Malamy, M. H., Fiandt, M., Szybalski, W. 1972. Electron microscopy of polar insertions in the *lac* operon of *Escherichia coli*. *Mol. Gen. Genet.* 119:207–22
103. Blattner, F. R., Fiandt, M., Haas, K. K., Twose, P. A., Szybalski, W. 1974. Deletions and insertions in the immunity region of coliphage λ: Revised measurement of the promoter-startpoint distance. *Virology* 62:458–71
104. Pfeifer, D., Kubai-Maroni, D., Habermann, P. 1977. Specific sites for integration of IS-elements within the transferase gene of the galactose operon. See Ref. 29
105. Saedler, H., Heiss, B. 1973. Multiple copies of the insertion DNA sequences IS*1* and IS*2* in the chromosome of *E. coli K-12*. *Mol. Gen. Genet.* 122:267–77
106. Saedler, H., Kubai, D., Nomura, M., Jaskunas, S. R. 1975. IS*1* and IS*2* mutations in the ribosomal protein genes of *E. coli K-12*. *Mol. Gen. Genet.* 141:85–89
107. Mosharrafa, E., Pilacinski, W., Zissler, J., Fiandt, M., Szybalski, W. 1976. Insertion sequence IS*2* near the gene for prophage λ excision. *Mol. Gen. Genet.* 147:103–10
108. Rak, B. 1976. gal mRNA initiated within IS*2*. I. Hybridization studies. *Mol. Gen. Genet.* 149:135–44
109. Malamy, M. H. 1966. Frameshift mutations in the lactose operon of *E. coli*. *Cold Spring Harbor Symp. Quant. Biol.* 31:189–201
110. Saedler, H., Starlinger, P. 1967. 0°-mutations in the galactose operon in *E. coli*. I. Genetic characterization. *Mol. Gen. Genet.* 100:178–89
111. Jordan, E., Saedler, H., Starlinger, P. 1967. Strong-polar mutations in the transferase gene of the galactose operon in *E. coli*. *Mol. Gen. Genet.* 100:296–306
112. Adhya, S. L., Shapiro, J. A. 1969. The galactose operon of *E. coli K-12*. I. Structural and pleiotropic mutations of the operon. *Genetics* 62:231–47
113. Jaskunas, S. R., Lindahl, L., Nomura, M. 1975. Isolation of polar insertion mutants and the direction of transcription of ribosomal protein genes in *E. coli*. *Nature* 256:183–87
114. Malamy, M. H. 1970. Some properties of insertion mutations in the *lac* operon. In *The Lactose Operon*, ed. J. R. Beckwith, D. Zipser, pp. 359–73. Cold Spring Harbor, NY: Cold Spring Harbor Lab.
115. Shapiro, J. A. 1969. Mutations caused by the insertion of genetic material into the galactose operon of *Escherichia coli*. *J. Mol. Biol.* 40:93–105
116. Michaelis, G., Saedler, H., Venkov, P., Starlinger, P. 1969. Two insertions in the galactose operon having different sizes but homologous DNA sequences. *Mol. Gen. Genet.* 104:371–77
117. Reif, H. J., Saedler, H. 1975. IS*1* is involved in deletion formation in the *gal* region of *E. coli K12 Mol. Gen. Genet.* 137:17–28
118. Reif, H. J., Saedler, H. 1977. IS*1*-dependent deletion formation in the *gal* region of *E. coli*. See Ref. 29
119. Foster, T. J. 1976. *R*-factor mediated tetracycline resistance in *Escherichia coli K-12*. Dominance of some tetracycline-sensitive mutants and relief of dominance by deletion. *Mol. Gen. Genet.* 143:339–44
120. Nevers, P., Reif, H. J., Saedler, H. 1977. A mutant of *Escherichia coli* defective in IS*1*-mediated deletion formation. See Ref. 29
121. Ahmed, A., Scraba, D. 1975. The nature of the *gal3* mutation of *Escherichia coli*. *Mol. Gen. Genet.* 136:233–42
122. Ahmed, A. 1975. Mechanism of reversion of the *gal3* mutation of *Escherichia coli*. *Mol. Gen. Genet.* 136:243–53
123. Ahmed, A., Johansen, E. 1975. Reversion of the gal3 mutation of *Escherichia coli*: Partial deletion of the insertion. *Mol. Gen. Genet.* 142:263–76
124. Saedler, H., Besemer, J., Kemper, B., Rosenwirth, B., Starlinger, P. 1972. Insertion mutations in the control region of the *gal* operon of *E. coli*. I. Biological characterization of the mutations. *Mol. Gen. Genet.* 115:258–65
125. Starlinger, P. 1977. Mutations caused by the integration of IS*1* and IS*2* into the *gal* operon of *E. coli*. See Ref. 29
126. Cohen, S. N., Kopecko, D. J. 1976. Structural evolution of bacterial plasmids: Role of translocating genetic elements and DNA sequence insertions. *Fed. Proc.* 34:2031–36

127. Cohen, S. N. 1976. Transposable genetic elements and plasmid evolution. *Nature* 263:731-38
128. Kondo, E., Mitsuhashi, S. 1964. Drug resistance of enteric bacteria. IV. Active transducing bacteriophage *P1 CM* produced by the combination of *R* factor with bacteriophage *P1*. *J. Bacteriol.* 88:1266-76
129. Chan, R. K., Botstein, D. 1972. Genetics of bacteriophage *P22* I. Isolation of prophage deletions which affect immunity to superinfection. *Virology* 49:257-67
130. Watanabe, T., Ogata, Y., Chan, R. K., Botstein, D. 1972. Specialized transduction of tetracycline resistance by phage *P22* in *Salmonella typhimurium*. I. Transduction of *R* factor *222* by phage *P22*. *Virology* 50:874-82
131. Scott, J. 1973. Phage *P1* cryptic. II. Location and regulation of prophage genes. *Virology* 53:327-36
132. Hedges, R. W., Jacob, A. E. 1974. Transposition of ampicillin resistance from *RP4* to other replicons. *Mol. Gen. Genet.* 132:31-40
133. Berg, D. E., Davies, J., Allet, B., Rochaix, J. D. 1975. Transposition of *R* factor genes to bacteriophage λ. *Proc. Natl. Acad. Sci. USA* 72:3628-32
134. Gottesman, M. M., Rosner, J. L. 1975. Acquisition of a determinant for chloramphenicol resistance by coliphage *lambda*. *Proc. Natl. Acad. Sci. USA* 72:5041-45
135. Heffron, F., Sublett, R., Hedges, R. W., Jacob, A. E., Falkow, S. 1975. Origin of the TEM β-lactamase gene found on plasmids. *J. Bacteriol.* 122:250-56
136. Heffron, F., Rubens, C., Falkow, S. 1975. Translocation of a plasmid DNA sequence which mediates ampicillin resistance: Molecular nature and specificity of insertion. *Proc. Natl. Acad. Sci. USA* 72:3623-27
137. Kleckner, N., Chan, R. K., Tye, B. K., Botstein, D. 1975. Mutagenesis by insertion of a drug-resistance element carrying an inverted repetition. *J. Mol. Biol.* 97:561-75
138. Kopecko, D. J., Cohen, S. N. 1975. Site-specific *recA*-independent recombination between bacterial plasmids: Involvement of palindromes at the recombinational loci. *Proc. Natl. Acad. Sci. USA* 72:1373-77
139. Tye, B. K., Chan, R. K., Botstein, D. 1974. Packaging of an oversize transducing genome by *Salmonella* phage *P22*. *J. Mol. Biol.* 85:485-500
140. Foster, T. J., Howe, T. G. B., Richmond, K. M. V. 1975. Translocation of the tetracycline resistance determinant from *R100-1* to the *Escherichia coli K-12* chromosome. *J. Bacteriol.* 124:1153-58
141. Barth, P. T., Datta, N., Hedges, R. W., Grinter, N. J. 1976. Transposition of a DNA sequence encoding trimethoprim and streptomycin resistances from *R483* to other replicons. *J. Bacteriol.* 125:800-10
142. Bennett, P. M., Richmond, M. H. 1976. The translocation of a discrete piece of DNA carrying an *amp* gene between replicons in *Escherichia coli*. *J. Bacteriol.* 126:1-6
143. Berg, D. E. 1977. Insertion and excision of the transposable kanamycin resistance determinant Tn5. See Ref. 29
144. Berg, D. E. 1977. Detection of transposable antibiotic resistance determinants using phage *lambda*. See Ref. 29
145. Botstein, D., Kleckner, N. 1977. Translocation and illegitimate recombination by the tetracycline resistance element. See Ref. 29
146. Heffron, F., Rubens, C., Falkow, S. 1977. Transposition of a plasmid DNA sequence which mediates ampicillin resistance: General description and epidemiological considerations. See Ref. 29
147. Heffron, F., Bedinger, P., Champoux, J., Falkow, S. 1977. Biochemical deletion mutations affecting transposition of TnA. See Ref. 29
148. Heffron, F., Bedinger, P., Champoux, J., Falkow, S. 1977. Deletions affecting the transposition of an antibiotic resistance gene. *Proc. Natl. Acad. Sci. USA* 74:702-6
149. MacHattie, L. A., Jackowski, J. B. 1977. Physical structure and deletion effects of the chloramphenicol-resistance element *Tn9* in phage *lambda*. See Ref. 29
150. Robinson, M. K., Bennett, P. M., Richmond, M. H. 1977. The inhibition of TnA translocation by TnA. *J. Bacteriol.* 129:407-14
151. Brevet, J., Nisen, P., Kopecko, D. J., Cohen, S. N. 1977. Promotion of insertions and deletions by translocating segments of DNA carrying antibiotic resistance genes. See Ref. 29
152. Kopecko, D. J., Brevet, J., Cohen, S. N. 1976. Involvement of multiple translocating DNA segments and recombinational hotspots in the structural evo-

lution of bacterial plasmids. *J. Mol. Biol.* 108:333–60
153. Rubens, C., Heffron, F., Falkow, S. 1976. Transposition of a plasmid deoxyribonucleic acid sequence that mediates ampicillin resistance: Independence from host *rec* functions and orientation of insertion. *J. Bacteriol.* 128:425–34
154. Berg, D. E., Botstein, D., Campbell, A., Novick, R., Starlinger, P. 1976. Nomenclature of transposable elements in prokaryotes. See Ref. 29
155. Beeftinck, F., Cunin, R., Glansdorff, N. 1974. Arginine gene duplications in recombination proficient and deficient strains of *Escherichia coli K12. Mol. Gen. Genet.* 132:241–53
156. Jamet-Vierny, C., Anagnostopoulos, C. 1975. Induction and transmission of a merodiploid condition near the terminal area of the chromosome of *Bacillus subtilis. Genetics* 81:437–58
157. Horiuchi, T., Horiuchi, S., Novick, A. 1963. The genetic basis of hyper-synthesis of β-galactosidase. *Genetics* 48:157–69
158. Rigby, P. W. J., Burleigh, B. D. Jr., Hartley, B. S. 1974. Gene duplication in experimental enzyme evolution. *Nature* 251:200–4
159. Folk, W. R., Berg, P. 1970. Characterization of altered forms of glycyl transfer ribonucleic acid synthetase and the effects of such alterations on aminoacyl transfer ribonucleic acid synthesis in vivo. *J. Bacteriol.* 102:204–12
160. Folk, W. R., Berg, P. 1971. Duplication of the structural gene for glycyl-transfer RNA synthetase in *Escherichia coli. J. Mol. Biol.* 58:595–610
161. Straus, D. S., Hoffmann, G. R. 1975. Selection for a large genetic duplication in *Salmonella typhimurium. Genetics* 80:227–37
162. Legrain, C., Stalon, V., Glansdorff, N., Gigot, D., Pierard, A., Crabeel, M. 1976. Structural and regulatory mutations allowing utilization of citrulline or carbamoylaspartate as a source of carbamoylphosphate in *Escherichia coli K-12. J. Bacteriol.* 128:39–48
163. Hill, C. W., Schiffer, D., Berg, P. 1969. Transduction of merodiploidy: Induced duplication of recipient genes. *J. Bacteriol.* 99:274–78
164. Hill, C. W., Fould, J., Soll, L., Berg, P. 1969. Instability of a missense suppressor resulting from a duplication of genetic material. *J. Mol. Biol.* 39:563–81
165. Basu, S. K., Margolin, P. 1973. The role of *recA* gene in the origin of chromosomal duplications in *Salmonella typhimurium. Abstr. Ann. Meet. Am. Soc. Microbiol.*, G 273
166. Anderson, R. P., Miller, C. G., Roth, J. R. 1976. Tandem duplications of the histidine operon observed following generalized transduction in *Salmonella typhimurium. J. Mol. Biol.* 105:201–18
167. Ravin, A. W., Takahashi, E. A. 1970. Merodiploid ribosomal loci arising by transformation and mutation in *Pneumococcus. J. Bacteriol.* 101:38–52
168. Ledbetter, M. L., Hotchkiss, R. D. 1975. Chromosomal basis of the merozygosity in a partially diploid mutant of *Pneumococcus. Genetics* 80:667–78
169. Kashmiri, S. V. S., Hotchkiss, R. D. 1975. Evidence of tandem duplication of genes in a merodiploid region of Pneumococcal mutants resistant to sulfonamide. *Genetics* 81:21–31
170. Jamet, C., Anagnostopoulos, C. 1969. Étude d'une mutation très faiblement transformable au locus de la thréonine désaminase de *Bacillus subtilis. Mol. Gen. Genet.* 105:225–42
171. Audit, C., Anagnostopoulos, C. 1975. Studies on the size of the diploid region in *Bacillus subtilis* merozygotes from strains carrying the *trpE26* mutation. *Mol. Gen. Genet.* 137:337–51
172. Trowsdale, J., Anagnostopoulos, C. 1975. Evidence for the translocation of a chromosome segment in *Bacillus subtilis* carrying the *trpE26* mutation. *J. Bacteriol.* 122:886–98
173. Trowsdale, J., Anagnostopoulos, C. 1976. Differences in the genetic structure of *Bacillus subtilis* strains carrying the *trpE26* mutation and strain *168.* J *J. Bacteriol.* 126:609–18
174. Anagnostopoulos, C., Trowsdale, J. 1976. Production of merodiploid clones in *Bacillus subtilis* strains. In *Microbiology—1976*, ed. D. Schlessinger, p. 44. Washington DC: Am. Soc. Microbiol.
175. Anagnostopoulos, C. 1976. Genetic analysis of *Bacillus subtilis* strains carrying chromosomal rearrangements. In *Modern Trends in Bacterial Transformation,* ed. A. Porcolés, R. López, M. Espinosa, pp. 211–30. Amsterdam: Elsevier North-Holland Biomed. Press
176. Stodolsky, M., Rae, M. E., Mullenbach, E. 1972. The addition of lac^+ chromosome fragments to the *E. coli proA-proB-lac* deletion *XIII* chromosomes. *Genetics* 70:495–510

177. Stodolsky, M. 1974. Recipient gene duplication during generalized transduction. *Genetics* 78:809–22
178. Weil, J., Terzaghi, B. 1970. The correlated occurrence of duplications and deletions in phage T4. *Virology* 42:234–37
179. Homyk, T. Jr., Weil, J. 1974. Deletion analysis of two nonessential regions of the T4 genome. *Virology* 61:505–23
180. Parma, D. H., Ingraham, L. J., Snyder, M. 1972. Tandem duplications of the *rII* region of bacteriophage *T4D. Genetics* 71:319–35
181. Parma, D. H., Snyder, M. 1973. The genetic constitution of tandem duplications of the *rII* of bacteriophage *T4D. Genetics* 73:161–83
182. Van de Vate, C., Symonds, N. 1974. A stable duplication as an intermediate in the selection of deletion mutants of phage T4. *Genet. Res.* 23:87–105
183. Van de Vate, C., Van den Ende, P., Symonds, N. 1974. A stable duplication of the *rII* region of bacteriophage T4. *Genet. Res.* 23:107–13
184. Bellett, A. J. D., Busse, H. G., Baldwin, R. L. 1971. Tandem genetic duplications in a derivative of phage λ. See Ref. 2, pp. 501–13
185. Busse, H. G., Baldwin, R. L. 1972. Tandem genetic duplications in a derivative of phage λ. II. Evidence for structure and location of endpoints. *J. Mol. Biol.* 65:401–12
186. Emmons, S. W. 1974. Bacteriophage λ derivative carrying two copies of the cohesive end site. *J. Mol. Biol.* 83:511–25
187. Emmons, S. W., MacCosham, V., Baldwin, R. L. 1975. Tandem genetic duplications in phage λ. III. The frequency of duplication mutants in two derivatives of phage λ is independent of known recombination systems. *J. Mol. Biol.* 91:133–46
188. Emmons, S. W., Thomas, J. O. 1975. Tandem genetic duplications in phage λ. IV. The locations of spontaneously arising tandem duplications. *J. Mol. Biol.* 91:147–52
189. Jackson, E. N., Yanofsky, C. 1973. Duplication-translocations of tryptophan operon genes in *Escherichia coli. J. Bacteriol.* 116:33–40
190. Yamagishi, H., Inokuchi, H., Ozeki, H. 1976. Excision and duplication of $su3^+$-transducing fragments carried by bacteriophage $\phi 80$. I. Novel structure of $\phi 80sus2psu3^+$ DNA molecules. *J. Virol.* 18:1016–23

191. Hill, C. W., Combriato, G. 1973. Genetic duplications induced at very high frequency by ultraviolet irradiation in *Escherichia coli. Mol. Gen. Genet.* 127:197–214
192. Straus, D. S. 1974. Induction by mutagens of tandem gene duplications in the *glyS* region of the *Escherichia coli* chromosome. *Genetics* 78:823–30
193. Andrésson, O. S., Magnusdottir, R. A., Eggertson, G. 1976. Deletions of ribosomal protein genes in *Escherichia coli.* Merodiploids heterozygous for resistance to streptomycin and spectinomycin. *Mol. Gen. Genet.* 144:127–30
194. Fraenkel, D. G., Banerjee, S. 1972. Deletion mapping of *zwf.* The gene for a constitutive glucose 6-phosphate dehydrogenase in *Escherichia coli. Genetics* 71:481–89
195. Castellazzi, M., Brachet, P., Eisen, H. 1972. Isolation and characterization of deletions in bacteriophage λ residing as prophage in *E. coli* K12. *Mol. Gen. Genet.* 117:211–18
196. Alper, M. D., Ames, B. N. 1975. Positive selection of mutants with deletions of the *gal-chl* region of the *Salmonella* chromosome as a screening procedure for mutagens that cause deletions. *J. Bacteriol.* 121:259–66
197. Saedler, H., Gullon, A., Fiethen, L., Starlinger, P. 1968. Negative control of the galactose operon in *E. coli. Mol. Gen. Genet.* 102:79–88
198. Schwartz, D. O., Beckwith, J. R. 1969. Mutagens which cause deletions in *Escherichia coli. Genetics* 61:371–76
199. Langridge, J., Campbell, J. H. 1969. Classification and intragenic position of mutations in the β-galactosidase gene of *E. coli. Mol. Gen. Genet.* 103:339–47
200. Ishii, Y., Kondo, S. 1972. Spontaneous and radiation-induced deletion mutations in *Escherichia coli* strains with different DNA repair capacities. *Mutat. Res.* 16:13–25
201. Yamamoto, K., Ishii, Y. 1974. 4-nitroquinoline 1-oxide-induced deletion mutations in *Escherichia coli* strains with different DNA repair capacities. *Mutat. Res.* 22:81–83
202. Ishii, Y., Kondo, S. 1975. Comparative analysis of deletion and base-change mutabilities of *Escherichia coli B* strains differing in DNA repair capacity (wild type, $uvrA^-$, $polA^-$, $recA^-$) by various mutagens. *Mutat. Res.* 27:27–44
203. Conkling, M. A., Grunau, J. A., Drake, J. W. 1976. Gamma-ray mutagenesis in bacteriophage T4. *Genetics* 82:565–75

204. Adams, A., Oishi, M. 1972. Genetic properties of arsenate-sensitive mutants of *Bacillus subtilis 168. Mol. Gen. Genet.* 118:295–310
205. Coukell, M. B., Yanofsky, C. 1970. Increased frequency of deletions in DNA polymerase mutants of *Escherichia coli. Nature* 228:633–35
206. Spudich, J. A., Horn, V., Yanofsky, C. 1970. On the production of deletions in the chromosome of *Escherichia coli. J. Mol. Biol.* 53:49–67
207. Coukell, M. B., Yanofsky, C. 1971. Influence of chromosome structure on the frequency of the *tonB trp* deletions in *Escherichia coli. J. Bacteriol.* 105:864–72
208. Ames, B. N., Hartman, P. E., Jacob, F., 1963. Chromosomal alterations affecting the regulation of histidine biosynthetic enzymes in *Salmonella. J. Mol. Biol.* 7:23–42
209. DeWitt, S. K., Adelberg, E. A. 1962. The occurrence of a genetic transposition in a strain of *Escherichia coli. Genetics* 47:577–85
210. Jacob, F., Wollman, E. L. 1961. *Sexuality and the Genetics of Bacteria,* p. 167. New York & London: Academic
211. Chow, L. 1977. The sequence arrangements of the *E. coli* chromosome and of putative insertion sequences, as revealed by electron microscope heteroduplex studies. See Ref. 29
212. Deonier, R. C., Hadley, R. G. 1976. Distribution of inverted IS-length sequences in the *E. coli K-12* genome. *Nature* 264:191–93
213. Fiandt, M., Hradecna, Z., Lozeron, H. A., Szybalski, W. 1971. Electron micrographic mapping of deletions, insertions, inversions and homologies in the DNAs of coliphages *lambda* and *phi 80.* See Ref. 2, pp. 329–54
214. Deleted in proof
215. Jaskunas, S. R., Nomura, M., Davies, J. 1974. Genetics of bacterial ribosomes. In *Ribosomes,* ed. M. Nomura, A. Tissiéres, P. Lengyel, pp. 333–68. Cold Spring Harbor, NY: Cold Spring Harbor Lab.
216. Wu, M., Davidson, N. 1975. Use of gene 32 protein staining of single-strand polynucleotides for gene mapping by electron microscopy: Application to the $\phi 80d_3$ *ilv su$^+$7* system. *Proc. Natl. Acad. Sci. USA* 72:4506–10
217. Lund, E., Dahlberg, J. E., Lindahl, L., Jaskunas, S. R., Dennis, P. P., Nomura, M. 1976. Transfer RNA genes between 16s and 23s rRNA genes in rRNA transcription units of *E. coli. Cell* 7:165–77
218. Jaskunas, S. R., Lindahl, L., Nomura, M. 1975. Identification of two copies of the gene for the elongation factor EF-Tu in *E. coli. Nature* 257:458–62
219. Kikuchi, A., Gorini, L. 1975. Similarity of genes *argF* and *argI. Nature* 256:621–24
220. Russell, R. L., Abelson, J. N., Landy, A., Gefter, M. L., Brenner, S., Smith, J. D. 1970. Duplicate genes for tyrosine transfer RNA in *Escherichia coli. J. Mol. Biol.* 47:1–13
221. Sanderson, K. E. 1976. Genetic relatedness in the family enterobacteriaceae. *Ann. Rev. Microbiol.* 30:327–49
222. Wallace, D. C., Horowitz, H. J. 1973. Genome size and evolution. *Chromosoma* 40:121–26
223. Hopwood, D. A. 1967. Genetic analysis and genome structure in *Streptomyces coelicolor. Bacteriol. Rev.* 31:373–403
224. Zipkas, D., Riley, M. 1975. Proposal concerning mechanism of evolution of the genome of *Escherichia coli. Proc. Natl. Acad. Sci. USA* 72:1354–58
225. Fincham, J. R. S., Sastry, G. R. K. 1974. Controlling elements in maize. *Ann. Rev. Genet.* 8:15–50
226. Hozumi, N., Tonegawa, S. 1976. Evidence for somatic rearrangement of immunoglobuline genes coding for variable and constant regions. *Proc. Natl. Acad. Sci. USA* 73:3828–32

IMMUNOGENETICS OF CELL SURFACE ANTIGENS OF MOUSE LEUKEMIA

✣3118

Lloyd J. Old and Elisabeth Stockert
Memorial Sloan-Kettering Cancer Center, New York, New York 10021

CONTENTS

INTRODUCTION	128
TECHNIQUES FOR THE IN VITRO ANALYSIS OF SURFACE ANTIGENS OF LEUKEMIA CELLS	129
Antibody Detection Systems	129
Preparation of Antisera	129
Qualitative and Quantitative Absorption Analysis In Vitro	132
Standard Test Cells	132
Other Techniques for Analyzing Cell Surface Antigens	134
CELL SURFACE ANTIGENS RELATED TO NATURALLY OCCURRING MURINE LEUKEMIA VIRUSES	134
Background	134
GCSA: Recognition and Initial Analysis	135
Expression of GCSA in $GCSA^+ \times GCSA^-$ Crosses	137
The G_{IX} Antigenic System	138
Genetics of G_{IX}	138
Relation of GCSA and G_{IX} to MuLV Structural Components	140
GCSA and G_{IX} Phenotype of Mouse Leukemia	140
GCSA and G_{IX} Induction by MuLV	140
Preleukemic Changes in Surface Alloantigens and MuLV-Related Antigens on AKR Thymocytes	142
X.1 System	142
$G_{(RADA1)}$ and $G_{(ERLD)}$	143
PC.1 Antigen: Relation to MuLV?	143
CELL SURFACE ANTIGENS OF THE TL SYSTEM	144
Background	144
Antigens Specified by the Tla Locus	145
Influence of Tla Haplotype on Expression of TL and H-2 Antigens by Thymocytes	147
Antigenic Modulation	147
TL Phenotype of Mouse Leukemia	148

Mapping of TL Antigens on the Surface of Thymocytes and Leukemia Cells	151
Preleukemic Expression of TL Antigens	153
TL and MuLV: Two Mechanisms of Leukemogenesis?	153
ML SURFACE ANTIGEN OF MOUSE LEUKEMIA	154
GENERAL COMMENTS	155

INTRODUCTION

This review deals with a description of the various categories of cell surface antigens that have now been defined on mouse leukemia. Because of the range of inbred mouse strains with known susceptibility to spontaneous or induced tumor development, analysis of cell surface antigens is most advanced with tumors of this species. Much of this work has been motivated by the search for antigens that are distinctive for malignant cells. Although initial evidence for such antigens came from experiments involving transplant rejection, serological techniques have proved far more powerful in sorting out the array of gene products that are expressed on the surface of tumor cells. As with so many areas of cancer research, the study of cancer illuminates much about normal cells, and this has certainly been true for the serological study of surface antigens of leukemia cells (1). A few examples suffice to illustrate this point. The surface markers that have come to be known as differentiation antigens, because their appearance relates to particular pathways of cellular differentiation, were initially recognized during an analysis of antisera to leukemia cells. The realization that other surface components of normal mouse cells were products of integrated viral genes came from the study of leukemias induced by murine leukemia virus (MuLV). The TL (thymus-leukemia) antigen, another surface marker of normal cells that was discovered during a study of mouse leukemia, is a product of a structural gene universally present in the mouse, but normally expressed in only certain strains. The surprising appearance of TL in the leukemias of strains that ordinarily do not express the antigen revealed the presence of the normally silent TL gene and represents the clearest example of genetic derepression or activation in malignant cells. Analysis of the TL system also gave rise to the recognition that surface markers of mammalian cells can undergo rapid and reversible changes following exposure to antibody comparable in some ways to those originally observed with protozoa, and antigenic modulation—as the phenomenon of TL antibody–induced suppression of TL came to be called—can now be viewed in relation to the fluidity of the surface membrane and the mobility of many of its constituents.

Our emphasis in this review is on surface antigens defined serologically. Because leukemia cells can be easily obtained in free cell suspension and are highly sensitive to cytotoxic antibody, they have been a favorite object of serological study. Mouse leukemias frequently arise in the thymus, and this permits comparative serological study of leukemia cells and normal thymocytes, two cell populations of similar derivation. Of the various surface antigens that have been recognized on mouse leukemia, those related to the MuLV-Gross and TL systems are among the most important with regard to natural leukemogenesis, and for this reason their discovery and analysis are the main topics of this review.

TECHNIQUES FOR THE IN VITRO ANALYSIS OF SURFACE ANTIGENS OF LEUKEMIA CELLS

Antibody Detection Systems

A variety of serological techniques have been devised to detect antibody to cell surface antigens. Nearly all depend on two properties of the attached antibody: (a) reaction with anti-immunoglobulin reagents labeled with visual (fluorescein, red blood cells, ferritin), radioactive (^{125}I, etc), or enzymatic (horseradish peroxidase) markers, or (b) ability to fix complement. Although fixation of complement can be detected by immune adherence, anticomplement reactions, or complement consumption assays, the consequence of complement fixation that has been most useful for analyzing surface antigens is cell lysis. This ability of antibody directed to components of the cell surface to kill target cells in the presence of complement is the basis for the cytotoxic test originally devised by Gorer & O'Gorman to demonstrate H-2 antigens on nucleated cells (2). Since its description, the cytotoxic test has been subject to considerable modification and improvement and has evolved into a powerful technique with exquisite sensitivity and reliability for the demonstration and quantitation of cell surface antigens. In fact, almost without exception, the surface antigens that have now been defined on nucleated cells of the mouse were initially discovered through application of this fundamentally very simple test.

Preparation of Antisera

The antisera that recognize the systems of surface antigens on thymocytes and leukemia cells of the mouse were raised by a variety of immunization procedures. Table 1 illustrates ten general categories of immune sera that have been useful in defining non-H-2 specificities on nucleated cells. Because antibody to products of the H-2 complex are usually the dominant ones formed during alloimmunization, the search for surface antigens determined by other loci has depended on the use of antisera that lack H-2 antibody or on serological test systems that make it irrelevant whether H-2 antibody is present or not. The recognition of TL and GCSA antigens was based on this latter procedure (Category 1, Table 1). In both cases, C57BL/6 (C57BL) mice were immunized with H-2 incompatible leukemia cells, resulting in antisera with high H-2 titers and, as subsequent analysis showed, anti-TL and anti-GCSA antibodies as well. Although absorption procedures, either in vivo or in vitro, could have been used to remove anti-H-2, this turned out to be unnecessary. Selected C57BL leukemias were found that expressed TL or GCSA and, by using these target cells in cytotoxic tests with the C57BL antisera, reactions due to H-2 antibodies were eliminated and the GCSA and TL specificities could be defined without interference by irrelevant alloantibodies. Of course, this method of antiserum analysis is possible only under conditions where the same antigen is expressed on the allogeneic cells used for immunization and on the syngeneic cells used for serological testing. The initial detection of the surface antigens related to Friend, Moloney, or Rauscher murine leukemia viruses (FMR complex) also came about through the application of this procedure of H-2 incompatible immunization (5). However, because these viruses induce leukemias with uniquely strong immunogenicity (6, 7), it was possible to produce cytotoxic antiserum by immunizing

Table 1 General description of immunization procedures for the production of cytotoxic antibody to cell surface antigens of the mouse

Category	Examples Antiserum preparation	Test cell	Cell surface antigen detected	Reference
1 H-2 incompatible immunization	C57BL anti-A strain spontaneous leukemia	C57BL X-ray leukemia	TL	3
	C57BL anti-AKR spontaneous leukemia	C57BL MuLV-Gross leukemia	GCSA	4
2 Syngeneic immunization	C57BL anti-MuLV-Rauscher leukemia	C57BL MuLV-Rauscher leukemia	FMR	6
3 Syngeneic immunization (rat)	(W/Fu × BN) F_1 anti-W/Fu MuLV leukemia absorbed in vitro	129 thymocyte	G_{IX}	8, 9
4 Congenic immunization	C57BL-TL$^+$ anti-C57BL X-ray leukemia	C57BL X-ray leukemia	TL.4	10
5 F_1 hybrid immunization	(BALB/c × C57BL) F_1 anti-BALB/c X-ray leukemia	BALB/c X-ray leukemia	X.1	11
6 H-2 compatible immunization	C3H anti-AKR thymocytes	AKR thymocyte	Thy-1 (formally θ)	12
	DBA/2 anti-BALB/c myeloma	BALB/c myeloma	PC.1	13
7 H-2 incompatible immunization: in vivo absorption	C57BL anti-AKR spontaneous leukemia absorbed in vivo in AKR	I strain thymocyte	Lyt-1 (formally Ly-A, Ly-1)	14
8 Heteroimmunization with mouse cells	Rabbit anti-mouse thymocytes absorbed in vivo in donor strain	thymocyte	MSLA	16
9 Heteroimmunization with MuLV structural components	Goat or rabbit anti-MuLV gp70, p30, p15	thymocyte or leukemia cell	gp70, p30, p15	20–23
10 Natural antibody to MuLV-related antigens	(C57BL-G_{IX}^+ × 129) F_1 normal serum	129 thymocyte	G_{IX}	27
	Swiss mouse normal serum	A strain X-ray leukemia	$G_{(RADA1)}$	Obata, Y., unpublished

mice with syngeneic FMR leukemias (Category 2). These FMR reagents proved of considerable value, not only because of high titer but also because they lack alloantibodies. For similar reasons, antisera prepared in inbred rats against syngeneic leukemias induced by MuLV-Gross have been widely used, and analysis of these rat antisera has resulted in the recognition of the G_{IX} antigenic system (Category 3).

Syngeneic immunization would appear to be the method of choice for producing antibody with specificity for leukemia cells. However, with the notable exception of FMR leukemias in the mouse, it has generally not been possible to detect specific antibody after syngeneic immunization with spontaneous, radiation-, or chemically induced mouse leukemias. Why this should be so is not known, but it may have to do with the inability of certain surface determinants to trigger a humoral immune response in the absence of additional, perhaps stronger, antigenic differences or be related to the absence of *Ir* genes necessary for antigen recognition. Experience with mice that are congenic for surface alloantigens, such as Thy-1 or Ly, may help to clarify this matter. We had expected that as the number of unrelated alloantigenic differences between donor and recipient were reduced, the immune response to a single surface determinant would be correspondingly increased. Although this proved to be true in several instances, immunization of congenic pairs exhibiting Thy-1 or single Ly differences results in little or no antibody. If, however, the congenic recipient is crossed to an unrelated mouse, the hybrid produces a good immune response to the determinant, and this result has been ascribed to heterozygosity at *Ir* loci. In this regard, one specificity of the TL system, TL.4, was recognized by immune sera produced in TL congenic mice (Category 4), but this cannot be considered a strictly congenic immunization, because C57BL and its congenic partner C57BL-TL$^+$ differ at more than the *Tla* locus. The most striking example of an immune response to a leukemia-associated antigen being determined by *Ir* genes relates to the X.1 system of BALB/c leukemias (Category 5). Syngeneic mice are incapable of forming anti-X.1, whereas hybrids with C57BL can produce the antibody. Genetic analysis shows that *H-2* linked *Ir* genes contributed by the C57BL partner are essential for X.1 recognition.

The definition of surface alloantigens characteristic of lymphocytes of T or B lineage and their malignant derivatives resulted from other methods of antisera preparation and analysis. Thy-1 and PC.1 alloantigens were detected and originally defined with antisera prepared in H-2–compatible mice (Category 6). The Ly series of T-cell alloantigens were recognized during analysis of antisera against leukemia cells prepared in H-2–incompatible mice. To remove H-2 antibody, these antisera were absorbed in vivo in normal mice of the strain in which the leukemia arose (Category 7). The absorbed sera retained antibody that reacted strongly with thymocytes and, in lower titers, with peripheral lymphocytes. Later tests proved that these antisera defined antigens that were restricted to T lymphocytes, and the two systems they identified were called Ly-A and Ly-B. Once the strain distribution of these systems was known, antisera could be prepared in H-2–compatible combinations that did not require in vivo absorption (15). These antisera, as well as others that have been defined subsequently, are proving of considerable value in distinguishing the various functionally distinct populations of T cells.

Heteroimmune sera prepared against mouse cells have been of only limited value in the analysis of cell surface antigens (Category 8). Properly absorbed, they can be useful markers for different cell populations in the mouse; absorbed heteroantisera with specificity for T lymphocytes (MSLA), B lymphocytes (MBLA), or plasma cells (MSPCA) have been described (16–19). As species antigens, they are not amenable to conventional genetic analysis, but for the study of interspecies hybrids in somatic cell genetics, they may find considerable use. In addition, a range of heteroantisera to individual structural proteins of MuLV are available and are now being used to study the expression of viral antigens on the cell surface (Category 9).

Since the original detection of antibody to MuLV-related antigens in normal mouse serum (24), naturally occurring antibody to MuLV structural antigens and to MuLV cell surface antigens has been found in a number of mouse strains, particularly those with low incidence of spontaneous leukemia (25, 26) (Category 10). Two such natural antibodies of the mouse are of particular interest. One is directed to the G_{IX} antigen which was originally defined by rat antiserum. Anti-G_{IX} had not previously been detected in inbred mice, and attempts at deliberate immunization of G_{IX}^- mice with G_{IX}^+ cells failed to induce antibody. However, certain $G_{IX}^+ \times G_{IX}^-$ or G_{IX}^+ hybrids form G_{IX} antibody, and this finding has also been explained on the basis of Ir genes necessary for G_{IX} recognition. The other naturally occurring antibody which defines a new system of MuLV-related antigens, $G_{(RADA1)}$, is found in the sera of random-bred Swiss mice (Category 10).

Qualitative and Quantitative Absorption Analysis In Vitro

Once an antiserum with sufficiently high titer for a target cell has been selected, the specificity of the cytotoxic antibody can be analyzed by absorption tests. With this technique, cells, cellular membranes, tissue homogenates, or isolated viral or cellular antigens can be tested for the presence of the relevant antigen by determining whether specific antibody for the target cell is removed or not from the antiserum. Absorption tests are invaluable for defining antigenic systems, particularly because of greater sensitivity of antigen detection than direct tests and because solid tissues such as liver, brain, and lung (where it would be impossible to obtain suitable target cells for direct tests) can be examined by absorption. Absorption tests to detect presence or absence of antigen are called qualitative absorptions. Absorption techniques can also be performed to quantitate the amount of surface antigen in a cell population. These quantitative absorption procedures have been useful in analyzing genotypic and phenotypic aspects involving cell surface antigens and in developing a method for mapping their spatial relations on the cell surface.

Standard Test Cells

Several lines of transplanted tumors have become prototype cells for the definition of certain cell surface antigens, and, for this reason, their surface antigenic structure is well characterized. The surface phenotype for some of these cells is given in Table 2. Even though many of these lines have been transplanted for a decade or so, their surface phenotype is remarkably stable. In a sense, these transplanted tumors represent the modern counterpart of the genetically undefined Ehrlich tumor and S-180 sarcoma so widely used by a former generation of cancer investigators. Although growing emphasis is being placed on using tumors of more recent origin in studies of cancer immunology, cells with well-characterized antigenic structure greatly facilitate the analysis of immune reactions to specific antigens and the definition of new antigenic systems. In fact, a good case could be made for the collective use of a limited number of these prototype cells by laboratories addressing various aspects of cell surface antigens, so that a comprehensive picture of the surface structure of a neoplastic mammalian cell could be constructed.

CELL SURFACE ANTIGENS OF MOUSE LEUKEMIA   133

Table 2 Surface phenotype of standard transplantable tumor cell lines of the mouse

Designation	Strain of origin	Tumor induction	H-2	TL 1 2 3 4	Thy-1	Lyt 1 2 3 4	PC.1	G_{IX}	GCSA	$G_{(RADA1)}$	gp70	MuLV p30	p15
ERLD	C57BL/6	X-ray leukemia	b	1 2 – 4	2	1 2 3 4	–	–	–	–	+	–	–
EδG2	C57BL/6	MuLV Gross leukemia	b	– – – –	2	1 – 3 4	–	+	+	+	+	+	+
EL4	C57BL	DMBA leukemia	b	– – – –	2	1 2 3 4	–	–	–	–	+	–	–
K36	AKR	Spontaneous leukemia	k	– – – –	–	– 2 – –	–	+	+	+	+	+	+
RLδ1	BALB/c	X-ray leukemia	d	1 2 – –	2	1 2 3 4	–	+	+	–	+	–	+
Meth A	BALB/c	Methylcholan-threne sarcoma	d	– – – –	–	– – – –	–	–	–	–	–	–	–
MOPC-70A	BALB/c	Mineral oil myeloma	d	– – – –	–	– – – –	1	–	+	–	–	–	+
ASL1	A	Spontaneous leukemia	a	1 2 3 –	2	1 2 3 4	–	–	–	–	+	–	–
RADA1	A	X-ray leukemia	a	1 2 3 –	2	– – 3 4	–	+	–	+	+	–	–

Other Techniques for Analyzing Cell Surface Antigens

After a new surface antigen has been serologically identified and distinguished from known systems by its strain and tissue distribution, the procedure for its further analysis is becoming rather straightforward. *Biochemical characterization* is possible by techniques that were originally developed to analyze the product of the *H-2* locus. *Relation to MuLV* can be assessed by absorption tests with known MuLV$^+$ and MuLV$^-$ cells and with structural components of MuLV, as well as by determining whether the antigen can be induced in cells by MuLV infection. *Chromosomal mapping* of genes determining surface antigens is facilitated by the extensive range of biochemical markers that distinguish inbred strains and the availability of recombinant inbred lines of mice. The techniques of somatic cell genetics will undoubtedly become increasingly useful for mapping genes specifying cell surface antigens, particularly those that are not amenable to conventional genetic analysis (e.g. species antigens and tumor-specific antigens). *Construction of congenic mice* on defined genetic background is now a standard approach to the immunogenetic analyses of serologically defined cell surface alloantigens, and congenic stocks for TL, G_{IX}, the Ly series, Thy-1 and PC.1, in addition to those related to the H-2 complex, are available.

Several years ago, there was considerable hope that the topographical relationship of cell surface antigens could be visualized by immunoelectronmicroscopy (1, 28). The introduction of a double labeling technique with two visually distinguishable markers made it possible to think about visual maps of antigens on the cell surface. However, the realization that the antibody probes themselves caused aggregation and rearrangement of the surface antigens led to the discontinuation of such studies and to the current search for methods that will give a picture of the antigenic mosaic as it exists in the unperturbed state. A method that may be of considerable value in determining the spacial relationship of cell surface antigens is the antibody-blocking technique, and the current status of these studies is discussed below.

CELL SURFACE ANTIGENS RELATED TO NATURALLY OCCURRING MURINE LEUKEMIA VIRUSES

Background

Murine leukemia viruses (MuLV) belong to a family of structurally related RNA viruses that are known to infect an extraordinary range of animals, from snakes and birds to higher apes. Although overt disease is a rare manifestation of infection by these viruses, they can under appropriate circumstances induce cancer, and it is this feature that has led to their being called *oncornaviruses*. In the mouse, three types of cancers (leukemia, sarcoma, and mammary adenocarcinoma) are closely associated with oncornaviruses, and there are indications that they may also play a role in autoimmune diseases as well. The presence of a DNA intermediary in the life cycle of oncornaviruses provides opportunity for close interaction with the genome of the host cell, and there is considerable evidence that genetic information related to these viruses resides in an integrated state in most, if not all, mice (29). The degree to which viral genes are expressed is determined by other factors, some clearly under the control of host genes. The picture that is emerging from the analysis of MuLV is one of extensive polymorphism of endogenous viruses, with several distinct classes of MuLV having been identified on the basis of host range, interference patterns, antigenicity, nucleic acid hybridization, and peptide mapping.

Serological study of MuLV-related antigens began after L. Gross demonstrated that leukemogenic virus could be isolated from mice of the high leukemic AKR strain (30). Subsequently, Friend, Moloney, Rauscher and others isolated MuLV variants with properties that distinguished them from Gross's original isolate (31). In these early studies, rabbit antisera to viral filtrates resulted in antibody of low titer, undoubtedly a consequence of little viral antigen in the immunizing inocula. Because of the difficulties in obtaining potent heteroantibody to MuLV, attention turned to the immune response of mice to surface antigens of MuLV-induced leukemias. It soon became evident that cytotoxic antisera could be easily raised to leukemias induced by Friend, Moloney, and Rauscher (FMR) viruses, whereas it proved more difficult to produce high titered antibody to leukemias induced by Gross virus. These early serological studies indicated that mouse leukemias could be placed into two groups on the basis of their surface antigens: (a) leukemias induced by FMR viruses sharing the FMR complex of antigens and (b) leukemias induced by Gross virus or occurring spontaneously in high leukemia incidence strains sharing G(Gross) antigen (4, 6, 32). As FMR antigens were restricted to leukemias induced by FMR viruses and could not be found in any other normal or neoplastic cell, it was concluded that these viruses played no role in the development of naturally occurring leukemia. The occurrence of G antigen in spontaneous and induced leukemias and in sarcomas of low leukemia incidence strains indicated widespread infection of mouse populations with MuLV and provided an explanation for the poor immunological reactivity of mice to MuLV-Gross-related antigens.

An important step in MuLV serology was taken when it was found that rat antisera to syngeneic MuLV-induced leukemias detected a far broader range of MuLV antigens than did mouse antisera (8). Rats, which lack endogenous MuLV, are highly susceptible to leukemia induction by MuLV. The transplantation behavior of these induced rat leukemias suggested strong immunogenicity; in order to maintain transplanted lines of certain rat leukemias, it was necessary to serially pass them in immunologically immature recipients. If, however, sufficiently large numbers of leukemia cells were transplanted to adult animals, progressively growing tumors resulted and the sera of these rats contained high levels of cytotoxic, neutralizing, and precipitating MuLV antibodies. Analyses of these rat antisera gave rise to the recognition of the G_{IX} antigenic system (9), the major group-specific core antigen of MuLV (8, 20), and the interspecies antigen shared by mammalian leukemia-sarcoma viruses (33). With the development of methods to concentrate and purify MuLV, heteroimmune sera prepared in rabbits and goats to intact virus and to isolated structural components have become valuable reagents to identify and quantitate virion antigens (22, 34).

GCSA: Recognition and Initial Analysis

The first attempts to detect cell surface antigens related to MuLV-Gross were carried out with antisera prepared in C3H mice by repeated sublethal inocula of syngeneic virus-induced leukemia cells (35). These antisera showed weak and variable reactivity, most likely a reflection of the high susceptibility of this strain to Gross virus leukemogenesis. For this reason, C57BL mice were chosen as a source of antibody, since it was known from the work of Gross that mice of this strain were exceptionally resistant to MuLV. From a large series of C57BL mice injected as newborns with Gross virus, a small number of leukemias were induced and, from one of the leukemic mice, a transplanted leukemia line designated E♂G2 was established. The C57BL antiserum showing highest titer against E♂G2 was prepared against a transplanted AKR spontaneous leukemia, K36 (Table 2 and 3). Absorption analysis of this antiserum showed that all leukemias induced by MuLV-Gross and all spontaneous leukemias occurring in mice of high leukemia incidence strains absorbed cytotoxic activity. In tests of normal young mice from different inbred strains, occurrence of antigen in spleen and other lymphoid tissues correlated closely with the incidence of spontaneous leukemia (Table 4). All high incidence strains, AKR, C58, PL, C3H/Figge, and F, were antigen positive, whereas low incidence strains, e.g. C57BL, A, and BALB/c, lacked antigen. In view of its obvious relationship to the leukemia virus originally described by Gross, the antigen was named G (Gross) cell surface antigen (GCSA) (4). It soon became clear that GCSA was not restricted to high incidence strains, but could also be found in normal and malignant tissues of low incidence strains. In fact, typing for GCSA provided some of the earliest evidence for the widespread infection of mouse population with MuLV. Spontaneous and X-ray–induced leukemias and chemically induced sarcomas of GCSA$^-$ strains were occasionally GCSA$^+$. Cell cultures derived from GCSA$^-$ mice frequently became GCSA$^+$ after in vitro passage. Age was also found to play a role in determining GCSA expression in certain GCSA$^-$ strains of mice. C3Hf/Bi, which are GCSA$^-$ at 2 months of age, became GCSA$^+$ later in life (36) and strains showing this age-related GCSA$^-$ → GCSA$^+$ change are referred to as *conversion strains* (Table 4). Electronmicroscopy revealed a good correlation between GCSA and occurrence of MuLV particles in both normal and tumor tissue, and much subsequent study has substantiated the early impression that GCSA is an almost invariable marker for MuLV replication in the mouse.

Table 3 Cell surface antigens related to MuLV-Gross defined by cytotoxic tests with antisera from mice or rats

Designation	Antiserum	Standard test cell	Relation to MuLV structural component	Reference
GCSA	C57BL anti-AKR spontaneous leukemia K36	E♂G2 leukemia	p15, p30	4, 46, 47
G_{IX}	1 (W/Fu × BN)F_1 anti-W/Fu MuLV induced leukemia (C58NT)D^a 2 (C57BL-G_{IX}^+ × 129)F_1 normal mouse serum	129 or C57BL-G_{IX}^+ thymocyte	gp70	9, 39, 45 27
X.1	(BALB/c × C57BL)F_1 anti-RL♂1	RL♂1 leukemia	?	11, 53
$G_{(RADA1)}$	Swiss mouse normal serum^b	RADA1 leukemia	gp70	Obata, Y., unpublished
$G_{(ERLD)}$	(C57BL × 129)F_1 normal serum	ERLD leukemia	?	

^{a} Absorbed in vitro with G_{IX}^- thymocytes.
^{b} Absorbed in vitro with G_{IX}^+ thymocytes.

Table 4 Strain distribution of cell surface antigens related to MuLV-Gross

	Inbred strains	GCSA	G_{IX}	X.1	$G_{(RADA1)}$	$G_{(ERLD)}$
High leukemia incidence strains	AKR, C58, C3H/Figge	+	+	+	+	+
Low leukemia incidence strains	A	−	+	−	−	+
	BALB/c	−	−	−	−	−
G_{IX} congenic strains	129	−	+	+	−	+
	129-G_{IX}^-	−	−	+	−	−
	C57BL	−	−	−	−	+
	C57BL-G_{IX}^+	−	+	−	−	+
Conversion strains	C3Hf/Bi young	−	−	−	−	+
	old	+	+	+	+	+
	NZB young	−	+	−	−	
	old	+	+	+	−	

Expression of GCSA in $GCSA^+ \times GCSA^-$ Crosses

In contrast to the codominant expression of other cell surface antigens, F_1 hybrids derived from $GCSA^+ \times GCSA^-$ crosses show either a $GCSA^-$ or $GCSA^+$ phenotype (Table 5). Certain $GCSA^-$ strains, such as C57BL, suppress GCSA in AKR or C58 hybrids, whereas other $GCSA^-$ strains do not (4). This effect of $GCSA^-$ strains on GCSA expression can now be ascribed to the *Fv-1* locus and its control over MuLV replication (37). The two alleles at *Fv-1*, *Fv-1^n* and *Fv-1^b*, determine the susceptibility of mouse cells to infection by N-tropic or B-tropic MuLV. AKR and other high leukemia strains share the *Fv-1^n* allele and leukemogenic virus from these strains have N-tropic properties. Low incidence strains have either the *Fv-1^n* or *Fv-1^b* allele and both N- and B-tropic MuLV have been isolated from these strains, although it is unclear whether these viruses have leukemogenic activity. AKR or C58 hybrids with Fv-1^n low incidence strains are $GCSA^+$, whereas hybrids with Fv-1^b low incidence strains are $GCSA^-$. An important exception to this rule has been found. The CBA strain is Fv-1^n, but crosses with AKR are $GCSA^-$. This absence of GCSA is surprising and deserves further study, especially since (AKR $\times$ CBA)F_1 mice are reported to have a low incidence of leukemia but a high level of MuLV (38).

Table 5 GCSA phenotype* of AKR or C58 crosses with $GCSA^-$ strains

$GCSA^+$ hybrids	$GCSA^-$ hybrids
129 $\times$ AKR†	A $\times$ AKR†
C3Hf/Bi $\times$ AKR†	BALB/c $\times$ AKR†
AKR $\times$ DBA/2	C57BL $\times$ AKR†
AKR $\times$ GR	C57BL-G_{IX}^+ $\times$ AKR
C57L $\times$ AKR	C57BL-H-2^k $\times$ AKR
129-G_{IX}^- $\times$ AKR	CBA $\times$ AKR†
129 $\times$ C58†	C57BL $\times$ C58
C58 $\times$ C3Hf/Bi	A $\times$ C58

*Based on absorption tests with spleen.
†Reciprocal crosses tested.

The G_{IX} Antigenic System

The G_{IX} antigen was recognized during the analysis of the polyvalent MuLV antisera produced in (W/Fu × BN)F_1 rats immunized with a W/Fu leukemia originally induced by MuLV (9) (Table 3). Initial tests with this antiserum gave rise to the impression that its specificity was similar to anti-GCSA, i.e. cytotoxic for $GCSA^+$ leukemias but not for $GCSA^-$ leukemias. Further analysis, however, indicated that its pattern of reactivity was distinct, and this was most evident in cytotoxic tests with normal thymocytes from different inbred mouse strains. Although thymocytes from the high leukemia incidence $GCSA^+$ strains were positive, thymocytes from some $GCSA^-$ strains (e.g. 129, A) were equally and, in some instances, more reactive. Because 129 thymocytes showed maximal sensitivity, they were selected as the prototype cell for defining the specificity of the rat antiserum in absorption tests, and this choice of target cells gave rise to the original designation of $G_{(129)}$ for this antigenic system. The later decision to rename it G_{IX} came from evidence that one of the two genes responsible for antigen expression resided in linkage group IX of the mouse. However, now that this genetic assignment must be questioned, the $G_{(129)}$ designation might be more appropriate. Nevertheless, to avoid unnecessary confusion, the designation G_{IX} will be preserved until a more definitive nomenclature can be devised.

Based on absorption tests with normal thymocytes, inbred mouse strains can be classified as G_{IX}^+ or G_{IX}^- (Table 4). Quantitative absorption analysis provided an explanation for the variable reactivity of thymocytes from different G_{IX}^+ strains detected in direct cytotoxic tests. G_{IX}^+ mouse strains are found to differ in the amount of G_{IX} expressed on thymocytes, and because absorption capacity follows a ratio of 3:2:1 these strains are designated G_{IX}^3, G_{IX}^2, or G_{IX}^1 (Table 6). As all cytotoxic activity for G_{IX}^3 cells can be removed by G_{IX}^1 or G_{IX}^2 thymocytes, and vice versa, the G_{IX} test system apparently detects a single determinant with variable expression. The thymocytes of F_1 hybrids from $G_{IX}^+ \times G_{IX}^-$ matings express 50% of the amount found in the G_{IX}^+ partner and crosses between $G_{IX}^3 \times G_{IX}^2$ or G_{IX}^1 strains also express 50% of the parental G_{IX} levels. In the $GCSA^+$ strains (AKR, C58), G_{IX} is found in spleen and other lymphatic tissue as well as thymus, and levels in the spleen are, in fact, higher than in the thymus. In $GCSA^-$ strains, the thymus is the only lymphoid tissue that expresses G_{IX}. Recent evidence indicates that antigen with G_{IX} specificity is found also in serum, on sperm, and in seminal vesicle fluid of G_{IX}^+ mice (27, 39).

Genetics of G_{IX}

The G_{IX} trait has been the subject of considerable genetic analysis (9, 40–43). The prototype G_{IX}^+ strain 129 has been the object of most study because the G_{IX}^3 thymocyte phenotype facilitates typing and because productive MuLV infection, which itself can cause G_{IX} appearance, does not occur in 129 mice. In backcross and F_2 populations with C57BL (the prototype G_{IX}^- strain), segregation ratios indicate that G_{IX} is specified by two unlinked genes, designated *Gv-1* and *Gv-2* (Gv = Gross virus related). Both *Gv-1* and *Gv-2* are required for G_{IX} expression, but *Gv-1*

Table 6 Strain distribution of the G_{IX}^3, G_{IX}^2, G_{IX}^1, and G_{IX}^- phenotype of mouse thymocytes

G_{IX}^+			G_{IX}^-	
G_{IX}^3	G_{IX}^2	G_{IX}^1		
129	AKR (and AKR.K, AKR-H-2^b)	SJL/J	C57BL (and C57BL-TL$^+$, C57BL-H-2^k, C57BL-Ly-1.1,	
C57BL-G_{IX}^+	AKR. T1ALD	DBA/2	C57BL-Ly-2.2, C57BL-Ly-2.1, Ly-3.1)	
CE	C58	C3H/An	C57BL/10	GR
	C3H/Figge	101	C57BL/Ka	MA
	C3H/He		C57BR	Swiss/NIH
	A (and A-TL$^-$, A-Thy-1.1)		C57L	SWR
	NZB*		129-G_{IX}^-	HSFS
	C57BR-G_{IX}^{+}M*		BALB/c (and BALB/c-T)	B10BR
	C57BL-G_{IX}^{+}M*		RF	HTG
			CBA (and CBA-T6)	HTH
			DBA/1	HTI
				C3Hf/Bi young†

*In quantitative absorption tests, these strains fall between G_{IX}^3 and G_{IX}^2 phenotype.
†Conversion strain.

behaves as a semidominant gene (50% G_{IX} expression in heterozygotes) and *Gv-2* as a dominant gene (full G_{IX} expression in heterozygotes). Comparable crosses with BALB/c (G_{IX}^-) mice give a one-gene ratio for G_{IX} and genetic analysis shows that this strain possesses the *Gv-2* allele. Thus, G_{IX}^- strains are either *Gv-1$^-$/Gv-2$^-$* (C57BL and CBA) or *Gv-1$^-$/Gv-2$^+$* (BALB/c, C57L, and the 129-G_{IX}^- congenic strain). An extensive search for the *Gv-1$^+$/Gv-2$^-$* genotype has not been productive, even in crosses that should have revealed them, and it has been suggested that this genetic constellation is lethal. The original linkage studies indicated that *Gv-1* was located on chromosome 17 (linkage group IX), 37 units from *H-2* (gene order centromere:*H-2*:*Gv-1*) and that *Gv-2* was located on chromosome 1, 34 units from *Hbb* (gene order *Gpi-1*:*Hbb*:*Gv-2*). However, when the G_{IX} trait of AKR mice (G_{IX}^2 strain) was studied, crosses with C57BL or BALB/c showed no linkage of *Gv-1* to *H-2* but indicated a chromosome 4 locus for *Gv-1,* approximately 19 units from the *Fv-1/Gpd-1* region. Two possibilities were considered to explain this discrepancy. Either *Gv-1* occupies different chromosomal sites in AKR and 129 mice or the assignment of *Gv-1* to chromosome 17 in 129 mice was incorrect. Because the *Fv-1* allele of C57BL and BALB/c (*Fv-1^b*) has a marked suppressive effect on MuLV production in crosses involving AKR (*Fv-1^n*) and this might influence the expression of G_{IX} in segregating populations, backcrosses of AKR to C57L (a G_{IX}^-:*Gv-1$^-$/Gv-2$^+$/Fv-1^n* strain) were analyzed. In this *Fv-1^n* homozygous cross, no linkage of *Gv-1* with chromosome 4 (marker Gpd-1) or with chromosome 17 (marker H-2) was noted. Similar findings have resulted from the further analysis of the *Gv-1/H-2* association of 129 mice. Linkage of *Gv-1* to *H-2* is noted when heterozygosity at the *H-2* locus is involved, but disappears when the crosses are homozygous for *H-2* (marker TL; < 2 map units from *H-2*). These results on the association of *Gv-1* with *Fv-1* and *H-2* must therefore be viewed as examples of pseudo- or quasi-linkage, in which certain combinations of unlinked genes appear to be inherited together. In light of this newer information, the initial assignment of *Gv-2* to chromosome 1 must also be reexamined.

Relation of GCSA and G_{IX} to MuLV Structural Components

The finding that G_{IX} and GCSA could be induced in rats by MuLV infection made it likely that MuLV genes directly code for these antigens (8, 9). Although the possibility was considered that G_{IX} and GCSA represented structural components of the virus incorporated into the cell surface during virus maturation, evidence at the time suggested that they were nonstructural products of MuLV. Immunoelectronmicroscopy with GCSA antibody showed labeling of the cell surface but no labeling of the envelope of budding or more mature viral particles (44). In contrast, both viral envelope and cell surface were labeled with broadly reactive anti-MuLV sera. In the case of G_{IX}, the antigen was classified as a nonvirion component because electronmicroscopy and infectivity studies revealed no evidence for MuLV in 129 thymocytes, the prototype G_{IX}^+ cell. With advances in the biochemical analysis of MuLV and new methods to define cell surface molecules, it has now been shown that both G_{IX} and GCSA are in fact viral structural components incorporated into the cell surface. G_{IX} is a type-specific antigen of the major envelope glycoprotein of MuLV, gp70 (39, 45). GCSA is related to the internal core proteins of MuLV, p30, and p15, which occur as glycosolated polyproteins on the surface of infected cells (46, 47). The failure of anti-GCSA to label intact virus can now be understood in relation to the internal location of the antigenically related structural components in the virion. Although GCSA has never been found in the absence of replicating MuLV, viral genes coding for G_{IX} can be expressed independently of other MuLV genes.

GCSA and G_{IX} Phenotype of Mouse Leukemia

Table 7 summarizes our accumulated evidence on expression of GCSA and G_{IX} in leukemias of various inbred and hybrid strains. Leukemias arising in the GCSA$^+$/G_{IX}^+ high leukemia incidence strains invariably express both antigens. In G_{IX}^3, G_{IX}^2, or G_{IX}^1 strains, where G_{IX} appears as a differentiation alloantigen of normal thymocytes, the leukemias are commonly G_{IX}^+, an expected finding in view of the T-cell origin of these leukemias. GCSA and G_{IX} may also occur in leukemias arising in strains that ordinarily do not express these antigens during normal life, revealing the presence of viral structural genes in these mice that are activated as a consequence of leukemogenesis. In some low leukemia strains (BALB/c), GCSA, and G_{IX} activation is common, whereas in others (C57BL), these antigens are only rarely found. In the 129 strain, GCSA activation has never been observed in a leukemia, and this correlates with the fact that MuLV has not been isolated from mice of this strain. (C57BL $\times$ AKR)F$_1$ mice resemble C57BL in the GCSA phenotype of their leukemias, indicating marked suppression of MuLV replication. (C57BL $\times$ A)F$_1$ resemble A mice in showing GCSA activation in a proportion of leukemias. The exceptional leukemia phenotype GCSA$^+$/G_{IX}^- may reflect activation of an MuLV that does not induce G_{IX}.

GCSA and G_{IX} Induction by MuLV

GCSA and G_{IX} can be induced in mouse fibroblasts by MuLV, and this consequence of viral infection has been useful in distinguishing various isolates of MuLV

Table 7 GCSA and G_{IX} phenotype of mouse leukemia

Strains	Thymocyte phenotype $GCSA/G_{IX}$	Leukemia induction	Leukemia cell phenotype			
			$GCSA^+/G_{IX}^+$	$GCSA^+/G_{IX}^-$	$GCSA^-/G_{IX}^+$	$GCSA^-/G_{IX}^-$
High leukemia incidence strains			Number of leukemias tested			
AKR, C58, PL, C3H/Figge	+/+	Spontaneous and urethan	>100			
Low leukemia incidence strains						
BALB/c, RF, DBA/1	-/-	Spontaneous and X-ray	3		2	2
A, A.BY, DBA/2	-/+	Spontaneous and X-ray	5	2	5	2
G_{IX} congenic strains						
129 (G_{IX}^+)	-/+	X-ray			9	
129-G_{IX}^-	-/-	X-ray			2	10
C57BL (G_{IX}^-)	-/-	X-ray	2	1		21
		Spontaneous				2
C57BL-G_{IX}^+	-/+	X-ray			26	
F_1 hybrids						
C57BL × A	-/- × -/+	X-ray	2	1	2	6
C57BL × AKR	-/- × +/+	X-ray			3	7

(48). Two G_{IX} phenotypes (G_{IX}^-, G_{IX}^+) and three GCSA phenotypes (GCSA$^-$, GCSA$^+$, GCSA^{2+}) can be scored in MuLV-infected fibroblasts by absorption of cytotoxic antibody, and five of the six possible G_{IX}/GCSA phenotypes have been found. (The missing phenotype is G_{IX}^+/GCSA$^-$.) With virus isolated from normal mouse tissues, N-tropic MuLV usually induces the G_{IX}^+/GCSA^{2+} phenotype, and B-tropic MuLV the G_{IX}^-/GCSA^{2+} phenotype, although N- and B-tropic MuLV isolates have been identified that do not follow this rule. G_{IX} and GCSA typing of fibroblasts infected with FMR viruses give results that parallel the phenotype of leukemias induced by these viruses—G_{IX}^-/GCSA$^+$. Xenotropic MuLV, a class of MuLV that cannot infect mouse cells but replicates well in certain heterologous cells, does not induce G_{IX}, and this is also true for the recently recognized amphotropic class of MuLV derived from wild mice (49). However, these viruses can be distinguished on the basis of GCSA typing and, in the case of xenotropic MuLV, this appears to correlate with their classification by nucleic acid hybridization and type-specific heteroantisera. Quantitative differences in GCSA levels induced by various MuLV undoubtedly reflect type-specific differences in the core protein of various MuLV, with all N- and B-tropic MuLV being most closely related (GCSA^{2+}), FMR viruses and MuLV isolated from wild mice and certain xenotropic MuLV more distantly related (GCSA$^+$), and AT124 and Caroli xenotropic MuLV being unrelated (GCSA$^-$). Presumably these latter viruses induce type-specific core protein of distinct antigenicity on the cell surface. Cell surface typing for G_{IX} and GCSA, as well as for other type-specific antigens specified by MuLV, may be of considerable value for analyzing MuLV recombinants. In this regard, the new MCF class of MuLV, originally recognized in preleukemic AKR mouse thymus, appears to represent a stable recombinant between a xenotropic and an N-tropic MuLV, having the properties of both classes by neutralization tests, host range, and interference patterns (50). MCF MuLV can induce G_{IX} antigen in infected cells, behaving therefore like an ecotropic virus and not a xenotropic MuLV in this respect.

Preleukemic Changes in Surface Alloantigens and MuLV-Related Antigens on AKR Thymocytes

The study of AKR mice during the preleukemic period (4–6 months) has revealed characteristic changes in the pattern of surface antigens expressed by thymocytes (51, 52). In contrast to the low H-2/high Thy-1 phenotype of normal thymocytes of young AKR mice, thymocytes from 6-month-old AKR show high H-2/low Thy-1 levels. In addition, expression of G_{IX}, GCSA, and other MuLV-related antigens on the surface of thymocytes is greatly amplified in preleukemic AKR mice. As a result of these changes, the surface phenotype of the preleukemic thymocyte comes to resemble the surface characteristics of AKR leukemia cells. Assays for MuLV reveal that the level of N-tropic MuLV remains constant during the 2- to 6-month period. However, virus with xenotropic properties appears in the thymus of 6-month-old animals, and the level of this virus correlates well with amplified MuLV-antigen expression (52). Further characterization of virus isolated from the preleukemic thymus led to the recognition of the MCF class of MuLV (discussed above).

X.1 System

This system was detected during the study of the transplantation behavior of X-ray-induced BALB/c leukemias (11). Although no evidence for resistance to transplants could be elicited in BALB/c mice, (BALB/c × C57BL)F_1 hybrids were found to reject large numbers of BALB/c leukemia cells. Considerable evidence points to the conclusion that this hybrid resistance to leukemic transplants is controlled by the *H-2*-linked *Ir* locus contributed by the C57BL partner. Serum from hybrids that had rejected repeated inocula of BALB/c leukemia RLσ1 defined an antigen, designated X.1, that was present on BALB/c, A, and AKR leukemias (Table 3). X.1 is unrelated to the GCSA or G_{IX} system as the standard G_{IX}^+/GCSA$^+$ leukemia EσG2 types X.1$^-$. X.1 is present in normal tissues of high leukemia strains, but unlike GCSA and G_{IX}, in very low levels. For this reason, mouse strains were typed for X.1 by the method of in vivo absorption (Table 4). All high leukemia strains are GCSA$^+$/G_{IX}^+/X.1$^+$, whereas low incidence GCSA$^-$ strains may be G_{IX}^+/X.1$^+$ (129 mice), G_{IX}^+/X.1$^-$ (A mice), G_{IX}^-/X.1$^+$ (129-G_{IX}^- mice), or G_{IX}^-/X.1$^-$ (BALB/c mice). X.1 antisera also detects a gp70 molecule (designated X-gp70) on the surface of leukemia cells that is distinguishable from G_{IX}-gp70 (53). However, as leukemia cells that type X-gp70$^+$/X.1$^-$ have been found, it appears likely that X-gp70 and X.1 represent distinct antigenic systems. Although the presence of X.1 in the normal tissues of high leukemia incidence strains and its appearance in leukemias of X.1$^-$ strains suggest an association with MuLV, more direct evidence that X.1 is an MuLV-related antigen (i.e. induction in MuLV-infected fibroblasts) is lacking.

$G_{(RADA1)}$ and $G_{(ERLD)}$

The serum of normal mice, particularly F_1 hybrids and random-bred Swiss mice, is proving to be an exceptional source of antibody capable of distinguishing new cell surface antigens related to MuLV. As discussed before, this may have to do with heterozygosity at Ir loci, permitting a broader range of immune reactions to antigens of endogenous and exogenous origin. We know that (C57BL-G_{IX}^+ X 129)F_1 mice produce G_{IX} antibody, and there is indication that X.1 antibody is found in normal (BALB/c X C57BL)F_1 mice. Antibody with GCSA specificity has not as yet been detected in normal mouse serum, but we expect that it too will be found in the appropriate hybrid. Several new systems of MuLV-related cell surface antigens have now been detected and two of these are sufficiently well analyzed to be briefly described. They are designated $G_{(RADA1)}$ and $G_{(ERLD)}$ to indicate their relation to MuLV-Gross and to the prototype leukemia cell lines that were used in their definition (Table 3). Random-bred Swiss mice are the source of $G_{(RADA1)}$ antibody, and (C57BL X 129)F_1 mice are the source of $G_{(ERLD)}$ antibody. Both $G_{(RADA1)}$ and $G_{(ERLD)}$ are found in normal and leukemic lymphoid tissues of strains with a high incidence of leukemia, and both can be induced in fibroblasts by infection with N-tropic MuLV. The distinct pattern of occurrence of these two antigens in the normal tissues of low incidence strains clearly distinguishes them from one another as well as distinguishing them from G_{IX}, GCSA, and X.1 (Table 4). Past studies with heteroantibody to MuLV-gp70 have indicated that all inbred mouse strains tested, with the exception of 129-G_{IX}^- and BALB/c, expressed gp70-like molecules (54). $G_{(ERLD)}$ has an equally widespread strain distribution and is similarly absent from 129-G_{IX}^- and BALB/c. We had originally assumed that $G_{(ERLD)}$ might be related to the 0-gp70 species defined by heteroantibody to gp70 on C57BL and other thymocytes (55), but as 129 thymocytes are $G_{(ERLD)}^+$/0-gp70$^-$, the two systems are serologically distinguishable. Present evidence indicates that the $G_{(RADA1)}$ antigen is related to a gp70 molecule on the surface of RADA1 cells that is not expressed by normal A strain cells. The relation of this gp70 to G_{IX}-gp70, X-gp70, and 0-gp70 remains to be determined.

PC.1 Antigen: Relation to MuLV?

The need to discuss this differentiation alloantigen arises from the suggestion that PC.1 is related to a class of MuLV that is widespread in mice. The PC.1 antigen was originally identified during a study of the surface antigens of myeloma cells (13). Antisera prepared against the BALB/c myeloma MOPC-70A in H-2–compatible DBA/2 mice were cytotoxic for BALB/c myeloma cells but not for normal thymocytes or thymic leukemias. Absorption tests showed that the antigen was present in normal BALB/c mice and that it had a distinctive tissue distribution in comparison with other known surface antigens. Because normal plasma cells as well as malignant plasma cells (myelomas) were marked by this antigen, it was called PC.1. Typing of mouse strains by absorption tests with normal spleen or liver distinguished PC.1$^+$ and PC.1$^-$ strains, and genetic studies indicated conventional single-gene specification (as yet not mapped) of the PC.1 system. After the PC.1 system was defined, it was observed that the serum of BALB/c mice, the prototype PC.1$^+$ strain, had cytotoxic antibody to BALB/c myelomas and that this BALB/c antibody appeared to be detecting an antigen with a strain and tissue distribution that was identical with PC.1 (56). Antibody of similar specificity was found in a number of other mouse strains, both PC.1$^+$ and PC.1$^-$. The suggestion was made that viral induction of PC.1 may be responsible for the widespread distribution of naturally occurring antibody to PC.1$^+$ myelomas and that the MuLV frequently found in myelomas is the likely candidate. If a PC.1 inducing virus does in fact exist, PC.1 might be expected to occur in myelomas arising in PC.1$^-$ strains, and one such instance was reported (56). However, we have tested 7 myelomas from PC.1$^-$ strains in the standard PC.1 typing system and they have all been PC.1$^-$. In addition, we have been unable to detect PC.1 antibody in normal BALB/c serum under the conditions of our cytotoxic test. If PC.1 is indeed virus-induced rather than the product of a conventional Mendelian gene, one might not expect to find an allelic product in PC.1$^-$ strains. Thus the definition of such an allele may be important in distinguishing these two alternatives.

CELL SURFACE ANTIGENS OF THE TL SYSTEM

Background

Recognition of the TL system of cell surface antigens came about during a study of radiation-induced leukemias of C57BL mice (3). Because resistance to transplants of these leukemias could not be induced in C57BL, it seemed unlikely that syngeneic immunization would give rise to antibody with specificity for the leukemia cells. For this reason, C57BL mice were immunized with radiation leukemia cells of (C57BL × A)F$_1$ origin, with the idea that radiation leukemias of the mouse share a common antigen. During rejection of the histoincompatible leukemia cells, antibody would be produced to alloantigenic differences and, if present, to leukemia-specific antigens as well. Reactions due to conventional alloantibodies could be eliminated by testing the resulting immune sera on syngeneic leukemia cells. In initial tests with these antisera prepared against F$_1$ leukemia cells, cytotoxic reactions were observed with several transplanted C57BL radiation leukemias, most notably ERLD, but not with any normal C57BL lymphoid cell. In examining the reactions of other C57BL antisera, highest titer for ERLD was found with an antiserum prepared against ASL1, a transplanted A strain spontaneous leukemia, and absorption tests with this antiserum and ERLD target cells formed the basis for the original analysis of the TL system. In C57BL mice, the antigen was strictly leukemia-specific; no normal tissue absorbed cytotoxic antibody. In A strain mice, absorption tests revealed the surprising fact that the antigen was found on normal thymocytes as well as on leukemia cells. Normal cells other than thymocytes lacked the antigen. Restriction of antigen to thymocytes and leukemia cells suggested the name thymus-leukemia or TL for this antigenic system.

We then showed that inbred mouse strains could be typed TL$^-$ (like C57BL) or TL$^+$ (like A strain) on the basis of absorption tests with normal thymocytes and that TL$^+$ leukemias can occur in TL$^-$ strains other than C57BL (57). In normal mice,

CELL SURFACE ANTIGENS OF MOUSE LEUKEMIA 145

TL is inherited as a Mendelian dominant trait, and linkage studies have placed the TL locus, designated *Tla,* on chromosome 17 < 2 units from the D end of the *H-2* complex (58). The generally accepted explanation for the anomalous appearance of TL^+ leukemias in mice of TL^- strains is that all mice possess the structural gene for TL, but that in TL^- strains, the gene is not normally expressed. As a consequence of leukemogenesis, the *Tla* locus is derepressed or activated, resulting in the appearance of the TL product on the surface of leukemia cells. According to this view, a regulatory gene controls expression of the TL structural gene, and allelism for the trait in normal mice is based on *expression* versus *nonexpression* alleles at the regulatory locus (1, 28). The possibility has been raised that the *Tla* locus on chromosome 17 represents the TL regulatory gene and that the TL structural gene, since it is present in all mice, could be elsewhere in the genome (59). Although this has not been formally excluded, current understanding of the TL system makes it most unlikely (60).

Antigens Specified by the Tla Locus

Initial serological study of TL indicated detection of a single antigenic determinant. However, subsequent tests have shown that the *Tla* locus specifies at least four serologically distinguishable determinants, TL.1, TL.2, TL.3, and TL.4 (Table 8). Antibodies detecting TL.2 and TL.3 were found in C57BL anti-ASL1 serum during analysis of this antiserum by the technique of in vivo absorption. Absorption in normal A mice eliminated all H-2 antibody but did not remove cytotoxic activity for TL^+ cells, such as ERLD. In direct tests with the absorbed antiserum, thymocytes from BALB/c mice, formerly classified TL^-, showed weak reactivity. Using this as a basis for absorption tests, it was found that thymocytes from all TL^+ strains expressed the new specificity and that several other strains, also previously typed TL^-, were positive. Formal demonstration that the new determinant, called *TL.2,* belonged to the TL system came from co-typing segregating populations from TL^-

Table 8 Antisera defining the four antigens of the TL system

Antiserum	Standard test cell	TL specificity detected
1 (C57BL × A-TL$^-$)F$_1$ anti-ASL1 (A strain spontaneous leukemia)	Strain A thymocyte	TL.1,2,3
	ERLD (C57BL X-ray leukemia)	TL.1
	BALB/c thymocyte	TL.2
2 (BALB/c × C3H/An)F$_1$ anti-ASL1 absorbed in vivo in (C57BL × A)F$_1$ mice bearing advanced transplants of ERLD	Strain A thymocyte	TL.3
3 (C57BL × A-TL$^-$)F$_1$ anti-ASL1 absorbed in vitro with ERLD	Strain A thymocyte	TL.3
4 C57BL-TL$^+$ anti-ERLD	ERLD	TL.4

× TL⁺ crosses. Recognition of TL.3 resulted from the analysis of (BALB/c × C3H/An)F₁ anti-ASL1 serum absorbed in vivo in (C57BL × A)F₁ mice growing ERLD leukemia. It was expected that all TL antibody would be eliminated from this antiserum when absorbed in this way, and, in fact, cytotoxic activity for ERLD was removed. However, strong cytotoxicity for thymocytes of A strain mice remained. Genetic tests similar to those performed with TL.2 showed that TL.3, as the antigen came to be called, was also specified by the *Tla* locus. With the recognition of TL.2 and TL.3, it became clear that the original typing system for TL (C57BL anti-ASL1 versus ERLD) was detecting a distinct specificity, TL.1, because ERLD cells lacked TL.3 and, for purposes of typing, TL.2. Thus, three phenotypes can now be distinguished in normal mice on the basis of thymocyte typing, TL⁻, TL.2, TL.1,2,3 and the corresponding genotypes have been designated *Tla^b*, *Tla^c*, and *Tla^a*.

The strain distribution of the three TL phenotypes based on TL typing in our laboratory is given in Table 9. The TL congenic stocks (C57BL-TL⁺ and A-TL⁻), which were derived from *H-2D/Tla* crossovers occurring in C57BL × A crosses, are proving valuable in mapping the region between *Tla* and *H-2D*. New specificities coded for by this region have been defined by skin graft rejection (61, 62) and by cytotoxic tests (63, 64). However, the occurrence of these antigens in tissues other than thymus clearly sets them apart from products of the *Tla* locus. TL.4, the most recently recognized TL specificity, has not been found on normal thymocytes of TL⁺ or TL⁻ strains, and in this regard, is the only antigen of the TL series that is restricted to leukemia cells. It is strongly represented on ERLD and other C57BL leukemia cells but has not been found on leukemias of A strain mice, and this accounts for the absence of anti-TL.4 in anti-ASL1 serum. The antiserum that detects TL.4 was produced in the TL congenic strain, C57BL-TL⁺, by immunization with ERLD. Because TL.4 does not occur on normal thymocytes, formal genetic proof that it is specified by the *Tla* locus has not been possible to obtain. However, evidence from analyses of H-2/TL-loss variants, modulation of TL.4 by TL antibody, and assignment of TL.4 to the TL.1/H-2D region of the cell surface by antibody blocking tests indicates that TL.4 clearly belongs to the TL system.

Table 9 Strain distribution of the TL.1,2,3, TL.2, and TL⁻ phenotype of mouse thymocytes

TL.1,2,3 (*Tla^a*)	TL.2 (*Tla^c*)	TL⁻ (*Tla^b*)	
A (and A-Thy-1.1)	129 (and 129-G$_{IX}$⁻)	C57BL (and C57BL-G$_{IX}$⁺, C57BL-G$_{IX}$⁺M,	
C57BL-TL⁺	BALB/c (and BALB/c-T)	C57BL-H-2^k, C57BL-Ly-1.1, C57BL-Ly-2.1,	
C57BR (and C57BR-G$_{IX}$⁺M)	DBA/2	C57BL-Ly-2.1,Ly-3.1)	
C58	C57L	C57BL/Ka	ASW
SJL/J	A.BY	A-TL⁻	HTH
HSFS	GR	BALB/H-2^b	HTG
PL	C × BD	C3H/An	101
NZB	C × BE	C3Hf/Bi	CE
HTI	C × BH	C3H/Figge	C × BG
AKR.K		CBA (and CBA-T6)	C × BI
		AKR (and AKR H-2^b)	C × BJ
		I	C × BK
		RF	

Influence of Tla Haplotype on Expression of TL and H-2 Antigens by Thymocytes

Quantitative absorption analysis has provided the means to study the expression of TL and H-2 on normal thymocytes from mice of various *Tla* haplotypes (65). Different Tla^a strains do not vary in the quantity of TL.1,2, or 3 found on thymocytes. Tla^c strains also do not show interstrain variation in TL.2, although the amount of TL.2 expressed by Tla^c thymocytes is lower than by Tla^a thymocytes. Tla^b X Tla^a or Tla^c heterozygotes express one half the quantity of individual TL specificities detected on thymocytes of Tla^a or Tla^c homozygotes. These findings are what one would expect from conventional gene dose effects, with no positive or negative interaction between *Tla* alleles. A different result occurs in Tla^a X Tla^c heterozygotes, where intermediate levels of TL.2, heterozygous levels of TL.3, but homozygous levels of TL.1 were found. The reason for a homozygous TL.1 phenotype in a mouse with a heterozygous genotype is unknown but could reflect interactions at the level of genetic regulation, antigen synthesis, or antigen insertion. The *Tla* haplotype also influences the expression of H-2D antigen on the thymocyte. The amount of H-2D antigen on TL$^-$ *(Tla^b)* thymocytes is significantly greater than on TL.1,2,3 or TL.2 (Tla^a or Tla^c thymocytes). In contrast, the level of H-2K and Thy-1 antigens is constant and unrelated to TL genotype. This inverse association between TL and H-2D may relate to the close spatial relation of these two molecules on the cell surface (see below). Absence of TL might relieve a steric hindrance to attachment of the H-2D antibody used to measure the amounts of H-2D by quantitative absorption tests. This idea receives support from the finding that TL$^+$ cells undergoing phenotypic suppression of TL by TL antibody (antigenic modulation) show a corresponding increase in the level of H-2D antigen (66).

Antigenic Modulation

This name was given to a phenomenon originally thought unique for TL antigens, but now considered likely to be of a more general significance. It refers to reversible changes in the sensitivity of cells to the cytotoxic action of TL antibody and complement brought about by prior exposure to TL antibody. Antigenic modulation was recognized while investigating the reason TL-immunized mice with high levels of TL antibody showed no resistance to transplants of syngeneic TL$^+$ leukemia cells (67). An explanation for this paradox became apparent when the leukemia cells were recovered from the immunized mice and found to be totally resistant to TL antibody and complement. As long as the cells were passed in animals with TL antibody, they remained TL$^-$, but when passed back to nonimmune mice, they reverted to TL$^+$ and high sensitivity to TL antibody. Passive transfer of TL antibody can also induce antigenic modulation, and, in this way, it was shown that normal thymocytes and leukemia cells of TL$^+$ mice (which are incapable of forming TL antibody) undergo modulation (68).

Antibody-mediated change in TL phenotype can also be brought about in vitro, and much that we know about antigenic modulation comes from such studies (66, 69). Loss of sensitivity to complement-mediated cytotoxicity occurs rapidly, in certain cases within 30 min after exposure to TL antibody. When TL antibody is removed, TL sensitivity returns after one to two cell divisions. The process is temperature-dependent and can be inhibited by certain metabolic antagonists, particularly actinomycin. Antibody to an individual specificity of the TL complex causes modulation of other TL antigens, and this is consistent with the current thinking that all TL specificities reside on a single molecule. Different TL^+ cell types modulate at different rates; leukemia cells undergo modulation more rapidly than normal thymocytes, a reflection most likely of the greater metabolic activity of leukemia cells.

A number of explanations have been proposed to account for antigenic modulation. With the recognition that antibody to certain cell surface antigens causes rearrangement of these molecules into patches and caps, there has been speculation that TL modulation represents a similar process. However, as monovalent Fab fragments of TL antibody can modulate TL but fail to cause patching and capping, antigenic modulation appears to represent a more subtle consequence of antibody attachment to surface molecules. Although there is some indication that the amount of TL remaining after modulation is reduced, considerable TL antigen can still be accounted for on the surface of modulated cells (70). Rather than antigen loss, current evidence points to antibody-induced conformational changes in TL sites that reduce the efficiency with which lytic complement components can be engaged. However, whatever the mechanism of antigenic modulation, it provides a remarkably effective mechanism for TL^+ leukemia cells to escape the consequence of their antigenicity.

TL Phenotype of Mouse Leukemia

Table 10 summarizes our accumulated information on the TL phenotypes of leukemias occurring in various mouse strains. The key feature of the TL system in regard to malignancy is the anomalous occurrence of TL^+ leukemias in strains with

Table 10 TL phenotype of mouse leukemia

	Tla haplotype	Thymocyte phenotype	Leukemia induction	Leukemia cell			
					Phenotype		
				1,2,3,4	1,2,3	1,2,4	1,2	–
TL congenic strains				Number of leukemias tested			
A (TL^+)	a	TL.1,2,3	Spontaneous	4			1
			X-ray	3			
A-TL^-	b	TL^-	X-ray		3		4
C57BL (TL^-)	b	TL^-	Spontaneous		1		3
			X-ray		45		8
C57BL-TL^+	a	TL.1,2,3	X-ray	4			
TL.2 strains							
129, BALB/c, DBA/2	c	TL.2	Spontaneous		2	2	1
			X-ray		9	13	6
			DMBA				6
TL^- high leukemia incidence strain							
AKR	b	TL^-	Spontaneous		10		78
			Urethan		10		2
F_1 hybrids							
C57BL × A	b × a	TL^- × TL.1,2,3	X-ray	7	6		1
C57BL × AKR	b × b	TL^- × TL^-	X-ray		9		4

the TL⁻ phenotype. The frequency of TL⁺ leukemias varies among inbred strains. For example, X-ray-induced leukemias of C57BL mice are frequently TL⁺, whereas most spontaneous leukemias of AKR mice are TL⁻. Urethan-induced AKR leukemias, however, are commonly TL⁺. This relation of TL appearance to different modes of leukemia induction requires more direct study.

Each of the TL specificities can be distinguished on the basis of appearance on normal thymocytes and on leukemia cells (Table 11). In normal mice, TL.1 is an antigen of normal differentiation, appearing on thymocytes of Tla^a (phenotype TL.1,2,3) strains. TL.2 is also an antigen of normal differentiation, appearing on normal thymocytes of Tla^a and Tla^c (phenotype TL.2) strains and is the only TL specificity that occurs alone on normal thymocytes. TL.1 and 2 are also antigens of leukemic differentiation, because they may occur anomalously on leukemias of Tla^b (phenotype TL⁻) strains and, in the case of TL.1, on the leukemias of Tla^c mice. It is this evidence that has led to the supposition that structural information for TL.1 and 2 is present in all mice. TL.3 is an antigen of normal differentiation only, its occurrence on leukemia cells being restricted to Tla^a strains where the antigen is a marker of normal thymocytes. Structural genes for TL.3 may not be universal, because TL.3 leukemias have never been found in mice of Tla^b or Tla^c haplotypes. An alternative explanation is that TL.3 structural genes may not be subject to derepression or activation during leukemogenesis. TL.4 is an antigen of leukemic differentiation only, having never been found on normal thymocytes of any strain. Its occurrence is restricted to leukemic mice of Tla^b and Tla^c haplotypes, suggesting that the structural gene for TL.4, like that for TL.3, is not present universally in mice.

The anomalous appearance of TL antigens on leukemias and the clearly distinctive behavior of individual TL components allow certain questions to be asked in regard to the relation of the Tla locus to leukemogenesis. The occurrence of TL⁻ leukemias of thymic derivation in both TL⁺ and TL⁻ strains indicates that TL activation is not an invariable consequence of T-cell leukemogenesis in the mouse.

Table 11 Conclusions derived from a comparison of the TL phenotypes of normal thymocytes and leukemia cells

Tla haplotype	Prototype strain	Phenotype		Conclusions
		Normal thymocyte	TL⁺ leukemia	
Tla^a	A, C57BL-TL⁺	TL.1,2,3	TL.1,2,3	Phenotype of TL⁺ leukemia resembles normal thymocyte. No evidence for appearance of anomalous TL components.
Tla^b	C57BL, A-TL⁻	TL⁻	TL.1,2,4	Invariable appearance of anomalous TL.1, TL.2, and TL.4 on TL⁺ leukemias.
Tla^c	BALB/c	TL.2	TL.1,2, or TL.1,2,4	No leukemia found with a TL.2 phenotype only, indicating that anomalous TL.1 appearance (but not TL.4) is an invariable consequence of leukemogenesis of the type involving TL activation in mice with the Tla^c haplotype.
$Tla^b \times Tla^a$	(C57BL × A)F₁	TL.1,2,3	TL.1,2,3, or TL.1,2,3,4	The TL phenotype TL.1,2,3,4 is found only in leukemias of (Tla^b or ^{c} × Tla^a) heterozygotes.
$Tla^b \times Tla^b$	(C57BL × AKR)F₁	TL⁻	TL.1,2,4	Otherwise the TL phenotype of F₁ leukemias is similar to those arising in Tla homozygotes.

With regard to the transformation pathway that leads to TL⁺ leukemias, one can ask whether anomalous TL antigens are invariably expressed. TL⁺ leukemias arising in mice of Tla^a strains are not revealing, because no anomalous TL products have been recognized in these mice. In Tla^c strains, however, where thymocytes are marked by TL.2, all TL⁺ leukemias are both TL.1 and TL.2 and never TL.2 alone, indicating that the structural gene for TL.1 is invariably activated during leukemogenesis involving this class of T cell. Given that there are two TL structural genes in a diploid cell, are both Tla loci activated in a TL⁺ leukemia arising in a TL⁻ mouse? Although quantitative absorption tests should be able to distinguish homozygous from heterozygous amounts of TL in such leukemias, studies of this sort have not given clear results. An indication that both loci need not be activated, however, comes from the analysis of leukemias occurring in $Tla^a \times Tla^b$ mice, where TL.4 can be used as the marker for activation of the Tla^b haplotype (Tables 10 and 11). Leukemias of both TL.1,2,3 and TL.1,2,3,4 phenotypes are found, indicating that the Tla^b locus need not be involved.

Another question relates to whether the product of the Tla locus is essential for the continued malignant behavior of TL⁺ leukemia cells. TL expression per se is apparently not critical, since the TL phenotype of TL⁺ leukemias can be suppressed by TL antibody (antigenic modulation) without the cell losing malignant characteristics. The isolation of a TL⁻ variant from an originally TL⁺ leukemia by in vitro immunoselection does not help in this regard because the leukemia came from a mouse of Tla^a genotype where Tla activation is not known to occur (71). TL variants can also be isolated by immunoselection in vivo (72) and Table 12 shows the results of our analysis of 3 TL⁺ leukemias of (C57BL × A)F₁ origin and the 6 H-2/TL-loss variants derived from them. The selection method, which has been extensively used to analyze H-2 variants (73), involved the passage of the F₁ leukemias into mice that are homozygous for the H-2ᵇ (C57BL) haplotype or for the H-2ᵃ (A strain) haplotype. In this way, stable variants that do not express one or the other H-2 haplotype escape immunological destruction and can be recovered from the original population of leukemia cells. A variety of explanations have been

Table 12 TL and H-2 phenotypes of three (C57BL × A)F₁ X-ray induced leukemias and the TL/H-2-loss variants derived from them by immunoselection in vivo

	TL.1	TL.3	TL.4	H-2ᵇ	H-2ᵃ
(C57BL × A)F₁ leukemia A	1	3	4	b	a
H-2ᵃ-loss variant	1	–	4	b	–
H-2ᵇ-loss variant	1	3	–	–	a
(C57BL × A)F₁ leukemia B	1	3	4	b	a
H-2ᵃ-loss variant	1	–	4	b	–
H-2ᵇ-loss variant	1	3	–	–	a
(C57BL × A)F₁ leukemia C	1	3	–	b	a
H-2ᵃ-loss variant	–	–	–	b	–
H-2ᵇ-loss variant	1	3	–	–	a

put forward to account for these antigenic loss variants, but at the moment their origin is unclear (74). Nonetheless, they have proved to be valuable in the study of TL for, as Table 12 shows, the process leading to the loss or inactivation of the *H-2* locus also leads to the loss of the closely linked *Tla* locus. Thus, H-2^a-loss variants lack products of the *Tla^a* locus (marker TL.3) and H-2^b-loss variants lack products on the *Tla^b* locus (marker TL.4). Parental variants of reciprocal type were obtained with all three leukemias, indicating that the presence of both *Tla* loci is not required to maintain the malignant phenotype. The H-2^b-loss variants of leukemia A and B show that the activated *Tla^b* locus is also unnecessary. With leukemia C, a TL$^-$ variant could be isolated from an originally TL$^+$ leukemia (H-2^a-loss variant), suggesting that TL is irrelevant as far as the malignant behavior of this leukemia is concerned. However, as there is no evidence for anomalous TL activation in this leukemia, it will be critical to show whether TL$^-$ variants can be isolated from TL$^+$ leukemias arising in *Tla^b* × *Tla^b* heterozygotes, such as (C57BL × AKR)F$_1$ mice.

Mapping of TL Antigens on the Surface of Thymocytes and Leukemia Cells

In 1968, our group proposed a map designating the relative positions of various alloantigens on the surface of mouse thymocytes (75). The method we used to develop this map was based on the idea that antibody to one antigenic site would inhibit or block the subsequent attachment of antibody to a second unrelated site if the two sites were sufficiently close to one another on the cell surface. The test (which has come to be called the *antibody blocking test*) is performed by exposing thymocytes to saturating levels of antibody directed to a particular surface alloantigen. After the cells are washed, they are tested for the amount of an unrelated surface antigen by quantitative absorption assays with antibody to the second specificity. Cells that are not exposed to antibody serve as the standard to measure the degree of blocking induced by attachment of antibody to the first specificity. Study of different antigenic systems by such pair-by-pair blocking tests gave rise to instances of no blocking, reciprocal blocking, or nonreciprocal blocking. The pattern that emerged from this analysis and forms the basis for the map is shown in Figure 1.

Based on ten distinct alloantigenic specificities, four regions of the map can be distinguished. Antigens defining region I, III, and IV are not within blocking range of one another. Thy-1 alloantigens appear to be proximal to antigens in region I and III, but the nonreciprocal nature of blocking suggests that its surface representation may be far greater than other alloantigens. This interpretation is consistent with the finding that the ratio of Thy-1 to H-2 antigenic sites on thymocytes, as estimated by the uptake of tritiated labeled alloantibody, is approximately 5 to 1 (76). At the time the map was constructed, it came as somewhat of a surprise that H-2D and H-2K were in separate blocking regions, but subsequent genetic, biochemical, and other studies have shown that these antigens do in fact reside on separate molecules. Two examples of surface association related to genetic linkage are found in region I. The *Tla* and *H-2D* loci are within 2 map units from one another, and this is reflected in the close proximity of their products on the cell surface. Lyt-2 and Lyt-3,

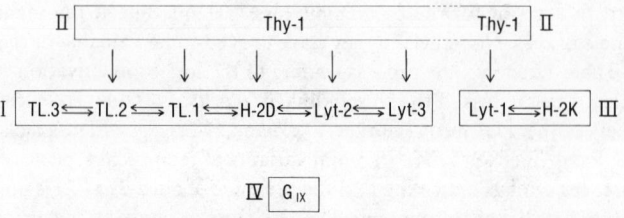

Figure 1 Spatial relation of alloantigens on the surface of mouse thymocytes as determined by antibody-blocking tests. *Arrow* indicates direction of antibody blocking (for example, antibody to the specificity at the arrow tail blocks attachment of antibody to the specificity at the arrowhead). *Arrows in both directions* (⇌) indicate reciprocal blocking. A *single arrow* (→) indicates nonreciprocal blocking. No blocking was found between alloantigens comprising region I and those in region III and IV.

determined by closely linked genes that have not as yet been resolved by crossing-over, are also closely linked on the cell surface. However, genetic linkage is not necessarily associated with surface linkage (e.g. H-2D/H-2K) and conversely, surface linkage does not necessarily reflect genetic linkage (e.g. H-2D/Lyt-2 and Lyt-1/H-2K). If current evidence indicating that the 3 TL specificities on normal thymocytes reside on a single molecule is correct, then their relative positions on the TL molecule is also resolved by the blocking test.

Blocking tests also permit questions to be asked about changes in the alloantigenic map following appearance of new tumor antigens on the cell surface. The TL system provides a unique opportunity to compare the surface site of TL on a leukemia cell that arose in a TL⁻ mouse with its known site on a TL⁺ thymocyte. For this purpose, the TL.1/H-2D surface association was explored with five TL⁺ leukemias (Table 13). In TL⁺ leukemias of TL⁻ mice, anomalous TL.1 mapped adjacent to H-2D. However, blocking is not reciprocal; H-2D antibody blocks TL sites, but TL antibody does not block H-2D sites. This can be interpreted to indicate that TL occupies the correct site (so that every TL molecule has an adjacent H-2D molecule), but not every H-2D site has an adjacent TL molecule, as if only one *Tla* locus had been activated during leukemogenesis. In TL⁺ leukemias of TL⁺ mice, nonreciprocal blocking was also observed, but, in the opposite direction; TL antibody blocked H-2D sites, whereas H-2D antibody did not block TL sites. The appearance of unrecognized anomalous TL components could explain this result, although other possibilities can be envisioned. Thus, the blocking technique may prove useful in mapping other tumor antigens, including viral antigens, that appear as a consequence of malignancy.

With the recognition that many of the protein constituents of the plasma membrane are highly mobile and that their distribution can be selectively altered by the attachment of antibody, doubt has been cast on the significance of the associations revealed by the blocking test. However, we now know that the temperature at which the test is performed (0°C) minimizes antibody-induced surface changes and that

Table 13 Spatial relation of TL.1 and H-2D antigens on the surface of normal thymocytes and TL$^+$ leukemia cells

Strain of origin (*Tla* haplotype)	Leukemia induction	Results of TL.1/H-2D blocking tests[a]	
		Normal thymocytes	Leukemia cells
C57BL (*Tla^b*)	X-ray (ERLD)	–	TL.1 ⟵ H-2D
C57BL (*Tla^b*)	X-ray	–	TL.1 ⟵ H-2D
AKR (*Tla^b*)	Spontaneous	–	TL.1 ⟵ H-2D
A (*Tla^a*)	Spontaneous (ASL1)	TL.1 ⇌ H-2D	TL.1 ⟶ H-2D
C57BL-TL$^+$ (*Tla^a*)	X-ray	TL.1 ⇌ H-2D	TL.1 ⟶ H-2D

[a] See Figure 1.

alloantigens, such as H-2 and Thy-1, in contrast to Ig and TL, undergo little surface rearrangement as a consequence of attached mouse antibody alone (77). Most important, a consistent pattern emerges from these blocking studies that indicates a preferential association of certain surface alloantigens which cannot be easily explained as an artifact of the technique. Clearly what is needed are other methods that can more directly assess the supramolecular organization of surface constituents on normal and malignant cells.

Preleukemic Expression of TL Antigens

Ever since the recognition of the TL system, it has been assumed that the anomalous appearance of TL antigen coincided with the transformation event and signaled the emergence of leukemia cells. However, the striking preleukemic changes in alloantigens and MuLV-related antigens in AKR thymus (51, 52) prompted us to examine this point directly in relation to X-ray leukemogenesis. In a study of C57BL mice that received fractionated whole body irradiation according to a schedule resulting in a high incidence of leukemia, TL$^+$ cells were found in the thymus long before the development of overt leukemia (78). Thus TL should be considered a marker for preleukemic changes occurring in a population of cells at high risk for leukemia transformation.

TL and MuLV: Two Mechanisms of Leukemogenesis?

A comparison of the GCSA/G$_{IX}$ phenotype and the TL phenotype of the leukemias that have now been studied (Table 7 and 10) distinguishes two major classes of mouse leukemia. One is exemplified by spontaneous leukemias of high incidence strains, such as AKR, which are primarily GCSA$^+$/G$_{IX}^+$/TL$^-$ and the other by X-ray-induced leukemias of low incidence TL$^-$ strains, such as C57BL, which are GCSA$^-$/G$_{IX}^-$/TL$^+$. An important question for future investigation is whether this difference reflects two distinct pathways of T-cell leukemogenesis, one a direct or indirect consequence of MuLV and the other not involving MuLV. There has been speculation that the *Tla* locus may represent the integration site of a defective

MuLV genome (28), but this appears unlikely in view of the similar biochemical characteristics of the TL products and the products of the closely linked *H-2* genes (79). Another possibility is that the *Tla* locus is activitated by a special class of MuLV involved in X-ray leukemogenesis. The fact that most C57BL leukemias lack GCSA, G_{IX}, and other MuLV structural antigens as well as infectious MuLV speaks against this idea (78). However, there is growing evidence that MuLV appears transiently during the preleukemic phase of X-ray leukemogenesis, and it could be at this time that *Tla* activation by MuLV might occur. If a TL-inducing MuLV can be identified, the relation between the TL-class and the MuLV-class of mouse leukemias would be clarified. In this regard, the frequent occurrence of trisomy 15 in spontaneous AKR leukemias (80) and X-ray-induced C57BL leukemias (81) suggests a common genetic basis for these leukemias.

ML SURFACE ANTIGEN OF MOUSE LEUKEMIA

Recognition of the ML system came about during the study of spontaneous leukemias of DBA/2 mice (82). Antisera were prepared in histoincompatible mice and were absorbed in vivo to remove alloantibody. The absorbed antisera retained cytotoxic activity for the immunizing leukemia and other DBA/2 leukemias. Absorption tests revealed that the antigen could not be detected in spontaneous or induced leukemias of any other mouse strain but was found in normal mammary tissue and spontaneous mammary tumors of mice infected with the mammary tumor virus (MTV). This restriction of antigen to leukemias of DBA/2 mice and MTV-infected cells prompted the designation ML for mammary-leukemia. The classical mode of MTV transmission is through the milk of infected mothers, and this permitted a more direct approach to the relation of MTV to ML antigen (32). When newborn mice of MTV^- strains were foster nursed on MTV^+ females, infection with MTV was accompanied by expression of ML in mammary tissue. Furthermore, in reciprocal crosses between MTV^+ and MTV^- strains, the mammary tissue of $(MTV^+ \times MTV^-)F_1$ mice was ML^+/MTV^+ and the mammary tissue of $(MTV^- \times MTV^+)F_1$ mice was ML^-/MTV^-, showing that maternal transmission of MTV led to ML antigen expression as well. Although there is little doubt that the ML antigen is coded directly or indirectly by MTV, the reason for ML appearance in DBA/2 leukemias is not clear. There are indications from molecular hybridization studies that genetic information related to MTV is present in all mouse strains (83), but that expression is restricted to certain strains (MTV^+ strains) and in these strains to certain tissues. Occurrence of ML antigen in leukemia cells could represent derepression of MTV genetic information in malignant lymphoid cells, although this leaves unanswered why this should occur only in DBA/2 mice and, as recently reported, in GR mice (84). Another possibility is that these mice carry a variant leukemogenic virus that arose through genetic interaction with MTV. Biochemical definition of the ML antigen, its relation to MTV structural components, and analysis of oncornaviruses from ML^+ leukemias should resolve some of these questions.

GENERAL COMMENTS

Several categories of surface antigens can now be recognized on mouse leukemia (Table 14). As leukemia has been the subject of the most intense scrutiny, we have a more comprehensive picture of the surface antigens of the leukemia cell and its normal counterpart, the thymocyte, than of any other cell type in the mouse. Serological study of nonlymphoid tumors of the mouse is at a very early stage and even rudimentary questions, such as expression of known alloantigens, have not been asked in relation to many classes of experimental cancer. Through the application of the serological techniques and approaches that have given us our present, albeit limited, view of surface antigens on leukemia cells, analysis of other tumor cell populations will undoubtedly uncover a large series of new differentiation alloantigens, as well as additional classes of tumor antigens that may have more immediate relevance to malignancy.

A diverse array of MuLV-related cell surface antigens has now been identified in mouse leukemia and there is every indication that the list will continue to grow. This complexity parallels comparable diversity of MuLV envelope antigens (85) and reflects the extensive MuLV polymorphism that has been recognized to exist in the mouse. None of the MuLV-related surface antigens can be considered to be transformation-specific, because they are also found on MuLV-infected cells that are not malignant. As it appears increasingly likely that cells transformed by avian and feline oncornaviruses express transformation-specific surface antigens that are unrelated to viral structural antigens (86, 87), search for this category of antigen on mouse leukemia needs to continue.

Knowledge of the TL and MuLV systems of antigens and the recognition that these antigens occur in leukemias and solid tumors of mice whose normal tissues do not express them have added considerable sophistication to our discussion of tumor-specific antigens and their genetic origin. They provide a secure basis for searching for other examples of distinctive tumor products that are determined by

Table 14 Categories of serologically demonstrable surface antigens on mouse leukemia

Conventional alloantigens	H-2D, H-2K
Differentiation alloantigens	Lyt-1,2,3,4, Thy-1, TL.1,2,3
MuLV-structural antigens	MuLV-gp70, p30, p15
MuLV-related antigens	GCSA, G_{IX}, X.1, $G_{(RADA1)}$, $G_{(ERLD)}$
Transformation-specific MuLV antigens	None defined
MTV-related antigens	ML
Derepression antigens	TL.1,2,4 (in TL$^-$ strains)
	GCSA, G_{IX}, $G_{(RADA1)}$, X.1 (in strains not expressing MuLV-related antigens in normal tissue)
Individually unique antigens	None defined
Species antigens	MSLA
Embryonic and fetal antigens	None defined

genes that are never expressed in normal life or expressed for only a limited period during some developmental phase. Much speculation has centered on the possibility that tumor-specific antigens may arise from derepressed fetal genes, but the evidence from transplantation and serological studies that has been presented to support this idea does not exclude other interpretations. For example, MuLV antigens, which are found so widely in mouse tumors, might be expressed during restricted periods of fetal development and so would mimic a fetal gene product. A serological definition of the surface antigens of embryonic and fetal cells will clarify such questions and permit us to know whether unique embryonic- or fetal-specific surface antigens not expressed by adult cells do in fact exist and, if so, whether these reappear on malignant cells. The analysis of mouse teratocarcinoma cells and preimplantation embryos shows much promise as a way to approach these important issues (88, 89).

Genes determining histocompatibility antigens, particularly those of the H-2 complex, have been considered another source of genetic information for tumor-specific antigens. Given the TL model, it is not difficult to imagine that structural genes for certain H-2 components are present in all mice but expressed in only some strains, and that anomalous appearance of H-2 antigens may occur as a consequence of malignancy. Suggestive evidence for this possibility comes from the study of chemically induced tumors, where immunization with normal tissues from an H-2 incompatible strain led to transplantation resistance against syngeneic tumor (90, 91). With the highly developed state of H-2 serology and biochemistry, identification of anomalous H-2 antigens on the tumor cell surface should be straightforward, and by analyzing a number of similarly but independently derived tumors, one could determine how frequently this phenotypic change occurs and whether a particular region of the H-2 complex is invariably involved.

The tumor antigens that have not as yet been defined serologically are the individually unique antigens that represent the classic tumor-specific transplantation antigens of cancer immunology (92). Since the discovery of these antigens in sarcomas induced by polycyclic hydrocarbons, individually distinct transplantation antigens have been detected in spontaneous mammary tumors (93, 94) and spontaneous reticulum cell sarcomas (95). The remarkable feature of these antigens is their polymorphism, even two tumors arising in the same inbred mouse being antigenically distinguishable. Their serological analysis is complicated by the frequent and unpredictable presence of MuLV and its antigens in these tumors, and past studies that did not take this fact into account are impossible to interpret. The genetic origin of these unique antigens in mouse tumors has been the subject of much discussion, and the commonly held view is that they represent the products of derepressed or mutated genes. Another possibility not involving genetic change relates their origin to a self-perpetuating phenotypic error in membrane synthesis or assembly (96). The view that they represent preexisting antigenic diversity in the normal cell population seems to have been excluded by the demonstration that transformed cell populations derived from the same clone have noncrossreacting antigenicity (97). Although antigens of this class have not been detected on leukemia cells, their presence may have been obscured by other systems of antigens, particularly those related to MuLV, and planned immunizations can now be devised to detect their existence.

CELL SURFACE ANTIGENS OF MOUSE LEUKEMIA   157

Literature Cited

1. Old, L. J., Boyse, E. A. 1973. Current enigmas in cancer research. *Harvey Lect.* 67:273–315
2. Gorer, P. A., O'Gorman, P. 1956. The cytotoxic activity of isoantibodies in mice. *Transplant. Bull.* 3:142–43
3. Old, L. J., Boyse, E. A., Stockert, E. 1963. Antigenic properties of experimental leukemias. I. Serological studies *in vitro* with spontaneous and radiation-induced leukemias. *J. Natl. Cancer Inst.* 31:977–86
4. Old, L. J., Boyse, E. A., Stockert, E. 1965. The G (Gross) leukemia antigen. *Cancer Res.* 25:813–19
5. Old, L. J., Boyse, E. A., Lilly, F. 1963. Formation of cytotoxic antibody against leukemias induced by Friend virus. *Cancer Res.* 23:1063–68
6. Old, L. J., Boyse, E. A., Stockert, E. 1964. Typing of mouse leukemias by serological methods. *Nature* 201:777–79
7. Klein, E., Klein, G. 1964. Antigenic properties of lymphomas induced by the Moloney agent. *J. Natl. Cancer Inst.* 32:547–68
8. Geering, G., Old, L. J., Boyse, E. A. 1966. Antigens of leukemias induced by naturally occurring murine leukemia virus: Their relation to the antigens of Gross virus and other murine leukemia viruses. *J. Exp. Med.* 124:753–72
9. Stockert, E., Old, L. J., Boyse, E. A. 1971. The G_{IX} System. A cell surface alloantigen associated with murine leukemia virus; implications regarding chromosomal integration of the viral genome. *J. Exp. Med.* 133:1334–55
10. Boyse, E. A., Stockert, E., Old, L. J. 1969. Properties of four antigens specified by the *Tla* locus: Similarities and differences. In *International Convocation on Immunology*, Buffalo, New York, ed. N.R. Rose, F. Milgrom, pp. 353–57. Basel & New York: Karger
11. Sato, H., Boyse, E. A., Aoki, T., Iritani, C., Old, L. J. 1973. Leukemia-associated transplantation antigens related to murine leukemia virus. *J. Exp. Med.* 138:593–606
12. Reif, A. E., Allen, J. M. V. 1964. The AKR thymic antigen and its distribution in leukemias and nervous tissues. *J. Exp. Med.* 120:413–33
13. Takahashi, T., Old, L. J., Boyse, E. A. 1970. Surface alloantigens of plasma cells. *J. Exp. Med.* 131:1325–41
14. Boyse, E. A., Miyazawa, M., Aoki, T., Old, L. J. 1968. Ly-A and Ly-B: Two systems of lymphocyte isoantigens in the mouse. *Proc. R. Soc. London Ser. B.* 170:175–93
15. Shen, F.-W., Boyse, E. A., Cantor, H. 1975. Preparation and use of Ly antisera. *Immunogenetics* 2:591–95
16. Shigeno, N., Arpels, C., Hämmerling, U., Boyse, E. A., Old, L. J. 1968. Preparation of lymphocyte-specific antibody from antilymphocyte serum. *Lancet* 2:320–23
17. Raff, M. C., Nase, S., Mitchison, N. A. 1971. Mouse-specific bone marrow-derived lymphocyte antigen as a marker for thymus-independent lymphocytes. *Nature* 230:50–51
18. Takahashi, T., Old, L. J., Hsu, C.-J., Boyse, E. A. 1971. A new differentiation antigen on plasma cells. *Eur. J. Immunol.* 1:478–82
19. Watanabe, T., Yagi, Y., Pressman, D. 1971. Antibody against neoplastic plasma cells. I. Specific surface antigens on mouse myeloma cells. *J. Immunol.* 106:1213–21
20. Gregoriades, A., Old, L. J. 1969. Isolation and some characteristics of a group-specific antigen of murine leukemia viruses. *Virology* 37:189–202
21. Strand, M., August, J. T. 1973. Structural proteins of oncogenic ribonucleic acid viruses. Interspec II, a new interspecies antigen. *J. Biol. Chem.* 248:5627–33
22. Fleissner, E., Ikeda, H., Tung, J.-S., Vitetta, E. S., Tress, E., Hardy, W. D. Jr., Stockert, E., Boyse, E. A., Pincus, T., O'Donnell, P. 1975. Characterization of murine leukemia virus-specific proteins. *Cold Spring Harbor Symp. Quant. Biol.* 39:1057–66
23. Hunsmann, G., Claviez, M., Moennig, V., Schwarz, H., Schäfer, W. 1976. Properties of mouse leukemia viruses. X. Occurrence of viral structural antigens on the cell surface as revealed by a cytotoxicity test. *Virology* 69:157–68
24. Aoki, T., Boyse, E. A., Old, L. J. 1966. Occurrence of natural antibody to the G (Gross) leukemia antigen in mice. *Cancer Res.* 26:1415–19
25. Hanna, M. G. Jr., Ihle, J. N., Batzing, B. L., Tennant, R. W., Schenley, C. K. 1975. Assessment of reactivities of natural antibodies to endogenous RNA tumor virus envelope antigens and virus-

induced cell surface antigens. *Cancer Res.* 35:164–71
26. Nowinski, R. C., Klein, P. A. 1975. Anomalous reactions of mouse alloantisera with cultured tumor cells. II. Cytotoxicity is caused by antibodies to leukemia viruses. *J. Immunol.* 115:1261–68
27. Obata, Y., Stockert, E., Boyse, E. A., Tung, J.-S., Litman, G. W. 1976. Spontaneous autoimmunization to G_{IX} cell surface antigen in hybrid mice. *J. Exp. Med.* 144:533–42
28. Boyse, E. A., Old, L. J. 1969. Some aspects of normal and abnormal cell surface genetics. *Ann. Rev. Genet.* 3:269–90
29. Rowe, W. P. 1977. Leukemia virus genomes in the chromosomal DNA of the mouse. *Harvey Lect.* 71: In press
30. Gross, L. 1951. "Spontaneous" leukemia developing in C3H mice following inoculation in infancy with AK leukemic extracts or AK embryos. *Proc. Soc. Exp. Biol. Med.* 76:27–32
31. Gross, L. 1970. In *Oncogenic Viruses.* Oxford, New York: Pergamon. 2nd ed.
32. Old, L. J., Boyse, E. A. 1965. Antigens of tumors and leukemias induced by viruses. *Fed. Proc.* 24:1009–17
33. Geering, G., Aoki, T., Old, L. J. 1970. Shared viral antigen of mammalian leukemia viruses. *Nature* 226:265–66
34. Bolognesi, D. P. 1974. Structural components of RNA tumor viruses. *Adv. Virus Res.* 19:315–59
35. Slettenmark-Wahren, B., Klein, E. 1962. Cytotoxic and neutralization tests with serum and lymph node cells of isologous mice with induced resistance against Gross lymphomas. *Cancer Res.* 22:947–54
36. Nowinski, R. C., Old, L. J., Boyse, E. A., de Harven, E., Geering, G. 1968. Group-specific viral antigens in the milk and tissues of mice naturally infected with mammary tumor virus or Gross leukemia virus. *Virology* 34:617–29
37. Lilly, F., Pincus, T. 1973. Genetic control of murine viral leukemogenesis. *Adv. Cancer Res.* 17:231–77
38. Barnes, R. D., Tuffrey, M. A., Crewe, P., Dawson, L., Brown, K., Joyner, J. 1976. Levels of C-type viral p30 antigens in lymphoma-resistant mice. *Cancer Res.* 36:3622–24
39. Obata, Y., Ikeda, H., Stockert, E., Boyse, E. A. 1975. Relation of G_{IX} antigen of thymocytes to envelope glycoprotein of murine leukemia virus. *J. Exp. Med.* 141:188–97

40. Stockert, E., Sato, H., Itakura, K., Boyse, E. A., Old, L. J., Hutton, J. J. 1972. Location of the second gene required for expression of the leukemia-associated mouse antigen G_{IX}. *Science* 178:862–63
41. Ikeda, H., Stockert, E., Rowe, W. P., Boyse, E. A., Lilly, F., Sato, H., Jacobs, S., Old, L. J. 1973. Relation of chromosome 4 (linkage group VIII) to murine leukemia virus-associated antigens of AKR mice. *J. Exp. Med.* 137:1103–7
42. Ikeda, H., Rowe, W. P., Boyse, E. A., Stockert, E., Sato, H., Jacobs, S. 1976. Relationship of infectious murine leukemia virus and virus-related antigens in genetic crosses between AKR and the *Fv-1* compatible strain C57L. *J. Exp. Med.* 143:32–46
43. Stockert, E., Boyse, E. A., Sato, H., Itakura, K. 1976. Heredity of the G_{IX} thymocyte antigen associated with murine leukemia virus: Segregation data simulating genetic linkage. *Proc. Natl. Acad. Sci. USA* 73:2077–81
44. Aoki, T., Boyse, E. A., Old, L. J., de Harven, E., Hämmerling, U., Wood, H. A. 1970. G (Gross) and H-2 cell surface antigens: Location on Gross leukemia cells by electron microscopy with visually labeled antibody. *Proc. Natl. Acad. Sci. USA* 65:569–76
45. Tung, J.-S., Vitetta, E. S., Fleissner, E., Boyse, E. A. 1975. Biochemical evidence linking the G_{IX} thymocyte surface antigen to the gp69/71 envelope glycoprotein of murine leukemia virus. *J. Exp. Med.* 141:198–205
46. Tung, J.-S., Yoshiki, T., Fleissner, E. 1976. A core polyprotein of murine leukemia virus on the surface of mouse leukemia cells. *Cell* 9:573–78
47. Snyder, H. W. Jr., Stockert, E., Fleissner, E. 1977. Characterization of molecular species carrying Gross cell surface antigen (GCSA). *J. Virol.* 23:302–14
48. O'Donnell, P. V., Stockert, E. 1976. Induction of G_{IX} antigen and Gross cell surface antigen after infection by ecotropic and xenotropic murine leukemia viruses *in vitro. J. Virol.* 20:545–54
49. Hartley, J. W., Rowe, W. P. 1976. Naturally occurring murine leukemia viruses in wild mice: Characterization of a new "amphotropic" class. *J. Virol.* 19:19–25
50. Hartley, J. W., Wolford, N. K., Old, L. J., Rowe, W. P. 1977. A new class of murine leukemia virus associated with development of spontaneous lym-

phomas. *Proc. Natl. Acad. Sci. USA* 74:789–92

51. Kawashima, K., Ikeda, H., Stockert, E., Takahashi, T., Old, L. J. 1976. Age-related changes in cell surface antigens of preleukemic AKR thymocytes. *J. Exp. Med.* 144:193–208

52. Kawashima, K., Ikeda, H., Hartley, J. W., Stockert, E., Rowe, W. P., Old, L. J. 1976. Changes in expression of murine leukemia virus antigens and production of xenotropic virus in the late preleukemic period in AKR mice. *Proc. Natl. Acad. Sci. USA* 73:4680–84

53. Tung, J.-S., Shen, F.-W., Fleissner, E., Boyse, E. A. 1976. X-gp70: A third molecular species of the envelope protein gp70 of murine leukemia virus, expressed on mouse lymphoid cells. *J. Exp. Med.* 143:969–74

54. Strand, M., Lilly, F., August, J. T. 1974. Host control of endogenous murine leukemia virus gene expression: Concentrations of viral proteins in high and low leukemia mouse strains. *Proc. Natl. Acad. Sci. USA* 71:3682–86

55. Tung, J.-S., Fleissner, E., Vitetta, E. S., Boyse, E. A. 1975. Expression of murine leukemia virus envelope glycoprotein gp69/71 on mouse thymocytes: Evidence for two structural variants distinguished by presence versus absence of G_{IX} antigen. *J. Exp. Med.* 142:518–23

56. Herberman, R. B., Aoki, T. 1972. Immune and natural antibodies to syngeneic murine plasma cell tumors. *J. Exp. Med.* 136:94–111

57. Boyse, E. A., Old, L. J., Stockert, E. 1965. The TL (thymus leukemia) antigen: A review. In *Immunopathology, IV Int. Symp.*, ed. P. Grabar, P. A. Miescher, pp. 23–40. Basel: Schwabe

58. Boyse, E. A., Old, L. J., Luell, S. 1964. Genetic determination of the TL (thymus-leukemia) antigen in the mouse. *Nature* 201:779

59. Schlesinger, M. 1970. How cells acquire antigens. *Prog. Exp. Tumor Res.* 13:28–83

60. Boyse, E. A., Old, L. J. 1971. A comment on the genetic data relating to expression of TL antigens. *Transplantation* 11:561–62

61. Boyse, E. A., Flaherty, L., Stockert, E., Old, L. J. 1972. Histo-incompatibility attributable to genes near *H-2* that are not revealed by hemagglutination or cytotoxicity tests. *Transplantation* 13:431–32

62. Flaherty, L., Wachtel, S. S. 1975. *H(Tla)* System: Identification of two new loci *H-31* and *H-32*, and alleles. *Immunogenetics* 2:81–85

63. Stanton, T. H., Boyse, E. A. 1976. A new serologically defined locus, *Qa-1*, in the *Tla*-region of the mouse. *Immunogenetics* 3:525–31

64. Flaherty, L. 1976. The *Tla* region of the mouse: Identification of a new serologically defined locus, *Qa-2*. *Immunogenetics* 3:533–39

65. Boyse, E. A., Stockert, E., Old, L. J. 1968. Isoantigens of the *H-2* and *Tla* loci of the mouse. Interactions affecting their representation on thymocytes. *J. Exp. Med.* 128:85–95

66. Old, L. J., Stockert, E., Boyse, E. A., Kim, J. H. 1968. Antigenic modulation: Loss of TL antigen from cells exposed to TL antibody. Study of the phenomenon *in vitro*. *J. Exp. Med.* 127:523–39

67. Boyse, E. A., Old, L. J., Luell, S. 1963. Antigenic properties of experimental leukemias. II. Immunological studies *in vivo* with C57BL/6 radiation-induced leukemias. *J. Natl. Cancer Inst.* 31:987–95

68. Boyse, E. A., Stockert, E., Old, L. J. 1967. Modification of the antigenic structure of the cell membrane by thymus-leukemia (TL) antibody. *Proc. Natl. Acad. Sci. USA* 58:954–57

69. Lamm, M. E., Boyse, E. A., Old, L. J., Lisowska-Bernstein, B., Stockert, E. 1968. Modulation of TL (thymus-leukemia) antigens by Fab-fragments of TL antibody. *J. Immunol.* 101:99–103

70. Stackpole, C. W., Jacobson, J. B. 1977. Antigenic modulation. In *Handbook of Cancer and Immunology*, ed. H. Waters, Vol. 1. New York: Garland. In press

71. Hyman, R., Stallings, V. 1976. Characterization of a TL⁻ variant of a homozygous TL⁺ mouse lymphoma. *Immunogenetics* 3:75–84

72. Boyse, E. A., Stockert, E., Iritani, C. A., Old, L. J. 1970. Implications of TL phenotype changes in an H-2-loss variant of a transplanted H-2ᵇ/H-2ᵃ leukemia. *Proc. Natl. Acad. Sci. USA* 65:933–38

73. Bjaring, B., Klein, G. 1968. Antigenic characterization of heterozygous mouse lymphomas after immunoselection *in vivo*. *J. Natl. Cancer Inst.* 41:1411–29

74. Hauschka, T. S., Hitt, S. A., Zumpft, M., Shows, T. B., Boyse, E. A. 1975. Immunoselective loss of parental H antigens by somatic reduction in an

H-2ª/H-2ᵇ hybrid mouse leukemia. *Transplant Proc.* 7:165–71
75. Boyse, E. A., Old, L. J., Stockert, E. 1968. An approach to the mapping of antigens on the cell surface. *Proc. Natl. Acad. Sci. USA* 60:886–93
76. Hämmerling, U., Eggers, H. J. 1970. Quantitative measurement of uptake of alloantibody on mouse lymphocytes. *Eur. J. Biochem.* 17:95–99
77. Stackpole, C. W. 1977. Topographical differentiation of the cell surface. *Prog. Surf. Membr. Sci.* 12: In press
78. Stockert, E., Old, L. J. 1977. Preleukemic expression of TL antigens in X-irradiated C57BL/6 mice. *J. Exp. Med.* 146:271–76
79. Snell, G. D., Dausset, J., Nathenson, S. 1976. Biochemical and structural properties of the cell membrane located alloantigens of the major histocompatibility complex. In *Histocompatibility,* pp. 275–321. New York: Academic
80. Dofuku, R., Biedler, J. L., Spengler, B. A., Old, L. J. 1975. Trisomy of chromosome 15 in spontaneous leukemia of AKR mice. *Proc. Natl. Acad. Sci. USA* 72:1515–17
81. Chang, T. D., Biedler, J. L., Stockert, E., Old, L. J. 1977. Trisomy of chromosome 15 in X-ray-induced mouse leukemia. *Proc. Am. Assoc. Cancer Res.* 18: 225
82. Stück, B., Boyse, E. A., Old, L. J., Carswell, E. A. 1964. ML—A new antigen found in leukaemias and mammary tumours of the mouse. *Nature* 203: 1033–34
83. Varmus, H. E., Quintrell, N., Medeiros, E., Bishop, J. M., Nowinski, R. C., Sarkar, N. H. 1973. Transcription of mouse mammary tumor virus genes in tissues from high and low tumor incidence mouse strains. *J. Mol. Biol.* 79:663–79
84. Hilgers, J., Haverman, J., Nusse, R., van Blitterswijk, W. J., Cleton, F. J., Hageman, P. C., van Nie, R., Calafat, J. 1975. Immunologic, virologic and genetic aspects of mammary tumor-virus-induced cell surface antigens: Presence of these antigens and the Thy-1.2 antigen on murine mammary gland and tumor cells. *J. Natl. Cancer Inst.* 54: 1323–33
85. Aoki, T., Huebner, R. J., Chang, K. S. S., Sturm, M. M., Liu, M. 1974. Diversity of envelope antigens on murine type-C RNA viruses. *J. Natl. Cancer Inst.* 52:1189–97
86. Stephenson, J. R., Essex, M., Hino, S., Hardy, W. D. Jr., Aaronson, S. A. 1977. Feline oncornavirus-associated cell membrane antigen (FOCMA). VII. Distinction between FOCMA and the major virion glycoprotein. *Proc. Natl. Acad. Sci. USA* 74:1219–23
87. Rohrschneider, L. R., Kurth, R., Bauer, H. 1975. Biochemical characterization of tumor-specific cell surface antigens on avian oncornavirus transformed cells. *Virology* 66:481–91
88. Artzt, K., Bennett, D., Jacob, F. 1974. Primitive teratocarcinoma cells express a differentiation antigen specified by a gene at the T-locus in the mouse. *Proc. Natl. Acad. Sci. USA* 71:811–14
89. Jacob, F. 1977. Mouse teratocarcinoma and embryonic antigens. *Immunol. Rev.* 33:3–32
90. Martin, W. J., Gipson, T. G., Rice, J. M. 1977. H-2ª-associated alloantigen expressed by several transplacentally-induced lung tumours of C3Hf mice. *Nature* 265:738–39
91. Parmiani, G., Invernizzi, G. 1975. Alien histocompatibility determinants on the cell surface of sarcomas induced by methylcholanthrene. I. *In vivo* studies. *Int. J. Cancer* 16:756–67
92. Old, L. J., Boyse, E. A. 1964. Immunology of experimental tumors. *Ann. Rev. Med.* 15:167–86
93. Vaage, J. 1968. Nonvirus-associated antigens in virus-induced mouse mammary tumors. *Cancer Res.* 28:2477–83
94. Morton, D. L., Miller, G. F., Wood, D. A. 1969. Demonstration of tumor-specific immunity against antigens unrelated to the mammary tumor virus in spontaneous mammary adenocarcinomas. *J. Natl. Cancer Inst.* 42:289–302
95. Carswell, E. A., Wanebo, H. J., Old, L. J., Boyse, E. A. 1970. Immunogenic properties of reticulum cell sarcomas of SJL/J mice. *J. Natl. Cancer Inst.* 44: 1281–88
96. Boyse, E. A., Old, L. J. 1970. The invitation to surveillance. In *Immune Surveillance,* ed. R. T. Smith, M. Landy, pp. 5–30. New York: Academic
97. Basombrio, M. A., Prehn, R. T. 1972. Antigenic diversity of tumors chemically induced within the progeny of a single cell. *Int. J. Cancer* 10:1–8

Ann. Rev. Genet. 1977. 11:161–82

GENETICS OF STRUCTURE AND FUNCTION OF BACTERIAL FLAGELLA

❖3119

Tetsuo Iino
Laboratory of Genetics, Faculty of Science, The University of Tokyo, Hongo, Tokyo 113, Japan

CONTENTS

INTRODUCTION 161
GENETIC ANALYSIS OF FLAGELLAR CHARACTERS 163
Flagella Genes and Selection of Their Mutants 163
Overall Maps of Flagella Genes 164
REGULATION OF FLAGELLAR FORMATION 165
Cell Cycle and Flagellation 165
Sequence of Flagellar Morphogenesis 166
Regulation of Flagellin Synthesis 168
Growth of Flagellar Filaments 168
POLYMORPHISM OF FLAGELLAR FILAMENTS 170
Fine Structure and Polymorphism of Flagellin 170
Polymorphism in Helical Shape of Flagellar Filaments 172
Flagellar Phase Variation 173
GENETIC CONTROL OF MOTILITY 175
Function of the mot Genes 175
Role of Flagellar Filaments on Bacterial Motility 176
SUMMARY 177

INTRODUCTION

Eight years ago I wrote a review article on the genetics of bacterial flagella (1). Since then, the knowledge of the structure and function of bacterial flagella has advanced considerably.

Owing to the success of the isolation of an "intact flagellum" (a flagellar filament joined with a hook and a basal body) from osmotically lysed bacterial cells, basal body structures, which had been ambiguous for a long time, were clarified (2–6).

In gram-negative bacteria, such as *Escherichia, Salmonella,* and *Pseudomonas,* a basal body is composed of four rings, a rod passing through the centers of the four rings, and a cylinder filled in the space between two outer rings (Figure 1). The four rings are termed L, P, S, and M; each is connected by its periphery with a lipopolysaccharide outer layer, a peptidoglycan layer, a periplasmic space, and a cytoplasmic membrane respectively. Thus the basal bodies are entirely mounted in the cell envelope. In gram-positive bacteria, only two rings are present in a basal body (4). At an end adjacent to the L ring, the rod connects longitudinally to the hook, which extends out of the cell body and is joined to the filament at the other end of its tubular structure.

These flagellar structures are composed principally of protein. Purification and amino acid analysis of hook protein (7–9) as well as filament protein, i.e. flagellin (10–19), has been established. The basal body of *Escherichia coli* was fractionated into at least seven component proteins with differing molecular weights (20, 21). The helical arrangement of flagellin molecules along the longitudinal rows in a filament was visualized on some bacterial strains, mainly with optical diffraction and filtering methods (22–30).

Flagella function as locomotive organelles of bacteria by propagating helical waves and by pushing bacterial bodies forward. Recent investigations have provided sound experimental evidence that the flagellar filaments rotate relative to the cell

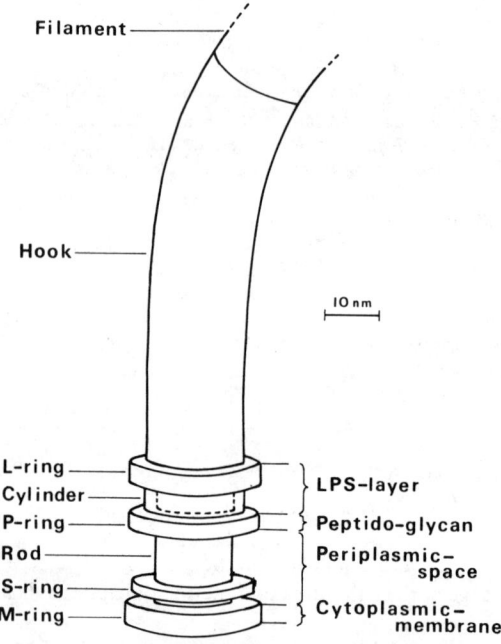

Figure 1 Model of the basal structure of a flagellum of *Salmonella typhimurium.*

body and that their motor machineries reside at their bases (31–36). The role of the basal structures in the locomotive function of flagella is thus gaining attention.

Genetic studies of bacterial flagella in recent years have complemented these researches on their structure and function. This review covers the progress in the knowledge of the structure and function of bacterial flagella since publication of my earlier review article (1). Taxis, a pronounced response of flagella, is to be covered in the review by Parkinson in this volume (37).

GENETIC ANALYSIS OF FLAGELLAR CHARACTERS

Flagella Genes and Selection of Their Mutants

The flagella genes whose functions were most clearly identified are the structural genes for flagellin, namely *H1* and *H2* in *Salmonella* (1), *H* in *Bacillus subtilis* (38) and *Pseudomonas aeruginosa* (39), and *hag* in *E. coli* (40). Their mutation results in the change of any one of the following phenotypes: efficiency of filament formation, shape of filaments, sensitivity to flagellotropic phage, and specificity of flagellar antigen (41–43).

Other flagella genes involved in flagellar formation are commonly given the gene symbol *fla* followed by a capital letter and/or a Roman numeral designating a cistron (1, 38–40, 44). The mutant phenotype generally used for genetic analysis of the *fla* genes is nonflagellate.

The flagella genes of the third category are termed *mot* (45–48). They control the locomotive function of flagella without affecting their overall structure. A mutation from mot^+ to mot^- results in flagellar paralysis. The genes for chemotaxis are termed *che* (37).

In addition to these genes essential for the structure and function of flagella, a modifier gene, *nml*, is present in some serotypes of *Salmonella* (49, 50). In the presence of nml^+, about half of the lysine residues of each flagellin molecule are methylated at the ϵ-position. This chemical modification sometimes is associated with the change of antigenicity of flagella (49).

Neither nonflagellate nor paralyzed mutants can spread and form swarms on semisolid medium (51), and both are resistant to flagellotropic phage (52–56). Therefore, by the combined use of semisolid medium and flagellotropic phage for selection, both types of nonmotile mutants are very efficiently isolated from a culture of motile bacterial strains.

Change of flagellar shape by mutation often associates with the change of motility of the bacteria without alteration in the sensitivity to flagellotropic phage (57). Consequently, flagellar shape mutants are detected among nonswarmers, e.g. straight mutants, or slow-swarmers, e.g. curly mutants, on semisolid medium, and they are differentiated from fla^- or mot^- by their sensitivity to flagellotropic phage (51).

An appropriate titer of antiserum prepared against a specific type of flagellar antigen inhibits motility of the bacteria of the same antigenic type but allows multiplication of their cells. The mutants of flagellar antigenic type are isolated as motile swarmers that escape from the inhibition by the antibody supplemented in semisolid medium (41).

Overall Maps of Flagella Genes

Genetic analyses of flagella genes were extensively carried out on *Salmonella,* mainly *S. typhimurium* and *S. abortusequi,* and on *E. coli.* The analyses on *Salmonella* were performed with P22-mediated transduction (41, 43, 58), and Hfr X F⁻ conjugation (46). On *E. coli,* F' X F⁻ conjugation (44, 59) and special transduction with λ-*fla* (60–63) as well as Pl-mediated transduction (47) were applied. Bacteriophage Mu-induced mutations have been known to prevent the expression of the genes that are transcribed later than the mutant genes in the same operon (64). Mu-induced flagellar mutations were successfully used for the disclosure of the organization of flagella genes into operons (65).

Distributions of the flagella genes on the chromosomes of *Salmonella* and *E. coli* are summarized in Figure 2. In both of these bacteria, the largest cluster of flagella genes resides at about the same map position, i.e. near the *his* operon (58, 59, 62, 66, 67). At the middle of the cluster, a flagellin gene, *H1* in *Salmonella* or *hag* in *E. coli,* is located. Besides, *fla, mot,* and *che* have been found in this cluster. Sixteen *fla* genes have been detected in *Salmonella* and 14 in *E. coli.* Two *mot* genes are present adjacent to each other in both bacteria. Near *mot,* a group of *che* genes was identified (37).

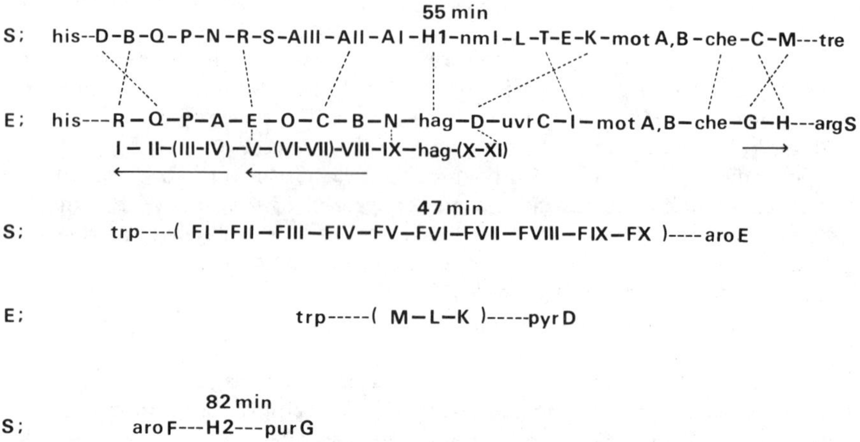

```
                              55 min
S;  his--D-B-Q-P-N-R-S-AIII-AII-AI-H1-nmI-L-T-E-K-motA,B-che-C-M---tre

E;  his---R-Q-P-A-E-O-C-B-N-hag-D-uvrC-I-motA,B-che-G-H---argS
        I-II-(III-IV)-V-(VI-VII)-VIII-IX-hag-(X-XI)              ⟶

                              47 min
S;       trp----( FI-FII-FIII-FIV-FV-FVI-FVII-FVIII-FIX-FX )----aroE


E;              trp-----( M-L-K )-----pyrD


              82 min
S;        aroF---H2---purG
```

Figure 2 Linkage map of the flagella genes of *Salmonella* (S) and *Escherichia coli* (E). his: histidine operon including A-I cistrons; trp: tryptophan operon including A-E cistrons; tre: trehalose; arg: arginine; aro: aromatic amino acids; pyr: pyrimidine; pur: purine; uvr: ultraviolet light sensitivity; nml: N-methyl lysine in flagellin; hag, H1, H2: flagellin; mot: motility; che: chemotaxis. Other symbols on which only cistron designations are given are *fla* (flagellation). An *arrow* indicates the direction of transcription of an operon. A *dotted line* indicates a pair of homologous cistrons. The chromosomal position of each linkage group is shown as "min" in the 138 min linkage map of *Salmonella typhimurium* (66). The arrangement of the genes in parentheses to their outside genes is not known. For the composition of cistrons in *che,* refer to Parkinson (37).

Functional homologies between the *fla* genes of *Salmonella* and *E. coli* in this cluster were examined by complementation tests between *fla*⁻ mutants in intergeneric transductional heterogenotes and conjugal hemizygotes of these bacteria (68; B. Stocker, personal communication). The relative positions of the functionally homologous genes on their chromosomes is not exactly the same in these two genera (Figure 2). This suggests the occurrence of inversion or translocation in the evolutionary history of these genera. In addition to these genes, *nml* is present in the same cluster of *Salmonella* (49, 50). Flagella genes of this cluster are separated into several operons (Figure 2) (59, 61, 67).

The second cluster of flagella genes in *Salmonella* and *E. coli* is located near the *trp* operon (44, 66, 69). In *Salmonella*, ten *fla* genes were identified in this cluster, while in *E. coli* three *fla* genes were found. This difference in number may simply reflect the extensiveness of the examined *fla* mutants of this cluster. The complementation test on transductional heterogenotes between *S. typhimurium* and *Shigella dysenteriae* indicated that in the latter bacterium at least a part of this *fla* cluster is present, but that the largest cluster near *his* is deleted (68). In diphasic *Salmonella*, an additional flagellin gene, *H2*, was mapped apart from the abovementioned two clusters. Its location is near *purG* (66, 70). *E. coli* entirely misses the *H2* locus. When *H2* is transduced from *Salmonella* to *E. coli*, it is translocated to the *hag* locus or more occasionally to a locus unrelated to flagella genes (71).

No advance has been made on mapping of the flagella genes in *B. subtilis* since Joys & Frankel (72) reported three nonallelic genes, *H, fla,* and *mot,* with transformation experiments. On *Proteus mirabilis*, eight *fla* and one *mot* were disclosed (73). Except for one *fla*, they were located in a cluster transduced by phage 34.13 simultaneously. Flagella genes of *Pseudomonas aeruginosa* were also mapped in a cluster, in which at least one *hag,* ten *fla,* and one *mot* were detected by phage F116-mediated transduction (T. Iino, unpublished information).

REGULATION OF FLAGELLAR FORMATION

Cell Cycle and Flagellation

The distribution of the number of flagella per cell is characteristic for each bacterial strain under a given cultural condition. During a cell cycle of peritrichously flagellate bacteria, the number of flagella doubles and upon cell division they distribute to daughter cells approximately evenly. This indicates that the formation of flagella is genetically regulated so as to be coupled with the cell cycle.

When light-density spores of *B. subtilis* are germinated at 46°C and grown at 37°C, they are well synchronized in both cell division and chromosome replication. In such bacteria, the doubling of the rate of flagellar formation was found to correspond to the time of replication of the *hisA1* gene (74). The synchronization of flagellar formation in a synchronized culture of *E. coli* was also reported (75). Several temperature-sensitive mutants of *E. coli* defective in DNA-replication, e.g. *fts* and *dna,* are retarded in flagellar formation at a nonpermissive temperature (76). Although these observations have not demonstrated that the responsible reaction step is at the initiation of flagellar formation, they strongly suggest that the initiation of flagellar formation is coupled with DNA replication. Whether the coupling is at the transcriptional level or, more indirectly, whether it involves a sequential induction system proceeding through a cell cycle or is associated with a specific state of the cell surface appearing at a stage in a cell cycle is left unanswered.

A regulatory factor presumably involved in the initiation of flagellation is adenosine 3',5'-cyclic phosphate (cAMP) and the protein specifically bound to it (CRP). Either cAMP-deficient mutants (cya^-) or CRP-deficient mutants (crp^-) of both *E. coli* and *S. typhimurium* are defective not only in sugar fermentation but also in flagellar formation (77–79). Thus the complex of cAMP and CRP has a pleiotropic regulatory function for both sugar fermentation and flagellation. Their function, however, is not invariably common to these phenomena, because suppressor mutants [constitutive flagella synthesis (*cfs*)] which restore flagella-forming ability in cya^- and crp^- were detected (78, 79). Sugar fermentation of these mutants remained defective in the presence of the suppressor. *cfs* is dominant over its wild allele. The mutant site of *cfs* was mapped in *flaI* in *E. coli* and *flaT* in *S. typhimurium* (78, 79). The step of flagellar formation sensitive to cAMP may be at a very early stage of flagellar formation because none of the precursor structures of flagellar bases were detected on cya^- cells. The most plausible explanation regarding the role of cAMP in flagellation is as follows: cAMP receptor protein together with cAMP modulates the wild allele of the gene in which the *cfs* mutation resides, and the gene in turn acts as a positive effector on the initial step of flagellar formation. Then, the *cfs* mutant gene may be regarded as the constitutive effector.

Sequence of Flagellar Morphogenesis

The intensive electron microscopical observation of the cell envelope fractions of various nonflagellate mutants enabled us to detect a series of precursor structures of flagella (69). The simplest precursor structure so far detected among the nonflagellate mutants of *Salmonella* is the complex of two inner rings and a rod (RIV particle). The next simplest is the complex of a RIV particle and a P ring (CAS particle). Then comes the complex of a CAS particle, a cylinder, and an L ring (BAB particle). A BAB particle is morphologically indistinguishable from a basal body of an intact flagellum. The precursor between BAB and the intact flagellum is the complex of a BAB particle and a hook (HOB particle). These precursor structures were cistron-specific. Summing up these results, the sequential process of flagellar morphogenesis is constructed as shown in Figure 3, and several *fla* cistrons were assigned to the resolved steps. Whether a *fla* cistron at a step is the structural gene of the forthcoming component or the positive regulator gene of the morphogenetic step is still not clear.

Among the fla^- mutants belonging to 13 cistrons listed at the left side of Figure 3, none of the precursor structures were detected electron microscopically. Some of these cistrons may be responsible for the synthesis of the components of RIV particles and the others for the regulation of the initiation of flagellar formation.

The existence of the complex of a rod and two inner rings as the first detectable structure suggests that the flagellar morphogenesis starts from the assembly of inner rings associated with the cell surface layers. Even if the genes specifically responsible for the formation of flagella are functional, flagellar formation is retarded by a mutation that causes a defect in the surface layers. Deep-rough mutants in *S. typhimurium* have no flagella (80). Flagellar formation in uridine diphosphoglucose pyrophosphorylase–deficient *galU*⁻ mutants of *E. coli* is remarkably depressed (81). The lipopolysaccharide layer is incomplete and some kinds of membrane-bound proteins are released in these mutants (80, 82). In a mutant of *B. subtilis,* production of exoenzymes and formation of flagella are pleiotropically affected presumably through certain alterations of the surface layers (83). Although the presence or absence of basal bodies has not been examined in these mutants, it is highly plausible that some components or structures of the surface layers are essential for the early stages of flagellar morphogenesis.

Because flagellar morphogenesis proceeds from the inner portion to the outer, the components of the latter structures constructed must be transported from the site of their synthesis in the cytoplasm to the outer surface layer. The intralayer protein transport systems may be responsible for this phenomenon, although they have not yet been identified.

In the sequential pathway of flagellar morphogenesis, three branches were found (Figure 3). Each of these branches resulted from a block of the main pathway by a *fla*⁻ mutation. The first and the second branches, as shown in Figure 3, indicate that (*a*) the formation of hooks can proceed to some extent even if outer rings are absent and (*b*) the L rings are formed inefficiently when cylinders are missing. The third branch consists of polyhooks or, in other words, superhooks. A polyhook is structurally homologous with a normal hook but exceedingly longer than that (8, 80, 82). Polyhooks have been detected in *flaE* mutants of *E. coli* (84) and *flaR* mutants of *S. typhimurium* (85). The mutant allele is recessive to its wild-type allele. In these mutants, few polyhooks bear flagellar filaments. The product of the

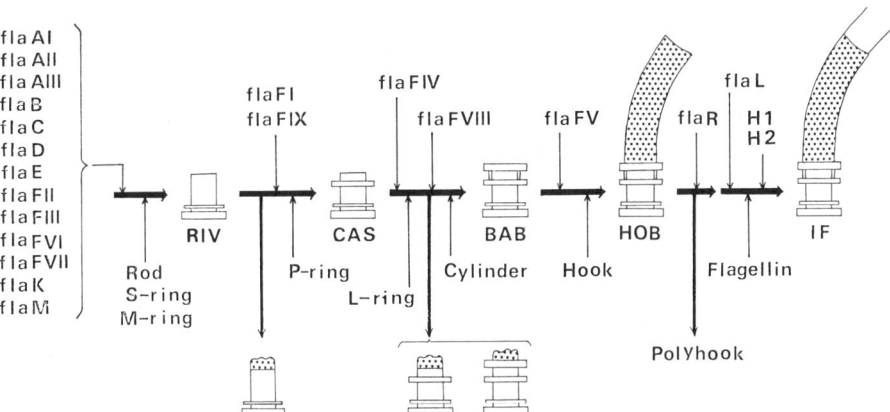

Figure 3 The process of flagellar morphogenesis in *Salmonella*. RIV: rod-inner ring complex; CAS: RIV-P ring complex; BAB: basal body; HOB: hook-basal body complex; IF: intact flagellum.

wild-type allele may control the length of the hooks and the initiation of filament growth, presumably acting as "stopper" for hook protein assembly and "initiator" for flagellin assembly at the tip of each hook.

Regulation of Flagellin Synthesis

The synthesis of flagellin, whose assembly occurs at the final step of flagellar morphogenesis, seems to be under the control of a complex regulatory system. So far, none of the proteins immunologically cross-reacting with flagellin have been detected in any *fla*⁻ mutants of *Salmonella* (1). A mutant assigned as *flaG* in *Salmonella* has been reported as the exception to this rule. However, it was later found to carry paralyzed flagellar filaments in less than 1% of the cells and its mutant site was mapped in *flaAII* (86).

In vitro synthesis of flagellin was carried out with a cell-free extract of *E. coli* directed by mRNA of *Salmonella* (86, 87). When RNA extracted from the flagellate strains was used, flagellin synthesis was demonstrated together with the synthesis of other proteins. When RNA of the nonflagellate mutants of various *fla* cistrons was used, no detectable flagellin synthesis occurred even though the overall synthesis of protein proceeded as efficiently as in the experiments with RNA of the flagellate strains. The same result was observed when a *cya* mutant, which has a primary defect in cAMP synthesis and consequently fails to produce flagella, was used as a source of mRNA (79). Extracts of *fla*⁻ mutant cells have no inhibitory effect on the synthesis of flagellin directed by mRNA of *fla*⁺ strains. Further, RNA-free extracts of *fla*⁺ strains do not promote the synthesis of flagellin by mRNA of *fla*⁻ strains (86).

These results mean that not only flagellin but also mRNA for flagellin are not synthesized in any *fla*⁻ mutant. In other words, when any one step of flagellar formation is genetically blocked, flagellin synthesis stops at the transcriptional level. The structural gene for flagellin constitutes an operon independent of any other known *fla* cistrons, and the *fla* cistrons are distributed among several discrete operons (65). Therefore, an interoperonic regulation system must be operating in the synthesis of flagellin mRNA. It is plausible that *flaL* of *Salmonella* functions as a "linker" between the induction system and the operons for flagellin synthesis (88). The *flaL* mutants can produce flagellar structures other than filaments, and synthesis of mRNA specific for both *H1* and *H2* is blocked simultaneously in the mutants. This is the case even in the deletion mutants of *flaL*. Further, some revertants of *flaL* partially recover the ability to synthesize flagellin, and the resulting partial revertant cells produce flagellar filaments shorter than those of the normal *fla*⁺ cells. It is possible that (*a*) flagellin monomers in an amount not detectable by the techniques applied so far work as an antirepressor of flagellin synthesis when they fail to polymerize in the cells of *fla*⁻ mutants, or (*b*) the product of *flaL* works as the antirepressor.

Growth of Flagellar Filaments

The process of assembly of flagellin to flagellar filaments has been clarified in detail at the molecular level by the in vitro polymerization experiments with *Salmonella* flagellin (89). A flagellin monomer binds to an end of an existing filament. The end where the monomer binds corresponds to the distal end of a flagellar filament on living bacteria. Next, the monomer is firmly incorporated into the filament accompanied by its conformational change. The incorporated monomer then acts as a part of the nucleus for polymerization of the next monomer to begin. Thus, a conformational change of flagellin molecules upon their assembly confers structural polarity to the flagellar filaments and also confines the assembling site to the distal end of each filament. These processes are regarded as self-assembly and proceed in the presence of appropriate concentrations of flagellin monomers and flagellar filaments under proper temperature, pH, and ionic strength (89). The growth of flagellar filaments in vivo may be essentially homologous as is that in vitro. The in vivo growth of a flagellar filament occurs by the polymerization of flagellin at the distal end of the filament (90–92), and under optimal conditions the maximum speed at the initial stage of the filament growth in vivo was the same as that in vitro (93).

Although polymerization of flagellin monomers in vitro is initiated without the presence of the filaments under conditions of high ionic strength (94), the filament must be present for polymerization to occur in the ordinary physiological environment. This suggests that the initiation of filament growth in vivo under ordinary physiological conditions occurs through a somewhat different process from that in vitro. In living bacteria, the proximal end of a filament is connected to the distal end cf a hook, and by the binding of flagellin at this end, the filament begins to grow. Thus the distal end of each hook serves as a nucleus for the polymerization of flagellin. In fact, isolated hooks were successfully used as a heteronucleus for the polymerization of flagellin in vitro, although the efficiency of the process was low compared with the polymerization in which fragments of flagellar filaments were used as the homonucleus (93). The requirement of a heteronucleus for the initiation of assembly of certain structural components seems to be an elaborate mechanism to confine the site of structure formation to a proper site in living organisms.

The rate of in vitro polymerization of flagellin monomers to flagellar filaments depends on the species and the amount of flagellin, and on the physicochemical condition of the environment (89). However, when these are fixed, the average rate is maintained at a constant, and filament growth proceeds constantly except for an occasional abrupt termination by an error occurring at the growing end of the filament (95). Consequently, it is experimentally possible to obtain filaments longer than 50 μm by in vitro polymerization. On the other hand, the average rate of in vivo elongation was found to decrease exponentially with the increase of filament length according to the following equation:

$$V = V_0 e^{-KL}$$

where V denotes the rate of filament growth at length L, V_0 the initial rate at $L = 0$, and K the constant characterizing the degree of decrease in rate per unit of length (93). This relationship is observed even among the filaments of various lengths growing on a single cell or a filament mechanically shortened by breakage. Therefore, the contribution of aging of a cell or of a flagellum-forming apparatus to the decrease in the growth rate of filaments is implausible. For the growth of a flagellar filament, flagellin monomers must be transported, probably by a sort of diffusion, from the cell body to the tip of the filament through its central canal (90–93). Consequently, the decrease in growth rate must be caused by the decrease in the efficiency of transportation with the increase in length of the filament. This means that the filament limits its own growth rate. The observed maximal length of flagellar filaments on living bacteria is explained by such autoregulation of the growth rate, although the possibility of a termination factor for filament elongation is not entirely excluded (93).

POLYMORPHISM OF FLAGELLAR FILAMENTS

Fine Structure and Polymorphism of Flagellin

Although flagellin molecules of various bacteria have common biochemical features as the component protein of flagellar filaments, they also exist as extensively polymorphic molecules in nature. Molecular weights of flagellin so far reported ranges from 33,000 to 60,000 among different bacterial species (14–17, 96).

Various mutants that differ in the spiral shape of flagellar filaments have been reported, and their mutant sites were all assigned to the structural genes for flagellin (1, 28, 42). In a straight flagellar mutant of *B. subtilis* an alanine residue in the polypeptide of normal flagellin is replaced with valine (97).

In some bacterial genera, such as *Salmonella,* many varieties of flagellar (H) antigen have been observed, and the difference of amino acid composition among different *Salmonella* serotypes has been demonstrated (1). The antigenic mutants of flagellin-i of *S. typhimurium* were attributed to either substitution of an amino acid or a small deletion in the flagellin (98). Even in flagellar filaments of the same shape and antigen type, their component flagellin molecules differ in primary structure between different strains of the same species (99).

Polymorphism of flagellin appears even in a clone of some bacteria. Flagellar phase variation in *Salmonella* is a remarkable example which will be described later. A single cell of *Vibrio parahaemolyticus* produces both polar and peritrichous flagella. Interestingly enough, flagellar filaments of these two types differ from each other in both shape and antigenicity, suggesting that they are composed of different types of flagellin (100, 101). Although mutants that produce only one type of flagella have been isolated, their genetic analysis has not yet been carried out.

This pronounced polymorphism of flagellin among bacteria suggests the existence of evolutionarily variable and conservative regions in the molecule. Genetic fine structure analysis of the *H1*-g cistron of *Salmonella* revealed that a quarter of the cistron from the terminus adjacent to *ah1* predominantly produces nonflagellate phenotype ("essential region" in Figure 4) (43). This region may correspond to a section of polypeptide chain essential for the specific conformation of flagellin. Evolutionarily, this region must have been conservative. Antigenic mutations occur in the remaining region ("antigenic region" in Figure 4). In the region primarily responsible for antigen-type determination, the sites of specific antigen-type determinants are arranged linearly. The antigenic region may correspond to an exposed part of the flagellin polypeptide. The relatively rare detection of nonflagellate mutations

GENETICS OF BACTERIAL FLAGELLA 171

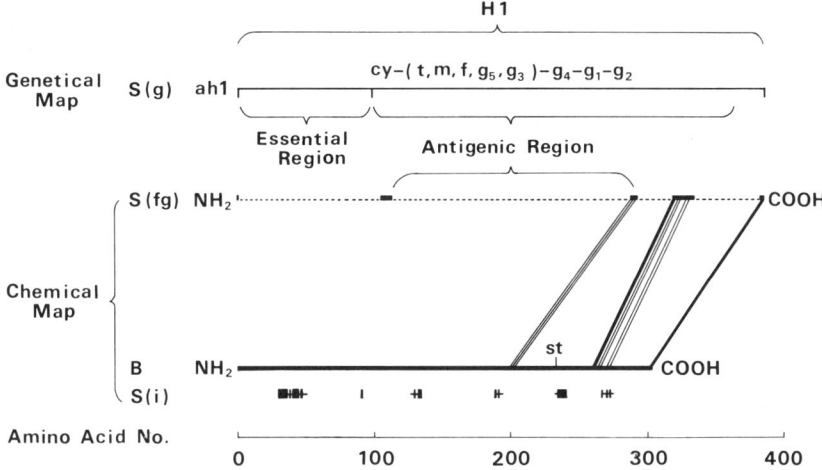

Figure 4 Comparison of the map of flagellin gene *H1-g*, shown as S(g), of *Salmonella* and the chemical maps of flagellin of *Salmonella adelaide*, S(fg), and *Bacillus subtilis* (B). Amino acid sequences of the regions indicated by dotted lines are not determined. The lines between S(fg) and B connect the identical amino acids of the homologous peptide segments. S(i) shows peptide segments of flagellin-i of *Salmonella typhimurium* whose homologous regions were identified in B. The identical amino acids between them are shown by vertical bars. st indicates straight mutant site; cy indicates curly mutant site.

in this region may mean that conformational alteration of flagellin caused by a mutational change in its surface is usually not so severe as to result in the loss of ability to polymerize. This region may be evolutionarily variable. These genetic results conform with the conclusion drawn from immunochemical studies on fg-type flagellin that all the antigenic specificities reside in the central region of flagellin polypeptide (Figure 4) (11, 12). The sites of amino acid substitution responsible for antigen-type mutations in the whole flagellin polypeptide are left for future investigations.

In *Salmonella*, partial amino acid sequences were determined on flagellin-i of *S. typhimurium*, and flagellin-fg of *S. adelaide* (12, 14). Flagellin-i and flagellin-fg are composed of 470 and 386 amino acids respectively. In the former, amino acid sequences were determined on 29 tryptic peptides of in total 217 amino acids. The order of these peptides in a flagellin polypeptide has not been determined yet. On flagellin-fg, the N-terminal amino acid, sequences of 37 amino acids of 4 tryptic peptides and the positions of these peptides in the polypeptide were determined.

It is premature to correlate this chemical information with the fine structure of the corresponding flagellin gene. However, the complete amino acid sequence of a flagellin molecule of *B. subtilis* strain 168, designated as flagellin-B for further discussion, was recently clarified (19), and the comparison of the sequence with the known partial amino acid sequences of *Salmonella* flagellin provides some interesting information. A flagellin-B molecule is composed of 304 amino acid residues. Out of the known 29 tryptic peptides of flagellin-i, 7 peptides correspond to homologous regions of flagellin-B (Figure 4). The largest homologous region, which corresponds to two consecutive peptides of flagellin-i, is assigned between the 30th and the 51st amino acids from the N terminus of flagellin-B. These two flagellin molecules are identical in 16 of the 22 amino acids in the region. The amino acids between the 233rd and the 241st of flagellin-B are identical in 7 of 9 amino acids with peptide-10 of flagellin-i. These two regions correspond to antigenically inert sections of flagellin-fg, and the former resides in a region corresponding to the "essential region" in the genetic map. In the homologous peptides assigned in the "antigenic region," flagellin-B and flagellin-i are identical in only 7 out of 16 amino acids. Thus as far as the identified homologous regions are concerned, the "essential region" seems to be more conservative than the antigenic region. Remarkable variation in molecular size among these three different flagellins indicates that numerous deletions or additions of genetic codes have occurred in the flagellin genes during the course of evolution. Distribution of the homologous peptide regions among flagellin-B, -i, and -fg suggests that the region responsible for such chromosomal variation is mainly in the antigenic region in the middle of the flagellin polypeptide (Figure 4).

As regards the sites of flagellar shape mutations, curly mutant sites in *Salmonella* were genetically mapped in a region proximal to the N terminus, and a straight mutant site of *B. subtilis* was chemically identified to be close to the C terminus. The antigenic region was located between them. These two separate regions on a flagellin polypeptide may jointly play an essential role for the determination of the mode of molecular assembly.

Polymorphism in Helical Shape of Flagellar Filaments

As described in the foregoing section, the genetic change of helical shape in flagellar filaments is attributed to a mutation at a specific site in the flagellin gene. The primacy of the type of flagellin for the determination of helical shape was also shown by in vitro assembly of flagellin into flagellar filaments (89). However, the helical shape manifested by a flagellar filament is not always restricted to only one form; transformation from one form to another occurs often. For example, the normal shaped filaments of *Salmonella* strain SJ25 undergo sequential transformation to coiled, semicoiled, and curly when the pH of the environmental solution is lowered from neutral to acidic (102). The transformation is also affected by ionic strength of the environment. The transformation from one form to another is discontinuous and the intermediate filaments appear as segmental chimera of two discrete forms. The comparable polymorphism is observed when the ratio of two different types of flagellin are changed in in vitro copolymerization of flagellin (103).

In order to explain the polymerization of a single kind of flagellin into a helical tubular structure and their discontinuous transformation, a hypothesis was presented that a single kind of flagellin can take two different conformations, namely R and T, as metastable states and that the molecules of each state lie along the longitudinal strands of a filament (89). Then, the various helical shapes appearing by the transformation are explained by the difference in the number of R and T strands in a filament. Another explanation assumes the presence of alternative bonding sites in a flagellin molecule (104). Whether it is attributed to the conformation or the bonding, the presence of at least two different states in a single kind of flagellin and their reversible transition from one to the other seem to be essential for the discontinuous transformation of the helix of flagellar filaments. If this is the case, the different flagellar shape mutants may produce flagellin differing not only in detailed conformation of its molecule but also in the stability of these states under ordinary physiological conditions. The polymorphous mutants detected on *Salmonella* may produce "flexible" flagellin which undergoes frequent transition between two states (42). Even though more than one form appears among a population of the polymorphous flagella, each individual flagellum comprises a single form except on rare occasions. For a flagellar filament to manifest a regular helical shape, flagellin molecules of the same state must be regularly arrayed along each of the longitudinal strands. This means that the terminus of an existing flagellar filament plays a decisive role in the choice of the state of polymerizing flagellin and successively determines the helical shape of the growing portion of the filament.

Flagellar Phase Variation

Flagellar phase variation in *Salmonella* is a dimorphic expression of flagellar characters in a clone. Many *Salmonella* species are characterized by their possession of a pair of nonallelic structural genes for flagellin designated *H1* and *H2*, whose expression is alternative in a clone. For example, in *S. typhimurium* the alternative expression of *H1-i* and *H2–1.2* occurs with the probability of about 10^{-3} per bacterial division. This phenomenon was first detected as the alternative expression of two discrete flagellar antigen types, and later observed on the shape of flagellar filaments (1). A bacterial strain that undergoes phase variation is termed a *diphasic strain* and a cell expressing *H1* or *H2* is said to be in *phase 1* or *phase 2* respectively. Early genetic studies on phase variation interpreted the process as follows: *H2* can exist in two different states, active and inactive; when *H2* is in the active state, the production of the phase-1 flagellin by *H1* is repressed, while *H2* carries out the production of phase-2 flagellin, and when *H2* changes to the inactive state the production of phase-1 flagellin, specified by *H1*, proceeds (1). As for the expression of each *H* gene, a closely linked factor *ah* was found to be responsible for activation of the adjoining *H* gene. Mutation of $ah1^+$ to $ah1^-$ or $ah2^+$ to $ah2^-$ results in the failure of production of phase-1 or phase-2 flagellin respectively (Table 1).

The process of phase variation has raised two questions associated with fundamental problems of genetics. The first is how the production of phase-1 flagellin is repressed in the *H2*-active state and the second is by what mechanism *H2* changes its state. Recent efforts to answer these questions have met with considerable success. Concerning the first question, a repressor gene, *rh1*, was found to exist and constitute an operon with *H2* and *ah2* (105). The genotype of diphasic cells is $ah2^+$-$H2^+$-$rh1^+$ (Table 1). *Ah2* controls the activity of both *H2* and *rh1*, and is regarded as the operator region. In the $ah2^-$ mutants the abilities of both phase-2 flagellin synthesis and inhibition of phase-1 flagellin synthesis are lost. Consequently, the phenotype of such mutants becomes stable phase 1. On the contrary, *rh1* remains

Table 1 Mutants of H1 or H2 operon and their phenotypes in *Salmonella typhimurium*

H1 operon		H2 operon				Phase variation	Phenotype in	
ah1	H1	vh2	ah2	H2	rh1		phase 1	phase 2
+	i	+	+	1.2	+	+	i	1.2
−	i	+	+	1.2	+	+	nf[a]	1.2
+	−	+	+	1.2	+	+	nf	1.2
+	i	−	+	1.2	+	s[b]	i	1.2
+	i	+	−	1.2	+	−	i	−
+	i	+	+	−	+	+	i	nf
+	i	+	+	1.2	−	+	i	i and 1.2

[a] nf = nonflagellate.
[b] s = stable in existing phase.

active in phase 2 of the $H2^-$ mutants, and neither phase 1 nor phase 2 flagellins are synthesized in this phase. Thus, the mutant clones reveal H-O variation. In the $rh1^-$ mutants, synthesis of both phase-1 and phase-2 flagellin proceeds in phase 2, and flagellar filaments are composed of both types of flagellin. In phase 1, they produce only phase-1 flagellin.

Studies on a cell-free system for in vitro protein synthesis directed by RNA of *Salmonella* were extended to the elucidation of the step blocked by the phase-1 repressor in phase-2 (106). Chromatographic analysis of the in vitro products showed the presence of synthesized flagellin corresponding to the phenotype of the cells from which the RNA was derived. That is, when RNA was extracted from the cells of the diphasic strain propagated from a single colony, expressing either phase 1 or phase 2, the in vitro synthesized flagellin was predominantly the same as that produced by the original colony. Translation of mRNA specific for phase-1 flagellin was not inhibited by the presence of mRNA specific for phase 2. This experimental evidence indicates that phase variation is due to the alternative synthesis of phase-specific mRNA and that the phase-1 repressor, i.e. the product of *rh1*, blocks the transcription of the *ah1-H1* operon (Figure 5).

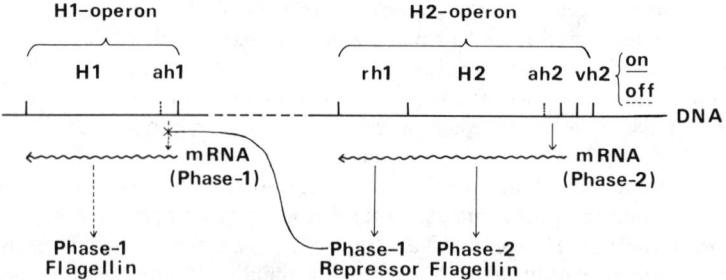

Figure 5 The regulation system operating on phase variation of *Salmonella*.

The repression of the *ah1-H1* operon by *rh1* is not complete throughout the growth phases of diphasic bacteria. Temporary derepression of phase-1 flagellar formation occurs in a fraction of the phase-2 cell population at late exponential and early stationary phase when a cell generation time exceeded 80 min (107). The resulting flagellar filaments carry a segment region composed of both phase-1 and phase-2 flagellin molecules. The copolymer segments are formed almost simultaneously in every growing flagellar filament of the derepressed cells for an average of about 8 min until the supply of phase-1 flagellin is exhausted after reestablishment of repression. The duration of detectable copolymer formation in a cell is of the same order as the half-life of flagellin mRNA (108). These phenomena show that a flagellum-forming apparatus can accept newly synthesized flagellin of phase 1 even after it has already started filament formation in phase 2. Thus, the flagellum-forming apparatus of each flagellum was found to be phase nonspecific. Formation of a homogeneous copolymer filament of two different kinds of flagellin was demonstrated also on the merodiploids of *E. coli* carrying two distinguishable *hag* loci (109).

As for the mechanism of the alternative activation and inactivation of the *ah2-H2-rh1* operon, presence of a chromosomal factor, *vh2*, which is closely linked to the operon was reported (1). Replacement of $vh2^+$ in a diphasic clone by $vh2^-$ of a monophasic clone reduced the rate of shifting from the existing state to the alternative one below 10^{-7} per bacterial division. Recently, molecular cloning techniques were applied to isolate the segment of *Salmonella* DNA which contain the *H2* region attached to λ (M. Simon, personal communication). When the cloned DNA molecules were denatured and then renatured, presumed heteroduplexes having a region similar to an "inversion bubble" were detected adjacent to *H2* on the molecules. The frequency of such molecules correlated with the ratio of phase-1 and phase-2 cells in a bacterial culture from which DNA was derived. Thus, the explanation was proposed that an inversion of DNA adjacent to *H2* is responsible for the flagellar phase variation in *Salmonella*. The correspondence of this inversion region with the genetically assigned stability controller, *vh2*, remains to be investigated. Destabilization of the *H2* state when the *H2* region of the phase-2 stable *Salmonella* is incorporated in the *E. coli* chromosome may be also associated with some sort of structural anomaly of the chromosomal region (71).

GENETIC CONTROL OF MOTILITY

Function of the mot Genes

Recently it has been established that the bacterial flagellum provides the thrust for cell movement by means of a rotary device at its base (31–36). This finding has led the studies of bacterial motility to focus on the rotary mechanism at the base of a flagellum. Two cistrons *motA* and *motB*, primarily responsible for motility of flagella, have been disclosed in both *Salmonella* (45, 46) and *E. coli* (48, 61). In *E. coli, motA* and *motB*, together with *cheA*, were found to be organized into an operon termed *Mocha* (61).

Neither electron microscopy of the flagellar basal structures nor SDS-acrylamide gel electrophoresis of their component proteins could detect differences between mot^+, $motA^-$, and $motB^-$ cells (21). However, the efforts to construct the λ-*mot* plasmids by molecular cloning techniques and to let them express *mot* functions in mot^- host bacteria helped identify the specific proteins corresponding to these two *mot* genes (48, 61). The molecular weights for *motA* and *motB* are 31,000 and 39,000 respectively. These proteins may be either the components of flagellar basal bodies contained in an amount too small to be detected in ordinary analytical procedures or regulatory factors not firmly bound to basal bodies. As intercistronic weak-complementation occurs in many combinations of $motA^-$ and $motB^-$ mutants (45–47), it is possible that the product proteins of these *mot* cistrons function as a structural complex. Although the actual function of these proteins in motility is left for further investigation, it is quite plausible that they participate in the link of the rotary apparatus of a basal body with the connecting membrane structure or with the machinery supplying energy for the rotation. It was shown in *E. coli* that chemoattractants or repellents induce a change of the membrane potential of motile mot^+ cells but not of a mot^- mutant cells of *E. coli* (110).

The third *mot* cistron, *motC*, reported in *Salmonella* (45), was later found to be identical with *flaAII* (58). Interestingly enough, a mutation in this cistron results in a defect of flagellar formation or paralysis of flagella depending on the site of mutation. Furthermore, the site of a never-tumbling che^- mutant was mapped in *flaAII* (111). Thus the product of *flaAII* is responsible for formation, rotation, and reversal of rotation. In nonflagellate mutants of *flaAII*, none of the flagellar structures were found to be associated with the membrane fraction (T. Suzuki and T. Iino, unpublished information). As mentioned in a foregoing section, the basal structure detectable in the earliest stage of flagellar morphogenesis is the rod–inner ring complex. Therefore, it may be assumed that this structure is the major portion of the rotary apparatus of a flagellum and that the product of *flaAII* either is a component of this complex structure or is associated with it.

The source of energy for the rotation of flagellar bases was inferred to be an intermediate in oxidative phosphorylation and not ATP directly (112), because, unlike their parents, mutants of *E. coli* and *S. typhimurium* that are blocked in the conversion of ATP to the intermediate of oxidative phsophorylation fail to swim anaerobically, even when they produce ATP. Further, carbonylcyanide *m*-chlorophenylhydrazone, which uncouples oxidative phosphorylation, completely inhibits motility even though ATP remains present. Similarly, anaerobic infection with flagellotropic phage χ is blocked in energy-transducing ATPase-defective ($uncA^-$) mutants and is restored by the addition of NO_3^-, which functions as a terminal electron acceptor for anaerobic respiration (113).

Role of Flagellar Filaments on Bacterial Motility

Rotation of the base of a flagellum lets the connecting filament rotate jointly and exerts the driving force for bacterial movement. When multiflagellate eubacteria swim translationally, a number of flagella on a cell form a bundle and rotate in unison (114, 115). For translational movement of such bacteria, the direction of rotation of the base must be the same as the handedness of the helix of the filament. The demonstration of basal rotation and identification of its direction were accomplished with an elaborate use of two kinds of flagella mutants, i.e. polyhook mutants and straight mutants of *E. coli* (32). Neither of these mutant cells can swim in liquid medium because they are missing normal filaments. However, when a cell of these mutants was tethered to a glass slide with antihook or antifilament antibodies, the cell body rotated alternatively clockwise (CW) and counterclockwise (CCW). This phenomenon indicates that, although the cell is nonmotile, its flagellar bases are actively rotating relative to the cell body. This experiment was further extended to cells with only one normal flagellar filament, and the frequency of rotation to each direction was measured (33). The direction of flagellar rotation was frequently CCW when it was seen from the side where the flagellum is attached, but abrupt stops of the rotation and successive CW rotation were also observed intermittently. For bacteria whose flagella rotate CCW, translational movements must be made when the helix is left-handed. Recent improvement of dark-field microscopy helped confirm the expectation (116) that the helix of normal-type flagellar filaments is left-handed. It was shown by dark-field microscopy that abrupt stop and tumbling of a normal flagellate cell during its translational movement corresponds to abrupt reversal of rotation, and flagellar filaments disperse around the cell at this stage (114, 115).

Information on the coordination between direction of basal rotation and handedness of the filament helix in translational movement explained a curious movement of the curly mutant cells, which continue to tumble in liquid medium and intermittently swim straight ahead. In contrast to normal flagellar filaments, curly filaments were found to be right-handed (116). Therefore, when flagellar bases of a mutant cell rotate CCW, its curly filaments cannot form a bundle and exert the necessary driving force for translational movement of the cell.

Reversal of the rotation of flagellar bases is an important event for the manifestation of tactic behavior, whose detail is described in the review article of Parkinson in this volume (37).

SUMMARY

Early genetic studies of bacterial flagella were focused mainly on their filament portion. It is now possible to describe the process of filament formation at the molecular level as the "self-assembly" of flagellin into filaments. The shape and length of the growing portion of a flagellar filament are determined by its preexisting portion. This may be taken as a typical example of the autoregulatory systems operating in molecular assembly in living organisms.

Fine structure analysis of flagellin is in progress both genetically and chemically. Structural and functional differentiation in the molecule is being investigated in relation to its assembly and antigenicity. In connection with the extensive polymorphism of flagellin, the completion of these investigations is expected to contribute to studies on molecular evolution. Investigation of the mechanism of phase variation disclosed a regulatory system operating at the transcriptional level between the operons in which the flagellin genes reside and demonstrated a possible contribution of a chromosomal inversion for the phase shift.

In addition to the above-mentioned studies on flagellar filaments, the studies on basal structures of flagella are progressing remarkably well. Through the elaborate use of nonmotile mutant cells, it was shown that the motility of a bacterium is activated by rotation of its flagellar bases. Two proteins specifically responsible for the rotation were identified by comparative chemical analysis of a motile bacterial strain and its paralyzed mutants. Studies on the energy transfer system for the basal rotation is also proceeding with appropriate mutants.

The structure of a flagellar basal body was clarified in detail on a number of bacterial species. An outline of the morphogenesis of flagellar basal structures was disclosed, in combination with the identification of the flagella genes responsible for the disclosed steps in the process. Detailed biochemical studies on the product of each step are in progress with the use of the cloned flagella genes.

Studies on the relation of flagellation to the cell cycle and cell envelope have been undertaken. They are expected to provide an excellent model for genetic regulation of cellular differentiation.

Literature Cited

1. Iino, T. 1969. Genetics and chemistry of bacterial flagella. *Bacteriol. Rev.* 33:454-75
2. Cohen-Bazire, G., London, J. 1967. Basal organelles of bacterial flagella. *J. Bacteriol.* 94:458-65
3. DePamphilis, M. L. 1971. Dissociation and reassembly of *Escherichia coli* outer membrane and of lipopolysaccharide, and their reassembly onto flagellar basal bodies. *J. Bacteriol.* 105:1184-99
4. DePamphilis, M. L., Adler, J. 1971. Fine structure and isolation of the hook-basal body complex of flagella from *Escherichia coli* and *Bacillus subtilis*. *J. Bacteriol.* 105:384-95
5. DePamphilis, M. L., Adler, J. 1971. Attachment of flagellar basal bodies to the cell envelope: Specific attachment to the outer, lipopolysaccharide membrane and the cytoplasmic membrane. *J. Bacteriol.* 105:396-407
6. Dimmitt, K., Simon, M. 1971. Purification and thermal stability of intact *Bacillus subtilis* flagella. *J. Bacteriol.* 105:369-75
7. Dimmitt, K., Simon, M. 1971. Purification and partial characterization of *Bacillus subtilis* flagellar hooks. *J. Bacteriol.* 108:282-86
8. Kagawa, H., Owaribe, K., Asakura, S., Takahashi, N. 1976. Flagellar hook protein from *Salmonella* SJ25. *J. Bacteriol.* 125:68-73
9. Raska, I., Mayer, F., Edelbluth, C., Schmitt, R. 1976. Structure of plain and complex flagellar hooks of *Pseudomonas rhodos*. *J. Bacteriol.* 125:679-88
10. Parish, C. R., Ada, G. L. 1969. Cleavage of bacterial flagellin with cyanogen bromide, chemical and physical properties of the protein fragments. *Biochem. J.* 113:489-99
11. Parish, C. R., Wistar, R. Jr., Ada, G. L. 1969. Cleavage of bacterial flagellin with cyanogen bromide, antigenic properties of the protein fragments. *Biochem. J.* 113:501-6
12. Davidson, B. E. 1971. The alignment of cyanogen bromide fragments from the flagellin of *Salmonella adelaide*. *Eur. J. Biochem.* 18:524-29
13. Tauschel, H. D. 1971. Der Geisselapparat von *Rhodopseudomonas palustris*. VI. Charakterisierung des Flagellins. *Arch. Mikrobiol.* 76:91-102
14. Joys, T. M., Rankis, V. 1972. The primary structure of the phase-1 flagellar protein of *Salmonella typhimurium*. *J. Biol. Chem.* 247:5180-93
15. DeLange, R. J., Chang, J. Y., Shaper, J. H., Martinez, R. J., Komatsu, S. K., Glazer, A. N. 1973. On the amino-acid sequence of flagellin from *Bacillus subtilis* 168: Comparison with other bacterial flagellins. *Proc. Natl. Acad. Sci. USA* 70:3428-31
16. Kondoh, H., Hotani, H. 1974. Flagellin from *Escherichia coli* K12: Polymeriza-

tion and molecular weight in comparison with *Salmonella* flagellins. *Biochim. Biophys. Acta* 336:117–39
17. Chang, J. Y., DeLange, R. J., Shaper, J. H., Glazer, A. N. 1976. Amino acid sequence of flagellin of *Bacillus subtilis* 168. I. Cyanogen bromide peptides. *J. Biol. Chem.* 251:695–700
18. Shaper, J. H., DeLange, R. J., Martinez, R. J., Glazer, A. N. 1976. Amino acid sequence of flagellin *Bacillus subtilis* 168. II. Tryptic peptides from maleylated flagellin. *J. Biol. Chem.* 251: 701–4
19. DeLange, R. J., Chang, J. Y., Shaper, J. H., Glazer, A. N. 1976. Amino acid sequence of flagellin of *Bacillus subtilis* 168. III. Tryptic peptides, N-bromosuccinimide peptides, and the complete amino acid sequence. *J. Biol. Chem.* 251:705–11
20. Simon, M., Hilmen, M., Silverman, M. 1975. The assembly and function of *E. coli* flagella. *Proc. Intersec. Congr. IAMS, 1st, Tokyo* 1:641–55
21. Hilmen, M., Simon, M. 1976. Motility and the structure of bacterial flagella. *Proc. Cold Spring Harbor Conf. Cell Proliferation, Cold Spring Harbor* 3:35–45
22. Champness, J. N., Lowy, J. 1967. The structure of bacterial flagella. *Symp. Fibrous Proteins, Australia,* pp. 106–14
23. Tauschel, H. D., Drews, G. 1970. Der Geisselapparat von *Rhodopseudomonas palustris.* III. Untersuchungen zur Feinstruktur der Geissel. *Cytobiologie* 2:87–107
24. Wakabayashi, K., Mitsui, T. 1970. An X-ray diffraction study of reconstituted straight *Salmonella* flagella. *J. Mol. Biol.* 53:567–70
25. Tauschel, H. D. 1971. Zur Feinstruktur von Spirillum serpens. *Cytobiologie* 3:25–36
26. Finch, J. T., Klug, A. 1972. The helical surface lattice of bacterial flagella. In *The Generation of Subcellular Structures,* ed. R. Markham, J. B. Bancroft, pp. 167–77. Amsterdam: North-Holland. 372 pp.
27. O'Brien, E. J., Bennett, P. M. 1972. Structure of straight flagella from a mutant *Salmonella. J. Mol. Biol.* 70: 133–52
28. Kondoh, H., Yanagida, M. 1975. Structure of straight flagellar filaments from a mutant of *Escherichia coli. J. Mol. Biol.* 96:641–52
29. Schmitt, R., Raska, I., Mayer, F. 1974. Plain and complex flagella of *Pseudomonas rhodos:* Analysis of fine structure and composition. *J. Bacteriol.* 117:844–57
30. Schmitt, R., Mayer, F., Lotz, W. 1975. Structure of the complex flagella of *Rhizobium lupini* H 13–3. *Proc. Intersec. Congr. IAMS, 1st, Tokyo* 1:691–95
31. Berg, H. C., Anderson, R. A. 1973. Bacteria swim by rotating their flagella filaments. *Nature* 245:380–82
32. Silverman, M., Simon, M. 1974. Flagellar rotation and the mechanism of bacterial motility. *Nature* 249:73–74
33. Larsen, S. H., Reader, R. W., Kort, E. N., Tso, W., Adler, J. 1974. Change in direction of flagellar rotation is the basis of the chemotactic response in *Escherichia coli. Nature* 249:74–77
34. Berg, H. C. 1974. Dynamic properties of bacterial flagellar motors. *Nature* 249:77–79
35. Berg, H. C. 1975. Bacterial behaviour. *Nature* 254:389–92
36. Berg, H. C. 1975. Flagellar rotation. *Proc. Intersec. Congr. IAMS, 1st, Tokyo* 1:665–69
37. Parkinson, J. S. 1977. Behavioral genetics in bacteria. *Ann. Rev. Genet.* 11: 397–414
38. Joys, T. M., Frankel, R. W. 1967. Genetic control of flagellation in *Bacillus subtilis. J. Bacteriol.* 94:32–37
39. Iino, T. 1969. Genetical studies of flagellation and motility in *Pseudomonas aeruginosa. Ann. Rep. Natl. Inst. Genet. Jpn.* 20:94
40. Armstrong, J. B., Adler, J. 1969. Location of genes for motility and chemotaxis on the *Escherichia coli* genetic map. *J. Bacteriol.* 97:156–61
41. Yamaguchi, S., Iino, T. 1969. Genetic determination of the antigenic specificity of flagellar protein in Salmonella. *J. Gen. Microbiol.* 55:59–74
42. Iino, T., Oguchi, T., Kuroiwa, T. 1974. Polymorphism in a flagellar-shape mutant of *Salmonella typhimurium. J. Gen. Microbiol.* 81:37–45
43. Horiguchi, T., Yamaguchi, S., Yao, K., Taira, T., Iino, T. 1975. Genetic analysis of *H1,* the structural gene for phase-1 flagellin in Salmonella. *J. Gen. Microbiol.* 91:139–49
44. Silverman, M., Simon, M. 1973. Genetic analysis of flagellar mutants in *Escherichia coli. J. Bacteriol.* 113:105–13
45. Enomoto, M. 1966. Genetic studies of paralyzed mutants in Salmonella. I. Genetic fine structure of the *mot* loci in *Salmonella typhimurium. Genetics* 54: 715–26
46. Enomoto, M. 1966. Genetic studies of paralyzed mutants in Salmonella. II.

Mapping of three *mot* loci by linkage analysis. *Genetics* 54:1069–76
47. Armstrong, J. B., Adler, J. 1967. Genetics of motility in *Escherichia coli:* Complementation of paralyzed mutants. *Genetics* 56:363–73
48. Silverman, M., Matsumura, P., Simon, M. 1976. The identification of the *mot* gene product with *Escherichia coli*-lambda hybrids. *Proc. Natl. Acad. Sci. USA* 73:3126–30
49. Stocker, B. A. D., McDonough, M. W., Ambler, R. P. 1961. A gene determining the presence or absence of ϵ-N-methyl-L-lysine in *Salmonella* flagellar protein. *Nature* 186:556–58
50. Konno, R., Fujita, H., Horiguchi, T., Yamaguchi, S. 1976. Precise position of the *nml* locus on the genetic map of Salmonella. *J. Gen. Microbiol.* 93:182–84
51. Iino, T., Enomoto, M. 1971. Motility. In *Methods in Microbiology 5A*, ed. J. R. Norris, D. W. Ribbons, pp. 145–63. New York: Academic. 450 pp.
52. Schade, S., Adler, J., Ris, H. 1967. How bacteriophage χ attacks motile bacteria. *J. Virol.* 1:559–609
53. Frankel, R. W., Joys, T. M. 1966. Adsorption specificity of bacteriophage PBS1. *J. Bacteriol.* 92:388–89
54. Appelbaum, P. C., Hugo, N., Coetzee, J. N. 1971. A flagellar phage for the Proteus-providence group. *J. Gen. Virol.* 13:153–62
55. Lotz, W., Acker, G., Schmitt, R. 1977. Bacteriophage 7-7-1 adsorbs to the complex flagella of *Rhizobium lupini* H 13-3. *J. Gen. Virol.* 34:9–17
56. Jollick, J. D., Wright, B. L. 1974. A flagella specific bacteriophage for *Caulobacter. J. Gen. Virol.* 22:197–205
57. Iino, T., Mitani, M. 1976. A mutant of *Salmonella* possessing straight flagella. *J. Gen. Microbiol.* 49:81–88
58. Yamaguchi, S., Iino, T., Horiguchi, T., Ohta, K. 1972. Genetic analysis of *fla* and *mot* cistrons closely linked to *H1* in *Salmonella abortusequi* and its derivatives. *J. Gen. Microbiol.* 70:59–75
59. Silverman, M., Simon, M. 1974. Positioning flagellar genes in *Escherichia coli* by deletion analysis. *J. Bacteriol.* 117:73–79
60. Silverman, M., Matsumura, P., Draper, R., Edwards, S., Simon, M. 1976. Expression of flagellar genes carried by bacteriophage lambda. *Nature* 261:248–50
61. Silverman, M., Simon, M. 1976. Operon controlling motility and chemotaxis in *E. coli. Nature* 264:577–80
62. Kondoh, H., Ozeki, H. 1976. Deletion and *amber* mutants of *fla* loci in *Escherichia coli* K12. *Genetics* 84:403–21
63. Komeda, Y., Shimada, K., Iino, T. 1977. Isolation of specialized lambda transducing bacteriophages for flagellar genes (*fla*) of *Escherichia coli* K12. *J. Virol.* 22:654–61
64. Nomura, M., Engbaek, F. 1972. Expression of ribosomal protein genes as analyzed by bacteriophage Mu induced mutations. *Proc. Natl. Acad. Sci. USA* 69:1526–30
65. Silverman, M., Simon, M. 1973. Genetic analysis of bacteriophage mu-induced flagellar mutants in *Escherichia coli. J. Bacteriol.* 116:114–22
66. Sanderson, K. E. 1972. Linkage map of *Salmonella typhimurium,* Edition IV. *Bacteriol. Rev.* 36:558–86
67. Hilmen, M., Silverman, M., Simon, M. 1974. The regulation of flagellar formation and function. *J. Supramol. Struct.* 2:360–71
68. Enomoto, M., Stocker, B. A. D. 1974. Transduction by phage P1*Kc* in *Salmonella typhimurium. Virology* 60:503–14
69. Suzuki, T., Iino, T., Yamaguchi, S., Horiguchi, T. 1976. Functions of genes of *flaF* cluster in the morphogenesis of flagellar basal structures in *Salmonella typhimurium. Jpn. J. Genet.* 51:441 (In Japanese)
70. Mäkelä, P. H. 1964. Genetic homologies between flagellar antigens of *Escherichia coli* and *Salmonella abony. J. Gen. Microbiol.* 35:503–10
71. Enomoto, M., Stocker, B. A. D. 1975. Integration, at *hag* or elsewhere, of *H2* (phase-2 flagellin) genes transduced from Salmonella to *Escherichia coli. Genetics* 81:595–614
72. Joys, T. M., Frankel, R. W. 1967. Genetic control of flagellation in *Bacillus subtilis. J. Bacteriol.* 94:32–37
73. Appelbaum, P. C., Prozesky, O. W. 1973. Transductional analysis of non-motile mutants in *Proteus mirabilis. J. Gen. Microbiol.* 77:89–97
74. Van Alstyne, D., Grant, G. F., Simon, M. 1969. Synthesis of bacterial flagella: Chromosomal synchrony and flagella synthesis. *J. Bacteriol.* 100:283–87
75. Kondoh, H., Ozeki, H. 1973. Evidences that flagella are formed at a stage of cell cycle in *E. coli* K12. *Jpn. J. Genet.* 48:426 (In Japanese)

76. Nishimura, A., Suzuki, H., Hirota, Y. 1975. Flagellar formation in *E. coli* is coupled with cell division in regulatory mechanism. *Jpn. J. Genet.* 50:484–85 (In Japanese)
77. Yokota, T., Gots, J. S. 1970. Requirement of adenosine 3'5'-cyclic phosphate for flagella formation in *Escherichia coli* and *Salmonella typhimurium*. *J. Bacteriol.* 103:513–16
78. Silverman, M., Simon, M. 1974. Characterization of *Escherichia coli* flagellar mutants that are insensitive to catabolite repression. *J. Bacteriol.* 120:1196–1203
79. Komeda, Y., Suzuki, H., Ishidsu, J., Iino, T. 1975. The role of cAMP in flagellation of *Salmonella typhimurium*. *Mol. Gen. Genet.* 142:289–98
80. Ames, G. F., Spudich, E. N., Nikaido, H. 1974. Protein composition of the outer membrane of *Salmonella typhimurium:* Effect of lipopolysaccharide mutations. *J. Bacteriol.* 117:406–16
81. Komeda, Y., Icho, T., Iino, T. 1977. Effects of *galU* mutation on flagellar formation in *Escherichia coli*. *J. Bacteriol.* 129:908–15
82. Fukasawa, T., Jokura, K., Kurahashi, K. 1963. Mutations in *Escherichia coli* that affect uridine diphosphate glucose pyrophosphorylase activity and galactose fermentation. *Biochim. Biophys. Acta* 74:608–20
83. Ayusawa, D., Yoneda, Y., Yamane, K., Maruo, B. 1975. Pleiotropic phenomena in autolytic enzyme(s) content, flagellation, and simultaneous hyperproduction of *Bacillus subtilis* mutant. *J. Bacteriol.* 124:459–69
84. Silverman, M. R., Simon, M. 1972. Flagellar assembly mutants in *Escherichia coli*. *J. Bacteriol.* 112:986–93
85. Patterson-Delafield, J., Martinez, R. J. 1973. A new *fla* gene in *Salmonella typhimurium*—flaR—and its mutant phenotype-superhooks. *Arch. Mikrobiol.* 90:107–20
86. Suzuki, H., Iino, T. 1975. Absence of messenger ribonucleic acid specific for flagellin in non-flagellate mutants of *Salmonella*. *J. Mol. Biol.* 95:549–56
87. Suzuki, H. 1975. Regulation of flagellin biosynthesis. *Proc. Intersec. Congr. IAMS 1st, Tokyo,* 1:656–64
88. Iino, T., Oguchi, T., Suzuki, T., Hirano, T. 1975. A regulator gene for flagellin synthesis in *Salmonella. Jpn. J. Genet.* 50:464 (In Japanese)
89. Asakura, S. 1970. Polymerization of flagellin and polymorphism of flagella. *Adv. Biophys.* 1:99–155
90. Iino, T. 1969. Polarity of flagellar growth in *Salmonella. J. Gen. Microbiol.* 56:227–39
91. Emerson, S., Tokuyasu, K., Simon, M. 1970. Bacterial flagella: Polarity of elongation. *Science* 169:190–92
92. Iino, T., Suzuki, H., Yamaguchi, S. 1972. Reconstitution of *Salmonella* flagella attached to cell bodies. *Nature New Biol.* 237:238–40
93. Iino, T. 1974. Assembly of Salmonella flagellin in vitro and in vivo. *J. Supramol. Struct.* 2:372–84
94. Wakabayashi, K., Hotani, H., Asakura, S. 1969. Polymerization of Salmonella flagellin in the presence of high concentrations of salts. *Biochim. Biophys. Acta* 175:195–203
95. Hotani, H., Asakura, S. 1974. Growth-saturation *in vitro* of *Salmonella* flagella. *J. Mol. Biol.* 86:285–300
96. Rhona, M. 1970. Properties of flagellin from *Bacillus megaterium* KM and its association with the cytoplasmic membrane. *Arch. Biochem. Biophys.* 139:97–103
97. Martinez, R. J., Ichiki, A. T., Lundh, N. P., Tronick, S. R. 1968. A single amino acid substitution responsible for altered flagellar morphology. *J. Mol. Biol.* 34:559–64
98. Joys, T. M., Martin, J. F. 1973. Identification of amino acid changes in serological mutants of the *i* flagellar antigen of *Salmonella typhimurium*. *Microbios* 7:71–73
99. Joys, T. M., Martin, J. F., Wilson, H. L., Rankis, V. 1974. Differences in the primary structure of the phase-1 flagellins of two strains of *Salmonella typhimurium*. *Biochim. Biophys. Acta* 351:301–5
100. Shinoda, S., Honda, T., Takeda, Y., Miwatani, T. 1974. Antigenic difference between polar monotrichous and peritrichous flagella of *Vibrio parahaemolyticus*. *J. Bacteriol.* 120:923–28
101. Miwatani, T., Shinoda, S., Yabuuchi, E., Honda, T., Takeda, Y. 1975. Existence of polar and peritrichous flagella and their antigenic difference in *Vibrio parahaemolyticus*. *Proc. Intersec. Congr. IAMS, 1st, Tokyo* 1:683–90
102. Kamiya, R., Asakura, S. 1977. Flagellar transformations at alkaline pH. *J. Mol. Biol.* 108:513–18
103. Asakura, S., Iino, T. 1972. Polymorphism of *Salmonella* flagella as investigated by means of *in vitro* colpolymeri-

zation of flagellins derived from various strains. *J. Mol. Biol.* 4:251–68
104. Calladine, C. R. 1976. Design requirements for the construction of bacterial flagella. *J. Theor. Biol.* 57:469–89
105. Fujita, H., Yamaguchi, S. 1973. Studies on H-O variants in *Salmonella* in relation to phase variation. *J. Gen. Microbiol.* 76:127–34
106. Suzuki, H., Iino, T. 1973. *In vitro* synthesis of phase-specific flagellin of *Salmonella*. *J. Mol. Biol.* 81:57–70
107. Iino, T., Oguchi, T., Hirano, T. 1975. Temporary expression of flagellar phase-1 in phase-2 clones of diphasic *Salmonella*. *J. Gen. Microbiol.* 89:265–76
108. Suzuki, H., Enomoto, M., Hirota, Y. 1974. Studies on control flagellin mRNA synthesis with *fla-ts* mutants of *E. coli. Ann. Rep. Natl. Inst. Genet. Jpn.* 24:14–15
109. Silverman, M., Simon, M. 1974. Assembly of hybrid flagellar filaments. *J. Bacteriol.* 118:750–52
110. Szmelcman, S., Adler, J. 1976. Change in membrane potential during bacterial chemotaxis. *Proc. Natl. Acad. Sci. USA* 73:4387–91
111. Collins, A. L. T., Stocker, B. A. D. 1976. *Salmonella typhimurium* mutants generally defective in chemotaxis. *J. Bacteriol.* 128:754–65
112. Larsen, S. H., Adler, J., Gargus, J. J., Hogg, R. W. 1974. Chemomechanical coupling without ATP: The source of energy for motility and chemotaxis in bacteria. *Proc. Natl. Acad. Sci. USA* 71:1239–43
113. Thipayathasana, P., Valentine, R. C. 1974. The requirement for energy transducing ATPase for anaerobic motility in *Escherichia coli. Biochim. Biophys. Acta* 347:464–68
114. Macnab, R. M. 1975. Smooth and tumbling motility in peritrichous bacteria. *Proc. Intersec. Congr. IAMS, 1st, Tokyo* 1:674–83
115. Macnab, R. M. 1976. Examination of bacterial flagellation by dark-field microscopy. *J. Clin. Microbiol.* 4:258–65
116. Shimada, K., Kamiya, R., Asakura, S. 1975. Left-handed to right-handed helix conversion in *Salmonella* flagella. *Nature* 254:332–34

SISTER CHROMATID EXCHANGE

❖3120

Sheldon Wolff
Laboratory of Radiobiology and Department of Anatomy, University of California, San Francisco, California 94143

CONTENTS

EARLY RESULTS	183
Sister Chromatid Exchange in Meiosis	183
Sister Chromatid Exchange in Somatic Cells	184
Isolabeling	186
Sister Chromatid Exchanges—Spontaneous or Induced?	186
RESULTS FROM NEW METHODS FOR DETECTING SISTER CHROMATID EXCHANGES	186
Reasons for Differential Staining	187
Harlequin Chromosomes and Chromosome Structure	189
Multiple sister chromatid exchanges	190
Autoradiographic image spread	190
Labeling for more than one replication cycle	190
The Question of Spontaneous Levels of Sister Chromatid Exchanges	191
Location of Sister Chromosome Exchanges in the Chromosome	193
Sister Chromatid Exchanges and Human Genetic Diseases	193
Sister Chromatid Exchanges as Indicators of Mutagenic Carcinogens	194
Lesions Responsible for Sister Chromatid Exchange Formation	196
CONCLUSION	197

EARLY RESULTS

Sister Chromatid Exchange in Meiosis

The question of whether or not sister strand crossing-over plays a role in genetic recombination has beguiled geneticists for years. According to the partial chiasmatype hypothesis (1, 2), crossing-over, which occurs between paired homologues at meiosis, entails the physical exchange of chromatids. The data in Drosophila (3) indicated that either of the chromatids from one of the homologues could exchange with either of the two chromatids from the other homologue, but that sister strand crossing-over did not occur. This conclusion had been anticipated in the study of ring/rod heterozygotes by Morgan (3a). Additionally, sister strand crossing-over did not seem to occur in somatic cells (4). Nonetheless, the cytological study of ring chromosomes in somatic cells of maize (5) showed the presence of double-sized dicentric rings that arose from sister chromatid exchanges (SCEs). Later, from the frequency of bridges present in anaphase I and II of maize heterozygous for a ring chromosome, it was found that sister strand crossing-over could occur in meiosis as well (6).

The lack of sister strand crossing-over in Drosophila was consistent with the fact that stable ring chromosomes have been found in this genus, and the difference between Drosophila and maize could be attributed to intrinsic qualitative differences between the two or to a quantitative difference in which the very much larger maize chromosomes would have a greater chance for exchange. The likelihood of the second possibility was diminished, however, by studies with large ring-X chromosomes in Drosophila [reviewed by Sandler (6a)] and by experiments with radioactively labeled large chromosomes in grasshoppers (7, 8), in which the exchanges seen in meiosis were directly related to chiasmata and were restricted to exchanges between homologues. The few SCEs seen were not related to the chiasmata or to crossing-over. These SCEs presumably occurred in the somatic gonial cells at the DNA replication preceding meiosis. SCEs have also been noticed in the unpaired sex chromosomes in metaphase I of mice (9). These, too, could have occurred at the division before meiosis.

It thus appears that, except for the case of maize heterozygous for a ring chromosome, the evidence that sister strand crossovers play a large role in genetic recombination occurring in meiosis is meager. In mitotic cells, however, it is well documented that sister chromatid exchange can occur not only in ring chromosomes but also in rod chromosomes that have been treated to distinguish the sister chromatids from one another.

SISTER CHROMATID EXCHANGE IN SOMATIC CELLS

The first direct observation of SCEs in rod chromosomes was made by Taylor (10), who previously had found that if sister chromatids were made to differ from one another in respect to their radioactivity, they could be distinguished from one another in autoradiograms (11). The procedure was to allow chromosomes to replicate once in the presence of [^{3}H] thymidine to produce chromosomes with equally labeled sister chromatids and then once again in its absence. After such a scheme of replication, the chromosomes contained one labeled chromatid and one unlabeled chromatid at any given point along their length. Because of the short path length of the β particle emitted from tritium, the autoradiographic resolution was such that silver grains in the autoradiographic emulsion were found mainly above the labeled chromatids. SCEs could be seen as places where radioactive material switched from one chromatid to its sister.

These studies, carried out before it was demonstrated that DNA replicated semiconservatively (12), showed that the label in chromosomes was distributed to daughter cells in a semiconservative manner. The original studies, which were carried out in plants, were soon extended to animal cells and shown to reflect a general phenomenon of eucaryotic chromosomes. Taylor soon went on to show that SCEs occurred with equal frequency in both cell cycles and that when an exchange occurred, whole DNA molecules seemed to exchange. Furthermore, from the ratio of events that occurred in the first cell cycle to those that occurred in the second cycle, he inferred that the subunits, which were segregating semiconservatively, were unlike, as would be expected if they were the complementary polynucleotide chains of the DNA double helix. This inference was made from studies in which the cells were not allowed to divide between the two replications but were made tetraploid by the use of colchicine to inhibit spindle formation after the first replication. In the tetraploid cells each original chromosome is represented by two daughter chromosomes and every SCE that occurs in the first replication cycle appears in both daughters. These are termed *twins*. On the other hand, any exchange that occurs in the second round of replication appears in only one of the pair of daughter chromosomes and is called a *single*. Taylor found twice as many singles as twins, which, because there were twice as many chromosomes present in the $4n$ cell as in the $2n$ cell, indicated that the frequency of induction of SCEs per chromosome was equal in the two cycles.

This 2:1 ratio is also expected if both of the subunits of the chromatid exchange and if they are unlike in respect to the way they rejoin. If only one of the subunits exchanges, then portions of the chromosomes would contain labeled and unlabeled sister chromatids at the metaphase following the first round of replication; however, this labeling pattern was not found. If the subunits were not unlike, but could rejoin randomly, then four possible combinations of exchange could occur. One fourth of the time, an exchange in the first replication cycle would result in the radioactive strands from each chromatid rejoining with the nonradioactive strands in the sister chromatids. This would form a twin observable in the tetraploid cell. Two fourths of the time, a radioactive strand would rejoin to a nonradioactive strand in only one of the two chromatids, leading to singles, and one fourth of the time, radioactive strands would join only to radioactive strands, leading to exchanges that would not be detectable. Thus, an exchange in the first division would lead to two singles to one twin. In the second division, where each chromosome contains one labeled and one unlabeled chromatid, all four types of exchange would lead to an exchange of label and would be recorded as singles. Thus, because there are twice as many chromosomes in the second-division cells, the results of exchange during both cycles would lead to ten singles to one twin, which is very different from the 2:1 ratio.

Nevertheless, some doubt existed about the autoradiographic data because discrepant ratios have been reported (13, 14). In addition, the determination of twins depended upon the correct pairing of daughter chromosomes in $4n$ cells that have four copies of each autosome. If an incorrect pairing is made, then two separate single exchanges that took place at approximately the same distance from the centromeres could be confused as a twin (15). Furthermore, although the autoradiographic resolution obtainable with tritium was quite good, it often was not good enough to distinguish between overlapping chromosomes or to determine small exchanges with certainty.

Isolabeling

Taylor's results showing the semiconservative distribution of label from chromosomes containing two subunits that were unlike, as might be expected if they were the complementary single polynucleotide chains of the DNA double helix, constituted some of the strongest evidence that the eucaryotic chromosome was not multistranded, but consisted of a single DNA double helix at any point along its length. Autoradiographic studies showed, however, that occasionally at the second division both sister chromatids appeared labeled over a part of their length, i.e. were isolabeled (16–18). This led to the suggestion that there was more than one DNA double helix at any given level along the chromosome.

Sister Chromatid Exchanges—Spontaneous or Induced?

Even though it is well known that radiation can break chromosomes and thus lead to the formation of chromosome and chromatid exchanges, the SCEs observed in autoradiographic studies were usually thought to be spontaneous events that occurred in all cells. The reasons for this were that the number of SCEs seemed to be independent of the amount of radioactive isotope and, therefore, of the amount of radiation present, and that increasing doses of X rays did not seem to increase the number of SCEs (19). It was pointed out, however, that the addition of exogenous radiation did seem to cause a slight increase in the number of SCEs, and it was postulated that unlike ordinary chromosome aberrations, the number of SCEs saturated with dose (20). Brewen & Peacock (21) showed that this was indeed the case in experiments with human ring-bearing cells. These investigators scored the formation of dicentric ring chromosomes, which are produced by sister chromatid exchange and do not require the presence of radioactive isotope for their visualization. They found a spontaneous level of sister chromatid exchange of about 0.12 per chromosome per cell cycle. Moreover, the addition of [^{3}H] thymidine increased the yield. At subsequent cell divisions, as the amount of tritium decreased, the level of dicentric rings stayed high, indicating a saturation. Other experiments by Gibson & Prescott (22) showed that if, indeed, the amount of tritium was decreased to very low levels, the number of SCEs observed in rod chromosomes did fall—again indicating that a large portion of the exchanges could be induced by the tritium.

RESULTS FROM NEW METHODS FOR DETECTING SISTER CHROMATID EXCHANGES

In spite of the uncertainties inherent in the use of autoradiography, Taylor's deductions have been confirmed by new methods that have simplified the study of sister chromatid exchange and allowed their observation with far greater clarity and resolution than is possible in autoradiography. The breakthrough came in 1972 with the publication of a paper by Zakharov & Egolina (23), who found that when they treated chromosomes of Chinese hamster cells with bromodeoxyuridine (BrdUrd) for two rounds of replication and subsequently stained the chromosomes with Giemsa, the two sister chromatids stained differently. After two rounds of replication one of the sister chromatids is unifilarly substituted (i.e. substituted in only one polynucleotide chain) with BrdUrd, whereas the other sister chromatid is bifilarly substituted. Zakharov & Egolina found that the unifilarly substituted chromatids stained darkly, and the bifilarly substituted ones stained lightly, thus enabling the visualization of SCEs. This work was quickly followed by the work of Latt (24), who found a very dramatic differential staining in human cells when they were treated with the fluorescent dye Hoechst 33258, and by the work of Ikushima & Wolff (25), who found that, in Chinese hamster cells, substitution of thymidine by iododeoxyuridine, as well as by BrdUrd, could cause the sister chromatids to stain differentially with Giemsa. It was subsequently shown that other fluorescent dyes such as acridine orange (26–28) and 4-6-diamidine-2-phenylindole (29) or combinations of fluorescent dyes and Giemsa staining [the fluorescent plus Giemsa (FPG) technique (26, 30, 31)] also enhanced the differential staining of sister chromatids to produce "harlequin" chromosomes. Treatment with hot salts at high pH has been found to enhance the staining when only Giemsa is used as the stain (32).

These techniques can also be used to distinguish chromatids that contain unsubstituted DNA molecules from their unifilarly substituted sisters, as are produced when cells are treated for one round of replication in the presence of the thymidine analogues followed by a round of replication in their absence in a manner analogous to the way the original experiments were performed with [^{3}H]thymidine.

With certain modifications the techniques also work for plant (33), fish (34), insect (35), and avian (36) chromosomes.

In all cases, cells are grown in the presence of a thymidine analogue and then stained. If they are stained with a fluorescent dye before being stained with Giemsa, they must be exposed to short-wave-length light, which causes the stain in the more heavily substituted chromatid to fade faster than in its sister chromatid. In order to make permanent preparations, the cells can then be stained with Giemsa or other stains (26, 37), often with a hot salt treatment between the two. Figure 1 shows a Chinese hamster ovary cell stained with a fluorescent dye, as well as a cell stained first with a fluorescent dye and then with Giemsa stain.

Reasons for Differential Staining

Zakharov & Egolina (23) noted that after Giemsa staining the pale chromatid was usually longer than its sister, and they postulated that protein synthesis that affected chromosome condensation and spiralization was delayed by the substitution of thymine by BrdUrd. Because proteins are more tightly bound to DNA containing BrdUrd than to unsubstituted DNA (38), Ikushima & Wolff (25) attributed the differential staining to a differential binding of protein to the DNA of chromatin. Studies of Chinese hamster ovary (CHO) chromosomes with the electron microscope by Korenberg & Ris (39) have shown that the primary effect of BrdUrd incorporation into chromosomes is exerted at the level of packing of the 25 nm fiber into the larger chromosomal unit. The bifilarly substituted chromatid is more open with looser gyres than is the unifilarly substituted chromatid. There is no visible effect on the 10 nm or 25 nm fibers. Korenberg & Ris, too, attribute the effect to nonhistone protein affecting the condensation of the chromosomes.

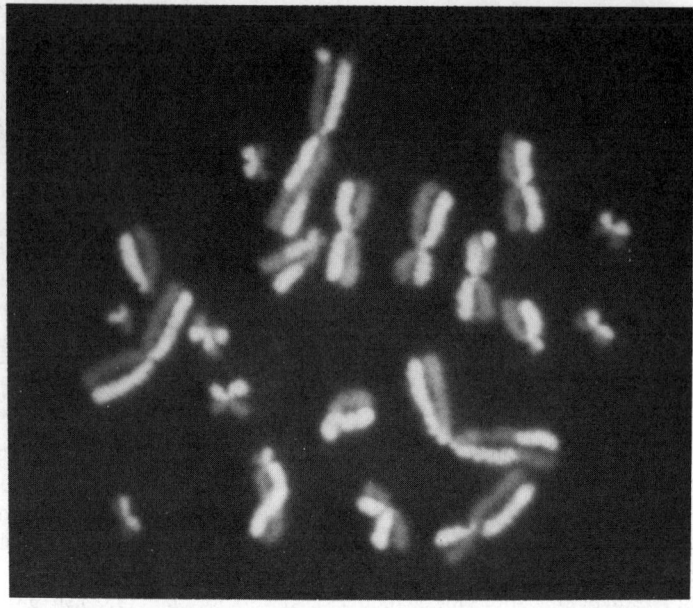

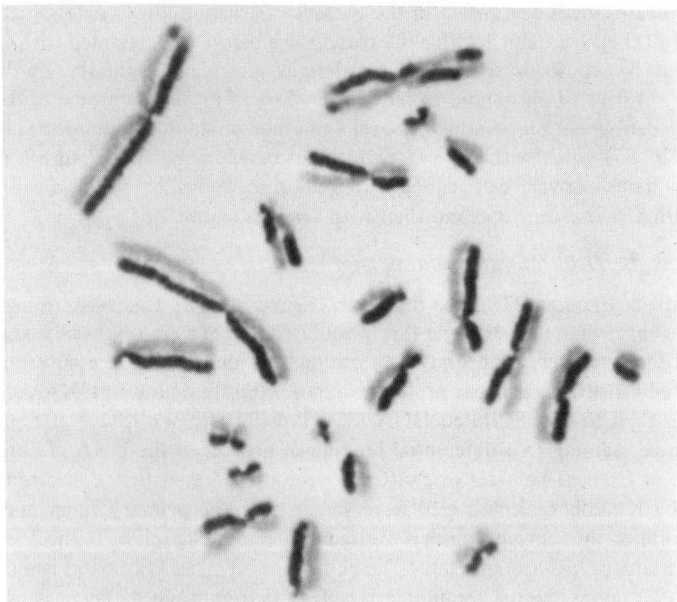

Figure 1 *Top*, Chinese hamster ovary (CHO) cell stained with a fluorescent dye. *Bottom*, CHO cell stained with a fluorescent dye and exposed to light before being stained with Giemsa.

Studies with the fluorescent dye Hoechst 33258 led Latt (24), however, to postulate that the difference in stainability was brought about by a BrdUrd-induced quenching of the fluorescence of the DNA-bound dye. Thus, both protein and DNA have been implicated in the differential staining.

Interestingly enough, the first step in the FPG technique, i.e. staining with a fluorescent dye, can be eliminated if the slides are exposed to very intense light. Experiments have been carried out (40) in which slides were exposed to the light from a high-pressure mercury burner and then simply stained in Giemsa; dramatic harlequinization results. Light can break disulfide bonds in protein (41), and further experiments with chemical agents that break disulfide bonds, such as dithiothreitol or β-mercaptoethanol, showed that incubation of cells in these chemicals in the dark prior to staining with Giemsa also leads to differential staining of the sister chromatids (40). These studies, too, implicate differential compaction of the chromosomes, influenced by the binding of nonhistone proteins, in the differential staining.

Harlequin Chromosomes and Chromosome Structure

Studies with harlequin chromosomes have by and large confirmed Taylor's deductions. They have provided a clear cytological demonstration of semiconservative replication of DNA and its subsequent distribution into chromatids. This was further confirmed in endoreduplicated cells, which are cells that undergo two rounds of replication without an intervening mitosis, so that the daughter chromosomes lie side by side. In these cells, the outer two of the four chromatids are bifilarly substituted with BrdUrd and so stain faintly, whereas the inner two chromatids are unifilarly substituted and stain darkly (28, 30, 42). This pattern indicates not only that the DNA replicates and segregates into the daughter chromatids in a semiconservative manner, but also that the newly made polynucleotide strands, which are coiled around their complementary strands, always segregate to the outside, as had been found in autoradiographs (14, 43, 44).

Studies of the numbers of twin and single exchanges either in endoreduplicated chromosomes (42, 45), where the paired daughter chromosomes preclude any ambiguity regarding the scoring of "false" twins, or in cells containing marker chromosomes that are present only once in diploid cells and so are present only twice in the tetraploids (42, 46) showed the ratio of singles to twins to be approximately 2:1, confirming the differences in polarity of the subunits previously reported (10, 47).

Although the data on SCE induction fit the concept that both polynucleotide strands of the DNA double helix in a chromatid are exchanged, it had been postulated that occasional single polynucleotide strand exchanges could occur to give rise to isolabeling in the second cell cycle (48–50). Such exchanges, however, would lead to heterolabeling (50), or harlequinization, in the first division after only a single round of replication in the presence of BrdUrd. No such heterolabeling was found in extensive studies in Chinese hamster (42), human (51), and Vicia (33) cells stained by the FPG technique, leading to the conclusion that single polynucleotide strand exchanges do not occur.

Although apparent isolabeling was found in autoradiograms, it has not been found in harlequinized chromosomes. In fact, studies with BrdUrd-substituted chromosomes have shown that isolabeling is an artifact of autoradiography that can be attributed to multiple SCEs (30), autoradiographic image spread (30), or labeling during more than one cycle of DNA synthesis (S. Wolff, J. Bodycote, and B. Rodin, unpublished data, 1977).

MULTIPLE SISTER CHROMATID EXCHANGES Occasionally harlequin chromosomes are found that contain several small SCEs within a short segment (24, 30, 32). These exchanges are so small that it is apparent that regions with a large number of these would appear isolabeled in autoradiograms. Indeed, in a study of FPG-stained chromosomes that had [^{3}H]BrdUrd incorporated during the first round of replication and nonradioactive BrdUrd during the second round, such regions of multiple small exhanges showed autoradiographic isolabeling (30).

AUTORADIOGRAPHIC IMAGE SPREAD It had been calculated that, because of the range of the β particle from ^{3}H-labeled chromosomes, enough particles originating in one chromatid could strike the emulsion over the sister chromatid to make it appear labeled (52). That such image spread, indeed, could cause apparent isolabeling, was found in the same FPG-stained autoradiographic preparations used to demonstrate multiple exchanges (30).

LABELING FOR MORE THAN ONE REPLICATION CYCLE If cells being prepared for autoradiography are incubated with [^{3}H]thymidine during the last part of one S period and then for another full S period before being incubated with nonradioactive thymidine for the next S period, the late-replicating regions of the chromosomes will have radioactive material in both sister chromatids and be isolabeled. In a carefully timed autoradiographic study with BrdUrd (S. Wolff, J. Bodycote, and B. Rodin, unpublished data, 1976), it was found that isolabeling occurred only in cells exposed to BrdUrd for more than two (but fewer than three) cycles. This isolabeling was confined to late-labeling regions such as the long arm of the X chromosome. In harlequin chromosomes the parental unsubstituted strands lead to dark staining and thus can be followed. The isolabeled regions were light staining, which showed that only bifilarly substituted DNA was present in the isolabeled regions. Such staining had also been noted by Crossen et al (53) in cells exposed to BrdUrd for more than two cycles.

In autoradiograms of cells that are exposed to [^{3}H]thymidine, the unsubstituted strands cannot be followed, so the distinction cannot be made.

In general, the number of SCEs is distributed randomly within the cells and within a given chromosome (28, 30, 54). That is, the number of cells or given chromosomes (for instance, chromosome number one) containing 0, 1, 2, 3, etc SCEs is distributed according to Poisson expectations. Furthermore, as has been previously found with autoradiographic studies (10, 17), the numbers of exchanges are positively correlated with chromosome length (25, 28, 33, 54–57). There are, however, some exceptions to these generalizations (see section on location of sister chromatid exchanges in the chromosome).

Another aspect of the new methods for observing differentially stained chromatids is that after three rounds of replication in the presence of BrdUrd, only one fourth of the chromatin is darkly stained, that is, only one half of the genome is harlequinized (28, 30, 58). This makes it possible to study SCEs formed in the third division. From the ratio of third division exchanges to those that occurred in the first and second divisions, Tice et al (58) were able to deduce that the rate of SCE formation was the same in all three divisions. They also confirmed that the subunits of the chromatids were restricted in the way that they could rejoin. Somewhat contrary results were published by Dutrillaux et al (28), who found fewer SCEs in the third division than in the others. The discrepancy might have been caused by the presence of less BrdUrd during the third division in the experiments of Dutrillaux et al.

Although the ratio of single to twin exchanges observed in tetraploid cells confirmed that the chromosomal subunits had different polarity, experiments on the types of ring chromosomes that were induced by X rays in G_1 cells showed that the polarity of the subunits could change along the length of the chromosome (59). The experiments were carried out by irradiating G_1 diploid Chinese hamster cells to induce ring chromosomes and then determining the types of rings found when the cells were immediately made tetraploid by the addition of Colcemid®. If the breaks that rejoined to form the rings were in regions of the chromosome that had identical polarity, then the tetraploid cells would contain pairs of monocentric rings. If, however, the breaks were in regions of opposite polarity, then single dicentric rings would be produced. The results of such an experiment with conventionally stained chromosomes were equivocal, however, because dicentric rings can arise not only from switches in polarity along the length of the chromosome, but also from an odd number of SCEs within monocentric rings (60). When the experiment was performed with harlequin chromosomes (59), however, the ambiguities no longer existed because those ring configurations that arose from sister chromatid exchange could be identified. The results indicated that in the Chinese hamster genome there is approximately one switch in polarity for every 10^9 normal 3'-5'-phosphodiester bonds.

In this experiment twin exchanges that occurred in the first cell cycle could be determined unambiguously in two different types of chromosomes, the rings themselves and symmetrical (mirror image) single dicentrics that arose from broken G_1 chromosomes in which the polynucleotide strands rejoined in a U-shaped configuration. In both the rings and these dicentrics a 2:1 ratio of singles to twins was observed.

The Question of Spontaneous Levels of Sister Chromatid Exchanges

The fact that the thymidine analogues used in the newer methods do not emit any radiation would, on the surface, indicate that SCEs observed with them represent truly spontaneous events. The situation, however, is not quite so simple as it seems, for it has long been known that incorporation of BrdUrd into mammalian chromosomes induces chromosome breaks and rearrangements (61, 62) just as does radiation. In fact, experiments with Chinese hamster (27, 30) and human (45, 63) cells have shown that at high concentrations of BrdUrd, the frequencies of SCEs increase as the BrdUrd concentration increases. In three of these experiments (30, 45, 63), the yield of SCEs seemed to fall slowly at low concentrations of BrdUrd, but in one experiment (27) the yield did not change as the concentration of BrdUrd was lowered from 3 to 0.3 μM.

Somewhat different estimates of the spontaneous level of SCEs were obtained in the two experiments with Chinese hamster cells (27, 30). The low-level plateau in the experiments by Kato (27) occurred at 2.3 SCEs per cell after two cell cycles. This reduces to 0.06 SCE per chromosome per cell cycle. If the ascending part of the curve in the experiments by Wolff & Perry (30) is extrapolated straight back to zero BrdUrd concentration, then a value of 0.15 SCE per chromosome per cell cycle is obtained, which is very close to the value obtained by Brewen & Peacock (21) for a human ring chromosome that was unsubstituted. Because the curve in Wolff & Perry's experiments rose steeply at very low levels of BrdUrd before leveling off, as does the curve obtained with [^{3}H]thymidine (22), and because the statistical errors associated with the points are such that the curve could be drawn from a zero SCE incidence at zero BrdUrd concentration, the possibility was raised that none of the SCEs might be spontaneous (30).

In addition, in vivo experiments on the induction of SCEs cast some doubt that the above figures represent a spontaneous level of SCE induction. When chicken embryos (36), mice (64–66), rats (66, 67), and fish (34) were injected with BrdUrd so that the chromosomes would become harlequinized, the number of SCEs observed per chromosome per cell cycle was lower in all of these in vivo studies than the number observed in tissue culture studies. The numbers were 0.035 SCE per macrochromosome per cell cycle in chicken embryos, 0.05 SCE per bone marrow chromosome per cell cycle and 0.033 SCE per spermatogonial chromosome per cell cycle in mice, 0.02 SCE per bone marrow chromosome per cell cycle in rats, and 0.06 SCE per gill or intestinal chromosome per cell cycle in fish.

These numbers, however, were derived primarily from the numbers of SCEs observed per cell and were not corrected for the amount of DNA present in the cell even though, as mentioned earlier, the number of SCEs is related to chromosome length and thus to DNA amount. The relation is not linear over all values of DNA content, however, as the small G group chromosomes of man seem to have too few SCEs (54) and the large number one chromosome of Chinese hamster has too many (25). Whether or not this is caused by chromosomal differences in BrdUrd concentration, i.e. differential incorporation in AT-rich and GC-rich chromosome regions, is unknown. Furthermore, in Drosophila cells, which have only one twentieth the DNA that a mammalian cell has, there is only 0.005 SCE per chromosome or 0.03 SCE per cell (35). If strict proportionality extends from mammalian cells, where 7–18 per cell seems to be the range with moderate amounts of BrdUrd (24, 27, 30, 31, 51, 54, 55, 57, 58, 68–71), then the rate of SCE formation in Drosophila seems to be very low, as might have been expected from the early results on sister strand crossing-over. Interestingly enough, *Vicia faba,* which has six to eight times as much DNA as mammalian cells, also seems to be under-represented (33). Twenty SCEs per cell have been observed, and although this reduces to 0.85 per chromosome per cell cycle, this number is really not too different from the numbers obtained in mammalian cells because in Vicia there are only 12 chromosomes. A similar apparent inflation in yield per chromosome can be seen in experiments with the Indian muntjac (56), which has only six large chromosomes and one dot chromosome per diploid cell. Here there were approximately 7.25 SCEs per cell, or 1.2 per chromosome.

Location of Sister Chromatid Exchanges in the Chromosome

Although SCEs seem to occur randomly in that they are distributed among the chromosomes according to Poisson expectations, there are several indications that their location within chromosomes might not be completely random. For instance, Latt (55) has found that the preponderance of SCEs occurs in the interband regions of G-banded chromosomes. Because the G bands are late replicating, and in this sense are like heterochromatin, these experiments raised questions about the relative amounts of sister chromatid exchange in euchromatin and heterochromatin. These were answered in studies with muntjac cells (56) that have an X chromosome containing a long thin heterochromatic "neck" region connecting its two euchromatic arms. Carrano & Wolff (56) showed that although there was one fifth the amount of DNA in this heterochromatic region as in a euchromatic region of the same length, there was a great deficit of SCEs in the heterochromatin. The junction between the heterochromatin and euchromatin, however, had a large excess of SCEs. A similar relation exists in kangaroo rat cells that have a GC-rich heterochromatin (72). The results with the muntjac suggested that all SCEs might occur where euchromatin abuts heterochromatin (56).

Although there are reports in *Microtus agrestis* that the heterochromatic part of the X chromosome contains either more (73) or only somewhat fewer (74) SCEs than does the euchromatin, work with Chinese hamster DON and *Microtus montanus* cells showed a deficit (75) of SCEs in heterochromatin, which is similar to the observations in the muntjac. In fact, the deficit in Chinese hamster DON cells was so great that it distorted the relation between SCEs and chromosome length (75).

Another type of nonrandomness had been noticed earlier in work with autoradiograms. There it was found that an excess of SCEs existed at the centromere in both Chinese hamster (19) and rat kangaroo (22) cells. This was confirmed in human (58) and mouse (76) cells treated with BrdUrd. Autoradiographic results with wallaby chromosomes (46) and other experiments with BrdUrd-substituted human chromosomes (54), however, indicated that the number at the centromeres was low. The difference between these two sets of results is as yet unresolved, although some of it could reflect a species difference in the heterochromatic constitution of the centromeres.

Sister Chromatid Exchanges and Human Genetic Diseases

Three human genetic diseases, Fanconi's anemia, Bloom's syndrome, and ataxia telangiectasia, are characterized by a high incidence of chromosome aberrations and a concomitant increased susceptibility to cancer. In cells from patients with two of these diseases, Fanconi's anemia (57, 68, 70, 71, 77) and ataxia telangiectasia (54, 68), the frequency of SCEs in untreated cells is normal. But in cells from patients with Bloom's syndrome there are 10–13 times more SCEs than normal (68). Normal numbers of SCEs also occur in cells from patients with the DNA-repair-deficient human disease xeroderma pigmentosum (51) as well as in cells from patients with Werner's syndrome (45). The xeroderma pigmentosum cells seem to be remarkably sensitive, however, to the induction of SCEs by chemical agents, which is contrary to the way in which chromosome aberrations in the same cells respond to many of these chemicals (78). In contrast, some Fanconi cells seem to be less sensitive to mitomycin C than are normal cells (57). Many of the mitomycin C–induced SCEs in these Fanconi cells are incomplete and thus contain a chromatid break at the point of the SCE. This is unlike the response of normal cells to X rays, where chromatid deletions are unrelated to SCEs (79).

Sister Chromatid Exchanges as Indicators of Mutagenic Carcinogens

SCEs have proved to be very sensitive indicators of the effects of chemical mutagens and carcinogens on eucaryotic chromosomes. The first study was carried out by Latt on human lymphocyte chromosomes treated with the bifunctional alkylating agent mitomycin C that cross-links DNA (63). Large numbers of SCEs were observed at chemical concentrations too low to cause an increase in ordinary chromosome aberrations, and it was suggested that the sensitivity of the system was such that it would have great utility as a test system for the effect of mutagens on mammalian chromosomes. Because of the ease in scoring SCEs and because of the dramatic effects, the work was soon followed up with other cell systems and other chemicals. A summary of this work is presented in Table 1.

Of particular interest is the extensive study carried out by Perry & Evans (69). They tested Chinese hamster ovary cells with 14 known or suspected mutagen/carcinogens and found that all the direct-acting mutagens gave dramatically increased yields of SCEs. With the exception of adriamycin and bleomycin, which are chemotherapeutic agents, no induction of chromosome aberrations was found at low concentrations that could increase the yield of SCEs by a factor of 10. The only chemicals that did not increase the yield of SCEs in this in vitro system were maleic hydrazide, which is not mutagenic in mammalian cells, and the chemotherapeutic agent cyclophosphamide, which requires metabolic activation to be effective. The latter type of chemical can be detected in sister chromatid exchange tests, however, either by treating whole animals, culturing their blood in the presence of BrdUrd (80), and counting the SCEs or by serially injecting animals with BrdUrd, treating with the test chemicals, and counting the SCEs in either bone marrow cells (64, 65) or spermatogonia (64). In such in vivo tests, cyclophosphamide was found to be activated and to produce a large increase in SCEs. If the studies are carried out on blood cells from large animals such as rabbits, the animal does not have to be sacrificed and serial samples can be taken from the same animal, which can also act as its own control (80). Under such conditions it is found that the increase in SCEs is transitory, first rising and then falling back to the control level.

Chemicals that require metabolic activation before being effective mutagens or carcinogens can also be tested in tissue culture if an activating system is incorporated into the cultures. Thus, Natarajan et al (81) and Stetka & Wolff (82), working with Chinese hamster ovary cells, found that the addition of an extract of rat liver

Table 1 Positive tests for chemicals that increase the yield of SCEs

Cells	Genetic constitution	Treatment mode	Chemical[a]	Reference
Animal				
Human				
Lymphocytes	normal	in vitro	mechlorethamine, MMC	Latt (63)
Lymphocytes	normal	in vitro	chlorambucil, QM	Solomon & Bobrow (93)
Lymphocytes	normal	in vitro	benzpyrene	Rudiger et al (96)
Lymphocytes	normal	in vitro	trenimon	Beek & Obe (97)
Lymphocytes	normal	in vitro	trenimon	Hayashi & Schmid (71)
Lymphocytes	Fanconi's anemia	in vitro	EMS, MMC	Latt et al (57)
Lymphocytes	normal	in vivo	AM	Perry & Evans (69)
Fibroblasts	normal	in vitro	trenimon	Hayashi & Schmid (71)
Fibroblasts	Fanconi's anemia	in vitro	MMC	Latt et al (57)
Fibroblasts	xeroderma pigmentosum	in vitro	DMS, EMS, ENU, MMC, MMS, MNNG, 4NQO	Wolff et al (78)
Chinese hamster				
Fibroblasts	normal	in vitro	MMC, 4NQO, proflavine	Kato (83)
Fibroblasts	normal	in vitro	MMC	Kato & Shimada (94)
Fibroblasts	normal	in vitro	AM, BLM, BPL, DEB, EMS, HN2, Hoechst 33258, MMC, MMS, MNNG, 4NQO, QM	Perry & Evans (69)
Fibroblasts	normal	in vitro	CP (with activation), EMS	Stetka & Wolff (82)
Fibroblasts	normal	in vitro	DEN, DMN (both with activation)	Natarajan et al (81)
Mouse				
Bone marrow	normal	in vivo	CP, triaziquon	Vogel & Bauknecht (65)
Bone marrow	normal	in vivo	CP	Allen & Latt (9)
Spermatogonia	normal	in vivo	CP	Allen & Latt (9)
Spermatogonia	normal	in vivo	MMC	Allen & Latt (64)
Fibroblasts	HGPRT[b]	in vitro	MMC	Lin & Alfi (29)
Rat				
Bone marrow	normal	in vitro	DMBA, TMBA	Ueda et al (98)
Rabbit				
Lymphocytes	normal	in vivo	CP, EMS, MMS	Stetka & Wolff (80)
Muntjac				
Fibroblasts	normal	in vitro	MMC	Huttner & Ruddle (99)
Plant				
Vicia faba	normal	in vivo	thiotepa	Kihlman (86)

[a] AM, adriamycin; BLM, bleomycin; BPL, β-propriolactone; CP, cyclophosphamide; DEB, diepoxybutane; DEN, diethylnitrosamine; DMBA, dimethylbenzanthracene; DMN, dimethylnitrosamine; DMS, dimethyl sulfate; EMS, ethylmethane sulfonate; ENU, ethyl nitrosourea; HN2, nitrogen mustard; MMC, mitomycin C; MMS, methylmethane sulfonate; MNNG, N-methyl-N'-nitro-N-nitrosoguanidine; 4NQO, 4-nitroquinoline-1-oxide; QM, quinicrine mustard; TMBA, trimethylbenzanthracene.
[b] Hypoxanthine-guanine phosphoribosyltransferase.

microsomes and a DPNH-generating system, which provides the mixed-function oxidases necessary to activate many chemicals, results in the activation of compounds that ordinarily would not induce SCEs.

This aspect of the work with SCEs not only has wide practicability, but also has ramifications leading to a better understanding of how SCEs are formed and what lesions are responsible for them. Thus, by a judicious choice of chemical and physical agents and by the use of cells defective in various metabolic processes, the studies of SCEs have been used to gather clues about the basic biology and biochemistry involved in SCE formation.

Because SCEs involve the exchange of partners, it has been suggested that SCE formation is dependent upon DNA repair processes (27, 50, 57, 63, 83–85), such as excision repair or post-replication repair. That the situation might not be quite this simple was indicated in experiments which showed that the "spontaneous" level of SCEs was normal in xeroderma pigmentosum cells that are defective in excision repair or post-replication repair (51, 77). Furthermore, when cells treated with alkylating agents were treated further with caffeine, which purportedly inhibits post-replication repair, the yields of SCEs were not increased (86). This stands in contrast to the results reported for caffeine and UV radiation (84).

Nonetheless, xeroderma pigmentosum cells are more sensitive to UV light (45) and to both UV-like and X-ray-like chemicals (78) than are normal cells. In fact, xeroderma pigmentosum cells provide the most sensitive mammalian system yet known for the detection of the effects of such chemicals on chromosomes (78).

Lesions Responsible for Sister Chromatid Exchange Formation

Although much information has been forthcoming on the mechanisms of SCE formation, the lesions that lead to SCEs are unknown. SCEs can be increased by several agents that attack DNA. Thus, it has been found that β particles from [^{3}H]thymidine (22), X rays (19, 48, 49), UV light (84, 85, 87), and chemical mutagens (Table 1) increase SCEs. In the cases of UV light (85) and chemicals (86), long-lived lesions are induced that lead to SCEs that are formed when the cells are in S. The lesions induced in G_2 do not give rise to exchange until after the cell has divided and proceeded through an S period (85, 86). That these long-lived lesions did not seem to be thymine dimers, which are induced very efficiently in DNA by UV light, was shown in experiments with repair-deficient cells. It was found that repair-deficient B14FAF Chinese hamster cells were as sensitive as normal Chinese hamster cells (85) and that in xeroderma pigmentosum cells with different repair capacities the sensitivity for SCE formation was not related in any simple way to the amount of repair (88). Thus the yield of SCEs seems to be independent of the residual amount of thymine dimers present in the cell. Xeroderma pigmentosum cells nevertheless, as noted earlier, are sensitive to chemical agents (78) [and to UV light (45)]. This is true both for UV-like chemical agents, whose damage xeroderma pigmentosum cells are unable to repair, and for X-ray-like chemical agents, whose damage they do seem able to repair (78). The work with the latter type of chemicals would indicate that although most of the lesions are repaired in xeroderma pigmentosum cells, there might be a small residual amount of unexcised damage that could lead to SCE formation (78). Although the lesion itself has not been characterized, experiments with ethylnitrosourea have led to speculations that guanine alkylated at the O-6 position might be responsible (89). Ethylnitrosourea, which induces a high yield of SCEs in xeroderma pigmentosum cells, also induces a relatively large amount of alkylation at the O-6 position of guanine as compared to that at N-7 position usually attacked. The same xeroderma pigmentosum cells cannot carry out excision repair of the guanine alkylated at the O-6 position.

After treatment with most monofunctional alkylating agents that react with guanine, the DNA becomes depurinated, which often leads to single-strand breaks (90, 91). This suggests that depurination following alkylation could cause SCEs. This suspicion is strengthened by the finding that visible light can increase the number of SCEs found in BrdUrd-substituted cells (25, 30). Visible light causes the photolysis of BrdUrd-containing DNA resulting in the formation of alkali-labile sites that become single-strand breaks when the DNA is exposed to high pH (92).

Whatever the nature of the lesions in the chromosome that ultimately lead to the formation of SCEs, because SCE induction differs in very many ways from the induction of ordinary chromosome aberrations, it is likely that the lesions will be quite different from those that lead to aberrations. SCEs, for instance, are induced at high frequencies by concentrations of chemicals that induce very few aberrations (63, 69, 93, 94), and they are not markedly increased by low doses of ionizing radiations, whereas aberrations are (69, 85). SCEs also saturate with increasing doses of ionizing radiation (19) and with [^{3}H]thymidine concentration (22), but aberrations do not. Furthermore, SCEs react differently to treatments with caffeine than do chromosome aberrations (84, 86), and they are not correlated in any consistent fashion with the increased aberrations seen in the human diseases, Bloom's syndrome (68), ataxia telangiectasia (54), and Fanconi's anemia (57), since the SCE levels are very high in Bloom's syndrome, normal in ataxia, and reportedly low in Fanconi's anemia.

It therefore appears that SCEs may be the result of fundamentally different cellular lesions and processes from those that cause chromosome aberrations (78, 95). The aberrations are associated with cell death, whereas SCEs do not seem to be so, thus making it likely that SCEs are more representative of events compatible with cell survival such as mutagenesis. It has been postulated that, although the majority of SCEs are probably genetically neutral because equal amounts of a sister chromatid are exchanged, some unequal exchange could occur leading to deletion, insertion, or frameshift mutagenesis (78). Nevertheless, the actual biological significance of sister chromatid exchange is still obscure.

CONCLUSION

The use of autoradiography for detecting sister chromatid exchange has been replaced by new methods, in which the sister chromatids are made to stain differentially. The ease of visualization of SCEs plus the added resolution obtained with the new techniques has already led to practical studies on the effects of chemicals on chromosomes and to basic studies on chromosome structure and the mechanisms involved in exchange formation. We now seem on the verge of obtaining answers to some of the questions that have puzzled geneticists for years.

ACKNOWLEDGMENT

This work was performed under the auspices of the US Energy Research and Development Administration.

Literature Cited

1. Janssens, F. A. 1909. Spermatogénése dans les Batraciens V., La théorie de la chiasmatypie, nouvelle interpretation des cinèses de maturation. *Cellule* 25: 387–411
2. Darlington, C. D. 1937. *Recent Advances in Cytology*. Philadelphia: Blakiston. 671 pp. 2nd ed.
3. Beadle, G. W., Emerson, S. 1935. Further studies of crossing over in attached-X chromosomes of *Drosophila melanogaster*. *Genetics* 20:192–206
3a. Morgan, L. V. 1933. A closed X chromosome in *Drosophila melanogaster*. *Genetics* 18:250–83
4. Stern, C. 1936. Somatic crossing over and segregation in Drosophila melanogaster. *Genetics* 21:625–730
5. McClintock, B. 1938. The production of homozygous deficient tissues with mutant characteristics by means of the aberrant mitotic behavior of ring-shaped chromosomes. *Genetics* 23: 315–76
6. Schwartz, D. 1953. Evidence for sister-strand crossing over in maize. *Genetics* 38:251–60
6a. Sandler, L. 1964. The meiotic mechanics of ring chromosomes in female *Drosophila melanogaster*. *Symp. Genes and Chromosome, Structure and Function. Natl. Cancer Inst. Monogr.* 18:243–73
7. Taylor, J. H. 1965. Distribution of tritium-labeled DNA among chromosomes during meiosis. I. Spermatogenesis in the grasshopper. *J. Cell Biol.* 25:57–67
8. Peacock, W. J. 1970. Replication, recombination, and chiasmata in *Goniaea Australasiae (Orthoptera: Acrididae)*. *Genetics* 65:593–617
9. Allen, J. W., Latt, S. A. 1976. In vivo BrdU-33258 Hoechst analysis of DNA replication kinetics and sister chromatid exchange formation in mouse somatic and meiotic cells. *Chromosoma* 58:325–40
10. Taylor, J. H. 1958. Sister chromatid exchanges in tritium labeled chromosomes. *Genetics* 43:515–29
11. Taylor, J. H., Woods, P. S., Hughes, W. L. 1967. The organization and duplication of chromosomes as revealed by autoradiographic studies using tritium-labeled thymidine. *Proc. Natl. Acad. Sci. USA* 43:122–38
12. Meselson, M., Stahl, F. W. 1958. The replication of DNA in *Escherichia coli*. *Proc. Natl. Acad. Sci. USA* 44:671–82
13. Sparvoli, E., Gay, H. 1973. Linear heterogeneity of Bellevalia mitotic chromosomes as evidenced by sister chromatid exchanges. *Chromosomes Today* 4:101–16
14. Walen, H. K. 1965. Spatial relationships in the replication of chromosomal DNA. *Genetics* 51:915–29
15. Heddle, J. A. 1969. The influence of false twins on the ratio of twin and single sister chromatid exchanges. *J. Theor. Biol.* 22:151–62
16. La Cour, L. F., Pelc, S. R. 1958. Effect of colchicine on the utilization of labelled thymidine during chromosomal reproduction. *Nature* 182:506–8
17. Peacock, W. J. 1963. Chromosome duplication and structure as determined by autoradiography *Proc. Natl. Acad. Sci. USA* 49:793–801
18. Deaven, L. L., Stubblefield, E. 1969. Segregation of chromsomal DNA in Chinese hamster fibroblasts *in vitro*. *Exp. Cell Res.* 55:132–35
19. Marin, G., Prescott, D. M. 1964. The frequency of sister chromatid exchanges following exposure to varying doses of H^3 thymidine or X-rays. *J. Cell Biol.* 21:159–67
20. Wolff, S. 1964. Are sister chromatid exchanges sister strand crossovers or radiation-induced exchanges? *Mutat. Res.* 1:337–43
21. Brewen, J. G., Peacock, W. J. 1969. The effect of tritiated thymidine on sister-chromatid exchange in a ring chromosome. *Mutat. Res.* 7:433–40
22. Gibson, D. A., Prescott, D. M. 1972. Induction of sister chromatid exchanges in chromosomes of rat kangaroo cells by tritium incorporated into DNA. *Exp. Cell Res.* 74:397–402
23. Zakharov, A. F., Egolina, N. A. 1972. Differential spiralisation along mammalian mitotic chromosomes. 1. BUdR-revealed differentiation in Chinese hamster chromosomes. *Chromosoma* 38: 341–65
24. Latt, S. A. 1973. Microfluorometric detection of deoxyribonucleic acid replication in human metaphase chromosomes. *Proc. Natl. Acad. Sci. USA* 70: 3395–99
25. Ikushima, T., Wolff, S. 1974. Sister chromatid exchanges induced by light flashes to 5-bromodeoxyuridine and 5-iododeoxyuridine substituted Chinese hamster chromosomes. *Exp. Cell Res.* 87:15–19

26. Perry, P., Wolff, S. 1974. New Giemsa method for the differential staining of sister chromatids. *Nature* 251:156-58
27. Kato, H. 1974. Spontaneous sister chromatid exchanges detected by a BrdU-labeling method. *Nature* 251:70-72
28. Dutrillaux, B., Fosse, A. M., Prieur, M., Lejeune, J. 1974. Analyses des échanges de chromatides dans les cellules somatiques humaines. *Chromosoma* 48:327-40
29. Lin, M. S., Alfi, O. S. 1976. Detection of sister chromatid exchanges 4'-6-diamidino-2-phenyl-indole fluorescence. *Chromosoma* 57:219-25
30. Wolff, S., Perry, P. 1974. Differential Giemsa staining of sister chromatids and the study of sister chromatid exchange without autoradiography. *Chromosoma* 48:341-53
31. Kim, M. A. 1974. Chromatidaustausch und Heterochromatinveranderungen menschlicher Chromosomen nach BUdR-Markierung. *Humangenetik* 25:179-88
32. Korenberg, J. R., Freedlender, E. F. 1974. Giemsa technique for the detection of sister chromatid exchanges. *Chromsoma* 48:355-60
33. Kihlman, B. A., Kronberg, D. 1975. Sister chromatid exchanges in *Vicia faba*. I. Demonstration by a modified fluorescent plus Giemsa (FPG) technique. *Chromosoma* 51:1-10
34. Kligerman, A. D., Bloom, S. E. 1976. Sister chromatid differentiation and exchanges in adult mudminnows (Umbra limi) after *in vivo* exposure to 5-bromodeoxyuridine. *Chromosoma* 56:101-9
35. Wienberg, J. 1977. BrdU-Giemsa-technique for the differentiation of sister chromatids in somatic cells of *Drosophila melanogaster*. *Mutat. Res.* In press
36. Bloom, S. E., Hsu, T. C. 1975. Differential fluorescence of sister chromatids in chicken embryos exposed to 5-bromodeoxyuridine. *Chromosoma* 51:261-67
37. Goto, K., Akematsu, T., Shimazu, H., Sugiyama, T. 1975. Simple differential Giemsa staining of sister chromatids after treatment with photosensitive dyes and exposure to light and the mechanism of staining. *Chromosoma* 53:223-30
38. David, J., Gordon, J. S., Rutter, W. J. 1974. Increased thermal stability of chromatin containing 5-bromodeoxy-uridine-substituted DNA. *Proc. Natl. Acad. Sci. USA* 71:2808-12
39. Korenberg, J. R., Ris, H. 1977. The effect of BrdU incorporation on metaphase chromosome structure. *J. Cell Biol.* In press
40. Wolff, S., Bodycote, J. 1977. The production of harlequin chromosomes with chemical and physical agents that disrupt protein structure. In *Human Molecular Cytogenetics*. New York: Academic. In press
41. Mousseron-Canet, M., Mani, J.-C. 1972. *Photochemistry and Molecular Reactions*. Jerusalem: Israel Program Sci. Transl. 268 pp.
42. Wolff, S., Perry, P. 1975. Insights on chromosome structure from sister chromatid exchanges and the lack of both isolabelling and heterolabelling as determined by the FPG technique. *Exp. Cell Res.* 93:23-30
43. Schwarzacher, H. G., Schnedl, W. 1965. Endoreduplication in human fibroblast cultures. *Cytogenetics* 4:1-18
44. Herreros, B., Giannelli, F. 1967. Spatial distribution of old and new chromatid subunits and frequency of chromatid exchanges in induced human lymphocyte endoreduplications. *Nature* 216:286-88
45. Bartram, C. R., Koske-Westphal, T., Passarge, E. 1976. Chromatid exchanges in ataxia telangiectasia, Bloom syndrome, Werner syndrome, and xeroderma pigmentosum. *Ann. Hum. Genet.* 40:79-86
46. Geard, C. R. 1974. Comparison of sister chromatid exchanges from three successive cell cycles in Wallabia bicolor chomosomes. *Mutat. Res.* 23:67-78
47. Brewen, J. G., Peacock, W. J. 1969. Restricted rejoining of chromosomal subunits in aberration formation: A test for subunit dissimilarity. *Proc. Natl. Acad. Sci. USA* 62:389-94
48. Gatti, M., Olivieri, G. 1973. The effect of X-rays on labelling pattern of M_1 and M_2 chromosomes in Chinese hamster cells. *Mutat. Res.* 17:101-12
49. Gatti, M., Pimpinelli, S., Olivieri, G. 1974. The frequency of distribution of isolabelling in Chinese hamster chromosomes after exposure to X-rays. *Mutat. Res.* 23:229-38
50. Bender, M. A., Griggs, H. G., Bedford, J. S. 1974. Recombinational DNA repair and sister chromatid exchanges. *Mutat. Res.* 24:117-23
51. Wolff, S., Bodycote, J., Thomas, G. H., Cleaver, J. E. 1975. Sister chromatid ex-

change in xeroderma pigmentosum cells that are defective in DNA excision-repair or post-replication repair. *Genetics* 81:349–55
52. Gibson, D. A., Prescott, D. M. 1973. Sister chromatid exchanges in isolabelling. *Exp. Cell Res.* 83:445–47
53. Crossen, P. E., Pathak, S., Arrighi, F. E. 1975. A high resolution study of the DNA replication patterns of Chinese hamster chromosomes using sister chromatid differential staining technique. *Chromosoma* 52:339–47
54. Galloway, S. M., Evans, H. J. 1975. Sister chromatid exchange in human chromosomes from normal individuals and patients with ataxia telangiectasia. *Cytogenet. Cell Genet.* 15:17–29
55. Latt, S. A. 1974. Localization of sister chromatid exchanges in human chromosomes. *Science* 185:74–76
56. Carrano, A. V., Wolff, S. 1975. Distribution of sister chromatid exchanges in the euchromatin and heterochromatin of the Indian muntjac. *Chromosoma* 53:361–69
57. Latt, S. A., Stetten, G., Juergens, L. A., Buchanan, G. R., Gerald, P. S. 1975. Induction by alkylating agents of sister chromatid exchanges and chromatid breaks in Fanconi's anemia. *Proc. Natl. Acad. Sci. USA* 72:4066–70
58. Tice, R., Chaillet, J., Schneider, E. L. 1975. Evidence derived from sister chromatid exchanges of restricted rejoining of chromatid subunits. *Nature* 256:642–44
59. Wolff, S., Lindsley, D. L., Peacock, W. J. 1976. Cytological evidence for switches in polarity of chromosomal DNA. *Proc. Natl. Acad. Sci. USA* 73:877–81
60. Peacock, W. J., Wolff, S., Lindsley, D. L. 1973. Continuity of chromosome subunits. *Chromosomes Today* 4:85–100
61. Hsu, T. C., Somers, C. E. 1962. Properties of L cells resistant to 5-bromodeoxyuridine. *Exp. Cell Res.* 26:404–10
62. Dewey, W. C., Humphrey, R. M. 1965. Increase in radiosensitivity to ionizing radiation related to replacement of thymidine in mammalian cells with 5-bromodeoxyuridine. *Radiat. Res.* 26:538–53
63. Latt, S. A. 1974. Sister chromatid exchanges, indices of human chromosome damage and repair: Detection by fluorescence and induction by mitomycin-C. *Proc. Natl. Acad. Sci. USA* 71:3162–66
64. Allen, J. W., Latt, S. A. 1976. Analysis of sister chromatid exchange formation in vivo in mouse spermatogonia as a new test system for environmental mutagens. *Nature* 260:449–51
65. Vogel, W., Bauknecht, T. 1976. Differential chromatid staining by *in vivo* treatment as a mutagenicity test system. *Nature* 260:448–49
66. Schneider, E. L., Chaillet, J. R., Tice, R. R. 1976. *In vivo* BUdR labeling of mammalian chromosomes. *Exp. Cell Res.* 100:396–99
67. Tice, R., Chaillet, J., Schneider, E. L. 1976. Demonstration of spontaneous sister chromatid exchanges *in vivo*. *Exp. Cell Res.* 102:426–29
68. Chaganti, R. S. K., Schonberg, S., German, J. 1974. A manyfold increase in sister chromatid exchanges in Bloom's syndrome lymphocytes. *Proc. Natl. Acad. Sci. USA* 71:4508–12
69. Perry, P., Evans, H. J. 1975. Cytological detection of mutagen-carcinogen exposure by sister chromatid exchange. *Nature* 258:121–25
70. Sperling, K., Wegner, R. D., Riehm, H., Obe, G. 1975. Frequency and distribution of sister chromatid exchanges in a case of Fanconi's anemia. *Humangenetik* 27:227–30
71. Hayashi, K., Schmid, W. 1975. The rate of sister chromatid exchanges parallel to spontaneous chromosome breakage in Fanconi's anemia and trenimon-induced aberrations in human lymphocytes and fibroblasts. *Humangenetik* 29:201–6
72. Bostock, C. J., Christie, S. 1976. Analysis of the frequency of sister chromatid exchange in different regions of chromosomes of the kangaroo rat (*Dipodomys ordii*). *Chromosoma* 56:275–87
73. Natarajan, A. R., Klášterská, I. 1975. Heterochromatin and sister chromatid exchanges in the chromosomes of *Microtus agrestis*. *Hereditas* 79:150–54
74. Pera, F., Mattias, P. 1976. Labelling of DNA and differential sister chromatid staining after BrdU treatment *in vivo*. *Chromosoma* 57:13–18
75. Hsu, T. C., Pathak, S. 1976. Differential rates of sister chromatid exchanges between euchromatin and heterochromatin. *Chromosoma* 58:269–73
76. Holmquist, G. P., Comings, D. E. 1975. Sister chromatid exchange and chromosome organization based on bromodeoxyuridine Giemsa-C-banding tech-

nique (TC-banding). *Chromosoma* 52: 245–59
77. Kato, H., Stich, H. F. 1976. Sister chromatid exchanges in ageing and repair-deficient human fibroblasts. *Nature* 260:447–48
78. Wolff, S., Rodin, B., Cleaver, J. E. 1977. Sister chromatid exchanges induced by mutagenic carcinogens in normal and xeroderma pigmentosum cells. *Nature* 265:347–48
79. Wolff, S., Bodycote, J. 1975. The induction of chromatid deletions in accord with the breakage-and-reunion hypothesis. *Mutat. Res.* 29:85–91
80. Stetka, D. G., Wolff, S. 1976. Sister chromatid exchange as an assay for genetic damage induced by mutagen-carcinogens. Part I. *In vivo* test for compounds requiring metabolic activation. *Mutat. Res.* 41:333–42
81. Natarajan, A. T., Tates, A. D., van Buul, P. P. W., Meijers, M., de Vogel, N. 1976. Cytogenetic effects of mutagens/carcinogens after activation in a microsomal system *in vitro*. I. Induction of chromosome aberrations and sister chromatid exchanges by diethylnitrosamine (DEN) and dimethylnitrosamine (DMN) in CHO cells in the presence of rat-liver microsomes. *Mutat. Res.* 37:83–90
82. Stetka, D. G., Wolff, S. 1976. Sister chromatid exchange as an assay for genetic damage induced by mutagen-carcinogens. Part II. *In vitro* test for compounds requiring metabolic activation. *Mutat. Res.* 41:343–50
83. Kato, H. 1974. Induction of sister chromatid exchanges by chemical mutagens and its possible relevance to DNA repair. *Exp. Cell Res.* 85:239–47
84. Kato, H. 1973. Induction of sister chromatid exchanges by UV light and its inhibition by caffeine. *Exp. Cell Res.* 82:383–90
85. Wolff, S., Bodycote, J., Painter, R. B. 1974. Sister chromatid exchanges induced in Chinese hamster cells by UV irradiation of different stages of the cell cycle: The necessity for cells to pass through S. *Mutat. Res.* 25:73–81
86. Kihlman, B. A. 1975. Sister chromatid exchanges in *Vicia faba*. II. Effects of thiotepa, caffeine and 8-ethoxycaffeine on the frequency of SCEs. *Chromosoma* 51:11–18
87. Rommelaere, J., Susskind, M., Errera, M. 1973. Chromosome and chromatid exchanges in Chinese hamster cells. *Chromosoma* 41:243–57
88. de Weerd-Kastelein, E. A., Keijzer, W., Rainaldi, G., Bootsma, D. 1977. Induction of sister chromatid exchanges in xeroderma pigmentosum cells following exposure to ultraviolet light. *Mutat. Res.* In press
89. Goth-Goldstein, R. 1977. Repair of DNA damaged by alkylating carcinogens is defective in xeroderma pigmentosum-derived fibroblasts. *Nature* 267:81–82
90. Strauss, B., Coyle, M., Robbins, M. 1968. Alkylation damage and its repair. *Cold Spring Harbor Symp. Quant. Biol.* 33:277–87
91. Brookes, P., Lawley, P. 1961. The reaction of mono- and di-functional alkylating agents with nucleic acids. *Biochem. J.* 80:496–503
92. Hutchinson, F. 1973. The lesions produced by ultraviolet light in DNA containing 5-bromouracil. *Q. Rev. Biophys.* 6:201–46
93. Solomon, E., Bobrow, M. 1975. Sister chromatid exchanges—A sensitive assay of agents damaging human chromosomes. *Mutat. Res.* 30:273–78
94. Kato, H., Shimada, H. 1975. Sister chromatid exchanges induced by mitomycin C: A new method of detecting DNA damage at chromosomal level. *Mutat. Res.* 28:459–64
95. Wolff, S. 1977. Chromosome effects induced by low levels of mutagens. *Research in Photobiology*, pp. 721–32. New York: Plenum
96. Rudiger, H. W., Kohl, F., Mangels, W., von Wichert, P., Bartram, C. R., Wohler, W., Passarge, E. 1976. Benzpyrene induces sister chromatid exchanges in cultured human lymphocytes. *Nature* 262:290–92
97. Beek, B., Obe, G. 1975. The human leukocyte test system. VI. The use of sister chromotid exchanges as possible indicators for mutagenic activities. *Humangenetik* 29:127–34
98. Ueda, N., Uenaka, H., Akematsu, T., Sugiyama, T. 1976. Parallel distribution of sister chromatid exchanges and chromosome aberrations. *Nature* 262:581–83
99. Huttner, K. M., Ruddle, F. H. 1976. Study of mitomycin C-induced chromosomal exchange. *Chromosoma* 56:1–13

THE GENETIC STRUCTURE OF RNA TUMOR VIRUSES

✦3121

Peter K. Vogt and Sylvia S. F. Hu

Department of Microbiology, University of Southern California, School of Medicine, Los Angeles, California 90033

CONTENTS

INTRODUCTION	203
PHYSICAL PROPERTIES OF THE GENOME	204
Virion RNA	204
Provirus	205
The 5' and 3' Termini	206
GENES OF RNA TUMOR VIRUSES	207
gag	207
pol	209
env	211
src	213
RECOMBINATION	216
GENE MAPPING	218
SYNTHESIS OF VIRAL RNA	221
CONCLUSION	222

INTRODUCTION

RNA tumor viruses constitute one of three subfamilies of the retroviridae; the other two are spumaviruses and lentiviruses (1–3). All retroviruses have an RNA genome and replicate via a DNA intermediate with the help of a virion RNA-dependent DNA polymerase. The recommended scientific term for RNA tumor viruses is oncovirinae or oncoviruses. We use the scientific and common terms interchangeably, but in the context of this paper restrict their meaning to include only the genus of type C oncoviruses, the only one for which significant genetic data are known. We first discuss the physical properties of the viral genome, then review the physiology of viral genes illustrated by appropriate mutants, and finally consider recombination, a genetic map, and viral RNA synthesis.

PHYSICAL PROPERTIES OF THE GENOME

Virion RNA

The RNA extracted from oncoviruses by sodium dodecyl sulfate and phenol contains a high molecular weight component which sediments at 60–70S and appears to carry all viral genetic information (4). Reported molecular weight figures for this 60–70S RNA vary with the virus and with the accuracy of the technique used; but most values lie between 4.5 and 7 $\times$ 10^6 daltons. Methods used in these molecular weight measurements include particle weight determination (5), equilibrium sedimentation (6), electron microscopy (7–9), and sedimentation in conjunction with gel electrophoresis (10). The 60–70S RNA can be dissociated into 30–40S pieces by heating or treatment with dimethylsulfoxide (11) suggesting that it is a complex, multimeric molecule whose components are held together by hydrogen bonding. The 30–40S RNA has a molecular weight of 2.2–3.5 $\times$ 10^6. The lower values are found with defective oncoviruses from which one or more genes are deleted. The higher values are characteristic of nondefective sarcoma viruses which have the largest genomes of RNA tumor viruses (6, 10, 12, 13). The 30–40S virion RNA has termini typical for messenger RNAs. The 3' ends consist of poly(A), about 200 residues long (14–18), the 5' ends are characterized by a capped structure in which a 7-methyl guanosine links 5' to 5' via a triphosphate to a second 2'-O-methylated nucleotide, represented as m^{7}G$^{5'}$ppp$^{5'}$NmpNp (19–21). The functions of the 3' terminal poly(A) and of the 5' capping are not known. The 30–40S virion RNA can also act as messenger both in vivo and in vitro and therefore is of the plus polarity (22–24). Several lines of evidence support the hypothesis that individual 30–40S RNA molecules contain a complete set of viral genes. Oligonucleotide analysis indicates that the genome of RNA tumor viruses has a unique sequence complexity of 3–4 $\times$ 10^6 daltons corresponding to a 30–40S molecule of single-stranded RNA (15, 16, 25–32). Hybridization kinetics of viral RNA to DNA have yielded less clear-cut data on genetic complexity, with values of 3 $\times$ 10^6 and 12 $\times$ 10^6 daltons, depending to some extent on the choice of the molecular weight standard (33, 34). The lower figure is the one generally accepted. The minimum size of infectious proviral DNA under conditions of single-hit infection also corresponds to the length of a 30–40S RNA (35–40). These studies thus show that the 60–70S RNA genome of oncoviruses is polyploid (41). Comparisons of the molecular weight of 60–70S and of 30–40S RNAs suggest diploidy in accord with electron microscopic observations. Such electron microscopic data, available mainly for mammalian oncoviruses, support the following general structure of the 60–70S virion RNA (9, 42–44). The two free ends of this linear complex contain poly(A) which can be visualized by tagging with relaxed circular SV40 DNA to which poly(dT) tails have been added with terminal transferase (43). A Y-shaped configuration, termed *dimer linkage structure* is seen in the center of the molecule and is believed to consist of the joined 5' ends of the two 30–40S RNA molecules. The 60–70S complex thus has an inverted polarity with a plane of two-fold symmetry going through the dimer linkage structure and 3'→5' polarity running in opposite directions in the two halves of the molecule. Dimer linkage is probably caused by base-pairing, and different viruses have dimer linkage structures of different stability. Since base-pairing cannot occur between parallel strands of nucleic acid, the two 5' ends could be held together either by a small linker RNA molecule or by regions of self-complementarity. As is discussed below, there is evidence for the latter. The failure of 30–40S monomeric RNA to reassociate into native 60–70S dimer complexes would be compatible with either possibility (58).

Besides the 60–70S genomic RNA the virion contains several classes of low molecular weight RNA (11, 45–47). These include ribosomal RNAs and 10–15 species of 4S RNA. The latter have structural and functional properties of tRNA, and some are hydrogen-bonded to the 60–70S complex (47–49). One of these serves as a primer for the initiation of DNA synthesis by reverse transcriptase (50–52, 58). The 4S primer RNA may be considered an essential component of the virion; other low molecular weight RNAs, especially the ribosomal species, are likely to represent accidental cytoplasmic contaminants incorporated into budding virus particles (53–55). The primer found in avian sarcoma and leukemia viruses is identical with cellular tRNA^{Trp}, and the primer of Moloney murine leukemia virus is tRNA^{Pro} (56–57). Primer tRNA binds tightly to the homologous 30–40S RNA. The melting temperature of this complex is higher than for that of other 4S RNA species bound to the viral genome (50, 58, 59). There is one main primer attachment site per molecule of 30–40S RNA; it is located close to the 5' end (60) extending from residues 103 to 118 (counted from the 5' terminus) in the avian sarcoma virus genome (61, 62). The tRNA is bound through a sequence beginning with the penultimate base at its 3' end and extending through the acceptor stem into loop IV (63). This mode of attachment disrupts the native conformation of tRNA. It appears that the 3' terminal adenosine of the primer is not paired with a complementary nucleotide on the 30–40S virion RNA (63). Other primer binding sites may exist as is suggested by multiple initiation of DNA synthesis in vitro, but such secondary sites are of doubtful importance for the in vivo transcription of viral RNA (64). Primer tRNA also binds strongly to the RNA-dependent DNA polymerase of the virion (65). Much of the DNA transcribed in vitro from the virion RNA is covalently linked to the 3' end of the primer (66).

Provirus

Vira DNA is first synthesized in the cytoplasm about 3 hr after infection (67–71a), at 6 hr the new provirus is moved into the nucleus, and by 9–10 hr virus-specific DNA is found at this cellular site, where it becomes integrated into high molecular weight DNA of the cell (69). The kinetics of this process suggests that free viral DNA is precursor of the integrated provirus (72). Unintegrated viral DNA occurs in the form of linear molecules and, in the nucleus, as closed circular double strands; open circular molecules have also been recovered (71a, 73, 74). The closed circular supercoiled viral DNA has a size corresponding to the 30–40S unit genome of the virus (69, 74–77). Both closed circular supercoiled and nonsupercoiled DNA are infectious in transfection experiments (37). The bulk of infectivity is found in a discrete, homogeneous fraction of linear DNA. The size of this component corresponds to one 30–40S unit genome (37–40). This infectious linear DNA consists of an intact full-length minus strand complementary to virion RNA. To this minus strand, fragments of plus strand of about 5×10^5 daltons are paired (70–74). In the case of mouse sarcoma virus the linear DNA has been studied with restriction endonucleases, and a map of genome fragments has been obtained (78). These techniques have also been applied to integrated provirus (79–80). For spleen necrosis virus, a member of the avian reticuloendotheliosis group, the data suggest that there is a preferred site of viral integration in the cellular genome. Avian sarcoma and leukosis viruses seem to have a limited number of integration sites, possibly determined by the endogenous provirus (81–82).

The 5' and 3' Termini

The majority of the DNA synthesized by the virion RNA-dependent DNA polymerase initiates at a position less than 200 nucleotides distant from the 5' terminus of the RNA genome (60, 62). One of the initial products of DNA synthesis is a small DNA species of discrete size, 101 nucleotides in avian sarcoma and 135 nucleotides in murine leukemia virus (83). During the early phase of the reaction this species constitutes a significant fraction of the total product, but later longer transcripts become prevalent (R. Junghans, 1977, personal communication). The small DNA species are initiated with the primer tRNA and constitute the sequence from the primer to the 5' end of the genome (61, 62, 83, 84, 85). These discrete, homogeneous DNA fragments from the 5' end have been isolated, purified, and in case of the avian sarcoma viruses they have been sequenced by the method of Maxam & Gilbert (86). The sequence of the 5' terminus has several interesting features that may shed light on important properties and functions of the genome. It contains a region that is complementary to the 3' end of eucaryotic 18S ribosomal RNA (87, 88) followed closely by the initiation codon AUG. These structures may represent a ribosome binding site on the viral genome and the initiation signal for the translation of *gag*, the first structural gene at the 5' end of the genome. About 20 nucleotides immediately adjacent to the 3' terminal poly(A) of the same avian sarcoma virus have also been sequenced (89). The sequence of these 3' terminal nucleotides is a repeat of the nucleotide sequence at the 5' end. This terminal redundancy of the viral genome is also demonstrable by hybridization and suggests a possible mechanism for reverse transcription (78, 90, 91). Initiation of DNA synthesis takes place at the tRNA primer and proceeds to the 5' end of the genome (92). Transcription of the remaining genomic RNA could be accomplished by circularization which would be facilitated by the terminal redundancy. In this process the RNase H activity of the reverse transcriptase may become active when the transcription complex reaches the 5' end of the template digesting the template from the 5' end to the primer attachment site. The exposed 5' complementary DNA could then hybridize to the repeat sequence adjacent to the 3' poly(A) of the same or of another genome, and transcription could proceed to the primer attachment site (91). In the course of circularization the redundant sequence may be transcribed only once. It has been suggested that the integration site in the cellular genome also contains this sequence and may be the source for the redundancy in progeny virus (91). This would provide a compelling argument for the requirement for integration in oncovirus infection. Several models have been proposed to explain the nature of the dimer linkage structure which holds the 5' ends of the two 30-40S unit genomes together in the 60-70S complex. One of these models postulates the presence of inverted repeat sequences in each strand (43). The sequence of the 5' end of avian sarcoma viruses can indeed be arranged into three hairpin loops with base-paired stems (61, 62). Such structures may then serve as bridges in dimer formation.

GENES OF RNA TUMOR VIRUSES

All RNA tumor viruses that can reproduce by themselves carry three genes coding for components of the virion: *gag* contains the information for four nonglycosylated internal structural proteins; *pol* directs the synthesis of the RNA-dependent DNA polymerase of the virion; and *env* is the gene for the envelope glycoproteins. Sarcoma viruses are often defective in one or several of these genes, but carry a specific gene, *src,* which codes for a transforming principle required for the initiation and maintenance of oncogenic transformation. In the following section we discuss these four oncoviral genes.

gag

In the avian leukosis and sarcoma viruses the four *gag* proteins are referred to as p27, p19, p15, and p12, where "p" stands for protein, and the number following it represents the first two digits of the molecular weight ($\times 10^{-3}$). In the murine leukemia and sarcoma viruses the *gag* proteins are p31, p15, p12, and p10 (93). The primary product of *gag* is a polyprotein that is cleaved proteolytically to yield the individual *gag* proteins found in the virion (94). The polyprotein precursor of the avian leukosis and sarcoma virus *gag* proteins has a molecular weight of 76,000 and is referred to as pr76. An analogous notation applies to intermediate cleavage products pr66, pr60, pr32, and pr14 which can be found in infected cells and which have been arranged in a pattern of sequential cleavages leading to p27, p19, p15, and p12 (95, 96). There is evidence that the *gag* protein p15 of avian sarcoma virus is a protease and may carry out the cleavage of the *gag* precursor (97). The prevalent *gag* precursors in murine leukemia virus infected cells are 80,000 and 65,000 daltons (98-101). In addition, large polyproteins of 150,000 to 200,000 daltons occur which carry immunological determinants of the virion DNA polymerase and *gag*-specific antigenic sites (99, 102). Such large "read through" polypeptides consisting of *gag* and of *pol* sequences are also present in avian leukosis and sarcoma virus infection (103; H. Oppermann and L. Levintow, 1976, personal communication). In murine leukemia virus-infected cells even larger virus-specific polyproteins of 250,000 to 350,000 daltons are demonstrable (102, 104). These can be precipitated with sera against *gag* proteins as well as with antipolymerase sera; they may represent translational "read throughs" of the entire viral genome. There is, however, no evidence that these very large polyproteins are the obligatory precursors for all viral proteins made in the cell; their function in infection is not known. In vitro systems of protein synthesis translate the genomic RNA of oncoviruses predominantly into *gag* proteins (24, 105). These occur in the form of precursor polypeptides of the expected size in the 60 to 80 × 10^3 dalton range (100, 105). Larger polypeptides of 180 × 10^3 daltons have also been detected, and could represent *gag-pol* "read throughs" (106). The virion RNA introduced into oocytes of *Xenopus laevis* is translated into polypeptides that belong again mainly to the *gag* family of precursors, intermediate cleavage products, and viral structural proteins (23). Parental virion RNA can also act as *gag* messenger immediately after infection and before reverse transcription in the natural host. No physiological need for this early production of *gag* proteins is apparent, but there is evidence that early virus-directed protein synthesis is essential in infection (22, 107).

gag stands for group-specific antigen. The prevalent antigenic sites of these proteins are shared among viruses of a group or species; that is, common antigenic determinants of the avian leukosis and sarcoma viruses and of the murine leukemia and sarcoma viruses respectively reside in *gag* proteins (108–114). These antigens shared within one viral species are also referred to as species-specific. In addition, *gag* proteins carry immunological determinants of both wider and narrower specificity. For instance, p31, p15, and p10 of the murine leukemia viruses contain broadly cross-reacting determinants which can also be detected in feline leukemia viruses and are referred to as interspecies antigens (115–123). Subgroup and type-specific antigenic determinants can be demonstrated on *gag* proteins by radioimmune competition assays and this narrow specificity is also reflected in the tryptic peptide analysis of these proteins. Thus p19 of avian leukosis has strong subgroup-specific determinants; to a lesser extent such determinants are also found on p15 and p12 (124–127). In the murine leukemia viruses, type- and subgroup-specific antigenic sites are found on p31, p15, and p12 (128–130).

As the *gag* proteins are derived from one precursor molecule, they should occur in equimolar amounts in the cell (131). The embryonic fibroblasts of Line 6 chickens are an apparent exception to this rule. They show endogenous expression of *gag* proteins p27 and p19, but not of p15 (132). A deletion in that part of the *gag* gene coding for p15 may explain this noncoordinate expression. If p15 is the specific protease required for the processing of the *gag* precursor, Line 6 cells should also be defective in precursor cleavage. The *gag* proteins are internal viral components. The p31 of murine leukemia viruses and p27 of the avian leukosis viruses form the icosahedral shell that surrounds a probably helical ribonucleoprotein containing p12 in the avian and p10 in the murine viruses (133–139). The avian p19 appears to be homologous to the mammalian p12; both are phosphorylated and show viral species-specific binding to virion RNA (140–144). Besides the virion structural and possibly proteolytic functions, the *gag* proteins may also play another role in the infected cells. Glycosylated *gag* protein precursor has been demonstrated to occur on the cell surface in murine leukemia virus infection (145). The effects of inserting modified internal viral proteins in this cellular site are not clear, but they can hardly be insignificant.

Temperature-sensitive mutants in *gag* are usually defective in virion assembly. *LA*3342, a temperature-sensitive mutant of avian sarcoma virus, produces only noninfectious, structurally aberrant virions at the nonpermissive temperature. The protein complement of these particles differs from that of infectious virions by the presence of new mutant proteins that appear to be intermediate cleavage products of the gag precursor (146–148). The cleavage of the pr76 gag precursor is inhibited at the nonpermissive temperature in LA3342-infected cells, but the pr76 made under these conditions can still be utilized for infectious virus synthesis after a shift to permissive conditions. One of the tryptic peptides of pr76 from LA3342 has an altered amino acid composition that may indicate the site of the mutation (148). Among the temperature-sensitive mutants of Rauscher murine leukemia virus there are two, namely ts25 and ts26, that show retarded cleavage of the gag precursor at the nonpermissive temperature (149). They differ from the avian examples in that they do not release noninfectious particles at the nonpermissive temperature but virus synthesis is resumed within a short time after downshift (150–152). There are a number of other temperature-sensitive late mutants both of Rauscher and Moloney leukemia virus; some of these may turn out to have lesions in the gag gene (152–154). Many mammalian sarcoma viruses cannot replicate without a helper virus, have a smaller RNA as compared to leukemia viruses, and in single infection transform the host cell but fail to synthesize virus structural proteins. It is probable that most of these agents have a deletion of at least a portion of the gag gene in addition to the envelope deletion that appears common to sarcoma viruses (155, 156).

pol

Oncoviruses carry in the virion an RNA-dependent DNA polymerase popularly known as *reverse transcriptase* (157, 158). There are approximately 70 molecules of reverse transcriptase per virion (159, 175). The enzyme transcribes the genomic RNA of the virus into DNA via an RNA-DNA hybrid (160, 161). This transcription occurs in the cytoplasm within the first 6 hr after infection (67, 70). Reverse transcriptase shows two different catalytic functions, DNA polymerase and RNase H activity (160, 162–164). Both DNA and RNA can serve as template for the polymerase activity (160, 165, 166). Similar to other DNA polymerases, reverse transcriptase requires a free 3' OH end which is provided by the tRNA primer attached to the genomic RNA of the virion (52, 57, 167). Synthesis of the new DNA polynucleotide chain is in the 5' to 3' direction (168). In the native enzyme RNase H activity of reverse transcriptase is a processive exonuclease that digests the ribonucleotide chains of RNA-DNA hybrids (169–173). RNase H and DNA polymerase activities reside in different catalytic sites of the enzyme molecule as is indicated by their differential susceptibility to heat inactivation and to various chemical inhibitors (174). Purified reverse transcriptase of avian myeloblastosis virus has a sedimentation coefficient of 7.5S corresponding to an approximate molecular weight of 1.7×10^5 (175–177). Electrophoresis in sodium dodecylsulfate polyacrylamide gels reveals that the holoenzyme of avian oncoviruses consists of two subunits of 9.5×10^4 daltons referred to as β and 6.5×10^4 daltons termed α (175, 178, 179). The α subunit is derived from β by proteolytic cleavage (180). Thus the viral genetic information necessary to code for reverse transcriptase is at most that needed for the synthesis of β, an RNA segment of about 0.9×10^6 daltons. Isolated α subunits of the avian myeloblastosis viral enzyme carry both polymerase and exonuclease activities, but their RNase H activity is distributive, and they show a lowered affinity for the template and fail to bind to tRNA primer (181, 182). Recently $\beta\beta$ dimers have also been obtained from virions of avian sarcoma virus B77 grown in duck cells. These $\beta\beta$ dimers are enzymatically active (182a, 182b). The reverse transcriptase isolated from murine oncoviruses is smaller (4 to 4.5S) than that from avian sources (173, 183). Only one size of polypeptide is seen in sodium dodecylsulfate polyacrylamide gels (8.4×10^4 daltons). The properties of this polypeptide resemble those of the β subunit from avian oncoviruses in its affinity for template RNA and the processive nature of the RNase H activity (184).

Both deletion and temperature-sensitive viral mutants of the *pol* gene have been isolated and indicate that the reverse transcriptase is a virus-coded enzyme. Temperature-sensitive mutants of *pol* fall into two physiological categories characterized by (*a*) temperature sensitivity of the enzyme and (*b*) temperature sensitivity of enzyme synthesis. To the first category belong the avian sarcoma virus mutants *LA*335, *LA* 336, and *LA*337 and the Rauscher murine leukemia virus mutant *ts*29. The sarcoma viruses show "coordinate" temperature sensitivity; that is, under nonpermissive conditions neither virus synthesis nor oncogenic transformation of the cell takes place (185). The murine leukemia virus *ts*29 fails to replicate or produce viral antigens under nonpermissive conditions; it also lacks leukemogenic activity (186–88). Temperature-shift experiments have shown that the temperature-sensitive phase of these mutants occurs early in infection (before 6–10 hr) and is transient. If the infection is allowed to proceed under permissive conditions for 10–12 hr, a change to nonpermissive conditions has no effect on transformation or virus synthesis. This duration of the temperature-sensitive phase coincides with the requirement for reverse transcription in oncovirus infection. Isolated virion polymerases of *LA* 335, *LA*336, and *LA*337, and of *ts*29 are more heat labile than wild-type enzyme (189–191). Both DNA synthesis and RNase H activity are affected although the latter is affected to a lesser extent. Similar to wild-type virion polymerase, the enzymes from *LA*336 and *ts*29 are partially protected against heat inactivation, if they are complexed with template. No such protection is seen with *LA*335 and *LA* 337, presumably because the binding of enzyme to template is affected by the temperature-sensitive lesion. From *LA*337 the α subunit of the polymerase has been purified and shown to contain the temperature-sensitive lesion (192). In all three avian sarcoma virus mutants intracellular viral DNA synthesis is greatly reduced at the elevated temperature (190). That the failure of these mutants to transform the host cell and to reproduce under nonpermissive conditions is indeed caused by the temperature-sensitive polymerase has been shown by genetic experiments (193). Revertants to wild-type biological properties also acquire wild-type enzyme. In recombination experiments there is complete linkage between the properties of the polymerase and the transforming and replicating ability of the virus. The second physiological category of temperature-sensitive mutants that seems affected in the *pol* gene is represented by the avian sarcoma virus mutant *LA*672 (194, 195). This virus is temperature-sensitive in the production of infectious progeny, but virus synthesized under permissive conditions can infect and transform cells at the elevated temperature. Noninfectious virus is released from cells under nonpermissive conditions; these noninfectious virions lack polymerase activity, which suggests that the defect in *LA*672 is in the synthesis of functional polymerase. The mutation in *LA*672 cannot be complemented by *LA*335 or *LA*337 and therefore is probably located in *pol*. However, if all polymerase were derived from a polyprotein produced by translational "read through" of *gag* and *pol*, a mutation in *gag* could affect processing of the polyprotein and would explain the properties of *LA*672.

Certain defective avian sarcoma viruses have a deletion that extends over all or most of the *env* gene, and in some cases also includes a portion of *pol*. The *env* deletions are discussed in the next section. The combined *env-pol* deletions occur in RSVα(-), a derivative of the Bryan high titer strain of Rous sarcoma virus, and in *NY*8α obtained from the Schmidt-Ruppin strain of Rous sarcoma virus. These mutants reproduce only with the aid of a helper virus acquiring both envelope glycoprotein and polymerase by phenotypic mixing (196, 197). Single, phenotypically mixed particles can infect and transform cells, but the progeny produced in this solitary infection is not infectious. Besides functional envelope glycoprotein, these progeny particles lack polymerase activity. The noninfectious particles are also free of material that cross-reacts immunologically with monospecific polymerase antiserum (159, 198). The size of the *env-pol* deletion in RSVα(-) and *NY*8α is not detectably larger than the *env* deletion seen in the same strains of RSV; therefore it includes probably only a small portion of *pol* (P. H. Duesberg and L.-H. Wang, 1976, private communication). Recently a new type of deletion mutant has been produced with RSV(-)α, containing *env* but still defective in *pol* (199). The main conclusions to be drawn from the genetic experiments with *pol* mutants are that (*a*) the polymerase is under genetic control of the virus genome and (*b*) that polymerase is needed early in infection for both oncogenic transformation and for virus reproduction.

env

The *env* gene codes for the envelope proteins of oncoviruses. In the avian leukosis and sarcoma viruses there are two such proteins of 85,000 and 37,000 daltons known as gp85 and gp37. They are both glycosylated (93). In murine leukemia viruses there is a major envelope glycoprotein gp69/71; a glycoprotein component in the 40,000 dalton region is not regularly seen. However, the murine leukemia viruses have a nonglycosylated envelope protein of 15,000 daltons (p15E) which is distinct from the p15 of the virion core and is probably coded for by the *env* gene (200). The glycoproteins can be visualized with the electron microscope on the surface of negatively stained virus particles where they protrude as knobs and spikes (201, 202). They are accessible to specific surface labels (203, 204). The larger glycoprotein of the avian leukosis viruses, gp85, has been shown to form the knob structure, the smaller one, gp37, the spike (205, 206). Gp37 and gp85 are linked as dimers by disulfide bonds (207). Disulfide linkage has also been found for gp69/71 and p15E of murine leukemia viruses (208, 209). Gp69/71 and p15E of the murine leukemia viruses and gp85 and gp37 of the avian leukosis viruses are derived by proteolytic cleavage and addition of fucose residues to the carbohydrate side chains of a glycosylated polyprotein precursor of 90,000 to 95,000 daltons (99, 103, 210–213). This glycosylated precursor seems to arise in turn from a 70,000 dalton nonglycosylated protein (104, 214). In the murine leukemia viruses p15E may be further cleaved to yield p12E; and in the avian leukosis and sarcoma viruses it is suspected that p10 is derived from gp37 by similar cleavage (210; D. P. Bolognesi, 1977, personal communication).

The envelope glycoproteins of oncoviruses determine type-specific properties of the virion: host range, interaction with neutralizing antibodies, and viral interference. In the avian leukosis and sarcoma viruses and the feline leukemia viruses these properties provide markers for virus classification into subgroups (215–217). The gp69/71 of murine leukemia viruses also mediates hemagglutination (218, 219). In general the *env*-determined type-specific properties of oncoviruses are genetically stable. However, a recent study on avian sarcoma virus B77 suggests that mutations in host range can occur at high frequency (220). This phenomenon deserves further study. Viral glycoprotein gp69/71 occurs not only in the cell and in the virion, but is also released into the tissue culture medium and into the plasma of animals as free glycoprotein (221–223). The major glycoproteins gp85 and gp69/71 of avian and of murine leukemia viruses have been purified and retain all of the properties which they show in situ (224, 225). They elicit the production of neutralizing antibody and absorb neutralizing antibody produced in response to injection of whole virus. They induce viral interference and mediate hemagglutination (226–228). Besides these prevalent type-specific properties the antigenic determinants of gp85 and gp69/71 include group-specific and interspecies-specific components (225, 229–231). The antigenic determinants in the p15E envelope protein of murine leukemia viruses are mainly group-specific. Although anti-p15E sera precipitate intact virus, their neutralizing activity is poor (200, 232, 233). In the avian sarcoma viruses the amount of carbohydrate in gp85 varies with virus subgroup; it is up to 50% in subgroup A and about 15% in subgroups B and C (234, 332). Since the carbohydrate side chains of the viral glycoproteins are synthesized under cellular control, one would not expect them to play a major role in determining specific viral properties. Indeed, a large proportion of carbohydrate can be enzymatically removed from gp69/71 without affecting immunological and interference-inducing capacities of the molecule; only the hemagglutinating ability is destroyed (235, 236). Monospecific sera against gp69/71, administered after infection, delay or prevent the development of leukemia in mice and can lead to a complete recovery of the animals from infection (237, 238). The mechanism of this intriguing therapeutic effect remains to be elucidated.

Two *ts* mutants of avian sarcoma viruses with a lesion in *env* have been isolated, *LA*30 and *PH*734. *LA*30 seems to have temperature-labile glycoproteins, a property that becomes evident if it is used as helper virus for the glycoprotein-defective Bryan high titer strain of Rous sarcoma virus. The temperature-sensitive phase occurs before penetration and therefore affects virus replication and cell transformation (F. Tato and J. A. Wyke, 1976, personal communication). *PH*734 appears defective in the synthesis of envelope glycoproteins. Virions produced under nonpermissive conditions contain reduced amounts of these surface components and are noninfectious. PH734 is also unable to complement the *env* deletion mutant RSV(–) at 41° (239). Major insights into the function of the envelope glycoproteins have come from deletion mutants that lack part or all of the *env* gene. The best-studied of this kind are the Bryan high titer strain of Rous sarcoma virus and the *NY*8 derivative of the Schmidt-Ruppin strain of Rous sarcoma virus (197, 240) (for review see 155). Both *NY*8 and Bryan high titer RSV can transform cells, but the progeny that is produced in single infection is noninfectious. These noninfectious particles lack functional glycoprotein. The Bryan high titer strain appears to carry some nonfunctional glycoprotein rudiment (whether this is present in the viral envelope or not is not known); the *NY*8 has no detectable glycoprotein at all (241, 242). No knobs or spikes are visible on the noninfectious viral particles (243, 244). As a result of this deficiency these viruses cannot enter cells on their own. If fused into cells with inactivated Sendai virus, they transform but then again produce only defective progeny (197, 245). The envelope defect can be complemented by wild-type virus growing in the same cell and supplying envelope glycoprotein to the deletion mutant by phenotypic mixing. The resultant infectious sarcoma viruses are called *pseudotypes* (246). The helper function of providing envelope glycoproteins can be fulfilled by exogenous oncoviruses infecting the same cell infected by the deletion mutant, or by an endogenous virus already integrated in the host genome (247–250). Endogenous helper activity for envelope deletions is correlated with the expression of the major glycoprotein in the uninfected cell (251, 252). The level of this endogenous expression does not seem to be affected by superinfection of the cell (253). The envelope deletion in *NY*8 has made possible the preparation of a specific complementary DNA representing the deletion and thus the *env* gene (254–256). Such *env*-specific DNA is obtained by transcribing the RNA of the nondefective Schmidt-Ruppin parental sarcoma virus in vitro with reverse transcriptase into single-stranded complementary DNA. This transcript is then hybridized to the RNA of the deletion mutant. The nonhybridized DNA is isolated and purified by repeated cycles of hybridization to *NY*8 RNA. This procedure yields a gene-specific probe that has been used to study genetic relatedness between different *env* genes and to follow the expression of *env*. Among chicken leukosis viruses *env* genes of subgroups A and C appear to be closely related; subgroups B, D, and E have diverged from A and C. The F and G specificities of *env* obtained from pheasants show little or no relatedness to the chicken leukosis viruses in their corresponding nucleic acid sequences. *env*-specific RNA is present in normal chicken cells that express endogenous virus helper activity for *env* deletion mutants (254). In mammalian cells transformed by avian sarcoma viruses *env*-specific RNA may be synthesized, but it does not leave the nucleus and is not translated (257).

src

The *src* gene is responsible for oncogenic transformation induced by RNA sarcoma viruses. Since the product of *src* has not yet been isolated, genetic approaches have been mainly used to characterize this portion of the viral genome. Numerous temperature-sensitive mutants in *src* have been isolated, most of them from nondefective strains of avian sarcoma virus and some from murine sarcoma virus (258–269). With very few exceptions these mutants have the same defect; they fail to induce and to maintain transformation at the nonpermissive temperature, while showing unimpaired or even increased replication under these conditions. Transformation is commonly measured by the formation of morphologically altered cell foci in tissue culture, but other manifestations of oncogenicity are also temperature-sensitive. The list of these is long and contains seemingly unrelated properties. Some of these are temperature-sensitive only in certain *src* mutants but not in others. These transformation-related properties include tumor formation in the chicken (262, 268, 270–273), transformation-related increase of cell ruffles (273), production of excess hyaluronic acid (267), transformation-related changes in glycolipid patterns (274), excretion of proteases (275, 276), increase in hexose uptake (261, 267, 277, 278), transformation-specific changes in the cytoskeleton (279–281), lowering of the cyclic AMP level in transformed cells (282–285), blockage of cellular differentiation (286–288), induction of anchorage-independent growth (colony formation in agar) (262, 268, 289), increased saturation density in culture (265, 290), the presence of "tumor specific cell surface antigens" (290), and the gain or loss of transformation-related cell surface proteins (291–296). If cell cultures transformed by *ts src* mutants are shifted from the nonpermissive to the permissive temperature, a reversion to the transformed state is observed. With most mutants retransformation is prevented by inhibitors of protein synthesis [for an exception see (267)], and at least in one case also by an inhibitor of glycosylation, suggesting that *src* codes for a transforming protein which is itself glycosylated or acts through a glycoprotein (261, 263, 296, 297). This hypothesis of a transforming protein is also supported by the observation that in some mammalian cells transformed by wild-type avian sarcoma viruses, inhibitors of protein synthesis can bring about a change from the transformed to normal cell morphology (280). However, other interpretations of the data not involving a virus-coded transforming protein are possible and have not been ruled out (155). Besides these common *ts* mutants in *src* (also referred to as Class T-1 mutants) a small number of mutants have been described that are temperature-sensitive only for the initiation but not for the maintenance of transformation and that replicate at the nonpermissive temperature (265, 266, 298). The existence of such mutants was not expected from our present knowledge of oncovirus infection: All early and transient viral gene functions should affect transformation and replication alike. It could be that these early transforming mutants are in fact also temperature-sensitive for replication, but that this property has been missed because of leakiness.

In nondefective sarcoma viruses all of which have been isolated from chicken tumors, spontaneous deletions of the *src* gene or of a portion of it occur (260, 299). The resulting transformation-defective (*td*) viruses can still replicate, but fail to transform. This observation and the fact that *ts src* mutants synthesize infectious progeny virus at the nonpermissive temperature show that the *src* gene is not required in virus replication. A similar deletion of *src* may also occur in defective sarcoma viruses but would result in a multiple deletion mutant that can neither replicate nor transform. Such a deletion would be difficult to detect. Little is known about the mechanism by which *td* deletions arise. Nondefective viruses do not segregate *td* deletions under conditions where infection of new cells is absent (300, 301). Reverse transcription may therefore be of importance for the generation of *td* viruses. The size of the deletion in independently isolated *td* viruses is quite uniform; complementation and recombination tests with *ts src* mutants indicate that it includes most if not all of the *src* gene (302, 303). Occasionally, however, smaller *td* deletions are seen, and these can recombine with some of the *ts src* mutants to give wild-type virus (304; S. Kawai, H. Hanafusa, and P. H. Duesberg, personal communication; M. M.-C. Lai, S. S. F. Hu, and P. K. Vogt, in preparation). Some avian *td* viruses can induce leukosis in fowl (305); whether this is a property of the *td* genome alone, or whether it requires the acquisition of additional genetic information from endogenous virus or from the cell is not known.

td viruses and their nondefective avian sarcoma virus progenitors have been used to obtain complementary single-stranded DNA representative of the deleted *src* gene. The procedure is analogous to the preparation of cDNA representative of the *env* gene: a single-stranded DNA transcript complementary to the RNA of the parental nondefective sarcoma virus is obtained and is hybridized to the RNA of the deletion mutant. The nonhybridized DNA is representative of the deletion and is referred to as $cDNA_{sarc}$. Complementary DNA_{sarc} prepared from the Prague strain of Rous sarcoma virus hybridizes to the RNA genome of all six avian sarcoma viruses tested (the virus strains are Prague, Schmidt-Ruppin, Bratislava 77, Carr-Zilber, Bryan, and Fujinami) but lacks homology to avian leukosis and *td* viruses, and to sarcoma viruses isolated from other vertebrate species (255, 306). Complementary DNA_{sarc} also anneals to normal cell DNA, but thermal denaturation studies indicate some mismatching in comparison to annealing with proviral DNA. The reaction with other avian cell DNAs decreases with increasing evolutionary distance from the chicken (255, 307). Complementary DNA_{sarc} sequences present in normal chicken cells are located in the microchromosomes, in contrast to the endogenous leukosis virus that resides on (a) macrochromosome(s) (308). Some homology to the $cDNA_{sarc}$ probe from avian sarcoma viruses has also been found in mammalian cells (D. Spector et al, 1977, private communication). The $cDNA_{sarc}$ can also be used to measure the amount of *src* deletions in populations of avian oncovirus genomes by hybridization (309). RNA sequences complementary to $cDNA_{sarc}$ occur in polysomal fractions of normal and chemically transformed cells of chicken and of quail. No clear-cut correlation between the amount of such RNA and the growth rate of the cells has been established (255; D. Spector, 1977, private communication). The normal cell sequences hybridizing with $cDNA_{sarc}$ appear to be cellular genetic elements. Their function in the cell is not known, but it may be regulatory. The conclusion to be drawn from these results is that *src* is derived from the cell, and that it has undergone further evolutionary change after incorporation into the viral genome. This hypothesis on the origin of *src* is also supported by studies with murine sarcoma viruses. Both Harvey and Kirsten murine sarcoma viruses have been isolated after passage of murine leukemia virus in rats. Their genomes consist of two sets of sequences, one homologous to murine leukemia virus, the other homologous to rat cell DNA (269, 310–313.) The latter sequences contain the sarcoma-specific portion of the genome. It is possible that these sarcoma-specific sequences represent, at least in part, endogenous viral information from rat cells, but this question requires further study (313–315). Moloney murine sarcoma virus, which is of mouse origin, does not contain the rat-related sequences seen in the Harvey and Kirsten strains but it too carries a specific genetic region not present in leukemia viruses (316). A sarcoma-specific probe prepared for representing this region detects similar sequences in normal mouse cells and, with lesser homology, in the cells of other mammals (317–319). Unlike the avian sarcoma viruses which contain closely related *src* genes, the sarcoma-specific sequences of the different murine sarcoma virus strains are not related and probably have different origins (310, 317, 318).

Leukemia viruses lack a *src* gene, and so far only *gag, pol,* and *env* have been identified in these viruses. A transforming ("*luk*") gene which has no effect on virus replication should be demonstrable with the same kind of conditional mutants that have been isolated in *src*. The development of in vitro assays for leukemic transformation will facilitate the search for such mutants.

RECOMBINATION

Mixed infection with different oncoviruses leads to the appearance of genetically stable progeny virus that unites markers from the two parental viruses by recombination (320, 321). The frequency of recombinants in mixed virus yields is rather high, exceeding 10% and often reaching 40%. Recombination between oncoviruses is not reassortment of genes located on independently segregating elements as was first suggested, but it represents crossing-over between two nucleic acid molecules. The experimental evidence in favor of crossing-over has been reviewed recently (29, 155). Therefore, it will suffice to point out that the existence of functionally distinct, independently segregating genetic elements in the oncovirus genome, as postulated by the reassortment hypothesis is inconsistent with the established structure of the unit genome which is a single molecule of RNA containing one copy of all viral genes. Furthermore, oligonucleotide mapping has provided direct evidence for crossovers in the genomes of recombinant oncoviruses (31, 32, 322–327).

Recombination occurs not only between oncoviruses that are experimentally introduced into the same cell, but also between exogenous virus and endogenous viruses that are already present and integrated in the host (328, 329). However, recombination with endogenous viral sequences appears to take place only if these sequences are transcribed in the cell, presumably because they need to be incorporated into heterozygote virus particles to participate in a recombination event (see below). Although recombination is of general occurrence among the avian oncoviruses tested, involves all genes, and can occur within a single gene (303, 329–332), there are some notable exceptions where recombination does not take place: Defective sarcoma viruses fail to recombine with their leukosis helper viruses to give nondefective sarcoma viral progeny (328, 333–335). An explanation for this observation can be offered by pointing out that both the sarcoma viruses (lacking *env* and possibly other genes) and the helper virus (lacking *src*) are deletion mutants. Since *env* and *src* are adjacent genes on the map of avian sarcoma viruses (and possibly of other sarcoma viruses as well) (30, 32, 322), a mating between an *env*⁻ and a *src*⁻ mutant would lead to opposing deletion loops in the *env-src* regions, preventing the formation of recombinants that have both *src* and *env* (336). However the replication-defective sarcoma viruses do not show a generalized inability to recombine. The avian sarcoma viruses $NY8\alpha$ and RSV(–)α which appear to have a small deletion in *pol* in addition to that in *env* can recombine with leukosis virus to acquire a functional polymerase (196, 197). Another example of recombination between a defective sarcoma virus and a leukosis virus is provided by the recovery of RAV-61 from pheasant cells infected by the Bryan high-titer strain of Rous sarcoma virus (337, 338). Recombination occurs also between defective and nondefective sarcoma viruses (339). Most of the work on crossover between oncovirinae has been done with avian sarcoma and leukosis viruses, but there are some reports of recombination between murine leukemia viruses (187, 340, 341). However, for the murine viruses the occurrence of genetic crossing-over, although reasonably expected by analogy, has not yet been directly demonstrated.

Several hypotheses attempting to explain the mechanism of crossing-over between oncoviruses have been put forward (41, 342). In these hypotheses heterozygotes play a key role in explaining the high frequency of crossing-over. Heterozygote particles carrying the two parental alleles of the same gene have been observed. Some evidence exists that such particles are not mere virus clumps, but this possibility has not been definitely ruled out (328, 343). The occurrence of heterozygotes is also in agreement with the diploid structure of the genome. The most recent hypothesis of recombination takes into account the current knowledge of genome structure and reverse transcription (E. Hunter, private communication, 1976). This heterozygote hypothesis of recombination is briefly summarized here. The first step after mixed infection is the formation of a heterozygote particle containing the two parental viral genomes. All recombinants are derived from such heterozygotes in a second round of infection during reverse transcription of the heterozygote genome. Recombinants have indeed not been found, if the second cycle of infection is prevented from taking place, even though the cells initially infected with the two parental viruses continue to produce progeny (344). It is thought that the intermediates of recombination in this second round of infection are circularized DNA transcripts of 30–40S unit genomes consisting of a full-length minus strand complementary to the virion RNA and shorter-length plus strands base-paired to the minus strand. Linear minus strands with short pieces of plus strands have been found in infected cells early after infection (73, 74). Because of the proximity of the parental DNA synthesis occurring in the same heterozygote particle, it is proposed that short positive strands of one parent can become aligned with the minus strand of the other parent, and that this process of "strand invasion" occurs at a relatively high frequency. This heterozygote hypothesis of recombination is also in agreement with the observation that the terminal markers of the genome, *src* and *gag*, appear to be linked, suggesting a circular intermediate in recombination (344, 345). According to this hypothesis all recombinants should show double crossover points, and many recombinants that have been mapped by oligonucleotide analysis do so. Those that show only one crossover may still contain a second one in an area in which the oligonucleotide maps of the two parental viruses are indistinguishable (32, 322–324). There are, however, some weak points in this hypothesis of recombination, which require ad hoc explanations. Electron microscopic and sedimentation analyses have failed to demonstrate the occurrence of heterodimer RNA in preparations of murine sarcoma and leukosis virus (44, 346). Such heterodimers consisting of the smaller sarcoma viral genome and the larger leukosis virus genome would be expected, if murine sarcoma and leukosis viruses recombined within the homologous portions of their genomes. In the absence of heterodimers one must assume that either there is no recombination between murine sarcoma and leukemia viruses, or that dimer linkage between sarcoma and leukemia viruses is less stable in the analytical methods used than is linkage between leukemia-leukemia and sarcoma-sarcoma genomes. Unfortunately, virion RNA dimers of avian oncoviruses have not yet been seen with the electron microscope, probably because they dissociate during the preparation procedures. A second problem for postulating heterozygotes as precursors of recombinants comes from the behavior of replication-defective avian sarcoma viruses. Since these can recombine with the helper virus in *pol* and possibly in *gag*, the common population of sarcoma virus pseudotypes which is derived from cells doubly infected by sarcoma and helper virus should contain a high frequency of heterozygotes consisting of sarcoma and leukosis virus unit genomes. Infection of new cells with such heterozygote particles should lead to complementation of the sarcoma virus replication defect, but this does not happen, indicating that either heterozygotes are not being formed, or only one of the two genomes carried by a virus particle becomes integrated in the cell. A third potential difficulty for the heterozygote hypothesis of recombination is created by the recent work on marker rescue (347). Restriction fragments and mechanically generated fragments of viral DNA can enter *ts* mutant-infected cells and recombine with the *ts* virus to yield wild type. According to the heterozygote hypothesis of recombination, the DNA fragments would have to integrate, be transcribed, and form part of a heterozygote as a prerequisite of recombination. This process seems intuitively less likely than direct crossover between DNA fragments and provirus. Unfortunately the efficiency of infection with DNA fragments is too low to encourage a search for heterozygote particles containing fragment size RNA.

GENE MAPPING

The gene order of avian sarcoma viruses is probably 5'-*gag-pol-env-src*-poly(A)-3'. Several lines of investigation support this genetic map. Although the data from any one alone are not convincing, taken together they form a strong case.

Three factor crosses involving nonconditional *src* and *env* markers plus a temperature-sensitive mutant have yielded the following qualitative relationships of linkage and nonlinkage: *env* is linked to *pol* and *src* to *gag*. *src* is not linked to *pol*, and *gag* is not linked to *env* (193, 195, 325, 344, 345). These results are compatible with a *gag-pol-env-src* sequence, provided the genome circularizes during recombination which would bring *src* in proximity to *gag*. This assumption seems reasonable in view of the physical findings of circular viral DNA in infected cells (37, 74). The results from three factor crosses could also be reconciled with a sequence reading *gag-src-pol-env*, but this alternative is ruled out by biochemical and electron microscopic data.

The most extensive information on the genetic map of avian leukosis and sarcoma viruses comes from oligonucleotide analysis of virion RNA (31, 32, 322–327). In this approach ^{32}P–labeled virion RNA fragments containing the poly(A) of the 3' end are isolated and then fractionated into several size classes from small pieces including only the poly(A) plus a short heteropolymeric sequence to long RNAs representing the whole genome. These fractions are digested exhaustively with RNase T$_1$, and oligonucleotides large enough to be unique in an RNA of 10×10^3 nucleotides are identified after two-dimensional separation. The complexity of such fingerprints is a function of fragment size indicating that the oligonucleotides occur at a fixed order in the linear genome and that there is no circular permutation. The position of the unique oligonucleotides with respect to the 3' end of the genome can therefore be determined by the minimal size of the RNA fragment in which a given oligonucleotide occurs. The smaller the size, the closer the oligonucleotide is to the 3' end. Twenty to thirty unique oligonucleotides can thus be arranged in linear order from the 3' to the 5' end of the RNA.

Biological functions can be assigned to this sequence of oligonucleotides by deletion and recombination mapping. Deletions of the *src* gene result in a loss of two to three oligonucleotides close to the 3' end, locating *src* at this position. However, all *src* deletion mutants retain a short heteropolymeric region adjacent to the terminal poly(A) tract, and the deletions represent therefore interstitial, not terminal, excisions of genetic material. These retained heteropolymeric sequences proximal to poly(A) include highly conserved elements in all avian RNA tumor viruses and are referred to as "C" (= constant) region (30, 348). A deletion of *env* is correlated with a loss of oligonucleotides close to the 5' end of *src,* suggesting that *env* and *src* are adjacent genes (322, 349).

Deletion mapping comparing several defective murine leukemia and sarcoma viruses suggests similarity to the avian oncovirus map (350). Recombination mapping is based on the fact that the oligonucleotide maps of parental viruses used in a cross are distinguishable, because they contain oligonucleotides that are characteristic for a specific viral type besides those oligonucleotides that are the same in different parents and represent conserved genetic sequences. Thus it is possible to determine which part of the map in a recombinant comes from one parent and which from the other. In order to locate *pol* and *gag,* three factor crosses have been carried out in which the *src* of one and the *env* of the other parent have been selected for, the third marker being a *ts* mutant in either *gag* or *pol.* All recombinants that have received the *ts* allele should have the corresponding segment of the oligonucleotide map derived from the *ts* parent, and all wild-type recombinants should have that genome segment derived from the wild-type parent. Such recombinant maps suggest that the *pol* gene is located in a highly conserved area close to *env*, and the *gag* gene occupies the 5' end of the genome (323–327).

The position of *gag* at the 5' end is also supported by translation of virion RNA in vitro which in the avian as well as in the murine RNA tumor viruses results mainly in the synthesis of *gag* precursor (24, 100, 105, 106). Since translation begins at the 5' end of the messenger RNA, and there is no evidence for internal initiation, *gag* is likely to be located at the 5' end.

As the *gag* proteins are derived from a polyprotein precursor, it is possible to determine their position in the precursor by pactamycin mapping. For the avian leukosis viruses this sequence of *gag* proteins reads p19 (p27–p12) p15 (96). In murine leukemia and sarcoma viruses a sequence of proteins within the *gag* gene has been suggested by identifying immunological determinants on *gag* precursors and intermediate cleavage products in murine leukemia virus infection, and by analyzing *gag* precursors from various cell lines infected by replication-defective sarcoma viruses. The latter have partial deletions, probably from the 3' end of the *gag* region. The sequence of *gag* proteins in the murine leukemia-sarcoma viruses reads p15-p12-p30-p10 (156). If homology between avian-murine proteins is p19-p12, p27-p30, p12-p10, p15-p15, the murine sequence would be a circular permutation of the avian *gag* sequence.

Recently, fragments of DNA from wild-type avian sarcoma virus prepared by Eco RI restriction endonuclease have been used successfully to rescue *ts* mutants by recombination (347). These studies have identified a fragment containing the *env* and the *pol* genes and suggest, in accord with the data discussed above, that these two genes are located adjacent to each other.

Results with heteroduplex mapping of virion RNA and complementary DNA are in general agreement with other mapping data. In the past, reverse transcription of the RNA tumor virus genome has yielded only short DNA segments (162). These short transcripts could still be used for heteroduplex mapping if they were labeled with visual tags, for instance short poly(dT) tails, to facilitate recognition of hybrid regions (351). Recently, improved conditions of transcription have resulted in the synthesis of long complementary DNA (85, 352, 353). Further orientation of electron microscope mounts has been achieved by marking the 3' poly(A) of the virion RNA with relaxed circular SV40 DNA to which poly(dT) tails have been added with terminal transferase (43). Heteroduplex analysis has defined the size and the location of the *src* deletion in transformation-defective derivatives of a Prague strain of Rous sarcoma virus (91). The results are in accord with the data obtained by oligonucleotide analysis (30, 32, 322) and gel electrophoresis (302), and locate *src* close to the 3' end of the genome. However, transformation-defective viruses from Prague strain Rous sarcoma virus subgroup C have a deletion that is slightly different in size from transformation defectives of subgroup B. A heteroduplex study comparing the RNA sequence homology of Moloney murine leukemia virus and the defective Moloney sarcoma virus indicates that several recombination or excision events at different sites have occurred in the sarcoma virus genome. Several regions of the leukemia virus genome are absent from sarcoma virus (354). The sarcoma virus in turn contains one sequence that is not shared with leukemia virus. This segment, which may contain the gene for transformation, lies near the 3' poly(A) end, a location comparable to that of the *src* gene on the avian sarcoma virus map.

SYNTHESIS OF VIRAL RNA

A comprehensive review of this area has been prepared recently by Fan (355). In the present paper we discuss only the main outlines of transcription and translation as they pertain to viral gene expression and regulation. Initiation of viral RNA synthesis depends on mitosis of the host cell (356, 357). In cells actively producing RNA tumor virus, 0.3 to 1% of the total RNA is virus-specific (253, 358–361). This RNA can be divided into messenger RNA and progeny virion RNA. The two appear to belong to different metabolic pools, because in the presence of actinomycin D, virion RNA is rapidly depleted from the cell, but messenger RNA is not diverted into virus; rather it continues to function in protein synthesis resulting in production of noninfectious particles defective in RNA (362). The overwhelming majority of intracellular viral RNA is of the same polarity as virion RNA, but, recently, small amounts of complementary RNA have been detected (363). The functional significance of this minus strand RNA is not known (364). Viral RNA is synthesized in the nucleus, probably by cellular RNA polymerase II, because the process is sensitive to inhibition by α-amanitin (365–367). The size of intracellular viral RNA is roughly 30–35S, the length of the whole genome, plus smaller species which sediment between 20–30S (358, 368, 369). Larger RNA (40–60S) has been detected but does not seem to be a precursor of the 35S molecules (355, 370). Viral messenger RNA defined by its EDTA-sensitive association with polysomes has been resolved into several size classes (360, 369, 371, 372). At least three classes 35S, 28S, and 21S of vital messenger RNA occur in chick embryo fibroblasts infected by avian sarcoma virus and have been characterized with respect to their gene content (S. Weiss, H. E. Varmus, and J. M. Bishop, 1977; W. Hayward and H. Hanafusa, 1977, personal communications). All three size classes contain the 3' end of the genome with its poly(A) and the highly conserved heteropolymeric C sequence. They extend for discrete distances toward the 5' terminus. Using complementary DNA probes specific for certain parts of the genome it was found that the 35S messenger RNA encompasses the sequences of the entire genome, including the *gag, pol, env,* and *src* genes. In 28S messenger RNA the 3' half of the genome is represented, i.e. *env* and *src,* besides the C region. In cells infected with nondefective avian sarcoma virus the 21S messenger RNA shows the same gene representation as 28S, but may actually consist of two different species, one carrying only *src* the other only *env*. The latter one could be derived from transformation-defective deletion mutants which are present in stocks of avian sarcoma viruses. Cells infected with the Bryan high titer strain of Rous sarcoma virus which is an *env* deletion mutant show only *src* sequences present in the 21S messenger RNA. In cells infected with transformation-defective viruses (*src* deletions), the 21S RNA contains only *env* genes. These observations are in accord with a model for viral gene expression which states that different genes are translated from different size messenger RNAs (373). The two species of 21S RNA would be translated into *src* and into *env* products. *Env* proteins would also be synthesized from 28S RNA, and the 35S RNA could direct synthesis of *gag* protein plus the occasional *gag-pol* "read-through" seen in in vitro translation systems and in infected cells. Thus in each class of messenger RNA the gene located at the 5' end would be translated; in the 35S species, translation would occasionally include the penultimate gene *pol* as well. This hypothesis is supported by the observation that all polysomes containing 35S messenger RNA can be precipitated with antibodies against *gag* proteins (373, 374). Furthermore, in vitro protein synthesis directed by 24–30S RNA yields non-*gag* proteins, and smaller poly(A)-containing RNA pieces from sarcoma virions are translated into what appear to be nonstructural viral polypeptides, possibly coded for by *src* (T. Pawson and A. E. Smith, 1977; T. Pawson and P. H. Duesberg, 1977; T. Hunter and K. Beemon, 1977, personal communications).

CONCLUSION

Oncoviruses contain only four genes, three of which are essential for virus replication and a fourth of which is required exclusively in sarcomagenic transformation. Viral genes for carcinoma or leukemic transformation have not been identified. A virus of such genetic simplicity is likely to depend on numerous host functions and genes, not only for its reproduction but also for the oncogenic transformation of the cell. Practically nothing is known about this participation of the cell in the virus life cycle, and pioneering work remains to be done in this area.

The field of RNA tumor virus genetics has made major advances in the past five to seven years. It is not inconceivable that entire RNA tumor virus genomes or at least large portions thereof will be sequenced within the next few years. Our understanding of dimer genome linkage, of RNA-DNA transcription, possibly of integration, and viral gene regulation will be immensely enhanced by such data. One might also predict that the function of the known viral genes will become better characterized through additional mutants, and that unrecognized viral genes, if any, will be identified. Continuation of genetic analysis in conjunction with biochemical techniques will provide insights into the mechanism of recombination between RNA tumor viruses. But will these advances bring about an understanding of oncogenic transformation? Not necessarily. First steps toward the solution of this central problem of tumor virology would be the isolation of a transforming protein and, more important, the design of a biological assay for it. Ultimately oncogenic transformation is a special problem of eucaryotic gene regulation. Substantial progress in this area now seems within reach through the new techniques of DNA cloning.

Acknowledgments

Work of the authors is supported by the National Cancer Institute, US Public Health Service, and the Leukemia Society of America.

Literature Cited

1. Fenner, F. 1976. The classification and nomenclature of viruses. Summary of results of meetings of the International Committee on Taxonomy of Viruses in Madrid, September 1975. *Virology* 71:371–78
2. Vogt, P. K. 1976. The oncovirinae—a definition of the group. In *WHO Center for Collection and Evaluation of Data on Comparative Virology*, ed. P. Thein, Rep. No. 1, pp. 327–39. Munich: WHO
3. Fenner, F. 1976. Classification and nomenclature of viruses. Second Report of the International Committee on Taxonomy of Viruses. *Intervirology* 7:8–102
4. Robinson, W. S., Pitkanen, A., Rubin, H. 1965. The nucleic acid of the Bryan strain of Rous sarcoma virus: Purification of the virus and isolation of the nucleic acid. *Proc. Natl. Acad. Sci. USA* 54:137–44
5. Bellamy, A. R., Gillies, S. C., Harvey, J. D. 1974. Molecular weight of two oncornavirus genomes: Derivation from particle molecular weights and RNA content. *J. Virol.* 13:88–93
6. Riggins, C. H., Bondurant, M., Mitchell, W. M. 1975. Physical properties of Moloney murine leukemia virus high-molecular-weight RNA: A two subunit structure. *J. Virol.* 16:1528–35
7. Mangel, W. F., Delius, H., Duesberg, P. H. 1974. Structure and molecular weight of the 60–70S RNA and the 30-40S RNA of the Rous sarcoma virus. *Proc. Natl. Acad. Sci. USA* 71:4541–45
8. Delius, H., Duesberg, P. H., Mangel, W. F. 1975. Electron microscope measurements of Rous sarcoma virus RNA. *Cold Spring Harbor Symp. Quant. Biol.* 39:835–43
9. Kung, H.-J., Bailey, J. M., Davidson, N., Nicolson, M. O., McAllister, R. M. 1975. Structure, subunit composition, and molecular weight of RD-114 RNA. *J. Virol.* 16:397–411
10. King, A. M. Q. 1976. High molecular weight RNAs from Rous sarcoma virus and Moloney murine leukemia virus contain two subunits. *J. Biol. Chem.* 251:141–49
11. Duesberg, P. H. 1968. Physical properties of Rous sarcoma virus RNA. *Proc. Natl. Acad. Sci. USA* 60:1511–18
12. Duesberg, P. H., Vogt, P. K. 1973. RNA species obtained from clonal lines of avian sarcoma and from avian leukosis virus. *Virology* 54:207–19
13. Maisel, J., Klement, V., Lai, M. M.-C., Ostertag, W., Duesberg, P. 1973. Ribonucleic acid components of murine sarcoma and leukemia viruses. *Proc. Natl. Acad. Sci. USA* 70:3536–40
14. Lai, M. M.-C., Duesberg, P. H. 1972. Adenylic acid-rich sequence in RNAs of Rous sarcoma virus and Rauscher mouse leukemia virus. *Nature* 235:383–86
14a. Stephenson, M. L., Scott, J. F., Zamecnik, P. C. 1973. Evidence that the polyadenylic acid segment of 35S RNA of avian myeloblastosis virus is located at the 3' OH terminus. *Biochem. Biophys. Res. Commun.* 55:8–15
15. Wang, L.-H., Duesberg, P. 1974. Properties and location of poly (A) in Rous sarcoma virus RNA. *J. Virol.* 14:1515–29
16. Quade, K. E., Smith, R. E., Nichols, J. L. 1974. Evidence for common nucleotide sequences in the RNA subunits comprising Rous sarcoma virus 70S RNA. *Virology* 61:287–91
17. King, A. M. Q., Wells, R. D. 1976. All intact subunit RNAs from Rous sarcoma virus contain poly(A). *J. Biol. Chem.* 251:150–52
18. Rho, H. M., Green, M. 1974. The homopoly adenylate and adjacent nucleotides at the 3'-terminus of 30–40S RNA subunits in the genome of murine sarcoma-leukemia virus. *Proc. Natl. Acad. Sci. USA* 71:2386–90
19. Rose, J. K., Haseltine, W. A., Baltimore, D. 1976. 5'-Terminus of Moloney murine leukemia virus 35S RNA is m^7G^5GmpCp. *J. Virol.* 20:324–41
20. Furuichi, Y., Shatkin, A. J., Stavnezer, E., Bishop, J. M. 1975. Blocked, methylated 5'-terminal sequence in avian sarcoma virus RNA. *Nature* 257:618–20
21. Keith, J., Fraenkel-Conrat, H. 1975. Identification of the 5' end of Rous sarcoma virus RNA. *Proc. Natl. Acad. Sci. USA* 72:3347–50
22. Gallis, B. M., Eisenman, R. N., Diggelmann, H. 1976. Synthesis of the precursor to avian RNA tumor virus internal structure proteins after early infection. *Virology* 74:302–13
23. Salden, M., Asselbergs, F., Bloemendal, H. 1976. Translation of oncogenic virus RNA in *Xenopus laevis* oocytes. *Nature* 259:696–99
24. Von der Helm, K., Duesberg, P. H. 1975. Translation of Rous sarcoma

virus RNA in a cell-free system from Ascites Krebs II cells. *Proc. Natl. Acad. Sci. USA* 72:614–18
25. Billeter, M. A., Parsons, J. T., Coffin, J. M. 1974. The nucleotide sequence complexity of avian tumor virus RNA *Proc. Natl. Acad. Sci. USA* 71:3560–64
26. Beemon, K., Duesberg, P. H., Vogt, P. K. 1974. Evidence for crossing-over between avian tumor viruses based on analysis of viral RNAs. *Proc. Natl. Acad. Sci. USA* 71:4254–58
27. Beemon, K. L., Faras, A. J., Haase, A. T., Duesberg, P. H., Maisel, J. E. 1976. Genomic complexities of murine leukemia and sarcoma, reticuloendotheliosis, and visna viruses. *J. Virol.* 17:525–37
28. Weissmann, C., Parsons, J. T., Coffin, J. W., Rymo, L., Billeter, M. A., Hofstetter, H. 1975. Studies on the structure and synthesis of Rous sarcoma virus RNA. *Cold Spring Harbor Symp. Quant. Biol.* 39:1043–56
29. Duesberg, P., Vogt, P. K., Beemon, K., Lai, M. M.-C. 1975. Avian RNA tumor viruses: Mechanism of recombination and complexity of the genome. *Cold Spring Harbor Symp. Quant. Biol.* 39:847–57
30. Wang, L.-H., Duesberg, P. H., Beemon, K., Vogt, P. K. 1975. Mapping RNase T_1-resistant oligonucleotides are near the poly (A) end and oligonucleotides common to sarcoma and transformation-defective viruses are at the poly (A) end. *J. Virol.* 16:1051–70
31. Coffin, J. M., Billeter, M. A. 1976. A physical map of the Rous sarcoma virus genome. *J. Mol. Biol.* 100:293–318
32. Joho, R. H., Billeter, M. A., Weissmann, C. 1975. Mapping of biological functions on RNA of avian tumor viruses: Location of regions required for transformation and determination of host range. *Proc. Natl. Acad. Sci. USA* 72:4772–76
33. Baluda, M. A., Shoyab, M., Markham, P. D., Evans, R. M., Drohan, W. N. 1975. Base sequence complexity of 35 S avian myeloblastosis virus RNA determined by molecular hybridization kinetics. *Cold Spring Harbor Symp. Quant. Biol.* 39:869–74
34. Taylor, J. M., Varmus, H. E., Faras, A. J., Levinson, W. E., Bishop, J. M. 1974. Evidence for nonrepetitive subunits in the genome of Rous sarcoma virus. *J. Mol. Biol.* 84:217–21
35. Montagnier, L., Vigier, P. 1972. Un intermediaire ADN infectieux et transformant du virus du sarcome de Rous dans les cellules de poule transformées par ce virus. *C. R. Acad. Sci. Ser. D* 274:1977–80
36. Cooper, G. M., Temin, H. M. 1974. Infectious Rous sarcoma virus and reticuloendotheliosis virus DNAs. *J. Virol.* 14:1132–41
37. Smotkin, D., Gianni, A. M., Rosenblatt, S., Weinberg, R. A. 1975. Infectious viral DNA of murine leukemia virus. *Proc. Natl. Acad. Sci. USA* 72:4910–13
38. Hill, M., Hillova, J. 1974. RNA and DNA forms of the genetic material of C-type viruses and the integrated state of the DNA form in the cellular chromosome. *Biochim. Biophys. Acta* 355:7–48
39. Levy, J. A., Kazan, P. A., Varmus, H. E. 1974. The importance of DNA size for successful transfection of chicken embryo fibroblasts. *Virology* 61:297–302
40. Smotkin, D., Yoshimura, F. K., Weinberg, R. A. 1976. Infectious linear, unintegrated DNA of Moloney murine leukemia virus. *J. Virol.* 20:621–26
41. Vogt, P. K. 1973. The genome of avian RNA tumor viruses: A discussion of four models. In *Possible Episomes in Eukaryotes,* ed. L. Silvestri, pp. 35–41. Amsterdam: North-Holland
42. Kung, H.-J., Hu, S., Bender, W., Bailey, J. M., Davidson, N., Nicolson, M. O., McAllister, R. M. 1976. RD-114, baboon, and woolly monkey viral RNA's compared in size and structure. *Cell* 7:609–20
43. Bender, W., Davidson, N. 1976. Mapping of poly(A) sequence in the electron microscope reveals unusual structure of type C oncornavirus RNA molecules. *Cell* 7:595–607
44. Dube, S., Kung, H.-J., Bender, W., Davidson, N., Ostertag, W. 1976. Size, subunit composition, and secondary structure of the Friend virus genome. *J. Virol.* 20:264–72
45. Erikson, R. L. 1969. Studies on the RNA from avian myeloblastosis virus. *Virology* 37:124–31
46. Montagnier, L., Goldé, A., Vigier, P. 1969. A possible subunit structure of Rous sarcoma virus RNA. *J. Gen. Virol.* 4:449–52
47. Faras, A. J., Garapin, A. C., Levinson, W. E., Bishop, J. M., Goodman, H. M. 1973. Characterization of the low-molecular-weight RNAs associated

with the 70S RNA of Rous sarcoma virus. *J. Virol.* 12:334–42
48. Erikson, E., Erikson, R. L. 1971. Association of 4S ribonucleic acid with oncornavirus ribonucleic acids. *J. Virol.* 8:254–56
49. Rosenthal, L. J., Zamecnik, P. C. 1973. Minor base composition of "70S-associated" 4S RNA from avian myeloblastosis virus. *Proc. Natl. Acad. Sci. USA* 70:865–69
50. Faras, A. J., Taylor, J. M., Levinson, W. E., Goodman, H. M., Bishop, J. M. 1973. RNA-directed DNA polymerase of Rous sarcoma virus: Initiation of synthesis with 70S viral RNA as template. *J. Mol. Biol.* 79:163–83
51. Dahlberg, J. E., Sawyer, R. C., Taylor, J. M., Faras, A. J., Levinson, W. E., Goodman, H. M., Bishop, J. M. 1974. Transcription of DNA from the 70S RNA of Rous sarcoma virus. I. Identification of a specific 4S RNA which serves as primer. *J. Virol.* 13:1126–33
52. Faras, A. J., Dahlberg, J. E., Sawyer, R. C., Harada, F., Taylor, J. M., Levinson, W. E., Bishop, J. M., Goodman, H. M. 1974. Transcription of DNA from the 70S RNA of Rous sarcoma virus. II. Structure of a 4S RNA primer. *J. Virol.* 13:1134–42
53. Bishop, J. M., Levinson, W. E., Quintrell, N., Sullivan, D., Fanshier, L., Jackson, J. 1970. The low molecular weight RNAs of Rous sarcoma virus. I. The 4S RNA. *Virology* 42:182–95
54. Bonar, R. A., Sverak, L., Bolognesi, D. P., Langlois, A. J., Beard, D., Beard, J. W. 1967. Ribonucleic acid components of BAI strain A (myeloblastosis) avian tumor virus. *Cancer Res.* 27:1138–57
55. Obara, T., Bolognesi, D. P., Bauer, H. 1971. Ribosomal RNA in avian leukosis virus particles. *Int. J. Cancer* 7:535–46
56. Harada, F., Sawyer, R. C., Dahlberg, J. E. 1975. A primer ribonucleic acid for initiation of in vitro Rous sarcoma virus deoxyribonucleic acid synthesis. *J. Biol. Chem.* 250:3487–97
57. Dahlberg, J. E., Harada, F., Sawyer, R. C. 1975. Structure and properties of an RNA primer for initiation of Rous sarcoma virus DNA synthesis in vitro. *Cold Spring Harbor Symp. Quant. Biol.* 39:925–32
58. Canaani, E., Duesberg, P. 1972. Role of subunits of 60 to 70S avian tumor virus ribonucleic acid in its template activity for the viral deoxyribonucleic acid polymerase. *J. Virol.* 10:23–31

59. Waters, L. C., Mullin, B. C., Bailiff, E. G., Popp, R. A. 1975. tRNA's associated with the 70S RNA of avian myeloblastosis virus. *J. Virol.* 16:1608–14
60. Taylor, J. M., Illmensee, R. 1975. Site on the RNA of an avian sarcoma virus at which primer is bound. *J. Virol.* 16:553–58
61. Haseltine, W. A., Maxam, A. M., Gilbert, W. 1977. Rous sarcoma virus genome is terminally redundant: The 5' sequence. *Proc. Natl. Acad. Sci. USA* 74:989–93
62. Shine, J., Czernilofsky, P., Friedrich, R., Goodman, H. M., Bishop, J. M. 1977. Nucleotide sequence at the 5' terminus of the avian sarcoma virus genome. *Proc. Natl. Acad. Sci. USA* 74:1473–77
63. Cordell, B., Stavnezer, E., Friedrich, R., Bishop, J. M., Goodman, H. M. 1976. Nucleotide sequence that binds primer for DNA synthesis to the avian sarcoma virus genome. *J. Virol.* 19:548–58
64. Cashion, L. M., Joho, R. H., Planitz, M. A., Billeter, M. A., Weissmann, C. 1976. Initiation sites of Rous sarcoma virus RNA-directed DNA synthesis in vitro. *Nature* 262:186–90
65. Panet, A., Haseltine, W. A., Baltimore, D., Peters, G., Harada, F., Dahlberg, J. E. 1975. Specific binding of tryptophan transfer RNA to avian myeloblastosis virus RNA-dependent DNA polymerase (reverse transcriptase). *Proc. Natl. Acad. Sci. USA* 72:2535–39
66. Taylor, J. M., Garfin, D. E., Levinson, W. E., Bishop, J. M., Goodman, H. M. 1974. Tumor virus ribonucleic acid directed deoxyribonucleic acid synthesis: Nucleotide sequence at the 5' terminus of nascent deoxyribonucleic acid. *Biochemistry* 13:3159–63
67. Varmus, H. E., Vogt, P. K., Bishop, J. M. 1973. Integration of deoxyribonucleic acid specific for Rous sarcoma virus after infection of permissive and nonpermissive hosts. *Proc. Natl. Acad. Sci. USA* 70:3067–71
68. Lovinger, G. G., Ling, H. P., Klein, R. A., Gilden, R. V., Hatanaka, M. 1974. Unintegrated murine leukemia viral DNA in newly infected cells. *Virology* 62:280–83
69. Varmus, H. E., Guntaka, R. V., Fan, W. J. W., Heasley, S., Bishop, J. M. 1974. Synthesis of viral DNA in the cytoplasm of duck embryo fibroblasts in enucleated cells after infection by avian

sarcoma virus. *Proc. Natl. Acad. Sci. USA* 71:3874–78
70. Lovinger, G. G., Klein, R. A., Ling, H. P., Gilden, R. V., Hatanaka, M. 1975. Kinetics of murine type C virus-specific DNA synthesis in newly infected cells. *J. Virol.* 16:824–31
71. Varmus, H. E., Shank, P. R. 1976. Unintegrated viral DNA is synthesized in the cytoplasm of avian sarcoma virus-transformed duck cells by viral DNA polymerase. *J. Virol.* 18:567–73
71a. Fritsch, E., Temin, H. M. 1977. Formation and structure of infectious DNA of spleen necrosis virus. *J. Virol.* 21:119–30
72. Ali, M., Baluda, M. A. 1974. Synthesis of avian oncornavirus DNA in infected chicken cells. *J. Virol.* 13:1005–13
73. Gianni, A. M., Weinberg, R. A. 1975. Partially single-stranded form of free Moloney viral DNA. *Nature* 255:646–48
74. Varmus, H. E., Guntaka, R. V., Deng, C. T., Bishop, J. M. 1975. DNA in permissive and nonpermissive cells. *Cold Spring Harbor Quant. Biol.* 39:987–96
75. Gianni, A. M., Smotkin, D., Weinberg, R. A. 1975. Murine leukemia virus: Detection of unintegrated double-stranded DNA forms of the provirus. *Proc. Natl. Acad. Sci. USA* 72:447–51
76. Gianni, A. M., Hutton, J. R., Smotkin, D., Weinberg, R. A. 1976. Proviral DNA of Moloney leukemia virus, purification and visualization. *Science* 191:569–71
77. Guntaka, R. V., Richards, O. C., Shank, P. R., Kung, H.-J., Davidson, N., Fritsch, E., Bishop, J. M., Varmus, H. E. 1976. Covalently closed circular DNA of avian sarcoma virus: Purification from nuclei of infected quail tumor cells and measurement by electron microscopy and gel electrophoresis. *J. Mol. Biol.* 106:337–58
78. Canaani, E., Dina, D., Duesberg, P. H. 1977. Cleavage map of linear mouse sarcoma virus DNA. *Proc. Natl. Acad. Sci. USA* 74:29–33
79. Kopecka, H., Hillova, J., Hill, M. 1976. Effect of restriction endonucleases on infectivity of Rous sarcoma virus DNA. *Nature* 262:72–74
80. Battula, N., Temin, H. M. 1977. Infectious DNA of spleen necrosis virus is integrated at a single site in the DNA of chronically infected chicken fibroblasts. *Proc. Natl. Acad. Sci. USA* 74:281–85
81. Khoury, A. T., Hanafusa, H. 1976. Synthesis and integration of viral DNA in chicken cells in different times after infection with various multiplicities of avian oncornavirus. *J. Virol.* 18:383–400
82. Shoyab, M., Dastoor, M. N., Baluda, M. A. 1976. Evidence for tandem integration of avian myeloblastosis virus DNA with endogenous provirus in leukemic chicken cells. *Proc. Natl. Acad. Sci. USA* 73:1749–53
83. Haseltine, W. A., Kleid, D. G., Panet, A., Rothenberg, E., Baltimore, D. 1976. Ordered transcription of RNA virus genomes. *J. Mol. Biol.* 106:109–31
84. Haseltine, W. A., Baltimore, D. 1976. In vitro replication of RNA tumor viruses. In *Animal Virology*, ed. D. Baltimore, A. H. Huang, C. F. Fox, 4:175–213. New York: Academic
85. Rothenberg, E., Baltimore, D. 1977. Increased length of DNA made by virions of murine leukemia virus at limiting magnesium ion concentration. *J. Virol.* 21:168–78
86. Maxam, A. M., Gilbert, W. 1977. A new method for sequencing DNA. *Proc. Natl. Acad. Sci. USA* 74:560–64
87. Shine, J., Dalgarno, L. 1974. Identical 3'-terminal octanucleotide sequence in 18S ribosomal ribonucleic acid from different eukaryotes. *Biochem. J.* 141:609–15
88. Eladari, M. E., Galibert, F. 1976. Sequence determination of the 3' terminal T1 oligonucleotide of 18S ribosomal RNA. *Nucleic Acid Res.* 3:2749–55
89. Schwartz, D. E., Zamecnik, P. C., Weith, H. L. 1977. Rous sarcoma virus genome is terminally redundant: The 3' sequence. *Proc. Natl. Acad. Sci. USA* 74:994–98
90. Collett, M. S., Dierks, P., Cahill, J. F., Faras, A. J., Parsons, J. T. 1977. Terminally repeated sequences in the avian sarcoma virus RNA genome. *Proc. Natl. Acad. USA.* In press
91. Junghans, R. P., Hu, S., Knight, C. A., Davidson, N. 1977. Heteroduplex analysis of avian RNA tumor viruses. *Proc. Natl. Acad. Sci. USA* 74:477–81
92. Taylor, J. M., Illmensee, R., Trusal, L., Summers, J. 1976. Transcription of avian sarcoma virus RNA. See Ref. 84, pp. 161–73
93. August, J. T., Bolognesi, D. P., Fleissner, E., Gilden, R. V., Nowinski, R. C. 1974. A proposed nomenclature for the virion proteins of oncogenic RNA viruses. *Virology* 60:595–601
94. Vogt, V. M., Eisenman, R. 1973. Identification of a large polypeptide precursor

to avian oncornavirus proteins. *Proc. Natl. Acad. Sci. USA* 70:1734–38
95. Eisenman, R., Vogt, V. M., Digglemann, H. 1975. Synthesis of avian RNA tumor virus structural proteins. *Cold Spring Harbor Symp. Quant. Biol.* 39:1067–75
96. Vogt, V. M., Eisenman, R., Diggelmann, H. 1975. Generation of avian myeloblastosis virus structural proteins by proteolytic cleavage of a precursor polypeptide. *J. Mol. Biol.* 96:471–94
97. Von der Helm, K. 1977. In vitro cleavage of Rous sarcoma viral polypeptide precursor into internal structural proteins in vitro involves protein p15. *Proc. Natl. Acad. Sci. USA* 74:911–15
98. Van Zaane, D., Dekker-Michielsen, M. J. A., Bloemers, H. P. J. 1976. Virus-specific precursor polypeptides in cells infected with Rauscher leukemia virus: Synthesis, identification and processing. *Virology* 75:113–29
99. Arcement, L. J., Karshin, W. L., Naso, R. B., Jamjoom, G., Arlinghaus, R. B. 1976. Biosynthesis of Rauscher leukemia viral proteins: Presence of p30 and envelope p15 sequences in precursor polypeptides. *Virology* 69:763–74
100. Gielkens, A. L. J., Van Zaane, D., Bloemers, H. P. J., Bloemendal, H. 1976. Synthesis of Rauscher murine leukemia virus-specific polypeptides in vitro. *Proc. Natl. Acad. Sci. USA* 73:356–60
101. Van Zaane, D., Gielkens, A. L. J., Dekker-Michielsen, M. J. A., Bloemers, H. P. J. 1975. Virus-specific precursor polypeptides in cells infected with Rauscher leukemia virus. *Virology* 67:544–52
102. Naso, R. B., Arcement, L. J., Arlinghaus, R. B. 1975. Biosynthesis of Rauscher leukemia viral proteins. *Cell* 4:31–36
103. Hayman, M. J. 1977. Two new viral polyproteins in chick embryo fibroblasts infected with avian sarcoma leukosis viruses. *Virology*. In press
104. Shapiro, S. Z., Strand, M., August, J. T. 1976. High molecular weight precursor polypeptides to structural proteins of Rauscher murine leukemia virus. *J. Mol. Biol.* 107:459–77
105. Pawson, T., Martin, G. S., Smith, A. E. 1976. Cell-free translation of virion RNA from nondefective and transformation defective Rous sarcoma viruses. *J. Virol.* 19:950–67
106. Kerr, I. M., Olshevsky, U., Lodish, H. F., Baltimore, D. 1976. Translation of murine leukemia virus RNA in cell-free systems from animal cells. *J. Virol.* 18:627–35
107. Salzberg, S., Robin, M. S., Green, M. 1977. A possible requirement for protein synthesis early in the infectious cycle of the murine sarcoma-leukemia virus. *Virology* 76:341–51
108. Huebner, R. J., Armstrong, D., Okuyan, M., Sarma, P. S., Turner, H. C. 1964. Specific complement-fixing viral antigens in hamster and guinea pig tumors induced by the Schmidt-Ruppin strain of avian sarcoma. *Proc. Natl. Acad. Sci. USA* 51:742–50
109. Bauer, H., Schäfer, W. 1966. Origin of group-specific antigen of chicken leukosis viruses. *Virology* 29:494–97
110. Bauer, H., Bolognesi, D. P. 1970. Polypeptides of avian RNA tumor viruses. II. Serological characterization. *Virology* 42:1113–26
111. Payne, F. E., Solomon, J. J., Purchase, H. G. 1966. Immunofluorescent studies of group-specific antigen of the avian sarcoma-leukosis viruses. *Proc. Natl. Acad. Sci. USA* 55:341–49
112. Kelloff, G., Vogt, P. K. 1966. Localization of avian tumor virus group-specific antigen in cell and virus. *Virology* 29:377–87
113. Schäfer, W., Anderer, A. F., Bauer, H., Pister, L. 1969. Studies on mouse leukemia viruses. I. Isolation and characterization of a group-specific antigen. *Virology* 38:387–94
114. Hartley, J. W., Rowe, W. P., Capps, W. E., Huebner, R. J. 1965. Complement fixation and tissue culture assays for mouse leukemia viruses. *Proc. Natl. Acad. Sci. USA* 53:931–38
115. Schäfer, W., Lange, J., Pister, L., Seifert, E., de Noronha, F., Schmidt, F. W. 1970. Eine Komplementbindungsreaktion zum nachweis der bei Leukamieviren verschiedener Sauger Vorkommenden gemeinsamen antigenen Komponente. *Z. Naturforsch.* 25:1029–36
116. Schäfer, W., Hunsmann, G., Moennig, V., de Noronha, F., Bolognesi, D. P., Green, R. W., Huper, G. 1975. Polypeptides of mammalian oncornaviruses. II. Characterization of a murine leukemia virus polypeptide (p15) bearing interspecies reactivity. *Virology* 63:48–59
117. Geering, G., Hardy, W. D. Jr., Old, L. J., DeHarven, E. 1968. Shared group-specific antigen of murine and feline leukemia viruses. *Virology* 36:678–707

118. Strand, M., Wilsnack, R., August, J. T., 1974. Structural proteins of mammalian oncogenic RNA viruses: Immunological characterization of the p15 polypeptide of Rauscher murine virus. *J. Virol.* 14:1575–83
119. Strand, M., August, J. T. 1974. Structural proteins of mammalian oncogenic RNA viruses: Multiple antigenic determinants of the major internal protein and envelope glycoprotein. *J. Virol.* 13:171–80
120. Gilden, R. V., Oroszlan, S., Huebner, R. J. 1971. Coexistence of intraspecies and interspecies specific antigenic determinants on the major structural polypeptide of mammalian C-type viruses. *Nature New Biol.* 231:107–9
121. Davis, J., Gilden, R. V., Oroszlan, S. 1976. Multiple species-specific and interspecific antigenic determinants of a mammalian type-C RNA virus internal protein. *Immunochemistry* 12:67–72
122. Barbacid, M., Stephenson, J. R., Aaronson, S. A. 1976. Structural polypeptides of mammalian type-C viruses: Isolation and immunologic characterization of a low molecular weight polypeptide p10. *J. Biol. Chem.* 251:4859–66
123. Parks, W. P., Scolnick, E. M. 1972. Radioimmunoassay of mammalian type-C viral proteins: Interspecies antigenic reactivities of the major internal polypeptide. *Proc. Natl. Acad. Sci. USA* 69:1766–70
124. Bolognesi, D. P., Ishizaki, R., Huper, G., Vanaman, T. C., Smith, R. E. 1975. Immunological properties of avian oncornavirus polypeptides. *Virology* 64: 349–57
125. Herman, A. C., Green, R. W., Bolognesi, D. P., Vanaman, T. C. 1975. Comparative chemical properties of avian oncornavirus polypeptides. *Virology* 64:339–48
126. Stephenson, J. R., Smith, E. J., Crittenden, L. B., Aaronson, S. A. 1975. Analysis of antigenic determinants of structural polypeptides of avian type-C tumor viruses. *J. Virol.* 16:27–33
127. Hayman, M. J., Vogt, P. K. 1976. Subgroup specific and antigenic determinants of avian RNA tumor virus structural proteins: Analysis of virus recombinants. *Virology* 73:372–80
128. Green, R. W., Bolognesi, D. P., Schäfer, W., Pister, L., Hunsmann, G., de Noronha, F. 1973. Polypeptides of mammalian oncornaviruses. I. Isolation and serological analysis of polypeptides from murine and feline viruses. *Virology* 56:565–79
129. Stephenson, J. R., Tronick, S. R., Aaronson, S. A. 1974. Analysis of typespecific antigenic determinants of two structural polypeptides of mouse RNA C-type viruses. *Virology* 58:1–8
130. Buchhagen, D. L., Stutman, O., Fleissner, E. 1975. Chromatographic separation and antigenic analysis of proteins of the oncornaviruses. IV. Biochemical typing of murine viral proteins. *J. Virol.* 15:1148–57
131. Chen, J. H., Hayward, W. S., Hanafusa, H. 1974. Avian tumor virus proteins and RNA in uninfected chicken embryo cells. *J. Virol.* 14:1419–29
132. Smith, E. J., Stephenson, J. R., Crittenden, L. B., Aaronson, S. A. 1976. Avian leukosis-sarcoma virus gene expression. *Virology* 70:493–501
133. Bolognesi, D. P., Huper, G., Green, R. W., Graf, T. 1974. Biochemical properties of oncornavirus polypeptides. *Biochem. Biophys. Acta* 355:220–35
134. Bolognesi, D. P., Luftig, R., Shaper, J. H. 1973. Localization of RNA tumor virus polypeptides. I. Isolation of further virus substructures. *Virology* 56: 549–64
135. Bolognesi, D. P., Gelderblom, H., Bauer, H., Mölling, K., Huper, G. 1972. Polypeptides of avian RNA tumor viruses. V. Analysis of the virus core. *Virology* 47:567–78
136. Davis, N. L., Rueckert, R. R. 1972. Properties of a ribonucleoprotein particle isolated from Nonidet P-40 treated Rous sarcoma virus. *J. Virol.* 10: 1010–20
137. Quigley, J. P., Rifkin, D. B., Compans, R. W. 1972. Isolation and characterization of ribonucleoprotein substructures from Rous sarcoma virus. *Virology* 50:65–75
138. Fleissner, E., Tress, E. 1973. Isolation of ribonucleoprotein structure from oncornaviruses. *J. Virol.* 12:1612–15
139. Stromberg, K., Hurley, N. E., Davis, N. L., Rueckert, R. R., Fleissner, E. 1974. Structural studies of avian myeloblastosis virus: Comparison of polypeptides in virion and core component by dodecyl sulfate polyacrylamide gel electrophoresis. *J. Virol.* 13:513–28
140. Pal, B. K., Roy-Burman, P. 1975. Phosphoproteins: Structural components of oncornavirus. *J. Virol.* 15:540–49
141. Pal, B. K., McAllister, R. M., Gardner, M. B., Roy-Burman, P. 1975. Comparative studies on the structural phospho-

proteins of mammalian type C viruses. *J. Virol.* 16:123–31
142. Lai, M. M.-C. 1976. Phosphoproteins of Rous sarcoma viruses. *Virology* 74:287–301
143. Sen, A., Sherr, C. J., Todaro, G. J. 1976. Specific binding of the type C viral core protein p12 with purified viral RNA. *Cell* 7:21–32
144. Sen, A., Todaro, G. J. 1977. An RNA-binding protein, p19, isolated from Rous sarcoma virus. *Cell* 10:91–100
145. Tung, J. S., Yoshiki, T., Fleissner, E. 1976. A core polyprotein of murine leukemia virus on the surface of mouse leukemia cells. *Cell* 9:573–78
146. Hunter, E., Hayman, M. J., Rongey, R. W., Vogt, P. K. 1976. An avian sarcoma virus mutant which is temperature-sensitive for virion assembly. *Virology* 69:35–49
147. Friis, R. R., Ogura, H., Gelderblom, H., Halpern, M. S. 1976. The defective maturation of viral progeny with a temperature-sensitive mutant of avian sarcoma virus. *Virology* 73:259–72
148. Rohrschneider, J. M., Diggelmann, H., Ogura, H., Friis, R. R., Bauer, H. 1976. Defective cleavage of a precursor polypeptide in temperature-sensitive mutant of avian sarcoma virus. *Virology* 75:177–87
149. Stephenson, J. R., Tronick, S. R., Aaronson, S. A. 1975. Murine leukemia virus mutants with temperature-sensitive defects in precursor polypeptide cleavage. *Cell* 6:543–48
150. Stephenson, J. R., Aaronson, S. A. 1973. Characterization of temperature-sensitive mutants of murine leukemia virus. *Virology* 54:53–59
151. Stephenson, J. R., Tronick, S. R., Aaronson, S. A. 1974. Temperature-sensitive mutants of murine leukemia virus. IV. Further physiological characterization and evidence for genetic recombination. *J. Virol.* 14:918–23
152. Yeger, H., Kalnins, V. E., Stephenson, J. R. 1976. Electron microscopy of mammalian type-C RNA viruses: Use of conditional lethal mutants in studies of virion maturation and assembly. *Virology* 74:459–69
153. Wong, P. K. Y., McCarter, J. A. 1974. Studies of two temperature-sensitive mutants of Moloney murine leukemia virus. *Virology* 58:396–408
154. Wong, P. K. Y., MacLeod, R. 1975. Studies on the budding process of a temperature-sensitive mutant of murine leukemia virus with a scanning electron microscope. *J. Virol.* 16:434–46
155. Vogt, P. K. 1977. The genetics of RNA tumor viruses. In *Comprehensive Virology,* ed. H. Fraenkel-Conrat, R. Wagner, Vol. IX. New York: Academic. In press
156. Barbacid, M., Stephenson, J. R., Aaronson, S. A. 1976. The *gag* gene of mammalian type-C RNA tumor viruses. *Nature* 262:554–59
157. Temin, H. M., Mizutani, S. 1970. RNA-dependent DNA polymerase in virions of Rous sarcoma virus. *Nature* 226:1211–13
158. Baltimore, D. 1970. Viral RNA-dependent DNA polymerase. *Nature* 226:1209–11
159. Panet, A., Baltimore, D. Hanafusa, T. 1975. Quantitation of avian RNA tumor virus reverse transcriptase by radioimmunoassay. *J. Virol.* 16:146–52
160. Temin, H. M., Baltimore, D. 1972. RNA-directed DNA synthesis and RNA tumor viruses. *Adv. Virus Res.* 17:129–86
161. Bishop, J. M., Varmus, H. E. 1975. Molecular biology of RNA tumor viruses. In *Cancer,* ed. F. F: Becker, 2:3–48. New York: Plenum
162. Green, M., Gerard, G. F. 1974. RNA-directed DNA polymerase—properties and functions in oncogenic RNA viruses and cells. *Prog. Nucleic Acid Res. Mol. Biol.* 14:187–334
163. Sarngadharan, M. G., Allaudeen, H. S., Gallo, R. C. 1976. Reverse transcriptase of RNA tumor viruses and animal cells. *Methods Cancer Res.* 12:3–47
164. Wu, A. M., Gallo, R. C. 1975. Reverse transcriptase. *CRC Crit. Rev. Biochem.* 3:289–347
165. Baltimore, D., Smoler, D. F. 1971. Primer requirement and template specificity of the DNA polymerase of RNA tumor viruses. *Proc. Natl. Acad. Sci. USA* 68:1507–11
166. Goodman, N. C., Spiegelman, S. 1971. Distinguishing reverse transcriptase of the RNA tumor virus from other known DNA polymerases. *Proc. Natl. Acad. Sci. USA* 68:2203–6
167. Verma, I. M., Meuth, N. L., Bromfeld, E., Manly, K. F., Baltimore, D. 1971. Covalently linked RNA-DNA molecule as initial product of RNA tumor virus DNA polymerase. *Nature New Biol.* 233:131–34
168. Smoler, D., Molineux, I., Baltimore, D. 1971. Direction of polymerization by the avian myeloblastosis virus deoxy-

ribonucleic acid polymerase. *J. Biol. Chem.* 246:7697–7700
169. Mölling, K., Bolognesi, D. P., Bauer, W., Busen, W., Plassmann, H. W., Hausen, P. 1971. Viral reverse transcriptase association with RNase specific for RNA-DNA hybrids. *Nature New Biol.* 234:240–43
170. Keller, W., Crouch, R. 1972. Degradation of DNA-RNA hybrids by ribonuclease H and DNA polymerases of cellular and viral origin. *Proc. Natl. Acad. Sci. USA* 69:3360–64
171. Baltimore, D., Smoler, D. 1972. Association of an endoribonuclease with the avian myeloblastosis virus deoxyribonucleic and polymerase. *J. Biol. Chem.* 247:7282–87
172. Grandgenett, D. P., Green, M. 1974. Different mode of action of ribonuclease H in purified α and β ribonucleic acid-directed deoxyribonucleic acid polymerase from avian myeloblastosis virus. *J. Biol. Chem.* 249:5148–52
173. Verma, I. M. 1975. Studies on reverse transcriptase of RNA tumor viruses. III. Properties of purified Moloney murine leukemia virus DNA polymerase and associated RNase H. *J. Virol.* 15:843–54
174. Verma, I. M. 1977. The reverse transcriptase. In *BBA Rev. Cancer* 473:1–38
175. Duesberg, P., Helm, K. V. D., Canaani, E. 1971. Properties of a soluble DNA polymerase isolated from Rous sarcoma virus. *Proc. Natl. Acad. Sci. USA* 68:747–51
176. Grandgenett, D. P., Gerard, G. F., Green, M. 1973. A single subunit from avian myeloblastosis virus with both RNA-directed DNA polymerase and ribonuclease H activity. *Proc. Natl. Acad. Sci. USA* 70:230–34
177. Verma, I. M., Meuth, N. L., Fan, H., Baltimore, D. 1974. Hamster leukemia virus: Lack of endogenous DNA synthesis and unique structure of its DNA polymerase. *J. Virol.* 13:1075–82
178. Gibson, W., Verma, I. M. 1974. Studies on the reverse transcriptase of RNA tumor viruses. Structural relatedness of two subunits of avian RNA tumor viruses. *Proc. Natl. Acad. Sci. USA* 71:4991–94
179. Kacian, D. L., Watson, K. F., Burny, A., Spiegelman, S. 1971. Purification of the DNA polymerase of avian myeloblastosis virus. *Biochem. Biophys. Acta* 246:365–83
180. Mölling, K. 1975. Reverse transcriptase and RNase H: Present in a murine virus and in both subunits of an avian virus. *Cold Spring Harbor Symp. Quant. Biol.* 39:969–74
181. Grandgenett, D. P., Rho, H. M. 1975. Binding properties of avian myeloblastosis virus DNA polymerase to nucleic acid affinity columns. *J. Virol.* 15:526–33
182. Grandgenett, D. P., Vohra, A. K., Faras, A. J. 1976. Different states of avian myeloblastosis virus DNA polymerase and their binding capacity to primer tRNA^{Trp}. *Virology* 75:26–32
182a. Hizi, A., Joklik, W. K. 1977. RNA dependent DNA polymerase of avian sarcoma virus B77. I. Isolation and partial characterization of the α_1, β_2, and $\alpha\beta$ forms of the enzyme. *J. Biol. Chem.* 252:2281–89
182b. Hizi, A., Leis, J. P., Joklik, W. K. 1977. RNA dependent DNA polymerase of avian sarcoma virus B77. II. Comparison of the catalytic properties of the α_1, β_2, and $\alpha\beta$ enzyme forms. *J. Biol. Chem.* 252:2290–95
183. Mölling, K. 1974. Characterization of reverse transcriptase and RNase from Friend murine leukemia virus. *Virology* 62:46–59
184. Mölling, K. 1976. Further characterization of the Friend murine leukemia virus reverse transcriptase-RNase H complex. *J. Virol.* 18:418–25
185. Linial, M., Mason, W. S. 1973. Characterization of two conditional early mutants of Rous sarcoma virus. *Virology* 53:258–73
186. Greenberger, J. S., Stephenson, J. R. 1975. Temperature-sensitive mutants of murine leukemia virus. V. Impaired leukemogenic activity in vivo. *Int. J. Cancer* 15:1009–15
187. Stephenson, J. R., Aaronson, S. A. 1974. Temperature-sensitive mutants of murine leukemia virus. III. Mutants defective in helper functions for sarcoma virus fixation. *Virology* 58:294–97
188. Baltimore, D., Verma, I. M., Drost, S. D., Mason, W. S. 1975. Temperature sensitive DNA polymerase from RSV mutants. In *Tumor Virus Host Cell Interactions*, ed. A. Kolber, pp. 267–72. New York: Plenum
189. Verma, I. M., Mason, W. S., Drost, S. D., Baltimore, D. 1974. DNA polymerase activity from two temperature-sensitive mutants of Rous sarcoma virus is thermolabile. *Nature* 251:27–31
190. Verma, I. M., Varmus, H. E., Hunter, E. 1976. Characterization of early temperature-sensitive mutants of avian sar-

coma viruses—Biological properties and thermal lability of the reverse transcriptase in vitro and synthesis of viral DNA in infected cells. *Virology* 74:16-29
191. Tronick, S. R., Stephenson, J. R., Verma, I. M., Aaronson, S. A. 1975. Thermolabile reverse transcriptase of a mammalian leukemia virus mutant temperature-sensitive in its replication and sarcoma virus helper functions. *J. Virol.* 16:1476-82
192. Verma, I. M. 1975. Studies on reverse transcriptase of RNA tumor viruses. I. Localization of thermolabile DNA polymerase and RNase H activities on one polypeptide. *J. Virol.* 15:121-26
193. Mason, W. S., Friis, R. R., Linial, M., Vogt, P. K. 1974. Determination of the defective function in two mutants of Rous sarcoma virus. *Virology* 61:559-74
194. Friis, R. R., Hunter, E. 1973. A temperature-sensitive mutant of Rous sarcoma virus that is defective for replication. *Virology* 53:479-83
195. Friis, R. R., Mason, W. S., Chen, Y. C., Halpern, M. S. 1975. A replication defective mutant of Rous sarcoma virus which fails to make a functional reverse transcriptase. *Virology* 64:49-62
196. Hanafusa, H., Hanafusa, T., 1968. Further studies on RSV production from transformed cells. *Virology* 34:630-36
197. Kawai, S., Hanafusa, H. 1973. Isolation of defective mutant of avian sarcoma virus. *Proc. Natl. Acad. Sci. USA* 70:3493-97
198. Nowinski, R. C., Watson, K. F., Yaniv, A., Spiegelman, S. 1972. Serological analysis of the deoxyribonucleic acid polymerase of avian oncornaviruses. II. Comparison of avian deoxyribonucleic acid polymerases. *J. Virol.* 10:959-64
199. Murphy, H. M. 1977. A new replication-defective variant of the Bryan high titer strain Rous sarcoma virus. *Virology.* In press
200. Ikeda, H., Hardy, W. Jr., Tress, E., Fleissner, E. 1975. Chromatographic separation and antigenic analysis of proteins of the oncornaviruses. V. Identification of a new murine viral protein, p15(E). *J. Virol.* 16:53-61
201. Nermut, M. V., Frank, H., Schäfer, W. 1972. Properties of mouse leukemia viruses. III. Electron microscopic appearance as revealed after conventional preparation techniques as well as freeze drying and freeze etching. *Virology* 49:345-58
202. Rifkin, D. B., Compans, R. W. 1971. Identification of the spike proteins of Rous sarcoma virus. *Virology* 46:485-89
203. Witte, O. N., Weissman, I. L., Kaplan, H. S. 1973. Structural characteristics of some murine RNA tumor viruses studied by lactoperoxidase iodination. *Proc. Natl. Acad. Sci. USA* 70:36-40
204. McLellan, W. L., August, J. T. 1976. Analysis of the envelope of Rauscher murine oncornavirus: In vitro labeling of glycopeptides. *J. Virol.* 20:627-36
205. Bolognesi, D. P., Bauer, H., Gelderblom, H., Huper, G. 1972. Polypeptides of avian RNA tumor viruses. IV. Components of the viral envelope. *Virology* 47:551-66
206. Bolognesi, D. P. 1974. Structural components of RNA tumor viruses. *Adv. Virus Res.* 19:315-60
207. Leamnson, R. N., Halpern, M. S. 1976. Subunit structure of the glycoprotein complex of avian tumor virus. *J. Virol.* 18:956-68
208. Leamnson, R. N., Shander, M. H. M., Halpern, M. S. 1976. A structural protein complex in Moloney leukemia virus. *Virology* 76:437-39
209. Witte, O. N., Tsukamoto-Adey, A., Weissman, I. L. 1977. Cellular maturation of oncornavirus glycoproteins: Topological arrangements of precursor and product forms in cellular membranes. *Virology* 76:539-53
210. Famulari, N. G., Buchhagen, D. L., Klenk, H.-D., Fleissner, E. 1976. Presence of murine leukemia virus envelope proteins gp70 and p15(E) in a common polyprotein of infected cells. *J. Virol.* 20:501-19
211. Naso, R. B., Arcement, L. J., Karshin, W. L., Jamjoom, G. A., Arlinghaus, R. B. 1976. Fucose-deficient glycoprotein precursor to Rauscher leukemia virus gp69-71. *Proc. Natl. Acad. Sci. USA* 73:2326-31
212. Arlinghaus, R. B., Naso, R. B., Jamjoom, G. A., Arcement, L. J., Karshin, W. L. 1976. Biosynthesis and processing of Rauscher leukemia viral precursor polyproteins. See Ref. 84, pp. 689-715
213. England, J. M., Bolognesi, D. P., Dietzschold, B., Halpern, M. S. 1977. Evidence that the precursor glycoprotein is cleaved to yield the major glycoprotein of avian tumor virus. *J. Virol.* 21:810-14
214. Halpern, M. S., Bolognesi, D. P., Lewandowski, L. J. 1974. Isolation of

the major viral glycoprotein and a putative precursor from cells transformed by avian sarcoma viruses. *Proc. Natl. Acad. Sci. USA* 71:2342–46
215. Vogt, P. K. 1970. Envelope classification of avian RNA tumor viruses. *Bibl. Haematol. Basel* 36:153–67
216. Sarma, P. S., Log, T. 1973. Subgroup classification of feline leukemia and sarcoma viruses by viral interference and neutralization tests. *Virology* 54:160–69
217. Sarma, P. S., Log, T., Jain, D., Hill, P. R., Huebner, R. J. 1975. Differential host range of viruses of feline leukemia-sarcoma complex. *Virology* 64:438–46
218. Witter, R., Frank, H., Moennig, V., Hunsmann, G., Lange, J., Schäfer, W. 1973. Properties of mouse leukemia viruses. IV. Hemagglutination assay and characterization of hemagglutinating surface components. *Virology* 54:330–45
219. Witter, R., Hunsmann, G., Lange, J., Schäfer, W. 1973. Properties of mouse leukemia viruses. V. Hemagglutination-inhibition and indirect hemagglutination tests. *Virology* 54:346–58
220. Zarling, D. A., Temin, H. M. 1976. High spontaneous mutation rate of an avian sarcoma virus. *J. Virol.* 17:74–84
221. Bolognesi, D. P., Langlois, A. J., Schäfer, W. 1975. Polypeptides of mammalian oncornaviruses. IV. Structural components of murine leukemia virus released as soluble antigens in cell culture. *Virology* 68:550–55
222. Strand, M., August, J. T. 1976. Oncornavirus envelope glycoprotein in serum of mice. *Virology* 75:130–44
223. Lerner, R. A., Wilson, C. B., del Villano, B. C., McConahey, P. J., Dixon, F. J. 1976. Endogenous oncornaviral gene expression in adult and fetal mice: Quantitative cystologic and physiologic studies of the major viral glycoprotein gp70. *J. Exp. Med.* 143:151–66
224. Strand, M., August, J. T. 1973. Structural proteins of oncogenic ribonucleic acid viruses. Interspec II, a new interspecies antigen. *J. Biol. Chem.* 248:5627–33
225. Moennig, V., Frank, H., Hunsmann, G., Schäfer, W. 1974. Properties of mouse leukemia viruses. VII. The major viral glycoprotein of Friend leukemia virus, isolation and physiochemical properties. *Virology* 61:100–11
226. Hunsmann, G., Moennig, V., Pister, L., Seifert, E., Schäfer, W. 1974. Properties of mouse leukemia viruses. VIII. The major viral glycoprotein of Friend leukemia virus. Seroimmunological, interfering and hemagglutinating capacities. *Virology* 62:307–18
227. Steeves, R. A., Strand, M., August, J. T. 1974. Structural proteins of mammalian oncogenic RNA viruses: Murine leukemia virus neutralization by antisera prepared against purified envelope glycoprotein. *J. Virol.* 14:187–89
228. Schäfer, W., Fischinger, P. J., Collins, J. J., Bolognesi, D. P. 1977. Role of carbohydrate in biological functions of Friend murine leukemia virus gp71. *J. Virol.* 21:35–40
229. Ihle, J. N., Denny, T. P., Bolognesi, D. P. 1976. Purification and serological characterization of the major envelope glycoprotein from AKR murine leukemia virus and its reactivity with autogenous immune sera from mice. *J. Virol.* 17:727–36
230. Rohrschneider, L., Bauer, H., Bolognesi, D. P. 1975. Group-specific antigenic determinants of the large envelope glycoprotein of avian oncornaviruses. *Virology* 67:234–41
231. Strand, M., August, J. T. 1974. Structural proteins of mammalian oncogenic RNA viruses: Multiple antigenic determinants of the major internal protein and envelope glycoprotein. *J. Virol.* 13:171–80
232. Schäfer, W., Hunsmann, G., Moennig V., de Noronha, F., Bolognesi, D. P. Green, R. W., Huper, G. 1975. Poly peptides of mammalian oncornaviruses II. Characterization of a murine leu kemia virus polypeptide (p15) bearin interspecies reactivity. *Virology* 63 48–59
233. Ihle, J. N., Hanna, M. G. Jr., Schäfer W., Hunsmann, G., Bolognesi, D. P. Huper, G. 1975. Polypeptides of mam malian oncornaviruses. III. Localiza tion of p15 and reactivity with natura antibody. *Virology* 63:60–67
234. Krantz, M. J., Lee, Y. C., Hung, P. P. 1976. Characterization and comparison of the major glycoprotein from three strains of Rous sarcoma virus. *Arch. Biochem. Biophys.* 174:66–73
235. Bolognesi, D. P., Collins, J. J., Leis, J. P., Moennig, V., Schäfer, W., Atkinson, P. H. 1975. Role of carbohydrate in determining the immunochemical properties of the major glycoprotein (gp71) of Friend murine leukemia virus. *J. Virol.* 16:1453–63
236. Schäfer, W., Fischinger, P. J., Collins, J. J., Bolognesi, D. P. 1977. Role of carbohydrate in biological functions of

Friend murine leukemia virus gp71. *J. Virol.* 21:35–40
237. Schäfer, W., Schwarz, H., Thiel, H.-J., Wecker, E., Bolognesi, D. P. 1976. Properties of mouse leukemia viruses. XIII. Serum therapy of virus-induced murine leukemias. *Virology* 75:401–18
238. Huebner, R. J., Gilden, R. V., Roni, R., Hill, R. W., Trimmer, R. W., Fish, D. C., Sass, B. 1976. Prevention of spontaneous leukemia in AKR mice in type-specific immuno-suppression of endogenous ecotropic virogenes. *Proc. Natl. Acad. Sci. USA* 73:4633–35
239. Mason, W. S., Yeater, C. 1977. A mutant of Rous sarcoma virus with a conditional defect in the determinants of viral host range. *Virology* 77:443–56
240. Hanafusa, H., Hanafusa, T., Rubin, H. 1963. The defectiveness of Rous sarcoma virus. *Proc. Natl. Acad. Sci. USA* 49:572–82
241. Scheele, C. M., Hanafusa, H. 1971. Proteins of helper-dependent RSV. *Virology* 45:401–10
242. Halpern, M. S., Bolognesi, D. P., Friis, R. R. 1976. Viral glycoprotein synthesis studied in an established line of Japanese quail embryo cells infected with the Bryan high-titer strain of Rous sarcoma virus. *J. Virol.* 18:504–10
243. DeGiuli, C., Kawai, S., Dales, S., Hanafusa, H. 1975. Absence of surface projections on some noninfectious forms of RSV. *Virology* 66:253–60
244. Ogura, H., Friis, R. 1975. Further evidence for the existence of a viral envelope protein defect in Bryan high-titer strain of Rous sarcoma virus. *J. Virol.* 16:443–46
245. Hanafusa, T., Miyamoto, T., Hanafusa, H. 1970. A type of chick embryo cell that fails to support formation of infectious RSV. *Virology* 40:55–64
246. Rubin, H. 1965. Genetic control of cellular susceptibility to pseudotypes of Rous sarcoma virus. *Virology* 26:270–76
247. Weiss, R. A. 1969. The host range of Bryan strain Rous sarcoma virus synthesized in the absence of helper virus. *J. Gen. Virol.* 5:511–28
248. Weiss, R. A. 1969. Interference and neutralization studies with Bryan strain Rous sarcoma virus synthesized in the absence of helper virus. *J. Gen. Virol.* 5:529–39
249. Hanafusa, T., Hanafusa, H., Miyamoto, T. 1970. Recovery of a new virus from apparently normal chick cells by infection with avian tumor viruses. *Proc. Natl. Acad. Sci. USA* 67:1797–1803
250. Hanafusa, H., Miyamoto, T., Hanafusa, T. 1970. A cell-associated factor essential for formation of an infectious form of Rous sarcoma virus. *Proc. Natl. Acad. Sci. USA* 66:314–21
251. Hanafusa, H., Aoki, T., Kawai, S., Miyamoto, T., Wilsnack, R. E. 1973. Presence of antigen common to avian tumor viral envelope antigen in normal chick embryo cells. *Virology* 56:22–32
252. Halpern, M. S., Bolognesi, D. P., Friis, R. R., Mason, W. S. 1975. Expression of the major viral glycoprotein of avian tumor virus in cells of *chf* (+) chicken embryos. *J. Virol.* 15:1131–40
253. Hayward, W. S., Hanafusa, H. 1976. Independent regulation of endogenous and exogenous avian RNA tumor virus genes. *Proc. Natl. Acad. Sci. USA* 73:2259–63
254. Hayward, W. S., Wang, S. Y., Urm, E., Hanafusa, H. 1976. Transcription of the avian RNA tumor virus glycoprotein gene in uninfected and infected cells. See Ref. 84, 4:21–35
255. Varmus, H. E., Stehelin, D., Spector, D., Tal, J., Fujita, F., Padgett, T., Roulland-Dussoix, D., Kung, H.-J., Bishop, J. M. 1976. Distribution and function of defined regions of avian tumor virus genomes in viruses and uninfected cells. See Ref. 84, pp. 339–58
256. Tal, J., Fujita, D. J., Kawai, S., Varmus, H. E., Bishop, M. J. 1977. Purification of DNA complementary to the *env* gene of avian sarcoma virus and analysis of relationship among the *env* genes of avian-leukosis-sarcoma viruses. *J. Virol.* 21:497–505
257. Bishop, J. M., Deng, D.-T., Mahy, B. J. W., Quintrell, N., Stavnezer, E., Varmus, H. E. 1976. Synthesis of viral RNA in cells infected by avian sarcoma viruses. See Ref. 84, pp. 1–20
258. Toyoshima, K., Vogt, P. K. 1969. Temperature-sensitive mutants of avian sarcoma virus. *Virology* 39:930–31
259. Martin, G. S. 1970. Rous sarcoma virus: A function required for the maintenance of the transformed state. *Nature* 227:1021–23
260. Martin, G. S., Duesberg, P. H. 1971. The *a* subunit in the RNA of transforming avian tumor viruses II. Spontaneous loss resulting in nontransforming variants. *Virology* 47:494–97
261. Kawai, S., Hanafusa, H. 1971. The effects of reciprocal changes in temperature on the transformed state of cells

infected with a Rous sarcoma virus mutant. *Virology* 46:470–79
262. Friis, R. R., Toyoshima, K., Vogt, P. K. 1971. Conditional lethal mutants of avian sarcoma viruses. I. Physiology of *ts* 75 and *ts* 149. *Virology* 43:375–89
263. Biquard, J.-M., Vigier, P. 1971. Isolement et étude d'un mutant conditionel transformante thermosensible. *C. R. Acad. Sci. Ser. D.* 271:2430–33
264. Biquard, J.-M., Vigier, P. 1972. Characteristics of a conditional mutant of Rous sarcoma virus defective in ability to transform cells at high temperature. *Virology* 47:444–55
265. Wyke, J. A., Linial, M. 1973. Temperature-sensitive avian sarcoma viruses: A physiological comparison of twenty mutants. *Virology* 53:152–61
266. Bookout, J. B., Sigel, M. M. 1975. Characterization of a conditional mutant of Rous sarcoma virus with alternations in early and late functions of cell transformation. *Virology* 67:474–86
267. Bader, J. P. 1972. Temperature-dependent transformation cells infected with a mutant of Bryan Rous sarcoma virus. *J. Virol.* 10:267–76
268. Balduzzi, P. C. 1976. Cooperative transformation studies with temperature-sensitive mutants of Rous sarcoma virus. *J. Virol.* 18:332–43
269. Scolnick, E. M., Goldberg, R. J., Parks, W. P. 1975. A biochemical and genetic analysis of mammalian RNA containing sarcoma viruses. *Cold Spring Harbor Symp. Quant. Biol.* 39:885–95
270. Toyoshima, K., Owada, M., Kozai, Y. 1973. Tumor producing capacity of temperature-sensitive mutants of avian sarcoma viruses in chicks. *Biken J.* 16:103–10
271. Burger, M. M., Martin, G. S. 1972. Agglutination of cells transformed by Rous sarcoma virus by wheat germ agglutinin and concanavalin A. *Nature New Biol.* 237:9–12
272. Biquard, J.-M. 1973. Agglutinability of Rous cells by concanavalin A: Study with a temperature-sensitive RSV mutant and inhibitors of macromolecular synthesis. *Intervirology* 1:220–23
273. Ambros, V. R., Chen, L. B., Buchanan, J. M. 1975. Surface ruffles as markers for studies of cell transformation by Rous sarcoma virus. *Proc. Natl. Acad. Sci. USA* 72:3144–48
274. Hakomori, S.-I., Wyke, J. A., Vogt, P. K. 1976. Glycolipids of chick embryo fibroblasts infected with temperature-sensitive mutants of avian sarcoma viruses. *Virology* 76:485–93
275. Unkeless, J. C., Tobia, A., Ossowski, L., Quigley, J. P., Rifkin, D. B., Reich, E. 1973. An enzymatic function associated with transformation of fibroblasts by oncogenic viruses. I. Chick embryo fibroblast cultures transformed by avian RNA tumor viruses. *J. Exp. Med.* 137:85–111
276. Chen, L. B., Buchanan, J. M. 1975. Plasminogen-independent fibrinolysis by proteases produced by transformed chick embryo fibroblasts. *Proc. Natl. Acad. Sci. USA* 72:1132–36
277. Martin, G. S., Venuta, S., Weber, M., Rubin, H. 1971. Temperature-dependent alterations in sugar transport in cells infected by a temperature-sensitive mutant of Rous sarcoma virus. *Proc. Natl. Acad. Sci. USA* 68:2739–41
278. Venuta, S., Rubin, H. 1973. Sugar transport in normal and Rous sarcoma virus-transformed chick-embryo fibroblasts. *Proc. Natl. Acad. Sci. USA* 70:653–57
279. Edelman, G. M., Yahara, I. 1976. Temperature-sensitive changes in surface modulating assemblies of fibroblasts transformed by mutants of Rous sarcoma virus. *Proc. Natl. Acad. Sci. USA* 73:2047–51
280. Ash, J. F., Vogt, P. K., Singer, S. J. 1976. The reversion from transformed to normal phenotype by inhibition of protein synthesis in normal rat kidney cells infected with a temperature-sensitive Rous sarcoma virus mutant. *Proc. Natl. Acad. Sci. USA* 73:3603–7
281. Wang, E., Goldberg, A. R. 1976. Changes in microfilament organization and surface topography upon transformation of chicken embryo fibroblasts with Rous sarcoma virus. *Proc. Natl. Acad. Sci. USA* 73:4065–69
282. Otten, J., Bader, J., Johnson, G. S., Pastan, I. 1972. A mutation in a Rous sarcoma virus gene that controls adenosine 3',5'-monophosphate levels and transformation. *J. Biol. Chem.* 247:1632–33
283. Anderson, W. B., Johnson, G. S., Pastan, I. 1973. Transformation of chick-embryo fibroblasts by wild-type and temperature-sensitive Rous sarcoma virus alters adenylate cyclase activity. *Proc. Natl. Acad. Sci. USA* 70:1055–59
284. Anderson, W. B., Lovelace, E., Pastan, I. 1973. Adenylate cyclase activity is decreased in chick embryo fibroblasts transformed by wild-type and temperature sensitive Schmidt-Ruppin Rous

sarcoma virus. *Biochem. Biophys. Res. Commun.* 52:1293-99
285. Yoshida, M., Owada, M., Toyoshima, K. 1975. Strain specificity of changes in adenylate cyclase activity in cells transformed by avian sarcoma viruses. *Virology* 63:68-76
286. Fiszman, M. Y., Fuchs, P. 1975. Temperature-sensitive expression of differentiation in transformed myeloblasts. *Nature* 254:429-31
287. Holtzer, H., Biehl, J., Yeoh, G., Meganathan, R., Kaji, A. 1975. Effect of oncogenic virus on muscle differentiation. *Proc. Natl. Acad. Sci. USA* 72:4051-55
288. Roby, K., Boettiger, D., Pacifici, M., Holtzer, H. 1976. Effects of Rous sarcoma virus on the synthesis programs of chondroblasts and retinal melanoblasts. *Am. J. Anat.* 147:401-5
289. Bauer, H., Kurth, R., Rohrschneider, L., Pauli, G., Friis, R. R., Gelderblom, H. 1975. The role of cell surface changes in RNA tumor virus transformed cells. *Cold Spring Harbor Symp. Quant. Biol.* 39:1181-85
290. Kurth, R., Friis, R. R., Wyke, J. A., Bauer, H. 1975. Expression of tumor-specific surface antigens on cells infected with temperature-sensitive mutants of avian sarcoma virus. *Virology* 64:400-8
291. Wickus, G. G., Robbins, P. W. 1973. Plasma membrane proteins of normal and Rous sarcoma virus-transformed chick embryo fibroblasts. *Nature New Biol.* 245:65-67
292. Stone, K. R., Smith, R.E., Joklik, W. K. 1974. Changes in membrane polypeptides that occur when chick embryo fibroblasts and NRK cells are transformed with avian sarcoma viruses. *Virology* 58:86-100
293. Vaheri, A., Ruoslahti, E. 1974. Disappearance of a major cell-type specific surface glycoprotein antigen (SF) after transformation of fibroblasts by Rous sarcoma virus. *Int. J. Cancer* 13:579-86
294. Robbins, P. W., Wickus, G. G., Branton, P. E., Gaffney, B. J., Hirschberg, C. B., Fuchs, P., Blumberg, P. M. 1975. The chick fibroblast cell surface after transformation by Rous sarcoma virus. *Cold Spring Harbor Symp. Quant. Biol.* 39:1173-80
295. Isaka, T., Yoshida, M., Owada, M., Toyoshima, K. 1975. Alterations in membrane polypeptides of chick embryo fibroblasts induced by transformation with avian sarcoma viruses. *Virology* 65:226-37
296. Hynes, R. O., Wyke, J. A. 1975. Alterations in surface proteins in chicken cells transformed by temperature-sensitive mutants of Rous sarcoma virus. *Virology* 64:492-504
297. Somers, K. D., May, J. T., Kit, S. 1973. Control of gene expression in rat cells transformed by a cold-sensitive murine sarcoma virus (MSV) mutant. *Intervirology* 1:176-84
298. Hunter, E., Vogt, P. K. 1976. Temperature-sensitive mutants of avian sarcoma viruses: Genetic recombination with wild type sarcoma virus and physiological analysis of multiple mutants. *Virology* 69:23-24
299. Vogt, P. K. 1971. Spontaneous segregation of nontransforming viruses from cloned sarcoma viruses. *Virology* 46:939-46
300. Duesberg, P. H., Vogt, P. K. 1973. RNA species obtained from clonal lines of avian sarcoma and avian leukosis virus. *Virology* 54:207-19
301. Hillova, J., Hill, M., Kalekine, M. 1976. Inability of the nondefective Rous sarcoma provirus to generate, upon transfection, a transformation-defective virus. *Virology* 74:540-53
302. Duesberg, P. H., Vogt, P. K. 1973. Gel electrophoresis of avian leukosis and sarcoma viral RNA in formamide: Comparison with other viral and cellular RNA species. *J. Virol.* 12:594-99
303. Bernstein, A., MacCormick, R., Martin, G. S. 1976. Transformation-defective mutants of avian sarcoma viruses: The genetic relationship between conditional and non-conditional mutants. *Virology* 70:206-9
304. Stone, K. R., Smith, R. E., Joklik, W. K. 1975. 35S a and b RNA subunits of avian RNA tumor virus strains cloned and passaged in chick and duck cells. *Cold Spring Harbor Symp. Quant. Biol.* 39:859-68
305. Biggs, P. M., Milne, B. S., Graf, T., Bauer, H. 1973. Oncogenicity of nontransforming mutants of avian sarcoma viruses. *J. Gen. Virol.* 18:399-403
306. Stehelin, D., Guntaka, R. V., Varmus, H. E., Bishop, J. M. 1976. Purification of DNA complementary to nucleotide sequences required for neoplastic transformation of fibroblasts by avian sarcoma viruses. *J. Mol. Biol.* 101:349-65
307. Stehelin, D., Varmus, H. E., Bishop, J. M., Vogt, P. K. 1976. DNA related to the transforming gene(s) of avian sar-

coma viruses is present in normal avian DNA. *Nature* 260:170–73
308. Padgett, T., Stubblefield, E., Varmus, H. E. 1977. Separation and sucrose gradients of chicken macrochromosomes containing the provirus of an endotheliosis RNA tumor virus and microchromosomes containing sequences related to the transforming gene of avian sarcoma virus. *Cell.* In press
309. Stehelin, D., Fujita, D. J., Padgett, T., Varmus, H. E., Bishop, J. M. 1977. Detection and enumeration of transformation defective strains of avian sarcoma virus with molecular hybridization. *Virology* 76:675–84
310. Maisel, J., Scolnick, E. M., Duesberg, P. 1975. Base sequence differences between the RNA components of Harvey sarcoma virus. *J. Virol.* 16:749–53
311. Scolnick, E. M., Howk, R. S., Anisowicz, A., Peebles, P. T., Scher, C. D., Parks, W. P. 1975. Separation of sarcoma virus-specific and leukemia virus specific genetic sequences of Moloney sarcoma virus. *Proc. Natl. Acad. Sci. USA* 72:4650–54
312. Roy-Burman, P., Klement, V. 1975. Derivation of mouse sarcoma virus (Kirsten) by acquisition of genes from heterologous host. *J. Gen. Virol.* 28:193–98
313. Anderson, G. R., Robbins, K. C. 1976. Rat sequences of the Kirsten and Harvey murine sarcoma virus genomes: Nature, origin and expression in rat tumor RNA. *J. Virol.* 17:335–51
314. Scolnick, E. M., Goldberg, R. J., Williams, D. 1976. Characterization of rat genetic sequences of Kirsten sarcoma virus: Distinct class of endogenous rat type-C viral sequences. *J. Virol.* 18:559–66
315. Scolnick, E. M., Williams, D., Maryak, J., Vass, W., Goldberg, R. J., Parks, W. P. 1976. Type-C particle-positive and type-C particle negative rat cell lines: Characterization of the coding capacity of endogenous sarcoma virus-specific mRNA. *J. Virol.* 20:570–82
316. Dina, D., Beemon, K., Duesberg, P. 1976. The 30 S Moloney sarcoma virus RNA contains leukemia virus nucleotide sequences. *Cell* 9:299–310
317. Frankel, A. E., Fischinger, P. J. 1976. Nucleotide sequences in mouse DNA and RNA specific for Moloney sarcoma virus. *Proc. Natl. Acad. Sci. USA* 73:3705–9
318. Frankel, A. E., Neubauer, R. L., Fischinger, P. J. 1976. Fractionation of DNA nucleotide transcripts from Moloney sarcoma virus and isolation of sarcoma virus-specific complementary DNA. *J. Virol.* 18:481–90
319. Frankel, A. E., Fischinger, P. J. 1977. Rate of divergence of cellular sequences homologous to segments of Moloney sarcoma virus. *J. Virol.* 21:153–60
320. Vogt, P. K. 1971. Genetically stable reassortment of markers during mixed infection with avian tumor viruses. *Virology* 46:947–52
321. Kawai, S., Hanafusa, H. 1972. Genetic recombination with avian tumor virus. *Virology* 49:37–44
322. Wang, L.-H., Duesberg, P. H., Kawai, S., Hanafusa, H. 1976. The location of envelope-specific and sarcoma-specific oligonucleotides on the RNA of Schmidt-Ruppin Rous sarcoma virus. *Proc. Natl. Acad. Sci. USA* 73:447–51
323. Wang, L.-H., Duesberg, P. H., Mellon, P., Vogt, P. K. 1976. Distribution of envelope-specific and sarcoma-specific nucleotide sequences from different parents in the RNAs of avian tumor virus recombinants. *Proc. Natl. Acad. Sci. USA* 73:1073–77
324. Joho, R. H., Stoll, E., Friis, R. R., Billeter, M. A., Weissmann, C. 1976. A partial genetic map of Rous sarcoma virus RNA: Location of polymerase, envelope and transformation marker. See Ref. 84, pp. 127–45
325. Duesberg, P. H., Wang, L.-H., Mellon, P., Mason, W. S., Vogt, P. K. 1976. Toward a complete genetic map of Rous sarcoma virus. See Ref. 84, pp. 107–25
326. Duesberg, P. H., Wang, L.-H., Mellon, P., Mason, W. S., Vogt, P. K. 1977. The genetic map of Rous sarcoma virus. In *Genetic Manipulation as it Affects the Cancer Problems,* ed. J. Schultz, Z. Brada. New York: Academic. In press
327. Wang, L.-H., Galehouse, D. M., Mellon, P., Duesberg, P. H., Mason, W. S., Vogt, P. K. 1976. Mapping oligonucleotides of Rous sarcoma virus RNA that segregate with polymerase and group-specific antigen markers in recombinants. *Proc. Natl. Acad. Sci. USA* 73:3952–56
328. Weiss, R. A., Mason, W. S., Vogt, P. K. 1973. Genetic recombinants and heterozygotes derived from endogenous and exogenous avian RNA tumor viruses. *Virology* 52:535–52
329. Hayward, W. S., Hanafusa, H. 1975. Recombination between endogenous and exogenous RNA tumor virus genes

as analyzed by nucleic acid hybridization. *J. Virol.* 15:1367–77
330. Wyke, J. A. 1973. Complementation of transforming functions by temperature-sensitive mutants of avian sarcoma virus. *Virology* 54:28–36
331. Kawai, S., Metroka, C. E., Hanafusa, H. 1972. Complementation of functions required for cell transformation by double infection with RSV mutants. *Virology* 49:302–4
332. Galehouse, D. M., Duesberg, P. H. 1976. Differences in the glycoproteins of avian tumor virus recombinants: Evidence for intragenic crossing over. See Ref. 84, pp. 227–36
333. Aaronson, S. A., Bassin, R. H., Weaver, C. 1972. Comparison of murine sarcoma viruses in nonproducer and S^+L^--transformed cells. *J. Virol.* 9: 701–4
334. Aaronson, S. A., Stephenson, J. R., Hino, S., Tronick, S. R. 1975. Differential expression of helper virus structural polypeptides in cells transformed by clonal isolates of woolly monkey sarcoma virus. *J. Virol.* 16:1117–23
335. Scolnick, E. M., Parks, W. P. 1973. Isolation and characterization of a primate sarcoma virus: Mechanism of rescue. *Int. J. Cancer* 12:138–47
336. Wyke, J. A. 1974. The genetics of C-type RNA tumor viruses. *Int. Rev. Cytol.* 38:67–109
337. Hanafusa, T., Hanafusa, H. 1973. Isolation of leukosis-type virus from pheasant embryo cells: Possible presence of viral genes in cells. *Virology* 51:247–51
338. Fujita, D. J., Chen, Y. C., Friis, R. R., Vogt, P. K. 1974. RNA tumor viruses of pheasants: Characterization of avian leukosis subgroups F and G. *Virology* 60:558–71
339. Kawai, S., Hanafusa, H. 1976. Recombination between temperature-sensitive mutant and a deletion mutant of Rous sarcoma virus. *J. Virol.* 19:389–97
340. Wong, P. K. Y., McCarter, J. A. 1973. Genetic studies of temperature-sensitive mutants of Moloney-murine leukemia virus. *Virology* 53:319–26
341. Fischinger, P. J., Nomura, S., Bolognesi, D. P. 1975. A novel murine oncornavirus with dual eco- and xenotropic properties. *Proc. Natl. Acad. Sci. USA* 72:5150–55
342. Cooper, P. D., Wyke, J. A. 1975. The genome of RNA tumor viruses: A functional requirement for a polyploid structure? *Cold Spring Harbor Symp. Quant. Biol.* 39:997–1004

343. Wyke, J. A., Bell, J. G., Beamand, J. A. 1975. Genetic recombination among temperature-sensitive mutants of Rous sarcoma virus. *Cold Spring Harbor Symp. Quant. Biol.* 39:897–905
344. Hunter, E., Vogt, P. K. 1976. Temperature-sensitive mutants of avian sarcoma viruses: Genetic recombination with wild type sarcoma virus and physiological analysis of multiple mutants. *Virology* 69:23–34
345. Hayman, M. J., Vogt, P. K. 1976. Subgroup specific antigenic determinants of avian RNA tumor virus structural proteins: Analysis of virus recombinants. *Virology* 73:372–80
346. Scolnick, E. M., Williams, D., Parks, W. P. 1976. Purification and characterization of viral RNA of a sarcoma virus isolated from a woolly monkey. *Nature* 264:809–11
347. Cooper, G. M., Castellot, S. B. 1977. Assay of non-infectious fragments of DNA of avian leukosis virus-infected cells by marker rescue. *J. Virol.* 22:300–7
348. Tal, J., Kung, H.-J. 1977. Characterization of DNA complementary to nucleotide sequences adjacent to poly(A) at the 3' terminus of the avian sarcoma virus genome. *Virology.* In press
349. Duesberg, P. H., Kawai, S., Wang, L.-H., Vogt, P. K., Murphy, H. M., Hanafusa, H. 1975. RNA of replication-defective strains of Rous sarcoma virus. *Proc. Natl. Acad. Sci. USA* 72:1569–73
350. Parks, W. P., Howk, R. S., Ansowicz, A., Scolnick, E. M. 1976. Deletion mapping of Moloney type C virus: Polypeptide and nucleic acid expression in different transforming virus isolates. *J. Virol.* 18:491–503
351. Hu, S., Davidson, N., Nicolson, M. O., McAllister, R. M. 1977. A heteroduplex study of the sequence relations between RD-114 and baboon viral RNAs. *J. Virol.* In press
352. Rothenberg, E., Baltimore, D. 1976. Synthesis of long, representative DNA copies of the murine RNA tumor virus genome. *J. Virol.* 17:168–74
353. Junghans, R. P., Duesberg, P. H., Knight, C. A. 1975. In vitro synthesis of full-length DNA transcripts of Rous sarcoma virus RNA by viral DNA polymerase. *Proc. Natl. Acad. Sci. USA* 72:4895–99
354. Hu, S., Davidson, N., Verma, I. M. 1977. A heteroduplex study of the sequence relationships between the RNAs

of M-MSV and M-MLV. *Cell.* In press
355. Fan, H. 1977. The expression of RNA tumor viruses at the levels of translation and transcription. *Curr. Top. Microbiol. Immunol.* In press
356. Humphries, E. H., Temin, H. M. 1972. Cell cycle–dependent activation of Rouse sarcoma virus-infected stationary chicken cells: Avian leukosis virus group-specific antigens and ribonucleic acid. *J. Virol.* 10:82–87
357. Humphries, E. H., Temin, H. M. 1974. Requirement for cell division for initiation of transcription of Rous sarcoma virus RNA. *J. Virol.* 14:531–46
358. Leong, J. A., Garapin, A. C., Jackson, N., Fanshier, L., Levinson, W. E., Bishop, J. M. 1972. Virus-specific ribonucleic acid in cells producing Rous sarcoma virus: Detection and characterization. *J. Virol.* 9:891–902
359. Jolicoeur, P., Baltimore, D. 1976. Effect of Fv-1 gene product on synthesis of N-tropic and B-tropic murine leukemia viral RNA. *Cell* 7:33–39
360. Schincariol, A. L., Joklik, W. K. 1973. Early synthesis of virus-specific RNA and DNA in cells rapidly transformed with Rous sarcoma virus. *Virology* 56:532–48
361. Hayward, W. S., Hanafusa, H. 1973. Detection of avian tumor virus RNA in uninfected chicken embryo cells. *J. Virol.* 11:157–67
362. Levin, J. G., Rosenak, M. J. 1976. Synthesis of murine leukemia virus proteins associated with virions assembled in actinomycin D-treated cells; Evidence for the persistence of viral messenger RNA. *Proc. Natl. Acad. Sci. USA* 73:1154–58
363. Coffin, J. M., Temin, H. M. 1972. Hybridization of Rous sarcoma virus deoxyribonucleic acid polymerase product and ribonucleic acids from chicken and rat cells infected with Rous sarcoma virus. *J. Virol.* 9:766–75
364. Stavnezer, E., Ringold, G., Varmus, H. E., Bishop, J. M. 1976. RNA complementary to the genome of RNA tumor viruses in virions and virus-producing cells. *J. Virol.* 20:342–47
365. Rymo, L., Parsons, J. T., Coffin, J. M., Weissmann, C. 1974. In vitro synthesis of Rous sarcoma virus is catalyzed by a DNA-dependent RNA polymerase. *Proc. Natl. Acad. Sci. USA* 71:2782–86
366. Dinowitz, M. 1975. Inhibition of Rous sarcoma virus by α-amanitin: Possible role of cell DNA-dependent RNA polymerase form II. *Virology* 66:1–9
367. Jacquet, M., Groner, Y., Monroy, G., Hurwitz, J. 1974. The in vitro synthesis of avian myeloblastosis viral RNA sequences. *Proc. Natl. Acad. Sci. USA* 71:3045–49
368. Tsuchida, N., Robin, M. S., Green, M. 1972. Viral RNA subunits in cells transformed by RNA tumor viruses. *Science* 176:1418–19
369. Fan, H., Baltimore, D. 1973. RNA metabolism of murine leukemia virus: Detection of virus-specific RNA sequences in infected and uninfected cells and identification of virus-specific messenger RNA. *J. Mol. Biol.* 80:93–117
370. Haseltine, W. A., Baltimore, D. 1976. Size of murine RNA tumor virus-specific nuclear RNA molecules. *J. Virol.* 19:331–45
371. Gielkens, A. L. J., Salden, M. H. L., Bloemendal, H. 1974. Virus-specific messenger RNA on free and membrane-bound polyribosomes from cells infected with Rauscher leukemia virus. *Proc. Natl. Acad. Sci. USA* 71:1093–97
372. Shanmugan, G., Bhaduri, S., Green, M. 1974. The virus-specific RNA species in free and membrane-bound polyribosomes of transformed cells replicating murine sarcoma-leukemia viruses. *Biochem. Biophys. Res. Commun.* 56:697–704
373. Mueller-Lantzsch, N., Fan, H. 1976. Monospecific immunoprecipitation of murine leukemia virus polyribosomes: Identification of p30 protein specific messenger RNA. *Cell* 9:579–88
374. Mueller-Lantzsch, N., Hatlen, L., Fan, H. 1976. Immunoprecipitation of murine leukemia virus-specific polyribosomes: Identification of virus-specific messenger RNA. See Ref. 84, pp. 37–53

REGULATION OF GENE EXPRESSION IN EUCARYOTES

❖3122

Bert W. O'Malley, Howard C. Towle, and Robert J. Schwartz

Department of Cell Biology, Baylor College of Medicine, Houston, Texas 77030

CONTENTS

INTRODUCTION	240
TRANSCRIPTIONAL PROPERTIES OF INTERPHASE CHROMOSOMES IN VITRO	240
Basic Elements of Eucaryotic Transcriptional Machinery	240
Measurements of Total Template Capacity	241
Measurement of Initiation Sites for RNA Polymerase on Chromatin	242
Comparison of Chromatin Transcription by Bacterial and Eucaryotic RNA Polymerase	244
TRANSCRIPTION OF SPECIFIC GENES FROM CHROMATIN	245
Analysis of Products Using Hybridization Techniques	245
Transcription of Ribosomal, 5S, and tRNA Genes	246
Transcription of Specific mRNA Sequences	248
Fidelity of In Vitro Chromatin Transcription of Specific mRNA Products	249
CHROMOSOMAL PROTEINS AS DETERMINANTS OF GENE EXPRESSION	251
Reconstitution of Chromatin	251
Role of Histones in Transcription	254
Role of Nonhistone Proteins in Transcription	255
The globin system	255
The ovalbumin system	256
The histone system	257
The Mechanism of Action of Chromosomal Proteins on the Transcription Process	258
Initiation of RNA synthesis	258
Selective DNase digestion	259
STEROID HORMONES AS INDUCERS OF GENE EXPRESSION	259
Hormone-Receptor Interactions with Chromatin	259
Activation of Gene Expression by Purified Hormone Receptors	262
Mechanism of Steroid-Receptor Action on Transcription	263
Differentiation and hormone action	264
The ovalbumin minichromosome	265

INTRODUCTION

An important aspect of gene expression during development and cellular differentiation occurs at the level of gene transcription. Only a fraction of the total genetic potential of any cell is ever expressed, and yet this fraction must be capable of changing to accommodate the processes of development and respond to normal physiological stresses in the organism. Thus, the regulation of gene expression is basic to the living cell. Despite the central importance of this issue, little is known regarding the molecular mechanisms involved in the process. This article focuses on one approach being used to study gene expression and its regulation in higher organisms: the in vitro transcription of isolated chromatin.

TRANSCRIPTIONAL PROPERTIES OF INTERPHASE CHROMOSOMES IN VITRO

Basic Elements of Eucaryotic Transcriptional Machinery

The DNA of eucaryotic cells is present in a complex nucleoprotein structure. When isolated from interphase cells, this complex is generally referred to as chromatin. It is important to recognize from the outset, however, that the definition of chromatin is merely operational. There are almost as many methods for the isolation of chromatin as there are laboratories doing it. These methods vary from very gentle procedures involving the lysing of nuclei, followed by centrifugation to separate soluble nuclear proteins from the nucleoprotein complex (1), to more stringent procedures involving salt washes of the chromatin material (2) or sedimentation through a high density sucrose cushion (3). The final composition of the chromatin isolated can vary substantially with regard to the proteins and RNA complexed to the DNA (4). The definition of chromatin can thus be more accurately based on a functional basis, rather than on structure or composition. For the case of in vitro transcription, chromatin can best be defined by examining the nature of the RNA products synthesized and comparing these products with their in vivo RNA counterparts. When examining results from different laboratories, the method of chromatin isolation should also be carefully scrutinized.

The DNA of eucaryotic cells contains an enormous amount of potential genetic information. At any one time in a particular cell type, only a small portion (2–15%) of the total genome is expressed (5–6). In addition to gene sequences which are present in a single or limited number of copies per haploid genome, there are sequence elements which are present in several hundred to hundreds of thousands of copies per nucleus (7). Since most proteins that have been examined are coded for by unique DNA elements (8–9), the possible relationship of repetitive DNA sequences to control of gene expression remains to be elucidated.

The proteins associated with DNA in chromatin are divided into two general classes. The histones are a set of low molecular weight, highly basic proteins present on a roughly equal weight basis with DNA. In most cells, there are generally only five different types of histones. This simplicity has argued against a role of the histones as specific regulators of gene expression. Recent findings have indicated that histones may be involved in a structural role in chromosomes, as is discussed in a later section of this article.

The second group of proteins associated with DNA contains a very diverse population known as the nonhistone proteins. Nonhistones are generally acidic in nature and vary markedly in molecular weight and relative frequency. This class of proteins is thought to contain specific controlling elements for gene transcription, as is discussed below.

The process of transcription in eucaryotic cells is carried out by three distinct DNA-dependent RNA polymerases [for reviews see (10, 11)]. The Class I or A RNA polymerases are nucleolar in origin and are responsible for the synthesis of ribosomal RNA. Heterogenous nuclear and messenger RNA are thought to be transcribed by the nucleoplasmic RNA polymerase II or B. RNA polymerase III or C is present only in small quantities, is nucleoplasmic in origin, and is responsible for the synthesis of 5S ribosomal RNA, tRNA, and possibly other low molecular weight RNA species. Thus, a gross level of control in eucaryotic cells is provided by the existence of multiple RNA polymerases which are specific for the transcription of different classes of RNA.

Measurements of Total Template Capacity

Over a decade ago, it was first demonstrated that isolated chromatin could serve as a template for the synthesis of RNA (12–14). In such studies, chromatin was incubated in the presence of exogenously added RNA polymerase, generally of bacterial origin, and the four ribonucleoside triphosphates, under conditions favorable for RNA synthesis. When compared with purified DNA of the same origin, an equivalent amount of chromatin was found to be less effective at supporting RNA synthesis than the deproteinized DNA (12, 14). This difference was postulated to reflect the fact that certain sequences were not available for transcription in the chromatin. These repressed sequences presumably corresponded to DNA sequences not actively transcribed in the tissue from which the chromatin was isolated. A more definitive demonstration of the restriction of certain gene sequences in chromatin was observed later in molecular hybridization studies (see below).

The level of RNA synthesized by a given amount of chromatin is generally referred to as its template capacity. Measurements of template capacity have been frequently used to monitor changes in the transcriptive level of a tissue. For instance, chromatin isolated from several steroid-responsive tissues increased in template capacity shortly following administration of hormone (15–20). Similarly, the transcriptional activity of isolated chromatin in a number of developmental systems correlated well with the developmental stage of the tissue from which the chromatin was isolated (21–24). Increases in template capacity in these studies have generally been interpreted to indicate an increase in the DNA sequences available for transcription in the cell.

The quantitation of template capacity should be considered only a relative estimate, because of the many technical difficulties involved in the assay. Most procedures for the isolation of chromatin yield a product with only limited solubility. This inherent insolubility of chromatin may play a role in its relative template activity (13, 25). In an attempt to circumvent this problem, many laboratories shear the chromatin to achieve a more soluble product with higher template capacity. The shearing process, however, results in nicks and breaks in the DNA which may act as artificial initiation sites for RNA synthesis. Furthermore, at least for histones, there is now reasonable evidence that shearing can result in migration of chromosomal proteins along the DNA (26). Thus, results obtained using sheared chromatin should be considered cautiously.

In addition to the problem of solubility, most chromatin preparations contain certain endogenous enzyme activities which can influence RNA synthesis. Endogenous ribonucleases and RNA polymerases are among these activities. While some workers have reported preparations of chromatin with no detectable RNase activity (27, 28), most chromatin preparations contain RNase capable of reducing RNA to fragments of 4 to 5S. The use of an RNase inhibitor, such as heparin, can lead to the recovery of larger RNA molecules (28, 29); however, the possibility of secondary effects, such as removal of chromosomal proteins, must be considered (29). The level of endogenous RNA polymerase in most chromatin preparations is generally quite low compared to exogenously added RNA polymerase. This may be due, in part, to the fact that most chromatin isolation procedures lead to the loss or inactivation of considerable RNA polymerase activity (3). The activity of endogenous RNA polymerase appears to be primarily due to a completion of RNA chains that were being synthesized at the time of chromatin isolation (29).

The interpretation of template capacity measurements can also be complicated by the complexity of the RNA reaction itself. The synthesis of an RNA chain involves many subreactions: binding of RNA polymerase to the template, initiation of RNA synthesis, elongation of the nascent chain, termination and release of the completed chain, and reinitiation. Changes in any of these steps can potentially affect measurements of total template capacity. Therefore, the correlation of template capacity with the availability of DNA sequences in the chromatin for transcription is somewhat tenuous. To overcome these problems, attempts have been made to specifically measure the level of initiation sites for RNA synthesis on chromatin.

Measurement of Initiation Sites for RNA Polymerase on Chromatin

Two methods for measuring the number of initiation sites on chromatin have been developed. Both methods attempt to utilize conditions under which each initiation site in chromatin is responsible for the synthesis of only one RNA molecule. Under these conditions, the number of initiation sites will be equal to the number of RNA chains synthesized. It should be noted that measurement of initiation sites by $[\gamma\text{-}^{32}P]$-ATP and GTP incorporation is very difficult because of the high level of protein kinase or polyphosphate kinase associated with chromatin and RNA polymerase.

Cedar & Felsenfeld (30) first measured the number of sites available for RNA chain initiation by *Escherichia coli* RNA polymerase in chromatin, utilizing a method originally developed for bacteriophage T7 DNA by Hyman & Davidson

(31). In this method, RNA polymerase was first preincubated with chromatin in the presence of three of the four ribonucleoside triphosphates to allow RNA chain initiation and limited elongation. Elongation was then allowed to proceed by the addition of the fourth nucleoside triphosphate together with a high concentration of ammonium sulfate to inhibit any secondary initiations. RNA chains were quantitated by determination of the total amount and size of RNA produced. The main drawback of this procedure is the uncertain mode of action of ammonium sulfate. While high concentrations of this compound certainly inhibit initiation, the effect on preexisting "initiation" complexes has not been well characterized. In this regard, only one initiation site is measured by this method on bacteriophage T7 DNA (31), whereas mapping studies have provided evidence for at least three strong promoter sites (32).

Tsai et al (33) adopted the method of Bautz & Bautz (34) and Chamberlin & Ring (35) for measuring chromatin initiation sites. This method involved first preincubating RNA polymerase and chromatin in the absence of any nucleoside triphosphates to allow formation of stable binary complexes. This process occurs through a series of random, nonspecific interactions between RNA polymerase and DNA. When such binding occurs at a specific initiation site, a transition takes place that leads to the formation of the stable complex between enzyme and DNA. This transition is thought to involve local unwinding of the DNA duplex in the region of the initiation site. After preincubation, RNA synthesis is initiated by the addition of the four ribonucleoside triphosphates together with the bacterial RNA polymerase inhibitor rifampicin and allowed to proceed to completion. At the concentration of rifampicin used, initiation from the stable binary complex was highly efficient, while secondary initiations were strongly inhibited (33, 36). Thus, the number of RNA chains synthesized was equivalent to the number of initiation sites at which RNA polymerase was bound. A possible drawback to this method is the potential for overestimation of initiation sites at high polymerase concentrations due to initiation at secondary or "weak" start sites.

When either method is used, a lower number of initiation sites were measured on chromatin than on purified DNA from the same source. Cedar & Felsenfeld (30) found about one site per 1200 nucleotide pairs on calf thymus DNA and one tenth that number on chromatin. Tsai et al (33) found a similar relationship between chick DNA (one site per 700 nucleotide pairs) and chick oviduct chromatin (one site per 40,000 nucleotide pairs). Thus, the restriction of DNA sequences first observed by template capacity measurements actually reflects a difference in the availability of DNA sites in the chromatin.

The rifampicin-nucleotide challenge assay has been utilized to follow the level of initiation sites on chick oviduct chromatin during estrogenic stimulation of the immature chick (37). The level of initiation sites in chromatin from unstimulated animals, 13,000 per haploid genome equivalent, was found to increase by 50% by 8 hr after estrogen administration. This level continued to increase to day 8 of stimulation and then declined to a plateau level approximately twofold higher than the unstimulated animal by day 14. The changes in initiation sites coincided with the pattern of increased growth and differentiation previously reported for estrogen stimulation of the chick oviduct (38). Chromatin isolated from chicks that had been stimulated with estrogen for 14 days and then withdrawn from hormone showed an increase in the level of initiation sites as early as 30 min after restimulation. Thus, changes in chromatin initiation sites appear to be one of the earliest measured responses to steroid hormone administration (39).

The measurement of initiation sites for *E. coli* RNA polymerase in chromatin from the estrogen-stimulated chick oviduct correlates fairly well with estimates of the total poly(A)-containing RNA sequence complexity of stimulated chick oviduct RNA (9). At present, however, it is impossible to demonstrate that the initiation sites measured in vitro correspond to actual in vivo sites for RNA chain initiation. Until this question can be resolved, measurements of chromatin initiation can be considered only as an approximation of the DNA sequences in chromatin available for transcription and as an analytic tool for quantitating the overall level of gene expression.

Comparison of Chromatin Transcription by Bacterial and Eucaryotic RNA Polymerase

Most of the studies discussed to this point have been carried out with bacterial RNA polymerase. This is due to the relative ease in obtaining sufficient quantities of highly purified enzyme from bacterial sources. The use of a bacterial enzyme to transcribe eucaryotic chromatin, however, is obviously subject to reasonable question. There is no reason a priori to assume that a bacterial RNA polymerase should be able to recognize eucaryotic regulatory signals.

Many studies comparing the transcriptive properties of eucaryotic and bacterial RNA polymerases on chromatin have been performed. Maryanka & Gould (27) found that bacterial RNA polymerase transcribed smaller RNA products from rat liver chromatin than the homologous enzyme. Keshgegian & Furth (40) found differences between calf thymus and *E. coli* RNA polymerases when examining the enzyme kinetics of the chromatin-directed reaction. The same workers also found a difference between bacterial and eucaryotic enzymes in the ratio of RNA chains initiated by ATP and GTP (41). Indirect studies such as these are difficult to interpret because of the complexities of the reaction and relative impurity of most eucaryotic RNA polymerase preparations. While these studies may indicate some differences in the properties of the enzymes, they do not give any information on the specificity of RNA synthesis.

Direct competition experiments between bacterial and eucaryotic RNA polymerase were first performed by Butterworth et al (42). In these studies, mixing experiments with rat liver RNA polymerase II and bacterial RNA polymerase demonstrated a lack of competition in the transcription of rat liver chromatin. However, these competition experiments were performed without any inhibitors of reinitiation. Thus, it is conceivable that any initiation site could be used sequentially by the bacterial and eucaryotic RNA polymerases without apparent competition. Meilhac & Chambon modified the technique for performing competition experiments by including an inhibitor of reinitiation-rifamycin AF/013 (43). This compound acts at a step prior to the formation of the first phosphodiester bond to inhibit both bacterial and eucaryotic RNA polymerase (42, 44). The competition experiments involved preincubating template with either one enzyme alone or both enzymes together, prior to the addition of rifamycin AF/013 and nucleoside triphosphate. When this technique was used, *E. coli* RNA polymerase and calf thymus RNA polymerase II did not compete for binding sites on either calf thymus or SV40 DNA (43, 45). Similar results were reported by Tsai et al (46) when comparing hen oviduct and *E. coli* RNA polymerase initiation sites on chick DNA. However, when chick oviduct chromatin was used as a template, these investigators observed a direct competition between the two enzymes. Thus, the presence of chromosomal proteins appeared to modify the initiation specificity of the two enzymes so that they bound to similar sites which conferred resistance to rifamycin AF/013. Recently, Cedar (47) reported that calf thymus chromatin can support 30 times fewer specific sites for the homologous RNA polymerase than for the bacterial enzyme. This method involved the use of high salt to inhibit reinitiation, as discussed earlier for measuring initiation sites for bacterial RNA polymerase (30). The explanation for the discrepancy between the two methods is unclear. One possibility is that the high concentration of ammonium sulfate removed a portion of the calf thymus RNA polymerase II bound to the chromatin, since the eucaryotic enzyme may be more easily dissociated from chromatin as compared to the procaryotic enzyme. Further characterization of the inhibitors used will be necessary to understand fully the differences observed.

Studies comparing the transcriptive properties of chromatin by measurement of total RNA synthesis are not capable of providing an unequivocal answer to the question of whether the bacterial enzyme can recognize the proper sites for initiation on chromatin. This question can be better explored by an examination of the nature of the RNA products transcribed, to be dealt with in the next section.

TRANSCRIPTION OF SPECIFIC GENES FROM CHROMATIN

Analysis of Products Using Hybridization Techniques

Examination of the total RNA products synthesized from chromatin can provide useful insights into the general properties of the process. However, the critical questions regarding chromatin transcription are: what specific RNA products are being synthesized and in what quantity? The answers to these types of questions require the use of molecular hybridization techniques by which specific RNA molecules can be individually monitored.

Pioneering work in this area was performed using filter hybridization techniques. In this method, total cellular DNA was immobilized on nitrocellulose filters and then allowed to hybridize with radiolabeled RNA. In most cases, specific RNA in chromatin transcripts was detected by competition experiments with in vivo RNA. The utilization of filter hybridization to study eucaryotic transcription is subject to certain limitations arising from the tremendous complexity of the eucaryotic genome and the presence of repetitive genome elements. Because of technical restraints on RNA concentration and incubation time, the method is only capable of detecting hybridization to reiterated gene sequences or hybridization by "highly abundant" sequences in the RNA transcripts (48). Thus, the hybridization can distinguish only between molecules from similar, but not necessarily identical, gene loci. Within these limitations, however, chromatin was found to support the transcription of only a subset of RNA sequences when compared to purified DNA (49–51). Furthermore, the transcripts of chromatin isolated from one organ were found in DNA-RNA competition hybridization reactions to be more like in vivo RNA isolated from the same organ than RNA from a different organ (50–52). Since the competition could reach completion, it was suggested that all sequences transcribed in vivo from repetitive genome elements were also present in the in vitro transcripts (53). Thus, at the level of sensitivity attainable in such experiments, chromatin appeared to maintain the specific restriction pattern of the tissue from which it originated.

The precise detection of specific RNA products required the isolation of highly purified, complementary DNA sequences which could be used in tracer quantities in RNA-excess hybridization. In a few cases in which the DNA of interest was present in repeated copies in the genome, this isolation could be performed from total cellular DNA. For instance, the genes for ribosomal RNA (54) and 5S RNA (55) have been substantially purified from Xenopus by density equilibrium centrifugation techniques. Studies on specific messenger RNA molecules were greatly facilitated by the discovery of RNA-dependent DNA polymerase (reverse transcriptase) from RNA tumor viruses. This enzyme was capable of synthesizing a highly radiolabeled complementary DNA (cDNA) copy of purified mRNA molecules. With such cDNA probes, it became possible to detect and estimate the concentration of specific mRNA sequences present in extremely low concentrations in total RNA populations.

Transcription of Ribosomal, 5S, and tRNA Genes

The many distinctions between the transcription of ribosomal, transfer and 5S RNA species, and messenger RNAs warrant separate discussion for these genes. The transcription of the former RNA species is carried out by different RNA polymerases (I and III) from those that carry out the transcription of mRNA (II). Furthermore, while genes coding for structural proteins are present in only a limited number of copies per genome, the ribosomal, transfer, and 5S RNA genes are present in reiterated copies generally clustered in a few specific sites in the genome. Thus, the mechanisms of transcriptional regulation for these different classes of genes may be very different.

Transcription of ribosomal RNA products from whole nuclei by endogenous RNA polymerase has been found to follow the in vivo pattern of transcription (56, 57). RNA was transcribed asymmetrically to yield a 45S ribosomal RNA precursor without transcription of "spacer" regions which were not normally transcribed in vivo. Transcription of these products from nuclei, however, represents completion of RNA chains initiated in vivo and does not indicate in vitro initiation specificity.

The availability of purified ribosomal DNA cistrons has made it possible to investigate whether the isolated DNA and the RNA polymerase alone are capable of faithfully reproducing the in vivo pattern of transcription. Using bacterial RNA polymerase, several groups have reported a transcriptional preference for the sense strand of ribosomal DNA (58, 59). While initiation was found to occur primarily from a specific site, it is now known that this site is not the correct initiation point for ribosomal RNA synthesis. Thus, the apparent specificity of the bacterial enzyme may be merely due to the presence of a nucleotide sequence arising by chance which acts as a strong promoter for bacterial RNA polymerase.

Analogous studies utilizing eucaryotic RNA polymerase have given rise to some conflicting results. Early studies of Roeder et al (60) failed to demonstrate a specific transcription of ribosomal DNA with homologous RNA polymerase I or II. Beebee & Butterworth (61) found some indications of specific transcription by RNA polymerase I; however, their studies were difficult to quantitate accurately. Only in the case of the yeast RNA polymerase I is there any evidence to support the hypothesis that the eucaryotic enzyme might be able to recognize intrinsic factors in the ribosomal DNA leading to specific transcription of ribosomal RNA (62–64). Interestingly, these studies indicate preferential transcription of ribosomal genes only when DNA of high integrity is used as template. Under such conditions, however, the transcription is symmetric. Final resolution of this question will require assessment of whether the in vitro initiation sites actually correspond to sites used in vivo.

Transcription of ribosomal RNA genes from chromatin has provided little encouraging results. Reeder (65) reported that *E. coli* RNA polymerase transcribed ribosomal and 5S cistrons in an aberrant fashion from Xenopus chromatin. Both strands of the ribosomal DNA, as well as spacer regions, were read by the bacterial enzyme. Analogous results were obtained when homologous RNA polymerase I and II were tested in the same system (66). These results suggested that factors other than chromosomal proteins and purified RNA polymerase are required for proper transcription of ribosomal RNA. Alternatively, the machinery for transcription may have been altered or damaged during preparation of chromatin or RNA polymerase.

In contrast to ribosomal RNA genes, the in vitro transcription of the products of RNA polymerase III—5S and tRNA—provide the best examples of faithful chromatin transcription presently known. Marzluff & Huang (1) have reported the transcription of tRNA and 5S RNA genes by endogenously bound RNA polymerase of mouse myeloma chromatin. The transcription was found to occur asymmetrically to produce RNA products of the same size as their in vivo counterparts. Furthermore, the synthesis represented to a large extent the in vitro initiation and completion of the RNA products. Recently, Roeder and co-workers found that the endogenous transcription of 5S RNA from Xenopus chromatin could be stimulated 10- to 50-fold by addition of purified RNA polymerase III from Xenopus (11, 67). The transcription by the exogenous RNA polymerase occurred in an asymmetric fashion to yield a product predominantly of the proper size. Addition of RNA polymerases I or II or *E. coli* RNA polymerase stimulated total RNA synthesis but did not selectively increase the transcription of 5S RNA sequences as observed with RNA polymerase III. This preferential and asymmetric transcription of 5S DNA genes, however, did not occur when naked Xenopus DNA was employed. Thus, the presence of both chromosomal proteins and purified RNA polymerase III are required to reproduce the in vivo pattern of 5S RNA synthesis. Further fractionation of this system may provide much useful information on the mechanism of transcription of this class of genes.

Transcription of Specific mRNA Sequences

The transcription of specific mRNA sequences from chromatin was first demonstrated by Axel et al (68) and Gilmour & Paul (69). Using highly radioactive cDNA to globin mRNA, these workers demonstrated that RNA transcripts of chromatin isolated from erythropoietic tissues contained globin-specific mRNA sequences. Chromatin isolated from tissues not active in globin synthesis, however, supported very little synthesis of globin sequences. Since these initial observations, similar findings have been reported for several other specific mRNA species, including ovalbumin mRNA (70), histone mRNA (71), and immunoglobulin κ light chain mRNA (72). Similarly, virus-specific RNA sequences were detected in the transcripts of chromatin from viral-infected or transformed tissues, but not normal cells (73–76). These studies supported the concept that chromatin is a restricted template and that this restriction is maintained in vitro. Furthermore, the restriction is such that it can be recognized by a bacterial RNA polymerase, although little information is provided on the accuracy of the transcription process.

The detection of specific mRNA sequences in chromatin transcripts has been complicated by the presence of endogenous mRNA sequences in the chromatin. Such endogenous mRNA is obviously indistinguishable from newly synthesized mRNA sequences by cDNA hybridization. Controls run in the absence of added RNA polymerase have indicated backgrounds ranging from about 10% for the transcription of ovalbumin mRNA from chick oviduct chromatin (70) to 50% for synthesis of globin mRNA from rabbit marrow chromatin (77). The presence of newly synthesized mRNA sequences in chromatin transcripts was confirmed by ^{32}P labeling of the synthesized RNA, followed by density equilibrium centrifugation to separate the mRNA-cDNA hybrid (77). Quantitative estimation was still imprecise, however, because of the low levels of ^{32}P incorporated into specific products in the presence of a high background.

A recent technical advance has been attained by the utilization of the 5-mercuri derivative of UTP as a substrate for RNA polymerase (78). Newly synthesized RNA containing mercury substitutions can be largely separated from endogenous RNA sequences by means of sulfhydryl-Sepharose affinity chromatography (72, 78–81). With this technology, fairly quantitative estimates of the levels of specific mRNA sequences in chromatin transcripts have been obtained (72, 79–81). However, it should be noted that even this technique is susceptible to certain procedural artifacts. Both RNA aggregation and RNA-dependent RNA synthesis can lead to retention of some endogenous contaminating mRNA on the sulfhydryl-Sepharose column. These problems can be counteracted by dilution and heat denaturation prior to application to the column. The nascent RNA should, of course, be susceptible to inhibition by actinomycin D.

The transcription of the ovalbumin gene from chick oviduct chromatin by *E. coli* RNA polymerase provides an interesting example of in vitro transcription of a specific mRNA gene (46, 70, 81). Chromatin isolated from the oviducts of chicks that received estrogen stimulation is capable of supporting the synthesis of ovalbumin mRNA sequences at a level of about 0.01% of the total RNA. The transcription has tissue specificity since chromatin isolated from chick spleen or reticulocyte is much less efficient as a template for ovalbumin mRNA synthesis. Furthermore, the transcription follows the expected pattern of hormone dependence. Chromatin isolated from unstimulated or hormone-withdrawn chick oviduct does not actively direct the synthesis of ovalbumin sequences. When deproteinized chick DNA was used as a template, the level of ovalbumin mRNA synthesized was about one one-hundredth of that seen with chromatin. This decreased level of transcription reflects the dilution of ovalbumin sequences by RNA species whose transcription is not restricted on the DNA. Thus, by several criteria the transcription of the ovalbumin gene follows the expected in vivo pattern. These studies, however, do not provide information on the accuracy of transcription. Bacterial RNA polymerase could transcribe any "available" DNA sequences in the chromatin in a random fashion. Attempts to examine this question are discussed in the next section.

Fidelity of In Vitro Chromatin Transcription of Specific mRNA Products

The fidelity of in vitro chromatin transcription encompasses two different problems. One question is the qualitative aspect of the accuracy of transcription: Is the transcribed RNA initiated and terminated at the proper sites on the DNA strand? The most direct approach to this problem would be to compare the nucleotide sequence of the 5'- and 3'-regions of the in vitro RNA with the actual in vivo product. This approach, of course, is not feasible because of the uncertain nature of the primary gene product. As a first approximation, the accuracy of transcription is sometimes monitored by examining the asymmetry of transcription. With the exception of a few anomolous systems, most genes are transcribed from only one strand, the coding or sense strand, of the DNA duplex (57, 83). Thus, the in vitro product should be transcribed asymmetrically if the RNA polymerase maintains the correct in vivo specificity. On the other hand, random transcription would lead to a highly symmetrical RNA product composed of transcripts from both nonsense and sense strands of the gene.

The second question of fidelity involves quantitative aspects of RNA synthesis. In vivo mRNA populations consist of members with widely varying individual frequencies which are, in part, due to differing transcription rates. Chromatin transcripts should follow the in vivo pattern, if proper regulatory signals are observed. Again, an unequivocal answer to this question cannot be attained at present because of uncertainties concerning the relative contributions of transcription rate and RNA turnover to the steady state level of cellular RNA. However, some aspects of this problem have proven amenable to investigation.

Quantitative estimations of specific chromatin products became reasonable with the recent techniques involving mercurated nucleotides. Using these techniques,

Towle et al (81) were able to examine the effect of varying the RNA polymerase-to-chromatin ratio on the transcription of the ovalbumin gene from chick oviduct chromatin. As the enzyme-to-DNA ratio in the reaction was lowered over a sixteen-fold range, the percentage of ovalbumin mRNA in the transcripts was found to increase sevenfold (81). This effect was not specific for the ovalbumin gene in oviduct chromatin alone; the transcription of globin mRNA from chick reticulocyte chromatin showed a similar change. On the other hand, the low level of nonspecific transcription of globin mRNA sequences from oviduct chromatin was not affected by changing the enzyme-to-DNA ratio. Likewise, the transcription of ovalbumin mRNA sequences from chick DNA, free of chromosomal protein, was not influenced by changing the enzyme-to-DNA ratio. Thus, a preferential transcription of certain genes relative to total RNA synthesis can occur from chromatin in vitro. Genes that are preferentially transcribed appear to coincide with genes that are preferentially transcribed in vivo. Since the preferential transcription does not occur from deproteinized DNA, the presence of chromosomal proteins must somehow be necessary for determining preferential transcription.

Recently, Biessmann et al (83) have examined Drosophila chromatin transcripts by hybridization to a cDNA probe to total nuclear poly A–containing RNA. In this study, the cDNA was fractionated into classes representing the high abundancy RNA and low abundancy RNA. The rate of hybridization of the chromatin transcripts to the high abundance cDNA probe was found to be about one hundred times greater than to the probe representing rarer sequences. Therefore, different sites in the chromatin are utilized by *E. coli* RNA polymerase with varying efficiencies. The observed differences show a quantitative relationship to the in vivo abundancies of the RNA product.

Preferential transcription of chromatin genes has thus been demonstrated both in the case of a specific RNA product and for total nuclear poly A–containing RNA. Studies on the asymmetry of chromatin transcription, however, have been slightly less encouraging. As discussed previously, genes for ribosomal RNA were transcribed symmetrically from chromatin by either homologous RNA polymerases or bacterial RNA polymerase (65, 66). Wilson et al (82) studied the asymmetry of transcription of rabbit marrow chromatin using an RNA probe synthesized from globin cDNA. Although a preference (2:1) for globin mRNA was found, both coding and noncoding strands of the globin gene were actively transcribed in vitro by either bacterial or homologous RNA polymerase. Beissmann et al (83) found sequences complementary to in vivo nuclear RNA in transcripts of Drosophila chromatin, indicating a considerable amount of transcription from the incorrect strand. Thus, in a number of systems, a significant degree of symmetric transcription of chromatin has been indicated.

A notable exception to these findings of apparently aberrant synthesis comes from studies on the chromatin of SV40 transformed cell lines. *E. coli* RNA polymerase transcripts of such chromatin were found to contain RNA complementary to only one strand of SV40 DNA (73, 74). The strand of DNA transcribed in vitro was the same as that coding for in vivo RNA products of the transformed cell line. Quantitation of the frequency of transcription from different regions of the SV40 genome demonstrated a pattern similar to that observed in the transformed cell (84). This pattern of restriction was not observed with purified DNA of the transformed cell line. Thus, in the presence of chromosomal proteins, *E. coli* RNA polymerase appeared to transcribe the viral gene sequences in a specific manner.

Recent studies on the transcription of chick oviduct chromatin have demonstrated an interesting effect of varying the RNA polymerase-to-chromatin ratio on the asymmetry of transcription (81). At high enzyme-to-DNA ratios (4 μg enzyme/ μg DNA), sequences homologous to both coding and noncoding strands of the ovalbumin gene were transcribed. As the ratio of enzyme-to-DNA was lowered, however, the transcription became increasingly asymmetric. At the lowest enzyme-to-DNA ratio tested (0.25 μg enzyme/μg DNA), approximately 95% of the ovalbumin gene transcripts were synthesized from the coding strand of the gene. Since most work on chromatin transcription has been performed at high enzyme-to-DNA ratios, this observation may explain some of the symmetry measured in various systems (82, 83). The effect of enzyme-to-DNA ratios on ovalbumin mRNA transcription was also observed with the eucaryotic RNA polymerase II from wheat germ.

The asymmetrical and preferential transcription of the ovalbumin gene from chromatin by *E. coli* RNA polymerase provides rather strong evidence that the bacterial enzyme can recognize certain regulatory features of the chromatin. This result, however, does not imply that the bacterial polymerase has the capacity to recognize eucaryotic promoters in an identical fashion with the eucaryotic enzyme. Similar conclusions have been derived from studies on the transcription of SV40 viral sequences from transformed cell chromatin and of total nuclear poly A– containing RNA from Drosophila chromatin. Furthermore, the specific pattern of transcription observed in these studies was not found on deproteinized DNA. Thus, chromosomal proteins appear to play a dominant role in regulating the specificity of the process. In the next section, we examine in more detail the specific effects of chromosomal proteins on in vitro chromatin transcription.

CHROMOSOMAL PROTEINS AS DETERMINANTS OF GENE EXPRESSION

Reconstitution of Chromatin

The control of transcriptional activity must be encoded by nucleotide sequences in DNA so that chromosomal activator or repressor proteins can read specific sequences and control the amount or quality of RNA synthesized. The interaction of chromosomal proteins may also have a structural basis in the packaging of active and inactive regions of the genome. Therefore, we consider the interaction of DNA with chromosomal proteins in relationship to transcriptional control.

Initial studies to identify the components responsible for the specific restriction of chromatin were pioneered by Paul & Gilmour (50, 86, 87) and Bonner (88), using reconstructed chromatin from isolated components. Gilmour & Paul (50, 86, 87) first implicated the nonhistone protein fraction of chromatin as a determinant for the specificity of RNA synthesis in eucaryotic systems. Essentially, chromatin was physically separated into three components: DNA, histones, and nonhistone proteins. Homologous and heterologous chromatins were reconstituted by a slow stepwise dialysis of required components from 6 M urea and 2 M NaCl into 0.2 M NaCl, after which the insoluble nucleoprotein was centrifuged and washed repeatedly with distilled water. Native and reconstituted chromatins were transcribed by bacterial RNA polymerase, and the nucleotide sequences in the synthesized RNA were assayed by hybridization competition experiments. Paul & Gilmour showed that RNA transcribed from reconstituted chromatin was indistinguishable from RNA synthesized from native chromatin, and that both were extensively similar to RNA isolated from the tissue (calf thymus) from which the components were derived. Although the nonhistone proteins used by these investigators contained variable amounts of nucleic acid, further experiments showed that the template properties of the reconstituted chromatins were dependent upon the source of the nonhistone fraction. Subsequently a number of workers were able to show by a variety of hybridization techniques that the synthesis of RNA was related to the origin of nonhistone protein (2, 53, 92).

In parallel experiments of this era, both Bonner's laboratory (88) and Huang & Huang (89) suggested that a particular species of protein-bound chromosomal RNA played an important role in DNA protein complex recognition in reconstitution. However, the nature of the chromosomal RNA remains in question since a series of reports have stated that chromosomal RNA is a mixture of degraded rRNA and tRNA (90, 91).

Recently, the conclusions derived from reconstitution experiments have been criticized on the grounds that the conditions of low RNA concentrations and relatively short incubation times (low Cot conditions) permit only the RNA transcribed from the repetitious DNA to form hybrids (see above). However, in more recent experiments in which RNA synthesized from chromatin was hybridized to unique-sequence DNA probes representing cytoplasmic messengers, the RNA products synthesized in vitro also appeared to be organ-specific. The putative role of the different types of nonhistone proteins in this process is discussed further in a later section.

Undoubtedly the technique of chromosomal protein reconstitution will be instrumental in elucidating the mechanism of gene control. However, because of the hybridization methods used, the fidelity of the reported reconstituted chromatins remains obscure. Furthermore, several reports have appeared over the past few years suggesting that chromatin protein components may be degraded as a result of treatment incurred during dissociation, fractionation, and reconstitution (93, 94). Thymus chromatin histones have been found to degrade under reconstitution buffer conditions of 5 M urea, 2 M NaCl, and 10 mM Tris, pH 8 (95, 96). Chae & Carter (94) reported that high molecular weight nonhistone proteins are degraded in liver chromatin even in the presence of sodium bisulfite, a standard inhibitor of chromatin-bound protease (97). While liver and thymus chromatin are known to have protease activity (97, 98), it is quite possible that chromatin isolated from other tissues or cultured cell lines may also contain proteases. Although a number of investigators have not observed degradation of proteins by polyacrylamide gel electrophoresis (99), these experiments raise questions concerning the integrity of the reconstituted chromosomal protein and create a degree of doubt about the validity of transcription experiments that used only repetitious sequence hybridization as the sole criterion for evaluating reconstitution.

It appears necessary to establish whether reconstituted chromatin is a satisfactory representative of "native" chromatin as isolated intact from cells. Several lines of evidence have suggested some degree of fidelity of chromatin reconstitution. First, thermal denaturation profiles were recorded by Spelsberg et al (95) to provide information about the physical state of the DNA in nucleoprotein. Such profiles would indicate the presence of any appreciable amount of uncomplexed DNA as a hyperchromic rise taking place in a temperature region between 45° and 50°C. Native and reconstituted thymus or liver chromatin demonstrated hyperchromicity data that were in agreement with reports on thermal denaturation (88). Second, reporter molecules (i. e. nitroaniline-labeled diammonium salt) with specificity for the minor groove of the DNA helix (100) were bound to native and reconstituted HeLa cell chromatin (99). The maximum number of bound reporter molecules, one per seven base-pairs, was similar in native and reconstituted chromatin. Third, circular dichroism was used as another probe to assess structural integrity of chromatin following reconstitution. Stein and co-workers (99) found that the CD spectra of a maximum ellipticity at 272 nm, the point of cross over at 258 nm, the point of negative ellipticity at 245 nm, as well as the peptide chromophore region at 220 nm were identical for both native and reconstituted chromatin. A number of studies have shown that the template activities of native and reconstituted chromatin are quantitatively identical (2, 88, 101). Tsai et al (102) showed that when native and reconstituted oviduct chromatin were transcribed under conditions in which reinitiation is inhibited, the transcriptional activity as well as the number of RNA polymerase initiation sites were identical in both chromatin preparations. Taken together, these reports suggest, but do not conclusively establish, the fidelity of chromatin reconstitution. Furthermore, discrepancies in the quality of the reconstituted product may arise because of differing technical protocols used in the reassociation process.

An extensive study of the association events that occur during the reconstitution of salt-urea dissociated chromatin was reported by Chae (103). Chae found that the reconstitution methods commonly used in conjunction with simple salt-urea dissociation could produce divergent effects on the reconstituted product. One might predict that under the conditions used by Bekhor et al (88) and Gilmour & Paul (86, 87), most of the nonhistone proteins and histones reassociated with DNA only in the last few hours of their experiment. The reassociation of histones and nonhistone proteins was more gradual in the procedure of Kleiman & Huang (96) than in the protocol of Bekhor et al (88). The chromatin recovered by Stein most likely lacked most of the nonhistones and a small amount of histone (103). In the reconstitution procedures, it is still not certain whether the nonhistone proteins rebind to DNA at the same time as histones or actually reassociate subsequent to formation of histone-DNA complexes.

Recently, Felsenfeld and co-workers (104) have devised a new method for reconstitution. In this procedure the concentration of salts is progressively lowered to 0.6 M, followed by dialysis to remove the urea and then the remaining salt. This method results in the reconstitution of chromatin which again contains the original nucleosome structure (see 105, 106).

Role of Histones in Transcription

Stedman & Stedman (107) first proposed that histones could be specific gene repressors. It was found that the addition of histones to DNA reduced the template capacity of the DNA (12, 108). Other results suggested that the apparent decrease in DNA template activity in the presence of histones reflected the precipitation of the nucleohistone complex (13, 109, 110). However, Hindley (111) was able to show that precipitation of the DNA was not in itself responsible for the decreased template activity. Furthermore, Johns (112) studied the binding of the histone H_2b fragments to DNA and found that while the N-terminal fragment, and to a lesser degree the C-terminal fragment, can bind to DNA and precipitate it, only the N-terminal fragment is able to depress the template activity of the DNA. The reduction in template activity was recently examined by the rifampicin-nucleotide challenge assay. Histone (1.2:1) reduced severely the DNA template capacity in comparison to chromatin (102). In agreement with Koslov & Georgiev (113), we also found that histones inhibited RNA chain elongation rate in DNA complex in comparison to native chromatin (R. J. Schwartz, unpublished observations).

The histones are bound to the DNA by both ionic and hydrophobic bonds (96, 113–115) into a basic unit of structure called *nu body* or *nucleosome*, which involves approximately 140 base-pairs of DNA complexed with two each of the four major histones (H_{2a}, H_{2b}, H_3, and H_4) (116). Nuclease digestion studies have supported a subunit or particulate structure for chromatin. Both endogenous (116, 117) and exogenous nuclease (118–122) appear to recognize a repeating nucleoprotein unit along the chromatin fiber and support the bead-on-string model of chromatin structure proposed by Olins & Olins (105).

The elucidation of a regular distribution of nu bodies in chromatin poses the question as to the possible role of this structure in transcription. Analysis of the organization of this subunit with respect to DNA revealed that a specific sequence of nucleotides was not necessary for nu body formation (123, 124). Felsenfeld's laboratory (125) examined the distribution of protein bound to the transcriptionally active globin gene of duck reticulocyte chromatin. They found that the globin gene sequence was distributed in both micrococcal nuclease-sensitive (open) and nuclease-resistant (covered) regions of reticulocyte chromatin. They also reported that the total open and covered DNA had similar sequence populations. These results were confirmed by Kuo et al (126) using cDNA to total poly (A)–containing RNA as a hybridization probe. Thus, current evidence suggests that histones are not arranged in a specific manner relative to active genes or total nucleotide sequences. It should be noted, however, that proteins can undergo a number of different post-translational modifications (127, 128). It is unclear whether the relative concentrations of acetylated, phosphorylated, and methylated histones vary in the regions of actively transcribed genes. Nevertheless, there is no direct evidence to support the hypothesis that histone modification is responsible for inducing transcription of specific genes.

Role of Nonhistone Proteins in Transcription

Several lines of indirect evidence have implicated nonhistone proteins as the potential candidates for regulating gene expression in eucaryotic cells (reviewed in 129–131). First, nonhistone proteins are present in increased levels in tissues which are active in RNA synthesis, while the level of histones remains constant (132). Second, the nonhistone protein fraction exhibits at least two to three orders of magnitude greater diversity than histones (133, 134). Third, nonhistone proteins possess both tissue specificity and DNA binding specificity (135–140). Fourth, certain classes of nonhistone proteins have been shown to stimulate RNA synthesis in vitro (2, 141, 142). Fifth, synthesis of specific classes of nonhistone proteins is associated with the induction of gene activity (143–145).

THE GLOBIN SYSTEM As discussed previously, chromatin from erythropoietic tissues supported the synthesis of globin mRNA sequences, but chromatin from brain or other tissues did not (68, 69). To study the effect of chromosomal proteins on this process, Barrett et al (146) performed reconstitution experiments analogous to the earlier work of Paul & Gilmour (50). Chromatin was reconstituted from the purified DNA, histones, and nonhistone of either reticulocytes or liver and then transcribed with bacterial RNA polymerase. The fractional yield of globin sequences for native reticulocyte chromatin was 0.019% in agreement with values reported by Axel et al (68) and Steggles et al (85). The reconstituted chromatin containing reticulocyte nonhistone had a similar template activity to the native chromatin and a fractional yield of globin sequences in the range of 0.007 to 0.02%. Importantly, there were no detectable globin sequences contaminating the reconstituted fraction prior to the addition of RNA polymerase. In contrast, no globin-specific sequences were detected in the transcripts of reticulocyte chromatin which was reconstituted with nonhistone proteins obtained from liver. Neither free DNA nor reconstituted erythrocyte nucleohistone, a poor template, yielded RNA with detectable globin sequences.

Gilmour & MacGillivray (147) also found that the dissociation and reconstitution of mouse fetal liver chromatin did not affect the capacity to act as a template for globin mRNA sequences. Furthermore, evidence has been presented to suggest that nonerythroid (mouse brain) chromatin can prime for the synthesis of globin mRNA following reconstitution in the presence of the nonhistone protein extracted from fetal liver chromatin (148). In order to remove endogenous RNA contamination, these investigators developed a two-step reconstitution procedure in which chromosomal proteins were initially dissociated and then separated from nucleic acid contamination by CsCl-urea gradients. Reconstitution of CsCl-purified proteins to purified DNA yielded a functionally active chromatin that was essentially devoid of endogenous RNA. Thus, Gilmour & MacGillivray (147) showed that expression of the globin gene is under the direct influence of the nonhistone component of erythroid chromatin, in agreement with Barrett et al (146).

Gilmour & MacGillivray (147) further applied these techniques to assay for globin gene "activator" activity in a number of subfractions of fetal liver nonhistone protein obtained by hydroxylapatite chromatography. The bulk of activity was present in a fraction that elutes in 0.05 M sodium phosphate. This fraction contains 12% of the chromosomal protein but essentially all of the globin gene stimulating activity. Isolation of a pure activator protein still remains a difficult problem because of the high degree of protein heterogeneity in this fraction. Chiu et al (149) have also developed a fractionation scheme to isolate chromosomal nonhistone protein involved in regulation of the globin gene. The nonhistone proteins, which are tightly bound to DNA and represent less than 5% of the total reticulocyte chromatin proteins, were found to be essential for the in vitro transcription of globin gene sequences by the reconstituted chromatin preparation.

THE OVALBUMIN SYSTEM The chick oviduct system is particularly suitable for investigating the function of nonhistone proteins in the control of gene expression. In the chick oviduct, the expression of the gene coding for ovalbumin mRNA is under hormonal control (150). The substantial difference in the concentration of ovalbumin gene sequences in the in vitro transcripts of estrogen-stimulated and hormonally withdrawn chick oviduct chromatin provided an approach to study the role of chromatin nonhistone proteins in the regulation of gene expression.

Reconstitution studies were performed using histones, extractable nonhistone proteins, and tightly bound nonhistone protein-DNA complex which were fractionated from chromatins isolated from estrogen-stimulated and hormone-withdrawn chick oviducts (102). The results demonstrated that 0.011% of the in vitro RNA transcripts produced from reconstituted estrogen-stimulated chromatin were ovalbumin mRNA sequences as compared to 0.0015% obtained for reconstituted hormone-withdrawn chromatin. Thus, the ratio of ovalbumin mRNA sequences in the in vitro transcripts of reconstituted stimulated and withdrawn chromatins was eight to one, identical with the levels observed in transcripts of native estrogen-stimulated and hormone-withdrawn chromatins. Furthermore, the number of initiation sites for RNA synthesis on reconstituted and native chromatins was indistinguishable. The specificity of transcription of isolated chromatins appeared to be conserved upon dissociation and fractionation of the chromatin proteins followed by reconstitution of these constituents to DNA.

The effect of chromatin proteins on gene expression was further examined by reconstituting components from different developmental stages. Reconstitution of extractable nonhistone proteins from estrogen-stimulated chromatin to histones and tightly bound nonhistone protein-DNA complexes from hormone-withdrawn chromatin resulted in the synthesis of a substantial amount of ovalbumin mRNA ($mRNA_{ov}$) sequences. Conversely, when extractable nonhistone proteins from withdrawn chromatin were reconstituted to histones and tightly bound nonhistone protein-DNA complexes from estrogen-stimulated chromatins, very low levels of mRNA$_{ov}$ sequences were detected. In contrast, interchange of the histones and tightly bound nonhistone protein-DNA complexes from hormone-withdrawn and estrogen-stimulated chromatins during reconstitution did not affect the level of mRNA$_{ov}$ sequences produced. Therefore, the extractable nonhistone proteins of chromatin appear to play a dominant role in regulating specific gene expression.

It was considered that the control elements residing in the nonhistone fraction could act either as a positive regulator(s) present in the nonhistone protein of the stimulated chick oviduct or as a negative regulatory protein present in the nonhistone of hormone-withdrawn chromatin (151). To differentiate between these two general modes of action, extractable nonhistone proteins were isolated from either withdrawn or stimulated chromatin and then reconstituted together with the full complement of their respective homologous components (152). The results indicated that extractable nonhistone proteins from stimulated chromatin were capable of activating the in vitro transcription of the ovalbumin gene when included in the reconstitution of withdrawn chromatin. On the other hand, addition of extractable nonhistone proteins from withdrawn chromatin to stimulated chromatin did not affect the synthesis of mRNA$_{ov}$ sequences. Therefore, the extractable nonhistone proteins from stimulated chromatin appear to contain a positive regulator(s) which controls the in vitro expression of the ovalbumin gene.

THE HISTONE SYSTEM In dividing cells, the synthesis of histones is confined to the S phase of the cell cycle (153–155). Several lines of evidence have suggested that nonhistone chromosomal proteins are important in the control of transcription during the cell cycle (101, 131). Variations in the composition and metabolism of nonhistone chromosomal proteins were consistent with a regulatory function for these proteins (156–158), while a series of chromatin reconstitution studies demonstrated that nonhistone chromosomal proteins are responsible for the increased transcriptional activity of S-phase chromatin in comparison to mitotic chromatin (101, 131).

Similar studies to those on the globin and ovalbumin genes have been carried out recently on the expression of the histone gene. Stein et al (71) utilized a [^{3}H]-labeled cDNA probe for histone mRNA to assay the presence of histone-specific sequences in RNA transcripts synthesized in vitro. Histone gene sequences are of a middle repetitive family and compose 1% of the total RNA transcript of native HeLa cell DNA (159). Stein showed that histone complexed to DNA can reduce the transcription of histone genes in a random and nonselective manner (159). When RNA transcripts from S-phase chromatin were examined for histone mRNA, approximately 8% of the transcripts contained hybridizable sequences (Cot$_{0.5}$ of 2.1 × 10^{-1}), while there were no detectable histone sequences in transcripts of G$_1$ phase chromatin hybridized up to Cot of 100. The technique of chromatin reconstitution demonstrated that the nonhistone chromosomal protein of the genome was responsible for the difference in the in vitro histone gene expression (71, 160).

The biochemical nature of the nonhistone proteins of S phase was examined by Kleinsmith et al (161). A nuclear phosphatase was used to dephosphorylate nonhistone protein from S-phase HeLa cells. Chromatin reconstituted with dephosphorylated proteins exhibited about a 50% reduction in the overall number of initiation sites available for transcription and a similar reduction in preferentiality of histone gene transcription in comparison to untreated nonhistone proteins. Although degradation was not ruled out during this experiment, the results tend to provide support for the theory that phosphorylation of protein is involved in regulating gene transcription.

The Mechanism of Action of Chromosomal Proteins on the Transcription Process

INITIATION OF RNA SYNTHESIS The importance of chromosomal proteins for specific gene transcription has been adequately stated. However, the mechanistic function of these proteins in relation to selective gene transcription has remained virtually unknown. Recently, a number of investigations have begun to elucidate some of the structural and functional aspects of chromatin proteins in regard to RNA chain initiation.

The most characteristic physical difference between chromatin and DNA was the finding that formation of the stable binary complex between RNA polymerase and chromatin was practically independent of temperature in comparison to DNA (33, 162, 163). In simple reconstitution experiments, both histones and nonhistone chromosomal proteins were found to be necessary to fully maintain this feature of chromatin transcription. Since the temperature of incubation is thought to be involved in the opening of the DNA strands during stable binary complex formation, chromatin may be present in a state in which conversion to the stable preinitiation complex is greatly facilitated (163). One possible explanation is that the initiation regions of chromatin are stretches of single-stranded DNA. This hypothesis is consistent with the results of Howk et al (164) and Groner et al (28), which demonstrated that the initial in vitro transcriptional products on chromatin using RNA polymerase II were RNA-DNA hybrids. Crick (165) has also proposed that in eucaryotic cells, RNA chain initiation may occur at single-stranded regions of DNA. The presence of large single-stranded regions accessible to RNA polymerase does not seem likely, however, because experiments failed to demonstrate any inhibition of chromatin transcription by either antibody to single-stranded DNA or single-strand specific nuclease (163). Recently, however, a large number of smaller single-stranded regions have been observed in cross-linked chromatin DNA (166).

It may also be reasonable to assume that chromosomal proteins present at initiation sites lower the activation energy required for opening of the DNA strands in forming the stable binary complex. Such proteins would be expected to be major determinants of the location of initiation sites for RNA polymerase and perhaps be composed of a variety of proteins capable of destabilizing DNA (167, 168). Competition experiments with *E. coli* RNA polymerase and hen oviduct RNA polymerase II indicate that these enzymes use the same region for initiation of RNA synthesis on chromatin. Since these polymerases have different template specificity on native DNA (34, 43), it seems likely that chromosomal proteins are exerting dominant influence in the selection of initiation sites on chromatin by both eucaryotic and procaryotic polymerases. Nevertheless, elucidation of the exact mechanisms of initiaation of eucaryotic unique gene expression will require definition of the primary transcript and the promoter itself.

SELECTIVE DNase DIGESTION   The enhanced susceptibility of actively transcribed genes to deoxyribonuclease has provided additional evidence to suggest that template-active regions of chromatin are structurally distinct from nontranscribable segments of the genome. Initial studies by Gottesfeld et al (169) showed that in a limit digest of chromatin, micrococcal nuclease sensitivity does not discriminate between active and inactive chromatin regions. However, when rat liver chromatin was fractionated by a limited digestion with DNase II, followed by a differential precipitation step in $MgCl_2$, the resulting soluble fraction contained about 11% of the total chromatin DNA and appeared to be the active gene region. This fraction contained nascent RNA chains, had a relatively high template activity, and was enriched in nonhistone proteins. The DNA of this fraction hybridized with a kinetic complexity that was about one tenth that observed for total DNA, indicating that it contained a subset of single-copy DNA sequences. Total cellular RNA preferentially hybridized with DNA from this fraction, supporting the notion that the fraction was active during in vivo transcription.

In experiments designed to examine specific gene sequences, Weintraub & Groudine (170) found that when 10% of the DNA of erythrocyte chromatin was made acid-soluble with pancreatic DNase I, the remaining 90% of the nuclear DNA failed to protect globin cDNA from nuclease digestion in a hybridization assay. Thus, similar to DNase II (171), DNase I appears to preferentially digest genes that are being actively transcribed. Weintraub also found the following: (a) fetal but not adult globin sequences are preferentially digested in fetal red cells; (b) adult but not fetal globin sequences are preferentially digested in adult red cells; (c) neither fetal nor adult globin sequences are preferentially digested in nuclei obtained from chick fibroblasts or chick brain; (d) when a cDNA probe to ovalbumin is used, ovalbumin DNA sequences are not preferentially digested in fibroblast, brain, or red cells. Garel & Axel (172) have confirmed recently the observations that DNase is capable of digesting transcriptionally active regions in chromatin.

In summary, digestion of nuclear chromatin by micrococcal nuclease has shown that histones cover both active and inactive transcription regions of DNA. In contrast, experiments utilizing DNase I digestion suggest that if histones cover active gene regions, the histone DNA complex must be in a conformation that is particularly sensitive to the nuclease. It is possible that the nonhistone proteins might selectively alter histone-DNA interaction to effect a structural alteration and thus create a transcriptionally active region.

STEROID HORMONES AS INDUCERS OF GENE EXPRESSION

Hormone-Receptor Interactions with Chromatin

Numerous studies have demonstrated that the action of steroid hormones is mediated at the level of genetic transcription (150, 173–175). It appears that the pattern of steroid hormone action may include the following events: First, steroids are passively taken up by the target cells and bind to a specific cytoplasmic receptor; second, the steroid-receptor complex is "activated" and translocated into the nuclear compartment; third, the receptor-hormone complex binds to "acceptor sites" on nuclear chromatin; and fourth, activation of the transcriptional process at chromatin "effector sites" induces the appearance of specific new RNA sequences. This general model now seems to hold for a wide variety of steroid hormones including estrogen (176–181), progesterone (182–184), aldosterone (185), glucocorticoids (186, 187), and androgens (188–190).

The relationship between nuclear receptor content and gene transcription has been studied in the chick oviduct model system. The kinetics of estrogenic stimulation of initiation sites for RNA polymerase on oviduct chromatin was shown to correlate with the changes in endogenous levels of nuclear estrogen receptor (192). The number of available initiation sites for RNA synthesis on chromatin was measured by the formation of rifampicin-resistant stable binary complexes (polymerase : DNA), as detailed previously. The endogenous levels of nuclear estrogen receptor were quantitated by [^{3}H]estrogen exchange assay developed by Anderson et al (191). It was found that upon withdrawal of estrogen from prestimulated chicks, both nuclear estrogen receptor levels and RNA initiation sites in chromatin declined in parallel with a similar half-time (39, 192). When estrogen was readministered to these chicks, the number of initiation sites increased twofold as early as 30 min and approached a maximal level (threefold) between 1–2 hr after hormone administration. During the same period of restimulation by estrogen, the number of estrogen receptor molecules bound to nuclei increased to a maximum at 20 min and then declined slightly at 1 hr to a steady state level which remained elevated compared to the number bound to nuclei in estrogen-withdrawn chicks. Simultaneous measurements of RNA chain length and RNA chain elongation rate demonstrated that these characteristics remained relatively constant throughout both estrogen withdrawal and secondary stimulation. The temporal correlation between the changes in nuclear estrogen receptor levels and the number of RNA initiation sites on chromatin in the same nuclei supported the notion that steroid hormone-receptor complexes act on chromatin to effect the observed changes in genetic transcriptive activity.

The mechanism by which biochemical information held by the steroid-receptor complex is transferred to the transcriptional apparatus has been a central problem in steroid hormone action. A number of groups have examined the properties of the nuclear acceptor sites for binding steroid hormone receptors (187, 193–200). Initial studies have examined the participatory role of DNA in the binding of receptors. Chromatin regions to which estrogen, aldosterone, and glucocorticoid receptors bind appear to be highly susceptible to digestion by DNase (187, 194–198, 200). These experiments suggest that receptors are binding directly to exposed regions of DNA. In agreement, Baxter et al (187) found that chromatin DNA with bound receptor was more resistant to DNase digestion. However, other studies utilizing the progesterone receptor have shown that nuclear digestion had little or no effect on receptors binding to oviduct chromatin (199). A number of reports have even demonstrated saturable binding of receptors to purified DNA (180, 187). However, there is no firm evidence for a specific acceptor on DNA as defined by a nucleotide sequence. This is adequately demonstrated by reports that showed that receptors bind equally to DNA from either eucaryotic or procaryotic sources (201–203).

If binding of a receptor to purified DNA is indeed relatively nonspecific, it seems rather unreasonable for DNA to be the sole determinant of receptor binding to chromatin. In mixing experiments performed in cell-free systems, it was demonstrated that steroid receptor complexes isolated from target tissue bind preferentially to nuclei and chromatin from the same target tissue (188, 204–207). These experiments suggested that chromatin proteins might participate in a cooperative fashion with DNA in defining nuclear acceptor sites. There are two possible candidates for such factors in the nucleus: histones or the nonhistone chromosomal proteins.

Spelsberg et al (204, 206) showed that histones are not acceptor molecules and cannot account for the tissue specificity of nuclear binding. More definitive experiments have implicated the nonhistone proteins as a prime component of the nuclear acceptor site. Puca et al (181) have detected an apparent high affinity interaction between the calf uterine estrogen receptor and a preparation of acid-extracted basic chromosomal proteins that are not histones. These proteins were covalently attached to Sepharose and used to remove estrogen receptor complexes from uterine extracts. These acceptor proteins were able to bind receptor most efficiently in the absence of DNA.

The acceptor role of nonhistone chromosomal proteins have been most heavily studied in the binding of the progesterone receptor to chick oviduct chromatin (204–206). The oviduct cytoplasmic progesterone-receptor complex in crude or pure state bound to chromatin much more effectively than either uncomplexed receptor or free progesterone alone. The capacity of oviduct chromatin to bind the progesterone-receptor complex was greater than that of nontarget-tissue chromatins such as spleen, heart, liver, or erythrocyte (206, 207). The gradual removal of histones from chromatin with salt and urea and subsequent reconstitution of the chromatin by sequential dialysis allowed preparation of "hybrid" chromatins in which histones from other tissues or species were substituted during the reconstitution process (206). Binding of progesterone receptor to homologous reconstituted chromatin was similar to its binding to the intact native oviduct chromatin. Moreover, the capacity to bind the steroid-receptor complex was retained by hybrid chromatins containing histones even from a nontarget tissue of a different species such as calf thymus. Dehistonized oviduct chromatin displayed slightly greater binding than dehistonized spleen chromatin. Treatment of the dissociated chromatin with ribonuclease before reconstitution did not appreciably affect the binding. However, if the nonhistone proteins were removed, the chromatin lost most of its capacity to bind the progesterone-receptor complex. These experiments indicated that histones and RNAs were not the acceptor molecules, whereas nonhistone proteins apparently played a major part in hormone-receptor interactions with nuclear acceptor sites (204, 205).

Nonhistone proteins imparted quantitative specificity to the chromatin acceptor sites. Reconstituted oviduct chromatin and intact oviduct chromatin bound the receptor in a quantitatively similar manner, whereas insertion of nonhistone proteins from a nontarget tissue markedly depressed binding. Conversely, insertion of oviduct nonhistone protein into erythrocyte chromatin bestowed binding capacity to this nontarget chromatin DNA. Additional experiments localized the "acceptor capacity" with a subfraction of the nuclear acidic proteins (206). Whether the postulated acceptor is analogous or related to an extractable factor from calf uterus as described by Puca (181) is uncertain. However, it seems probable that the actual chromatin-associated acceptor sites consist of a DNA backbone which is structurally modified by chromatin-associated nonhistone proteins.

Activation of Gene Expression by Purified Hormone Receptors

Experiments in intact cells have implied that steroids may have rapid effects on transcription (39, 183). A number of investigators have attempted to detect such changes by the addition of cytosol receptor preparations to cell-free chromatin or nuclear transcription systems.

Raynaud-Jammet & Baulieu (208) and Arnaud et al (209) have reported an effect of estrogen on uterine RNA polymerase in a crude cell-free system containing receptors and nuclei. More detailed observations have been made by Mohla et al (210) in the rat uterine system. Similarly Davies & Griffiths (211) have incubated cytosol fractions containing androgen receptor with prostatic nuclei and demonstrated an enhancement of total nuclear polymerase activity. These results tend to corroborate the proposition that steroid hormones can exert effects on nuclear gene transcription. Clearly, such systems are complex in that uncontrolled variables could render a number of interpretations. In fact, it has been reported that crude cytosol receptor fractions of eucaryotic cells can often spuriously stimulate RNA synthesis by a template-independent process which is unrelated to hormone receptors (212).

In order to study the molecular mechanism by which hormone-receptor complexes regulate genetic transcription, this laboratory utilized a cell-free system in which all parameters of transcription (hormone, receptor, chromatin, RNA polymerase, cofactors, and RNA precursors) were pure and subject to experimental manipulation (183). The rifampicin challenge assay described above was used to quantitate the number of stable preinitiation complexes formed between RNA polymerase and chromatin prior to and following an in vitro exposure of the template to progesterone-receptor complexes.

To test directly the effect of progesterone receptor on transcription in vitro, a reconstituted cell-free system was employed which contained cytoplasmic progesterone-receptor complexes, highly purified by affinity chromatography, *E. coli* RNA polymerase, and chromatin prepared from hormonally withdrawn chick oviducts (213). Purified progesterone-receptor complex stimulated transcription of oviduct chromatin in vitro by promoting an increase of 3000 to 5000 additional sites for RNA chain initiation (183). These data showed that progesterone receptor can directly increase the number of RNA polymerase binding and initiation sites in the chromatin template in the absence of a detectable change in either the rate of RNA chain propagation or the size of the RNA product. The incorporation of [γ-^{32}P]-GTP into the 5'-termini of RNA substantiated our observation that the progesterone receptor stimulates initiation of transcription in the cell-free assay by increasing the number of chromatin initiation sites available to RNA polymerase (183). The kinetics of progesterone-receptor stimulation of RNA synthesis in chromatin revealed a $T_{1/2}$ of 15 min for this effect to occur. This value was identical with the optimal time required for binding of receptor to chromatin under similar conditions (214). The concentration of receptor required for half-maximal stimulation of RNA chain initiation was $\sim 5 \times 10^{-9}$ M. This value agreed closely with our previously reported estimates of the affinity ($K_d \sim 5 \times 10^{-9}$ M) of the progesterone-receptor complex for oviduct chromatin (193, 199). The stimulatory effect of the purified progesterone receptor appeared to be relatively specific for oviduct chromatin in comparison to nontarget-tissue chromatins or chick DNA. The data presented here show that the steroid hormone-receptor complex can directly regulate gene transcription in vitro in a manner that mimics the events observed in vivo in target cells.

It has been suggested by Yamamoto & Alberts (215) that added receptors may alter the gross structure of chromatin, and that the transcriptional changes detected in vitro may be a consequence of structural changes rather than reflecting the specific primary response per se. Since progesterone administered in vivo can rapidly induce synthesis of ovalbumin mRNA in chicks withdrawn from hormone, it was conceivable that purified progesterone receptor could induce transcription of the ovalbumin gene in isolated oviduct chromatin. Such a demonstration would strongly reaffirm the notion that the in vitro receptor-chromatin interaction mimics the events in vivo in a rather specific manner. To test this proposition, bulk amounts of RNA were synthesized from withdrawn chromatin incubated alone and from withdrawn chromatin incubated in the presence of purified progesterone-receptor complex (1×10^{-8} M). Both RNA preparations were assayed for the presence of complementary sequences to ovalbumin [^{3}H]cDNA. The RNA synthesized in the presence of receptor-hormone complex contained a tenfold enrichment of mRNA$_{ov}$ sequences as compared to that found in untreated withdrawn chromatin controls (216). It thus appears that a steroid-receptor complex may act directly on chromatin to enhance the number of initiation sites for RNA synthesis, thereby promoting the synthesis of specific mRNAs for induced proteins.

Mechanism of Steroid-Receptor Action on Transcription

The mechanism of steroid hormone action may be related to the distinctive properties of the progesterone receptor. Our current working hypothesis for progesterone action in chick oviduct has been summarized elsewhere (173). The progesterone receptor is a dimer composed of A and B subunits that have different and unique properties (217, 218). The intact dimer (6S) is located in the cytoplasm of the target cell in the absence of hormone stimulation and translocates to the nuclear compartment on administration of progesterone (217). Both the A and B subunits bind a molecule of hormone. The B subunit binds to the nonhistone protein-DNA complexes of oviduct chromatin but not to pure DNA, whereas the A subunit binds to pure DNA but poorly to chromatin (205, 217). Accordingly, these observations have led to the suggestion that the A subunit could be the actual gene regulatory protein. In the absence of the B component of the dimer, the A subunit should encounter difficulty in locating the specific initiation sites (genes) it is to regulate, and the B subunit alone should be totally inactive as a transcriptional stimulant. Consistent with these predictions, purified A subunit protein (184) was capable of stimulating transcription on withdrawn chromatin but only at much higher concentrations (approximately 10- to 50-fold) than that required for the intact dimer (193). The isolated B subunit (219) was totally ineffective in stimulating transcription from oviduct chromatin at any concentration tested (184). These observations are consistent with a model in which one part of the progesterone receptor dimer (B subunit) acts as a binding site specifier to localize the dimer in certain regions of chromatin, whereas the other subunit (A) may destabilize a portion of the chromatin DNA so that new sites are available to RNA polymerase for initiation of RNA synthesis. To finally prove or disprove this model, a more simplified system will be required.

DIFFERENTIATION AND HORMONE ACTION A model hypothesis for the interrelationships of differentiation, specific gene restriction, and hormone induction is depicted in Figure 1. For illustrative purposes, we might consider oviduct genes to be assigned to one of three classes (N, O, and P). At an earlier stage of differentiation (Stage I), the genes of each of these classes would be inactive. As the tissue differentiates to the mature stage (Stage II), the distinction between these genes sets would become apparent. Class N would remain permanently repressed and would repre-

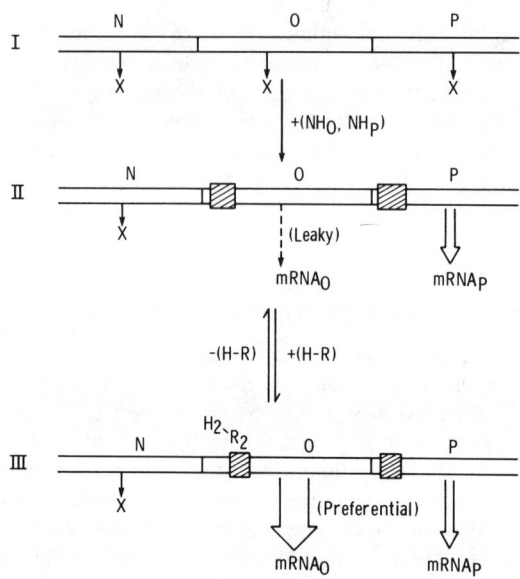

Figure 1 The role of nonhistone proteins in differentiation and acute hormonal production.

sent genes that need never be expressed in that particular tissue. The signal for repression is still unclear but could be simply a lack of availability of the appropriate nonhistone proteins required for expression. Class P genes represent those that are not under fine regulation. By virtue of a specific subclass of nonhistone (P), they exist in a conformation available to polymerase and are transcribed randomly. These sequences might represent those genes coding for cell structural proteins, metabolic enzymes, or other "housekeeping" proteins. Their low rate of transcription might be adequate to provide a sufficient level of product for daily maintenance functions. Class O genes would also be marked by a specific subclass of nonhistones (O). These genes exist in the differentiated state but are generally unavailable to polymerase for transcription in the absence of inducer (Stage II). They might be considered "leaky" in that an occasional transcript is synthesized but the rate of expression would be much less than random. In contrast, when inducer is provided, the rate of transcription of class O genes would increase to a level that greatly exceeds random (Stage II). One example of such inducers would be steroid hormone-receptor complex (H-R). The hormone-receptor would convert class O genes to a physical state in which these genes would be "preferentially" transcribed by nuclear polymerases.

THE OVALBUMIN MINICHROMOSOME In order to better understand the mechanism by which the steroid hormone-receptor complex activates the ovalbumin gene, we are presently involved in the isolation of the DNA sequences in the chick genome that contain the ovalbumin structural gene together with nearby possible regulatory sites. With the complete ovalbumin cDNA being used as a template, a double-stranded ovalbumin DNA (1850 nucleotide pairs) has been synthesized in vitro (220, 221). Amplification of this pure double-stranded DNA in bacterial plasmids has yielded large quantitites of the coding portion of the ovalbumin gene.

The natural ovalbumin gene has now been substantially purified from total chick DNA by affinity chromatography using ovalbumin mRNA and cDNA coupled covalently to phosphocellulose in separate columns (222). After repeated chromatography, a 10,000-fold purification of both the coding ($mRNA_{ov}$ column) and anticoding ($cDNA_{ov}$ column) ovalbumin DNA strands was obtained. The purified strands were further purified 15- to 20-fold by affinity chromatography over a SH-Sepharose column after hybridization with mercurated RNA synthesized in vitro from the structural gene (223). This overall procedure results in a preparation which is 150,000–200,000-fold purified and is 40–50% pure on a DNA strand basis. Furthermore, since the total DNA was originally sheared to a mean length of 5000 base pairs and no DNA degradation occurred during the isolation procedure, these DNA preparations contain sequences adjacent to the coding portion of the ovalbumin gene (223).

It is hoped that sequences located in tandem to the 3' end of the structural gene play an important regulatory role in the expression of this gene. Amplification of these DNA preparations with bacterial plasmids will now enable use of large quantities of the ovalbumin gene to purify putative chromosomal regulatory proteins by affinity chromatography. The eventual goal will be to reconstitute a "minichromosome" containing the regulatory and structural elements of the ovalbumin gene (DNA plus nonhistone proteins) and to study the interaction with RNA polymerase between this "minichromosome" and pure steroid hormone-receptor complexes. Such studies should lead then to a definitive description of the molecular mechanism of steroid hormone action and the regulation of gene expression in eucaryotes.

Literature Cited

1. Marzluff, W. F. Jr., Huang, R. C. C. 1975. Chromatin directed transcription of 5S and tRNA genes. *Proc. Natl. Acad. Sci. USA* 72:1082–86
2. Spelsberg, T. C., Hnilica, L. S. 1971. Proteins of chromatin in template restriction. I. RNA synthesis in vitro. *Biochim. Biophys. Acta* 228:202–11
3. Bonner, J., Chalkley, G. R., Dahmus, M. E., Fambrough, D. M., Fujimura, F., Huang, R. C. C., Huberman, J., Jensen, R., Marushige, K., Ohlenbusch, J., Olivera, B., Widholm, J. 1968. Isolation and characterization of chromosomal nucleoproteins. *Methods Enzymol.* 12B: 3–84
4. De Pomerai, D. I., Chesterton, C. J., Butterworth, P. H. W. 1974. Preparation of chromatin. Variation in the template properties of chromatin dependent on the method of preparation. *Eur. J. Biochem.* 46:461–71
5. Grouse, L., Chilton, M. D., McCarthy, B. J. 1972. Hybridization of ribonucleic acid with unique sequences of mouse deoxyribonucleic acid. *Biochemistry* 11: 798–805
6. Liarakos, C. D., Rosen, J. M., O'Malley, B. W. 1973. Effect of estrogen on gene expression in the chick oviduct. II. Transcription of chick tritiated unique deoxyribonucleic acid as measured by hybridization in ribonucleic acid excess. *Biochemistry* 12:2809–16
7. Britten, R. J., Kohne, D. E. 1968. Repeated sequences in DNA. *Science* 161:529–40
8. Bishop, J. O., Morton, J. G., Rosbash, M., Richardson, M. 1974. Three abundance classes of HeLa cell messenger RNA. *Nature* 250:199–204
9. Monahan, J. J., Harris, S. E., O'Malley, B. W. 1976. Effect of estrogen on gene expression in the chick oviduct. Effect of estrogen on the sequence and population complexity of chick oviduct poly-(A)-containing RNA. *J. Biol. Chem.* 251:3738–48
10. Chambon, P. 1975. Eukaryotic nuclear-RNA polymerases. *Ann. Rev. Biochem.* 44:613–38
11. Roeder, R. G., Golomb, M. W., Jaehning, J. A., Ng, S. Y., Parker, C. S., Schwartz, L. B., Sklar, V. E. F., Weinmann, R. 1977. In *Receptors and Hormone Action,* ed. B. W. O'Malley, L. Birnbaumer. New York: Academic. In press
12. Huang, R. C. C., Bonner, J. 1962. Histone, a suppressor of chromosomal RNA synthesis. *Proc. Natl. Acad. Sci. USA* 48:1216–22
13. Sonnenberg, B. P., Zubay, G. 1965. Nucleohistone as a primer for RNA synthesis. *Proc. Natl. Acad. Sci. USA* 54: 415–20
14. Marushige, K., Bonner, J. 1966. Template properties of liver chromatin. *J. Mol. Biol.* 15:160–74
15. Dahmus, M. E., Bonner, J. 1965. Increased template activity of liver chromatin, a result of hydrocortisone administration. *Proc. Natl. Acad. Sci. USA* 54:1370–75
16. Barker, K. L., Warren, J. C. 1966. Template capacity of uterine chromatin: Control by estradiol. *Proc. Natl. Acad. Sci. USA* 56:1298–1302
17. Breuer, C. B., Florini, J. R. 1966. Effects of ammonium sulfate, growth hormone, and testosterone propionate on ribonucleic acid polymerase and chromatin activities in rat skeletal muscle. *Biochemistry* 5:3857–65
18. Teng, C. S., Hamilton, T. H. 1968. The role of chromatin in estrogen action in the uterus: I. The control of template capacity and chemical composition and the binding of ^{3}H-estradiol-17β. *Proc. Natl. Acad. Sci. USA* 60:1410–17
19. Spelsberg, T. C., Steggles, A. W., O'Malley, B. W. 1971. Changes in chromatin composition and hormone binding during chick oviduct development. *Biochim. Biophys. Acta* 254: 129–34
20. Couch, R. M., Anderson, K. M. 1973. Rat ventral prostate chromatin: Effect of androgens on its chemical composition, physical properties, and template activity. *Biochemistry* 12:3114–21

21. Flickinger, R. A., Coward, S. J., Miyagi, M., Moser, C., Rollins, E. 1965. The ability of DNA and chromatin of developing frog embryos to prime for RNA polymerase-dependent RNA synthesis. *Proc. Natl. Acad. Sci. USA* 53:783–90
22. Marushige, K., Ozaki, H. 1967. Properties of isolated chromatin from sea urchin embryo. *Dev. Biol.* 16:474–88
23. Johnson, A. W., Hnilica, L. S. 1970. In vitro RNA synthesis and nuclear proteins of isolated sea urchin embryo nuclei. *Biochim. Biophys. Acta* 224:518–30
24. Chiu, N., Baserga, R. 1975. Changes in template activity and structure of nuclei from WI-38 cells in the prereplicative phase. *Biochemistry* 14:3126–32
25. Johns, E. W., Hoare, T. A. 1970. Histones and gene control. *Nature* 226:650–51
26. Doenecke, D., McCarthy, B. J. 1976. Movement of histones in chromatin induced by shearing. *Eur. J. Biochem.* 64:405–9
27. Maryanka, D., Gould, H. 1973. Transcription of rat liver chromatin with homologous enzyme. *Proc. Natl. Acad. Sci. USA* 70:1161–65
28. Groner, Y., Monroy, G., Jacquet, M., Hurwitz, J. 1975. Chromatin as a template for RNA synthesis in vitro. *Proc. Natl. Acad. Sci. USA* 72:194–99
29. Cox, R. F. 1973. Transcription of high-molecular weight RNA from hen oviduct chromatin by bacterial and endogenous form-B RNA polymerases. *Eur. J. Biochem.* 39:49–61
30. Cedar, H., Felsenfeld, G. 1973. Transcription of chromatin in vitro. *J. Mol. Biol.* 77:237–54
31. Hyman, R. W., Davidson, N. 1970. Kinetics of the in vitro inhibition of transcription by actinomycin. *J. Mol. Biol.* 50:421–38
32. Dunn, J. J., Studier, F. W. 1973. T7 early RNAs are generated by site-specific cleavages. *Proc. Natl. Acad. Sci. USA* 70:1559–63
33. Tsai, M. J., Schwartz, R. J., Tsai, S. Y., O'Malley, B. W. 1975. Effects of estrogen on gene expression in the chick oviduct. IV. Initiation of RNA synthesis on DNA and chromatin. *J. Biol. Chem.* 250:5165–74
34. Bautz, E. K. F., Bautz, F. A. 1970. Initiation of RNA synthesis: the function of σ in the binding of RNA polymerase to promoter sites. *Nature* 226:1219–22
35. Chamberlin, M. J., Ring, J. 1972. Studies of the binding of *Escherichia coli* RNA polymerase to DNA. V. T7 RNA chain initiation by enzyme-DNA complexes. *J. Mol. Biol.* 70:221–37
36. Mangel, W. F., Chamberlin, M. J. 1974. Studies of ribonucleic acid chain initiation by *Escherichia coli* ribonucleic acid polymerase bound to T7 deoxyribonucleic acid. I. An assay for the rate and extent of ribonucleic acid chain initiation. *J. Biol. Chem.* 249:2995–3001
37. Schwartz, R. J., Tsai, M. J., Tsai, S. Y., O'Malley, B. W. 1975. Effect of estrogen on gene expression in the chick oviduct. V. Changes in the number of RNA polymerase binding and initiation sites in chromatin. *J. Biol. Chem.* 250:5175–82
38. Kohler, P. O., Grimley, P. M., O'Malley, B. W. 1969. Estrogen-induced cytodifferentiation of the ovalbumin-secreting glands of the chick oviduct. *J. Cell Biol.* 40:8–27
39. Tsai, S. Y., Tsai, M. J., Schwartz, R. J., Kalimi, M., Clark, J. H., O'Malley, B. W. 1975. Effects of estrogen on gene expression in chick oviduct: Nuclear receptor levels and initiation of transcription. *Proc. Natl. Acad. Sci. USA* 72:4228–32
40. Keshgegian, A. A., Furth, J. J. 1972. Comparison of transcription of chromatin by calf thymus and *E. coli* RNA polymerases. *Biochem. Biophys. Res. Commun.* 48:757–63
41. Keshgegian, A. A., Garibian, G. S., Furth, J. J. 1973. Transcription of chromatin. Initial and terminal nucleotides of ribonucleic acid synthesized by calf thymus and *Escherichia coli* ribonucleic acid polymerases. *Biochemistry* 12:4337–42
42. Butterworth, P. H. W., Cox, R. F., Chesterton, C. J. 1971. Transcription of mammalian chromatin by mammalian DNA-dependent RNA polymerases. *Eur. J. Biochem.* 23:229–41
43. Meilhac, M., Chambon, P. 1973. Animal DNA-dependent RNA polymerases. Initiation sites on calf-thymus DNA. *Eur. J. Biochem.* 35:454–63
44. Meilhac, M., Tysper, Z., Chambon, P. 1972. Animal DNA-dependent RNA polymerases. 4. Studies on inhibition by rifamycin derivatives. *Eur. J. Biochem.* 28:291–300
45. Mandel, J. L., Chambon, P. 1974. Animal DNA-dependent RNA polymerases. Analysis of the RNAs synthesized on Simian Virus 40 superhelical DNA by mammalian RNA polyme-

rases AI and B. *Eur. J. Biochem.* 41:379–95
46. Tsai, M. J., Towle, H. C., Harris, S. E., O'Malley, B. W. 1976. Effect of estrogen on gene expression in the chick oviduct. Comparative aspects of RNA chain initiation in chromatin using homologous versus *Escherichia coli* RNA polymerase. *J. Biol. Chem.* 251:1960–68
47. Cedar, H. 1975. Transcription of DNA and chromatin with calf thymus RNA polymerase B in vitro. *J. Mol. Biol.* 95:257–69
48. McCarthy, B. J., Church, R. B. 1970. The specificity of molecular hybridization reactions. *Ann. Rev. Biochem.* 39:131–50
49. Paul, J., Gilmour, R. S. 1966. Template activity of DNA is restricted in chromatin. *J. Mol. Biol.* 16:242–44
50. Paul, J., Gilmour, R. S. 1968. Organ-specific restriction of transcription in mammalian chromatin. *J. Mol. Biol.* 34:305–16
51. Smith, K. D., Church, R. B., McCarthy, B. J. 1969. Template specificity of isolated chromatin. *Biochemistry* 8:4271–77
52. Tan, C. H., Miyagi, M. 1970. Specificity of transcription of chromatin in vitro. *J. Mol. Biol.* 50:641–53
53. Bacheler, L. T., Smith, K. D. 1976. Transcription of isolated mouse liver chromatin. *Biochemistry* 15:3281–90
54. Dawid, I. B., Brown, D. D., Reeder, R. H. 1970. Composition and structure of chromosomal and amplified ribosomal DNA's of *Xenopus laevis. J. Mol. Biol.* 51:341–60
55. Brown, D. D., Wensink, P. C., Jordan, E. 1971. Purification and some characteristics of 5S DNA from *Xenopus laevis. Proc. Natl. Acad. Sci. USA* 68:3175–79
56. Zylber, E. A., Penman, S. 1971. Products of RNA polymerases in HeLa cell nuclei. *Proc. Natl. Acad. Sci. USA* 68:2861–65
57. Reeder, R. H., Roeder, R. G. 1972. Ribosomal RNA synthesis in isolated nuclei *J. Mol. Biol.* 67:433–41
58. Reeder, R. H., Brown, D. D. 1970. Transcription of the ribosomal RNA genes of an amphibian by the RNA polymerase of a bacterium. *J. Mol. Biol.* 51:361–77
59. Hecht, R. M., Birnstiel, M. L. 1972. Integrity of the DNA template, a prerequisite for the faithful transcription of *Xenopus* rDNA in vitro. *Eur. J. Biochem.* 29:489–99
60. Roeder, R. G., Reeder, R. H., Brown, D. D. 1970. Multiple forms of RNA polymerase in *Xenopus laevis:* Their relationship to RNA synthesis in vivo and their fidelity of transcription in vitro. *Cold Spring Harbor Symp. Quant. Biol.* 35:727–35
61. Beebee, T. J. C., Butterworth, P. H. W. 1974. Template specificities of *Xenopus laevis* RNA polymerases: Selective transcription of ribosomal cistrons by RNA polymerase A. *Eur. J. Biochem.* 45:395–406
62. Cramer, J. H., Sebastian, J., Rownd, R. H., Halvorson, H. O. 1974. Transcription of *Saccharomyces cerevisiae* ribosomal DNA in vivo and in vitro. *Proc. Natl. Acad. Sci. USA* 71:2188–92
63. Van Keulen, H., Planta, R. J., Retel, J. 1975. Structure and transcription specificity of yeast RNA polymerase A. *Biochim. Biophys. Acta* 395:179–90
64. Hager, G., Holland, M., Valenzuela, P., Weinberg, F., Rutter, W. J. 1976. In *RNA Polymerase,* ed. R. Losick, M. Chamberlin, pp. 745–62. Cold Spring Harbor, NY: Cold Spring Harbor Lab. 899 pp.
65. Reeder, R. H. 1973. Transcription of chromatin by bacterial RNA polymerase. *J. Mol. Biol.* 80:229–41
66. Honjo, T., Reeder, R. H. 1974. Transcription of *Xenopus* chromatin by homologous ribonucleic acid polymerase: Aberrant synthesis of ribosomal and 5S ribonucleic acid. *Biochemistry* 13:1896–99
67. Parker, C. S., Roeder, R. G. 1977. Selective and accurate transcription of the *Xenopus laevis* 5S RNA genes in isolated chromatin by purified RNA polymerase III. *Proc. Natl. Acad. Sci. USA* 74:44–48
68. Axel, R., Cedar, H., Felsenfeld, G. 1973. Synthesis of globin ribonucleic acid from duck reticulocyte chromatin in vitro. *Proc. Natl. Acad. Sci. USA* 70:2029–32
69. Gilmour, R. S., Paul, J. 1973. Tissue-specific transcription of the globin gene in isolated chromatin. *Proc. Natl. Acad. Sci. USA* 70:3440–42
70. Harris, S. E., Schwartz, R. J., Tsai, M. J., O'Malley, B. W., Roy, A. K. 1976. Effect of estrogen on gene expression in the chick oviduct. In vitro transcription of the ovalbumin gene in chromatin. *J. Biol. Chem.* 251:524–29

71. Stein, G., Park, W., Thrall, C., Mans, R., Stein, J. 1975. Regulation of cell cycle stage-specific transcription of histone genes from chromatin by non-histone chromosomal proteins. *Nature* 257:764–67
72. Smith, M. M., Haung, R. C. C. 1976. Transcription in vitro of immunoglobulin kappa light chain genes in isolated mouse myeloma nuclei and chromatin. *Proc. Natl. Acad. Sci. USA* 73:775–79
73. Astrin, S. M. 1973. In vitro transcription of Simian Virus 40 sequences in SV3T3 chromatin. *Proc. Natl. Acad. Sci. USA* 70:2304–8
74. Shih, T. Y., Khoury, G., Martin, M. A. 1973. In vitro transcription of the viral-specific sequences present in the chromatin of cells transformed by Simian Virus 40. *Proc. Natl. Acad. Sci. USA* 70:3506–10
75. Jacquet, M., Groner, Y., Monroy, G., Hurwitz, J. 1974. The in vitro synthesis of avian myeloblastosis viral RNA sequences. *Proc. Natl. Acad. Sci. USA* 71:3045–49
76. Janowski, M., Baugnet-Mahieu, L., Sassen, A. 1974. Murine leukemia virus RNA transcription from chromatin of normal and infected BALB/c spleen. *Nature* 251:347–50
77. Wilson, G. N., Steggles, A. W., Kantor, J. A., Nienhuis, A. W., Anderson, W. F. 1975. Cell-free transcription of mammalian chromatin: Quantitative measurement of newly synthesized globin messenger RNA sequences. *J. Biol. Chem.* 250:8604–13
78. Dale, R. M. K., Ward, D. C. 1975. Mercurated polynucleotides: New probes for hybridization and selective polymer fractionation. *Biochemistry* 14:2458–69
79. Crouse, G. F., Fodor, E. J. B., Doty, P. 1976. In vitro transcription of chromatin in the presence of a mercurated nucleotide. *Proc. Natl. Acad. Sci. USA* 73:1564–67
80. Beebee, T. J. C., Butterworth, P. H. W. 1976. The use of mercurated nucleoside triphosphates as a probe in transcription studies in vitro. *Eur. J. Biochem.* 66:543–50
81. Towle, H. C., Tsai, M. J., Tsai, S. Y., O'Malley, B. W. 1977. Effect of estrogen on gene expression in the chick oviduct. Preferential initiation and asymmetrical transcription of specific chromatin genes. *J. Biol. Chem.* 252:2396–2404
82. Wilson, G. N., Steggles, A. W., Nienhuis, A. W. 1975. Strand-selective transcription of globin genes in rabbit erythroid cells and chromatin. *Proc. Natl. Acad. Sci. USA* 72:4835–39
83. Biessmann, H., Gjerset, R. A., Levy, B., McCarthy, B. J. 1976. Fidelity of chromatin transcription in vitro. *Biochemistry* 15:4356–63
84. Astrin, S. M. 1975. Mapping of the SV40 specific sequences transcribed in vitro from chromatin of SV40 transformed cells. *Biochemistry* 14:2700–4
85. Steggles, A. W., Wilson, G. N., Kantor, J. A., Picciano, D. J., Falvey, A. K., Anderson, W. F. 1974. Cell-free transcription of mammalian chromatin: Transcription of globin messenger RNA sequences from bone-marrow chromatin with mammalian RNA polymerase. *Proc. Natl. Acad. Sci. USA* 71:1219–23
86. Gilmour, R. S., Paul, J. 1969. RNA transcribed from reconstituted nucleoprotein is similar to natural RNA. *J. Mol. Biol.* 40:137–39
87. Gilmour, R. S., Paul, J. 1970. Role of non-histone components in determining organ specificity of rabbit chromatins. *FEBS Lett.* 9:242–44
88. Bekhor, I., Kung, G. M., Bonner, J. 1969. Sequence-specific interaction of DNA and chromosomal protein. *J. Mol. Biol.* 39:351–64
89. Huang, R. C. C., Huang, P. C. 1969. Effect of protein-bound RNA associated with chick embryo chromatin on template specificity of the chromatin. *J. Mol. Biol.* 39:365–80
90. Von Heyden, H. W., Zachau, H. G. 1971. Characterization of RNA in fractions of calf thymus chromatin. *Biochim. Biophys. Acta* 232:651–60
91. Scharpe, A., Van Parijs, R. 1974. A study of the RNA associated with chromosomal proteins in chromatin from *Pisum sativum* L. embryos. *Biochim. Biophys. Acta* 353:45–62
92. Kamiyama, M., Wang, T. Y. 1971. Activated transcription from rat liver chromatin by non-histone proteins. *Biochim. Biophys. Acta* 228:563–76
93. Bekhor, I., Lapeyre, J. N., Kim, J. 1974. Fractionation of nonhistone chromosomal proteins isolated from rabbit liver and submandibular salivary glands. *Arch. Biochim. Biophys.* 161:1–10
94. Chae, C. B., Carter, D. B. 1974. Degradation of chromosomal proteins during dissociation and reconstitution of chromatin. *Biochem. Biophys. Res. Commun.* 57:740–46

95. Spelsberg, T. C., Hnilica, L. S., Ansevin, A. T. 1971. Proteins of chromatin in template restriction. III. The macromolecules in specific restriction of the chromatin DNA. *Biochim. Biophys. Acta* 228:550–62
96. Kleiman, L., Huang, R. C. C. 1972. Reconstitution of chromatin. The sequential binding of histones to DNA in the presence of salt and urea. *J. Mol. Biol.* 64:1–8
97. Bartley, J., Chalkley, R. 1970. Further studies of a thymus nucleohistone-associated protease. *J. Biol. Chem.* 245:4286–92
98. Garrels, J. I., Elgin, S. C. R., Bonner, J. 1972. A histone protease of rat liver chromatin. *Biochem. Biophys. Res. Commun.* 46:545–51
99. Stein, G. S., Mans, R. J., Gabbay, E. J., Stein, J. L., Davis, J., Adawadkar, P. D. 1975. Evidence for fidelity of chromatin reconstitution. *Biochemistry* 14:1859–66
100. Gabbay, E. J., Sanford, K. 1974. Dissymmetric recognition of the helical sense of deoxyribonucleic acid and evidence for binding of reporter molecules from the minor groove of DNA. *Bioorg. Chem.* 3:91–102
101. Stein, G. S., Farber, J. 1972. Role of nonhistone chromosomal proteins in the restriction of mitotic chromatin template activity. *Proc. Natl. Acad. Sci. USA* 69:2918–21
102. Tsai, S. Y., Harris, S. E., Tsai, M. J., O'Malley, B. W. 1976. Effects of estrogen on gene expression in chick oviduct. XII. The role of chromatin proteins in regulating transcription of the ovalbumin gene. *J. Biol. Chem.* 251:4713–21
103. Chae, C. B. 1975. Reconstitution of chromatin: Mode of reassociation of chromosomal proteins. *Biochemistry* 14:900–6
104. Camerini-Otero, R. D., Sollner-Webb, B., Felsenfeld, G. 1976. The organization of histones and DNA in chromatin: Evidence for an arginine-rich histone kernel. *Cell* 8:333–47
105. Olins, A. L., Olins, D. E. 1974. Spheroid chromatin units (ν bodies). *Science* 183:330–32
106. Elgin, S. C. R., Weintraub, H. 1975. Chromosomal proteins and chromatin structure. *Ann. Rev. Biochem.* 44:725–74
107. Stedman, Edgar, Stedman, Ellen, 1950. Cell specificity of histones. *Nature* 166:780–81
108. Butler, J. A. V., Chipperfield, A. R. 1967. Inhibition of RNA polymerase by histones. *Nature* 215:1188–89
109. Roy, A. K., Zubay, G. 1966. RNA synthesis stimulated by sonicated nucleohistone. *Biochim. Biophys. Acta* 129:403–5
110. Moskowitz, G. J., Ogawa, Y., Starbuck, W. C., Rusch, H. 1969. Comparative effects of RNA polymerase of the whole GAR histone and its peptides containing clusters of basic amino acids. *Biochem. Biophys. Res. Commun.* 35:741–46
111. Hindley, J. 1963. The relative ability of reconstituted nucleohistones to allow DNA-dependent RNA synthesis. *Biochem. Biophys. Res. Commun.* 12:175–79
112. Johns, E. W. 1972. Histones, chromatin structure and RNA synthesis. *Nature New Biol.* 237:87–88
113. Koslov, Y. V., Georgiev, G. P. 1970. Mechanism of inhibitory action of histones on DNA template activity in vitro. *Nature* 228:245–47
114. Bartley, J. A., Chalkley, R. 1972. The binding of deoxyribonucleic acid and histone in native nucleohistone. *J. Biol. Chem.* 247:3647–55
115. Kornberg, R., Thomas, J. O. 1974. Chromatin structure: Oligomers of the histones. *Science* 184:865–68
116. Hewish, D. R., Burgoyne, L. A. 1973. Chromatin substructure. The digestion of chromatin DNA at regularly spaced sites by a nuclear deoxyribonuclease. *Biochem. Biophys. Res. Commun.* 52:504–10
117. Burgoyne, L. A., Hewish, D. R., Mobbs, J. 1974. Mammalian chromatin substructure studies with the calcium-magnesium endonuclease and two dimensional polyacrylamide gel electrophoresis. *Biochem. J.* 143:67–72
118. Clark, R. J., Felsenfeld, G. 1971. Structure of chromatin. *Nature New Biol.* 229:101–6
119. Clark, R. J., Felsenfeld, G. 1974. Chemical probes of chromatin structure. *Biochemistry* 13:3622–28
120. Axel, R., Melchior, W., Sollner-Webb, B., Felsenfeld, G. 1974. Specific sites of interaction between histones and DNA in chromatin. *Proc. Natl. Acad. Sci. USA* 71:4101–5
121. Rill, R., Van Holde, K.E. 1973. Properties of nuclease-resistant fragments of calf thymus chromatin. *J. Biol. Chem.* 248:1080–83

122. Noll, M. 1974. Subunit structure of chromatin. *Nature* 251:249–51
123. Polisky, B., McCarthy, B. 1975. Location of histones on simian virus 40 DNA. *Proc. Natl. Acad. Sci. USA* 72: 2895–99
124. Griffith, J. D. 1975. Chromatin structure: Deduced from a minichromosome. *Science* 187:1202–3
125. Axel, R., Cedar, H., Felsenfeld, G. 1975. The structure of the globin gene in chromatin. *Biochemistry* 14:2489–95
126. Kuo, M. T., Sahasrabuddhe, C. G., Saunders, G. F. 1976. Presence of messenger specifying sequences in the DNA of chromatin subunits. *Proc. Natl. Acad. Sci. USA* 73:1572–75
127. Allfrey, V. G. 1971. In *Histones and Nucleohistones,* ed. D. M. P. Phillips, p. 241. London: Plenum. 305 pp.
128. Hnilica, L. S. 1972. *The Structure and Biological Functions of Histones,* p. 79. Cleveland, Ohio: CRC Press
129. Simpson, R. J. 1973. Structure and function of chromatin. *Adv. Enzymol.* 38:41–108
130. MacGillivray, A. J., Rickwood, D. 1974. The role of chromosomal proteins as gene regulators. In *Biochemistry of Cell Differentiation,* ed. J. Paul, pp. 301–61. London: Butterworths
131. Stein, G. S., Spelsberg, T. C., Kleinsmith, L. J. 1974. Nonhistone chromosomal proteins and gene regulation. *Science* 183:817–24
132. Dingman, C. W., Sporn, M. B. 1964. Studies on chromatin. I. Isolation and characterization of nuclear complexes of deoxyribonucleic acid, ribonucleic and protein from embryonic and adult tissue of the chicken. *J. Biol. Chem.* 239:3483–92
133. Orrick, L. R., Olson, M. O. J., Busch, H. 1973. Composition of nucleolar proteins of normal rat liver and Novikoff hepatoma ascites cells by two dimensional polyacrylamide gel electrophoresis. *Proc. Natl. Acad. Sci. USA* 70: 1316–20
134. Peterson, J. L., McConkey, E. H. 1976. Nonhistone chromosomal proteins from Hela cells. A survey by high resolution, two dimensional electrophoresis. *J. Biol. Chem.* 251:548–54
135. Wilhelm, J. A., Ansevin, A. T., Johnson, A. W., Hnilica, L. S. 1972. Proteins of chromatin in genetic restriction. IV. Comparison of histone and nonhistone proteins of rat liver nucleolar and extra nuclear chromatin. *Biochim. Biophys. Acta* 272:220–30
136. Elgin, S., Bonner, J. 1970. Limited heterogeneity of the major nonhistone chromosomal proteins. *Biochemistry* 9:4440–47
137. MacGillivray, A. J., Carroll, D., Paul, J. 1971. The heterogeneity of the nonhistone chromatin proteins from mouse tissues. *FEBS Lett.* 13:204–8
138. Sevall, J. S., Cockburn, A., Savage, M., Bonner, J. 1975. DNA-protein interactions of the rat liver non-histone chromosomal protein. *Biochemistry* 14: 782–88
139. Rickwood, D., Riches, P. G., MacGillivray, A. J. 1973. Studies of the *in vitro* phosphorylation of chromatin non-histone proteins in isolated nuclei. *Biochim. Biophys. Acta* 299:162–71
140. Chytil, F., Spelsberg, T. C. 1971. Tissue differences in antigenic properties of non-histone protein-DNA complexes. *Nature New Biol.* 233:215–18
141. Teng, C. S., Teng, C. T., Allfrey, V. G. 1971. Studies of nuclear acidic proteins, evidence for their phosphorylation, tissue specificity, selective binding to deoxyribonucleic acid and stimulatory effects on transcription. *J. Biol. Chem.* 246:3597–3609
142. Kostraba, N. C., Montagna, R. A., Wang, T. Y. 1975. Study of the loosely bound non-histone chromatin proteins. *J. Biol. Chem.* 250:1548–55
143. Teng, C. S., Hamilton, T. H. 1969. Role of chromatin in estrogen action in the uterus. II. Hormone-induced synthesis of nonhistone acidic protein which restore histone inhibited DNA-dependent RNA synthesis. *Proc. Natl. Acad. Sci. USA* 63:465–72
144. Enea, V., Allfrey, V. G. 1973. Selective synthesis of liver nuclear acidic proteins following glucagon administration in vivo. *Nature* 242:265–67
145. Salas, J., Green, H. 1971. Proteins binding to DNA and their relation to growth in cultured mammalian cells. *Nature New Biol.* 229:165–69
146. Barrett, T., Maryanka, D., Hamlyn, P. H., Gould, H. J. 1974. Nonhistone proteins control gene expression in reconstituted chromatin. *Proc. Natl. Acad. Sci. USA* 71:5057–61
147. Gilmour, R. S., MacGillivray, A. J. 1976. The role of non-histone proteins in the regulation of globin gene activity. In *the Molecular Biology of Hormone Action,* ed. J. Papaconstantinou, pp. 15–30. New York: Academic
148. MacGillivray, A. J., Cameron, A., Krauze, R. J., Rickwood, D., Paul, J.

1972. The nonhistone proteins of chromatin. Their isolation and composition in a number of tissues. *Biochim. Biophys. Acta* 277:384–402
149. Chiu, J. F., Tsai, Y. H., Sakuma, K., Hnilica, L. S. 1975. Regulation of in vitro mRNA transcription by a fraction of chromosomal proteins. *J. Biol. Chem.* 250:9431–33
150. O'Malley, B. W., Means, A. R. 1974. Female steroid hormones and target cell nuclei. *Science* 183:610–20
151. Kostraba, N. C., Wang, T. Y. 1975. Inhibition of transcription *in vitro* by a non-histone protein isolated from Ehrlich ascites tumor chromatin. *J. Biol. Chem.* 250:8938–42
152. Tsai, S. Y., Tsai, M. J., Harris, S. E., O'Malley, B. W. 1976. Effects of estrogen on gene expression in the chick oviduct. XIV. Control of ovalbumin gene expression by nonhistone proteins. *J. Biol. Chem.* 251:6475–78
153. Levy, R., Levy, S., Rosenberg, S. A., Simpson, R. T. 1973. Selective stimulation of nonhistone chromatin protein synthesis in lymphoid cells by phytohemagglutinin. *Biochemistry* 12:224–28
154. Rovera, G., Baserga, R. 1971. Early changes in the synthesis of acidic nuclear proteins in human diploid fibroblasts stimulated to synthesize by changing the medium. *J. Cell. Physiol.* 77:201–12
155. Stein, G. S., Matthews, D. E. 1973. Nonhistone chromosomal protein synthesis: Utilization of pre-existing and newly transcribed messenger RNA's. *Science* 181:71–73
156. Stein, G. S., Borun, T. W. 1972. The synthesis of acidic chromosomal proteins during the cell cycle. *J. Cell Biol.* 52:292–307
157. Borun, T. W., Stein, G. S., 1972. The synthesis of acidic chromosomal proteins during the cell cycle of HeLa S_3 cells. *J. Cell Biol.* 52:308–15
158. Stein, G. S., Chaudhuri, S. C., Baserga, R. 1972. Gene activation in WI-38 fibroblasts stimulated to proliferate. Role of nonhistone chromosomal proteins. *J. Biol. Chem.* 247:3918–22
159. Stein, J. L., Reed, K., Stein, G. S. 1976. Effect of histones and nonhistones chromosomal proteins on the transcription of histone genes from HeLa S_3 cell DNA. *Biochemistry* 15:3291–95
160. Park, W. D., Stein, J. L., Stein, G. S. 1976. Activation of in vitro histone gene transcription from HeLa S_3 chromatin by s-phase nonhistone chromosomal proteins. *Biochemistry* 15:3297–3300
161. Kleinsmith, L. J., Stein, J. Stein, G. 1976. Dephosphorylation of nonhistone proteins specifically alters the pattern of gene transcription in reconstituted chromatin. *Proc. Natl. Acad. Sci. USA* 73:1174–78
162. Masaaki, H., Tsai, M. J., O'Malley, B. W. 1976. Effect of estrogen on gene expression in the chick oviduct. VII. Kinetics of initiation of *in vitro* transcription on chromatin. *J. Biol. Chem.* 251:1137–46
163. Tsai, M. J., Tsai, S. Y., Towle, H. C., O'Malley, B. W. 1976. Effect of estrogen on gene expression in the chick oviduct. XIII. Studies on the initiation of RNA synthesis on chromatin in vitro. *J. Biol. Chem.* 251:5565–74
164. Howk, R. S., Williams, D. R., Haberman, A. B., Parks, W. P., Scolnick, E. M. 1974. A comparison of the transcription of mouse cell chromatin by homologous RNA polymerase I and II. *Cell* 3:15–22
165. Crick, F. 1971. General model for the chromosomes of higher organisms. *Nature* 234:25–27
166. Hanson, C. V., Shen, C. J., Hearst, J. E. 1976. Cross-linking of DNA in situ as a probe for chromatin structure. *Science* 193:62–64
167. Thomas, T. L., Patel, G. L. 1976. DNA unwinding component of the nonhistone chromatin proteins. *Proc. Natl. Acad. Sci. USA* 73:4364–68
168. Herrick, G., Alberts, B. 1976. Purification and physical characterization of nucleic acid helix-unwinding proteins from calf thymus. *J. Biol. Chem.* 251: 2124–32
169. Gottesfeld, J. M., Garrard, W. T., Bagi, G., Wilson, R. F., Bonner, J. 1974. Partial purification of the template-active fraction of chromatin: A preliminary report. *Proc. Natl. Acad. Sci. USA* 71: 2193–97
170. Weintraub, H., Groudine, M. 1976. Chromosomal subunits in active genes have an altered conformation. *Science* 193:848–56
171. Gottesfeld, J. M., Murphy, R. F., Bonner, J. 1975. Structure of transcriptionally active chromatin. *Proc. Natl. Acad. Sci. USA* 72:4404–8
172. Garel, A., Axel, R. 1976. Selective digestion of transcriptionally active ovalbumin genes from oviduct nuclei. *Proc. Natl. Acad. Sci. USA* 73:3966–70

173. O'Malley, B. W., Schrader, W. T. 1976. The receptors of steroid hormones. *Sci. Am.* 234:232-43
174. Jensen, E. V., DeSombre, E. R. 1972. Mechanism of action of the female sex hormones. *Ann. Rev. Biochem.* 41:203-30
175. Gorski, J., Toft, D., Shyamala, G., Smith, D., Notides, A. 1968. Hormone receptors: Studies on the interaction of estrogen with the uterus. *Recent Prog. Horm. Res.* 24:45-80
176. Jensen, E. V., Suzuki, T., Kawashima, T., Stumpf, W. E., Jungblut, P. W., DeSombre, E. R. 1968. A two step mechansim for the interaction of estradiol with rat uterus. *Proc. Natl. Acad. Sci. USA* 59:632-38
177. O'Malley, B. W., McGuire, W. L., Kohler, P. O., Korenman, S. G. 1969. Studies on the mechanism of steroid hormone regulation of synthesis of specific proteins. *Recent Prog. Horm. Res.* 25:105-60
178. Gorski, J., Nicolette, J. A. 1963. Early estrogen effects on newly synthesized RNA and phospholipid in subcellular fractions of rat uteri. *Arch. Biochem. Biophys.* 103:418-23
179. Hamilton, T. H., Widnell, C. C., Tata, J. R. 1965. Sequential stimulation by oestrogen of nuclear RNA synthesis and DNA-dependent RNA polymerase activities in rat uterus. *Biochim. Biophys. Acta* 108:168-72
180. King, R. J. B., Gordon, J. 1972. Involvement of DNA in the acceptor mechanism for uterine oestradiol receptor. *Nature New Biol.* 240:185-87
181. Puca, G. A., Sica, V., Nola, E. 1974. Identification of a high affinity nuclear acceptor site for estrogen receptor of calf uterus. *Proc. Natl. Acad. Sci. USA* 71:979-83
182. O'Malley, B. W., McGuire, W. L. 1968. Studies on the mechanisms of action of progesterone in regulation of the synthesis of specific protein. *J. Clin. Invest.* 47:654-64
183. Schwartz, R. J., Kuhn, R. W., Buller, R. E., Schrader, W. T., O'Malley, B. W. 1976. Progesterone-binding components of chick oviduct. XI. *In vitro* effects of purified hormone receptor complexes on the initiation of RNA synthesis in chromatin. *J. Biol. Chem.* 251:5166-77
184. Buller, R. E., Schwartz, R. J., Schrader, W. T., O'Malley, B. W. 1976. Progesterone-binding components of chick oviduct. XII. *In vitro* effect of receptor subunits on gene transcription. *J. Biol. Chem.* 251:5178-86
185. Edelman, I. S., Fimognari, G. M. 1968. On the biochemical mechanism of action of aldosterone. *Recent Prog. Horm. Res.* 24:1-33
186. Kenny, F. T., Kull, F. J. 1963. Hydrocortisone stimulated synthesis of nuclear RNA in enzyme induction. *Proc. Natl. Acad. Sci. USA* 50:493-99
187. Baxter, J. D., Rousseau, G. G., Benson, M. C., Garcea, R. L., Ito, J., Tomkins, G. M. 1972. Role of DNA and specific cytoplasm receptors in glucocorticoid action. *Proc. Natl. Acad. Sci. USA* 69:1892-96
188. Fang, S., Liao, S. 1971. Androgen receptors: Steroid and tissue specific retention of a 17β hydroxy-5α-androstan-3-one protein complex by the cell nuclei of ventral prostate. *J. Biol. Chem.* 246:16-24
189. Mainwaring, W. I. P., Peterken, B. M. 1971. A reconstituted cell free system for the specific transfer of steroid receptor complexes into nuclear chromatin isolated from rat ventral prostate gland. *Biochem. J.* 125:285-95
190. Bullock, L. P., Bardin, C. W. 1974. Androgen receptor in mouse kidney: A study of male, female and androgen-insensitive (tfm/y) mice. *Endocrinology* 94:746-56
191. Anderson, J., Clark, J. H., Peck, E. J. Jr. 1972. Oestrogen and nuclear binding sites. Determination of specific sites by [³H] oestradiol exchange. *Biochem. J.* 126:561-67
192. Kalimi, M., Tsai, S. Y., Tsai, M. J., Clark, J. H., O'Malley, B. W. 1976. Effect of estrogen on gene expression in the chick oviduct. VIII. Correlation between nuclear-bound estrogen receptor and chromatin initiation sites for transcription. *J. Biol. Chem.* 251:516-23
193. Buller, R. E., Schrader, W. T., O'Malley, B. W. 1975. Progesterone-binding components of chick oviduct. IX. The kinetics of nuclear binding. *J. Biol. Chem.* 250:809-18
194. Marver, D., Stewart, J., Funder, J. W., Feldman, D., Edelman, I. S. 1974. Renal aldosterone receptors: Studies with [³H] aldosterone and the antimineral corticoid [³H] spirolactone (SC-26304). *Proc. Natl. Acad. Sci. USA* 71:1432-35
195. Gehring, U., Mohit, B., Tomkins, G. M. 1972. Glucocorticoid action on hybrid clones derived from cultured myeloma and lymphoma cell lines. *Proc. Natl. Acad. Sci. USA* 69:3124-27

196. Chamness, G. C., Jennings, A. W., McGuire, W. L. 1974. Estrogen receptor binding to isolated nuclei. A nonsaturable process. *Biochemistry* 13:327–31
197. Higgins, S. J., Rousseau, G. G., Baxter, J. D., Tomkins, G. M. 1973. Nature of nuclear acceptor sites for glucocorticoids and estrogen receptor complexes. *J. Biol. Chem.* 248:5873–79
198. Harris, G. S. 1971. Nature of oestrogen specific binding sites in the nuclei of mouse uteri. *Nature New Biol.* 321:246–48
199. Buller, R. E., Toft, D. O., Schrader, W. T., O'Malley, B. W. 1975. Progesterone binding components of chick oviduct. VII. Receptor activation and hormone dependent binding to purified nuclei. *J. Biol. Chem.* 250:801–8
200. Beato, M., Kalimi, M., Konstam, M., Feigelson, P. 1973. Interaction of glucocorticords with rat liver nuclei. II. Studies on the nature of the cytosol transfer factor and the nuclear acceptor site. *Biochemistry* 12:3372–78
201. Yamamoto, K. R., Alberts, B. 1974. On the specificity of the binding of the estradiol receptor protein to deoxyribonucleic acid. *J. Biol. Chem.* 249:7076–86
202. Yamamoto, K. R. 1974. Characterization of the 4S and 5S forms of the estradiol receptor protein and their interactions with deoxyribonucleic acid. *J. Biol. Chem.* 249:7068–75
203. Yamamoto, K. R., Alberts, B. M. 1972. In vitro conversion of estradiol-receptor protein to its nuclear form: Dependence on hormone and DNA. *Proc. Natl. Acad. Sci. USA* 69:2105–9
204. Spelsberg, T. C., Steggles, A. W., O'Malley, B. W. 1971. Progesterone-binding components of chick oviduct. III. Chromatin acceptor sites. *J. Biol. Chem.* 246:4188–97
205. O'Malley, B. W., Spelsberg, T. C., Schrader, W. T., Chytil, F., Steggles, A. W. 1972. Mechanism of interaction of a hormone-receptor complex with the genome of a eukaryotic target cell. *Nature* 235:141–44
206. Spelsberg, T. C., Steggles, A. W., Chytil, F., O'Malley, B. W. 1972. Progesterone binding components of chick oviduct. V. Exchange of progesterone-binding capacity from target to nontarget tissue chromatins. *J. Biol. Chem.* 247:1368–74
207. Steggles, A. W., Spelsberg, T. C., Glasser, S. R., O'Malley, B. W. 1971. Soluble complexes between steroid hormones and target tissue receptors bind specifically to target-tissue chromatin. *Proc. Natl. Acad. Sci. USA* 68:1479–82
208. Raynaud-Jammet, C., Baulieu, E. E. 1969. Action de l'oestradiol *in vitro:* Augmentation de la biosynthèse d'acide ribonucleique dans les noyaux uterins. *C. R. Acad. Sci. Ser. D.* 268:3211–14
209. Arnaud, M., Beziat, Y., Guilleux, J. C., Hough, A., Hough, D., Mousseron-Canet, M. 1971. Les recepteurs de l'oestradiol dans l'uterus cle genisse stimulation de la biosynthese de RNA *in vitro. Biochim. Biophys. Acta* 232:117–31
210. Mohla, S., DeSombre, E. R., Jensen, E. V. 1972. Tissue-specific stimulation of RNA synthesis by transformed estradiol-receptor complex. *Biochem. Biophys. Res. Commun.* 46:661-67
211. Davies, P., Griffiths, K. 1973. Stimulation of ribonucleic acid polymerase activity *in vitro* by prostatic steroid-protein receptor complexes. *Biochem. J.* 136:611–22
212. Buller, R. E., Schwartz, R. J., O'Malley, B. W. 1976. Steroid hormone receptor fraction stimulation of RNA synthesis: A caution. *Biochem. Biophys. Res. Commun.* 69:106–13
213. Kuhn, R. W., Schrader, W. T., Smith, R. G., O'Malley, B. W. 1975. Progesterone binding components of chick oviduct. X. Purification by affinity chromatography. *J. Biol. Chem.* 250:4220–28
214. Jaffe, R. C., Socher, S. H., O'Malley, B. W. 1975. An analysis of the binding of the chick oviduct progesterone-receptor to chromatin. *Biochim. Biophys. Acta* 399:403–19
215. Yamamoto, K. R., Alberts, B. M. 1976. Steroid receptors: Elements for modulation of eukaryotic transcription. *Ann. Rev. Biochem.* 45:721–46
216. Schwartz, R. J., Schrader, W. T., O'Malley, B. W. 1976. Mechanism of steroid hormone action: *In vitro* control of gene expression in chick oviduct chromatin by purified steroid receptor complexes. In *The Juvenile Hormones,* ed. L. I. Gilbert, pp. 530–56. New York: Plenum
217. Schrader, W. T., Heuer, S. S., O'Malley, B. W. 1975. Progesterone receptors of chick oviduct: Identification of 6S receptor dimers. *Biol. Reprod.* 12:134–42
218. Schrader, W. T., O'Malley, B. W. 1972. Progesterone-binding components of chick oviduct IV. Characterization of purified subunits. *J. Biol. Chem.* 247:51–59

219. Schrader, W. T., Kuhn, R. W., O'Malley, B. W. 1977. Progesterone-binding components of chick oviduct. XIII. Receptor B subunit protein purified to apparent homogeneity from laying hen oviducts. *J. Biol. Chem.* 252:299-307
220. Monahan, J. J., Harris, S. E., Woo, S. L. C., Robberson, D. L., O'Malley, B. W. 1976. The synthesis and properties of the complete complementary DNA transcript of ovalbumin mRNA. *Biochemistry* 15:223-33
221. Monahan, J. J., McReynolds, L. A., O'Malley, B. W. 1976. The ovalbumin gene. II. *In vitro* enzymatic synthesis and characterization. *J. Biol. Chem.* 251:7355-62
222. Woo, S. L. C., Smith, R. G., Means, A. R., O'Malley, B. W. 1976. The ovalbumin gene. I. Partial purification of the coding strand. *J. Biol. Chem.* 251: 3868-74
223. Woo, S. L. C., Monahan, J. J., O'Malley, B. W. 1977. The ovalbumin gene. V. Purification of the anticoding strand. *J. Biol. Chem.* In press

INTERACTIONS BETWEEN HOST AND VIRAL GENOMES IN MOUSE LEUKEMIA

❖3123

Richard Steeves and Frank Lilly
Department of Developmental Biology and Cancer and Department of Genetics,
Albert Einstein College of Medicine, Bronx, New York 10461

CONTENTS

INTRODUCTION	277
MENDELIAN INHERITANCE OF THE MuLV GENOME	279
INFLUENCE OF GENES THAT AFFECT NORMAL DEVELOPMENT	280
Altered Leukemia Incidence Associated with Anemia-Inducing Genes	280
Increased Leukemia Incidence Associated with the Hairless Gene	280
RECEPTORS FOR VIRUS	281
BLOCKED INTEGRATION OF VIRAL GENOME	281
The Fv-1 Gene	281
Genes That Interact with Fv-1	282
CONTROL OF SFFV REPLICATION	283
RESISTANCE IN VITRO TO FRIEND VIRUS	284
Fv-3: Resistance of Lymphocytes to Mitogen-Induced Proliferation	284
Fv-4: Resistance of the G Mouse Strain	285
OUTGROWTH OF TUMOR CELLS	285
HOST GENES DEFINED BY OTHER TYPE C VIRUSES	286
CHANGES IN THE VIRUS	287
Conversion from N or B Tropism to NB Tropism	287
Evolution of Nonleukemogenic, Fibrotropic Type C Viruses	287
Isolation of Mink Cell Focus-Inducing Viruses	288
Evolution of Viruses That Contain "Sarc" and SFFV-Specific Sequences	288
CONCLUSIONS	289

INTRODUCTION

The incidence of spontaneous or induced leukemia in mice is influenced by a complex array of interactions between host and viral genomes. In spite of our ignorance about the mechanisms of action of host genes that affect viral leukemogenesis, it is useful to create a temporary framework for approaching the problem. Genes are described here with the full expectation that the framework will have to change as more information becomes available. Therefore we review the effects of chromosomally integrated murine leukemia virus (MuLV) genomes, host genes that affect response to MuLV by influencing cellular differentiation, genes that alter various stages of MuLV growth cycle, genes that control the outgrowth of tumor cells, and viral genes that affect the tropism or pathogenicity of the virus.

Viruses that induce leukemia in mice have been called type C viruses (1), but not all type C viruses are leukemogenic. Type C viruses that are infectious in hosts of the species in which they arose are said to be ecotropic, while those infectious only in hosts of other species are xenotropic (2). Wild mice harbor xenotropic and/or amphotropic viruses (3–5), the latter infectious both in non-murine species and in N-type mice (see section on blocked integration of viral genome), whereas inbred mouse strains harbor xenotropic and/or ecotropic viruses. There have been many isolations of different ecotropic MuLV strains since Gross's discovery of MuLV in 1951 (6). Most of these viruses induce thymus-dependent lymphomas months after inoculation of neonatal mice (7), but some isolates induce nonthymic leukemias within only a few weeks of infection. For example, the Abelson virus induces a B-cell lymphosarcoma (8), and the spleen focus-forming virus (SFFV) induces erythroleukemia (9). Both viruses induce rapidly progressive diseases, and SFFV is highly pathogenic even in adult mice of susceptible strains.

Like the sarcoma-inducing viruses of mice (MuSV), SFFV and Abelson virus are defective for replication (9, 10). The virus replication cycle involves a series of steps: attachment, penetration, and uncoating; synthesis and integration of DNA from the viral RNA template; messenger and viral RNA synthesis; viral protein and glycoprotein (envelope) synthesis; virion assembly (budding) and release. The replication-defective viruses lack a portion of the MuLV genome which includes the determinant of the major envelope protein, and they can complete their replication cycle only if this protein is supplied by an accompanying helper virus which replicates within the same cell. This "rescue" of defective virus by helper virus can be compared to the phenomenon of phenotypic mixing, in which cells mixedly infected with two competent viruses can generate new virions consisting of a genome of one enveloped in a coat coded by the other. Phenotypically mixed particles do not breed true upon single infection of cells, but defective viruses retain their helper-derived properties at each replication cycle because of the absolute requirement for coinfection with helper virus. If, however, a different helper virus is substituted for the original one, the defective virus will then show the acquired phenotypic properties of this new helper and will be called a *pseudotype*.

SFFV is present in preparations of both the Friend (11) and Rauscher (12) strains of MuLV (9). It is not known whether or not the SFFV is identical in these two virus preparations, but their native helper viruses differ in antigenicity and in host range (13). Partly because of the more limited host range of the original Friend helper virus, the Friend (FV) complex became a favorite tool for geneticists during their early studies on virus-induced leukemia in mice (14–16).

MENDELIAN INHERITANCE OF THE MULV GENOME

AKR mice, which show a high incidence of spontaneous leukemia at 8–12 months of age, also show high levels of infectious, ecotropic MuLV in their tissues from the first week after birth (17). Expression of this infectious virus is dominant in F_1 hybrids of AKR with virus-negative strains of mice. However, the levels of virus expression in these F_1 hybrids may be strongly influenced by other host genes (e.g. *Fv-1*) to be described later. Studies of backcrosses of these virus-positive F_1 mice to the virus-negative parental strain indicate that two unlinked dominant genes of the AKR govern the capacity to express this virus (18, 19). One of these genes, *Akv-1*, has been mapped to a position linked to the albino and *Gpi-1* loci on chromosome 7 (20); the independently segregating *Akv-2* locus remains unmapped.

Recent studies have demonstrated that the *Akv-1* and *Akv-2* genes of AKR mice consist of chromosomally integrated ecotropic MuLV genomes. Probes consisting of DNA transcripts (cDNA) of the viral RNA hybridize with cellular DNA from AKR mice and from crosses with mice of virus-free strains in a pattern consistent with the number of *Akv* alleles present in each case (21). Furthermore, viral cDNA hybridizes with a segment of cellular DNA that has been mapped to the same chromosomal position as the *Akv-1* locus (22). Chromosomal integration of a reverse transcript of the viral genome is known to be a possible and perhaps a necessary event in the replicative cycle of MuLV in somatic cells, but mice infected with exogenous MuLV at birth do not transmit the virus to their offspring by a Mendelian mechanism (23, 24), implying that sex cells of these mice do not become infected by the virus under these conditions. This restriction of cells of the germ line can be overcome by infecting mice with ecotropic (Moloney) MuLV at the 4–8 cell preimplantation stage (25). It is likely that incubation of such embryos with virus leads to infection of only a fraction of the blastomeres. Nonetheless, some of the infected embryos, transplanted to the uterine horns of pseudopregnant mothers, eventually reached adulthood, developed leukemia, and transmitted the virus to their progeny (in crosses with uninfected mice) according to Mendelian expectations (26). Thus infection of mice at an early stage of embryonic development, before any differentiation into "target" and "nontarget" (germ line) cells had taken place, led to germ line integration of the virus with the host genome—the conversion of an exogenous virus into an endogenous virus. The question remains whether the integration of viral genomes at the *Akv-1* and *Akv-2* loci occurred in the recent or distant past.

Chromosomally integrated ecotropic viral genomes have been demonstrated in mice of several strains in addition to AKR. These include not only other high-leukemia strains such as C58 and C3H/Fg, but also a number of low-leukemia strains such as BALB/c and DBA/2 (27). Existing evidence indicates that integration has occurred at several other chromosomal sites in addition to the *Akv-1* and *Akv-2* loci, but it is not clear just how large is the number of possible sites.

Not only ecotropic, but also xenotropic, viral genomes may be chromosomally integrated and therefore transmitted in a Mendelian manner. Studies of BALB/c mice have shown that both ecotropic and xenotropic MuLV genomes are present and transmitted independently in backcrosses to mice of virus-free strains (28). Endogenous xenotropic MuLV genomes are not ordinarily expressed as infectious virus in mice of most strains, although mice of the NZB strain show high levels of this virus from early in life (2) and AKR thymic tissues show significant levels from about 6 months of age (29). In addition, tissues from mice free of detectable infectious MuLV frequently show significant levels of certain proteins encoded by the MuLV genome (30). Whether these proteins result from expression of incomplete viral genomes or incomplete expression of competent viral genomes is not known.

INFLUENCE OF GENES THAT AFFECT NORMAL DEVELOPMENT

Altered Leukemia Incidence Associated with Anemia-Inducing Genes

All three of the hereditary anemias studied to date are associated with a marked loss of susceptibility to Friend SFFV. The pleiotropism at the dominant spotting, or W (31), and steel, or Sl (32) loci can lead not only to macrocytic anemia and decreased fertility but also to a lack of hair pigmentation, so mutant alleles were most often detected and maintained in dark-coated mice, especially those of the C57BL family which are resistant to SFFV as a result of their genotype at other loci (see below). Therefore, before the effect on host susceptibility of various alleles at these loci could be tested, it was necessary to isolate and characterize a host-adapted FV strain that could induce spleen focus formation in adult C57BL and WB mice (33). With two such isolates it was found that $+/W$, $+/Sl^T$, and Sl^T/Sl^T mice were as susceptible as $+/+$ littermates, while $+/W^v$ and $+/Sl^d$ were significantly more resistant, and the compound mutants W/W^v and Sl/Sl^d were completely refractory to spleen focus formation by SFFV (34, 35). Thus, although the Sl locus influences the hemopoietic environment whereas the W locus affects the progenitor cells themselves (36), these genetic loci control steps in both normal differentiation and in neoplastic transformation, and they provide powerful tools for future definition of the target cell for SFFV in vivo.

Anemic Sl/Sl^d mice also show an increased risk of spontaneous lymphocytic leukemia (37), though it is not clear whether this is due to the known pleiotropic effects of the mutant steel genes, such as anemia, or through other, as yet unknown, effects.

The recessive gene "flexed" (f) induces a transitory siderocytic anemia by affecting chiefly the intermediate generation erythrocytes, which stem largely from fetal liver hematopoiesis (38). Axelrad has observed that mutant f/f mice had a significantly lower susceptibility to FV than wild-type F/F homozygotes (39).

Increased Leukemia Incidence Associated with the Hairless Gene

Hairless (hr/hr) mice of the HRS/J strain spontaneously develop leukemia with an incidence of 45% at 8 to 10 months of age and 72% at 18 months, whereas only 1% of heterozygotes ($+/hr$) develop leukemia by 8–10 months, and 20% by 18 months (40). Although initial studies of type C ecotropic virus titers in the spleens of weanling mice (40) or in whole embryos (41) failed to reveal a difference between mutant and normal mice, more recent studies with tail extracts from 6-month-old mice revealed an average of 13 times more type C virus (as measured by the XC assay) in hairless than in normal mice (42). Whether the *hr* gene provides a factor that enhances virus expression or the wild-type allele inhibits virus expression through some negative regulatory control is not known, but the latter alternative would explain the similarity in virus titers of heterozygotes and wild-type homozygotes, and it might have interesting implications for the prevention of leukemia.

RECEPTORS FOR VIRUS

Various classes of MuLVs appear to have adapted to the utilization of cell surface molecules (virus receptors) in order to facilitate penetration into host cells. In the case of virus/host cell interactions in certain other viral systems [e.g. avian leukosis viruses (43)], host genetic polymorphism with respect to the cell surface molecule in question has permitted identification of the gene(s) involved on the basis of the presence or absence of the virus receptor function (see below). However, among the inbred mouse strains tested there is no known polymorphism for MuLV receptors.

Although ecotropic strains of MuLV are to some extent infectious in cells of virtually all strains of mice, they are infectious in cells of few other species. In the case of hamster cells, data exist that indicate that resistance to ecotropic MuLV is due to a lack of functional virus receptors. Hamster cells are fully capable of supporting the replication of vesicular stomatitis virus (VSV), but the pseudotype VSV(MuLV) (VSV genome in a MuLV coat) is not infectious for these cells (44). This pseudotype is fully infectious for murine cells, however (45).

Another indication of the existence of receptors for MuLV on mouse cells comes from the observation that virus-infected, nonproducer cell lines that express the gp70 virus envelope antigen are resistant to superinfection with ecotropic MuLV, but when this viral protein ceases to be expressed the cells regain susceptibility to superinfection with the virus (P. Roe and H. Freedman, unpublished observations).

The resistance of mouse cells to infection by xenotropic strains of MuLV appears to be due at least in part to the absence of appropriate receptors (46, 47). Cells of nonmurine species susceptible to xenotropic MuLV could be infected by pseudotype virus consisting of murine sarcoma virus in a xenotropic MuLV coat, but most mouse cells could not be infected by this pseudotype (46).

It is not difficult to conceive of evolutionary mechanisms by which ecotropic viruses might have adapted themselves to using as a cellular receptor a molecule whose homologues, if any, in other species do not possess the necessary properties to support this function. By contrast, it is more difficult to conceive of an evolutionary mechanism by which xenotropic virus might have achieved a nearly opposite result, i.e. that of being incapable of penetration into cells of its presumed natural host but highly penetrant into cells of many other species.

BLOCKED INTEGRATION OF VIRAL GENOME

The Fv-1 Gene

The major determinant of exogenous infection of mouse cells with naturally occurring ecotropic MuLV is at the *Fv-1* locus. This locus, on chromosome 4 (48), has two alleles, $Fv\text{-}1^n$ and $Fv\text{-}1^b$, and mouse strains are said to be N-type or B-type, respectively, depending on their $Fv\text{-}1$ genotype (49). An MuLV not restricted in N-type mice is said to be N-tropic, and similarly, virus not restricted in B-type mice is B-tropic. The degree of restriction of N-tropic virus in B-type mice or of B-tropic virus in N-type mice is generally 50- to 1000-fold. In F_1 crosses between N-type and B-type mice, resistance is dominant, and both N-tropic and B-tropic MuLV are restricted (49). However, both N- and B-tropic viruses can be adapted by repeated forced passage in $Fv\text{-}1$-restrictive mice to become NB-tropic; that is, the adapted virus is no longer sensitive to the $Fv\text{-}1$ type of its host (as described in the section on changes in the virus).

The $Fv\text{-}1$ gene has been studied principally in three systems: by direct titration of MuLV in mouse embryo fibroblasts in culture with the XC cell assay (50), by an immunofluorescence focus assay (50a), and by indirect titration of Friend helper virus for SFFV, where the complete FV complex is measured by the spleen focus assay in mice (51). In all titration systems the dose-response relationships are similar: High-titer, one-hit curves are obtained in $Fv\text{-}1$-susceptible hosts, and low-titer, two-hit curves are usually observed in $Fv\text{-}1$-restrictive hosts. It is not clear, in the latter instance, why workers in two laboratories have been unable to duplicate the two-hit curves (52, 53).

The molecular basis of the $Fv\text{-}1$ effect on virus titration patterns remains unknown, although it is currently being studied intensively. Recent biochemical evidence indicates that the host cell restricts the virus at a late stage in the virus replication cycle (54, 55) probably at some step between the synthesis of viral DNA and its integration into the host genome (56, 57). A soluble extract has been isolated from resistant cells that confers $Fv\text{-}1$-type resistance to otherwise susceptible cells in culture (58, 59). It has been hypothesized that the $Fv\text{-}1$ gene product interacts with a specific but unidentified viral gene product as its target, thereby preventing viral integration (60).

Studies by W. P. Rowe and J. W. Hartley (personal communication) indicate that there are probably more than two alleles at the $Fv\text{-}1$ locus. A third allele, named $Fv\text{-}1^{nr}$, characterizes four mouse strains (RF, 129, NZB, and NZW) previously said to be N-type which are four-fold or more resistant to N-tropic MuLV than to other N-type strains. In crosses between CBA and 129 mice, $Fv\text{-}1^{nr}$ (resistance) was dominant over $Fv\text{-}1^n$, and it segregated with the $Gpd\text{-}1$ locus, a marker closely linked with $Fv\text{-}1$. Also, after N-tropic or B-tropic MuLV was forced-passed in RF mice, it became more infectious in cells of other $Fv\text{-}1^{nr}$ strains.

Genes That Interact with Fv-1

The "$Fv\text{-}1$ story" is complicated further by the presence of genes that modify the $Fv\text{-}1$ restriction of certain B-tropic viruses. The Srv gene (61), whose dominant allele occurs in C57BL mice, modifies the patterns (from two-hit to one-hit) and elevation of titration curves of a particular B-tropic virus, radiation leukemia virus, assayed in $Fv\text{-}1$-restrictive mouse embryo cells. However, neither the titration pattern of the Tennant subclass of B-tropic virus nor that of N-tropic viruses is affected by Srv. A different gene has been found (62) which also modifies the titration pattern (from two-hit to one-hit) but decreases the infectivity of the Tennant (B-tropic) pseudotype (and not the native N-tropic strain) of Friend SFFV in DBA/2 and other N-type mice. This unnamed gene is recessive for resistance, but it is not expressed in $Fv\text{-}1^b$ homozygotes (e.g. in congenic DBA/2.Fv-1^b mice). The mechanism of action of these genes is not known, and neither gene has been mapped, but they are not closely linked to $Fv\text{-}1$.

CONTROL OF SFFV REPLICATION

Linked to the dilute locus on chromosome 9, the $Fv\text{-}2$ gene exerts a major effect on the response of mice to FV (63) and also to Rauscher virus. Mice of the C57-C58 family of strains possess the recessive $Fv\text{-}2^r$ allele for resistance, while all other laboratory strains tested have the dominant $Fv\text{-}2^s$ allele for susceptibility. Two major systems have been used in identifying this gene: the spleen weight assay (64) and the spleen focus assay (65). Recently a test for $Fv\text{-}2$ resistance in vitro has been developed, based on thymidine incorporation into spleen cells (66). In vivo, $Fv\text{-}2^r$ homozygotes are absolutely resistant to the virus (14, 15, 67, 68). However, the $Fv\text{-}2$ gene does not affect susceptibility to the helper component of FV or to other strains of MuLV, whatever their tropism; neither lymphatic leukemogenesis in vivo nor XC plaque formation in vitro is inhibited by $Fv\text{-}2$ resistance. Therefore, the major effect of $Fv\text{-}2$ is to restrict SFFV infection.

Studies of the mechanism of this resistance have been hampered by the fact that the most widely available $Fv\text{-}2^r$ mouse is the C57BL, and this mouse has the curious property of having collected into its genome the alleles for resistance to viral leukemogenesis at many different loci. Therefore, detailed studies of $Fv\text{-}2$ have awaited the availability of a congenic strain. Such strains have been developed in three laboratories: Axelrad (68) developed the B6.S strain, congenic with C57BL/6 but susceptible at $Fv\text{-}2$. Normal-appearing foci developed in FV-infected B6.S mice, but the number of foci was diminished in comparison with most other mouse strains carrying the $Fv\text{-}2^s$ gene. Odaka (69) developed the DDD.Fv-2^r strain, congenic with virus-susceptible DDD mice. Our attempts to produce a DBA/2.Fv-2^r strain that was truly congenic with DBA/2 led to an unexpected conclusion: that the $Fv\text{-}2^r$ gene (or a gene very closely linked to it) is lethal at an early embryonic age in mice homozygous for it *unless* they also possess another unidentified gene (present in C57BL but not in DBA/2 mice) which protects them from this lethality (70).

Our conclusion is based on the observation that, after 13 backcross generations of DBA/2 × C57BL with DBA/2 mice, intercrossing heterozygous $Fv\text{-}2^s/Fv\text{-}2^r$ backcross mice yielded homozygous $Fv\text{-}2^s/Fv\text{-}2^s$ and heterozygous $Fv\text{-}2^s/Fv\text{-}2^r$ mice in a 1:2 ratio, but no $Fv\text{-}2^r/Fv\text{-}2^r$ homozygotes. Since at no time during the backcross breeding of $Fv\text{-}2^s/Fv\text{-}2^r$ heterozygotes with $Fv\text{-}2^s/Fv\text{-}2^s$ DBA/2 mice were $Fv\text{-}2^r/Fv\text{-}2^r$ homozygotes produced, any genes from the C57BL ancestor that might be needed for survival of $Fv\text{-}2^r/Fv\text{-}2^r$ homozygotes could have been lost. It is interesting to speculate why Odaka did not encounter this situation during his production of the DDD.Fv-2^r strain: Either the DDD strain had the "protector" gene missing from DBA/2 mice, or Odaka's use of the cross-intercross system (rather than the continuous backcross system which we used) made him select $Fv-2^r/Fv-2^r$ homozygotes which also carried the C57BL "protector" gene in each generation.

Fortunately, we obtained homozygous, partially congenic DBA/2.Fv-2^r mice after only two backcross generations to DBA/2 mice. This strain, since it was viable, had obtained the "protector" gene and probably about 12% of the rest of its genome from C57BL mice, but was suitable for preliminary studies on the mechanism of action of the $Fv-2$ gene. Our published results (71) indicate that the gene acts primarily by inhibiting SFFV replication and by blocking spleen colony formation by SFFV-infected cells. However, more recent observations have shown that the latter parameter (spleen colony formation) is dependent on the former (SFFV replication) in unirradiated hosts (72), and we are currently seeking a mechanism whereby the $Fv-2^r$ gene suppresses SFFV replication, mindful that such a mechanism, once found, might lead to a potentially lethal defect in the absence of the C57BL "protector" gene.

RESISTANCE IN VITRO TO FRIEND VIRUS

Fv-3: Resistance of Lymphocytes to Mitogen-Induced Proliferation

Mice of the C57-C58 family are resistant not only to the erythroleukemic effects of FV, by virtue of the $Fv-2^r$ gene, but also to most bone marrow allografts (73). Recently Kumar & Bennett suggested that a gene responsible at least in part for the latter resistance may also confer resistance to FV (74, 75). The evidence to date is based on a simple cell proliferation assay in vitro for both T- and B-cell mitogens, such as Concanavalin A, dextran sulfate, or lipopolysaccharide. Normal lymphocytes of all mouse strains tested would respond to these mitogens in vitro, but after infection with NB-tropic FV, lymphocytes from mice genetically susceptible to FV (e.g. BALB/c, C3H, 129) were unable to respond to the mitogens, whereas lymphocytes of mice resistant to FV (e.g. C58, C57BL) responded well in the presence or absence of the virus. Suppression of the mitogenic response of T or B cells in the presence of FV is thought to be mediated by T-suppressor cells, whose population size and/or function is controlled in turn by the so-called M cell, a bone marrow-dependent immune cell responsible for the rejection of bone marrow allografts. Resistant C57BL mice treated with the bone-seeking isotope, ^{89}Sr, were selectively depleted of M cells, and, perhaps because of increased T-suppressor cell function, were rendered susceptible to FV-induced leukemia (76).

The gene controlling the capacity of lymphocytes to respond to mitogens in vitro in the presence of NB-tropic FV has been tentatively designated $Fv-3$, since the recessive phenotype of B10.D2 mice segregated independently of resistance at the $Fv-2$ locus among progeny of backcrosses with DBA/2 mice (V. Kumar and M. Bennett, 1977, personal communication). In the same backcross, however, there was a complete correlation between resistance of lymphocytes in vitro and resistance to virus-induced depression of the humoral response to sheep erythrocytes. Just as Dent (77) asked whether the virus-induced immunodepression associated with viral leukemia is a cause or a consequence of the oncogenic process, so one might ask whether the lack of immunodepression associated with genetically controlled host resistance to FV is a cause or a consequence of this resistance. Nonetheless, it is tempting to speculate that *Fv-3* may determine whether or not the infected host is immunodepressed by FV. In any case the precise role of this gene in determining resistance in vivo to Friend virus and to bone marrow allografts remains to be elucidated.

Fv-4: Resistance of the G Mouse Strain

A new mouse strain, G, was selected from a non-inbred mouse colony in Japan on the basis of its resistance to N-tropic FV (78). Tests of the susceptibility of the offspring of G X DBA/2 and (G X BALB/c) X DBA/2 matings revealed that the resistance of the G strain was controlled by one gene, dominant for resistance, which segregated independently of *Fv-1* and the coat color genes, *b, d,* and *c* (79). More recently Kai et al (80) observed with the XC test that cultured G embryo cells were resistant, though not absolutely, to all type C viruses tested, including N-tropic AKR virus, N- and NB-tropic FV, NB-tropic Rauscher virus complex, B-tropic WN 1802B virus, NB-tropic Moloney leukemia and sarcoma viruses, and N-tropic Kirsten sarcoma virus. Clearly, the *Fv-4* gene affects a broader range of MuLV strains of differing host range than *Fv-1* does, and hence *Fv-4* masks the *Fv-1* phenotype in G mice.

OUTGROWTH OF TUMOR CELLS

One of the earliest discoveries of a gene involved in susceptibility to viral leukemogenesis was *Rgv-1,* a function that maps within or in close association with the major histocompatibility complex, *H-2,* on chromosome 17 (81). The initial studies of *H-2* association with leukemia were in systems of lymphatic leukemogenesis with the Gross (82) and B/T-L (83) viruses. These studies were rapidly followed by others in different viral systems, including FV erythroleukemogenesis (84), mammary tumorigenesis (85), and sarcomagenesis (86), in each of which a comparable influence of *H-2* was demonstrable.

It is clear that the resistance conferred by certain *H-2* haplotypes is relative rather than absolute. This is most obvious in studies with FV, where an additional clue to the mechanism of the resistance is the finding that *H-2* type influences the capacity of the mouse to recover from the initial stages of the disease syndrome (84). At low doses of virus the *H-2* influence may result in all-or-none differences in the occurrence of the disease. This observation suggests that the resistance may be mediated by an immunologic mechanism. [It is noteworthy that preliminary observations concerning lung cancer in man suggest that the human major histocompatibility complex, *HLA,* is not associated with the *occurrence* of the disease but may be a factor in the *prognosis* for response to therapy for the disease (87)].

An obvious candidate for identity with *Rgv-1* would be one of the *H-2*–linked immune response genes (88) since *Rgv-1* maps in the general vicinity of these genes (81). Increased immune responsiveness, either cellular or humoral, to viral antigens might well be expressed as increased resistance to the disease. The H-2–linked control of immune responsiveness to tumor cells bearing the X.1 antigen is an important observation that strengthens this interpretation (89).

A recent and perhaps surprising finding has been that, in addition to Rgv-1, which maps proximally toward the K region of H-2, another gene, Rfv-1, which maps distally in the D region, is the major determinant of H-2–associated resistance to Friend virus erythroleukemia (90). Since no immune response genes are known to be located in the immediate vicinity of Rfv-1, there is every reason to seek another sort of immunologic explanation of its mechanism.

The recent observation of H-2 restriction in the host response to lymphocytic choriomeningitis virus (91), which has also been found to occur in the immune response to FV infection, may provide a key to the elucidation of this problem. Mice immunized with syngeneic, FV-induced tumor cells often respond by producing T-killer cells. These killer cells have two different requirements for their cytotoxic activity. Potential target cells must express simultaneously both antigens induced by the virus and H-2 antigens identical with those of the immunizing cell (92). The basis for this requirement for dual recognition remains to be clarified, but preliminary findings have suggested a hypothetical mechanism which might bring together the two phenomena of H-2 restriction and the H-2–associated influence on viral leukemogenesis.

Several lines of evidence suggest that certain viral proteins may become physically associated on the cell surface with H-$2K$ and/or H-$2D$ molecules. In the FV system, evidence for the formation of such complexes includes (*a*) partial and selective cocapping of certain H-2 antigenic determinants by treatment of infected cells with anti-FV antibodies in the absence of complement (93) and (*b*) the selective inclusion of certain H-2 antigens into FV particles during virion assembly at the cell membrane (94).

This H-2/viral protein complex, which is not necessarily a stable structure, appears to provide a new antigenic determinant, not present on either component alone, which is capable of triggering the induction of cytotoxic T lymphocytes (95). Since only certain H-$2K$ and H-$2D$ molecules appear to possess this property of associating with FV proteins to form complex structures, only infected cells bearing these H-$2K$ or H-$2D$ molecules can induce cytotoxic lymphocytes. Mice possessing these molecules should be relatively resistant to the FV disease by comparison with mice possessing only H-$2K$ and H-$2D$ molecules incapable of forming the complexes.

It is not yet possible to state with certainty whether or not this phenomenon of T-cell cytotoxicity based on H-2/viral protein complex formation is indeed the basis of either the Rgv-1 or the Rfv-1 effect, but the hypothesis is an attractive one.

HOST GENES DEFINED BY OTHER TYPE C VIRUSES

Several host genes have been described by investigators working with type C viruses other than FV, and some of these genes may be the same as those described above. In one case, Toth et al (96) have described two genes, Rv-1 and Rv-2, which confer resistance in DBA/1 and C57BL/10 mice, respectively, against Rauscher virus–induced leukemia. In crosses with susceptible BALB/c mice, $Rv-1$ is dominant for resistance against the helper virus, while $Rv-2$ is recessive for resistance to SFFV in the Rauscher virus complex. It would be reasonable to surmise that $Rv-1$ and $Rv-2$ are the same as $Fv-1$ and $Fv-2$, respectively, on the hypothesis that the stock of Rauscher virus used was B-tropic, although most characterized stocks of Rauscher virus are NB-tropic.

Another case is the elegant work of Chieco-Bianchi and his collaborators (86, 97, 98) on host genes that confer resistance to murine sarcoma virus (MuSV). In these studies the genes identified were grouped into those that affect the early occurrence of virus-induced tumors and those that affect late-occurring tumors. While the former group appear to be unique to the control of host response to MuSV, the latter group are those ($Fv-1$ and $H-2$-associated) that control the host response to MuLV.

CHANGES IN THE VIRUS

Conversion from N or B Tropism to NB Tropism

NB-tropic MuLV is not known to occur in nature, but both N-tropic and B-tropic viruses can be adapted to NB tropism by repeated passage at high doses in $Fv-1$-restrictive hosts. The adaptation, which requires from 2 to 5 serial passages in vivo (16, 99) and often more in cell culture (100, 101), is permanent, for NB-tropic virus does not revert to its original tropism when it is returned to hosts of the permissive $Fv-1$ type. NB-tropic viruses differ from mixtures of N-tropic and B-tropic viruses in that only the former are not restricted in $Fv-1^n/Fv-1^b$ heterozygotes. This last observation suggests instead that NB-tropic virus has lost some factor(s) or gene product(s) involved in the heterozygote's recognition and restriction of N-tropic and B-tropic virus. Further support for the concept of recognition factors in the virion has come from experiments with phenotypically mixed MuLV particles which were dually restricted. Rein et al (60) observed that progeny virus from cells mixedly infected with both N-tropic and B-tropic MuLV was sensitive to $Fv-1$ restriction in both N- and B-type cells, but it retained high infectivity for an unusual line of mouse cells (SC-1) which does not exhibit any $Fv-1$ restriction. The conversion to NB tropism has recently been associated with an altered electrophoretic mobility of the major capsid protein, p30, but whether or not the association is causal is not clear (102).

Evolution of Nonleukemogenic, Fibrotropic Type C Viruses

Several strains of leukemogenic virus have been observed to lose their leukemogenic capacity after prolonged growth in mouse embryo fibroblasts in culture (103–111). Although the virus is attenuated in terms of its leukemogenicity, it is highly infectious in mouse embryo fibroblasts and is therefore said to be fibrotropic. When this was first observed with radiation leukemia virus (RadLV), it was interpreted as the result of "modulation" or selection in the intracellular environment of the fibroblast (111). Then Haas & Hilgers (112) found that RadLV propagated on lymphoma cells in vitro lacked fibrotropic activity but retained its leukemogenic (thymotropic) activity, and they suggested that RadLV may be composed of a heterologous mixture of two different B-tropic viruses, one thymotropic and the other fibrotropic. However, Declève et al showed recently that RadLV contains no detectable B-tropic, fibrotropic viral component (113), and they have concluded that the thymotropic RadLV induces the expression of an endogenous B-tropic fibrotropic virus in vitro. Whether this "induction" involves complete de novo induction of the virus or a complementation-rescue mechanism is not known.

An important corollary of these findings is that a tissue-specific restriction system governs the replication of murine ecotropic type C viruses in addition to the genes described above. Mouse thymic lymphocytes in vivo may selectively synthesize thymotropic, leukemogenic viruses, whereas mouse fibroblasts in vitro tend to preferentially synthesize fibrotropic viruses.

Isolation of Mink Cell Focus-Inducing Viruses

A new class of MuLV was recently detected in thymuses of leukemic and late preleukemic AKR mice, in lymphomas of NIH Swiss mice carrying the *Akv-1* or *Akv-2* genes, and in the thymus of a preleukemic C58 mouse (114). The viruses induce focal areas of morphologic alteration in a mink lung cell line; they have the host range of both xenotropic and N-tropic (ecotropic) MuLV, but, unlike amphotropic viruses, they are neutralized by antisera to both ecotropic and xenotropic MuLV, and are interfered with by both viruses. Mink cell focus-inducing (MCF) viruses are thought to represent a particular type of genetic recombinant which emerges during the preleukemic period in mouse strains with high ecotropic virus titers. If they play a significant role in the etiology of spontaneous or radiation-induced lymphomas, they might account for several previously noted discrepancies between ecotropic virus titer and leukemogenicity.

Evolution of Viruses That Contain "Sarc" and SFFV-Specific Sequences

Recombinant viruses of a different type arise rarely, but once formed they are highly visible because of their rapid oncogenic properties. The sarcoma-inducing, fibroblast-transforming, replication-defective viruses of mice (MuSV) apparently arose by recombination between portions of murine type C viral genomes and distinct sequences that are sarcoma-specific, as recently indicated from the elegant nucleic acid hybridization studies of Scolnick and others (115–119). The Harvey and Kirsten strains, isolated by passage of MuLV through rats, are recombinants between murine type C viruses and a special class of endogenous rat type C viral gene sequences (115–117), whereas the Moloney strain, derived by passage of Moloney MuLV through BALB/c mice, seems to be a recombinant between a portion of the Moloney MuLV genome and a distinct set of sequences which are partially homologous to sequences present in normal mouse cellular DNA (118, 119).

A similar approach toward understanding the origin of SFFV is now feasible with the development of SFFV-infected nonproducer cell lines that are free of detectable helper virus (120). Preliminary findings indicate that the SFFV genome includes genetic sequences homologous to a portion of the Friend helper virus genome and SFFV-specific sequences not present in the helper virus. The latter sequences, also present in the Rauscher virus genome, are partially homologous to three separate strains of xenotropic MuLV but not to several cloned ecotropic MuLVs. Thus, the Friend and Rauscher strains of SFFV seem to represent recombinants between a portion of the ecotropic helper MuLV present in the original virus isolates and a class of endogenous mouse type C viral sequences which are related to sequences contained in xenotropic MuLV (121).

CONCLUSIONS

It is clear that the pathogenesis of leukemia in mice is a complex process and that a number of preconditions must be satisfied in order for it to occur and run its course to the death of the host animal. By analogy with the sequence of events (transformations) involved in a metabolic pathway, the scheme

$$A \xrightarrow{a} B \xrightarrow{b} C \xrightarrow{c} \longrightarrow \longrightarrow X,$$

which represents a series of molecules (products of one enzymatic reaction and substrates for another), might also represent a sequence of cells progressing, through a series of alterations, from a normal one (A) to a highly malignant one (X) (122). By extension of this analogy, interactions between host and viral genomes in lymphocytic cells appear to contribute to the evolution of viral leukemia in mice through various developmental stages.

Two major subdivisions of this evolutionary process are pre- and post-transformational events. The transforming event itself remains completely mysterious in lymphatic leukemia and is only faintly glimpsed at present in certain other type C virus–induced diseases. Pretransformational events include (a) the occurrence of appropriate target cells [Availability of target cells in the FV disease is influenced by genes that produce hereditary anemias, and in thymus-dependent lymphocytic leukemia it is reasonable to suppose that the nude (nu) gene, which prevents the development of a functional thymus (123), might similarly influence the availability of the usual target cells for these MuLVs.]; (b) the occurrence of the MuLV genome [In addition to exogenous sources (e.g. from another cell within the same individual), MuLV genomes of various sorts are contained within the host genomes of many if not most mice.]; (c) the presence of cell surface receptors for the attachment and penetration of infectious virions [Unidentified genes specify molecules that are necessary for cell-to-cell transmission of virus within an individual from rare spontaneously producing cells (124).]; (d) viral replication within cells [Genes such as $Fv-1$, and apparently $Fv-2$, regulate as yet poorly defined events in the replication of various classes of MuLV.]

Following transformation of one or more target cells, extensive cellular replication must occur in order to produce a clinically recognizable disease. Host immune responses, regulated by genetic factors such as $H-2$, to viral gene products or to altered host-cell molecules may interfere with this outgrowth of tumor cell clones.

It has long been clear that cells may be infected with type C viruses and still not possess any demonstrable malignant properties. Is this because (a) the cell infected is not an appropriate target for the transforming mechanism of the virus, or (b) the virus used is one that lacks a transforming mechanism? Evidence for both alternatives has emerged from recent studies suggesting that only thymotropic viruses give rise to lymphatic leukemia.

If such a thymotropic virus is the basis of the disease, and if the virus is endogenous in all cells of at least some mice, why then does the spontaneous disease appear to be monoclonal in origin and to occur only relatively late in life? One answer to this question could be that in many cases the viral genome is under suppressive control by other host genetic factors, and the late emergence of the disease represents the breakdown of suppression with aging. Another recently available hypothesis adds the further factor that perhaps the leukemogenic virus does *not* preexist as such in the host genome, but rather that it emerges as the result of an unusual recombinational event between different preexisting viral genomes in a particular target cell (114).

Such recombinational events can occur only in cells in which two different viral genomes are being replicated. In high leukemia AKR mice, many cells actively produce infectious ecotropic MuLV throughout the life span of the host, although these cells are not necessarily thymic target cells. The second virus, xenotropic MuLV, appears to be under strong suppressive regulation and is detected only in the thymus and only later in life ($\sim$ 6 months of age). It seems likely that only a minute fraction of thymic cells, even in older AKR mice, express both viral genomes simultaneously and can, therefore, generate recombinants.

Given the complex origin of thymotropic recombinant MuLV, the question remains of the nature of the presumed transforming capacity of this virus. Transforming genes of other mostly defective strains of type C virus (e.g. MuSV, SFFV) appear to consist of host genetic sequences attached to truncated ecotropic MuLV genomes (121). No direct evidence for such a transforming gene exists for viruses associated with thymus-dependent lymphocytic leukemia. It is tempting to speculate that, if such a transforming gene were acquired by these MuLV, then the capacity of the virus to be transmitted from cell to cell would be lost. Even so, the new virus might serve as a vehicle for transporting the host sequences from one chromosomal site to another. It has been suggested (121) that this host material might consist of genes involved in the control of cell replication, and these genes, whose expression is regulated in their original chromosomal location, might function constitutively in the new site, thereby leading to the loss of growth control which is a prime characteristic of malignant cells.

A case, imperfect but provocative, can be made for the hypothesis that all leukemias in mice are in some way associated with type C viruses. Leukemogenic viruses have been isolated from tissues of not only spontaneous cases of the disease but also cases induced by X rays (125, 126) and chemical substances (127–129). Such environmental agents are probably very important factors in leukemia in natural populations. Although the establishment of high-leukemia strains of mice and the identification of leukemogenic type C viruses represent momentous steps in the direction of elucidating the disease process, until we also know how chemical and physical agents from the environment interact with host and viral genomes, leukemogenesis in mice will remain only partially understood.

Acknowledgments

We are grateful for prepublication information from Drs. M. Bennett, A. Declève, R. Jaenisch, W. Rowe, and D. Troxler. Our research has been supported by grant Nos. CA-14529, CA-19873, and CA-19931 and by contract NO1 CP NCI 71017 from the National Cancer Institute, Department of Health, Education, and Welfare, Bethesda, Maryland.

Literature Cited

1. Bernhard, W., Guérin, M. 1958. Présence de particules d'aspect virusal dans les tissus tumoraux de souris atteintes de leucémie spontanée. *C. R. Acad. Sci.* 247:1802–5
2. Levy, J. A. 1973. Xenotropic viruses: Murine leukemia viruses associated with NIH Swiss, NZB, and other strains. *Science* 182:1151–53
3. Rasheed, S., Gardner, M. B., Chan, E. 1976. Amphotropic host range of naturally occurring wild mouse leukemia viruses. *J. Virol.* 19:13–18
4. Hartley, J. W., Rowe, W. P. 1976. Naturally occurring mouse leukemia viruses in wild mice: Characterization of a new "amphotropic" class. *J. Virol.* 19:19–25
5. Bryant, M. L., Klement, V. 1976. Clonal heterogeneity of wild mouse leukemia viruses: Host range and antigenicity. *Virology* 73:532–36
6. Gross, L. 1970. *Oncogenic Viruses*, pp. 470–601, New York: Pergamon. 991 pp. 2nd ed.
7. Siegler, R., Rich, M. A. 1966. Pathogenesis of murine leukemia. *Natl. Cancer Inst. Monogr.* 22:525–47
8. Abelson, H. T., Rabstein, L. S. 1970. Lymphosarcoma: Virus-induced thymic-independent disease in mice. *Cancer Res.* 30:2213–22
9. Steeves, R. A. 1975. Spleen focus-forming virus in Friend and Rauscher leukemia virus preparations. *J. Natl. Cancer Inst.* 54:289–97
10. Scher, C. D., Siegler, R. 1975. Direct transformation of 3T3 cells by Abelson murine leukemia virus. *Nature* 253:729–31
11. Friend, C. 1957. Cell-free transmission in adult Swiss mice of a disease having the character of a leukemia. *J. Exp. Med.* 105:307–18
12. Rauscher, F. J. 1962. A virus-induced disease of mice characterized by erythropoiesis and lymphoid leukemia. *J. Natl. Cancer Inst.* 29:515–43
13. Eckner, R. J., Steeves, R. A. 1972. A classification of the murine leukemia viruses. Neutralization of pseudotypes of Friend spleen focus-forming virus by type-specific murine antisera. *J. Exp. Med.* 136:832–50
14. Odaka, T., Yamamoto, T. 1962. Inheritance of susceptibility to Friend mouse leukemia virus. *Jpn. J. Exp. Med.* 32:405–13
15. Axelrad, A. A. 1966. Genetic control of susceptibility to Friend leukemia virus in mice: Studies with the spleen focus assay method. *Natl. Cancer Inst. Monogr.* 22:619–29
16. Lilly, F. 1967. Susceptibility to two strains of Friend leukemia virus in mice. *Science* 155:461–62
17. Rowe, W. P., Pincus, T. 1972. Quantitative studies of naturally occurring murine leukemia virus infection of AKR mice. *J. Exp. Med.* 135:429–36
18. Rowe, W. P. 1972. Studies of genetic transmission of murine leukemia virus by AKR mice. I. Crosses with $Fv-1^n$ strains in mice. *J. Exp. Med.* 136:1272–85
19. Rowe, W. P., Hartley, J. W. 1972. Studies of genetic transmission of murine leukemia virus by AKR mice. II. Crosses with $Fv-1^b$ strains of mice. *J. Exp. Med.* 136:1286–1301
20. Rowe, W. P., Hartley, J. W., Bremner, T. 1972. Genetic mapping of a murine leukemia virus-inducing locus of AKR mice. *Science* 178:860–62
21. Lowy, D. R., Chattopadhyay, S. K., Teich, N. M., Rowe, W. P., Levine, A. S. 1974. AKR murine leukemia virus genome: Frequency of sequences in DNA of high-, low-, and non-virus-yielding mouse strains. *Proc. Natl. Acad. Sci. USA* 71:3555–59
22. Chattopadhyay, S. K., Rowe, W. P., Teich, N. M., Lowy, D. R. 1975. Definitive evidence that murine C-type virus inducing locus $Akv-1$ is viral genetic material. *Proc. Natl. Acad. Sci. USA* 72:906–10

23. Law, L. W., Moloney, J. B. 1961. Studies of congenital transmission of a leukemia virus in mice. *Proc. Soc. Exp. Biol. Med.* 108:715–23
24. Buffett, R. F., Grace, J. T., DiBerardino, L. A., Mirand, E. A. 1969. Vertical transmission of murine leukemia virus. *Cancer Res.* 29:588–95
25. Jaenisch, R., Fan, H., Croker, B. 1975. Infection of preimplantation mouse embryos and of newborn mice with leukemia virus: Tissue distribution of viral DNA and RNA and leukemogenesis in the adult animal. *Proc. Natl. Acad. Sci. USA* 72:4008–12
26. Jaenisch, R. 1976. Germ line integration and Mendelian transmission of the exogenous Moloney leukemia virus. *Proc. Natl. Acad. Sci. USA* 73:1260–64
27. Aaronson, S. A., Stephenson, J. R. 1976. Endogenous type-C RNA viruses of mammalian cells. *Biochim. Biophys. Acta* 458:323–54
28. Aaronson, S. A., Stephenson, J. R. 1973. Independent segregation of loci for activation of biologically distinguishable RNA C-type viruses in mouse cells. *Proc. Natl. Acad. Sci. USA* 70:2055–58
29. Kawashima, K., Ikeda, H., Stockert, E., Takahashi, T., Old, L. J. 1976. Age-related changes in cell surface antigens of preleukemic AKR thymocytes. *J. Exp. Med.* 144:193–208
30. Strand, M., Lilly, F., August, J. T. 1974. Host control of endogenous murine leukemia virus gene expression: Concentrations of viral proteins in high and low leukemia mouse strains. *Proc. Natl. Acad. Sci. USA* 71:3682–86
31. Russell, E. S. 1949. Analysis of pleiotropism at the W-locus in the mouse: Relationship between the effects of W and W^v substitution on hair pigmentation and on erythrocytes. *Genetics* 34:708–23
32. Sarvella, P. A., Russell, L. B. 1956. Steel, a new dominant gene in the house mouse. *J. Hered.* 47:123–28
33. Steeves, R. A., Mirand, E. A., Bulba, A., Trudel, P. J. 1970. Spleen foci and polycythenia in C57BL mice infected with host-adapted Friend leukemia virus. *Int. J. Cancer* 5:346–56
34. Steeves, R. A., Bennett, M., Mirand, E. A., Cudkowicz, G. 1968. Genetic control by the W locus of susceptibility to (Friend) spleen focus-forming virus. *Nature* 218:372–74
35. Bennett, M., Steeves, R. A., Cudkowicz, G., Mirand, E. A., Russell, L. B. 1968. Mutant Sl alleles of mice affect susceptibility to Friend spleen focus-forming virus. *Science* 162:564–65
36. McCulloch, E. A., Siminovitch, L., Till, J. E., Russell, E. S., Bernstein, S. E. 1965. The cellular basis of the genetically determined hemopoietic defect in anemic mice of genotype Sl/Sl^d. *Blood* 26:399–410
37. Murphy, E. D. 1977. Effects of mutant steel alleles on leukemogenesis and lifespan in the mouse. *J. Natl. Cancer Inst.* 58:107–10
38. Grüneberg, H. 1942. The anemia of flexed-tail mice (*Mus musculus L.*). I. Static and dynamic hematology. *J. Genet.* 43:45–68
39. Axelrad, A. 1968. Genetic and cellular basis of susceptibility or resistance to Friend leukemia virus infection in mice. *Can. Cancer Conf.* 8:313–43
40. Meier, H., Myers, D. D., Huebner, R. J. 1969. Genetic control by the *hr* locus of susceptibility and resistance to leukemia. *Proc. Natl. Acad. Sci. USA* 63:759–66
41. Hackett, A. J., Manning, J. S., Owens, R. B. 1971. In vitro production of murine leukemia virus by cells differing in a single allele. *J. Natl. Cancer Inst.* 46:1335–42
42. Heiniger, H. J., Huebner, R. J., Meier, H. 1976. Effect of allelic substitutions at the hairless locus on endogenous ecotropic murine leukemia virus titers and leukemogenesis. *J. Natl. Cancer Inst.* 56:1073–74
43. Vogt, P. K., Moscovici, C., Duff, R. 1968. Host resistance and the biological analysis of avian tumor virus infections. *Can. Cancer Conf.* 8:286–312
44. Huang, A. S., Besmer, P., Chu, L., Baltimore, D. 1973. Growth of pseudotypes of vesicular stomatitis virus with N-tropic murine leukemia virus coats in cells resistant to N-tropic viruses. *J. Virol.* 12:659–62
45. Závada, J. 1972. Pseudotypes of vesicular stomatitis virus with the coat of murine leukemia and of avian myeloblastosis virus. *J. Gen. Virol.* 15:183–91
46. Ishimoto, A., Hartley, J. W., Rowe, W. P. 1977. Detection and quantitation of phenotypically mixed viruses between ecotropic and xenotropic murine leukemia viruses. *Virology*. In press
47. Fischinger, P. J., Nomura, S., Blevins, C. S., Bolognesi, D. P. 1975. Two levels of restriction by mouse or cat cells of murine sarcoma virus coated by en-

dogenous xenotropic oncornavirus. *J. Gen. Virol.* 29:51–62
48. Rowe, W. P., Humphrey, J. B., Lilly, F. 1973. A major genetic locus affecting resistance to infection with murine leukemia viruses. III. Assignment of the *Fv-1* locus to linkage group VIII of the mouse. *J. Exp. Med.* 137:850–53
49. Pincus, T., Hartley, J. W., Rowe, W. P. 1971. A major genetic locus affecting resistance to infection with murine leukemia viruses. I. Tissue culture studies of naturally occurring viruses. *J. Exp. Med.* 133:1219–33
50. Pincus, T., Hartley, J. W., Rowe, W. P. 1975. A major genetic locus affecting resistance to infection with murine leukemia viruses. IV. Dose-response relationships in *Fv-1*-sensitive and resistant cell cultures. *Virology* 65:333–42
50a. Declève, A., Niwa, O., Gelmann, E., Kaplan, H. S. 1975. Replication kinetics of N- and B-tropic murine leukemia viruses on permissive and non-permissive cells *in vitro. Virology* 65:320–32
51. Steeves, R. A., Eckner, R. J. 1970. Host-induced changes in infectivity of Friend spleen focus-forming virus. *J. Natl. Cancer Inst.* 44:587–94
52. Jolicoeur, P., Baltimore, D. 1975. Effect of the *Fv-1* locus on the titration of murine leukemia viruses. *J. Virol.* 16:1593–98
53. Schuh, V., Blackstein, M. E., Axelrad, A. A. 1976. Inherited resistance to N- and B-tropic murine leukemia viruses in vitro: Titration patterns in strains SIM and SIM.R congenic at the *Fv-1* locus. *J. Virol.* 18:473–80
54. Sveda, M. M., Fields, B. N., Soeiro, R. 1974. Host restriction of Friend leukemia virus: Fate of input virion RNA. *Cell* 2:271–77
55. Jolicoeur, P., Baltimore, D. 1976. Effect of *Fv-1* gene on synthesis of N-tropic and B-tropic murine leukemia viral RNA. *Cell* 7:33–39
56. Jolicoeur, P., Baltimore, D. 1976. Effect of *Fv-1* gene product on proviral DNA formation and integration in cells infected with murine leukemia viruses. *Proc. Natl. Acad. Sci. USA* 73:2236–40
57. Sveda, M. M., Soeiro, R. 1976. Host restriction of Friend leukemia virus: Synthesis and integration of the provirus. *Proc. Natl. Acad. Sci. USA* 73:2356–60
58. Tennant, R. W., Schluter, B., Yang, W. K., Brown, A. 1974. Reciprocal inhibition of mouse leukemia virus infection by *Fv-1* allele cell extracts. *Proc. Natl. Acad. Sci. USA* 71:4241–45
59. Tennant, R. W., Schluter, B., Myer, F. E., Otten, J. A., Yang, W. K., Brown, A. 1976. Genetic evidence for a product of the *Fv-1* locus that transfers resistance to mouse leukemia viruses. *J. Virol.* 20:589–96
60. Rein, A., Kashmiri, S. V., Bassin, R. H., Gerwin, B., Duran-Troise, G. 1976. Phenotypic mixing between N- and B-tropic murine leukemia viruses: Infectious particles with dual sensitivity to *Fv-1* restriction. *Cell* 7:373–79
61. Declève, A., Niwa, O., Kojola, J., Kaplan, H. S. 1976. New gene locus modifying susceptibility to certain B-tropic murine leukemia viruses. *Proc. Natl. Acad. Sci. USA* 73:585–90
62. Lilly, F., Steeves, R. A. 1973. B-tropic Friend virus: A host-range pseudotype of spleen focus-forming virus (SFFV). *Virology* 55:363–70
63. Lilly, F. 1970. *Fv-2*: Identification and location of a second gene governing the spleen focus response to Friend leukemia virus in mice. *J. Natl. Cancer Inst.* 45:163–69
64. Rowe, W. P., Brodsky, I. 1959. A graded-response assay for the Friend leukemia virus. *J. Natl. Cancer Inst.* 23:1239–48
65. Axelrad, A. A., Steeves, R. A. 1964. Assay for Friend leukemia virus: Rapid quantitative method based on enumeration of macroscopic spleen foci in mice. *Virology* 24:513–18
66. Mitani, S., Wakabayashi, K. 1975. The effect of *Fv-2* gene on the incorporation of thymidine in the mouse spleen. *Eur. J. Cancer* 11:649–55
67. Odaka, T., Yamamoto, T. 1965. Inheritance of susceptibility to Friend mouse leukemia virus. II. Spleen foci method applied to test the susceptibility of crossbred progeny between a sensitive and a resistant strain. *Jpn. J. Exp. Med.* 35:311–14
68. Axelrad, A. A., Ware, M., van der Gaag, H. C. 1972. Host cell susceptibility and resistance to murine leukemia viruses and their genetic control. In *RNA Viruses and Host Genome in Oncogenesis*, ed. P. Emmelot, P. Bentvelzen, pp. 239–54. Amsterdam: North-Holland
69. Odaka, T. 1970. Inheritance of susceptibility to Friend mouse leukemia virus. VII. Establishment of a resistant strain. *Int. J. Cancer* 6:18–23

70. Blank, K. J., Lilly, F. 1976. Lethality of the Fv-2 resistance allele in mice with a DBA/2 genetic background. *Genetics* 83:8(Suppl.)
71. Blank, K., Steeves, R. A., Lilly, F. 1976. The Fv-2^r resistance gene in mice: Its effect on spleen colony formation by Friend virus-transformed cells. *J. Natl. Cancer Inst.* 57:925–30
72. Steeves, R. A., Steinheider, G., Lilly, F. 1977. Spleen colony formation in mice by Friend leukemia cells: A result of donor or host cell proliferation? *Proc. Am. Assoc. Cancer Res.* 18:71
73. Cudkowicz, G., Bennett, M. 1971. Peculiar immunobiology of bone marrow allografts. I. Graft rejection by irradiated responder mice. *J. Exp. Med.* 134:83–102
74. Kumar, V., Bennett, M. 1976. Mechanisms of genetic resistance to Friend virus leukemia in mice. II. Resistance of mitogen-responsive lymphocytes mediated by marrow-dependent cells. *J. Exp. Med.* 144:713–27
75. Kumar, V., Caruso, T., Bennett, M. 1976. Mechanism of genetic resistance to Friend virus leukemia. III. Susceptibility of mitogen-responsive lymphocytes mediated by T cells. *J. Exp. Med.* 144:728–40
76. Kumar, V., Bennett, M., Eckner, R. J. 1974. Mechanisms of genetic resistance to Friend virus leukemia in mice. I. Role of ^{89}Sr-sensitive effector cells responsible for rejection of bone marrow allografts. *J. Exp. Med.* 139:1093–1109
77. Dent, P. B. 1972. Immunodepression by oncogenic viruses. *Prog. Med. Virol.* 14:1–35
78. Suzuki, S., Matsubara, S. 1975. Isolation of Friend leukemia virus resistant line from non-inbred mouse colony. *Jpn. J. Exp. Med.* 45:467–71
79. Suzuki, S. 1975. Fv-4: A new gene affecting the splenomegaly induction by Friend leukemia virus. *Jpn. J. Exp. Med.* 45:473–78
80. Kai, K., Ikeda, H., Yussa, Y., Suzuki, S., Odaka, T. 1976. Mouse strain resistant to N-, B-, and NB-tropic murine leukemia viruses. *J. Virol.* 20:436–40
81. Lilly, F. 1970. The role of genetics in Gross virus leukemogenesis. In *Comparative Leukemia Research, 1969*, ed. R. M. Dutcher, 36:213–20. Basel: Karger. 799 pp.
82. Lilly, F., Boyse, E. A., Old, L. J. 1964. Genetic basis of susceptibility to viral leukemogenesis. *Lancet* ii:1207–9
83. Tennant, J. R., Snell, G. D. 1968. The H-2 locus and viral leukemogenesis as studied in congenic strains of mice. *J. Natl. Cancer Inst.* 41:597–604
84. Lilly, F. 1968. The effect of histocompatibility-2 type on response to the Friend leukemia virus in mice. *J. Exp. Med.* 127:465–73
85. Mühlbock, O., Dux, A. 1974. Histocompatibility genes (the H-2 complex) and susceptibility to mammary tumor virus in mice. *J. Natl. Cancer Inst.* 53:993–96
86. Colombatti, A., Chieco-Bianchi, L., De Rossi, A., D'Andrea, E., Collavo, D. 1977. Genetics of murine sarcoma virus (MSV)-induced tumors in AKR mice: Evidence that late progressing and early regressing tumors are controlled by different genes. *Int. J. Cancer* 19:565–75
87. Dellon, A. L., Rogentine, G. N., Chretien, P. B. 1975. Prolonged survival in bronchogenic carcinoma associated with HL-A antigens W-19 and HL-A5: A preliminary report. *J. Natl. Cancer Inst.* 54:1283–86
88. McDevitt, H. O., Benacerraf, B. 1969. Genetic control of specific immune responses. *Adv. Immunol.* 11:31–74
89. Sato, H., Boyse, E. A., Aoki, T., Iritani, C., Old, L. J. 1973. Leukemia-associated transplantation antigens related to murine leukemia virus. The X.1 system: Immune response controlled by a locus linked to H-2. *J. Exp. Med.* 138:593–606
90. Chesebro, B., Wehrly, K., Stimpfling, J. H. 1974. Host genetic control of recovery from Friend leukemia virus-induced splenomegaly. Mapping of a gene within the major histocompatibility complex. *J. Exp. Med.* 140:1457–67
91. Doherty, P. C., Zinkernagel, R. M. 1975. H-2 compatibility is required for T cell mediated lysis of target cells infected with lymphocytic choriomeningitis virus. *J. Exp. Med.* 141:502–7
92. Blank, K. J., Freedman, H. A., Lilly, F. 1976. T-lymphocyte response to Friend virus-induced tumor cell lines in mice of strains congenic at H-2. *Nature* 260:250–52
93. Bubbers, J. E., Steeves, R. A., Lilly, F. 1976. Evidence for a physical association between Friend virus-induced and histocompatibility antigens on leukemia cell surfaces. *Proc. Am. Assoc. Cancer Res.* 17:93
94. Bubbers, J. E., Lilly, F. 1977. Selective incorporation of H-2 antigenic determi-

nants into Friend virus particles. *Nature* 266:458-59
95. Blank, K. J., Bubbers, J. E., Lilly, F. 1977. H-2/viral protein interaction at the cell membrane as the basis for H-2-restricted T-lymphocyte immunity. In *Regulation of the Immune System: Genes and the Cells in Which They Function*, ed. E. Sercarz, L. Herzenberg, C. Fox. New York: Academic
96. Toth, F. D., Vaczi, L., Balogh, M. 1973. Inheritance of susceptibility and resistance to Rauscher leukemia virus. *Acta Microbiol. Acad. Sci. Hung.* 20:183-89
97. Colombatti, A., Collavo, D., Biasi, G., Chieco-Bianchi, L. 1975. Genetic control of oncogenesis by murine sarcoma virus Moloney pseudotype. I. Genetics of resistance in AKR mice. *Int. J. Cancer* 16:427-34
98. Colombatti, A., Collavo, D., Biasi, G., Chieco-Bianchi, L. 1975. Genetic control of oncogenesis by murine sarcoma virus Moloney pseudotype. II. A dominant epistatic susceptiblity gene. *Int. J. Cancer* 16:435-41
99. Fieldsteel, A. H., Dawson, P. J., Bostick, W. L. 1961. Quantitative aspects of Friend leukemia virus in various murine hosts. *Proc. Soc. Exp. Biol. Med.* 108: 826-29
100. Lilly, F., Pincus, T. 1973. Genetic control of murine viral leukemogenesis. *Adv. Cancer Res.* 17:231-77
101. Yoshikura, H. 1975. Adaptation of N-tropic Friend leukemia virus and its murine sarcoma virus pseudotype to non-permissive B-type C57BL/6 mouse cell line. *J. Gen. Virol.* 29:1-9
102. Hopkins, N., Schindler, J., Hynes, R. 1977. Six NB-tropic murine leukemia viruses derived from a B-tropic virus of BALB/c have altered p30. *J. Virol.* 21:309-18
103. Ginsburg, H., Sachs, L. 1962. Leukemia induction in mice by Moloney virus from long- and short-term tissue cultures, and attempts to detect a leukemogenic virus in cultures from X-ray-induced leukemia. *J. Natl. Cancer Inst.* 28:1391-1410
104. Manaker, R. A., Jensen, E. M., Korol, W. 1964. Long-term propagation of a murine leukemia virus in an established cell line. *J. Natl. Cancer Inst.* 33:363-71
105. Barski, G., Youn, J. K. 1966. Protective effect of specific immunization in Rauscher leukemia. *Natl. Cancer Inst. Monogr.* 22:659-70
106. Sinkovics, J. G., Bertin, B. A., Howe, C. D. 1966. Occurrence of low leukemogenic but immunizing mouse leukemia virus in tissue culture. *Natl. Cancer Inst. Monogr.* 22:349-67
107. Wright, B. S., Lasfargues, J. C. 1966. Attenuation of the Rauscher murine leukemia virus through serial passages in tissue culture. *Natl. Cancer Inst. Monogr.* 22:685-700
108. Somers, K. D., Kirsten, W. H. 1968. Long-term propagation of a murine erythroblastosis virus in vitro. *J. Natl. Cancer Inst.* 40:1053-65
109. Yoshikura, H., Hirokawa, Y., Ikawa, Y., Sugano, H. 1969. Infectious but non-leukemogenic Friend leukemia virus obtained after prolonged cultivation in vitro. *Int. J. Cancer* 4:636-40
110. Schlom, J., Moloney, J. B., Groupe, V. 1971. Evidence for the rapid decrease in leukemogenic potential of Rauscher leukemia virus in cell culture. *Cancer Res.* 31:260-64
111. Lieberman, M., Niwa, O., Declève, A., Kaplan, H. S. 1973. Continuous propagation of radiation leukemia virus on a C57BL mouse embryo fibroblast line, with attenuation of leukemogenic activity. *Proc. Natl. Acad. Sci. USA* 70: 1250-53
112. Haas, M., Hilgers, J. 1975. In vitro infection of lymphoid cells by thymotropic radiation leukemia virus grown in vitro. *Proc. Natl. Acad. Sci. USA* 72:3546-50
113. Declève, A., Lieberman, M., Ihle, J. N., Kaplan, H. S. 1976. Biological and serological characterization of radiation leukemia virus. *Proc. Natl. Acad. Sci. USA* 73:4675-79
114. Hartley, J. W., Walford, N. K., Old, L. J., Rowe, W. P. 1977. A new class of murine leukemia virus associated with development of spontaneous leukemia. *Proc. Natl. Acad. Sci. USA* 74:789-92
115. Scolnick, E. M., Rands, E., Williams, D., Parks, W. P. 1973. Studies on the nucleic acid sequences of Kirsten sarcoma virus: A model for formation of a mammalian RNA-containing sarcoma virus. *J. Virol.* 12:458-63
116. Scolnick, E. M., Parks, W. P. 1974. Harvey sarcoma virus: A second murine type C sarcoma virus with rat genetic information. *J. Virol.* 13:1211-19
117. Scolnick, E. M., Goldberg, R. J., Williams, D. 1976. Characterization of rat genetic sequences of Kirsten sarcoma virus: Distinct class of endogenous rat type C viral sequences. *J. Virol.* 18: 559-66

118. Scolnick, E. M., Howk, R. S., Anisowicz, A., Peebles, P. T., Scher, C. D., Parks, W. P. 1975. Separation of sarcoma virus-specific and leukemia virus-specific genetic sequences of Moloney sarcoma virus. *Proc. Natl. Acad. Sci. USA* 72:4650–54
119. Frankel, A. E., Neubauer, R. L., Fischinger, P. J. 1976. Fractionation of DNA nucleotide transcripts from Moloney sarcoma virus and isolation of sarcoma virus-specific complementary DNA. *J. Virol.* 18:481–90
120. Troxler, D., Parks, W. P., Vass, W. C., Scolnick, E. M. 1977. Isolation of a fibroblast nonproducer cell line containing the Friend strain of the spleen focus-forming virus. *Virology* 76:602–15
121. Troxler, D. H., Boyars, J. K., Parks, W. P., Scolnick, E. M. 1977. The Friend strain of spleen focus-forming virus: A recombinant between mouse type-C ecotropic viral sequences and sequences related to xenotropic virus. *J. Virol.* 22:361–372
122. Foulds, L. 1958. The natural history of cancer. *J. Chronic Dis.* 8:2–37
123. Pantelouris, E. M. 1968. Absence of thymus in a mouse mutant. *Nature* 217:370–71
124. Rowe, W. P. 1973. Genetic factors in the natural history of murine leukemia virus infection: G. H. A. Clowes memorial lecture. *Cancer Res.* 33:3061–68
125. Gross, L. 1958. Serial cell-free passage of a radiation-activated mouse leukemia agent. *Proc. Soc. Exp. Biol. Med.* 100:102–5
126. Lieberman, M., Kaplan, H. S. 1959. Leukemogenic activity of filtrates from radiation-induced lymphoid tumors of mice. *Science* 130:387–88
127. Haran-Ghera, N. 1967. A leukemogenic filtrable agent from chemically induced lymphoid leukemia in C57BL mice. *Proc. Soc. Exp. Biol. Med.* 124:697–99
128. Igel, H. J., Huebner, R. J., Turner, H. D., Kotin, P., Falk, H. L. 1969. Mouse leukemia virus activation by chemical carcinogens. *Science* 166:1624–26
129. Ball, J. K., McCarter, J. A. 1971. Repeated demonstration of a mouse leukemia virus after treatment with chemical carcinogens. *J. Natl. Cancer Inst.* 46:751–62

GENETICS OF BACTERIAL RIBOSOMES[1]

❖3124

Masayasu Nomura and Edward A. Morgan
Departments of Genetics and Biochemistry, Institute for Enzyme Research, University of Wisconsin, Madison, Wisconsin 53706

S. Richard Jaskunas
Department of Chemistry, Indiana University, Bloomington, Indiana 47401

CONTENTS

INTRODUCTION	298
GENES FOR RIBOSOMAL RNA	299
Redundancy of rRNA Genes	299
Number and Location of rRNA Operons in E. coli	300
Structure of rRNA Operons and Presence of tRNA Genes in rRNA Operons	304
Heterogeneity of rRNA Operons	306
Chromosome Rearrangements Using rRNA Operon Homology	307
rRNA PROCESSING	308
LOCATION AND ORGANIZATION OF GENES FOR RIBOSOMAL PROTEINS	309
Str-Spc Cluster	309
Identification of genes	309
Order of genes on the λfus2 genome	311
Ribosomal protein transcription units in the str-spc cluster	312
Rif Cluster	314
Identification of genes on λrif^{d}18	314
Organization of r-protein genes on λrif^{d}18	315
Other Regions	316
rpsR (S18)	316
rpsB (S2) and tsf (EF-Ts)	316
rpsT (S20)	316
rpsO (S15) and rplU (L21)	317
rpsF (S6)	317

[1]This is paper number 2148 from the Laboratory of Genetics.

ISOLATION OF MUTATIONS IN RIBOSOMAL PROTEIN GENES 317
   Antibiotic-Resistant Mutants of E. coli 318
      Streptomycin resistance 318
      Spectinomycin resistance 318
      Erythromycin resistance 318
      Neamine resistance 318
      Neomycin-kanamycin resistance 319
   Suppressors of Conditionally Lethal Mutants 319
      Suppressors of drug-dependent strains 319
      Suppressors of temperature-sensitive tRNA synthetase mutants ... 319
   Temperature-Sensitive Mutants 319
PHYSIOLOGICAL EFFECTS OF MUTANT RIBOSOMAL PROTEINS 320
   Mutations Affecting Ribosomal Assembly 320
   Mutations Affecting Translational Fidelity 321
   Mutations Affecting Other Cellular Processes 323
GENETICS OF TRANSLATION FACTORS 324
REGULATION OF THE BIOSYNTHESIS OF RIBOSOMES 325
   Regulation of r-Protein Synthesis 325
   Regulation of rRNA Synthesis 326
   Gene Dosage Effects and the Mechanism of Regulation 328
   Coordinate Control of rRNA and r-Protein Gene Expression 330
   Analysis of Gene Expression In Vitro 331
CONCLUDING REMARKS: RIBOSOMAL GENE EXPRESSION AND OTHER
      CELLULAR ACTIVITIES RELATED TO GROWTH 332

INTRODUCTION

Genetic studies of bacterial ribosomes aim at two goals. The first is to identify all the structural genes for ribosomal components, elucidate their organization, including their promoters and any other regulatory genes needed for expression, and to understand the mechanisms of regulation of ribosome biosynthesis in relation to cellular growth. The second is to isolate mutants with alterations in some of the ribosomal components and to analyze alterations of ribosomal structure and function, as well as physiological alterations of mutant cells in vivo. This latter analysis will provide information on the structure and function of ribosomes in general as well as the physiological roles of the products of mutated genes.

Several important advances have recently been made in the study of the genetics of ribosomal components in *Escherichia coli.* The most dramatic ones have concerned the organization of ribosomal-protein (r-protein) and rRNA genes. Our understanding of the organization and regulation of these genes in bacteria has lagged far behind our understanding of many operons involved in intermediary metabolism. A major problem in studying the organization of r-protein genes is that the structure of ribosomes is complex. In addition, they are required for viability. Thus, many types of mutations, such as nonsense mutations and deletions, used to study classical bacterial operons could not be used to study r-protein genes because they would be lethal. Furthermore, studies on rRNA genes have been hampered because of the presence of many gene copies on the chromosome. These obstacles have been recently overcome with the use of specialized transducing phages and small plasmids that carry r-protein and rRNA genes from the *E. coli* chromosome. The availability of these phages and plasmids and the resulting ability to isolate DNA carrying these genes for in vitro experiments has made it possible to apply several powerful techniques to analyze the organization and regulation of these genes.

This review describes recent advances in the field, with somewhat more emphasis on the subjects related to the first goal mentioned above, namely, gene organization and regulation of gene expression. A brief review of some recent results, discussing specifically new strategies developed, is also given in reference (1). For more detailed general reviews of earlier observations, readers should consult references 2-7. Although ribosome genetics has become intertwined with RNA polymerase genetics, we have not made any attempt to review the latter field. The review by Scaife (8) should be consulted for a recent review of RNA polymerase genetics.

GENES FOR RIBOSOMAL RNA
Redundancy of rRNA Genes

Genes for rRNAs in bacteria are unusual because they are present in multiple copies (a minimum of seven in *E. coli*, see below) per haploid chromosome. Although some protein genes such as the gene for EF-Tu (see below) are now known to exist in more than one copy, redundancy of protein genes is rare in bacteria. Redundancy of rRNA genes must have evolved in order to supply a large quantity of rRNA for fast-growing cells. Simple calculations using various growth parameters (e.g. 9-11) obtained with *E. coli* cells indicate that one rRNA transcription unit ("rRNA operon") can supply only a small fraction of the rRNA required by fast-growing *E. coli* cells dividing every 20 min, even if the rRNA operon is used at or near full efficiency; that is, if the operon is transcribed by RNA polymerase at the maximum density that can be accommodated on DNA. Therefore, at least several rRNA operons must be present to allow the high rRNA synthesis needed for fast growth. In contrast, r-protein genes exist in single copies per haploid chromosome. The difference between rRNA genes and r-protein genes is, of course, that there is "translational amplification" for r-protein genes to make the final gene products (r-proteins), while the transcription products of rRNA genes have to be utilized directly. Problems of coordination between rRNA synthesis and r-protein are discussed later in this article. (It is not known whether there are any elements corresponding to operators in rRNA transcription units or in r-protein transcription units. Therefore, we cannot call these units *rRNA operons* or *r-protein operons* in a strict sense. However, for the sake of convenience, we call these transcription units *operons* in this article.)

The redundancy of rRNA genes poses a number of problems. As is described below, the basic structure of rRNA operons is as follows: promoter-16S rRNA gene-spacer tRNA gene(s)-23S rRNA gene-5S RNA gene-"distal tRNA gene(s)." We can ask whether and how these basic arrangements differ among different rRNA operons. We can also ask whether and how rRNAs (and tRNAs) produced from one rRNA operon are chemically and functionally different from those produced from other operons, whether all rRNA operons function in all growth conditions, and whether there are any recombinational events between rRNA operons as a result of their sequence homology.

Number and Location of rRNA Operons in E. coli

Early studies measured the proportion of total *E. coli* DNA saturated by hybridization with rRNA and estimated that there are about five to ten copies of rRNA genes per haploid chromosome (reviewed in 2–4). Since it was difficult to obtain mutations in the structural genes for rRNAs, earlier studies to localize the rRNA genes were done using two approaches. The first was to examine saturation levels for rRNA hybridized to DNA prepared from various *E. coli* strains that are merodiploid for a small region of the chromosome. An increase in saturation levels was taken to indicate the presence of rRNA genes on the F'-factor carried by pertinent strains (12–15). Since the number of rRNA operons is at least seven (see below), an increase in the number of rRNA gene copies by only one in merodiploids would have been very difficult to detect with this approach.

The second approach was to use sequence differences in rRNAs in interspecies or intergeneric crosses. Although it was difficult to map 16S or 23S rRNA genes by this approach, it was successfully used to map 5S RNA genes by Jarry & Rosset (16–18). Using fingerprinting techniques and exploiting the naturally occurring sequence heterogeneity within 5S RNA of *E. coli* strains K12 and B they estimated that there are six to seven 5S RNA genes in *E. coli*, as oligonucleotides present in presumably only one 5S RNA gene copy in *E. coli* occur at a molar fraction of 0.15 in the bulk 5S RNA isolated from cells. By genetic analysis of 5S RNA sequence heterogeneity, the authors were able to map four 5S RNA genes, and called them *cqsA*, *cqsB*, *cqsC*, and *cqsD*. As expected from our belief in the general structure of rRNA operons in *E. coli*, subsequent work demonstrated the presence of 16S and 23S rRNA genes at these locations. They correspond to *rrnA*, *rrnC*, *rrnF*, and *rrnD*, respectively, in Figure 1 (see also Table 1). The chromosomal locations of known rRNA operons, incorporating new data to be described below and including subsequently discovered rRNA operons, are given in Figure 1. These rRNA operons as well as various genes associated with the rRNA operons are summarized in Table 1. [Confusion in nomenclature of rRNA operons has resulted from the use of different designations by different authors (see Table 1). To avoid further confusion, the operon and gene designations in Table 1 and Figure 1 are referred to in this review in lieu of designations used by other authors for what are, to the best of our knowledge, the same operons or genes.]

More recent studies to identify and map individual rRNA operons were done by isolating segments of the *E. coli* chromosome on transducing phages or on plasmids, and by analyzing them using various techniques including electronmicrographic heteroduplex techniques and RNA-DNA hybridization techniques.

The first pioneering work in this direction was done by Davidson and his co-workers, using F'-factor F14 and a transducing phage $\phi 80d_3 ilv^+ su7$ (19, 20). This phage was isolated using *su7* as a selection marker and was found to carry an rRNA

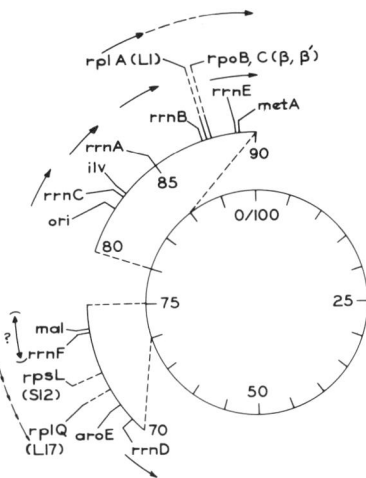

Figure 1 The locations of rRNA operons and the two r-protein gene clusters on the *E. coli* genetic map. The locations of *rrnA* through *rrnF* (cf Table 1) are shown; the *arrows* indicate the direction of transcription of these operons. In addition, the locations and the directions of transcription of r-protein gene clusters at the *str-spc* region (72 min), the r-protein gene cluster at the *rif* region (88 min), and the genes for RNA polymerase subunits β and β' (88 min) are indicated. The origin of DNA replication ("ori" at 82 min) is also shown.

operon, which is close to *ilvA* and is presumably *rrnC*. The F'-factor F14 was found to carry two rRNA operons, one presumably corresponding to *rrnC* [or *rrnB*, or a "recombinant" rRNA operon formed between *rrnC* and *rrnB*, see reference (19)] and the other corresponding to *rrnA*, which is located about 80 kb (1 kb is 1000 base-pairs), clockwise from *ilvA*.

Subsequently, other specialized transducing phages have been isolated which carry complete or partial rRNA operons. Several specialized transducing phages for *aroE* were isolated and found to carry a portion of or a complete rRNA operon, *rrnD* (21). A specialized transducing phage isolated by Jørgensen, called λ*ilv*5, carries genes of the *ilv* operon, and also a complete rRNA operon (22), confirming the presence of an rRNA operon near *ilv*. The presence of an rRNA operon (*rrnB*) near *rif* at 88 min was demonstrated (23; see Figure 2) by detection of an rRNA operon on two independently isolated transducing phages carrying the *rif* gene, ϕ 80*rif*^{r} (24) and λ*rif*^{d}18 (25). Finally, isolation and analysis of transducing phages carrying *purD* and/or *metA* proved the presence of an rRNA operon (*rrnE*) between *purD* and *metA* at 89 min (26). The presence of an rRNA operon in this region had been suggested by Hill and his co-workers on other grounds (27, 28; see below).

An extensive effort has recently been made to isolate rRNA operons on plasmids (29). Approximately 2000 recombinant plasmids were constructed by Clarke & Carbon (30). These plasmids consist of *E. coli* DNA which was mechanically

Table 1 Chromosomal locations and encoded genes of known rRNA operons in *E. coli* K12[a]

rRNA operon designation used in this review	Chromosomal location	Known genes in operon	Other designations that have been used (references in parentheses)	Other references
rrnA	85 min	*rrsA*, (*tilA*, *talA*), *rrlA*, *rrfA*	*cqsA* (18, 32)	19, 20, 29, 42, 44
rrnB	88 min	*rrsB*, *tgtB*, *rrlB*, *rrfB*	*rrnB*$_1$ (19), *rrnE* (32), *cqsE* (32)	40, 29, 38, 39, 42, 44
rrnC	83 min	*rrsC*, *tgtC*, *rrlC*, *rrfC*, *tasC*	*rrnB* (19, 32), *cqsB* (18), *rrnB*$_2$ (19)	39, 22, 29, 42, 44
rrnD	71 min	*rrsD*, (*tilD*, *talD*), *rrlD*, *rrfD*	*cqsD* (18)	21, 29, 42, 44
rrnE	89 min	*rrsE*, *tgtE*, *rrlE*, *rrfE*	*rrnD* (28)	26, 29, 42
rrnF	74 min	*rrsF*, (tRNA gene?), *rrlF*, *rrfF*	*cqsC* (18), *rrnC* (18, 32)	
Group I	unmapped	*rrs*_, (*til*_, *tal*_), *rrl*_, *rrf*_, *tas*_	—	29, 42, 44
Group VI	unmapped	*rrs*_, *tgt*_, *rrl*_, *rrf*_	—	29, 42, 44

[a] In this summary, *rrnA* (and *rrnB*, etc) is a designation referring to a transcription unit (rRNA operon), each of which contains certain genes that code for the designated gene product: *rrs* (16S rRNA), *rrl* (23S rRNA), *rrf* (5S RNA), *tgt* (tRNA$_2^{Glu}$), *til* (tRNA$_1^{Ile}$), *tal* (tRNA$_1^{Ala}$), or *tas* (tRNA$_1^{Asp}$). The designations Group I and Group VI are rRNA operons that have been isolated and shown to have chromosomal locations different from *rrnA* through *rrnE* (29). However, they have not yet been mapped, and therefore, their *rrn* designations have been left unspecified. The Group I or Group VI operon may prove to be identical with *rrnF*. tRNA genes of *rrnF* have not been analyzed. Only genes for which substantial evidence exists for their inclusion in rRNA operons are identified. (As described in the text, there is some evidence for the existence of some other tRNA genes at the distal ends of rRNA operons.) It should be noted that Champney & Kushner (328) have made a proposal for nomenclature of rRNA genes. Their designations for 16S and 23S rRNA genes (*rrs* and *rrl*, respectively) have been adopted here, but that for 5S RNA (which is also *rrl*) has not been adopted. We propose to use *rrf* for 5S RNA genes to distinguish them from 23S rRNA genes (*rrl*). The gene for tRNA$_2^{Glu}$ was previously designated *gltT* (e.g. 40), and the two tRNA^{Asp} genes were called *aspT* and *aspU* (42) in accordance with the previous custom for naming tRNA genes. However, tRNA gene products coming from different rRNA operons may have minor sequence differences and it appears that the best way to distinguish these redundant genes is to specify the rRNA operons with which tRNA genes are associated. Therefore, we propose to designate these tRNA genes starting with *t* followed by two letters related to pertinent amino acids, and then by a capital letter specifying the appropriate rRNA operon. Any new tRNA genes found will be named according to the present manner, if they are associated with rRNA operons, but will be named according to the previous manner (three letter codes related to amino acids followed by *T*, *U*, *V*, etc), if they are not associated with rRNA operons. The nomenclature proposed here has been adopted after consultation with B. Low and B. Bachmann.

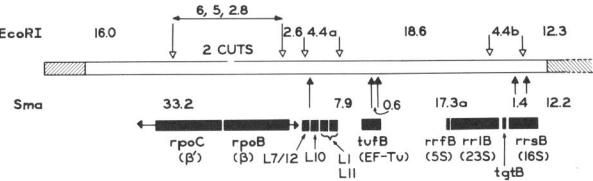

Figure 2 Location of genes for rRNA, ribosomal proteins, RNA polymerase subunits, and elongation factors on $\lambda rif^d 18$. A schematic diagram of the left arm of $\lambda rif^d 18$ is given. The *open bar* represents bacterial DNA and the *shaded bar* represents λ DNA. The *open arrows* on top represent DNA restriction endonuclease *Eco*RI-sensitive sites and the *closed arrows* beneath the bar represent *Sma*-sensitive sites. The sizes of restriction DNA fragments are indicated in %λ units: 100 %λ units are 46.5 kilobases. The sizes and locations of the genes are represented by *black bars*. Genes for r-proteins are indicated by the names of the r-proteins. With other genes, the gene products are indicated in parentheses except *tgtB* which is a structural gene for tRNA$_2^{Glu}$. Approximate sizes of the genes are estimated from the molecular weights of the gene products, protein, and RNA.

sheared and inserted into the *Eco*RI sensitive site of colicin E1 factor by the poly dA·dT "connector" method (31). This plasmid bank was screened by Kenerley et al (29) for plasmids that hybridize *E. coli* rRNA, using in situ hybridization of rRNA to plasmids in agarose gels. Sixteen plasmids were identified that carry rRNA genes. By a combination of heteroduplex, biochemical, and genetic techniques, these plasmids were classified into six groups on the basis of their chromosomal origins. Homology with known transducing phage DNAs and genetic mapping have assigned locations on the *E. coli* chromosome (corresponding to *rrnB*, *rrnC*, and *rrnD*) to three of the six groups. In addition, the fourth plasmid group was found to carry a 5S RNA gene that codes for 5S RNA molecules with a sequence specified by the *cqsA* gene (16–18) mentioned above. Therefore, this plasmid group probably carries *rrnA* at 85 min. The chromosomal locations for the remaining two plasmid groups have not been determined. Neither of the two corresponds to *rrnE*, which was isolated on $\lambda metA 20$ (26). Apparently, *rrnE* was not found in the plasmid collection. However, it is possible that one of the unmapped plasmid groups may correspond to the rRNA operon (called *cqsC* previously, now called *rrnF*) mapped by Jarry & Rosset (18) and isolated recently by Vola et al on an F' plasmid (32). Therefore, at least one of these plasmid groups must derive from an as yet unmapped rRNA operon not identical with *rrnA* through *rrnF*.

In summary, at least seven different rRNA operons have been isolated and characterized, and, therefore, the minimum number of rRNA operons per haploid chromosome of *E. coli* is seven. The location of six rRNA operons is known (Figure 1). Since the original estimate made by RNA-DNA hybridization data as well as the data on the stoichiometry of minor oligonucleotides [obtained by Jarry & Rosset (16–18), see above] was six to seven, it appears that we have isolated most if not all of the rRNA operons in *E. coli*.

Structure of rRNA Operons and Presence of tRNA Genes in rRNA Operons

Initially, kinetic experiments suggested that rRNA genes were organized into several transcription units (rRNA operons) with the structure: promoter-16S rRNA gene (*rrs*)-23S rRNA gene (*rrl*)-5S RNA gene (*rrf*) [see earlier reviews (2, 3, 33)]. This suggested structure has been largely proven by recent experiments which involved two different approaches. One approach involved analysis of 30S precursor ribosomal RNA accumulated in an RNase III-deficient *E. coli* mutant in the presence of chloramphenicol. The accumulated 30S precursor ribosomal RNA was shown to contain the 16S, 23S (34–37; reviewed in 33), and 5S rRNA sequences (37). Thus, these three rRNA molecules are cotranscribed and subsequently processed into mature forms. This approach was useful in demonstrating cotranscription of the three genes, but analysis of the gene order and fine structures of rRNA operons using this approach was not easy. This is partly because the 30S precursor RNA is a mixture of different RNA molecules with considerable sequence heterogeneity (see below).

The second approach involved isolation of DNA carrying rRNA operons, as described in the previous section, and analysis of gene organization using various techniques, such as electronmicrographic heteroduplex techniques, examination of DNA fragments obtained with various restriction enzymes, and use of RNA-DNA hybridization techniques to identify gene locations. In addition, DNA was analyzed that had rRNA operons with deletions to various extents either from the promoter side or from the distal side (29, 38). With these methods, the gene order in transcription units was directly proven to be 16S rRNA gene (*rrs*)-23S rRNA gene (*rrl*)-5S RNA gene (*rrf*) (19, 20, 38–40; cf Figure 2). In addition, the presence of tRNA genes in the rRNA operons was discovered and studied in detail using this second approach.

Davidson and his co-workers (19, 20) examined heteroduplex structures formed between *rrnC* and *rrnA* on F14, or between *rrnC* on $\phi 80d_3ilv^+su7$ and *rrnA* on F14, and found that the length of homology of DNAs from different rRNA operons is on the order of 5.3 kb. This is sufficient to accommodate mature 16S, 23S, and 5S rRNA species, as well as known rRNA precursor-specific sequences. In addition, the heteroduplex structures indicated that between genes for 16S and 23S rRNA there was a region with the length of about 0.25 kb that was nonhomologous in heteroduplexes formed between *rrnC* and *rrnA*. This region is termed the *spacer region*.

The presence of tRNA genes associated with rRNA operons was first indicated in the analysis of RNA synthesized in UV-irradiated *E. coli* cells infected with λ rif^d18 and $\phi 80d_3ilv^+su7$ transducing phages (39). The tRNA synthesized was identified as $tRNA_2^{Glu}$ (39). This gene is located between the 16S and 23S rRNA genes as proven by hybridization of radioactive $tRNA_2^{Glu}$ to various DNA fragments obtained after digestion of λrif^d18 DNA with *Eco*RI (39; cf. Figure 2) as well as to DNA from mutants of λrif^d18 which have deletions covering the rRNA operon to various extents (38). Similar experiments were repeated using the $\phi 80 rif^r$

transducing phage, and a spacer tRNA carried by this transducing phage was identified as tRNA$_1^{Ile}$. Heteroduplex analysis between $\lambda rif^d 18$ and $\phi 80$-rif^r had indicated that the rRNA operons carried by these two phages had nonhomologous spacer regions (39), and hence the nonidentity of the spacer tRNA genes carried by these two phages was consistent with the nonidentity of the spacers revealed in the heteroduplex analysis. [There is evidence indicating that $\phi 80rif^r$ carries a "hybrid" rRNA operon formed by recombination of two rRNA operons (see below). This explains why the two phages carrying bacterial DNA derived from the same *rif* region have different spacer tRNAs.] The presence of a tRNA gene in a spacer region was also demonstrated by electronmicrographic techniques (41).

Subsequent more extensive analysis of the transducing phages and ColE1 hybrid plasmids that carry rRNA operons indicated that there are probably only two classes of "spacer tRNA" arrangements in *E. coli* (42), which correspond to the two classes of "spacer regions" observed in the analysis of heteroduplexes (29). Of the seven rRNA operons with different chromosomal locations (see above), four have been shown to carry only tRNA$_2^{Glu}$ gene(s) in their spacers and three have been shown to carry genes for both tRNA$_1^{Ile}$ and tRNA$_{1B}^{Ala}$ (42; see Table 1). Although we have no proof that the seven different rRNA operons so far isolated include all the rRNA operons in *E. coli*, it is probably safe to conclude that the results on spacer tRNAs shown in Table 1 reflect the general pattern of spacer tRNA genes in *E. coli*.

The above conclusion on spacer tRNA genes is consistent with the data of Lund & Dahlberg (43), who showed formation of tRNA$_2^{Glu}$, tRNA$_{1B}^{Ala}$, and tRNA$_1^{Ile}$ from 30S precursor rRNA after incubation with crude bacterial extracts in vitro. In addition, the existence of tRNA$_2^{Glu}$ in the spacer region of $\phi 80d_3 ilv^+ su7$ was shown by analysis of tRNA-DNA hybridization (E. A. Morgan and M. Nomura, unpublished experiments). [As noted above, in earlier experiments (39), the synthesis of only tRNA$_1^{Ile}$, but not tRNA$_{1B}^{Ala}$, was detected in UV-irradiated, $\phi 80rif^r$-infected cells, and hence the presence of the tRNA$_{1B}^{Ala}$ gene in the spacer was not recognized. The failure to observe the synthesis of tRNA$_{1B}^{Ala}$ in the above experiments was probably due to poor maturation of precursor rRNAs in the UV-irradiated cells under the conditions used for the experiments. Later experiments using tRNA-DNA hybridization techniques demonstrated the presence of both alanine and isoleucine spacer tRNA genes on $\phi 80rif^r$ (E. A. Morgan and M. Nomura, unpublished experiments)].

Although only three spacer tRNA genes have been identified (in the two types of spacers) so far, it appears that several more tRNA genes are associated with rRNA operons at the 3' end of the rRNA operons. For example, genes for tRNA$_1^{Asp}$ were discovered very close to the distal end of 5S RNA in two rRNA operons carried by ColE1 hybrid plasmids (42, 44; see Table 1). In preliminary experiments, it was shown that deletion of the promoter region (for rRNA genes) abolishes the expression of not only rRNA genes and spacer tRNA genes, but also the expression of tRNA$_1^{Asp}$ genes associated with the rRNA operons carried by plasmids. This suggests that the tRNA$_1^{Asp}$ genes are cotranscribed with the rRNA genes (T. Ikemura, E. A. Morgan, and M. Nomura, unpublished experiments; cited in 42, 44). However, not all rRNA operons seem to have tRNA genes associated with them at the distal ends of the rRNA operons (42). Therefore, the significance of this apparent association of tRNA genes with some rRNA operons at the distal ends is not clear.

The association of several tRNA genes with rRNA operons has been established. It should be noted that the tRNA genes identified so far code for major "abundant" tRNAs and that these tRNAs constitute a signficant fraction of total amounts of tRNA in *E. coli*. Because these tRNA genes are in the same transcription units, they are coregulated with rRNA genes. This explains at least in part the common features in the regulation of both rRNA synthesis and the synthesis of the bulk of tRNA observed in earlier work. However, the physiological significance of association of certain tRNA genes (but apparently not other tRNA genes) with rRNA genes is not entirely clear. Initially, the possibility was considered that each rRNA operon in *E. coli* has its own unique spacer tRNA genes and that these tRNA genes might serve to maintain the specified number of rRNA operons optimum for cellular growth (39). This is clearly not the case. Another possibility we should consider seriously is that the presence of tRNA genes in rRNA operons plays an important role in the regulation of their own expression. As discussed in a later section, the presence of uncharged tRNAs inhibits the accumulation of rRNA, presumably through synthesis of guanosine tetraphosphate. For example, amino acids (such as glutamic acid, alanine, and aspartic acid) that are used to charge tRNAs encoded by rRNA operons are located in major metabolic pathways and their synthesis depends on the supply of corresponding keto acids in the TCA cycle and the glycolytic pathway. Perhaps tRNAs for these amino acids may be the ones most commonly used for "stringent control" of synthesis of ribosomal components (see below). Production of these tRNAs simultaneously with rRNAs through cotranscription and by the use of common processing pathways might be a sensitive way to feedback-regulate the transcription of rRNA operons (see below for further discussion).

Heterogeneity of rRNA Operons

As described above, the seven rRNA operons so far identified are obviously not uniform with respect to the tRNA genes associated with the operons. Fellner presented some experimental evidence indicating that different rRNA operons code for 16S and 23S rRNA molecules with minor differences in their sequences (45). The sequence heterogeneity of 5S RNA was well studied (16–18). However, the sequence heterogeneity in larger rRNAs has been difficult to analyze because of the complexity of oligonucleotides in enzyme digests used for sequencing the RNAs. Even in the case of 5S RNA molecules where sequence heterogeneity was established definitively at several sites, separation of individual "pure" 5S RNA molecules was very difficult and comparison of complete sequences of each molecular species was not possible. However, such comparison has now become possible. As mentioned above, six different rRNA operons are now isolated on ColE1 hybrid plasmids (29), and strains carrying these plasmids overproduce RNA transcripts coded for by individual rRNA operons carried by the plasmids (44; see below). The results obtained so far confirmed the heterogeneity of 5S RNA, and in addition, suggested that different 5S RNAs can be equally used for making ribosomes without any obvious effects on cellular growth (44).

The question whether heterogeneity exists in the "promoters" of the rRNA operons has been asked. Although analysis of transcription of rRNA operons in vitro suggested the possibility of the presence of functionally significant heterogeneity in the promoters (46), analyses of rRNA synthesis in vivo failed to detect any significant differences in the expressions of different rRNA operons under various growth conditions (44, 47). With the available systems mentioned above, further experiments could be designed to answer this question.

Chromosome Rearrangements Using rRNA Operon Homology

Since there are at least seven rRNA operons with extensive sequence homology, it is possible that recombination between two different rRNA operons takes place in *E. coli* cells. There are several different conceivable ways in which this sort of "unequal crossing-over" might take place. For example, crossing-over between two neighboring operons may result in either (*a*) looping out and excision of a circular DNA, when the orientations of the two are the same as in the case of *rrnB* and *rrnE* (see Figure 1) or (*b*) inversion of the segment of the chromosome between the two rRNA operons, when the orientations of the two are different as in the case of *rrnC* and *rrnD* in Figure 1. In the former case (*a*), the result could be deletion (the excised circular DNA lost), exchange of spacer tRNA genes (the excised circular DNA inserted back into the "original" operon, but the second recombination taking place at a site within the operon that is different from the original site), or translocation (the excised circular DNA inserted into the third rRNA operon). Some of these rearrangements may well be detrimental to cells. We have some evidence that at least such recombinational events *can* take place, but we do not know how frequently such events are taking place in "normal" *E. coli* cells. For example, we have already noted that independently isolated phages, $\lambda rif^d 18$ and $\phi 80 rif^r$, which carry bacterial DNA from the same chromosomal location, have rRNA operons distinguishable by the type of spacer they carried. A similar situation was also observed with the two phage pairs $\phi 80d_3 ilv^+ su7$ and $\lambda ilv5$, both of which carry DNA from the *ilv* gene region. Heteroduplex studies between these transducing phage DNAs and plasmid DNAs indicated that two of these phages ($\phi 80 rif^r$ and $\lambda ilv5$) carry hybrid rRNA operons produced by recombination between two rRNA operons located at distant sites on the *E. coli* chromosome, probably during phage excision. Such phages may have been selected if the location of the original prophages and the gene used for selection were widely separated and recombination between two rRNA operons "decreased" this separation.

Another example of chromosomal rearrangements caused by such unequal crossing-over is the duplication of the *glyT-purD* region studied by Hill & Combriato (27, 28). This duplication was detected because of the presence of two different cell types among mutants with *glyT* missense suppressor, a fast-growing type and a very slowly growing type. Apparently, the glycine tRNA, which recognizes the GGA codon, is produced from a single structural gene, *glyT*$^+$, and the mutation of *glyT*$^+$ to the *glyT* suppressor gene on the normal haploid chromosome is harmful to cellular growth. The fast-growing cell type was shown to have duplications including the *glyT* region and has both the mutant *glyT* gene and the wild-type *glyT*$^+$ gene. Hill and co-workers discovered that a large proportion of these duplications extended from near *rrnB* to between *purD* and *metA*, and proposed that such duplications were produced by unequal crossing-over between the two rRNA operons surrounding the *glyT*$^+$ gene, for example, between *rrnB* and a possible rRNA operon between *purD* and *metA* (*rrnE*, as subsequently proven, see above). They suggested possible models to explain the mechanism for formation of the observed duplications (27).

If recombination between the two adjacent rRNA operons takes place with reasonable frequency, this could result in spacer tRNA genes being exchanged between the two operons without any other noticeable chromosomal alterations. Therefore, comparison of spacer tRNA genes in an rRNA operon at certain chromosomal locations among various different *E. coli* strains or closely related bacteria might give an indication as to how frequently such recombinational events occur. It remains to be determined what the frequency of recombination between rRNA operons might be, and whether *E. coli* has a mechanism to reduce the frequency of "harmful" recombination between rRNA operons.

rRNA PROCESSING

As described above, each rRNA operon is transcribed in the order 16S-tRNA-23S-5S, with tRNAs at the ends of some transcripts. This precursor molecule must be processed to give mature RNA species. Many of these maturational events have been previously reviewed (3, 48). Recent progress has been primarily in understanding the enzymes involved in this maturation, and in recognizing that tRNA species are also in the transcript.

A mutation has been identified that greatly reduces or eliminates the level of RNase III (49). This mutation has been mapped near 55 min on the *E. coli* chromosome (50, 51). Cells carrying this mutation grow more slowly than normal cells, especially at high temperatures (51–53). In these cells, a large rRNA precursor ("30S precursor RNA") can be detected, especially in the presence of chloramphenicol, which may be identical with the primary rRNA transcript (34–37). In vitro studies (34, 36, 37, 54, 55) indicate that the 30S precursor is cleaved by purified RNase III to give the well-characterized precursors to 16S and 23S rRNA (for review see 3; and see below). In addition, spacer tRNAs have also been processed in vitro from the 30S precursor (43). The 5S RNA precursor produced in vitro by RNase III cleavage is larger than the precursor 5S RNA found in vivo (37, 55). It is not clear whether the RNase III mutants studied are completely devoid of RNase III activity. However, alternate pathways for rRNA processing apparently exist in RNase III mutants, as rRNA precursors with sizes between 30S precursor RNA and mature 23S or 16S rRNA are formed in these cells which differ in size from those found in normal cells (56). From both in vivo and in vitro evidence, it appears that RNase III is clearly involved in rRNA processing in normal cells.

As tRNAs are found in rRNA transcripts, enzymes that process tRNA may be involved in rRNA processing in normal cells, or may provide alternate processing

GENETICS OF BACTERIAL RIBOSOMES 309

pathways in RNase III⁻ mutants. The relationship between enzymes involved in rRNA processing and tRNA processing is not clear, but may prove to be interesting. Several enzymes involved in tRNA processing have been identified (57–63) [for reviews, see (48, 64, 65)]. Temperature-sensitive mutations in RNase P (58, 63), which is involved in tRNA processing, result in accumulation of precursors of all known spacer tRNAs at restrictive temperatures (66; T. Ikemura, personal communication). This indicates that RNase P participates in processing of rRNA operon transcripts to produce spacer tRNAs. It is possible that the alternative pathways in RNase III mutants that cleave the primary transcripts involve the endonucleolytic activity of RNase P. RNase II, which may be involved in tRNA processing (59–62), has been reported to be involved in the processing of a 16S rRNA precursor (67, 68) but this has been disputed (69).

In normal cells, the 30S precursor does not accumulate to detectable levels. However, smaller precursors to 16S, 23S, and 5S rRNA can be detected and have been well characterized [see (3, 48) for reviews]. The precursor to 16S rRNA (p16S) has approximately 200 nucleotides not found on 16S rRNA, some located at both the 5' and the 3' ends. A nuclease has been isolated from *E. coli* that can cleave at the 3' end of p16S to give the normal 3' terminus of mature 16S (70). A mutation has also been identified that is defective in processing the 5' end of p16S rRNA (71). Meyhack et al (72) have partially purified an enzyme, which they call *RNase M*, which they believe is a new enzyme capable of converting p16S to a product like mature 16S rRNA, using precursor ribosomal particles as a substrate.

The 5S RNA precursor of *Bacillus subtilis*, which accumulates in the presence of chloramphenicol and is considerably larger than the precursor found in *E. coli* under similar conditions, has been sequenced (73). An enzyme has been purified that is capable of cleaving this precursor to give mature 5S rRNA (74).

LOCATION AND ORGANIZATION OF GENES FOR RIBOSOMAL PROTEINS

In this chapter we discuss the locations of r-protein genes mapped so far and their organization. In particular, we discuss r-protein genes in two chromosomal regions, the *str-spc* region and the *rif* region. Detailed mapping of genes and analysis of their organization into operons were done mainly using transducing phages carrying these chromosomal regions. The use of various methods to isolate r-protein mutants and the properties of individual mutants obtained by these methods is discussed in the following sections.

Str-Spc Cluster

IDENTIFICATION OF GENES Approximately half the genes for r-proteins in *E. coli* are located in a cluster near 72 min on the recalibrated genetic map (75). One of the first indications that such a cluster existed came from the discovery that the *ramA* gene and also the genes for resistance to streptomycin (*strA*), spectinomycin (*spcA*), erythromycin (*eryA* and *eryB*), and neamine (*neaA* and *neaB*) were clustered at this region [see (2, 4) for reviews]. These genes are now known to be structural genes for r-proteins (see below). The *fus* gene, which codes for EF-G

(76–78), was also mapped near *strA*. We refer to this cluster as the *str-spc* cluster or region.

Further evidence that many r-protein genes were located at this region came from studies using interspecies or intergeneric crosses (for review see 79). These experiments suggested that there were at least 20 genes for 30S and 50S r-proteins of *E. coli* near *strA*.

Much of the recent progress on the organization of r-protein genes in the *str-spc* region has come from experiments using specialized λ transducing phages carrying DNA from this region on the *E. coli* chromosome (80–92). Four phages were isolated that carry different lengths of DNA from this region: λ*trkA*, λ*spc*1, λ*spc*2, and λ*fus*2 (80, 88). The physical structures of these phages have been determined (85, 86, 92), and the results are shown in Figure 3.

λ*fus*2 carries the largest substitution of bacterial DNA, 27×10^6 daltons or about 44,000 base-pairs (44 kb) (85). It seems to have all the DNA between *aroE* and *strA*. λ*fus*3, which has been used to isolate polar mutants in r-protein transcription units ("r-protein operons"), appears to be the same as λ*fus*2 except it carries a str^s allele instead of a str^r allele (88).

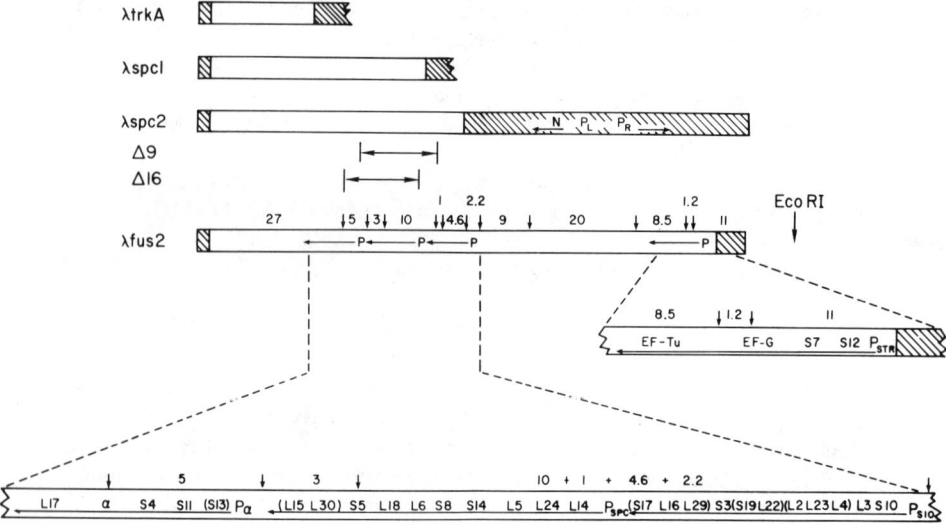

Figure 3 Location of genes for r-proteins, RNA polymerase subunit α, and elongation factors on λ*fus*2. A schematic diagram of the whole λ*fus*2 genome is given. The *open bar* represents bacterial DNA and the *shaded bar* represents λ DNA. Bacterial DNA segments carried by λ*trkA* and λ*spc*1 as well as the whole genome of λ*spc*2 are also indicated. The DNA restriction endonuclease (*Eco*RI) sensitive sites as well as the sizes of restriction DNA fragments are indicated as in Figure 2. The genes are identified by their protein products. p_α, p_{spc}, p_{S10}, and p_{str} represent the approximate positions of bacterial promoters. p_L and p_R are the two major λ promoters. The *arrows* under the genes give the directions of transcription. Groups of genes in which the order is unknown are given in parentheses. Information obtained using both in vivo and in vitro approaches is summarized. For further explanations, see the text.

The r-protein genes on λfus2 and the other phages were identified by both in vivo and in vitro techniques. In the in vivo technique, ultraviolet light (UV)-irradiated bacteria were infected with the phages and the proteins whose synthesis was stimulated were identified (80, 82, 88). The UV irradiation of the bacterial host substantially reduced the expression of the host genes. Stimulation of the synthesis of a gene product was taken as evidence that the gene was carried by the phage. In the in vitro technique, proteins were identified that were synthesized in a DNA-dependent protein synthesizing system by using transducing phage DNA as template (80, 82, 86, 92). These in vitro experiments have verified the assumption that the proteins synthesized in UV-irradiated cells corresponded to genes carried by the phages, rather than to host genes turned on by an indirect effect of the phage on the host.

The in vivo and in vitro synthesized proteins were identified by several techniques including co-electrophoresis with mature r-proteins, fingerprint analysis, and precipitation with antisera against purified r-proteins (80, 82, 83, 86, 88, 92). In addition, it was demonstrated that most of the r-proteins synthesized in vitro could assemble into ribosome particles (80, 86, 92).

The conclusions were that genes for the following 27 r-proteins are carried on λfus2 (88, 92): S3, S4, S5, S7, S8, S10, S11, S12, S13, S14, S17, S19, L2, L3, L4, L5, L6, L14, L15, L16, L17, L18, L22, L23, L24, L29, and L30. In addition, the phage was found to carry genes for EF-Tu, EF-G, and the α subunit of RNA polymerase (82, 83).

These results are generally in agreement with those obtained by interspecies or intergeneric bacterial crosses (79). In addition, genes for several of the proteins whose genes were found on λfus2 have also been mapped at the *str-spc* region through the isolation of mutants [see (2, 4) and below]. A mutation in the gene for the α subunit of RNA polymerase, which restricts the growth of phage P2 and is called *gro,* has also been mapped near *spcA* (84, 96, 97). No mutants in r-protein genes have been mapped at this region whose genes were not found on λfus2, with the possible exception of S6 (93–95; and see below).

A cluster of r-protein genes has also been found on *Bacillus subtilis* genome near *cysA*. Genes coding for resistance to several antibiotics that affect protein synthesis have been mapped in this region [see (2, 7) for reviews]. These genes are also linked to genes for EF-G (98, 99), EF-Tu (99), rRNA (100), and the β subunit of RNA polymerase, *rif* (101–103). In *E. coli,* the *rif* gene is not located at the *str-spc* region. However, it is clustered with genes for r-proteins, EF-Tu, and rRNA (see below). Thus, the clustering of genes for ribosome components, translation factors, and RNA polymerase subunits may be a general phenomenon in bacteria.

ORDER OF GENES ON THE λfus 2 GENOME Several techniques have been used to determine the order of the genes on λfus2. One approach has been to compare the genes present on phages that carry different amounts of the bacterial DNA on λfus2 (88, 92). This is a form of deletion mapping.

Another technique has been to identify the proteins synthesized in a DNA-dependent protein synthesizing system primed with purified restriction fragments

(86, 92). This is a particularly powerful technique since it does not seem to require the presence of a promoter on the fragment (92). In principle, it is possible to determine the order of all the genes by using overlapping fragments generated by different restriction enzymes. By using *Eco*RI, *Hind*III, and *Sma* restriction enzymes to generate fragments of known order on the λ*fus*2 genome, it was possible to determine the location of most of the r-protein genes (92).

An analogous technique has been to clone fragments from λ*fus*2 with a λ-cloning vector and identify the proteins synthesized in UV-irradiated cells after infection with the hybrid phage (90, 91). Analysis of the r-protein genes carried by these fragments was consistent with the results of the in vitro experiments. In addition, these experiments provided some information on the location of the promoters for the r-protein operons (see below).

The order of most of the genes in the Spc operon has been determined by analyzing a set of 20 mutants of λ*spc*1 and λ*fus*3 that carry polar insertions that reduced the expression of the *spcA* gene (89). The results of these experiments are discussed in the following section.

All of the results on the order of the r-protein genes on λ*fus*2 are summarized in Figure 3. Several observations can be made: 1. The *str-spc* cluster is actually two clusters. The *spc* cluster contains genes for 25 r-proteins plus the α gene, and consists of three operons (Figure 3; see below). The *str* cluster contains genes for S7, S12, EF-G, and EF-Tu. These two clusters are separated by 14×10^3 base-pairs. No genes have been mapped in this region. 2. Genes for 30S and 50S r-proteins are intermingled. 3. There is no obvious correlation between the order of genes and the assembly map of the 30S subunit (104, 105). With respect to the 50S subunit, there is insufficient information available on its assembly to determine whether it could be correlated with the order of genes. 4. The α gene appears to lie between the genes for L17 and S4, i.e. intermingled with r-protein genes. 5. The genes for EF-Tu and EF-G are clustered with r-protein genes.

RIBOSOMAL PROTEIN TRANSCRIPTION UNITS IN THE *str-spc* CLUSTER The directions of transcription of the r-protein genes on λ*fus*2 were determined by separating the strands of λ*spc*1 DNA and λ̄*fus*2 DNA and hybridizing with in vivo labeled RNA (81). The results indicated that all the genes were transcribed leftward on the λ*fus*2 map, which is counterclockwise on the *E. coli* chromosome. The directions of transcription of the *spcA* and *strA* genes were also established by determining the relative order of several mutations in these genes and the positions of the mutations in the amino acid sequences of the proteins (106, 107). The results also indicated that these genes were transcribed counterclockwise.

Analyses of several polar mutants of λ*spc*1, λ*spc*2, and λ*fus*3 have indicated that the ribosome genes on λ*fus*2 are organized into at least four operons (81–83, 88, 89), which have been named α operon, Spc operon, S10 operon, and Str operon (see Figure 3).

The Spc (81, 89) and Str (82) operons were identified by the isolation of several mutants of λ*spc*1 and λ*fus*3 that carried insertions of IS-like elements. They were isolated by selecting mutants of lysogens in which the expression of the *spc*[s] or *str*[s] gene on the phage genome had been reduced. The locations of the insertions on the

GENETICS OF BACTERIAL RIBOSOMES 313

phage genomes were determined by heteroduplex analysis. To determine which other genes had been affected by the insertions, the expressions of the genes from the mutant and parent phages in UV-irradiated cells were compared.

The IS2 insertion in λspc1-I16 (108) reduced the expression of ten r-protein genes, including the spcA gene, to about 10% of their level of expression from λspc1 (81, 89). The genes whose expressions were reduced are clustered immediately to the left of the insertion, "downstream" in the direction of transcription. IS2 is known to reduce the expression of distal genes when it is inserted into transcription units in its polar orientation (for review see 109), which was its orientation in λspc1-I16 (108). These observations were interpreted as indicating that these ten genes are organized into an operon. By testing mutants of λspc1 and λfus3 that had insertions at different sites in the Spc operon, it was possible to determine the order of all the genes that are between the spcA gene and the promoter (89). Information on the order of genes in the Spc operon has also come from in vitro experiments with purified restriction fragments and the DNA from the phages carrying insertions (92). The conclusion from all these experiments was that the orders of genes in the Spc operon is as follows: promoter-rplN(L14)-rplX(L24)-rplE(L5)-rpsN(S14)-rpsH(S8)-rplF(L6)-rplR(L18)- rpsE(S5)-[rplO(L15), rpmD(L30)] (Figure 3). The order of the genes for L15 and L30 is not known.

Similar analyses of polar insertion mutants and deletion mutants of λfus3 that inactivated the str^s gene indicated it is in a transcription unit with genes for S7, EF-G, and EF-Tu (82). The order of genes is given in Figure 3 (91, 92; L. Post and M. Nomura, unpublished experiments).

The genes for S13, S11, S4, α, and L17 are "downstream" in the direction of transcription from the Spc operon (see Figure 3). Insertions in the Spc operon did not reduce their expression to the same extent as the genes in the Spc operon (89). This observation suggests that these genes can be expressed from another promoter (or promoters) that is located to the left of the Spc operon. Evidence for the cotranscription of the gene for α with genes for S11, S4, and L17 has come from the analysis of the expression of those genes on λspc2 and a deletion mutant, λspc2-Δ9 (83). The Δ9 deletion appeared to have deleted their common bacterial promoter and fused them onto the λ p_L promoter, resulting in their expression being controlled by the λ repressor. Thus there appears to be no promoter between the left end of the Δ9 deletion and the L17 gene; these genes appear to be in a separate operon that has been named the α operon. The order of genes seems to be as follows: promoter-rpsM(S13)-rpsK(S11)-rpsD(S4)-rpoA(α)-rplQ(L17) (83, 86). There is no direct evidence that the S13 gene is in the α operon. However, it does not seem to be part of the Spc operon (81, 89). Thus, it is either in the α operon or a separate operon.

The remaining cluster of r-protein genes on λfus2, which are located to the right of the Spc operon, also seems to be organized into a transcription unit called the S10 operon (30). The following is the order of the genes in this operon: promoter-rpsJ(S10)-rplC(L3)-[rplB(L2), rplD(L4), rplW(L23)]-[rpsS(S19), rplV(L22)]-rpsC(S3)-[rpsQ(S17), rplP(L16), rpmC(L29)] (88, 92; Figure 3).

The promoters for the four identified r-protein operons on λfus2 have been localized to the following EcoRI fragments: 5% for the α operon, 10% for the Spc operon, 2.2% for the S10 operon, and 11% for the Str operon. Part of the evidence for these conclusions has come from analysis of the expression of r-protein genes on several EcoRI fragments of λfus2 that have been cloned with a λ-cloning vector (90, 91). The ability to express the bacterial genes in λ-immune cells when there was no transcription from the major phage promoters was taken as evidence that the genes were being expressed from a bacterial promoter on one of the cloned fragments (91).

Earlier analysis of Mu-induced mutations in the *str-spc* cluster had suggested that the *strA* and *spcA* genes are cotranscribed clockwise on the *E. coli* chromosome (110). These predictions have not been confirmed by the recent analysis of the r-protein transcription units on λfus2. It seems likely that the phenotypes of many of the Mu-induced mutants resulted from deletions rather than insertion of the Mu genome into r-protein genes (111, 112). When efforts were made to avoid deletions, analysis of Mu-induced mutants suggested that the *strA* and *spcA* genes are in separate operons (112).

The genes in the Spc cluster appear to be organized into at least three operons. Since all the genes are transcribed in the same direction, it is possible that there is a single promoter for all the protein genes in the Spc cluster, and that there are subpromoters within a larger transcription unit. However, the order of the induction of individual r-proteins after a nutritional shift-up (113) could be correlated with the order of the genes in each of the three operons in the Spc cluster (92). This suggests each of the promoters for these operons is used during a nutritional shift-up, and therefore, there are three discrete operons in the Spc cluster.

This investigation of the transcription units on λfus2 has led to the expected conclusion that the r-protein genes in the *str-spc* cluster are organized into r-protein operons, which presumably at least partially accounts for their coordinate regulation. The genes for EF-G and EF-Tu also appear to be part of an r-protein operon, which presumably means that their expression is coordinately regulated with these r-protein genes.

The most unexpected finding concerning the genes on λfus2 was that the gene for α is intermingled with r-protein genes and appears to be cotranscribed with them. This suggests that the synthesis of α is coordinated with the synthesis of ribosomes. Iwakura et al (114) concluded that the synthesis of α was coordinated with the synthesis of the β and β' subunits of RNA polymerase. Hence the observation that the α gene is in an r-protein transcription unit suggests that the biosynthesis of ribosomes and RNA polymerase may be coordinately regulated. This possibility is discussed further in a later section.

Rif Cluster

IDENTIFICATION OF GENES ON λrif^a18 Another cluster of r-protein genes on the *E. coli* chromosome has been discovered near *rif*, the structural gene for the β subunit of RNA polymerase (8). The first indication of this cluster came from the isolation of a new relaxed mutant in which the mutation, called *relC*, was mapped near *rif* rather than at the classical *relA* locus (115). The relaxed phenotype of the mutant seemed to result from a mutation in the L11 gene (116).

GENETICS OF BACTERIAL RIBOSOMES   315

Transducing phages carrying DNA from the *rif* region have been used to investigate the r-protein genes in this region. Most of the experiments have been done with λ*rif*^{d}18, which was isolated by Kirschbaum & Konrad (25). It has been shown to code for the β and β' subunits of RNA polymerase (25, 117). The r-protein genes carried by this phage have been identified and mapped by the same techniques that were employed to study the r-protein genes on λ*fus*2.

Infection of UV-irradiated cells with λ*rif*^{d}18 was found to stimulate the synthesis of L1, L7/L12, L10, and L11 (23, 118). [L7 and L12 have the same sequence except that the N-terminal serine is acetylated on L7 but not on L12 (119, 120).] These five proteins have also been synthesized in vitro using λ*rif*^{d}18 DNA as template (23, 40, 121). Thus, it was concluded that the four genes for these 50S r-proteins are on the λ*rif*^{d}18 genome and on the *E. coli* chromosome near *rif* (23, 118).

Infection with λ*rif*^{d}18 also appeared to stimulate the synthesis of a protein with the electrophoretic mobility of L8 (23, 118). Pettersson et al (122) found that a complex of L7 and L10 migrated to the same position as L8 on the two-dimensional gels that were used to identify the "L8" coded by λ*rif*^{d}18. Comparison of the one-dimensional electrophoresis of the leucine-containing tryptic peptides of L7, L12, L10, and the "L8" synthesized from λ*rif*^{d}18 has suggested that the "L8" was composed at least in part of L7/L12 (S. R. Jaskunas and M. Nomura, unpublished experiments). It was not clear whether it was a complex of L7 and L10. Thus, there probably is not a separate gene for an L8 protein on λ*rif*^{d}18. In addition, λ*rif*^{d}18 was also found to carry a gene for EF-Tu (*tufB*) (another gene for EF-Tu is located in the *str-spc* region, see above) (82), and an rRNA operon (see above).

ORGANIZATION OF r-PROTEIN GENES ON λ*rif*^{d}18   The locations of the genes for β, β', L1, L7/L12, L10, L11, and EF-Tu on the λ*rif*^{d}18 genome were determined by analysis of the proteins synthesized in an in vitro system using purified restriction fragments as templates (40). The results are summarized in Figure 2. The genes for the r-proteins were found in a cluster close to the genes for β and β'. The order from right to left on the λ*rif*^{d}18 genome is as follows: *rrnB-tufB*-[*rplA*(L1), *rplK* (L11)]-*rplJ*(L10)-*rplL*(L7/L12)-*rpoB*(β)-*rpoC*(β') (29, 40).

Previous results indicated the β and β' genes are cotranscribed (123; for review see 8). There is no information available on the transcription units for the r-protein gene on λ*rif*^{d}18. However, all the genes for rRNA, r-proteins, β, β', and EF-Tu are transcribed leftward on the λ*rif*^{d}18 genome (124), which is clockwise on the *E. coli* chromosome (40, 125; see Figure 1). Thus, since the α gene seems to be cotranscribed with r-protein genes, it is possible that the β and β' genes are cotranscribed with the adjacent cluster of r-protein genes. In general, there are a greater number of ribosomes in the cell than RNA polymerase molecules (8). Thus, if the β and β' genes are cotranscribed with r-protein genes, there would have to be some mechanism to permit a greater synthesis of r-proteins than β and β'.

The cluster of r-protein genes on λ*rif*^{d}18 can be correlated with the in vitro assembly of the 50S subunit. The binding of L7/L12 to 50S "core" particles depended on the presence of L10, and the binding of L10 seemed to be stimulated by the addition of L11 (126). L7/L12 and L10 are now known to form a complex in solution (122). They also may be complexed together in the ribosome. Conceivably, the proximity of the genes for these proteins aids in the formation of the complex as soon as the proteins are synthesized so that they are assembled into the ribosome as a preformed complex rather than one protein at a time.

Other Regions

Six r-protein genes have now been mapped at regions of the chromosome other than at the *str-spc* or *rif* regions. These are genes for S2, S6, S15, S18, S20, and L21, which were found at five different regions. There is no evidence for a large cluster of r-protein genes at any of these sites. It remains to be seen whether the other r-protein genes are scattered or whether any of them will be clustered.

rpsR (S18) The mapping of this gene was made possible by the discovery of a mutant that had an S18 with an altered electrophoretic mobility (127, 128). From chemical analysis, the mutation was shown to be in the structural gene of S18 (128). The altered electrophoretic mobility of the mutant protein was used to map the gene for S18 at 94 min between *cycA* and *pyrB* (129).

rpsB (S2) AND *tsf* (EF-Ts) Nashimoto & Uchida isolated several cold-resistant revertants of a cold-sensitive spc^r mutant (130). Some of them contained an altered S2, which resulted from a mutation that was mapped near *polC* at 3.5 min. Experiments using intergeneric crosses also have suggested that the S2 gene is in this region (131).

Transducing phages were isolated carrying DNA from this region (132, 133). They were found to stimulate the synthesis of S2 and EF-Ts after infection of UV-irradiated cells. From the comparison of several different transducing phages, it was concluded that the structural genes for S2 and EF-Ts are between *tonA* and *polC*, and that the gene order is *tonA*, *rpsB* (S2), *tsf* (EF-Ts), *polC* (132).

Friesen et al (133) also reported that λ*polCdapD*-9 stimulated the synthesis of a protein that appeared to be the σ subunit of RNA polymerase. However, the observations made by Nakamura et al (134) as well as by Harris et al (135) suggest that a σ gene is near *argG*. Thus, the identity of the σ-like protein whose synthesis was stimulated by λ*polCdapD*-9 is not certain.

rpsT (S20) Böck, Wittmann, and their collaborators have isolated several mutants with altered r-proteins by selecting suppressors of the temperature-sensitive phenotype of alanine- or valine-tRNA synthetase mutants (136–141). Some of the suppressor mutations seem to be in the S20 gene (136, 137). These mutations were mapped at about 0.5 min on the genetic map (138, 139). Takata also mapped the S20 gene in this region by use of intergeneric crosses (131).

Transducing phages carrying DNA from the region were isolated and found to stimulate the synthesis of S20 after infection of UV-irradiated cells (142, 143). No other r-protein genes seemed to be carried by these phages. The S20 gene is located between *thr* and *dapB* (142, 143), and probably counterclockwise to *ileS* (138).

GENETICS OF BACTERIAL RIBOSOMES   317

In contrast to these results, the gene for S20 (which is homologous to *E. coli* S20) in *B. subtilis* was found in the *str-spc* cluster (144). For this organism, there are no examples yet of an r-protein gene that is not in the *str-spc* cluster.

rpsO (S15) AND *rplU* (L21)   The technique using intergeneric crosses has been used to locate the genes for S15 and L21 (145). *E. coli* episomes covering the *argG* region but not the *str-spc* region were transferred to *Serratia marcescens*. The ribosomes in these diploids contained the *E. coli* S15 and L21 r-proteins. It is not known whether the S15 and L21 genes are adjacent.

rpsF (S6)   Isono & Kitakawa have recently reported that the S6 gene is at 97 min (146). This conclusion resulted from the analysis of a mutant isolated by Isono and his collaborators that contained an S6 with an altered electrophoretic mobility (147, 148). Strains that were diploid for the 97–98 min region contained both the wild-type and mutant forms of S6. Hence, the authors concluded that the mutant gene resulting in the altered S6 is the structural gene for S6 (146).

Other reports have suggested the S6 gene may be near *aroE* (93–95). No evidence was found for an S6 gene on λ*fus*2 (88, 92). Of course, it could be counterclockwise from *aroE* and not carried by λ*fus*2. The mapping of the S6 gene at 97 min should be considered tentative until the chemical nature of the altered S6 in several mutants can be investigated further and the mutations responsible for the alterations can be mapped.

In conclusion, genes for about 37 r-proteins have now been mapped; 27 in the *str-spc* cluster, four in the *rif* cluster, and these six that are scattered. The experiments using interspecies or intergeneric crosses suggested that a few more r-protein genes may be found at the *str-spc* region. Since they are not found on λ*fus*2, the data suggest that they may be located between *strA* and *crp* (75). Many mutants have been isolated that have altered r-proteins (see below). Some of these mutants contain alterations in r-proteins whose genes have not been mapped. Thus, we can expect that many of these genes will be mapped soon.

ISOLATION OF MUTATIONS IN RIBOSOMAL PROTEIN GENES

Selecting mutants that are resistant to antibiotics that inhibit protein synthesis has been the classical way to obtain mutations in r-protein genes. Recently, several other techniques have been used to obtain mutants containing altered r-proteins. Strains containing a mutated r-protein gene have frequently been identified by looking for those that contain an r-protein with altered electrophoretic or chromatographic properties. Analyzing r-proteins by two-dimensional polyacrylamide gel electrophoresis (149) is a sensitive and convenient method to do this screening. The application of all these techniques has resulted in the isolation of many mutants containing alterations in 30 different r-proteins. Some of these mutations have already given us new information on the location of r-protein genes. However, most of them have not been thoroughly characterized or exploited.

This section reviews the isolation of mutations in r-protein genes. The next section deals with the physiological effects of some of the mutant r-proteins. Cold-sensitive mutants and suppressors of cold-sensitive r-protein mutants are reviewed below in the section on ribosome assembly.

Antibiotic-Resistant Mutants of E. coli

Only a cursory review of these mutants is given here because most of the information on antibiotic-resistant mutants has been reviewed previously (2, 4, 153).

STREPTOMYCIN RESISTANCE Several types of *E. coli* mutants can be isolated that are resistant to high concentrations of streptomycin, including some that are streptomycin-dependent (2, 4). All str^r and most str^d mutations are in the *strA* gene, which codes for S12 (150–153). Breckenridge & Gorini distinguished four classes of str^r mutants in *E. coli* (154, 155) on the basis of their restriction of the efficiency of suppression by nonsense suppressors. Wittmann and his co-workers have identified the amino acid substitutions in S12 from these four classes (156) and also from several str^d mutants (157–159). Four classes of Str-R mutants have also been distinguished on the basis of their effect on the efficiency of plating by MS2 and T7 in male strains (160). There is no information available as to whether these four classes can be correlated with the four classes distinguished by restriction of suppression by nonsense suppressors.

SPECTINOMYCIN RESISTANCE Resistance to high concentrations of spectinomycin results from mutation in the *spcA* gene, which codes for S5 (2, 4, 161–163). Many spc^r mutants are cold-sensitive (164), and some may be sucrose-dependent (165).

Mutations in the S5 gene have also been isolated by selecting suppressors for streptomycin dependence (166, 167) or a temperature-sensitive alanyl-tRNA synthetase mutant (137). The amino acid replacements in these three kinds of mutant S5 proteins were found at three different regions of the protein (106, 107, 159, 168–170).

A spectinomycin-dependent strain has been isolated (171). It appears to contain one mutation in the *spcA* gene and another in a gene near *rif*.

ERYTHROMYCIN RESISTANCE Two classes of Ery-R mutations that map near *spcA* have been distinguished (172–175). One class has an altered L4 (*eryA*) and the other has an altered L22 (*eryB*) (172). The genes for these two r-proteins are on λ*fus*2 (see Figure 3). Thus, *eryA* and *eryB* are probably the genes for L4 and L22, respectively.

Pardo & Rosset have isolated a third class of erythromycin-resistant mutants, whose genetic locus, called *eryC,* was mapped at 83 min (175, 176). The structures and functions of the ribosomes were abnormal. However, an altered r-protein has not yet been identified.

NEAMINE RESISTANCE Bollen and his co-workers have isolated and characterized two classes of neamine-resistant mutants, *neaA* and *neaB* (177). The *neaA* gene was mapped near *spcA,* and chemical analysis has established that it codes for S17, (178, 179). The *neaB* mutants contained two mutations, one in the S5 gene and one in the S12 gene (159, 180). The mutation in the S12 gene conferred Str-D by itself and the S5 mutation suppressed the Str-D phenotype. The amino acid changes in the S12 and S5 from *neaB* mutants were similar to the changes in S12 from conventional *str*^{d} strains and the mutant S5 that can suppress Str-D (159).

NEOMYCIN-KANAMYCIN RESISTANCE Mutants of *E. coli*, called *nek*, have been isolated that were resistant to both neomycin and kanamycin (93, 181). The mutation was mapped between *aroE* and *spcA* (93). This mutant contains an altered S6, which was found to have only two glu residues at the C-terminus rather than the usual four to six (94). Nevertheless, it is still not clear which gene or genes have been mutated in the *nek* mutant. One problem is that the location of the S6 gene is not clear (see above). Also, the heterogeneity in the number of glu residues at the C-terminus of S6 suggests it may be affected by post-translational modification.

Suppressors of Conditionally Lethal Mutants

SUPPRESSORS OF DRUG-DEPENDENT STRAINS The first conditionally lethal mutants to be used in this way were *str*^{d} strains. Mutations in either the S4 or S5 gene have been isolated by selecting revertants of the Str-D phenotype (153, 166, 167, 182–184). A new streptomycin-dependent mutant that has an altered S8 has been used to isolate many more mutations in r-protein genes (185). Out of 120 Str-independent "revertants," 13 had changes in a 30S r-protein and 15 had changes in a 50S r-protein. "Revertants" of a spectinomycin-dependent mutant have been obtained with an altered S4, L5, or L1 (171).

SUPPRESSORS OF TEMPERATURE-SENSITIVE tRNA SYNTHETASE MUTANTS Böck and his collaborators discovered that some of the revertants of a temperature-sensitive alanyl-tRNA synthetase mutant had an altered S5 or S20 (136–141). These investigations led to the mapping of the S20 gene near *thr* (see above). Similarly, some of the "revertants" of a temperature-sensitive valyl-tRNA synthetase mutant had an altered S8 or S20 (140). This S8 mutation rendered the strain cold-sensitive. Some of the "revertants" of this strain that were no longer cold-sensitive contained an altered L30 (186).

It appears to be generally possible to obtain mutations in r-protein genes by selecting suppressors of aminoacyl-tRNA synthetase mutants as long as the synthetase has a small residual activity at the nonpermissive temperature (140). Buckel et al have suggested that the mechanism of suppression by r-protein mutations is due to a reduction in the rate of polypeptide synthesis that would keep it in balance with the rate of charging of tRNAs (141).

Temperature-Sensitive Mutants

Several groups have initiated efforts to isolate temperature-sensitive mutations in r-protein genes by localized mutagenesis of the *str-spc* regions in *E. coli* (95, 187–190) and *B. subtilis* (98, 99). The most frequent approach for *E. coli* has been to transduce an *aroE*$^-$ recipient to *aroE*$^+$ with a mutagenized P1 lysate and then to screen by replica plating to find mutants that are *ts*. Mutants of *E. coli* with an altered S17 (190), L18 (187), and L24 (190) have been obtained with this technique.

Uchida and his co-workers (95, 189) and Bollen and his co-workers (190) have attempted to isolate amber mutations in r-protein genes with this technique by using a recipient for P1 transduction that carried a temperature-sensitive amber suppressor gene. Nashimoto et al isolated one mutant that appeared to have an amber mutation in the S6 gene (95).

Isono and his collaborators have used a more general approach to isolating temperature-sensitive mutations in r-protein genes. They isolated 5000 random temperature-sensitive mutants (147). r-Proteins from 650 of these mutants have been analyzed by two-dimensional gel electrophoresis (147, 148). A total of 101 (about 15%) had an alteration in one of 30 different r-proteins. One advantage of this approach is that it may be possible to isolate mutations in r-protein genes that have not been mapped. Among these have been found mutants with alterations in S15, S16, L19, or L33.

Miller and his co-workers have isolated many temperature-sensitive mutations that map near rif (125). Some of these mutations appear to be in two of the 50S r-protein genes that are carried by $\lambda rif^{d}18$ (191).

Several years ago, Flaks et al isolated a temperature-sensitive mutant that had a 50S r-protein with an altered electrophoretic mobility (192). The temperature-sensitive mutation and the r-protein mutation were separable. Watson et al have shown that the altered r-protein is L7/L12 (118). There are two to four copies of L7 + L12 per ribosome (193, 194), while most other r-proteins are present in only one copy per ribosome (194). Thus, there must be two to four times more L7 + L12 synthesized than other r-proteins. One possible mechanism for this would be to have several copies of the gene for L7/L12. However, since all the L7 and all the L12 appeared to be affected by the mutation, there appears to be only one gene for L7/L12.

Finally, a temperature-sensitive mutant has been isolated that at first appeared to contain a temperature-sensitive S1 (195). However, more recent experiments have indicated that the mutation is probably not in the S1 gene, but is in the gene for asparaginyl-tRNA synthetase (M. Yamamoto, M. Nomura, H. Ohsawa, and B. Maruo, unpublished experiments).

PHYSIOLOGICAL EFFECTS OF MUTANT RIBOSOMAL PROTEINS

Mutations Affecting Ribosomal Assembly

The genetics of ribosome assembly has been extensively reviewed (2–5). The results of the earlier work can be summarized as follows. Cold-sensitive mutants of *E. coli* (164, 196–199) and *Salmonella typhimurium* (200, 201) have been isolated in which the assembly of only 50S or both 30S and 50S subunits appeared defective at the nonpermissive temperature. Some of these mutations were mapped at the *str-spc* region (164, 197, 200–203), and some are known to have alterations in S4 (202, 203) or S5 (164). However, many of these mutations in *E. coli* were mapped outside the *str-spc* region (197–199). It is not known whether they are in previously unmapped r-protein genes or in other genes that directly or indirectly affect ribosome assembly (204).

Recently, a few more *E. coli* mutants have been isolated in which ribosome assembly is abnormal. Some of the S5 suppressors of the *alaS* mutant and also the S8 suppressor of a *valS* mutant resulted in a cold-sensitive phenotype (136, 140, 186) and defective ribosome assembly. Most of the S20 suppressors of the *alaS* mutant were not cold-sensitive (137, 139). However, ribosome assembly was abnormal. Temperature-sensitive mutants have also been isolated in which ribosome assembly appeared defective at the nonpermissive temperature (190, 205).

What do these assembly mutants reveal about the in vivo pathways for the assembly of each subunit? Are the results consistent with the assembly maps determined with in vitro techniques? Many of the assembly mutants have alterations in r-proteins that bind directly and specifically to rRNA.

For example, there are assembly mutants that have an alteration in S4 (202, 203), S8 (186), S17 (190), and S20 (137, 139), all of which bind directly to 16S rRNA (105, 206). Also, there is a temperature-sensitive assembly mutant that has an altered L24 (190), which binds to 23S rRNA (206). Thus, the observation that changes in these proteins result in abnormal assembly is consistent with the importance of these proteins for in vitro assembly.

The results from studies on the cold-sensitive S5 mutant (Spc-49) were generally consistent with the in vitro assembly map for 30S. This mutant accumulated a 21S particle whose r-protein composition was similar to a 21S intermediate for in vitro assembly (for review see 105). Suppressors of the cold-sensitive phenotype and the assembly defect of a mutant, Spc-49 (164), had alteration in S2, S3, or S5 (130). From the in vitro assembly map, it is not clear how alteration in these three proteins would have such an important effect on assembly, since they are added late during assembly (104, 105). However, the three interact with each other in the in vitro assembly. Nashimoto & Uchida suggested that these three proteins may form a subassembly complex that activates the 21S intermediate particles (130).

Perhaps the most interesting observation on Spc-49 is that the assembly of 50S particles was also defective, although only a 30S r-protein appeared to have been altered (164). The S2 and S3 suppressor mutations also suppressed the defect in 50S assembly in Spc-49 (130). The assembly mutants with altered S4 (202, 203), S8 (186), and S20 (137, 139) also seemed to have a defect in 50S assembly. Furthermore, the cold-sensitivity of the S8 mutant apparently could be suppressed by mutations in a gene for a 50S r-protein (L30) (186). These observations suggest that 50S assembly may be coupled to 30S assembly. However, there are two assembly mutants that only appeared to have a defect in 30S assembly (190, 205). Thus, it is still not clear whether there is any obligate coupling of 50S and 30S assembly in vivo.

Mutations Affecting Translational Fidelity

The effect of streptomycin and *strA* and *ramA* mutations on translational fidelity has been extensively reviewed by Gorini (207). He and his co-workers discovered that str^r mutations decreased the efficiency with which nonsense codons could be decoded by suppressor tRNAs in vivo, a phenomenon referred to as *restriction of genotypic suppression*. Translational leakiness, which is the reading of nonsense codons by wild-type tRNAs, was also restricted by str^r mutations. The *ramA* mutation in S4 had the opposite effect. It increased translational leakiness, and the efficiency of suppression. The effect of the *ramA* mutation was similar to the effect of streptomycin, which is known to induce misreading in vitro and in vivo. Gorini's interpretation of these observations was that there is an inherent ambiguity in the codon-anticodon interaction such that there is some low-level misreading of some and perhaps all codons, including nonsense codons. Streptomycin and the *ramA* mutation increase the level of misreading, and the str^r mutations decrease it. His interpretation of str^d is that the mutations are so restrictive they are lethal unless the restriction is relaxed by the addition of streptomycin or a *ramA* mutation, which is known to relieve the requirement for streptomycin of most str^d strains. He visualized that the ribosome screens incoming tRNAs for the correct codon-anticodon interaction. Ninio has suggested the ribosome screening process is a reflection of the kinetics of the binding of tRNA to the ribosome-mRNA complex (208).

S5 mutations that relieve a Str-D phenotype are similar to *ramA* mutations; that is, they increase translational ambiguity (209, 210). Mutations in the S5 gene that have this effect have been called *ramC* (210). These observations and the previous ones provide genetic evidence that S4, S5, and S12 modulate the interaction of aminoacyl-tRNA with the codon.

Another protein that seems to be involved is S17. The *neaA* mutation that has an altered S17 was similar to *strA* mutations in that it restricted genotypic and phenotypic suppression (178, 211). The introduction of a *ramA* mutation relieved the restriction (211), just as it does for *strA* mutants.

Several aspects of Gorini's theory have been confirmed by Yates et al with biochemical experiments (212). They investigated the synthesis of the λ *O* protein directed by λ DNA in a cell-free system. Two sizes of the *O* polypeptide were synthesized by wild-type ribosomes. The larger resulted from read-through at a UGA termination codon. However, there was no read-through at this codon when ribosomes from a restrictive str^r mutant (SM3) were used. This mutant probably falls in the most restrictive class of str^r mutants (212, 213). Suppression of an *O* amber mutation by amber suppressor tRNA in vitro was also greatly reduced when these restrictive ribosomes were used (212).

Zengel et al have examined the rate of protein synthesis in the SM3 mutant (214). This mutant, as do several other str^r mutants, had a reduced growth rate compared to the parent. The polypeptide chain growth rate in the mutant was found to be reduced. This indicates the str^r mutation affected protein chain elongation. Presumably, the altered S12 interfered with the binding of other tRNAs to their codons just as it appeared to interfere with the binding of suppressor tRNAs to nonsense codons.

The observations of Yates et al (212) and Zengel et al (214) provide some possible explanations for the many pleiotropic effects of some str^r mutations [for example, see (215, 216)]. One possibility is that read-through synthesis of certain proteins may be important for various processes. For example, Qβ phage read-through proteins are probably essential for phage growth (217), and failure of growth of Qβ phage in some str^r mutants (218) might be related to the observed restriction of the read-through (212) by the str^r mutations. The reduced rate of protein chain elongation could be used to explain many pleiotropic effects. The str^r mutation in SM3 was also found to affect the specificity of initiation of translation in vitro (213; see also 219). Thus, restrictive str^r mutations affect initiation, elongation, and termination of translation.

Mutations Affecting Other Cellular Processes

We consider two phenomena here. The first is the intraction of str^r and rif^t mutations on T7 growth in male strains of *E. coli*. The other is the effect of some spc^r mutations on the maintenance of episomes.

T7 can grow on female strains but not wild-type male strains of *E. coli* (220). The mechanism by which T7 growth is restricted in male strains is not clear. Different workers have attributed the block to inhibition of transcription (221) or translation (222) of T7 genes, or a leakage of nucleotides after infection of T7 (223, 224). The isolation of most spontaneous str^r derivatives of male strains resulted in an increase in the efficiency of plating T7 (160). The str^r mutations appeared to increase early transcription of the T7 genome by the host RNA polymerase, as if they affected the transcription properties of the polymerase (221). Thus, Chakrabarti & Gorini looked for and found some spontaneous rif^t mutations that reversed the effect of these str^r mutations on T7 growth in male strains (221). They have also discovered that the presence of certain $strA$ and rif mutations in the same strain conferred a temperature-sensitive phenotype on the strain even though each mutation alone did not (225). These surprising observations suggest there may be some functional interdependence of ribosomes and RNA polymerase.

Another interesting phenomenon is the effect of spc^r mutations on the replication of episomes. Several years ago, Jacob et al isolated mutants that could not support F replication at 42°C (226). Some of these had a mutation on the chromosome that was mapped close to $spcA$ (227). This mutation may have been in the $spcA$ gene because Yamagata & Uchida found that many spc^r mutants of a strain carrying an F' *lac* also could not maintain the episome (228). They also started with a strain carrying a temperature-sensitive F' *lac* and isolated mutants that could maintain the episome at 42°C (229). One of these had a mutation in the *str-spc* region. The original temperature-sensitive phenotype of this strain could be recovered by isolating spontaneous spc^r mutants. Although the original mutated gene has not been identified, the authors concluded that it was probably a ribosomal mutation because of its location and its interaction with $spcA$ mutations. They also suggested that one possible explanation for these observations is that ribosomes or r-proteins interact with the bacterial membrane.

To explore this possibility, Miyoshi & Yamagata have looked for and found spc^r mutants that had simultaneously become sucrose-dependent (165). The two phenotypes appeared to be due to a single mutation that may be in the $spcA$ gene. A major protein was found missing from the cytoplasmic membrane of these mutants (230). The loss of this protein seemed to be due to the spc^r mutation since revertants that had become spc^s had regained the protein.

GENETICS OF TRANSLATION FACTORS

One of the important results from the recent work with transducing phages has been the mapping of genes for EF-Tu (82), EF-Ts (132, 133), and also the gene for EF-G (82), which had been previously mapped. A temperature-sensitive mutation in the EF-G gene (231) was used to select the λ*fus* transducing phages (88). Subsequent genetic and biochemical analyses confirmed that λ*fus*2 carried the gene for EF-G (82, 92). It was physically mapped close to *strA* (91, 92), which confirmed the previous genetic analysis (77, 78).

Furthermore, analysis of the genes of λ*fus*2 and λ*rif*^{d}18 indicated that both carried a gene for EF-Tu (82). The one on λ*fus*2 was named *tufA* and the one on λ*rif*^{d}18 was named *tufB*. Thus, there are at least two genes for EF-Tu in *E. coli*. It is not known whether there is any microheterogeneity between the two genes that could have functional significance. The two genes have been physically mapped on the λ*fus*2 (92) and λ*rif*^{d}18 (40) genomes and the results are given in Figures 2 and 3. As indicated above, the *tufA* gene seems to be in a transcription unit with the genes for EF-G, S7, and S12 (82).

A gene for EF-Ts (*tsf*) was found on the λ*polC* phages that also carried the S2 gene (132, 133; see above). It is not known whether they are cotranscribed.

Two temperature-sensitive mutants of *E. coli* have been isolated that appeared to have mutations in the EF-Ts (232, 233) and EF-Tu gene (234). One was the HAK88 mutant that required tRNA for growth at 42°C (232). The EF-Ts from this mutant appeared to be temperature-sensitive as judged by several tests (233, 235, 236). However, the λ*polC* phages were not able to transduce this strain to temperature resistance (132, 133), indicating the mutant contained a temperature-sensitive mutation in a gene not carried by λ*polC*. Of course, it could also have a mutation in the EF-Ts gene. Pedersen et al discovered that this strain had a mutation in the *tufB* gene that resulted in the EF-Tu from this gene having a different electrophoretic mobility than the EF-Tu from *tufA* (237). However, transductants containing only this mutation were not temperature-sensitive, indicating it also was not solely responsible for the temperature-sensitive phenotype of HAK88. The mutant reported by Lupker et al (234) with an apparent defect in EF-Tu has not been characterized further.

Dubnau et al have isolated three temperature-sensitive mutants of *B. subtilis* that seemed to have a mutation in the EF-G gene (99). Another mutant seemed to have a mutation in an EF-Tu gene. All four mutations were mapped very close to *strA*, suggesting that genes for EF-G and EF-Tu are clustered near *strA* in *B. subtilis* just as they are in *E. coli*.

In general, the synthesis of translation factors EF-G, EF-Ts, and EF-Tu appears to be coordinated with the synthesis of ribosomes. For example, the synthesis of these factors increases with increasing growth rate (238–241). Also, the ratio of both EF-G and EF-Ts to ribosomes is about 1 (238–240). However, in the case of EF-Tu, the ratio is 8–14, depending on the growth rate (241). (This means that the synthesis of EF-Tu and ribosomes is *not strictly* coordinated.) In general, the synthesis of each of these factors also appears to be under stringent control (242–244; see below). The cotranscription of *fus* and *tufA* with r-protein genes presumably at least partially accounts for this coordination (82). It is possible that the *tsf* (132, 133) and *tufB* (40) genes are also cotranscribed with r-protein genes.

The greater number of EF-Tu molecules than ribosomes is probably partially due to the presence of more than one gene. It may also be partially due to a greater activity of the *tufA* than r-protein genes, which has been observed (237, 245).

A temperature-sensitive mutant has recently been isolated that appeared to have a thermolabile IF3 (246). The temperature-sensitive phenotype of the mutant was mapped near 38 min. A transducing phage was isolated that carried this region of the chromosome (247). Infection of UV-irradiated cells with the phage stimulated the synthesis of a protein that was the size of IF3 and could be precipitated with antisera to IF3. Thus, a gene for IF3 appears to be near 38 min. However, it is not clear whether the apparent mutation in the IF3 gene of the mutant is responsible for the temperature-sensitive phenotype.

REGULATION OF THE BIOSYNTHESIS OF RIBOSOMES

In bacteria, the biosynthesis of ribosomes appears to be regulated. First, most ribosomal components exist in single copies per ribosome [for example, reference (194)], and there does not appear to be a significant pool of free ribosomal components (249, 250). All the ribosomes appear to be homogeneous in vivo, and the stoichiometric relationship among ribosomal components does not appear to change in different environments (251). Therefore, it appears that all the ribosomal components are synthesized coordinately and stoichiometrically. Second, under a variety of nutritional environments, the number of ribosomes per genome is approximately proportional to the growth rate; that is, the protein synthesis efficiency per ribosome particle remains approximately the same and bacterial cells appear to change the rate of protein synthesis, in response to changes of nutritional conditions, by changing the rate of synthesis of ribosomes rather than their efficiency in protein synthesis (9, 252). Third, during amino acid starvation, stringent (rel^+) bacteria, but not relaxed (rel^-) mutants, stop the accumulation of rRNA. This is called stringent control of rRNA synthesis (253).

There have been many studies related to the regulation of ribosome biosynthesis. A comprehensive discussion of these studies is beyond the scope of this review. We limit our discussion to some recent developments. For more comprehensive discussions of the subject, especially on many earlier studies, the reader should refer to several recent reviews (9, 254–256).

Regulation of r-Protein Synthesis

Recent progress in techniques for the separation of individual r-proteins has made possible analysis of r-protein synthesis rates under a variety of conditions with respect to individual proteins. Thus, as mentioned above, coordinated regulation of the synthesis of most r-proteins has been rigorously demonstrated (251). In addition, changes in the differential synthesis rate of r-proteins (called "α_r"; α_r = rate of synthesis of r-proteins/rate of synthesis of total protein) in various nutritional conditions were shown to be due to changes in the amounts of r-protein mRNA (257–259). Measurements of r-protein mRNA have become possible because of the availability of transducing phage DNAs carrying various r-protein genes (see above).

The increase in α_r values during nutritional shift-up has been examined (113). As noted above, the order of the "induction" of individual proteins after the shift-up could be correlated with the order of genes in the three operons in the Spc region (92, 113). This suggests that such induction involves an increase in the frequency of initiation of transcription.

Earlier studies indicated that amino acid starvation does not dramatically affect the rate of mRNA synthesis (e.g. 248, 260–263). Therefore, it was thought that amino acid starvation affects only rRNA and tRNA synthesis. Partly because of this and partly because even rel^- cells can adjust their ribosome synthesis according to nutritional environment and in response to nutritional shift-up or shift-down (see below), it was not clear whether the stringent control system plays a major role in the regulation of ribosome biosynthesis [see, for example, the discussion by Maaløe, (264)]. However, recent experiments demonstrated that the synthesis of r-proteins is also under stringent control (265), and this control is exerted at the level of transcription of r-protein genes (266). Subsequently, the synthesis of chain elongation factors such as EF-G, EF-Ts, and EF-Tu was also shown to be influenced by the stringent control system (242–244). Thus, it is now clear that the regulation of not only r-protein gene expression but also several genes related to translation (and perhaps transcription, see above) is very similar to the regulation of rRNA genes, and that the stringent control system involving guanosine tetraphosphate (ppGpp or "MSI") may play an important role in the regulation of expression of all of these genes related to ribosomes. [However, the question whether ppGpp is *directly* involved in the regulation of the expression of ribosomal genes has not been established (see below).]

Regulation of rRNA Synthesis

We discuss briefly several problems related to stringent control of rRNA synthesis. In vitro synthesis of ppGpp (and pppGpp) (267–270) has revealed the molecular mechanism involved in the formation of ppGpp during amino acid starvation in rel^+ strains. The $relA^-$ strains do not produce the functional stringent factor (a nucleotide pyrophosphotransferase) which was shown to be required for this in vitro reaction. Correlation between the level of ppGpp and the synthesis rate of rRNA is striking (reviewed in 254–256) and may explain the regulation of ribosome synthesis under various conditions. Studies on *spoT* mutants (see below) suggest that it is ppGpp, not pppGpp, which participates in the regulation of ribosome synthesis (271–273).

However, there are several instances where this correlation appears to break down. For example, in a mutant of *B. subtilis* with a temperature-sensitive EF-G growing at a subrestrictive temperature, rRNA synthesis is not inhibited during amino acid starvation even though the cells accumulate ppGpp (98). This observation may be related to another case observed with *E. coli* mutants with temperature-sensitive EF-G, where rRNA synthesis was inhibited at high temperatures without accumulation of ppGpp (189). These observations suggest that EF-G may play a role in rRNA synthesis and that perhaps the intact EF-G must be present for ppGpp to exert the inhibitory effect. Other factors with similar roles, such as EF-Tu, were also proposed in the past [for example, references (274, 275)]. Other examples of the breakdown of the correlation are RNA synthesis during carbon source shift-down (276–278) and that after temperature shift-up (279). In the latter case, Gallant and co-workers (279) found that when *E. coli* cells growing at 23°C are shifted to 40°C, rapid accumulation of ppGpp takes place, yet the rate of accumulation of rRNA increases, rather than decreases. Therefore, it appears that a high level of ppGpp is not a sufficient condition for the inhibition of rRNA synthesis. Thus, the search for the *additional* missing factor(s) or the *real* factor(s) responsible for regulation is being actively pursued by several investigators [see, for example, (275, 280–282)].

It has been known that *relA*$^-$ mutants are still able to regulate rRNA synthesis according to changes in nutritional conditions. For example, both *relA*$^+$ and *relA*$^-$ mutants still accumulate ppGpp and restrict rRNA synthesis normally during carbon source shift-down [276–278; for reviews, see (254–256)]. Therefore, the mechanism of accumulation of ppGpp is apparently different in carbon source shift-down from that in amino acid starvation. A current explanation for this phenomenon is that most of the *relA*$^-$ mutants are leaky and synthesize a small amount of ppGpp, that the amount of ppGpp is determined by both the rate of its synthesis and the rate of its degradation, and that the carbon source shift-down causes inhibition of the degradation, leading to accumulation of ppGpp synthesized by residual activity in *relA*$^-$ mutants. Such an explanation has also received support from the analysis of *spoT* mutants.

The *spoT* mutation was first discovered in a strain of *E. coli* which accumulates only ppGpp but not pppGpp during amino acid starvation (271, 272). The mutation was found to slow down the degradation of ppGpp (271–273). Subsequently, several *spoT*$^-$ mutants were isolated as "suppressor mutants" starting from *relC*$^-$ strains (273); inhibition of degradation of ppGpp apparently increases the level of ppGpp in *relC*$^-$ cells, resulting in a pseudo-stringent phenotype. The similarity in the pattern of accumulation of guanosine polyphosphate compounds during carbon source shift-down to that in *spoT* mutants suggests that the shift-down affects the degradation of ppGpp like *spoT* mutations. Studies using *spoT* mutants indicated that the pathway of synthesis and degradation of ppGpp in vivo is probably GTP + ATP → pppGpp → ppGpp → X → [273, 283; but see the previous reports with different suggestions (271, 272)]. However, the *spoT* gene product has not been identified and the exact reactions involved in ppGpp degradation as well as the presumed signals to regulate the degradation step during carbon source shift-down are entirely unknown. In addition, it is still possible that some other, as yet unknown, mechanism of production of ppGpp exists and is responsible for the synthesis of basal levels of ppGpp in *relA*$^-$ cells, as observed after carbon source shift-down.

We have already mentioned *relC* mutations, which involve an alteration in r-protein L11 (see above) and result in a relaxed phenotype similar to that caused by *relA* mutations. This mutation presumably interferes in some way with the ribosome-dependent production of guanosine polyphosphate compounds by the stringent factor. Since many other ribosomal components probably participate in this reaction, we expect that other types of *rel* mutations involving other r-proteins may occur.

The question whether the regulation of rRNA accumulation (by the stringent control system or during shift-up or shift-down) really takes place at the initiation of transcription or involves RNA degradation was a matter of controversy [see reviews (9, 254)]. Recent analysis of pulse-labeled RNA during amino acid starvation ascribed a definite but small role of degradation in the regulation of RNA accumulation (284, 285). In addition, a comparison of the rates of synthesis and the rates of accumulation of rRNA under a variety of nutritional conditions also showed that the change in the rate of accumulation is largely due to the regulation of the rate of synthesis (259). However, breakdown of nascent rRNA does take place, especially at low growth rates (259). The role of RNA degradation in coordinate regulation of rRNA and r-protein accumulation is further discussed below.

Gene Dosage Effects and the Mechanism of Regulation

There are two possibilities one has to consider regarding the regulation of transcription of rRNA operons (and that of several r-protein operons). 1. The number of rRNA operons (or "promoters" of rRNA operons) is in excess and the number of "special" RNA polymerase molecules capable of transcription of rRNA operons, or of some positive regulatory element(s), is limiting ("polymerase limiting model"). In this case, regulation would be accomplished by regulating the number of "special" polymerases or the amount of positive regulatory element(s). 2. The number of functional RNA polymerases is in excess, and the "capacity" of rRNA operons (or the "capacity" of their promoters) is limiting ("promoter limiting model"). In this case, regulation could be done by changing the number of functional rRNA operons or the "capacity" of their promoters. Some negative regulatory elements, like a hypothetical repressor in conjunction with some effectors, e.g. ppGpp, could be used to alter the capacity of the promoters.

A model proposed by Travers (286, 287) assumes, in its simplest form, the first possibility. According to this model, there are several different "forms" of RNA polymerase in cells and the level of ppGpp decreases the amount of a special RNA polymerase capable of transcription of rRNA operons. A simple prediction of this model is that the increase in gene dosage would not affect the rate of rRNA transcription.

In earlier experiments, Pedersen (284) attempted to detect gene dosage effects in rRNA synthesis using a synchronized culture of *E. coli*. The failure to detect the expected gene dosage effect was thought to be consistent with the polymerase limiting model. However, the expected gene dosage effect was rather small and might have been masked by experimental errors inherent in the hybridization technique used.

More recently, two approaches have been used to increase rRNA gene dosage much further above the seven to ten copies normally occurring in *E. coli*. The induction, and subsequent replication, of a specialized transducing phage carrying an rRNA operon appeared to result in an increase in rRNA accumulation (22). More extensive studies on gene dosage effects were done using hybrid ColE1 plasmids carrying rRNA operons, which are described in the previous sections (44). Approximately tenfold increases in rRNA gene dosage were achieved after amplification of hybrid ColE1-rDNA plasmids by exposure of cells to chloramphenicol or amino acid starvation, and a twofold increase was observed in exponentially growing cells without any treatment (44). Gene dosage effects were examined by analyzing the accumulation of spacer tRNAs and 5S RNA, rather than analyzing pulse-labeled rRNA by RNA-DNA hybridization methods. This approach has an advantage, because spacer tRNAs (and 5S RNA) are much more stable than free 16S or 23S rRNA, and measurements of tRNA accumulation are more accurate than measurements of rRNA by RNA-DNA hybridization methods. It was observed that both during chloramphenicol treatment and after recovery from amino acid starvation (treatments that increase rRNA gene dosage, as mentioned above), the rate of transcription of rRNA operons relative to the transcription of "ordinary" tRNA genes (usually tRNA^{Val} gene as a reference) was about six times higher in the cells carrying rRNA operons on plasmids than in control cells carrying plasmids without rRNA operons; under these conditions of high rRNA gene dosages, synthesis from the plasmid-borne rRNA operon represented about 80–90% of the rRNA synthesized. However, direct comparison of the absolute rate of transcription of rRNA operons demonstrated that it is about threefold higher than the rate of rRNA transcription in the control cells; the transcription rate from *individual* tRNA and rRNA promoters in the experimental cells was depressed by about 50%. This decrease is due either to depletion of RNA polymerase and/or factors required for stable RNA synthesis (see below), or to specific (but only partial) inhibition as a result of overproduction of (uncharged) spacer tRNA and other tRNAs associated with rRNA operons (see above). Even with normal exponentially growing (plasmid-carrying) cells, where gene dosages were about twofold higher than control cells, the absolute increase in the rate of transcription of rRNA operons (relative to control cells) appeared to be higher. In this case, the decrease of transcription rate from *individual* tRNA and rRNA promoters was very slight. No significant difference in growth rate was observed between the experimental and the control strains in either exponential growth or growth after recovery from amino acid starvation. Furthermore, no difference was observed in the *accumulation* of 16S and 23S rRNAs, and hence in the synthesis of ribosomes, between the experimental and control strains. [Note that the gene dosage effect was *not* observed in the accumulation of rRNA in this system, and hence the results are different from the reported results on the gene dosage effects after transducing phage induction (22), as mentioned above.]

The clear gene dosage effects observed in the above experiments seems to be more consistent with the "promoter limiting model" than the "polymerase limiting model." In fact, is is known that *E. coli* cells contain a large excess (two- to fivefold) of nontranscribing RNA polymerase molecules (9, 289–291). According to the promoter limiting model, the free RNA polymerase molecules are functional and available for cellular requirements, including the transcription of rRNA operons.

However, other earlier experiments lead to a different conclusion, namely, the excess RNA polymerase molecules are inactive and the amount of functional RNA polymerase in the pool is probably very small (292, 292a). Further studies are required to clarify these problems.

Coordinate Control of rRNA and r-Protein Gene Expression

As discussed in previous sections, several operons have been identified that carry a number of genes for r-proteins, protein chain elongation factors, and certain RNA polymerase subunits. In addition, genes for all three rRNAs have been found to be organized into polygenic rRNA operons. The coordinate control of certain components of the transcriptional and translational machinery is therefore assured, at least at the transcriptional level. However, the mechanism(s) by which the expression of separate operons is coordinated is not yet clear. In the case of coordination of individual r-protein operons, promoters (or any other structures involved in the regulation of initiation) for these r-protein operons may be the same or similar, and probably respond to the same regulatory signals, leading to the coordination of transcription.

We have already mentioned coordination of rRNA synthesis and r-protein synthesis in connection with stringent control of production of both rRNA and r-proteins. Three possibilities can be considered for the possible mechanisms involved in the coordination. The first possibility is that rRNA synthesis is the primary target of regulatory mechanisms and the regulation of r-protein operon expression is a secondary consequence of the regulation of rRNA synthesis. A specific model that has been suggested previously and belongs to this category is that some r-proteins are autogenous repressors [see, for example, discussion in (266, 288)]. According to this model, overproduction of rRNA synthesis should remove (hypothetical) r-protein repressors and stimulate r-protein synthesis. No such stimulation was observed in the ColE1-hybrid plasmid system described above (A. M. Fallon and M. Nomura, unpublished experiments) either in measuring pulse-labeled r-proteins or in measuring accumulation of ribosomes, despite the large increase in the transcription of rRNA operons. In addition, specific inhibition of $\lambda fus2$ DNA-directed in vitro r-protein synthesis by ppGpp was demonstrated under conditions where there were no rRNA genes and where no detectable rRNA synthesis took place (293). Therefore, this autogenous r-protein repressor model does not seem likely, although in earlier work, some experimental results were interpreted on the basis of this autogenous repressor model (266, 288).

A second possible way to achieve the coordination is to regulate the r-protein genes and to use either r-protein itself or some products synthesized from r-protein operons to regulate rRNA synthesis. A specific model proposed by Maaløe (264) assumes such a mechanism, and in addition, assumes that the synthesis of r-proteins is passively regulated, that is, the rate of r-protein synthesis is determined by the amount of nontranscribing RNA polymerase available in the pool. This in turn is determined by the activities of many other genes concerned with "general metabolism" and regulated by nutritional environmental factors. As discussed above, there appear to be large amounts of functional free RNA polymerases in the pool. In addition, r-protein mRNA synthesis appears to be directly regulated by stringent control, rather than indirectly regulated by rRNA gene expression (see above). The proposed passive control does not appear to explain these results without invoking futher complicated assumptions. The idea of a coupling mechanism that uses some r-proteins as inducer(s) for rRNA synthesis is interesting, but also perhaps unlikely, because rRNA synthesis can take place in the absence of r-protein synthesis. However, it should be noted that under these conditions (rRNA synthesis in the presence of protein synthesis inhibitors such as chloramphenicol, or rRNA synthesis during amino acid starvation in rel^- strains), there is always some breakdown of preexisting ribosomes (294), liberating free r-proteins to the pool. Thus, the proposed mechanism cannot be excluded. In addition, the synthesis of r-protein mRNA takes place under these conditions and is always coordinate with the synthesis of rRNA. Thus one could consider the possibility of r-protein mRNA (or some transcripts from r-protein operons) as positive regulatory elements for rRNA synthesis. If such hypothetical regulatory elements are products of r-protein operons in the *str-spc* region, an increase in dosages of these operons would be expected to have stimulatory effects on the transcription of rRNA operons. No convincing experiments either with positive or negative results have yet been performed.

Finally, the third possibility is that both rRNA and r-protein operon expressions are regulated, and exact coordination is achieved either by a balance of transcriptional and translational efficiencies inherent in the DNA and mRNA structures themselves, or by degradation of products synthesized in excess, or by both. The ability of cells to degrade rRNA was observed under a variety of conditions (259, 284, 295, 296). In the gene dosage experiments described above (44), the data indicated rapid degradation of as much as 50% of the rRNAs synthesized in exponentially growing cells in a glucose-peptone medium without any apparent harmful effects on growth. The presence of such high degradation activity in normally growing cells suggests that the degradation of free excess rRNA may play an important role in achieving the coordination of accumulation of rRNA and r-proteins.

Analysis of Gene Expression In Vitro

As described above, ppGpp has been thought to play an important role in the regulation of rRNA synthesis. In order to understand the mechanism involved in the ppGpp-mediated inhibition of rRNA synthesis, many in vitro experiments were performed either using "pure" transcription systems (22, 46, 274, 275, 297–300) or using crude bacterial extracts (280, 301–303), or "permeabilized" cells that may be a system very close to the real in vivo state (304, 305). Preferential inhibition of rRNA synthesis by ppGpp was observed in a number of laboratories (22, 46, 274, 275, 280, 297, 299, 300, 302) but not in others (298, 301, 303–305). Some reported the preferential inhibition in pure transcription systems, but others observed it only in crude systems containing many uncharacterized proteins. Thus, the question whether ppGpp has a direct effect on rRNA synthesis has not been settled (see the discussion on in vivo experiments described above). Similar in vitro studies were also done on the expression of r-protein genes in DNA-directed protein synthesizing systems, and preferential inhibition of synthesis of r-proteins (and other translational factors) has been observed (22, 293, 306).

From the analysis of in vitro transcription studies, several investigators suggested that there are multiple RNA polymerase binding sites ("promoters") for each rRNA operon, and that this might explain a high efficiency of transcription of rRNA operons (300, 307, 308). In one study, the presence of as many as 30 promoters was suggested (308). It should be noted that the proposed promoter region to accommodate 30 RNA polymerase molecules would occupy a large segment of DNA (perhaps as large as 2 kb). As mentioned above, the heteroduplex studies showed only 5.2 kb long regions of homology between various rRNA operons and this length is barely enough to code for the known RNA transcripts from rRNA operons. Thus, if true, a multiple promoter model would raise a serious question on heterogeneity of rRNA promoters (see above).

In vitro studies have not yielded consistent information on regulation. However, many defined homogeneous DNA templates carrying important ribosomal genes and their promoters are now available. This, combined with several new technical developments such as DNA cloning and DNA sequencing, promises useful in vitro approaches for obtaining detailed molecular mechanisms involved in the regulation of ribosomal gene expression.

CONCLUDING REMARKS: RIBOSOMAL GENE EXPRESSION AND OTHER CELLULAR ACTIVITIES RELATED TO GROWTH

As mentioned above, perhaps one of the most unexpected recent findings has been that genes for the α, β, and β' subunits of RNA polymerase are clustered with r-protein genes (1, 8, 23, 40, 83, 118). In the case of the α gene, there is strong genetic evidence that the α gene is cotranscribed with r-protein genes. In the case of the β and β' genes, the question whether these genes are cotranscribed with neighboring r-protein genes has not been settled. It is known that the direction of transcription of the β and β' genes and the adjacent r-protein genes is the same (124), suggesting the possibility of cotranscription of these genes. A recent analysis of pulse-labeled mRNA synthesized in various growth media (309) suggests that the rate of transcription of the β and β' genes is coordinate with, but only about one fourth to one fifth as high as, transcription of the adjacent r-protein genes. However, as there are conflicting reports on the amounts of α, β, and β' proteins in several different nutritional environments during steady-state growth, and during nutritional shift-up or shift-down (289, 290, 310–312; for review see 8), it is not clear whether the differential synthesis rate of RNA polymerase subunits is coordinate with that of r-proteins in various growth conditions. In addition, the synthesis of these polymerase subunit proteins does not appear to be stringently controlled in vivo (243, 244, 313). Thus, even though the gene for α appears to be cotranscribed and therefore coregulated with r-proteins, and the genes for β and β' may be possibly cotranscribed or coregulated with other r-proteins at the primary transcription level, additional and perhaps more complex regulatory mechanisms appear to be superimposed on this primary transcriptional coregulation. The nature of such regulatory mechanisms is unknown.

Another interesting recent development is related to the location of tRNA genes within rRNA operons, either in the spacer region or distal to the rRNA genes (38, 39, 42–44). Although the amounts of tRNA per genome were once thought to be invariant with growth rates (252, 264), more recent measurements show that the amounts vary with growth rates like other ribosomal components (9, 10, 314, 314a). In addition, the synthesis of tRNAs is stringently controlled (263, 315, 316). Therefore, the general picture here is consistent with recent genetic discoveries. However, as discussed above, the reason only certain tRNA genes are associated with rRNA operons is not entirely clear. We should now perhaps distinguish two kinds of tRNAs, one whose genes are associated with rRNA operons and the other, whose genes are apparently not associated with rRNA operons. Close examination of the two classes of tRNAs with respect to possible differences in the physiological roles of tRNA in bacteria or the metabolic roles of their cognate amino acids might suggest why only certain tRNAs are associated with rRNA operons. We have already discussed above the possible significance of the association of certain tRNA genes with rRNA operons in connection with regulation of ribosomal gene expression. Further experiments should be able to test the validity of such models.

Another example of links between ribosomal gene expression and other cellular activities is related to the possible functions of ppGpp. Besides the suggested role of ppGpp in the regulation of ribosomal genes, several experiments, both in vivo and in vitro, indicated that ppGpp is involved in the regulation of certain other non-ribosomal genes (281, 302, 317–322, 322a), such as the *his* operon in *S. typhimurium* (318). Thus, Ames and his co-workers have suggested that ppGpp is a general signal molecule in a control system that senses an amino acid deficiency and redirects various cellular activities in response (318). The response of synthesis of many other proteins, including various aminoacyl-tRNA synthetases, to a condition of amino acid deficiency is under current investigation by various groups (243, 244, 323–325).

Our final comments on recent results in ribosome genetics are on the direction of transcription of ribosomal genes. The *E. coli* genetic map is shown in Figure 1, and the directions of transcription are indicated for the two major r-protein gene clusters and five rRNA operons. The origin of replication of the *E. coli* genome is known to be at about 82 min, and the replication is bidirectional (326, 327). It is a striking coincidence that the direction of transcription of all the ribosomal genes so far analyzed is the same as that of DNA replication. As mentioned above, these ribosomal genes are transcriptionally very active and the density of actively transcribing RNA polymerase may be very high. One could therefore speculate that the observed organization of these genes may have evolved to prevent physical collision of the DNA replication machinery with actively transcribing RNA polymerase molecules.

It is now clear that the regulatory system for ribosomal gene expression is interconnected with other major regulatory systems, such as those governing the synthesis of transcriptional machinery, and functions as a part of a more complex system governing overall regulation of cellular growth. This is perhaps not surprising in view of the fact that ribosomes constitute a major fraction of the cellular mass and play a crucial role in growth. In the past, through use of data obtained in many physiological experiments, regulation of ribosome biosynthesis was analyzed in relation to synthesis of other macromolecules and to cellular growth (252, 264; see also 11). Recent progress in genetic analysis reemphasizes the importance of such a viewpoint. In addition, developments of the many powerful new approaches already discussed promise a real possibility of deeper understanding of the problems involved in the control of bacterial growth.

Acknowledgments

We acknowledge support from the National Institute of General Medical Sciences (GM-20427 administered by M.N.; GM-24347 administered by S.R.J.) and from the National Science Foundation (GB-31086 administered by M.N.). We thank our many colleagues for sending preprints describing their unpublished work, and Dr. M. Susman for reading the manuscript. We would like to take this occasion to express our appreciation of the great contribution to this field made by the late Dr. Luigi Gorini.

Literature Cited

1. Nomura, M. 1976. Organization of bacterial genes for ribosomal components: Studies using novel approaches. *Cell* 9:633–44
2. Jaskunas, S. R., Nomura, M., Davies, J. 1974. Genetics of bacterial ribosomes. In *Ribosomes*, ed. M. Nomura, A. Tissières, P. Lengyel, pp. 333–68. Cold Spring Harbor, NY: Cold Spring Harbor Lab. 930 pp.
3. Pace, N. R. 1973. Structure and synthesis of the ribosomal ribonucleic acid of prokaryotes. *Bacteriol. Rev.* 37:562–603
4. Davies, J., Nomura, M. 1972. The genetics of bacterial ribosomes. *Ann. Rev. Genet.* 6:203–34
5. Nomura, M. 1970. Bacterial ribosome. *Bacteriol. Rev.* 34:228–77
6. Nomura, M., Tissières, A., Lengyel, P., eds. 1974. *Ribosomes*. Cold Spring Harbor, New York: Cold Spring Harbor Lab. 930 pp.
7. Smith, I. 1977. Genetics of the translational apparatus. In *Molecular Mechanisms of Protein Biosynthesis*, ed. H. Weissbach, S. Pestka, pp. 627–700. New York: Academic. 720 pp.
8. Scaife, J. 1976. Bacterial RNA polymerases: The genetics and control of their synthesis. In *RNA Polymerase*, ed. R. Losick, M. Chamberlin, pp. 207–26. Cold Spring Harbor, NY: Cold Spring Harbor Lab. 899 pp.
9. Kjeldgaard, N. O., Gausing, K. 1974. Regulation of biosynthesis of ribosomes. See Ref. 2, pp. 369–92
10. Dennis, P. P., Bremer, H. 1974. Macromolecular composition during steady-state growth of *Escherichia coli* B/r. *J. Bacteriol.* 119:270–81
11. Bremer, H. 1975. Parameters affecting the rate of synthesis of ribosomes and RNA polymerase in bacteria. *J. Theor. Biol.* 53:115–24
12. Yu, M. T., Vermeulen, C. W., Atwood, K. C. 1970. Location of the genes for 16S and 23S ribosomal RNA in the genetic map of *Escherichia coli*. *Proc. Natl. Acad. Sci. USA* 67:27–31
13. Birnbaum, L. S., Kaplan, S. 1971. Localization of a portion of the ribosomal RNA genes in *Escherichia coli*. *Proc. Natl. Acad. Sci. USA* 68:925–29
14. Unger, M., Birnbaum, L. S., Kaplan, S. 1972. Location of the ribosomal RNA cistron of *Escherichia coli:* A second site. *Mol. Gen. Genet.* 119:377–80
15. Gorelic, L. 1970. Chromosomal location of ribosomal RNA cistrons in *Escherichia coli*. *Mol. Gen. Genet.* 106:323–27
16. Jarry, B., Rosset, R. 1971. Heterogeneity of 5S RNA cistrons in *Escherichia coli*. *Mol. Gen. Genet.* 113:43–50
17. Jarry, B., Rosset, R. 1973. Localization of some 5S RNA cistrons on *Escherichia coli* chromosome. *Mol. Gen. Genet.* 121:151–62
18. Jarry, B., Rosset, R. 1973. Further mapping of 5S RNA cistrons in *Escherichia coli*. *Mol. Gen. Genet.* 126:29–35

19. Deonier, R. C., Ohtsubo, E., Lee, H. J., Davidson, N. 1974. Electron microscope heteroduplex studies of sequence relations among plasmids of *Escherichia coli*. VII. Mapping the ribosomal RNA genes of plasmid F14. *J. Mol. Biol.* 89:619–29
20. Ohtsubo, E., Lee, H. J., Deonier, R. C., Davidson, N. 1974. Electron microscope heteroduplex studies of sequence relations among plasmids of *Escherichia coli*. VI. Mapping of F14 sequences homologous to φ80d*metBJF* and φ80d *argECBH* bacteriophages. *J. Mol. Biol.* 89:599–618
21. Jorgensen, P. 1976. A ribosomal RNA gene of *Escherichia coli (rrnD)* on λd*aroE* specialized transducing phages. *Mol. Gen. Genet.* 146:303–7
22. Jorgensen, P., Fiil, N. P. 1976. Ribosomal RNA synthesis in vitro. In *Alfred Benzon Symposium IX: Control of Ribosome Synthesis*, ed. N. O. Kjeldgaard, O. Maaløe, pp. 370–82. New York: Academic. 466 pp.
23. Lindahl, L., Jaskunas, S. R., Dennis, P. P., Nomura, M. 1975. Cluster of genes in *Escherichia coli* for ribosomal proteins, ribosomal RNA and RNA polymerase subunits. *Proc. Natl. Acad. Sci. USA* 72:2743–47
24. Konrad, E. B., Kirschbaum, J. B., Austin, S. 1973. Isolation and characterization of φ80 transducing bacteriophage for a ribonucleic acid polymerase gene. *J. Bacteriol.* 116:511–16
25. Kirschbaum, J. B., Konrad, E. B. 1973. Isolation of specialized lambda transducing bacteriophage carrying the beta subunit gene for *Escherichia coli* ribonucleic acid polymerase. *J. Bacteriol.* 116:517–26
26. Yamamoto, M., Nomura, M. 1976. Isolation of λ transducing phages carrying rRNA genes at the *metA-purD* region of the *Escherichia coli* chromosome. *FEBS Lett.* 72:256–61
27. Hill, C. W., Combriato, G. 1973. Genetic duplications induced at very high frequency by ultraviolet irradiation in *Escherichia coli*. *Mol. Gen. Genet.* 127:197–214
28. Hill, C. W. 1977. Chromosomal rearrangements resulting from recombination between ribosomal RNA genes. In *DNA Insertion Elements, Plasmids and Episomes*, ed. A. Bukhari, J. Shapiro, S. Adhya. Cold Spring Harbor, NY: Cold Spring Harbor Lab. In press
29. Kenerley, M. E., Morgan, E. A., Post, L., Lindahl, L., Nomura, M. 1977. Characterization of hybrid plasmids carrying individual ribosomal RNA transcription units of *Escherichia coli*. *J. Bacteriol.* In press
30. Clarke, L., Carbon, J. 1976. A colony bank containing synthetic Col E1 hybrid plasmids representative of the entire *E. coli* genome. *Cell* 9:91–99
31. Lobban, P. E., Kaiser, A. D. 1973. Enzymatic end-to-end joining of DNA molecules. *J. Mol. Biol.* 78:453–71
32. Vola, C., Jarry, B., Rosset, R. 1977. Linkage of 5S RNA and 16S and 23S RNA genes on the *E. coli* chromosome. *Mol. Gen. Genet.* 153:337–41
33. Schlessinger, D. 1974. Ribosome formation in *Escherichia coli*. See Ref. 2, pp. 393–416
34. Dunn, J. J., Studier, F. W. 1973. T7 early RNAs and *Escherichia coli* ribosomal RNAs are cut from large precursor RNAs in vivo by ribonuclease III. *Proc. Natl. Acad. Sci. USA* 70:3296–3300
35. Nikolaev, N., Silengo, L., Schlessinger, D. 1973. Synthesis of a large precursor to ribosomal RNA in a mutant of *Escherichia coli*. *Proc. Natl. Acad. Sci. USA* 70:3361–65
36. Nikolaev, N., Schlessinger, D., Wellauer, P. K. 1974. 30 S preribosomal RNA of *Escherichia coli* and products of cleavage by ribonuclease III: Length and molecular weight. *J. Mol. Biol.* 86:741–47
37. Ginsburg, D., Steitz, J. A. 1975. The 30 S ribosomal precursor RNA from *Escherichia coli*. A primary transcript containing 23 S, 16 S, and 5 S sequences. *J. Biol. Chem.* 250:5647–54
38. Yamamoto, M., Lindahl, L., Nomura, M. 1976. Synthesis of ribosomal RNA in *E. coli*: Analysis using deletion mutants of a λ transducing phage carrying ribosomal RNA genes. *Cell* 7:179–90
39. Lund, E., Dahlberg, J. E., Lindahl, L., Jaskunas, S. R., Dennis, P. P., Nomura, M. 1976. Transfer RNA genes between 16S and 23S rRNA genes in rRNA transcription units of *E. coli*. *Cell* 7:165–77
40. Lindahl, L., Yamamoto, M., Nomura, M., Kirschbaum, J. B., Allet, B., Rochaix, J.-D. 1977. Mapping of a cluster of genes for components of the transcriptional and translational machineries of *Escherichia coli*. *J. Mol. Biol.* 109:23–47
41. Wu, M., Davidson, N. 1975. Use of gene 32 protein staining of single-strand polynucleotides for gene mapping by

electron microscopy: Application to the $\phi 80d_3 ilvsu^{+}7$ system. *Proc. Natl. Acad. Sci. USA* 72:4506–10
42. Morgan, E. A., Ikemura, T., Nomura, M. 1977. Identification of spacer tRNA genes in individual ribosomal RNA transcription units of *Escherichia coli. Proc. Natl. Acad. Sci. USA* 74:2710–14
43. Lund, E., Dahlberg, J. E. 1977. Spacer transfer RNAs in ribosomal RNA transcripts of *E. coli:* Processing of 30S ribosomal RNA in vitro. *Cell* 11:247–62
44. Ikemura, T., Nomura, M. 1977. Expression of spacer tRNA genes in ribosomal RNA transcription units carried by hybrid Col E1 plasmids in *Escherichia coli. Cell* 11:779–93
45. Fellner, P. 1974. Structure of the 16S and 23S ribosomal RNAs. See Ref. 2, pp. 169–91
46. Oostra, B. A., van Ooyen, A. J. J., Gruber, M. 1977. In vitro transcription of three different ribosomal RNA cistrons of *E. coli.* Heterogeneity of control regions. *Mol. Gen. Genet.* 152:1–6
47. Morgan, E. A., Kaplan, S. 1976. Coordinate regulation of the individual ribosomal RNA operons in *Escherichia coli. Biochem. Biophys. Res. Commun.* 68:969–74
48. Perry, R. P. 1976. Processing of RNA. *Ann. Rev. Biochem.* 45:605–29
49. Kindler, P., Keil, T. U., Hofschneider, P. H. 1973. Isolation and characterization of a ribonuclease III deficient mutant of *Escherichia coli. Mol. Gen. Genet.* 126:53–59
50. Studier, F. W. 1975. Genetic mapping of a mutation that causes ribonuclease III deficiency in *Escherichia coli. J. Bacteriol.* 124:307–16
51. Apirion, D., Watson, N. 1975. Mapping and characterization of a mutation in *Escherichia coli* that reduces the level of ribonuclease III specific for double-stranded ribonucleic acid. *J. Bacteriol.* 124:317–24
52. Apirion, D., Neil, J., Watson, N. 1976. Revertants from RNase III negative mutants of *Escherichia coli. Mol. Gen. Genet.* 149:201–10
53. Apirion, D., Neil, J., Watson, N. 1976. Consequences of losing ribonuclease III on the *Escherichia coli* cell. *Mol. Gen. Genet.* 144:185–90
54. Nikolaev, N., Silengo, L., Schlessinger, D. 1973. A role for ribonculease III in processing of ribosomal ribonucleic acid and messenger ribonucleic acid precursors in *Escherichia coli. J. Biol. Chem.* 248:7967–69
55. Hayes, F., Vasseur, M., Nikolaev, N., Schlessinger, D., Sri Widada, J., Krol, A., Branlant, C. 1975. Structure of a 30 S preribosomal RNA of *E. coli. FEBS Lett.* 56:85–91
56. Gegenheimer, P., Watson, N., Apirion, D. 1977. Multiple pathways for the primary processing of ribosomal RNA in *Escherichia coli. J. Biol. Chem.* 252:3064–73
57. Robertson, H. D., Altman, S., Smith, J. D. 1972. Purification and properties of a specific *Escherichia coli* ribonuclease which cleaves a tyrosine transfer ribonucleic acid precursor. *J. Biol. Chem.* 247:5243–51
58. Schedl, P., Primakoff, P. 1973. Mutants of *Escherichia coli* thermosensitive for the synthesis of transfer RNA. *Proc. Natl. Acad. Sci. USA* 70:2091–95
59. Schedl, P., Roberts, J., Primakoff, P. 1976. In vitro processing of *E. coli* tRNA precursors. *Cell* 8:581–94
60. Bikoff, E., LaRue, B. F., Gefter, M. L. 1975. In vitro synthesis of transfer RNA. II. Identification of required enzymatic activities. *J. Biol. Chem.* 250:6248–55
61. Seidman, J. G., Schmidt, F. J., Foss, K., McClain, W. H. 1975. A mutant of *Escherichia coli* defective in removing 3' terminal nucleotides from some transfer RNA precursor molecules. *Cell* 5:389–400
62. Kitamura, N., Ikeda, H., Yamada, Y., Ishikura, H. 1977. Processing by ribonuclease II of the tRNA^{Tyr} precursor of *Escherichia coli* synthesized in vitro. *Eur. J. Biochem.* 73:297–306
63. Sakano, H., Yamada, S., Ikemura, T., Shimura, Y., Ozeki, H. 1974. Temperature sensitive mutants of *Escherichia coli* for tRNA synthesis. *Nucleic Acids Res.* 1:355–71
64. Altman, S. 1975. Biosynthesis of transfer RNA in *Escherichia coli. Cell* 4:21–29
65. Smith, J. D. 1976. Transcription and processing of transfer RNA and precursors. *Prog. Nucleic Acid Res. Mol. Biol.* 16:25–73
66. Ikemura, T., Shimura, Y., Sakano, H., Ozeki, H. 1975. Precursor molecules of *Escherichia coli* transfer RNAs accumulated in a temperature-sensitive mutant. *J. Mol. Biol.* 96:69–86
67. Yuki, A. 1971. Tentative identification of a "maturation enzyme" for precursor

16 s ribosomal RNA in *Escherichia coli*. *J. Mol. Biol.* 62:321–29
68. Corte, G., Schlessinger, D., Longo, D., Venkov, P. 1971. Transformation of 17 s to 16 s ribosomal RNA using ribonuclease II of *Escherichia coli*. *J. Mol. Biol.* 60:325–38
69. Weatherford, S. C., Rosen, L., Gorelic, L., Apirion, D. 1972. *Escherichia coli* strains with thermolabile ribonuclease II activity. *J. Biol. Chem.* 247:5404–8
70. Hayes, F., Vasseur, M. 1976. Processing of the 17-S *Escherichia coli* precursor RNA in the 27-S pre-ribosomal particle. *Eur. J. Biochem.* 61:433–42
71. Dahlberg, A. E., Tokimatsu, H., Zahalak, M., Reynolds, F., Calvert, P., Rabson, A. B., Dahlberg, J. E. 1977. Processing of the 17S precursor ribosomal RNA. In *Nucleic Acid-Protein Recognition*, ed. H. J. Vogel, pp. 509–17. New York: Academic. 587 pp.
72. Meyhack, B., Meyhack, I., Apirion, D. 1974. Processing of precursor particles containing 17S rRNA in a cell free system. *FEBS Lett.* 49:215–19
73. Sogin, M. L., Pace, N. R., Rosenberg, M., Weissman, S. M. 1976. Nucleotide sequence of a 5 S ribosomal RNA precursor from *Bacillus subtilis*. *J. Biol. Chem.* 251:3480–88
74. Sogin, M. L., Pace, B., Pace, N. R. 1977. Partial purification and properties of a ribosomal RNA maturation endonuclease from *Bacillus subtilis*. *J. Biol. Chem.* 252:1350–58
75. Bachmann, B. J., Low, K. B., Taylor, A. L. 1976. Recalibrated linkage map of *Escherichia coli* K-12. *Bacteriol. Rev.* 40:116–67
76. Bernadi, A., Leder, P. 1970. Protein biosynthesis in *Escherichia coli*. Purification and characteristics of a mutant G factor. *J. Biol. Chem.* 245:4263–68
77. Kuwano, M., Schlessinger, D., Rinaldi, G., Felicetti, L., Tocchini-Valentini, G. P. 1971. G factor mutants of *Escherichia coli:* Map location and properties. *Biochem. Biophys. Res. Commun.* 42:441–44
78. Tanaka, N., Kawano, G., Kinoshita, T. 1971. Chromosomal location of a fusidic acid resistant marker in *Escherichia coli*. *Biochem. Biophys. Res. Commun.* 42:564–67
79. Sypherd, P. S., Osawa, S. 1974. Ribosome genetics revealed by hybrid bacteria. See Ref. 2, pp. 669–78
80. Jaskunas, S. R., Lindahl, L., Nomura, M. 1975. Specialized transducing phages for ribosomal protein genes of *Escherichia coli*. *Proc. Natl. Acad. Sci. USA* 72:6–10
81. Jaskunas, S. R., Lindahl, L., Nomura, M. 1975. Isolation of polar insertion mutations and the direction of transcription of ribosomal protein genes in *E. coli*. *Nature* 256:183–87
82. Jaskunas, S. R., Lindahl, L., Nomura, M., Burgess, R. R. 1975. Identification of two copies of the gene for the elongation factor EF-Tu in *E. coli*. *Nature* 257:458–62
83. Jaskunas, S. R., Burgess, R. R., Nomura, M. 1975. Identification of a gene for the α subunit of RNA polymerase at the *str-spc* region of the *Escherichia coli* chromosome. *Proc. Natl. Acad. Sci. USA* 72:5036–40
84. Jaskunas, S. R., Burgess, R. R., Lindahl, L., Nomura, M. 1976. Two clusters of genes for RNA polymerase and ribosomal components in *Escherichia coli*. See Ref. 8, pp. 539–52
85. Fiandt, M., Szybalski, W., Blattner, F. R., Jaskunas, S. R., Lindahl, L., Nomura, M. 1976. Organization of ribosomal protein genes in *Escherichia coli*. I. Physical structure of DNA from transducing λ phages carrying genes from the *aroE-str* region. *J. Mol. Biol.* 106:817–35
86. Lindahl, L., Zengel, J., Nomura, M. 1976. Organization of ribosomal protein genes in *Escherichia coli*. II. Mapping of ribosomal protein genes by in vitro synthesis of ribosomal proteins using DNA fragments of a transducing phage as templates. *J. Mol. Biol.* 106:837–55
87. Lindahl, L., Post, L., Nomura, M. 1976. DNA-dependent in vitro synthesis of ribosomal proteins, protein elongation factors, and RNA polymerase subunit α: Inhibition by ppGpp. *Cell* 9:439–48
88. Jaskunas, S. R., Fallon, A. M., Nomura, M. 1977. Identification and organization of ribosomal protein genes of *Escherichia coli* carried by λ*fus*2 transducing phage. *J. Biol. Chem.* In press
89. Jaskunas, S. R., Nomura, M. 1977. Organization of ribosomal protein genes of *Escherichia coli* as analyzed by polar insertion mutations. *J. Biol. Chem.* In press
90. Williams, B. G., Blattner, F. R., Jaskunas, S. R., Nomura, M. 1977. Insertion of DNA carrying ribosomal protein genes of *Escherichia coli* into

Charon vector phages. *J. Biol. Chem.* In press
91. Jaskunas, S. R., Fallon, A. M., Nomura, M., Williams, B. G., Blattner, F. R. 1977. Expression of ribosomal protein genes cloned in Charon vector phages and identification of their promoters. *J. Biol. Chem.* In press
92. Lindahl, L., Post, L., Zengel, J., Gilbert, S. F., Strycharz, W. A., Nomura, M. 1977. Mapping of ribosomal protein genes by in vitro protein synthesis using DNA fragments of $\lambda fus3$ transducing phage DNA as templates. *J. Biol. Chem.* In press
93. Brown, M. E., Apirion, D. 1974. Mapping a cluster of ribosomal genes in *Escherichia coli. Mol. Gen. Genet.* 133:317–27
94. Hitz, H., Schäfer, D., Wittmann-Liebold, B. 1975. Primary structure of ribosomal protein S6 from the wild type and a mutant of *Escherichia coli. FEBS Lett.* 56:259–62
95. Nashimoto, H., Tsugawa, A., Uchida, H. 1975. Isolation of temperature-sensitive mutations in the *aroE-strA* region of the *Escherichia coli* chromosome. *Proc. 1975 Mol. Biol. Meet. Japan,* p. 12
96. Sunshine, M. G., Sauer, B. 1975. A bacterial mutation blocking P2 phage late gene expression. *Proc. Natl. Acad. Sci. USA* 72:2770–74
97. Fujiki, H., Palm, P., Zillig, W., Calendar, R., Sunshine, M. 1976. Identification of a mutation within the structural gene for the α subunit of DNA-dependent RNA polymerase of *E. coli. Mol. Gen. Genet.* 145:19–22
98. Kimura, A., Muto, A., Osawa, S. 1974. Control of stable RNA synthesis in a temperature-sensitive mutant of elongation factor G of *Bacillus subtilis. Mol. Gen. Genet.* 130:203–14
99. Dubnau, E., Pifko, S., Sloma, A., Cabane, K., Smith, I. 1976. Conditional mutations in the translational apparatus of *Bacillus subtilis. Mol. Gen. Genet.* 147:1–12
100. Chow, L. T., Davidson, N. 1973. Electron microscope mapping of the distribution of ribosomal genes of the *Bacillus subtilis* chromosome. *J. Mol. Biol.* 75:265–79
101. Harford, N., Sueoka, N. 1970. Chromosomal location of antibiotic resistance markers in *Bacillus subtilis. J. Mol. Biol.* 51:267–86
102. Haworth, S. R., Brown, L. R. 1973. Genetic analysis of ribonucleic acid polymerase mutants of *Bacillus subtilis. J. Bacteriol.* 114:103–13
103. Sonenshein, A. L., Cami, B., Brevet, J., Cote, R. 1974. Isolation and characterization of rifampicin-resistant and streptolydigin-resistant mutants of *Bacillus subtilis* with altered sporulation properties. *J. Bacteriol.* 120:253–65
104. Held, W. A., Ballou, B., Mizushima, S., Nomura, M. 1974. Assembly mapping of 30 S ribosomal proteins from *Escherichia coli:* Further studies. *J. Biol. Chem.* 249:3103–11
105. Nomura, M., Held, W. A. 1974. Reconstitution of ribosomes: Studies of ribosome structure, function and assembly. See Ref. 2, pp. 193–223
106. Wittmann, H. G., Yaguchi, M., Piepersberg, W., Böck, A. 1975. Direction of transcription of two ribosomal protein genes in *Escherichia coli. J. Mol. Biol.* 98:827–29
107. Piepersberg, W., Böck, A., Yaguchi, M., Wittmann, H. G. 1975. Genetic position and amino acid replacements of several mutations in ribosomal protein S5 from *Escherichia coli. Mol. Gen. Genet.* 143:43–52
108. Saedler, H., Kubai, D. F., Nomura, M., Jaskunas, S. R. 1975. IS1 and IS2 mutations in the ribosomal protein genes of *E. coli* K-12. *Mol. Gen. Genet.* 141:85–89
109. Starlinger, P., Saedler, H. 1977. IS-elements in microorganisms. In *Current Topics in Microbiology and Immunology.* In press
110. Nomura, M., Engbaek, F. 1972. Expression of ribosomal protein genes as analyzed by bacteriophage Mu-induced mutations. *Proc. Natl. Acad. Sci. USA* 69:1526–30
111. Howe, M. M., Bade, E. G. 1975. Molecular biology of bacteriophage Mu. *Science* 190:624–32
112. Cabezón, T., Faelen, M., De Wilde, M., Bollen, A., Thomas, R. 1975. Expression of ribosomal protein genes in *Escherichia coli. Mol. Gen. Genet.* 137:125–29
113. Dennis, P. P. 1974. Synthesis of individual ribosomal proteins in *Escherichia coli B/r. J. Mol. Biol.* 89:223–32
114. Iwakura, Y., Ito, K., Ishihama, A. 1974. Biosynthesis of RNA polymerase in *Escherichia coli.* I. Control of RNA polymerase content at various growth rates. *Mol. Gen. Genet.* 133:1–23
115. Friesen, J. D., Fiil, N. P., Parker, J. M., Haseltine, W. A. 1974. A new relaxed mutant of *Escherichia coli* with an al-

tered 50S ribosomal subunit. *Proc. Natl. Acad. Sci. USA* 71:3465-69
116. Parker, J., Watson, R. J., Friesen, J. D., Fiil, N. P. 1976. A relaxed mutant with an altered ribosomal protein L11. *Mol. Gen. Genet.* 144:111-14
117. Kirschbaum, J. B., Scaife, J. 1974. Evidence for a λ transducing phage carrying the genes for the β and β' subunits of *Escherichia coli* RNA polymerase. *Mol. Gen. Genet.* 132:193-201
118. Watson, R. J., Parker, J., Fiil, N. P., Flaks, J. G., Friesen, J. D. 1975. New chromosomal location for structural genes of ribosomal proteins. *Proc. Natl. Acad. Sci. USA* 72:2765-69
119. Terhorst, C., Möller, W., Laursen, R., Wittmann-Liebold, B. 1972. Amino acid sequence of a 50S ribosomal protein involved in both EFG and EFT dependent GTP-hydrolysis. *FEBS Lett.* 28:325-28
120. Terhorst, C., Möller, W., Laursen, R., Wittmann-Liebold, B. 1973. The primary structure of an acidic protein from 50-S ribosomes of *Escherichia coli* which is involved in GTP hydrolysis dependent on elongation factors G and T. *Eur. J. Biochem.* 34:138-52
121. Chu, F., Kung, H.-F., Caldwell, P., Weissbach, H., Brot, N. 1976. DNA dependent synthesis of protein L12 from *Escherichia coli* ribosomes, in vitro. *Proc. Natl. Acad. Sci. USA* 73:3156-59
122. Pettersson, I., Hardy, S. J. S., Liljas, A. 1976. The ribosomal protein L8 is a complex of L7/L12 and L10. *FEBS Lett.* 64:135-38
123. Errington, L., Glass, R. E., Hayward, R. S., Scaife, J. G. 1974. Structure and orientation of an RNA polymerase operon in *Escherichia coli. Nature* 249:519-22
124. Lindahl, L., Nomura, M. 1976. Analysis of organization of *Escherichia coli* ribosome and RNA polymerase genes using biochemical methods. See Ref. 22, pp. 206-17
125. Claeys, I. V., Miller, J. H., Kirschbaum, J. B., Nasi, S., Molholt, B., Gross, G., Fields, D. A., Bautz, E. K. F. 1976. Altered RNA polymerases resulting from temperature sensitive mutations in the *rif* region of the *E. coli* chromosome. See Ref. 22, pp. 56-65
126. Highland, J. H., Howard, G. A. 1975. Assembly of ribosomal proteins L7, L10, L11, and L12 on the 50 S subunit of *Escherichia coli. J. Biol. Chem.* 250:831-34
127. Bollen, A., Faelen, M., Lecocq, J. P., Herzog, A., Zengel, J., Kahan, L., Nomura, M. 1973. The structural gene for the ribosomal protein S18 in *Escherichia coli.* I. Genetic studies on a mutant having an alteration in the protein S18. *J. Mol. Biol.* 76:463-72
128. Kahan, L., Zengel, J., Nomura, M., Bollen, A., Herzog, A. 1973. The structural gene for the ribosomal protein S18 in *Escherichia coli.* II. Chemical studies on the protein S18 having an altered electrophoretic mobility. *J. Mol. Biol.* 76:473-83
129. De Wilde, M., Michel, F., Broman, K. 1974. The structural gene for ribosomal protein S18 in *Escherichia coli.* III. Mapping outside the ribosomal protein gene cluster at minute 84 on the genome. *Mol. Gen. Genet.* 133:329-33
130. Nashimoto, H., Uchida, H. 1975. Late steps in the assembly of 30 S ribosomal proteins in vivo in a spectinomycin-resistant mutant of *Escherichia coli. J. Mol. Biol.* 96:443-53
131. Takata, R. 1976. Genetic studies of the ribosomal proteins in *Escherichia coli.* IX. Mapping of the ribosomal proteins, S2 and S20, by intergeneric mating experiments between *Serratia marcescens* and *Escherichia coli* K12. *Mol. Gen. Genet.* 146:233-38
132. Yamamoto, M., Strycharz, W. A., Nomura, M. 1976. Identification of genes for elongation factor Ts and ribosomal protein S2 in *E. coli. Cell* 8:129-38
133. Friesen, J. D., Parker, J., Watson, R. J., Bendiak, D., Reeh, S. V., Pedersen, S., Fiil, N. P. 1976. A transducing bacteriophage λ carrying the structural gene for elongation factor Ts. *Mol. Gen. Genet.* 148:93-98
134. Nakamura, Y., Osawa, T., Yura, T. 1977. Chromosomal location of a structural gene for the RNA polymerase sigma factor in *Escherichia coli. Proc. Natl. Acad. Sci. USA* 74:1831-35
135. Harris, J. D., Martinez, I. I., Calendar, R. 1977. A gene from *Escherichia coli* affecting the sigma subunit of RNA polymerase. *Proc. Natl. Acad. Sci. USA* 74:1836-40
136. Buckel, P., Ruffler, D., Piepersberg, W., Böck, A. 1972. RNA overproducing revertants of an alanyl-tRNA synthetase mutant of *Escherichia coli. Mol. Gen. Genet.* 119:323-35
137. Wittmann, H. G., Stöffler, G., Piepersberg, W., Buckel, P., Ruffler, D., Böck, A. 1974. Altered S5 and S20 ribosomal

proteins in revertants of an alanyl-tRNA synthetase mutant of *Escherichia coli. Mol. Gen. Genet.* 134:225–36
138. Ruffler, D., Buckel, P., Piepersberg, W., Böck, A. 1974. Alanyl-tRNA synthetase of *Escherichia coli:* Genetic analysis of the structural gene and of suppressor mutations. *Mol. Gen. Genet.* 134:313–23
139. Böck, A., Ruffler, D., Piepersberg, W., Wittmann, H. G. 1974. Genetic analysis of an alteration of ribosomal protein S20 in revertants of an alanyl-tRNA-synthetase mutant of *Escherichia coli. Mol. Gen. Genet.* 134:325–32
140. Wittmann, H. G., Stöffler, G., Geyl, D., Böck, A. 1975. Alteration of ribosomal proteins in revertants of a valyl-tRNA synthetase mutant of *Escherichia coli. Mol. Gen. Genet.* 141:317–29
141. Buckel, P., Piepersberg, W., Böck, A. 1976. Suppression of temperature-sensitive aminoacyl-tRNA synthetase mutations by ribosomal mutations: A possible mechanism. *Mol. Gen. Genet.* 149:51–61
142. Friesen, J. D., Parker, S., Watson, R. J., Fiil, N. P., Pedersen, S. 1976. Isolation of a transducing phage carrying *rpsT,* the structural gene for ribosomal protein S20. *Mol. Gen. Genet.* 144:115–18
143. Buckel, P. 1976. Identity of a gene responsible for suppression of aminoacyl-tRNA synthetase mutations with *rpsT,* the structural gene for ribosomal protein S20. *Mol. Gen. Genet.* 149:225–28
144. Osawa, S. 1976. Gene locus of a 30s ribosomal protein S20 of *Bacillus subtilis. Mol. Gen. Genet.* 144:49–51
145. Takata, R., Kobata, K. 1976. Genetic studies of the ribosomal proteins in *Escherichia coli.* X. Mapping of the ribosomal proteins, L21 and S15, by intergeneric mating experiments between *Serratia marcescens* and *Escherichia coli* K12. *Mol. Gen. Genet.* 149:159–65
146. Isono, K., Kitakawa, M. 1977. A new ribosomal protein locus in *Escherichia coli:* The gene for protein S6 maps at 97 min. *Mol. Gen. Genet.* 153:115–20
147. Isono, K., Krauss, J., Hirota, Y. 1976. Isolation and characterization of temperature-sensitive mutants of *Escherichia coli* with altered ribosomal proteins. *Mol. Gen. Genet.* 149:297–302
148. Isono, K., Cumberlidge, A. G., Isono, S., Hirota, Y. 1977. Further temperature-sensitive mutants of *Escherichia coli* with altered ribosomal proteins. *Mol. Gen. Genet.* 152:239–43

149. Kaltschmidt, E., Wittmann, H. G. 1970. Ribosomal proteins. VII. Two-dimensional polyacrylamide gel electrophoresis for fingerprinting of ribosomal proteins. *Anal. Biochem.* 36:401–12
150. Traub, P., Nomura, M. 1968. Streptomycin resistance mutation in *Escherichia coli:* Altered ribosomal protein. *Science* 160:198–99
151. Ozaki, M., Mizushima, S., Nomura, M. 1969. Identification and functional characterization of the protein controlled by the streptomycin-resistant locus in *E. coli. Nature* 222:333–39
152. Birge, E. A., Kurland, C. G. 1969. Altered ribosomal protein in streptomycin-dependent *Escherichia coli. Science* 166:1282–84
153. Wittman, H. G., Wittmann-Liebold, B. 1974. Chemical structure of bacterial ribosomal proteins. See Ref. 2, pp. 115–40
154. Breckenridge, L., Gorini, L. 1970. Genetic analysis of streptomycin resistance in *Escherichia coli. Genetics* 65:9–25
155. Gorini, L. 1971. Ribosomal discrimination of tRNAs. *Nature New Biol.* 234:261–64
156. Funatsu, G., Wittmann, H. G. 1972. Ribosomal proteins. XXXIII. Location of amino-acid replacements in protein S12 isolated from *Escherichia coli* mutants resistant to streptomycin. *J. Mol. Biol.* 68:547–50
157. Itoh, T., Wittmann, H. G. 1973. Amino acid replacements in proteins S5 and S12 of two *Escherichia coli* revertants from streptomycin dependence to independence. *Mol. Gen. Genet.* 127:19–32
158. van Acken, U. 1975. Proteinchemical studies on ribosomal proteins S4 and S12 from *ram* (ribosomal ambiguity) mutants of *Escherichia coli. Mol. Gen. Genet.* 140:61–68
159. Yaguchi, M., Wittmann, H. G., Cabezón, T., De Wilde, M., Villarroel, R., Herzog, A., Bollen, A. 1975. Cooperative control of translational fidelity by ribosomal proteins in *Escherichia coli.* II. Localization of amino acid replacements in proteins S5 and S12 altered in double mutants resistant to neamine. *Mol. Gen. Genet.* 142:35–43
160. Chakrabarti, S., Gorini, L. 1975. Growth of bacteriophages MS2 and T7 on streptomycin-resistant mutants of *Escherichia coli. J. Bacteriol.* 121:670–74
161. Bollen, A., Davies, J., Ozaki, M., Mizushima, S. 1969. Ribosomal protein conferring sensitivity to the antibiotic spec-

tinomycin in *Escherichia coli. Science* 165:85–86
162. Dekio, S., Takata, R. 1969. Genetic studies of the ribosomal proteins in *Escherichia coli.* II. Altered 30s ribosomal protein component specific to spectinomycin resistant mutants. *Mol. Gen. Genet.* 105:219–24
163. Bollen, A., Herzog, A. 1970. The ribosomal protein altered in spectinomycin resistant *Escherichia coli. FEBS Lett.* 6:69–72
164. Nashimoto, H., Nomura, M. 1970. Structure and function of bacterial ribosomes. XI. Dependence of 50S ribosomal assembly on simultaneous assembly of 30S subunits. *Proc. Natl. Acad. Sci. USA* 67:1440–47
165. Miyoshi, Y., Yamagata, H. 1976. Sucrose-dependent spectinomycin-resistant mutants of *Escherichia coli. J. Bacteriol.* 125:142–48
166. Kreider, G., Brownstein, B. L. 1972. Ribosomal proteins involved in the suppression of streptomycin dependence in *Escherichia coli. J. Bacteriol.* 109:780–85
167. Hasenbank, R., Guthrie, C., Stöffler, G., Wittmann, H. G., Rosen, L., Apirion, D. 1973. Electrophoretic and immunological studies on ribosomal proteins of 100 *Escherichia coli* revertants from streptomycin dependence. *Mol. Gen. Genet.* 127:1–18
168. Funatsu, G., Schiltz, E., Wittmann, H. G. 1971. Ribosomal proteins. XXVII. Localization of the amino acid exchanges in protein S5 from two *Escherichia coli* mutants resistant to spectinomycin. *Mol. Gen. Genet.* 114:106–11
169. Funatsu, G., Nierhaus, K., Wittmann-Liebold, B. 1972. Ribosomal proteins. XXII. Studies on the altered protein S5 from a spectinomycin-resistant mutant of *Escherichia coli. J. Mol. Biol.* 64:201–9
170. De Wilde, M., Wittmann-Liebold, B. 1973. Localization of the amino-acid exchange in protein S5 from an *Escherichia coli* mutant resistant to spectinomycin. *Mol. Gen. Genet.* 127:273–76
171. Dabbs, E. R. 1977. A spectinomycin dependent mutant of *Escherichia coli. Mol. Gen. Genet.* 151:261–67
172. Wittmann, H. G., Stöffler, G., Apirion, D., Rosen, L., Tanaka, K., Tamaki, M., Takata, R., Dekio, S., Otaka, E., Osawa, S. 1973. Biochemical and genetic studies on two different types of erythromycin resistant mutants of *Escherichia coli* with altered ribosomal proteins. *Mol. Gen. Genet.* 127:175–89
173. Dekio, S., Takata, R., Osawa, S., Tanaka, K., Tamaki, M. 1970. Genetic studies of the ribosomal proteins in *Escherichia coli.* IV. Pattern of the alteration of ribosomal components in mutants resistant to spectinomycin and erythromycin in different strains of *Escherichia coli. Mol. Gen. Genet.* 107:39–49
174. Otaka, E., Teraoka, H., Tamaki, M., Tanaka, K., Osawa, S. 1970. Ribosomes from erythromycin-resistant mutants of *Escherichia coli* Q13. *J. Mol. Biol.* 48:499–510
175. Pardo, D., Rosset, R. 1974. Genetic studies of erythromycin resistant mutants of *Escherichia coli. Mol. Gen. Genet.* 135:257–68
176. Pardo, D., Rosset, R. 1977. A new ribosomal mutation which affects the two ribosomal subunits in *Escherichia coli. Mol. Gen. Genet.* 153:199–204
177. Cannon, M., Cabezón, T., Bollen, A. 1974. Mapping of neamine resistance: Identification of two genetic loci, *nea* A and *nea* B. *Mol. Gen. Genet.* 130:321–26
178. Bollen, A., Cabezón, T., De Wilde, M., Villarroel, R., Herzog, A. 1975. Alteration of ribosomal protein S17 by mutation linked to neamine resistance in *Escherichia coli.* I. General properties of *nea*A mutants. *J. Mol. Biol.* 99:795–806
179. Yaguchi, M., Wittmann, H. G., Cabezón, T., De Wilde, M., Villarroel, R., Herzog, A., Bollen, A. 1976. Alteration of ribosomal protein S17 by mutation linked to neamine resistance in *Escherichia coli.* II. Localization of the amino acid replacement in protein S17 from a *nea*A mutant. *J. Mol. Biol.* 104:617–20
180. De Wilde, M., Cabezón, T., Villarroel, R., Herzog, A., Bollen, A. 1975. Cooperative control of translational fidelity by ribosomal proteins in *Escherichia coli.* I. Properties of ribosomal mutants whose resistance to neamine is the cumulative effect of two distinct mutations. *Mol. Gen. Genet.* 142:19–33
181. Apirion, D., Schlessinger, D. 1968. Coresistance to neamycin and kanamycin by mutations in an *Escherichia coli* locus that affects ribosomes. *J. Bacteriol.* 96:768–76
182. Deusser, E., Stöffler, G., Wittmann, H. G., Apirion, D. 1970. Ribosomal proteins. XVI. Altered S4 proteins in *Es-

cherichia coli revertants from streptomycin dependence to independence. *Mol. Gen. Genet.* 109:298–302
183. Birge, E. A., Kurland, C. G. 1970. Reversion of a streptomycin-dependent strain of *Escherichia coli*. *Mol. Gen. Genet.* 109:356–69
184. Wittmann, H. G., Apirion, D. 1975. Analysis of ribosomal proteins in streptomycin resistant and dependent mutants isolated from streptomycin independent *Escherichia coli* strains. *Mol. Gen. Genet.* 141:331–41
185. Dabbs, E. R., Wittmann, H. G. 1976. A strain of *E. coli* which gives rise to mutations in a large number of ribosomal proteins. *Mol. Gen. Genet.* 149:303–9
186. Geyl, D., Böck, A., Wittmann, H. G. 1977. Cold-sensitive growth of a mutant of *Escherichia coli* with an altered ribosomal protein S8: Analysis of revertants. *Mol. Gen. Genet.* 152:259–66
187. Berger, I., Geyl, D., Böck, A., Stöffler, G., Wittmann, H. G. 1975. Localized mutagenesis of the *aroE-strA* section of the *Escherichia coli* chromosome coding for ribosomal proteins. *Mol. Gen. Genet.* 141:207–11
188. Kushner, S. R., Maples, V. F., Champney, W. S. 1977. Conditionally lethal ribosomal protein mutants: Characterization of a locus required for modification of 50S subunit proteins. *Proc. Natl. Acad. Sci. USA* 74:467–71
189. Nashimoto, H., Tsugawa, A., Uchida, H. 1976. Temperature-sensitive mutations in the *aroE-strA* region of the *Escherichia coli* chromosome affecting the synthesis of ribonucleic acid. *Proc. 1976 Mol. Biol. Meet. Jpn.*, p. 62
190. Cabezón, T., Delcuve, G., Herzog, A., Petre, J., Bollen, A. 1977. Characterization of conditionally lethal mutations affecting ribosomal proteins in *Escherichia coli*. *Biochem. Soc. Trans.* In press
191. Molholt, B. 1976. Extragenic suppression of two ribosomal protein cistrons lying near the *rif* locus in *Escherichia coli*. *J. Bacteriol.* 126:563–67
192. Flaks, J. G., Leboy, P. S., Birge, E. A., Kurland, C. G. 1966. Mutations and genetics concerned with the ribosomes. *Cold Spring Harbor Symp. Quant. Biol.* 31:623–31
193. Thammana, P., Kurland, C. G., Deusser, E., Weber, J., Maschler, R., Stöffler, G., Wittmann, H. G. 1973. Structural and functional evidence for a repeated 50S subunit ribosomal protein. *Nature New Biol.* 242:47–49

194. Hardy, S. J. S. 1975. The stoichiometry of the ribosomal proteins of *Escherichia coli*. *Mol. Gen. Genet.* 140:253–74
195. Ohsawa, H., Maruo, B. 1976. Restoration by ribosomal protein S1 of the defective translation in a temperature-sensitive mutant of *Escherichia coli* K-12: Characterization and genetic studies. *J. Bacteriol.* 127:1157–66
196. Guthrie, C., Nashimoto, H., Nomura, M. 1969. Structure and function of *E. coli* ribosomes. VIII. Cold-sensitive mutants defective in ribosome assembly. *Proc. Natl. Acad. Sci. USA* 63:384–91
197. Guthrie, C., Nashimoto, H., Nomura, M. 1969. Studies on the assembly of ribosomes in vivo. *Cold Spring Harbor Symp. Quant. Biol.* 34:69–75
198. Bryant, R. E., Sypherd, P. S. 1974. Genetic analysis of cold-sensitive ribosome maturation mutants of *Escherichia coli*. *J. Bacteriol.* 117:1082–92
199. Berman, D., Budzilowicz, C., Chang, F. N. 1973. Accumulation of different precursor ribonucleoprotein particles by various cold sensitive, antibiotic resistant mutants. *Biochem. Biophys. Res. Commun.* 54:991–97
200. Tai, P.-C., Kessler, D. P., Ingraham, J. 1969. Cold-sensitive mutations in *Salmonella typhimurium* which affect ribosome synthesis. *J. Bacteriol.* 97:1298–1304
201. Tyler, B., Ingraham, J. L. 1973. Studies on ribosomal mutants of *Salmonella typhimurium* LT-2. *Mol. Gen. Genet.* 122:197–214
202. Lewandowski, L. J., Brownstein, B. L. 1969. Characterization of a 43 S ribonucleoprotein component of a mutant of *Escherichia coli*. *J. Mol. Biol.* 41:277–90
203. Kreider, G., Brownstein, B. L. 1971. A mutation suppressing streptomycin dependence. II. An altered protein on the 30 S ribosomal subunit. *J. Mol. Biol.* 61:135–42
204. Bryant, R. E., Fujisawa, T., Sypherd, P. S. 1974. Ribosomal proteins and ribonucleic acids of ribosome maturation mutants of *Escherichia coli*. *Biochemistry* 13:2110–14
205. Rosset, R., Vola, C., Feunteun, J., Monier, R. 1971. A thermosensitive mutant defective in ribosomal 30S subunit assembly. *FEBS Lett.* 18:127–29
206. Zimmermann, R. A. 1974. RNA-protein interactions in the ribosome. See Ref. 2, pp. 225–69
207. Gorini, L. 1974. Streptomycin and mis-

208. Ninio, J. 1974. A semi-quantitative treatment of missense and nonsense suppression in the *strA* and *ram* ribosomal mutants of *Escherichia coli*; Evaluation of some molecular parameters of translation in vivo. *J. Mol. Biol.* 84:297–313
209. Piepersberg, W., Böck, A., Wittmann, H. G. 1975. Effect of different mutations in ribosomal protein S5 of *Escherichia coli* on translational fidelity. *Mol. Gen. Genet.* 140:91–100
210. Cabezón, T., Herzog, A., De Wilde, M., Villarroel, R., Bollen, A. 1976. Cooperative control of translational fidelity by ribosomal proteins in *Escherichia coli*. III. A *ram* mutation in the structural gene for protein S5 (*rpx* E). *Mol. Gen. Genet.* 144:59–62
211. Topisirovic, L., Villarroel, R., De Wilde, M., Herzog, A., Cabezón, T., Bollen, A. 1977. Translational fidelity in *Escherichia coli*: Contrasting role of *nea*A and *ram*A gene products in the ribosome functioning. *Mol. Gen. Genet.* 151:17–26
212. Yates, J. L., Gette, W. R., Furth, M. E., Nomura, M. 1977. Effects of ribosomal mutations on the read-through of a chain termination signal: Studies on the synthesis of bacteriophage λ *O* gene protein in vitro. *Proc. Natl. Acad. Sci. USA* 74:689–93
213. Gette, W. R. 1976. *Studies on the cistron selectivity of the 30 S ribosomal subunit of Escherichia coli in the translation of natural messenger RNA.* PhD dissertation. Univ. of Wisconsin, Madison
214. Zengel, J. M., Young, R., Dennis, P. P., Nomura, M. 1977. Role of ribosomal protein S12 in peptide chain elongation; Analysis of pleiotropic streptomycin-resistant mutants of *Escherichia coli*. *J. Bacteriol.* 129:1320–29
215. Couturier, M., Desmet, L., Thomas, R. 1964. High pleiotrophy of streptomycin mutations in *Escherichia coli*. *Biochem. Biophys. Res. Commun.* 16:244–48
216. Simonian, M. H., Mosteller, R. D. 1976. Increased loss of duplicated genes in streptomycin-resistant (*strA*) mutants of *Escherichia coli* K-12. *J. Bacteriol.* 125:382–84
217. Hofstetter, H., Monstein, H.-J., Weissmann, C. 1974. The readthrough protein A_1 is essential for the formation of viable Qβ particles. *Biochim. Biophys. Acta.* 374:238–51
218. Engelberg-Kulka, H., Dekel, L., Israeli-Reches, M. 1977. Streptomycin-resistant *Escherichia coli* mutant temperature sensitive for the production of Qβ-infective particles. *J. Virol.* 21:1–6
219. Held, W. A., Gette, W. R., Nomura, M. 1974. Role of 16S ribosomal ribonucleic acid and the 30S ribosomal protein S12 in the initiation of natural messenger ribonucleic acid translation. *Biochemistry* 13:2115–22
220. Hyman, R. W., Brunovskis, I., Summers, W. C. 1974. A biochemical comparison of the related bacteriophages T7, φI, φII, W31, H, and T3. *Virology* 57:189–206
221. Chakrabarti, S. L., Gorini, L. 1975. A link between streptomycin and rifampicin mutation. *Proc. Natl. Acad. Sci. USA* 72:2084–87
222. Morrison, T. G., Malamy, M. H. 1971. T7 translational control mechanisms and their inhibition by F factors. *Nature New Biol.* 231:37–41
223. Britton, J. R., Haselkorn, R. 1975. Permeability lesions in male *Escherichia coli* infected with bacteriophage T7. *Proc. Natl. Acad. Sci. USA* 72:2222–26
224. Condit, R. C. 1975. F factor-mediated inhibition of bacteriophage T7 growth: Increased membrane permeability and decreased ATP levels following T7 infection of male *Escherichia coli*. *J. Mol. Biol.* 98:45–56
225. Chakrabarti, S. L., Gorini, L. 1977. Interaction between mutations of ribosomes and RNA polymerase: A pair of *strA* and *rif* mutants individually temperature-insensitive but temperature-sensitive in combination. *Proc. Natl. Acad. Sci. USA* 74:1157–61
226. Jacob, F., Brenner, S., Cuzin, F. 1963. On the regulation of DNA replication in bacteria. *Cold Spring Harbor Symp. Quant. Biol.* 28:329–48
227. Hirota, Y., Ryter, A., Jacob, F. 1968. Thermosensitive mutants of *E. coli* affected in the processes of DNA synthesis and cellular division. *Cold Spring Harbor Symp. Quant. Biol.* 33:677–93
228. Yamagata, H., Uchida, H. 1972. Spectinomycin resistance mutations affecting the stability of sex-factors in *Escherichia coli*. *J. Mol. Biol.* 67:533–35
229. Yamagata, H., Uchida, H. 1972. Chromosomal mutations affecting the stability of sex-factors in *Escherichia coli*. *J. Mol. Biol.* 63:281–94
230. Mizuno, T., Yamada, H., Yamagata, H., Mizushima, S. 1976. Coordinated alteration in ribosomes and cytoplasmic

(reading of the genetic code. See Ref. 2, pp. 791–803)

membrane in sucrose-dependent, spectinomycin-resistant mutants of *Escherichia coli*. *J. Bacteriol.* 125:524–30
231. Tocchini-Valentini, G. P., Mattoccia, E. 1968. A mutant of *E. coli* with an altered supernatant factor. *Proc. Natl. Acad. Sci. USA* 61:146–51
232. Kuwano, M., Endo, H., Yamamoto, M. 1972. Temperature-sensitive mutation in regulation of ribonucleic acid synthesis in *Escherichia coli*. *J. Bacteriol.* 112:1150–56
233. Kuwano, M., Ono, M., Yamamoto, M., Endo, H., Kamiya, T., Hori, K. 1973. Elongation factor T altered in a temperature-sensitive *Escherichia coli* mutant. *Nature New Biol.* 244:107–9
234. Lupker, J. H., Verschoor, G. J., De Rooij, F. W. M., Rörsch, A., Bosch, L. 1974. An *Escherichia coli* mutant with an altered elongation factor Tu. *Proc. Natl. Acad. Sci. USA* 71:460–63
235. Kuwano, M., Endo, H., Kamiya, T., Hori, K. 1974. A mutant of *Escherichia coli* blocked in peptide elongation: Altered elongation factor Ts. *J. Mol. Biol.* 86:689–98
236. Hori, K., Harada, K., Kuwano, M. 1974. Function of bacteriophage Qβ replicase containing an altered subunit IV. *J. Mol. Biol.* 86:699–708
237. Pedersen, S., Blumenthal, R. M., Reeh, S., Parker, J., Lemaux, P., Laursen, R. A., Nagarkatti, S., Friesen, J. D. 1976. A mutant of *Escherichia coli* with an altered elongation factor Tu. *Proc. Natl. Acad. Sci. USA* 73:1698–1701
238. Gordon, J. 1970. Regulation of the in vivo synthesis of the polypeptide chain elongation factors in *Escherichia coli*. *Biochemistry* 9:912–17
239. Gordon, J., Weissbach, H. 1970. Immunochemical distinction between the *Escherichia coli* polypeptide chain elongation factors T_u and T_s. *Biochemistry* 9:4233–36
240. Krauss, S. W., Leder, P. 1975. Regulation of initiation and elongation factor levels in *Escherichia coli* as assessed by a quantitative immunoassay. *J. Biol. Chem.* 250:3752–58
241. Furano, A. V. 1975. Content of elongation factor Tu in *Escherichia coli*. *Proc. Natl. Acad. Sci. USA* 72:4780–84
242. Furano, A. V., Wittel, F. P. 1976. Syntheses of elongation factors Tu and G are under stringent control in *Escherichia coli*. *J. Biol. Chem.* 251:898–901
243. Blumenthal, R. M., Lemaux, P. G., Neidhardt, F. C., Dennis, P. P. 1976. The effects of the *rel*A gene on the synthesis of aminoacyl-tRNA synthetases and other transcription and translation proteins in *Escherichia coli* B. *Mol. Gen. Genet.* 149:291–96
244. Reeh, S., Pedersen, S., Friesen, J. D. 1976. Biosynthetic regulation of individual proteins in *rel*A$^+$ and *rel*A strains of *Escherichia coli* during amino acid starvation. *Mol. Gen. Genet.* 149:279–89
245. Pedersen, S., Reeh, S. V., Parker, J., Watson, R. J., Friesen, J. D., Fiil, N. P. 1976. Analysis of the proteins synthesized in ultraviolet light-irradiated *Escherichia coli* following infection with bacteriophages λ*drif*^{d} 18 and λ*dfus*-3. *Mol. Gen. Genet.* 144:339–43
246. Springer, M., Graffe, M., Grunberg-Manago, M. 1977. Characterization of an *E. coli* mutant with a thermolabile initiation factor IF3 activity. *Mol. Gen. Genet.* 151:17–26
247. Springer, M., Graffe, M., Hennecke, H. 1977. Specialized transducing phage for initiation factor IF3 gene in *E. coli*. *Proc. Natl. Acad. Sci. USA.* In press
248. Nierlich, D. P. 1968. Amino acid control over RNA synthesis: A reevaluation. *Proc. Natl. Acad. Sci. USA* 60:1345–52
249. Gausing, K. 1974. Ribosomal protein in *E. coli:* Rate of synthesis and pool size at different growth rates. *Mol. Gen. Genet.* 129:61–75
250. Lindahl, L. 1975. Intermediates and time kinetics of the in vivo assembly of *Escherichia coli* ribosomes. *J. Mol. Biol.* 92:15–37
251. Dennis, P. P. 1974. In vivo stability, maturation and relative differential synthesis rates of individual ribosomal proteins in *Escherichia coli* B/r. *J. Mol. Biol.* 88:25–41
252. Maaloe, O., Kjeldgaard, N. O. 1966. *Control of Macromolecular Synthesis.* New York: Benjamin. 284 pp.
253. Stent, G. S., Brenner, S. 1961. A genetic locus for the regulation of ribonucleic acid synthesis. *Proc. Natl. Acad. Sci. USA* 47:2005–14
254. Gallant, J., Lazzarini, R. A. 1976. The regulation of ribosomal RNA synthesis and degradation in bacteria. In *Protein Synthesis: A Series of Advances,* ed. E. McConkey, 2:309–59. New York & Basel: Dekker. 386 pp.
255. Cashel, M., Gallant, J. 1974. Cellular regulation of guanosine tetraphosphate and guanosine pentaphosphate. See Ref. 2, pp. 733–45

256. Cashel, M. 1975. Regulation of bacterial ppGpp and pppGpp. *Ann. Rev. Microbiol.* 29:301–18
257. Dennis, P. P., Nomura, M. 1975. Regulation of the expression of ribosomal protein genes in *Escherichia coli. J. Mol. Biol.* 97:61–76
258. Dennis, P. P., Nomura, M. 1976. Regulation of the expression of ribosomal protein genes in *Escherichia coli.* See Ref. 22, pp. 304–18
259. Gausing, K. 1976. Synthesis of rRNA and r-protein mRNA in *E. coli* at different growth rates. See Ref. 22, pp. 292–303
260. Forchhammer, J., Kjeldgaard, N. O. 1968. Regulation of messenger RNA synthesis in *Escherichia coli. J. Mol. Biol.* 37:245–55
261. Morris, D. W., Kjeldgaard, N. O. 1968. Evidence for the *non*-co-ordinate regulation of ribonucleic acid synthesis in stringent strains of *Escherichia coli. J. Mol. Biol.* 31:145–48
262. Edlin, G., Stent, G. S., Baker, R. F., Yanofsky, C. 1968. Synthesis of a specific messenger RNA during amino acid starvation of *Escherichia coli. J. Mol. Biol.* 37:257–68
263. Edlin, G., Broda, P. 1968. Physiology and genetics of the "ribonucleic acid control" locus in *Escherichia coli. Bacteriol. Rev.* 32:206–26
264. Maaløe, O. 1969. An analysis of bacterial growth. *Dev. Biol. Suppl.* 3:33–58
265. Dennis, P. P., Nomura, M. 1974. Stringent control of ribosomal protein gene expression in *Escherichia coli. Proc. Natl. Acad. Sci. USA* 71:3819–23
266. Dennis, P. P., Nomura, M. 1975. Stringent control of the transcriptional activities of ribosomal protein genes in *E. coli. Nature* 255:460–65
267. Haseltine, W. A., Block, R., Gilbert, W., Weber, K. 1972. MSI and MSII made on ribosome in idling step of protein synthesis. *Nature* 238:381–84
268. Haseltine, W. A., Block, R. 1973. Synthesis of guanosine tetra- and pentaphosphate requires the presence of a codon-specific, uncharged transfer ribonucleic acid in the acceptor site of ribosomes. *Proc. Natl. Acad. Sci. USA* 70:1564–68
269. Pedersen, F. S., Lund, E., Kjeldgaard, N. O. 1973. Codon specific, tRNA dependent in vitro synthesis of ppGpp and pppGpp. *Nature New Biol.* 243:13–15
270. Block, R., Haseltine, W. A. 1974. In vitro synthesis of ppGpp and pppGpp. See Ref. 2, pp. 747–61
271. Laffler, T., Gallant, J. 1974. *spoT,* a new genetic locus involved in the stringent response in *E. coli. Cell* 1:27–30
272. Stamminger, G., Lazzarini, R. A. 1974. Altered metabolism of the guanosine tetraphosphate, ppGpp, in mutants of *E. coli. Cell* 1:85–90
273. Fiil, N. P., Willumsen, B. M., Friesen, J. D., von Meyenburg, K. 1977. Interaction of alleles of the *relA, relC* and *spoT* genes in *Escherichia coli:* Analysis of the interconversion of GTP, ppGpp and pppGpp. *Mol. Gen. Genet.* 150:87–101
274. Travers, A. A., Kamen, R. I., Schleif, R. F. 1970. Factor necessary for ribosomal RNA synthesis. *Nature* 228:748–51
275. Travers, A. 1973. Control of ribosomal RNA synthesis in vitro. *Nature* 244:15–18
276. Winslow, R. M. 1971. A consequence of the *rel* gene during a glucose to lactate downshift in *Escherichia coli. J. Biol. Chem.* 246:4872–77
277. Hansen, M. T., Pato, M. L., Molin, S., Fiil, N. P., von Meyenburg, K. 1975. Simple downshift and resulting lack of correlation between ppGpp pool size and ribonucleic acid accumulation. *J. Bacteriol.* 122:585–91
278. Lazzarini, R. A., Cashel, M., Gallant, J. 1971. On the regulation of guanosine tetraphosphate levels in stringent and relaxed strains of *Escherichia coli. J. Biol. Chem.* 246:4381–85
279. Gallant, J., Palmer, L., Pao, C. C. 1977. Anomalous synthesis of ppGpp in growing cells. *Cell* 11:181–85
280. Block, R. 1976. Synthesis of ribosomal RNA in a partially purified extract from *Escherichia coli.* See Ref. 22, pp. 226–38
281. Aboud, M., Pastan, I. 1975. Activation of transcription by guanosine 5'-diphosphate, 3'-diphosphate, transfer ribonucleic acid, and a novel protein from *Escherichia coli. J. Biol. Chem.* 250:2189–95
282. Gallant, J., Shell, L., Bittner, R. 1976. A novel nucleotide implicated in the response of *E. coli* to energy source downshift. *Cell* 7:75–84
283. Kari, C., Török, I., Travers, A. 1977. ppGpp cycle in *Escherichia coli. Mol. Gen. Genet.* 150:249–56
284. Pedersen, S. 1976. Stability of nascent ribosomal RNA in *Escherichia coli.* See Ref. 22, pp. 345–52
285. Dennis, P. P. 1977. Influence of the stringent control system on the transcription of ribosomal ribonucleic acid

and ribosomal protein genes in *Escherichia coli. J. Bacteriol.* 129:580–88
286. Travers, A. 1976. Modulation of RNA polymerase specificity by ppGpp. *Mol. Gen. Genet.* 147:225–32
287. Travers, A. 1976. RNA polymerase specificity and the control of growth. *Nature* 263:641–46
288. Dennis, P. P. 1976. Effects of chloramphenicol on the transcription activities of ribosomal RNA and ribosomal protein genes in *Escherichia coli. J. Mol. Biol.* 108:535–46
289. Matzura, H., Hansen, B. S., Zeuthen, J. 1973. Biosynthesis of the β and β' subunits of RNA polymerase in *Escherichia coli. J. Mol. Biol.* 74:9–20
290. Iwakura, Y., Ito, K., Ishihama, A. 1974. Biosynthesis of RNA polymerase in *Escherichia coli.* I. Control of RNA polymerase content at various growth rates. *Mol. Gen. Genet.* 133:1–23
291. Rünzi, W., Matzura, H. 1976. In vivo distribution of ribonucleic acid polymerase between cytoplasm and nucleoid in *Escherichia coli. J. Bacteriol.* 125:1237–39
292. Dalbow, D. G., Bremer, H. 1975. Metabolic regulation of β-galactosidase synthesis in *Escherichia coli:* A test for constitutive ribosome synthesis. *Biochem. J.* 150:1–8
292a. Bremer, H., Dalbow, D. G. 1975. Regulatory state of ribosomal genes and physiological changes in the concentration of free ribonucleic acid polymerase in *Escherichia coli. Biochem. J.* 150:9–12
293. Lindahl, L., Post, L., Nomura, M. 1976. DNA-dependent in vitro synthesis of ribosomal proteins, protein elongation factors and RNA polymerase subunit α: Inhibition by ppGpp. *Cell* 9:439–48
294. Nomura, M., Watson, J. D. 1959. Ribonucleoprotein particles within chloromycetin-inhibited *Escherichia coli. J. Mol. Biol.* 1:204–17
295. Erlich, H., Gallant, J., Lazzarini, R. A. 1975. Synthesis and turnover of ribosomal ribonucleic acid in guanine-starved cells of *Escherichia coli. J. Biol. Chem.* 250:3057–61
296. Norris, T. E., Koch, A. L. 1972. Effect of growth rate on the relative rates of synthesis of messenger, ribosomal, and transfer RNA in *Escherichia coli. J. Mol. Biol.* 64:633–49
297. van Ooyen, A. J. J., de Boer, H. A., Ab, G., Gruber, M. 1975. Specific inhibition of ribosomal RNA synthesis in vitro by guanosine 3' diphosphate, 5' diphosphate. *Nature* 254:530–31
298. Haseltine, W. A. 1972. In vitro transcription of *Escherichia coli* ribosomal RNA genes. *Nature* 235:329–33
299. van Ooyen, A. J. J., Gruber, M., Jorgensen, P. 1976. The mechanism of action of ppGpp on rRNA synthesis in vitro. *Cell* 8:123–28
300. Travers, A., Baralle, F. E. 1976. In vitro transcription of *E. coli* ribosomal RNA. See Ref. 22, pp. 241–49
301. Murooka, Y., Lazzarini, R. A. 1973. In vitro synthesis of ribosomal ribonucleic acid by a deoxyribonucleic acid-protein complex isolated from *Escherichia coli. J. Biol. Chem.* 248:6248–50
302. Reiness, G., Yang, H.-L., Zubay, G., Cashel, M. 1975. Effects of guanosine tetraphosphate on cell-free synthesis of *Escherichia coli* ribosomal RNA and other gene products. *Proc. Natl. Acad. Sci. USA* 72:2881–85
303. Muto, A. 1975. Preferential ribosomal RNA synthesis in the lysate of *Escherichia coli. Mol. Gen. Genet.* 138:1–10
304. Lazzarini, R. A., Johnson, L. D. 1973. Regulation of ribosomal RNA synthesis in cold-shocked *E. coli. Nature New Biol.* 243:17–20
305. Atherly, A. 1974. Ribonucleic acid regulation in permeabilized cells of *Escherichia coli* capable of ribonucleic acid and protein synthesis. *J. Bacteriol.* 118:1186–89
306. Morrissey, J. J., Cupp, L. E., Weissbach, H., Brot, N. 1976. Synthesis of ribosomal proteins L7L12 in relaxed and stringent strains of *Escherichia coli. J. Biol. Chem.* 251:5516–21
307. Venetianer, P., Sümegi, J., Udvardy, A. 1976. Properties of ribosomal RNA promoters. See Ref. 22, pp. 252–65
308. Mueller, K., Oebbecke, C., Förster, G. 1977. Capacity of ribosomal RNA promoters of *E. coli* to bind RNA polymerase. *Cell* 10:121–30
309. Dennis, P. P. 1977. Transcription patterns of adjacent segments on the chromosome of *Escherichia coli* containing genes coding for four 50S ribosomal proteins and the β and β' subunits of RNA polymerase. *J. Mol. Biol.* In press
310. Dalbow, D. G. 1973. Synthesis of RNA polymerase in *Escherichia coli* B/r growing at different rates. *J. Mol. Biol.* 75:181–84
311. Iwakura, Y., Ishihama, A. 1975. Biosynthesis of RNA polymerase in *Es-

cherichia coli. II. Control of RNA polymerase synthesis during nutritional shift up and down. *Mol. Gen. Genet.* 142:67–84
312. Engbaek, F., Gross, C., Burgess, R. R. 1976. Quantitation of RNA polymerase subunits in *Escherichia coli* during exponential growth and after bacteriophage T4 infection. *Mol. Gen. Genet.* 143:291–95
313. Maher, D. L., Dennis, P. P. 1977. In vivo transcription of *E. coli* genes coding for rRNA, ribosomal proteins and subunits of RNA polymerase: Influence of the stringent control system. *Mol. Gen. Genet.* In press
314. Rosset, R., Julien, J., Monier, R. 1966. Ribonucleic acid composition of bacteria as a function of growth rate. *J. Mol. Biol.* 18:308–20
314a. Dennis, P. P. 1972. Regulation of ribosomal and transfer ribonucleic acid synthesis in *Escherichia coli* B/r. *J. Biol. Chem.* 247:2842–45
315. Ikemura, T., Dahlberg, J. E. 1973. Small ribonucleic acids of *Escherichia coli.* II. Noncoordinate accumulation during stringent control. *J. Biol. Chem.* 248:5033–41
316. Primakoff, P., Berg, P. 1970. Stringent control of transcription of phage ϕ80p su_3. *Cold Spring Harbor Symp. Quant. Biol.* 35:391–96
317. Yang, H.-L., Zubay, G., Urm, E., Reinens, G., Cashel, M. 1974. Effects of guanosine tetraphosphate, guanosine pentaphosphate, and β-γ methylenylguanosine pentaphosphate on gene expression of *Escherichia coli* in vitro. *Proc. Natl. Acad. Sci. USA* 71:63–67
318. Stephens, J. C., Artz, S. W., Ames, B. N. 1975. Guanosine 5'-diphosphate 3'-diphosphate (ppGpp): Positive effector for histidine operon transcription and general signal for amino-acid deficiency. *Proc. Natl. Acad. Sci. USA* 72:4389–93
319. Pedersen, S., Kjeldgaard, N. O. 1976. A *rel* gene control of *lac* gene transcription in vivo. See Ref. 22, pp. 164–65
320. Schumacher, G., Ehring, R. 1973. RNA-directed cell-free synthesis of the galactose enzymes of *Escherichia coli. Mol. Gen. Genet.* 124:329–44
321. Kung, H.-F., Brot, N., Spears, C., Chen, B., Weissbach, H. 1974. Studies on the in vitro transcription and translation of the *lac* operon. *Arch. Biochem. Biophys.* 160:168–74
322. de Crombrugghe, B., Chen, B., Gottesman, M., Pastan, I., Varmus, H. E., Emmer, M., Perlman, R. L. 1971. Regulation of *lac* mRNA synthesis in a soluble cell-free system. *Nature New Biol.* 230:37–40
322a. Zubay, G., Gielow, L., Englesberg, E. 1971. Cell-free studies on the regulation of the arabinose operon. *Nature New Biol.* 233:164–65
323. Neidhardt, F. C. 1976. Regulation of aminoacyl-tRNA synthetases. See Ref. 22, pp. 342–44
324. Furano, A. V., Wittel, F. P. 1976. Effect of the *relA* gene on the synthesis of individual proteins in vivo. *Cell* 8:115–22
325. Laffler, T., Gallant, J. A. 1974. Stringent control of protein synthesis in *E. coli. Cell* 3:47–49
326. Bird, R. E., Louarn, J., Martuscelli, J., Caro, L. 1972. Origin and sequence of chromosome replication in *Escherichia coli. J. Mol. Biol.* 70:549–66
327. McKenna, W. G., Masters, M. 1972. Biochemical evidence for the bidirection replication of DNA in *Escherichia coli. Nature* 240:536–39
328. Champney, W. S., Kushner, S. R. 1976. A proposal for a uniform nomenclature for the genetics of bacterial protein synthesis. *Mol. Gen. Genet.* 147:145–51

GENETICS OF CELLULAR DIFFERENTIATION: STABLE NUCLEAR DIFFERENTIATION IN EUCARYOTIC UNICELLS[1]

♦3125

T. M. Sonneborn

Department of Biology, Indiana University, Bloomington, Indiana 47401

CONTENTS

I NUCLEAR CYTOLOGY	350
II THE A SYSTEM OF MATING TYPE DETERMINATION AND INHERITANCE IN SPECIES WITH ONLY TWO MATING TYPES	352
The A System in Paramecium primaurelia	356
The A System in Paramecium pentaurelia	356
The A system in Paramecium multimicronucleatum, Syngen 2	357
III THE A SYSTEM OF MATING TYPE DETERMINATION AND INHERITANCE IN SPECIES WITH MULTIPLE MATING TYPES	357
The A System in Tetrahymena canadensis	357
The A System in Tetrahymena thermophila	357
The A System in Stylonychia mytilus	359
IV THE B SYSTEM OF MATING TYPE DETERMINATION AND INHERITANCE IN PARAMECIUM TETRAURELIA AND PARAMECIUM SEPTAURELIA	359
V MACRONUCLEAR DIFFERENTIATION FOR OTHER TRAITS	362
VI MISCELLANEOUS CONCLUDING COMMENTS	363

This review offers a number of examples of an important, possibly general, class of genetic phenomena that have long seemed difficult to assimilate into the corpus of genetic principles. However, observations that may be related to them have recently been explained by reasonable molecular mechanisms. It therefore seems timely to review the still unexplained observations in the hope that this will stimulate efforts to explain them in molecular terms. Some of the observations in this class are old,

[1]Department of Biology Contribution Number 1057.

but they, as well as recent and current work, are not widely known and have not been included in recent reviews (1–5) of what seem to the present author likely to be similar or related phenomena. Therefore, this review calls attention to older as well as to recent and current work. Parts of the material in this review are dealt with in other reviews (6–21). Because the terminology and cytogenetics of the organisms, certain ciliated Protozoa, are unfamiliar but essential for understanding the rest of the review, they are summarized in the next section. When Tetrahymena and Paramecium are referred to without species designations, they refer to species of the *Tetrahymena pyriformis* complex (22) and the *Paramecium aurelia* complex (23). Other species of the same genera have somewhat different nuclear cytology.

I NUCLEAR CYTOLOGY

Ciliates contain two visibly different kinds of nuclei, large DNA-rich somatic macronuclei and small diploid germinal micronuclei. Micronuclei divide mitotically at cell division, undergo meiosis before their haploid products participate in fertilization, and give rise to macronuclei after fertilization. Micronuclei have relatively little or no effect on the phenotype. Viable amicronucleate clones occur in many species. Tetrahymena has one micronucleus per cell; Paramecium has two. The micronuclei of Tetrahymena have five pairs of chromosomes; those of Paramecium have from about 30 to more than 60 pairs in different stocks of the same species. Nevertheless, the amount of DNA in a haploid set is roughly the same in both, about the same as in Drosophila.

Two fertilization processes occur in Paramecium, conjugation and autogamy; in Tetrahymena, only conjugation occurs. Between meiosis and fertilization, one of the haploid nuclei undergoes a mitotic division to produce two genotypically identical gamete nuclei. Consequently, at autogamy, the diploid synkaryon formed by the union of the two gamete nuclei in a single unpaired cell is homozygous at all loci. Conjugation, on the other hand, involves reciprocal cross-fertilization (like a pair of reciprocal crosses). One gamete nucleus (the so-called male nucleus) of each mate migrates into the other mate and there unites with the stationary (female) nucleus to form a synkaryon. This results in genotypic identity of the synkarya in the two conjugants of each pair: If the genotypes of the two gamete nuclei in one mate are symbolized A and A and the two in the other mate B and B, then both synkarya will be AB. The genotypically identical clones produced by the two conjugants of a mated pair form a genetic unit referred to as a *synclone*.

After fertilization, either conjugation or autogamy, the synkaryon divides mitotically twice to produce four genotypically identical diploid nuclei. In Tetrahymena, one disintegrates and disappears, one stays a micronucleus, and two develop into macronuclei. In Paramecium, none disintegrates, two remaining micronuclei and two developing into macronuclei. These differentiations among the products of the synkaryon are determined by their positions in the cell (24, 25). At the first postzygotic cell division, in both genera, the micronuclei divide, but the new macronuclei do not; instead, they segregate, one going to each of the two products of this cell division. Thereafter, at each cell division, both micronuclei and macronuclei divide.

A subclone containing macronuclei descended from the same original new macronucleus is referred to as a *caryonide*. It is a genetic unit of key importance in the present review. Normally, there are thus two *sister caryonides* from each fertilized cell and each caryonide is traceable to a cell produced at the first postzygotic cell division.

Macronuclear development consists of several rounds of DNA replication, half or more occurring in the first postzygotic cell cycle and the rest during the second cell cycle. The fully developed G1 macronucleus of mature Tetrahymena has about 45 times as much DNA as a haploid chromosome set; that of Paramecium has more than 800 times as much as its haploid set. Unlike the situation in certain more complex ciliates, at least 90% of the DNA sequences in the micronucleus, both unique and repetitive sequences, are also represented by many copies in the macronucleus. How the DNA of the macronucleus is organized remains unknown. It has been variously postulated to be organized in diploid or haploid subunits, free intact chromosomes, and free intact chromosomal fragments such as replicons or single genes. Whatever the organization may be, and in spite of the imprecision of its amitotic division, the total amount of DNA per macronucleus seems to be regulated so as to compensate for inequalities of its division. Whether regulation extends to the quantity of genomes, individual chromosomes, replicons, or genes, remains unknown.

The macronucleus controls the cell phenotype and is essential for life. Amacronucleate cells quickly cease to grow and soon die.

The postzygotic cell contains, in addition to micronuclei and developing new macronuclei, the old macronucleus of the prezygotic mother cell. In Tetrahymena, the maternal macronucleus disappears before the first postzygotic cell division. In Paramecium, it breaks up into about 35 fragments during conjugation and autogamy. DNA synthesis in the fragments, but not RNA or protein synthesis, is inhibited by the presence of a postzygotic macronucleus. The fate of the fragments depends on nutritive and other conditions. If the fertilized cell is kept under nonnutrient conditions, the fragments are gradually resorbed in the course of several days; but if it is in growth medium, the fragments persist for six or seven cell generations and then rapidly disintegrate and disappear. Prior to their disappearance, they are randomly distributed to the cell progeny at cell division.

Under some genetic or environmental conditions, the prezygotic macronucleus has a different fate. The common prerequisite for this fate, which is very important in the genetic analyses reported in this review, is the absence from a cell of a postzygotic macronucleus or the presence of a grossly defective one. Under these conditions, the prezygotic macronucleus persists and continues to function instead of disappearing. The process is called *macronuclear retention* in Tetrahymena because the prezygotic macronucleus remains functional; in Paramecium, it is called *macronuclear regeneration* because fragments into which the prezygotic macronucleus breaks down undergo DNA synthesis and regenerate into functional macronuclei.

Macronuclear regeneration can begin at the first postzygotic cell division if the two newly developing postzygotic macronuclei fail to segregate, or at a later division if the macronucleus misdivides so that a cell fails to get a product of its division.

So long as a single intact fragment is present, it is capable of regenerating. Throughout macronuclear regeneration, the fragments grow until their *total* volume (and DNA content) per cell equals that of a mature macronucleus. Regenerating fragments are distributed randomly to daughter cells at cell division and do not divide until there is only one in a cell; by this time, it has reached normal macronuclear size and DNA content.

Regenerating or regenerated fragments develop from *prezygotic parental* macronuclei; *zygotic* macronuclei develop from a product of the synkaryon. Cell lineages that bear descendants of the prezygotic macronucleus are continuations of the parental caryonide; those with descendants of a zygotic macronucleus constitute a new caryonide. Misdivision of the zygotic macronucleus results in descent of both parental and zygotic caryonides from the same fertilized cell. This provides a critical test of an important point in the genetic analysis.

II THE A SYSTEM OF MATING TYPE DETERMINATION AND INHERITANCE IN SPECIES WITH ONLY TWO MATING TYPES

All of the genetic systems to be reviewed in this paper share basic features with the first such system to be delineated, analyzed, and interpreted, namely the so-called A system of mating type determination and inheritance in *Paramecium primaurelia*, formerly *P. aurelia*, variety or syngen 1 (23). Essentially the same system operates, but has been less investigated, in several other species of the *P. aurelia* complex (*P. triaurelia, P. pentaurelia, P. novaurelia, P. undecaurelia,* and *P. quadecaurelia*) and in a mutant of syngen 2 of *P. multimicronucleatum*. Modifications of the A system, due to multiple mating types, are dealt with in Section III. Because of its fundamental importance for all that follows, the system in *P. primaurelia* is set forth rather fully here although much of the work was done long ago. Points established in the early papers were summarized by Kimball (11); only later work is individually referenced here.

Mating types in the 14 species of the *P. aurelia* complex were originally designated by a series of pairs of numbers, in each species one odd and one even number. Currently they are designated O (for odd numbered) and E (for even numbered) because of evidence that the O's of different species are homologous, as are the E's.

The A System in Paramecium primaurelia

For later reference, each of the 21 kinds of observations in the A system in *P. primaurelia* is numbered, as follows.

   1. Both O and E as a rule reproduce true to type during asexual reproduction. Unstable lineages producing both O and E cells are relatively rare in *P. primaurelia*.
   2. Mating type frequently changes at sexual reproduction. The mating types of parent and offspring are not correlated at either autogamy or conjugation.
   3. Both mating types are produced in the same frequencies (under the same cultural conditions) at successive autogamies, starting with either mating type, even when selecting at each autogamy offspring of the same mating type as that of the initial parent. Since only one autogamy (see Section I) is required to result in

homozygosity at all loci, there can be no genotypic difference between the germinal micronuclei in the two mating types.

4. Crosses of O X E yield the same results for the progeny of each mate as does autogamy in either O or E.

5. The unit of inheritance of mating type is neither the synclone nor the clone, but the caryonide (see Section I). The two caryonides from an autogamous or conjugant cell are often of different mating type. This implies a macronuclear difference between the two mating types.

6. If the total frequency of type O in a group of caryonides is p and the total frequency of type E is $(1 - p)$, then the two sister caryonides of a clone are both O or both E or one of each in the frequencies p^2, $(1 - p)^2$, and $2p(1 - p)$, respectively. Likewise, the frequencies of various combinations of types among the four caryonides from a pair of conjugants are given by expansion of the binomial $[p + (1 - p)]^4$. Thus, each caryonide is independently and randomly determined for mating type.

7. In certain stocks, about 20% of the fertilized cells form three or four, rarely more, new macronuclei instead of the normal two. In the expected frequencies, mating type segregates (and caryonides arise) at the second instead of the first postzygotic cell division, and rarely at the third cell division. This confirms the macronuclear basis of the mating type difference.

8. Caryonides lacking micronuclei, but of course possessing macronuclei, may be either O or E.

9. When the two zygotic macronuclei in a fertilized cell are induced to fuse (by prolonged starvation), the resulting caryonide is often a selfer in which both O and E cells are continually produced (12), in agreement with expected frequent fusion of oppositely determined developing macronuclei.

10. External conditions applied prior to the first postzygotic cell division greatly influence the probabilities of occurrence of the two kinds of caryonides. Within the range 10° to 35°C, the frequency of type E caryonides in certain stocks of *P. primaurelia* equals $18.4 + 1.96\ t$ where t is degrees centigrade (26, 27).

11. Temperature is effective only during the first postfertilization cell cycle, i.e. only while zygotic macronuclei are arising and developing from products of the synkaryon. Early studies on *P. triaurelia* reported erratic effects of high temperature during conjugation as well as during the first cell cycle. Temperature effects after the first cell cycle have never been reported. The stages at which the temperature-sensitive period begins and ends have not been more precisely ascertained.

12. Exposure to $CaCl_2$ during the first postfertilization cell cycle is reported (27) to decrease the frequency of type E in nonselfing caryonides, but the data involve recognized difficulties that remain to be resolved.

13. Once macronuclei are determined for mating type, the determination appears to be completely stable and unalterable in all descendants of that macronucleus. This is shown not only by the constancy of mating type in the ordinary caryonide, but also by its constancy through macronuclear regeneration (see Section I), even when the macronuclear fragments destined to regenerate are, during the temperature-sensitive first postfertilization cell cycle, in the same cell with oppositely determined zygotic macronuclei. After the zygotic macronuclei and the regenerated fragments

of the prezygotic macronucleus pass into different cells, the lineages containing descendants of regenerated fragments always retain the parental mating type while the lineages containing descendants of a zygotic macronucleus often have the other mating type.

14. In other words, the mating type determined by regenerated macronuclear fragments is totally unaffected by temperature. Mating type is thus determined by an apparently irreversible differentiation occurring only in newly developing zygotic macronuclei.

15. Hallet (27, 28) found that, after exposure to $CaCl_2$ during (and *only* during) the sensitive period, varying percentages (0 to 25 in different experiments, average 8.4%) of the subclones initiated with a cell produced at the first cell division (i.e. a *presumptive* caryonide) are selfers and that these selfers give rise in the course of the next few cell generations to sublines all of which are nonselfers, some being pure type O, others pure type E. Examination after exposure to $CaCl_2$ regularly shows two zygotic macronuclei developing during the first cell cycle and only one per cell after the first cell division. So the results are not to be explained by development of multiple new macronuclei instead of just two (see item 7 above) or by macronuclear regeneration due to failure of segregation of the two new macronuclei at the *first* cell division. However, the observations do not exclude the possibility of macronuclear regeneration at later cell divisions, which could account for all of the results. Hallet interprets the results as due to delayed differentiation of the zygotic macronuclei. Before accepting this interpretation, macronuclear regeneration should be excluded by experiments using genetic markers and/or by adequate cytological observations.

16. The change from type E to type O at sexual reproduction is slow; cells of the new clone remain type E for variable numbers (4 to 6) of cell generations, then all become type O. Change occurs much more rapidly (1–2 cell generations) in the reverse direction, from type O to type E. The same unidirectional long phenomic lag occurs in all species of the *P. aurelia* complex and is one of the criteria for identifying homologies between the mating types in different species. In no case does it conflict with homologies based on mating reactions. The great difference in phenomic lags in the two directions is one of several facts consistent with Butzel's (29) hypothesis that a mating type O substance is a precursor of the mating type E substance.

17. Persistent selfing caryonides, containing both O and E cells, occur with different frequencies in different stocks of *P. primaurelia;* in some stocks, their frequency is 3% or less. Usually any isolated cell of either mating type tends to reproduce true to type for a few cell generations and then again yields progeny cells of both mating types.

18. Alleles at one locus, mt^o and mt^+, control the possibilities for macronuclear determination: homozygotes for mt^o cannot be determined for mating type E, all caryonides being type *O;* both homozygotes and heterozygotes for mt^+ can be determined for either type O or type E. All stocks collected in nature are mt^+/mt^+ except that those collected from one natural source were mt^o/mt^o. Wild type is thus prevailingly mt^+/mt^+. Alleles indistinguishable from mt^o have been isolated

from an mt^+/mt^+ stock after UV mutagenesis and after temperature shocks during sexual processes (29).

19. All attempts in *P. primaurelia* to obtain mutants restricted to mating type E, i.e. unable to produce caryonides of type O, have failed (29).

20. Genes at other loci affect (under the same temperature conditions) the probabilities of determination for the alternative mating types in homozygotes for mt^+ (29).

21. Haploids developed from wild-type (mt^+/mt^+) parents can yield caryonides of either mating type.

What more do the preceding 21 items tell about the underlying genetic mechanisms in the A system of mating type determination and inheritance in *P. primaurelia*? Items 3, 4, 18, and 21 together show that the same entirely homozygous micronuclear genotype can exist in both mating types and that the genetic information for both mating types is in each haploid wild-type genome. Items 5, 6, 7, 8, 9, 10, 11, 13, and 14 together show that (*a*) *macro*nuclei differ irreversibly in the two mating types; (*b*) this difference arises as new zygotic macronuclei develop from products of the synkaryon during the first cell cycle after fertilization; (*c*) each new zygotic macronucleus is independently differentiated or determined, even the two arising in the same fertilized cell; and (*d*) the probabilities of differentiation of new zygotic macronuclei for the two alternative mating types vary with the temperature (and perhaps other environmental conditions) prevailing only during the sensitive first cell-cycle, not later. Items 18 and 20 show that alleles at one locus control the capacity of macronuclei to be differentiated for type E and that genes at other loci affect (as does temperature) the probabilities of macronuclear determination for the two mating types.

The major remaining questions about the underlying genetic mechanisms are the following: the nature of the difference between macronuclei determined for type O and those determined for type E; the mechanism by which this differentiation is brought about; why it is normally or always created only in the first cell cycle; how the macronuclear difference is perpetuated during asexual reproduction; and why it is irreversible. None of these basic questions can at present be answered. One can only imagine possible molecular mechanisms for some of them and perhaps eliminate certain of these possibilities.

The possibility that the two kinds of macronuclei differ with respect to a mechanism such as that in the well-known example of perpetuated difference in the inactivity versus activity (repression versus derepression) of a gene cluster for β-galactosidase in *Escherichia coli* seems unlikely. One would, for example, expect fusion of two oppositely determined macronuclei to result in both component macronuclei becoming alike unless they were physically separated in the fused nucleus. They did not become alike; the descendant macronucleus remained heterogeneous (item 9). In the case of natural selfing caryonides, the assumption of physical separation into two subnuclei would be entirely gratuitous; yet again heterogeneity is maintained (item 17). These observations indicate that the same difference which distinguishes different macronuclei can also exist among different components of the same (highly compounded) macronucleus without either being altered by the presence of the other. Hence interpretations based on structural rather than mere physiological difference seem more reasonable.

Two kinds of structural difference have been suggested: chromosomal and genic. The simplest chromosomal possibility (suggested by known sex chromosome mechanisms) is that one member of a pair of micronuclear chromosomes is either not incorporated into some macronuclei (thereby determining them for one mating type) or is eliminated from them during their development in the first cell cycle. Chromosome loss seems to be ruled out by item 21 and (if the mt chromosome is the one subject to loss) also by the fact that heterozygotes (mt^o/mt^+) are with high frequency type E, not at least the 50% type O expected if the chromosome-bearing mt^+ is lost half of the time.

The gene mutation hypothesis, as first put forth (18), postulated that micronuclei were homozygous for a mutable mt allele determining type O, and that this gene mutated to an allele determining type E at a rate proportional to temperature (item 10) only during the early development of zygotic macronuclei (item 11), after the model of mutable genes mutating at certain developmental stages (30). In later discussions, the words *mutation* and *mutable gene* were dropped and replaced by *alternative stable genic states*, the active state determining type E and the inactive state type O. There is, however, no substantial difference between the two terminologies. Both the unidirectional phenomic lag (item 16) and the recessiveness of the mt^o allele (item 18) favor the assignment of inactivity to the stable O state, activity to the stable E state. Presumably constitutive genes at one or more other loci determine type O, and O would be converted to E when the mt^+ allele is active (29).

These mutational or genic-state interpretations have recently become credible in principle as knowledge of molecular genetics has led to readily conceivable molecular models for them (1–4; 31–33). They are referred to more particularly in Section VI.

The A System in Paramecium pentaurelia

The A system in this species (34, 35) is unique in that no caryonides are pure for type E. Caryonides are either pure O or selfers. Selfer caryonides are of two types. Some segregate pure O sublines, the rest being selfers. These caryonides seem to start with mixed macronuclei of which most components are determined for O. Other selfer caryonides are "pure" selfers. When grown slowly by limiting the food supply, the cells regularly express type E; when grown rapidly in excess food, they produce cells of both types. As rapidly grown cultures of these selfers become mating reactive (i.e. when the food supply approaches exhaustion), some E cells and later some O cells appear. These cells are not conversions of E cells to O; they are O as soon as they become reactive, but in about one hour they change to E in the absence of cell division and even in isolation (precluding contact with E cells). Any cell isolated from such selfing cultures can produce progeny of both mating types. The frequency of these selfers increases with temperature during the sensitive first cell cycle after fertilization. Hence, the selfers are in this respect like pure E caryonides in *P. primaurelia*. Their failure to be pure E seems to be due to reduced efficiency of action of an *mt* gene in its stable active state.

The A System in Paramecium multimicronucleatum, Syngen 2

Wild type in this species has a circadian rhythm of expression of the two mating types. Under proper nutritive conditions, all the cells are one mating type part of each day and the other mating type another part of the day. A recessive gene, c, abolishes the rhythm, and mating type then follows the A system: It is caryonidally inherited and, at macronuclear regeneration, remains unchanged (36).

III THE A SYSTEM OF MATING TYPE DETERMINATION AND INHERITANCE IN SPECIES WITH MULTIPLE MATING TYPES

The A System in Tetrahymena canadensis

In the A system of this species (37), formerly *T. pyriformis,* syngen 7 (22), homozygotes for mt^A can produce only two mating types, designated types II and IV. Crosses between these two types produce only the same two parental types. Inheritance is caryonidal. The frequency of determination of macronuclei for type IV increases with temperature. In Tetrahymena, the new macronuclei begin to develop before conjugants separate; therefore the cells were kept at the different temperatures throughout conjugation as well as during the first cell cycle. By analogy with *P. primaurelia,* types II and IV would correspond to types O and E, respectively, so far as temperature effect is concerned.

An allele of mt^A, designated mt^B, yields, in homozygotes, two other mating types, III and V, again with caryonidal inheritance. The frequency of macronuclear determination for type III increases with temperature during the sensitive period. So, types V and III correspond, in temperature effect, to types O and E, respectively, of *P. primaurelia.* In both species, the system in homozygotes appears to be based on a pair of alternative stable genic states. In *T. canadensis,* presumably determination of types II and V consists of inducing stable inactivity of mt^A and mt^B, respectively; determination of types IV and III consists of inducing stable activity of mt^A and mt^B, respectively.

What about the heterozygote mt^A/mt^B? Again inheritance is caryonidal, but now all four mating types (II to V) can be produced. Analysis of the temperature effects on heterozygotes led to the conclusions (*a*) that temperature also affects which allele will be expressed and (*b*) that the two temperature effects are not independent. The two temperature effects on heterozygotes interact in a complex pattern; their interpretation remains obscure.

At least one other mating type (I) is known in this species; but it has not been analyzed genetically. Possibly another allele similarly controls another pair of mating types of which one has not been found. Since relatively few wild stocks have been studied, there may be still more such alleles.

The A System in Tetrahymena thermophila

Suppose that unequal crossing-over brought the mt^A and mt^B alleles of *T. canadensis* into contiguous positions in one chromosome. A homozygote for such a chromosome would show caryonidal inheritance of multiple mating types with a complex pattern of macronuclear determination in response to temperature. This is in fact the sort of situation observed (6, 13, 14, 20, 38–47; D. L. Nanney, E. B. Meyer, and S. S. Ho Chen, submitted for publication) in *T. thermophila,* formerly *T. pyriformis,* syngen 1 (22). There is only one known (presumably complex) locus for the seven mating types. No known allele is associated with *only* two alternative mating types. Homozygotes for the most restrictive allele (mt^A) can be any of five types (all but types IV and VII); another allele (mt^B) permits six (all but type I); the heterozygote permits all seven (42). Homozygotes for a third allele, mt^C, like those for mt^A, can be any type except types IV and VII; but the five possible types occur with different frequencies in those two homozygotes (45). Genes at other loci also affect the frequencies of the mating types.

Regardless of micronuclear genotype, inheritance of mating type is caryonidal (40, 43). There is no correlation at sexual reproduction between the mating types of parent clone and offspring clone or between one caryonide and its sister caryonide from the same conjugant. Macronuclei are randomly differentiated for mating type.

Stability of macronuclear differentiation for mating type is demonstrated as in *P. primaurelia* by persistence of mating type through conjugation when macronuclear retention (see Section I) occurs, i.e. when the functional macronucleus is derived from the prezygotic parental macronucleus instead of from a product of the synkaryon.

As in *P. primaurelia,* the probabilities of macronuclear differentiation for the various mating type alternatives vary with the temperature prevailing during the sensitive period of development of postzygotic macronuclei. The data on temperature effects (46) have been rationalized (D. L. Nanney, E. B. Meyer, and S. S. Ho Chen, submitted for publication) on the following scheme. The *mt* locus is assumed to consist of three tandem subloci each of which can exist in either of two stable states, e.g. "on" or "off." Temperature affects separately the probability of each sublocus being stabilized in one or the other state. Each combination of states at the three subloci results in a different mating type, except that one combination (e.g. all in the "on" state) is "forbidden," thus accounting for the missing eighth mating type. The four pairs of mating types (including the forbidden type) due to opposite states at all three loci are inferred to be III and (VIII), II and IV, V and VI, I and VII. This ingenious theoretical analysis awaits confirmation.

Although many caryonides appear to be pure for a single mating type, selfing caryonides are common. Most of them are ditypic caryonides, but caryonides including three or more mating types occur rarely. Ditypic selfing caryonides have been extensively studied and discussed (6, 13, 14, 20, 38, 39, 44), particularly in relation to the subunit of macronuclear determination. The following are the main facts: 1. Any two types permitted by the genotype may occur within a caryonide. 2. In the course of asexual reproduction, sublines arise that are phenotypically pure for each of the two mating types (*phenotypic assortment*). 3. Sooner or later, the sum of the frequencies with which phenotypically pure sublines assort appears to stabilize at about 0.0113 per cell division. 4. The "output" ratio, i.e. the ratio of the frequencies of sublines of the two assorting pure types, varies greatly among caryonides. 5. Usually, however, sublines of one type tend to be much more common than sublines of the other type. This occurs even when the two types have to be determined by different alleles in the heterozygote (mt^A/mt^B), e.g. when one type is I (restricted to the mt^A allele) and the other is type IV (restricted to the mt^B allele). 6. Certain mating types tend to be the majority type far more than expected from their frequency of occurrence in monotypic caryonides.

The interpretation of these facts is by no means yet completely clear. The most important points for present purposes are the following: 1. A single macronucleus can be heterogeneous with respect to determination for mating type. 2. The heterogeneity of determination can persist in descendant macronuclei through hundreds of cell generations in some sublines. 3. After long maintenance of heterogeneity, macronuclei can segregate their heterogeneous components so that homogeneous daughter macronuclei arise and reproduce thereafter true to type. These facts strongly confirm the conclusion drawn above for certain kinds of selfers in *P. primaurelia,* namely, that the stability of macronuclear differentiation has a structural basis—perhaps in a localized chromosomal region or a gene, perhaps the *mt* gene itself—and is not to be understood in terms of a feedback or steady state physiological mechanism (47). Otherwise the long maintenance of heterogeneity and eventual assortment of diverse components cannot readily be accounted for.

The A System in Stylonychia mytilus

In one syngen (species), 48 mating types have been reported (48). Each haploid genome can yield several mating types. The evidence indicates macronuclear differentiation with the frequencies of determination of alternative types being affected by temperature during the sensitive period of macronuclear development from a micronucleus.

IV THE B SYSTEM OF MATING TYPE DETERMINATION AND INHERITANCE IN *PARAMECIUM TETRAURELIA* AND *PARAMECIUM SEPTAURELIA*

The *pattern of inheritance* of mating types in species with the B system differs strikingly from that in species with the A system; but mating type *determination* in the two systems is fundamentally the same except for one feature. In species with the B system, mating type usually is inherited without change through sexual (as well as asexual) reproduction. Type O parents produce type O clones from most fertilized cells regardless of whether produced at autogamy or conjugation, and type E parents usually produce type E clones at both sexual processes. Inheritance is clonal: sister caryonides from the same fertilized cell are usually alike in mating type. Temperature during the first cell cycle normally has no effect on mating type determination (but see below, this section). Nevertheless, mating type is determined at that time via macronuclear differentiations which develop even when the macronuclei arise, as they normally do, from micronuclei of identical genotype.

The one basic unique feature of the B system is the existence of a decisive cytoplasmic difference between O and E cells created and maintained by action of their macronuclei. This cytoplasmic difference then operates in each sexual generation so as to determine new zygotic macronuclei (in the first cell cycle) for O and

E, respectively, and for production of their distinctive cytoplasmic differential. Thus a nucleocytoplasmic feedback cycle results (*a*) in the maintenance of the same mating type and cytoplasmic state through sexual reproduction and (*b*) in agreement in these respects between sister caryonides.

These basic features of the B system were first discovered and studied (9, 12, 18, 21, 25, 49–53) in *P. tetraurelia;* additional features occur in *P. septaurelia* (54–56). Very little has been published (21) on other species in which the B system is known to occur (*P. biaurelia, P. sexaurelia, P. octaurelia, P. decaurelia,* and *P. dodecaurelia*). Unless otherwise stated, the following account is of work done on *P. tetraurelia* although, insofar as is known, essentially the same relations hold for the other species that have the B system.

The presence of a cytoplasmic element in the B system became evident from comparison of the genetics of mating types at conjugation (O X E) when mates do and do not experience transfer of cytoplasm (18, 49, 52). The latter, the usual case, yields the results set forth above; that is, the O parent usually produces an O clone, the E parent an E clone. When cytoplasm is visibly transferred between mates, usually both mates produce E clones; sometimes, however, both produce O clones or both produce selfer clones; sometimes only one mate (either one) produces a selfer clone. The frequencies of O, E, and S clones after cytoplasmic transfer may be affected by the composition of the culture medium (9). Cytoplasmic determination of mating type was shown more directly (50) by removing cytoplasm from E cells and injecting it into O cells just before the sensitive period of macronuclear origin from a product of the synkaryon: Fifty percent of the progeny clones were type E.

When a selfer clone develops from a fertilized cell, the clone commonly consists of two pure caryonides, one type O, the other E (49). The existence of sister caryonides differing in mating type suggests macronuclear differentiation; macronuclear regeneration experiments proved it (25). Mating type O persists through macronuclear regeneration even in the presence of cytoplasm that determines the newly arisen zygotic macronuclei for type E. As in the A system, macronuclear differentiation is stable.

Macronuclei are stably differentiated not only for mating type, but also for production of the corresponding cytoplasmic determiner. This was shown by following through the next autogamy the two diverse lines of descent (the O and E sublines) mentioned in the preceding paragraph. Although they both descended from a cell with E-determining cytoplasm, results at the next autogamy showed that this E cytoplasm was perpetuated only in the type E line of descent, not in the type O line of descent. Autogamy in the former yielded type E clones; in the latter, type O clones (25). The same results occur without macronuclear regeneration, when O and E sister caryonides have been produced from the same fertilized cell.

Stable differentiations for control of mating type and for production of cytoplasmic factors are regularly correlated in caryonides pure for one mating type, either O or E; the correlation, however, breaks down in selfer caryonides (49). Such caryonides contain both O and E cells. Unlike E cells of pure E caryonides that normally produce E progeny at conjugation, the E cells in selfer caryonides usually produce O progeny at conjugation. Most of the E cells thus seem to lack the E cytoplasmic state, but at least some have it, for they produce E progeny at conjugation. The proportion that produces E progeny, while greatly reduced at all temperatures, is greater at high than at low temperatures, the temperature being effective only before the first cell division, as in the A system. Thus at least some of the E cells are capable of producing either O or E progeny at conjugation, depending on external conditions. This indicates that cells of type E in selfing caryonides may possess simply a reduced and usually ineffective concentration of an E cytoplasmic factor.

More striking than the seemingly quantitative uncoupling of the two aspects of macronuclear differentiation in selfing caryonides of *P. tetraurelia* is their qualitative uncoupling in certain genotypes of *P. septaurelia* (54). Genotype mt^+/mt^+, like the similarly designated genotype in other species, can be mating type O or E, in both cases regularly having the corresponding cytoplasmic state. Homozygotes for the allele mt^o are always mating type O, but they always possess the E cytoplasmic state! They always produce E progeny when mt^+ is introduced into them. Moreover, heterozygotes (mt^o/mt^+) also always have the E cytoplasmic state regardless of whether they are mating type O or E. The allele mt^o is thus dominant for cytoplasmic state E, but recessive for restriction to mating type O. The E cytoplasmic state is stably inducible in mt^+/mt^+, but stably constitutive in mt^o/mt^+ and mt^o/mt^o. The latter genotype renders new zygotic macronuclei incapable of differentiating for determination of mating type E in spite of the presence of adequate concentrations of E cytoplasmic factor and the stable continued maintenance of it by these macronuclei and their descendants.

Another mutation, *n*, unlinked to *mt*, also restricts homozygotes of *P. septaurelia* to type O but has no effect on cytoplasmic state (54). Such O caryonides can have either the O or the E cytoplasmic state as detected by the mating type developed when the n^+ allele is reintroduced. In *n/n* as in n^+/n^+ lineages, both alternative cytoplasmic states are inherited through sexual reproduction. After each fertilization new macronuclei are differentiated for different cytoplasmic states depending on whether the parent had the O or E state, although both parent and progeny are restricted to mating type O. Thus, in this mutant, as in mt^o, cytoplasmic state can be uncoupled from mating type.

In *P. tetraurelia*, three unlinked mutations—mtA^o, mtB^o, mtC^o—have been found (53) to be like mutation *n* in *P. septaurelia*. Homozygotes for any one of these genes are restricted to mating type O but can have either O or E cytoplasm. When all three dominant alleles (mtA^+, mtB^+, and mtC^+) are present, the caryonides can be either mating type O or E depending upon whether the parent cell had O or E cytoplasm.

Homozygotes for another recessive mutation, mtB^s, are selfers when the homozygote has E cytoplasm (53). Most cells in such clones are type O. Selfing also occurs in the double heterozygotes, $mtA^o/mtA^+, mtC^o/mtA^+$; $mtB^o/mtB^+, mtC^o/mtC^+$; and $mtA^o/mtA^+, mtB^s/mtB^+$; but in these selfers most cells are type E.

In *P. septaurelia*, about 15% of wild-type (mt^+/mt^+, n^+/n^+) caryonides are selfers containing both O and E cells. As the clones age, there is a progression toward stable E; that is, selfing clones that are O when young become selfers and eventually pure E, or they are selfers when young and later become pure E. Shifts in the same direction occur in younger subcultures as they progressively starve.

Even individual cells may change in a few hours from O to "bisexual" to E (56). All of the results on selfers are consistent with Butzel's (29) hypothesis that O is a precursor of E.

The mutations in *P. tetraurelia* and *P. septaurelia* show that mating types depend on the action of genes at more than one locus. One of these loci in *P. septaurelia* is also involved in production of a cytoplasmic product that differentiates new zygotic macronuclei for mating type. Taub (54) suggests that this differentiation consists simply of an "on" or "off" state of a gene(s) for mating type E. If production of the cytoplasmic factor (in mt^+/mt^+, n^+/n^+) could be blocked by mutation or rendered constitutive instead of inducible, one would expect restriction to type O or to type E depending on whether the factor represses or derepresses the E gene(s). Search for mutants restricting to type E has until recently been unsuccessful; according to Y. Brygoo (manuscript submitted for publication), two such mutations—both lethal—have been obtained in Beisson's laboratory.

V MACRONUCLEAR DIFFERENTIATION FOR OTHER TRAITS

Working with *P. caudatum* and *P. aurelia* before mating types were known, Jollos (57) reported many examples of "enduring modifications" (Dauermodifikationen) held to be induced by exposure to various agents during the period immediately after conjugation and to persist during asexual reproduction, but to disappear at the next or a subsequent fertilization. Most of these Dauermodifikationen—resistance to arsenic, to calcium salts, or to high temperature, and alterations in cell size—showed various complications in transmission and were not analyzed in a way that could reveal a macronuclear basis, if it existed. But one was analzyed; it was a case of a 50% increase in growth rate following long culture at 31°C. This Dauermodifikation, obtained in stock h of *P. aurelia*, showed typical caryonidal inheritance. The two growth rates shifted back and forth at autogamy and conjugation, parents having either growth rate producing some daughter caryonides with the "normal" rate and some with the increased rate. Jollos therefore concluded that the basis of the difference lay in the macronuclei.

Recently, modifications of resistance to $CaCl_2$ and to extreme temperatures, induced by exposure to these agents, and variations of growth rate induced by exposure to different temperatures have been reinvestigated in *P. primaurelia* by Génermont (10, 58, 59). For these traits, he observed temperature effects during the standard sensitive period and his analysis of variance indicated that the system of determination and inheritance is essentially the same as the B system for mating types, although this species has the A system for mating types. He also calls attention (10) to three studies on *P. caudatum* that appear to show caryonidal inheritance (macronuclear differentiation) for cell size, resistance to high temperature, and the ratio of macronuclear to micronuclear DNA ("endopolyploidy" level).

Macronuclear differentiation appears to be involved to some extent in the determination and inheritance of serotypes, i.e. of ciliary immobilization antigens, in *P. primaurelia* and *P. tetraurelia*. In *P. primaurelia*, heterozygotes for certain allelic antigens usually express both antigens, sometimes only one or only the other one.

These variations seem to be caryonidal (60). In *P. tetraurelia,* cells of serotype B, when exposed to 31°C during the first post fertilization cell cycle, almost always produced clones of serotype A except if they underwent macronuclear regeneration, in which case they nearly always produced a clone of serotype B (61). This indicates rather stable macronuclear differentiation for serotype; high temperature during the sensitive period was much more effective in shifting expression to another serotype gene in zygotic macronuclei than in regenerating fragments. Other serotype shifts have also been correlated with the period of development of new zygotic macronuclei (62, 63). Shifts can, however, also be brought about at other times, but the possible shifts within any one caryonide may be only a part of the whole range of shifts possible for a particular homozygous genotype, different parts of the range being characteristic of different caryonides (62).

Macronuclear differentiation for phase of a circadian rhythm of mating type change has been indicated in studies on syngen 2 of *P. multimicronucleatum* (36).

The most fully investigated example of macronuclear differentiation, aside from mating types, concerns the capacity of cells of a stock of *P. tetraurelia* to respond to external stimuli (e.g. picric acid) by discharging trichocysts (64; T. M. Sonneborn and M. V. Schneller, in preparation). In all essential respects, the system of determination and inheritance is exactly the same as the B system of mating type determination. The cytoplasmic determiner of macronuclear differentiation for trichocyst discharge (D) is distinct from the cytoplasmic determiner of macronuclear differentiations for mating type E. Nondischarge (N) caryonides, like the O mating type, lack the cytoplasmic determiner. The existence of the cytoplasmic determiner has been shown both by normal transfer between mates and by microinjection of cytoplasm from D to N cells during the sensitive first cell cycle. There is no genotypic difference between N and D lines of descent within this stock.

Two additional features are shown by the N-D system: 1. The probability of shift from N to D is markedly affected (only during the sensitive period) not only by temperature, but also by the presence versus absence of food and by other conditions. 2. Under certain conditions it is possible to obtain caryonides of intermediate character containing, in addition to N and D cells, cells that discharge greatly differing percentages of their several thousand trichocysts. These intermediates are formally comparable to selfing caryonides but are much more susceptible to detailed analysis because of the readily scored quantitative gradations in cell character. The data are consistent with the assumption that intermediate caryonides are due to activation in the original ancestral macronucleus of only a fraction of the copies of a gene for trichocyst discharge. The amitotic divisions of such a partially activated macronucleus then would lead to the observed intracaryonidal variations.

VI MISCELLANEOUS CONCLUDING COMMENTS

1. The two unique and great analytical advantages of all of these nuclear differentiations in ciliates are their origin at the same readily synchronized sensitive period, the period when new zygotic macronuclei are developing, and the existence of many copies of the genome within a single nucleus, the macronucleus, that distributes these copies at amitotic divisions to daughter nuclei. The first advantage has been

exploited in all of the examples. The second one is of special value in the study of quantitative characters where gene dosage effects prevail, as in grades of resistance to temperature or chemical agents (10) and in the number of dischargeable trichocysts (see preceding paragraph).

2. The technique of microinjection, shown to be effective in transferring the cytoplasmic determiner of macronuclei in the B mating type system and the trichocyst-discharge system, opens the way to isolation and characterization of these determiners. This should be an important first step toward discovering how they activate genes and whether, as suspected, they are the products of the genes they activate.

3. The problem of ultimate interest is the nature of the stable alternative nuclear differentiations taking place during the development of macronuclei from micronuclei. The relative inertness of micronuclei (65–67) and the gain of activity when they develop into macronuclei invite comparison with the protein changes occurring in an inert sperm nucleus after it enters an egg. More specifically, the gene for trichocyst discharge becomes demonstrably active once and for all during macronuclear differentiation, but a gene for mating type does not overtly express itself for up to 100 or more cell generations in stocks that have a clonal period of sexual immaturity. Does this mean that the relevant gene was not activated during macronuclear differentiation, but only put into an activable state? Recent evidence for an immaturity substance transferable by microinjection in *P. caudatum* (68) opens the possibility that a mating type gene could be activated at the time of macronuclear differentiation, but that its overt expression during immaturity could be blocked by the immaturity substance.

The apparent maintenance of multiple active and inactive copies of the same gene within a single macronucleus and within its descendant macronuclei, as in selfing caryonides and in caryonides intermediate for the trichocyst discharge trait, argues for a structural difference between the active and inactive genes or between the associated chromosomal regions. If possible, a cytoplasm-free nucleus from an active cell should be injected into an inactive cell. If it becomes inactive, would injected cytoplasm from an active cell reactivate it? A more direct and potentially more informative line of attack would be to carry out experiments designed to test various molecular models that have been suggested to distinguish stable active from stable inactive genes: molecular mechanisms of heterochromatinization (1–3), methylation (4, 5) insertion factors (31), and reversible orientation of promoters or other DNA sequences (32, 33).

4. The attractiveness of the ciliate systems for sustained efforts to solve the problems of nuclear differentiation is heightened by evidence for stable nuclear differentiation in vertebrate development (69). The abundance and accessibility of such systems in ciliates makes them materials of choice for further attacks on nuclear differentiation, clearly a general problem of developmental genetics (70).

ACKNOWLEDGMENT

The author is indebted to D. L. Nanney for critical reading of the manuscript.

Literature Cited

1. Lyon, M. F. 1972. X-chromosome inactivation and developmental patterns in mammals. *Biol. Rev.* 47:1–35
2. Cook, P. R. 1974. On the inheritance of differentiated traits. *Biol. Rev.* 49:51–84
3. Cattanach, B. M. 1975. Control of chromosome inactivation. *Ann. Rev. Genet.* 9:1–18
4. Holliday, R., Pugh, J. E. 1975. DNA modification mechanisms and gene activity during development. *Science* 187:226–32
5. Sager, R., Kitchin, R. 1975. Selective silencing of eukaryotic DNA. *Science* 189:426–33
6. Allen, S., Gibson, I. 1973. Genetics of Tetrahymena. In *Biology of Tetrahymena*, ed. A. M. Elliott, pp. 307–73. Stroudsburg, Pa.: Dowden, Hutchinson & Ross. 508 pp.
7. Beale, G. H. 1954. *The Genetics of Paramecium aurelia*. London: Cambridge Univ. Press. 178 pp.
8. Beale, G. H. 1964. The genetic control of cell surfaces. *Recent Prog. Surf. Sci.* 2:261–351
9. Butzel, H. M. Jr. 1974. Mating type determination and development in Paramecium aurelia. In *Paramecium, A Current Survey*, ed. W. J. van Wagtendonk, pp. 91–130. New York: Elsevier. 499 pp.
10. Génermont, J. 1975. Le polymorphisme chez les protozoaires ciliés. *Mém. Soc. Zool. France* No. 37, pp. 205–33
11. Kimball, R. F. 1943. Mating types in the ciliate Protozoa. *Q. Rev. Biol.* 18:30–45
12. Nanney, D. L. 1954. Mating type determination in *Paramecium aurelia*. In *Sex in Microorganisms*, ed. D. H. Wenrich, I. F. Lewis, J. R. Raper, pp. 266–83. Washington DC: Am. Assoc. Adv. Sci. 362 pp.
12a. Nanney, D. L. 1957. The role of the cytoplasm in heredity. In *The Chemical Basis of Heredity*, ed. W. D. McElroy, B. Glass, pp. 134–64. Baltimore: Johns Hopkins Univ. Press. 548 pp.
12b. Nanney, D. L. 1958. Epigenetic factors affecting mating type expression in certain ciliates. *Cold Spring Harbor Symp. Quant. Biol.* 23:327–35
13. Nanney, D. L. 1964. Macronuclear differentiation and subnuclear assortment in ciliates. In *The Role of Chromosomes in Development*, ed. M. Locke, pp. 253–73. New York: Academic
14. Nanney, D. L. 1968. Ciliate genetics: Patterns and programs of gene action. *Ann. Rev. Genet.* 2:121–40
15. Preer, J. R. Jr. 1957. Genetics of the Protozoa. *Ann. Rev. Microbiol.* 11:419–38
16. Preer, J. R. Jr. 1968. Genetics of the Protozoa. In *Research in Protozoology*, ed. T. T. Chen, 3:130–278. New York: Pergamon. 744 pp.
17. Sonneborn, T. M. 1942. Inheritance in ciliate Protozoa. *Am. Nat.* 76:46–62
18. Sonneborn, T. M. 1947. Recent advances in the genetics of Paramecium and Euplotes. *Adv. Genet.* 1:263–358
19. Sonneborn, T. M. 1957. Breeding systems, reproductive methods, and species problems in Protozoa. In *The Species Problem*, ed. E. Mayr, Publ. No. 50, Am. Assoc. Adv. Sci., Washington DC
20. Sonneborn, T. M. 1975. *Tetrahymena pyriformis*. In *Handbook of Genetics*, ed. R. C. King, 2:433–67. New York: Plenum. 631 pp.
21. Sonneborn, T. M. 1975. *Paramecium aurelia*. See Ref. 20, pp. 469–594
22. Nanney, D. L., McCoy, J. W. 1976. Characterization of the species of the *Tetrahymena pyriformis* complex. *Trans. Am. Microsc. Soc.* 95:664–82
23. Sonneborn, T. M. 1975. The *Paramecium aurelia* complex of fourteen sibling species. *Trans. Am. Microsc. Soc.* 94:155–78
24. Nanney, D. L. 1953. Nucleo-cytoplasmic interaction during conjugation in Tetrahymena. *Biol. Bull.* 105:133–48
25. Sonneborn, T. M. 1954. Patterns of nucleo-cytoplasmic integration in Paramecium. *Caryologia* Suppl., pp. 307–25
26. Sonneborn, T. M. 1942. Inheritance of an environmental effect in *Paramecium aurelia*, variety 1, and its significance. *Genetics* 27:169
27. Hallet, M.-M. 1975. Differentiation of the macronuclear anlagen of *Paramecium aurelia* (syng. 1) in the presence of calcium chloride: Analysis of the variability of the disturbances. *Acta Protozool.* 14:281–90
28. Hallet, M.-M. 1972. Effet du chlorure de calcium sur le déterminisme génétique du type sexuel chez *Paramecium aurelia* (Souche 60, syng. 1). *Protistologica* 8:387–96
29. Butzel, H. M. Jr. 1955. Mating type mutations in variety 1 of *Paramecium*

aurelia, and their bearing upon the problem of mating type determination. *Genetics* 40:321–30
30. Demerec, M. 1941. Unstable genes in Drosophila. *Cold Spring Harbor Symp. Quant. Biol.* 9:145–49
31. Starlinger, P., Saedler, H. 1972. Insertion mutations in microorganisms. *Biochemie* 54:177–85
32. Zieg, J., Silverman, M., Hilmen, M., Simon, M. 1977. Recombinational switch for gene expression. *Science* 196:170–72
33. Hicks, J. B., Herskowitz, I. 1977. Interconversion of yeast mating types. II. Restoration of mating ability to sterile mutants in homothallic and heterothallic strains. *Genetics* 85:373–93
34. Bleyman, L. K. 1967. Determination and inheritance of mating type in *Paramecium aurelia*, syngen 5. *Genetics* 56:49–59
35. Bleyman, L. K. 1967. Selfing in *Paramecium aurelia*, syngen 5: Persistent instability of mating type expression. *J. Exp. Zool.* 165:139–46
36. Barnett, A. 1966. A circadian rhythm of mating type reversals in *Paramecium multimicronucleatum*, syngen 2, and its genetic control. *J. Cell. Physiol.* 67:239–70
37. Phillips, R. B. 1969. Mating type inheritance in syngen 7 of *Tetrahymena pyriformis:* Intra- and interallelic interactions. *Genetics* 63:349–59
38. Allen, S. L., Nanney, D. L. 1958. An analysis of nuclear differentiation in the selfers of Tetrahymena. *Am. Nat.* 92:139–60
39. Bleyman, L. K., Simon, E. M. 1968. Clonal analysis of nuclear differentiation in Tetrahymena. *Dev. Biol.* 18:217–31
40. Nanney, D. L., Caughey, P. A. 1953. Mating type determination in *Tetrahymena pyriformis. Proc. Natl. Acad. Sci. USA* 39:1057–63
41. Nanney, D. L., Caughey, P. A. 1955. An unstable nuclear condition in *Tetrahymena pyriformis. Genetics* 40:388–98
42. Nanney, D. L., Caughey, P. A., Tefankjian, A. 1955. The genetic control of mating type potentialities in *Tetrahymena pyriformis. Genetics* 40:668–80
43. Nanney, D. L., 1956. Caryonidal inheritance and nuclear differentiation. *Am. Nat.* 90:291–307
44. Nanney, D. L., Allen, S. L., 1959. Intranuclear coordination in Tetrahymena. *Physiol. Zool.* 32:221–29
45. Nanney, D. L. 1959. Genetic factors affecting mating type frequencies in variety 1 of *Tetrahymena pyriformis. Genetics* 44:1173–84
46. Nanney, D. L. 1960. Temperature effects on nuclear differentiation in variety 1 of *Tetrahymena pyriformis. Physiol. Zool.* 33:146–51
47. Nanney, D. L. 1963. Aspects of mutual exclusion in Tetrahymena. In *Biological Organization at Cellular and Supracellular Levels*, ed. R. J. C. Harris, pp. 91–109. New York: Academic. 261 pp.
48. Ammermann, D. 1965. Cytologische und genetische Untersuchungen an dem Ciliaten *Stylonychia mytilus* Ehrenberg. *Arch. Protistenkd.* 108:109–52
49. Nanney, D. L. 1957. Mating type inheritance at conjugation in variety 4 of *Paramecium aurelia. J. Protozool.* 4:89–95
50. Koizumi, S. 1971. The cytoplasmic factor that fixes macronuclear mating type determination in *Paramecium aurelia*, syngen 4. *Genetics* 68(Suppl):s34
51. Butzel, H. M. Jr. 1968. Mating type determination in stock 51, syngen 4 of *Paramecium aurelia* grown in axenic culture. *J. Protozool.* 15:284–90
52. Butzel, H. M. Jr. 1973. Abnormalities in nuclear behavior and mating type determination in cytoplasmically bridged exconjugants of doublet *Paramecium aurelia. J. Protozool.* 20:140–43
53. Byrne, B. C. 1973. Mutational analysis of mating type inheritance in syngen 4 of *Paramecium aurelia. Genetics* 74:63–80
54. Taub, S. R. 1963. The genetic control of mating type differentiation in Paramecium. *Genetics* 48:815–34
55. Taub, S. R. 1966. Regular changes in mating type composition in selfing cultures and in mating type potentiality in selfing caryonides of *Paramecium aurelia. Genetics* 54:173–89
56. Taub, S. R. 1966. Unidirectional mating type changes in individual cells from selfing cultures of *Paramecium aurelia. J. Exp. Zool.* 163:141–50
57. Jollos, V. 1921. Experimentelle Protistenstudien. I. Untersuchungen über Variabilität und Vererbung bei Infusorien. *Arch. Protistenkd.* 43:1–222
58. Génermont, J. 1961. Déterminants génétiques macronucléaires et cytoplasmiques controlant la résistance au chlorure de calcium chez *Paramecium aurelia* (souche 90, variété 1). *Ann. Génét.* 3(1):1–8

59. Génermont, J. 1966. Le déterminisme génétique de la vitesse de multiplication chez *Paramecium aurelia* (syng. 1). *Protistologica* 2:45–51
60. Capdeville, Y. 1971. Allelic modulation in *Paramecium aurelia* heterozygotes. *Mol. Gen. Genet.* 112:306–16
61. Preer, J. R. Jr., Bray, M., Koizumi, S. 1963. The role of cytoplasm and nucleus in the determination of serotype in Paramecium. *Proc. XI Int. Congr. Genetics*, p. 189
62. Skaar, P. D. 1956. Past history and pattern of serotype transformation in *Paramecium aurelia*. *Exp. Cell Res.* 10:646–56
63. Dryl, S. 1965. Antigenic transformation in relation to nutritional conditions and the interautogamous cycle in *Paramecium aurelia*. *Exp. Cell Res.* 37:569–81
64. Sonneborn, T. M. 1977. Local differentiations of the cell surface of ciliates: Their determination, effects, and genetics. In *Cell Surface Reviews*, Vol. 4, *The Synthesis, Assembly and Turnover of Cell Surface Components*, ed. G. Poste, G. L. Nicolson. In press
65. Sonneborn, T. M. 1946. Inert nuclei: Inactivity of micronuclear genes in variety 4 of *Paramecium Aurelia*. *Genetics* 31:231
66. Sonneborn, T. M. 1954. Is gene K active in the micronucleus of *Paramecium aurelia?* *Microb. Genet. Bull.* 11:25–26
67. Pasternak, J. 1967. Differential genic activity in *Paramecium aurelia*. *J. Exp. Zool.* 165:395–418
68. Miwa, I., Haga, N., Hiwatashi, K. 1975. Immaturity substances: Material basis for immaturity in Paramecium. *J. Cell Sci.* 19:369–78
69. Brothers, A. J. 1976. Stable nuclear activation dependent on a protein synthesized during oogenesis. *Nature* 260:112–15
70. Briggs, R. 1977. Genetics of cell type determination. In *Cell Interactions in Differentiation*, ed. L. Saxén, L. Weiss. New York: Academic. In press

GENETIC RECOMBINATION IN BACTERIA

❖3126

A. Eisenstark

Division of Biological Sciences, University of Missouri-Columbia, Columbia, Missouri 65201

CONTENTS

INTRODUCTION	369
PROPERTIES OF recA MUTANTS	370
Induction Model for Synthesis of Product(s)	371
The RecA Gene Product	373
Possible Relation to DNA Supercoil and Membrane	373
PROPERTIES OF recB AND recC MUTANTS	374
STAGES IN RECOMBINATION	376
GENETIC DUPLICATIONS AND DELETIONS	377
HOST INVOLVEMENT IN PLASMID AND PHAGE RECOMBINATION	380
Plasmid Recombination	380
Recombinational Genes in Phage	381
RELATION BETWEEN MUTATION AND RECOMBINATION	382
PERTURBATION (IN VIVO) OF RECOMBINATION BY PHYSICAL AND CHEMICAL AGENTS	384
rec MUTANTS IN NON-ENTERIC BACTERIA	386
CONCLUSIONS	386

INTRODUCTION

The current flow of information on bacterial recombination promises a clearer understanding of the sequence of molecular events in the crossover process. This review discusses some recent additions to a continuum of important observations of bacterial and phage recombination, a subject that has received periodic analysis (23, 31, 33, 34, 59, 65, 72, 81, 91, 146, 174, 201, 211).

This review emphasizes the pivotal role of the RecA gene product. The resolution of its structure and function, long complicated by its bewildering pleiotropic expressions, may soon be accomplished. An induction model (67) may explain some of the DNA metabolism phenotypes that *recA* shares with *lex,* but the model still leaves the puzzle of the unique *recombinational* function of RecA. The recombinational stages at which RecA and RecBC activities are needed are now identified. A miscellany of observations dealing with recombination is cited here, especially as they may relate to evolutionary processes at the DNA level.

Finally, the capability of perturbing the crossover process in vivo with chemical and physical agents is explored for clues of the various steps of recombination.

PROPERTIES OF *recA* MUTANTS

The extensive, pleiotropic controls of the RecA gene are striking and baffling. In addition to its critical role in recombination, the varied expressions of a *recA* mutation include high radiation sensitivity (129, 235); reduction in UV mutagenesis (229); lack of prophage induction, occurring either spontaneously or upon treatment (153, 158); reckless degradation of DNA (52, 134); lack of UV reactivation (Weigle reactivation) of irradiated phage (103, 153, 210); lack of prophage reactivation (153, 157, 158); growth abnormalities (26, 52, 145, 210); uncoupling of cell division and DNA synthesis (22, 58, 66, 95, 155, 209, 210, 232); loss of respiration control after UV (210); increased photoprotection to UV (210); lack of repair of double-strand DNA breaks (214); loss of repair of synergistic damage by peroxide and near-UV radiation (74); loss of ability to cleave phage repressor (144, 181, 213); and loss of colicin E1 expression, even if the plasmid gene is present in the cell (94).

The RecA pleiotropism may have obscured an understanding of its role in recombination. It now appears that most (if not all) of its nonrecombination properties may involve special *recA* product binding to single-stranded DNA to inhibit *exoV* activity (67, 68, 123, 141), which, in turn, may modulate other DNA metabolic functions. Although the search for this binding protein led to an interesting model for Lex and RecA repair functions (68), the model does not fully explain the role of *recA*$^+$ in recombination and other pleiotropic activities.

A proposed explanation (153) distinguishes between two separate functions of the RecA product, one inducible and the other constitutive. The constitutive function deals with recombination, spontaneous DNA degradation (exoV control), and host cell reactivation; the inducible functions may explain prophage development (158), synthesis of protein to achieve error-prone repair (190–192, 229), filament formation (22), inhibition of UV-provoked exoV activity (171), and UV reactivation (103, 153).

To explain this complex *recA* pleiotropy, Morand et al (153) propose that *recA*$^+$ is a gene essential to many aspects of DNA metabolism. The RecA protein may interact with DNA (alone or complexed with other proteins). Mutation at different sites within the cistron may yield different phenotypes (126), including those previously described for *lexB* (20, 153, 159), *tif* (58, 225), and *zab* (27).

The inability of the cell to undergo genetic recombination is indeed a special feature of certain *recA* mutants [see special cases (70, 126, 127)]. Their stringency is striking; recombination in such *recA* mutants has been estimated to occur at a maximum of 10^{-6} per mating (69). In our own experience with P22 transduction of *recA* recipients of *Salmonella typhimurium* LT2, we never observe recombinants (49). Perhaps even more striking is the noninducibility of one of the *S. typhimurium* prophages. When supernatant samples from a wild-type *S. typhimurium* LT2 culture are plated on the Boyd Q1 strain of *S. typhimurium*, thousands of plaques per milliliter are observed. However, if that strain carries the *recA* mutation, no plaques are observed upon plating of supernatant samples (49), even when cells have been exposed to a mutagen. It is obvious that the *recA* gene product in *S. typhimurium* LT2 rigidly controls prophage repression as well as recombination.

Additional regulatory roles of *recA* are the coupling of DNA synthesis with cell division (*recA* mutants lose this coupling) (22, 155, 232), and the coupling of DNA synthesis with respiration (210). These may involve intricate binding of the RecA product to DNA, other proteins, and membrane. Comparison of the properties of *lexA* (90' on new genetic map) (9) with *recA* (58') has been useful in sorting out the specific recombinational roles of *recA*. The *lexA* mutation does not duplicate RecA phenotypic losses in recombination proficiency, rate of DNA breakdown without UV damage, DNA synthesis by a single ^{32}P decay (214,) UV reactivation (103), spontaneous prophage inducibility, cleavage of prophage repressor (181, 213), integration of mu-1 phage (24), induction of colicin synthesis (94), or λ Fec⁻ growth (153). Both *lexA* and *recA* share such properties as radiation and drug sensitivity, noninducibility of prophage, and high rate of DNA degradation with UV (153, 160).

Induction Model for Synthesis of Product(s)

Since induction of phage production in lysogenic cells requires *intra*chromosomal recombination, *inter*chromosomal crossover might also be subject to analogous inducible regulation (66, 68, 123, 171, 190), although the system may be very different in detail.

The studies of McEntee and co-workers (139, 141) show that the *recA*⁺ protein can be induced by UV, thymine starvation, and nalidixic acid, thus supporting contentions of Sedgwick (191) that repair and recombination are inducible. The Gudas & Pardee model (67) is based on the observation that a protein (protein X), important in repair and recombination, is inducibly synthesized upon DNA damage. (Although it appears that protein X is the RecA product, this review uses the designation *protein X* when citing authors who themselves use the term.) Gudas & Pardee stated that the protein is synthesized when a repressor is removed from the operator site; this repressor is the gene product of LexA (90 min on the *Escherichia coli* chromosome). This protein binds to single-stranded DNA to prevent DNA degradation by the RecBC-coded exoV nuclease. According to this model, the RecA gene product, together with exoV-fragmented DNA inducers, are required to remove the LexA-coded repressor from the operator, possibly via protease action (144, 181, 213). This model has the attractive feature of a feedback loop for the regulation of DNA repair.

According to this model, a *lexA* product tenaciously binds to the operator to prevent protein X synthesis. Thus, *lexA* and *recA* mutants would yield those phenotypes characteristic of the absence of protein X. However, there is one important difference—*lexA* mutants are almost normal in recombination (152). This would indicate that, while protein X may be needed to stop action of exoV, it is not necessary for recombination, at least not in the same configuration nor complexed with the same components. The RecA repair phenotypes can be uncoupled from the recombination malfunctions via a second mutation in a *recA* strain (63, 161, 162). In addition to *lexA* production of a faulty repressor, there should be some *lexA* mutants that give *no* repressor molecules, or *tif* mutants with ineffective operator regions, where the protein would be synthesized constitutively. Constitutive synthesis is observed in *tsl, tsl recA,* and *tif* and *lexB* mutants (153). *LexB* alleles are less sensitive to UV and X ray than *recA*. Also, UV fails to induce λ in *lexB* mutants, but thymine starvation still induces λ. The characteristics and the mapping of *lexB, zab, tif,* and *recA* mutations support the view that they may be in the same gene (28, 126, 153). McEntee (138), using three-factor crosses, has mapped these mutants between *srlC* and *alaS* (9).

Morand et al (153) present two possible models to explain *recA* action. Model one states that $recA^+$ regulates the expression of a number of other genes. Different mutants would lead to different regulatory activities and to different phenotypes, as proposed by Gudas & Pardee (67). Model two states that *recA* is a gene essential to coordinated DNA metabolism. The *recA* protein interacts (alone or complexed with other proteins) with the DNA. Different mutations alter the RecA protein differently, and each of the phenotypic characteristics observed is a consequence of the specific alterations. The authors prefer this second model on the basis that mutations may affect different parts of the *recA* protein, each one being responsible for either induced or constitutive functions, but some mutations may interfere with both.

In addition to separating constitutive from inducible functions, there may be more than a single induction for the expression of the bewildering number of RecA phenotypes. Bridges et al (22) conclude that the induction of cell division in *E. coli* B_{s-12} is not part of the coordinate induction system usually associated with RecA. The cell division signaling system seems to "march to a different drummer" than the signaling system for DNA repair functions. If model two of Morand et al is valid, then the mutations within *recA* that alter induction capabilities may affect at least two separate systems. Prophage induction and error-prone DNA repair are apparently controlled by the same RecA-inducing signal, which can function constitutively via mutation (158).

Little & Hanawalt (123) have examined various ways that cells can be treated to make DNA degradation fragments to see which of these treatments lead to induction of protein X synthesis. They found that not all DNA degradation results in induction of protein X; the inducer may be a special DNA fragment, not necessarily dependent on exoV degradation of DNA. This fragment might either have a unique, specific structure or be present at a strategic location in the cell. The authors confirm that X ray, nalidixic acid, or thymine starvation result in protein X synthesis in wild-type strains (67). Also, they found that those DNA fragments produced by host-controlled restriction of phage λ do not induce protein X. Bleomycin degradation of DNA (even in *recBC* mutants) induces protein X (67). Their conclusion is that there is not a causal relationship between the production of DNA fragments and the induction of protein X.

In repair of DNA damage, DNA intermediates are competitors with prophage operators for the repressor binding (21), perhaps a clue as to a common repressor.

The RecA Gene Product

McEntee and colleagues have recently purified and identified the RecA protein of *E. coli* (138–141). This was accomplished by first isolating a plaque-forming λ phage carrying the RecA locus (140). By inducing a lysogen in which λ is integrated into the *srl* (sorbitol utilization), a locus adjacent to *recA*$^+$, they obtained increased yields of RecA protein (141). It is a soluble protein of 43,000 mol wt, found in heteropolymer of 150,000 mol wt after mild detergent lysis of host cells (141). This complexing may be a key to a number of activities of DNA repair, replication, and recombination, and may clarify *recA* pleiotropy. The RecA protein isolated has a molecular weight similar to protein X (67), and indeed, appears to be the same protein. About 80,000 molecules of protein X were reported to be in a fully induced cell (68). Both the RecA protein isolated by McEntee and protein X may be induced upon inhibition of DNA synthesis at the replication fork (68), followed by DNA degradation [see (123) for exception].

One can only speculate on how RecA protein structure is geared to account for the constitutive and inducible properties of *recA, lex, tif,* and related mutants; also, it is not known whether the RecA protein can act as a protease to cleave prophage repressors, as suggested in *E. coli* (143a, 181) and *S. typhimurium* (116, 213). The structural relation of the inducer to the RecA protein may be important in this regulation.

Possible Relation to DNA Supercoil and Membrane

The absolute dependence upon RecA product for recombination (see exceptions in reference 126) suggests that it is mandatory for this cell function. Tomizawa & Okawa (214) suggested that *recA*$^+$ might have a role in maintenance of chromosome structure, crucial for recombination but not for repair or replication. This hypothesis arose from observations of *recA* hypersensitivity to ^{32}P decay double-strand DNA breaks, suggesting that a rigid structure might be needed to keep the two strands together during repair; *recA* cells might lack this rigidity.

More recently, a number of investigators have characterized the superstructure of DNA in bacteria (60, 78, 169, 172). In a model described by Pettijohn (169) an RNA molecule as the core of a folded chromosome holds each of ca 140 loops in place; RNAse can unfold the chromosome. It takes many single-strand nicks to relax the DNA completely; therefore, relaxation in one of the loops is not transmitted to the other loops. Apparently, without external assault, there are only a few nicks per chromosome; thus, only a few of the 140 postulated loops are relaxed at any one time in the cell.

Although the argument that *recA* strains might be defective in chromosome structure is not very strong, there is indication that the RecA product may interact with supercoiled DNA. Holloman et al (88) propose that the energy required for the final uptake of the donor material is bound in the supercoil. If supercoiling were defective, recombination might be thwarted. A nuclease ($recC^+$) step follows $recA^+$ action. When supercoiled DNA from φX174 phage is mixed in vitro with fragments of single-strand DNA, recombinants are obtained in $recA$ spheroplasts (87). However, if the two parental components are added separately to cells without in vitro mixing, recombination occurs in $recA^+$ but not in $recA$ spheroplasts. The $recA^+$ product is apparently needed in vivo to consummate quickly the single-strand exchange with one of the two strands in the supercoil. A single strand apparently invades a double-stranded molecule, yielding a triple-stranded joint molecule in which one strand of the recipient molecule is subsequently cleaved. The Holloman model might account for differences in recombination for genes at different map positions in the chromosomes since the energies needed at precisely the time of recombination may be quite different; that is, supercoiling may not be identical throughout the chromosome, not even within short regions of the chromosome.

It is now known that an enzyme, DNA gyrase, can introduce superhelical turns, and is probably involved in λ integrative recombination (57). It would be interesting if this has a relation to a rec gene.

There is some argument that defective supercoiling is not an RecA trait. If the RecA product maintains the intricate superstructure of the cell's chromosome, it does not do this for the plasmid DNA that it harbors. Studies with P1 phage (182) and other plasmids (109) in $recA$ cells show that these plasmids are normal in supercoiling, recombination, and replication; the $recA^+$ product is obviously not essential for these plasmid activities.

The basis for RecA membrane involvement is even less solid, although a fraction of the isolated product binds to membrane (67, 68, 123, 141). Chipchase (31) has proposed that a protein complex, perhaps in the membrane, regulates recombination by controlling access of the necessary nucleases and polymerases to the chromosome. This is based on an earlier model of DNA-membrane binding in the crossover process (214). The idea receives support from Resnick (179) who describes rejoining of double-strand breaks via a membrane mechanism in *Micrococcus radiodurans,* an organism that has the extraordinary capability of repairing such breaks following radiation. He emphasizes that the process may be mechanical, to keep both parts of the broken duplex together until the crossover is accomplished. Study of membranes and attachment to supercoiled DNA, in DNA repair, using radiation-sensitive mutants, may be a way of probing the membrane role in recombination and other activities of the RecA product.

PROPERTIES OF $recB$ AND $recC$ MUTANTS

The exoV nuclease is capable of several DNA catalytic functions (121, 130). Its specific contribution to genetic recombination could be to create single-strand DNA gaps and ends (13, 14, 143), to create DNA fragments of very specific structure to act as inducers in the synthesis of protein X (67, 123), and/or as a "trimmer" of excess single-strand DNA after joint heteroduplex molecules are hydrogen bonded (130). In addition, RecBC may have DNA metabolic roles that are not directly related to recombination.

As noted below, the RecBC activity is probably involved in a late step in recombination (19), since crosses between two *lacZ* mutants can give rise to *lacZ* enzyme without exoV activity although no stable recombinant results (see section on stages in recombination.)

Enzymes similar to the exoV of *E. coli* not only are in other bacteria (147, 226) but also are coded for by phage genes (64, 101). The *red* gene in λ promotes transduction in *recB* hosts (64, 224), thus assuming one of the functions of the *recB* gene.

RecBC activity results from a complex of exoV polypeptide together with still another component (α subunit), coded for by a separate gene to form part of this enzymatic team (117, 118). The enzyme shortens duplex DNA by first binding both strands at a terminus. While attached to one strand, it unwinds the duplex and cleaves fragments from the other. After degrading several thousand nucleotides, the nuclease action is switched to the other strand (100). This activity is recycled until the DNA is finally reduced to oligonucleotides.

The RecBC enzyme may form complexes with each of the DNA polymerases (80, 100), promoting ATP-stimulated DNA synthesis in a reaction preferring DNA in duplex form. The complex, in extracts, converts double-stranded DNA into active templates. As noted for RecA protein, a DNA-membrane-protein complex may regulate recombination by controlling the access of the necessary nucleases and polymerases to the DNA (31).

Many models (143, 150) conceive of an unwinding function to generate single-stranded DNA for the initiation of recombination. At least two enzymes other than exoV, however, seem to be involved in the duplex unwinding in *E. coli* (1). One of them resembles exoV in some functions; it invades double-stranded regions and separates the strands by zipper-like action. This enzyme is apparently not *exoV*, but cooperates with it in the recombinational process.

The DNAse activity of *exoV* may have a secondary function in the restriction of foreign DNA (15, 194) (it can rip apart noncircular P22 or λ DNA), or perhaps in the termination of a recombination region by the placement of a judicious nick in the recombining DNA molecule. Also, exoV may degrade early markers of Hfr donor in conjugation (126).

DNA degradation is usually assigned to the *exoV*. However, over 10% degradation always occurs even in strains that lack this enzyme (234). This degradation may be a role of exoVIII or exoI, and may account for the low residual recombinational level in *recB* or *recC* crosses.

Modification of exoV activity via additional mutations has been useful in the identification and characterization of other nucleases that may have recombinational roles (59, 124, 125). A pseudo/revertant of *recB*, *sbcA*, allows for excess exonuclease VIII synthesis (110) and restores recombination capacity, as well as full postreplication repair capacity (195). A third mutation, *recF*, (in addition to *recB sbcA*), can again make it recombinationless (183). However, a *recF* mutation alone in wild type has almost normal recombination capacity. The *sbcA* gene codes for a product since, in heterozygotes, sbc^+ is dominant (123). There is another suppressor, *sbcB*, which indirectly suppresses *recBC* mutations by loss of exoI activity

(110). Recombination ability of *recB* and *recC* can be restored by *rac* (128) and *ror* (218) mutations.

Because of the retarded growth of *recB* and *recC* cultures (26, 34), it has been uncertain whether each *rec* cell is retarded, or whether occasional daughter cells are inviable. Recent experiments support the latter (145) since *rec* cultures in log phase consist of two components in the population, dividing and nondividing cells (145). Dividing cells support normal burst sizes of phage T4, they are normal in oxygen consumption, and retain β-galactosidase within the cell. The growth abnormalities of the culture are apparently due to the nondividing rec cells; these cells do not support T4 multiplication, they release β-galactosidase into the medium, and they eventually lyse.

STAGES IN RECOMBINATION

The interesting experiments of Birge & Low (19) established a stage in recombination at which the RecBC product is active. Upon conjugation of two parental cells, each with a *lacZ* defect, synthesis of the enzyme can be detected, even if the recipient is *recB* or *recC*. This indicates that, in a *recB* or *recC* recipient, recombination via *recA*$^+$ pathway has already progressed to the stage of a covalently bonded hybrid DNA segment so that a proper *lacZ* messenger can be transcribed. However, if the recipient is *recA*, no enzyme is detected. Apparently, in the latter case, the two parental *lacZ* segments have not yet joined; therefore, transcription does not occur.

Although this and other experiments (173) establish that RecA action precedes that of RecBC, we know little about the other steps.

Preceding RecA action, parental molecules may achieve the appropriate geographic positions, and must undergo some localized changes in preparation for the exchange. Little is known about this synapsis, although there is a hint from *Bacillus subtilis* studies (104) that an enzyme may bring two parental DNA molecules into register, a first step in recombination.

With regard to the importance of a naturally nicked DNA as a first step in the recombination process, Sobell (199) proposes that a restriction-like endonuclease produces a DNA nick that leads to synapsis. As noted below, chemical and physical assaults, many of which produce nicked DNA, can perturb the recombinational process. Chipchase (31) envisions an early step in recombination that does not involve DNA nicks initially but depends upon a heteroduplex formed via DNA melting proteins. This allows for strand assimilation followed by heteroduplex correction and strand crossover, perhaps under the control of the DNA membrane complex. This could explain a wide range of genetic data, including the relationship of mutation to recombination. Also, the unpairing of only a few nucleotides at a time in the process of hybrid formation might take place preceding a nick, as formulated by Meselson (142).

Studies with radioactive and heavy isotopes (168) verify that hydrogen-bonded joints are made before RecA product is needed, and that covalent linkages are made before RecBC action. However, the joint (hydrogen-bonded) molecules still have more single-strand breaks in $recA^-$ than in $recA^+$. An early role of RecA gene action is further supported by evidence (235) that its product needs to have been in operation before DNA can be repaired by either *recB, lexA, uvrD* products or the chloramphenicol-controlled repair steps of *uvrD, lexA, recB,* or *recF* gene actions. These results indicate that, if the RecA product has a multiple role, it functions early, and perhaps coordinately, in both repair and recombination.

How does this information fit into Holloman & Radding's (87) observation that supercoiled RF1 DNA from ϕX174 can recombine with single-stranded fragments in vitro? These complexes produce recombinants in both $recA^+$ and $recA$ spheroplasts. However, if RF1 and the single-stranded fragment are added separately, no recombinants are formed in *recA* cells. This further suggests that the RecA product, although not needed for synapsis, is needed in vivo for embracing the single-stranded fragments into the supercoil of the double strand. At a slow rate, the embracement apparently can take place in vitro without a RecA product.

Another possible role of RecA protein in recombination may be to bind single-stranded DNA at the junction of two parental strands to demarcate that DNA which exoV is allowed to digest. In the absence of exoV, the resulting untrimmed tails of single-stranded DNA may exert a destabilizing effect at the junctures, or perhaps, inhibit the localized structure from returning to the proper supercoiled state. Despite the Birge & Low information (19) that joint molecules had already been made, even in the absence of exoV, the unstable DNA condition could lead to a reversal, with DNA repair, but without recombination, to yield a viable cell.

Obviously, not only do the precise recombinational roles of RecA and RecBC remain a mystery, but other steps in this highly integrated process are still unknown, despite numerous models (31, 173, 179, 221). The awareness that other Rec genes, not yet fully explored (33, 34, 55), may have roles in recombination, adds to the obscurity.

GENETIC DUPLICATIONS AND DELETIONS

Crossover between different DNA segments within the same parental cell, even within the same chromosome, may be a common and spontaneous occurrence; such intrachromosomal recombination often results in multiple copies of the same gene, as well as in deletions and inversions. Genetic duplications may result from crossover mechanisms similar to those in biparental recombination. Of special interest are the duplications that occupy tandem locations. Although the $recA^+$ requirement for gene duplication has been described in many cases (11, 206–208), there are other cases of gene duplication in phage λ (50, 54), in *E. coli* (11), and in plasmids (36) that do not depend upon Rec genes.

The increase in frequency of tandem duplications by UV and other mutagens supports the view that they are formed by *intrachromosomal* recombination (86). As noted below, mutagens have long been known to increase the frequency of *biparental chromosomal* matings. Recent studies by Straus and colleagues (206–208) of tandem duplications in the glycine region in *S. typhimurium* deal with specific mutagens that induce these repeats, methods for selecting large genetic duplications [as much as one third of *S. typhimurium* genome (7)], and the large overlaps in duplicated regions. In *Pneumococcus* (113), it is postulated that UV stimulates crossover, leading to segregation of a duplicate region when daughter cells are formed.

The frequency of tandem duplications may vary greatly in different regions of the chromosome (82), with some regions or sites containing "hot-spots." Tandem duplications might be utilized to identify crossover hot-spots or "join points" in chromosomes (J. Roth, personal communication). Via transduction mapping, a join point should be cotransducible with a donor chromosomal marker (such as his^+).

In the case of duplication-translocation of tryptophan operon genes (98), unlike Straus's duplications (206), the length of the duplicated segment and the position to which it is translocated differ in each of seven cases studied. Since such duplication-translocation events may occur at a frequency comparable with point mutations, an important role in evolution (98) is noted.

The study of Anderson et al (7) further supports the view that duplications occur spontaneously, and are selected naturally, in Darwinian fashion. When certain *his* mutants were transduced with his^+ donors, mosaic colonies arose frequently. Upon analysis, Anderson et al found that the recipient originally had two mutant *his* copies. The his^+ transducing fragment recombined with one of the two his^- copies. Since there is a high frequency of intrachromosomal crossover in duplicate regions, with some resulting deletions, the daughter cells within a transduced colony would have three genotypes, his^+, his^-, his^+his^-, and thus the colony would have a mosaic appearance.

As additional cases are studied (J. Roth, personal communication), it becomes apparent that such tandem duplications in *S. typhimurium* are extremely common (ca 10^{-3} cells are duplicated for *his*), large (some include 25% of chromosome), and $recA^+$-dependent for formation. Further careful analysis of mosaic clones shows that many of the unstable his^+ transductants have a segregation pattern that suggests that exchanges between daughter chromosomes (sister chromatids) may be a common feature of bacterial growth.

Straus and his group have distinguished between two classes of gene duplication (208). In some cases, the duplicate regions appear to have unique end points whereas in others, the end points vary. The precise end point duplication was noted when a malate selection technique was used. The ability of Straus et al to get tandem duplications under these stressed conditions further supports the view that organisms with duplicate genes may have a selective advantage under similar environmental selective pressures. Duplicate regions may be easily deleted again by recombination. This is an example of Darwinian fine tuning, with duplications and inversions occurring at a relatively high frequency. Studies utilizing chemostats to obtain hypersynthesis of enzymes via duplication support this viewpoint (90). Also, there may be selection pressures for inversions that result in reduced recombination (212).

The observation that *recC* can influence the sectoring of mosaic Lac and Mal colonies following UV treatment suggests that these genes may be duplicated via recombination on the chromosome (208). Strains of *E. coli* with duplicate copies of a lac^- operon may undergo intrachromosomal recombination to yield lac^+ papillae on colonies (30, 55, 105, 106). From these strains, Freifelder (55) isolated new and interesting colonies with no lac^+ papillae, and thus defective in Rec genes. Novel Rec mutants have also been isolated by a similar papillae selection technique, utilizing *gal* meridiploids (200). Interestingly, some of the lac^+ isolates that yield excess arrays of papillae have a hyperrecombinational phenotype (55, 105). A known mutant (*xseA*), deficient in exonuclease VII (30), produces increased intrachromosomal *lac* recombination. In fact, if it is a double mutant with *polA*, the recombination frequency is even higher than with each mutant alone.

In another study in which the *lac* duplication was not tandem, not only did recombination depend on rec^+ genes, but formation of recombinants occurred primarily as the cells entered stationary phase (237).

Rigby et al (180) present an interesting discussion of the significance of gene duplication in experimental enzyme evolution; the ideas relate directly to the observations of frequent spontaneous tandem duplication (197), and to models of crossing-over that include the pairing, nicking, and crossing-over of palindromic loops (43, 108, 177, 197, 199, 221).

Observations of recipient gene duplication during generalized transduction of *E. coli* with P1 phage (84, 205) might also be explained by crossover of daughter chromosomes.

Not only do recent reports of gene duplication in bacteria and phage represent interesting examples, they also describe clever methods for selecting and scoring tandem duplicates (207), and methods for measuring amplified gene expression, especially in the genes that code for glycyl-tRNA synthetase (53, 206), arginine (61), and tryptophan (98). As another example of gene amplification, the presence of hybrids of *sbcB* plasmids with a duplicated region results in a seven-fold increase of exoI in a wild-type strain and a fifteen-fold increase in a *recA* strain (219). The hybrid plasmid with a duplicate region is likely to be more stable in the latter. Still another case of increase of gene product upon tandem duplication is that of a phage λ coat protein (176).

On the surface, it may seem that deletions are the flip side of duplications; that is, they occur by the same crossover mechanisms, analogous to lysogenization and induction. This may be true in some cases, but there are examples of differences. Tandem duplications in the rII region of T4 phage were found to be accompanied by a deleted region (167). Deletions in *S. typhimurium* are frequent (40), occur often in hot-spot positions on the chromosome (42), and the deleted segments are sometimes nonrandom in size and position; that is, they are "ditto" deletions (97, 189). They could result from insertion sequences at specific places (177). Relatively short deletions can be recovered in haploid organisms, apparently formed by recombination between unrelated DNA sequences; they occur in much lower frequency than the RecA-mediated recombination (208).

Straus suggests that the duplications and deletions that he has studied may be formed by fundamentally different molecular mechanisms (206). His conclusion is based on the observations that, with a battery of mutagens, only a few induce deletions, whereas numerous mutagens seem to be effective in producing tandem duplications. He noted that the frequency of gene duplications may be increased by some (but not all) mutagens (82, 206). UV, nitrous acid, hydroxylamine, ethylmethanesulfonate, and nitrosoguanidine increase duplication of the glycine region in *E. coli,* but other mutagens do not (82, 83). There is obviously some specificity for mutagen-induced deletions. Alper & Ames (4) found that, among a large number of mutagens tested, nitrous acid is especially effective, and only three other mutagens (nitrogen mustard, mitomycin C, and fast neutrons) yield any deletions. A particular scoring scheme, depending upon survival values, may eliminate nonviable deletions. However, this would still not explain mutagen specificity.

HOST INVOLVEMENT IN PLASMID AND PHAGE RECOMBINATION

In addition to crossover between bacterial chromosomes, recombinational events that involve plasmid and phage DNA show various degrees of interdependence with the Rec genes of the host.

Plasmid Recombination

When plasmids recombine among themselves in vivo (6, 36, 37, 96, 99), the $recA^+$ gene of the host is not always required. The *recA* gene is needed, however, for integration of the F' plasmid into the chromosome (166). Recombination intermediate molecules of colicin DNA are observed in wild-type and *recBC* mutants, but not in *recA* (173). Also, the $recA^+$ gene is needed for λ *adv-1* oligomers to form in lengths greater than 2X, and even the 2X lengths are reduced in number in *recA* mutants (85).

In some cases, the presence of a plasmid may influence some of the cell's Rec-associated properties (131, 132, 151, 154, 215, 216) and some plasmid properties may differ depending upon whether the cell has *rec* or *lex* mutations (151). Specifically, the R-Utrecht plasmid protects against UV damage in wild-type but not in *recA* mutants of *E. coli.* The presence of the plasmid also increases the UV mutagenicity of the cell. For certain phenotypes, a cell containing the R-Utrecht plasmid resembles a "hyper" $recA^+$ cell. This is also true of two plasmids that do not produce colicins (94).

The protective effect of R-Utrecht on wild-type, but not on *recA* derivatives, suggests that some UV-induction process may be involved. Swenson (210) argues that this protective effect may be linked to cell respiration. It is now becoming quite apparent that the switching off of respiration may depend upon a protein, regulated by $recA^+$ and lex^+ genes. The entire matter is still puzzling, however, since the presence of plasmid R46, which confers enhanced mutagenesis and UV resistance to the host (hyper recA?), does not affect protein X synthesis; increased protein X synthesis might be anticipated to accompany the enhanced mutagenesis and DNA repair (123).

When two distinct plasmids recombine in a *recA* mutant to form a large plasmid, the mechanism may depend upon short, palindromic DNA regions with short segments of homology in each of the plasmids. The same mechanism can, in reverse, be utilized for the development of two smaller plasmids from the larger one (108).

In these RecA-independent recombinations, a segment of DNA is inverted in the recombinant; one plasmid inserts into the other at a unique site. Theoretically, complementarity of only a few bases is sufficient to enhance considerably the probability of recombination (178). Plasmids from bacteria that have been independently isolated from recent hospital cases still have regions of DNA homology (6, 99) with plasmids from laboratory strains, indicating conservation of specific base sequences. The conserved palindrome may be a mechanism that allows for units to split, join, and/or assemble into new groupings, including the development of Hfr male strains of *E. coli* (24).

Inverted DNA sequences (palindromes) are not an exclusive feature of plasmids; there may be wide distribution of such sequences in the *E. coli* genome (43, 175). The importance of these may become more apparent, not only in bacterial recombination between two parental cells, but also in tandem duplication, deletion, and inversion formation, and perhaps in the crossover processes of transduction, lysogeny, and prophage induction. The work of Zieg et al (238) presents insights on "flip-flop" inversions that may be a model for many genetic phenomena. A number of "site-specific" recombinational events, previously assumed not to involve DNA homology, may indeed involve short, palindromic matching sequences (54). Whether phage mu-1 insertion is another representative remains a puzzle (24). Plasmids could have had an evolutionary role in the development of biological switches to activate or deactivate groups of genes by insertion, inversion, and excision. Palindromic structures may also be regions where promotor, operator, and other protein-DNA interacting sites are present (36). Many of these concepts of movement of genetic components ("jumping genes") via recombinational mechanisms stem from McClintock's (137) early observations, but are now gaining support from recent experiments in both procaryotes and eucaryotes (24). Such roving has long been postulated to explain the diversity of antibodies (198). As noted above, in addition to insertion and deletion switches, regulation may also be achieved by inversion of a DNA fragment. Such an inversion switch has been postulated (238) as a way of determining which of two flagellar antigens would be expressed in *S. typhimurium*. Via recombination, the DNA piece can be swung around in either direction.

It would be interesting to determine whether IS palindromes are present in those cases where P22 phage of *Salmonella* transduce plasmid genes into chromosomes (47), and in *E. coli* and *S. typhimurium* hybrid crosses (48, 188).

Recombinational Genes in Phage

Rec-equivalent genes are carried on chromosomes of some phages. In many cases, the host-equivalent Rec genes are shut off upon infection and the phage genes are turned on to accomplish recombination and other DNA metabolism functions. For example, the RecBC$^+$ DNAse of *E. coli* is markedly reduced upon infection with double-stranded T2, T3, T4, T5, T6, T7, λ, ϕ80, and P1, but not with single-stranded f1, or RNA phage Qβ (187).

In λ, particularly, RecA- and RecBC-equivalent genes have received special attention. The *gam* protein functions (64, 101, 196, 217, 224) by inhibiting exoV activity. One of the *red* genes in λ may be an analogue of *recF* (exoVIII) (59).

Certain mutations (chi mutations) act as hot spots of recombinational activity mediated by the Rec genes of the host (111, 202, 203).

A number of genes in T7 are involved in recombination (102, 114, 186). Gene 3 is apparently an endonuclease; gene 4, a DNA replication protein; gene 5, a DNA polymerase; and gene 6, an exonuclease (102) that may degrade a DNA strand distal to the whisker. In vitro studies (114) demonstrate a physical transfer of T7 fragments from one DNA molecule to another.

Phage T4, of course, has many genes involved in recombination (71, 146) including the unwinding protein 32, and gene products that can inhibit exoV. Such genes are also found in T7, λ, and other phages (12).

Phage T1 also has genes similar to those in λ and T4. Gene 4 seems to correspond to the *red* gene of λ (32). Phage T1 is capable of growing in rec^- hosts, although in some cases there may be reduced recombination.

Phage P22 of *S. typhimurium* has some special features with regard to recombination. An antirepressor function has been identified (the *ant* gene) that could be a protease to split the phage P22 repressor (116). The cleavage of the repressor does not take place in RecA cells; therefore, the *ant* gene may have to interact with the $recA^+$ product to mediate this cleavage. The repressor of P22 is sensitive to UV (213). This similarity between the antirepression of phage P22 prophage and the postulated antirepression of protein X (67) should be noted. Another interesting aspect of phage P22 is that recombination is mandatory for its multiplication (223); either the host rec^+ or phage erf^+ equivalent must be operative.

Phage *mu-1* also has some special recombination features. Faelen & Toussaint (51) propose a new tool to transpose and localize bacterial genes by use of phage *mu-1*. They show that transposition of chromosomal genes to the F ' occurs either after infection with *mu-1* or after induction of *mu-1* prophage. Also, at least one entire *mu-1* genome is transferred to the F ' episome. Moreover, it can be used for short-distance mapping. Another advantage is that the mapping can be done in rec^- cells. Although *mu-1* has genes to promote integration of its DNA into random positions of host DNA, the crossover sequence in the viral DNA is fixed (24). Apparently, it has no *rec*-equivalent genes (24) for its own vegetative recombination system.

In the case of phage f1, there is a RecA dependence of recombination involving breakage and rejoining of DNA molecules with the formation of long segments of heteroduplex DNA (239).

RELATION BETWEEN MUTATION AND RECOMBINATION

This relationship has been a lingering concern of geneticists. New discussions arose when it was found that mutants that are incapable of DNA postreplication repair (*recA* and *lex*) also lose the capacity for mutation by UV (Witkin effect) (228, 229). The $recA^+$ and lex^+ products lead to "error-prone" repair following DNA damage by UV; the absence of these products in *recA* and *lex* forces cells to utilize an "error-proof" (PolA?) repair system. However, not all mutation occurs via errors in postreplication repair; *recA* and *lex* are readily mutated by nitrous acid via a ligase-dependent pathway and by 2AP via replication error (18). Mutation by 5BU is more obscure; in *E. coli,* as with 2AP, repair genes are not required (230), but 5BU mutation of λ phage requires functional RecA or Lex, and Red genes (170). Frameshift-specific mutagens do not require RecA function for mutagenesis (44).

The stimulation of mutation and recombination by UV or UV-mimicking drugs need not necessarily be confined to dimer production. In addition to nondimer photoproducts, UV also stimulates DNA degradation; a degradation product may be an inducer of the error-prone enzyme (67, 72). For example, carcinogens such as 3-methyl chloranthrene could act as a mutagen in mammalian cells by inducing a protein similar to Gudas and Pardee's protein X.

Throughout this review, there has been an attempt to distinguish between the postreplication repair and the recombination functions of RecA. The Witkin effect (229) need not involve any crossover events. Indeed, there are suppressor mutants of *recA* that are "error-prone" while still incapable of recombination (44, 63, 161, 162); thus, the postreplication repair and the recombination properties are not under identical controls (153). Also, *rnm* suppressors of *lexA* restore UV resistance but not UV mutability (220).

However, the crossover process itself may lead to mutational errors. A specific verification of an association (but not a necessary dependence) between recombination and mutation is an increased yield of mutants upon certain crosses of λ (236), as well as an enrichment of recombinants when the selection is for mutants. Both mutagenesis and recombination of λ are stimulated by UV irradiation of the host. Mutagenesis could occur either at the exchange site by "slippage" of bases in the hybrid overlap region or by a recombinational polymerase that produces an error.

The relation between *mut* (mutator) and *rec* genes deserves attention. If there is an association between *recA* and mutation, some mutator sites might be expected to neighbor (or be within) the RecA or LexB cistron. Note that the MutS (39) and Tif (58, 229) sites are neighbors of (or within?) RecA (9). Some *tif* mutants have high spontaneous mutation rates (58). The MutR locus is near RecBC, as is *thyA* which has interesting mutator properties upon thymine starvation (89).

An early argument for a recombination-mutation correlation was the observation of increased phage mutations that accompanied UV reactivation (210). However, this increase of phage mutagenesis does not occur in *recA* or *lexA* cells (103). Again, it could be repair rather than the crossover event that is correlated with mutagenesis. The frequent correlation between mutation and recombination might involve a common enzymatic correction step. Recombination may arise as a result of repair processes normally active to preserve the integrity of the genetic message during replication of the chromosome; an unequal crossover could yield a mutant (41). Gene conversion may represent such an unequal crossover (31). In the process of recombination, the heteroduplex molecules with mismatched bases (3, 17, 221, 227) are likely to produce localized distortion of DNA. The mismatch repair that follows may increase recombination frequency of very close markers. An enzyme such as one isolated from wheat smut could take care of distortions, similar to enzymes that remove pyrimidine dimers (3). A mutation in pneumococcus (*hex*) apparently results in simultaneous inability to remove mismatched bases and high mutability (164). Recombination in T4 phage crosses via mismatch repair requires the host *rec* genes (227).

It should be noted that *dam* of *E. coli*, which is considered a mutator gene, makes the cell hyper in recombination capacity (133). Also, as already noted, mutants defective in exonuclease VII (*xseA*) (30), and especially the double mutant *xseA pol*^{ex1}*A* have hyperrecombinational properties, but change in mutation frequency has not been reported in these strains.

Another example of the relation of recombination and mutation is found with DNA-delay mutants of phage T4 (115, 163); this delay promotes not only recombination but also base substitution and addition-deletion mutation.

PERTURBATION (IN VIVO) OF RECOMBINATION BY PHYSICAL AND CHEMICAL AGENTS

Small doses of radiation or drugs can stimulate recombination (2, 10, 16, 35, 46, 56, 77, 79, 92, 93, 120). Soon after conjugation in *E. coli* was discovered, Clark et al (35) observed that UV and hydrogen peroxide could indeed increase recombinant frequencies. Both H_2O_2 (5) and UV (92, 93, 119, 120) induce DNA breaks and free ends, which are conducive to recombination. The early observations were verified by Garen & Zinder's (56) study of the increase in transductant numbers upon radiation of the donor phage. More recently, insights on the crossover mechanism have been gained through studies of such stimulation. Some examples have already been presented of stimulation of *intrachromosomal* recombination, leading to increased duplications, deletions, and inversions.

A number of DNA alterations via natural processes, as well as via external assault, are recombinogenic. Single-stranded DNA ends and single-stranded stretches have received particular attention (10, 13, 14, 38, 179). These can be increased in number by several routes. Via mutation, excision repair-deficient *hcr* mutants following irradiation enhance the number of phage recombinants as compared to *hcr*$^+$ hosts. Helling (79) verified the role of UvrA and UvrB nucleases in this stimulation. Also, as noted above, recombination is abnormally high with DNA delay T4 mutants and X mutants of λ (202–204). Clark (34) reviewed the effect of various other DNA mutations on recombination and states that most of these do not increase the frequency of crossover, including mutations in polymerase II and exonuclease III where increased breaks might be expected. In one case, DnaB mutants show a decrease in genetic recombination. T4 mutants deficient in ligase show increased recombination, presumably through increased opportunity for exchange of DNA fragments. This is discussed by Chipchase (31) who points out that in eucaryotes inhibition of replication during S phase also increases recombination.

If single-strand gaps or the opening of the helix leads to recombination, then transcription should stimulate recombination (34); indeed, it appears that it does (45).

As for external assaults, an understanding is developing as to how these agents, by damaging DNA, perturb recombination frequencies (45, 46, 79, 119, 120, 184). The comparative roles of dimers (29), DNA gaps, DNA single-stranded ends (143), and other DNA states in the stimulation of recombination are now clarified as a result of studies by Lin & Howard-Flanders (119, 120). They obtained the greatest increase in recombination upon UV irradiation of excision-deficient lysogens, and concluded that, *with DNA replication,* recombination increase can be attributed to dimers in phage DNA; these dimers lead to gaps in daughter strands. However, nondimer photoproducts increase recombination even in the *absence of DNA replication.* DNA protein cross-links or intrastrand DNA cross-links (231) might be the stimulatory damage.

The principal photoproducts induced in DNA by 254 nm UV include cyclobutane dimers, cytosine-thymine adducts, cytosine hydrate, dihydrothymine, and DNA-protein cross-links. The active product may be one of these, or a product yet to be described. It is unlikely that the active product is cytosine hydrate, since the effect is not reversed by incubation of phages of 37°, a procedure known to reverse hydrates, but having no effect upon 254 nm induced recombination.

Lin & Howard-Flanders (119, 120) observed that DNA-cross-linking agents produce different results. When a λ lysogen is superinfected with λ phage treated with psoralen and light, exchanges with the prophage occur readily if replication is permitted. However, when replication of the DNA containing the cross-links is blocked, exchanges are produced abundantly in wild-type lysogens, but not in lysogens that are also excision deficient mutants. In the absence of DNA replication, the chromosome structure necessary for crossover is apparently created by the action of excision enzymes on the cross-links. The psoralen photocross-links and monoadducts initiate recombination only following the action of excision enzymes; there is no stimulation in UvrA or UvrB strains. The excision enzymes may release one arm of each cross-link, producing a gap with free-strand ends.

It is interesting to note that nalidixic acid does not interfere with recombination; however, hydroxyphenylazopyrimidine, another inhibitor of replicative DNA synthesis, does (25).

Little (122) points out that the fraction of phage λ recombinants in a particular cross is increased severalfold if 5BU is substituted for thymine. A similar enhancement results from a period of thymine starvation. This suggests the need for caution in interpreting phage recombination experiments involving use of BU. In a *recA* host, where only the *red* system of λ is operative, BU stimulates recombination although to a lesser extent than a rec^+ control.

As one examines the Gudas-Pardee model, a question can be raised as to whether fluctuating quantities of protein X would stimulate or inhibit recombination. Adenine is apparently an inducer of protein X, and uracil may be an inhibitor. Gillen & Clark (59) note that a uracil analogue inhibits polymerase III enzyme (responsible for chromosome synthesis initiation) and also inhibits recombination.

One way that recombination might be manipulated is by controlling access of necessary polymerases and nucleases to the chromosomes (31).

As discussed above, the mutational process may also lead to recombination (many mutagens are also recombinogens); this could be attributed to common enzymatic correction steps. Recombination, as well as mutation, may be a secondary consequence of the normal repair processes to preserve the integrity of the chromosome during replication. The observation that interspecies *S. typhimurium* × *E. coli* crosses are enhanced via mutagens (47) and the presence of a mutator gene (148, 149) further support this relationship.

Goldfarb et al (62) use a protein from *E. coli* male strains to stimulate recombination in "young" merozygotes. It would appear that this is some step after DNA transfer but before integration of donor DNA into the recipient genome, perhaps at a synaptic stage.

Drexler & Kylberg (46) observed that, in a transductional system, not all markers are transduced equally upon stimulation with UV. They point out that the *gal* marker is the only one that they observed to show this stimulation.

It should be noted that in many of the experiments in which an increase in recombination per survivor is noted, an overdose of the agent may cause an actual reduction in the number of recombinants per survivor (112). Drake (44) indicates that if the DNA gaps are over 1000 bases in length, one would anticipate such reductions. Of particular interest in our laboratory is the observation that small doses of H_2O_2 can drastically reduce the recombination frequency in phage T4 crosses and *E. coli* conjugation (112).

rec MUTANTS IN NON-ENTERIC BACTERIA

Although most of the recombinational studies have been performed with enteric bacteria, the special features of recombination in other microorganisms (e.g. transforming system) may assist in unraveling the puzzle of precise molecular events. Recombination-deficient mutants of *B. subtilis* (73, 135, 136, 185, 233), *Haemophilus influenzae* (107, 164), and *Pseudomonas aeruginosa* (147) have now been characterized. A RecBC enzyme has been identified in *P. aeruginosa* (147). A number of *rec* genes in *B. subtilis* have been carefully mapped (73); several correspond in activity to those in *E. coli*. A *rec* mutant in *Micrococcus radiodurans* (155, 156) drastically reduces transformation frequency and radiation resistance, but it is uncertain as to whether it resembles *recA* or *recB* of *E. coli*. The wild type is exceptionally resistant to radiation, but there is no evidence that this mutation gives it hyperrecombination capacity.

Especially interesting is a description of nonspecific pairing of DNA molecules by recombination enzymes of *B. subtilis* (165). Köhnlein & Hutchinson (104), observing a four-stranded DNA, postulate that a Rec gene product is responsible for a separation of the four-stranded state, since the four-strand is found only in *rec*⁻ mutant. Ohi et al (165) discuss a mechanism whereby an enzyme can bring the two parental DNA molecules of *B. subtilis* close to each other.

Staphylococcus aureus rec mutants have been isolated (222). Plasmids of *S. aureus* require Rec to recombine efficiently with each other, although there is a residuum of interplasmid recombination in the absence of Rec product. Transduction of plasmid genes to *rec* recipients is normal; thus the Rec product is not needed (96).

CONCLUSIONS

Recent studies are highlighted by the isolation and characterization of a RecA protein (141). This protein appears to be the same as the inducible protein X isolated previously (68). Although the pleiotropy of *recA* mutants remains a complex puzzle, it now appears that mutation in different parts of the cistron leads to different phenotypic expressions, some of which are inducible and others of which are constitutive (153). The *lexA*⁺ product apparently acts at the operator region within a RecA cistron. Also, specific DNA fragments may play a role in the induction of RecA protein. RecA activity precedes RecBC activity in the recombinational process.

Studies of tandem duplications, deletions, inversions, and plasmid and phage activities fortify the view that intrachromosomal rearrangements (roving genes) may be natural and frequent, and influence cell populations via Darwinian selective and evolutionary patterns. Particularly exciting are studies (238) that reveal inversions with a "flip-flop" role in which two different proteins are synthesized when the DNA fragments are in the two different positions. Many genetic and regulatory phenomena could be explained by this recombinational process that leads to inversions.

The capability of perturbing recombinational frequencies in bacteria in vivo via chemicals, UV, and mutation aids in the dissection of the crossover process.

These recent advances lead to the optimistic view that the baffling puzzle of genetic recombination in bacteria, as well as the intricate relationship between recombination, mutation, DNA repair, and cell division, may be greatly clarified in the near future.

ACKNOWLEDGMENT

I should like to express my sincere thanks to Mrs. Jean Carr for her assistance in the typing and assembly of materials for this manuscript, to Mrs. Catherine Huckins for her assistance in reference searches, and to several colleagues who kindly sent me preprints of manuscripts. Experiments of the author are supported by PHS Grant No. 2-RO1-FDOO658.

Literature Cited

1. Abdel-Monem, M., Durwald, H., Berling, H. H. 1976. Enzymatic unwinding of DNA. 2. Chain separation by an ATP-dependent DNA unwinding enzyme. *Eur. J. Biochem.* 65:441–49
2. Adye, J. 1962. Effect of nitrous acid on transduction of *Salmonella* phage P22. *Virology* 18:627–32
3. Ahmad, A., Holloman, W. K., Holliday, R. 1975. Nuclease that preferentially inactivates DNA containing mismatched bases. *Nature* 258:54–56
4. Alper, M. D., Ames, B. N. 1975. Positive selection of mutants with deletions of Gal-Chl region of *Salmonella* chromosome as a screening procedure for mutagens that cause deletions. *J. Bacteriol.* 121:259–66
5. Ananthaswamy, H. N., Eisenstark, A. 1977. Repair of hydrogen peroxide-induced single strand breaks in *Escherichia coli* deoxyribonucleic acid. *J. Bacteriol.* 130:187–191
6. Anderson, E. S. 1974. Recombination between unrelated bacterial plasmids. *Ann. Microbiol. Paris* 125A(3):251–59
7. Anderson, R. P., Miller, C. G., Roth, J. 1976. Tandem duplication of the histidine operon observed following generalized transduction in *S. typhimurium. J. Mol. Biol.* 105:201–18
8. Audit, C., Anagnostopoulos, C. 1975. Studies on the size of the diploid region in *Bacillus subtilis* merozygotes from strains carrying the *trpE26* mutation. *Mol. Gen. Genet.* 137:337–51
9. Bachmann, B. J., Low, K. B., Taylor, A. L. 1976. Recalibrated linkage map of *Escherichia coli* K-12. *Bacteriol. Rev.* 40:116–67
10. Baker, R. M., Haynes, R. H. 1967. UV-induced enhancement of recombination among Lambda bacteriophages in UV-sensitive host bacteria. *Mol. Gen. Genet.* 100:166–77
11. Beeftinck, F., Cunin, R., Glansdorff, N.

1974. Arginine gene duplications in recombination proficient and deficient strains of *Escherichia coli* K-12. *Mol. Gen. Genet.* 132:241–53
12. Behme, M. T., Lilley, G. D., Ebisuzaki, K. 1976. Postinfection control by bacteriophage T4 of *Escherichia coli* recBC nuclease activity. *J. Virol.* 18:20–25
13. Benbow, R. M., Zuccarelli, A. J., Sinsheimer, R. L. 1974. A role for single-strand breaks in bacteriophage ϕX174 genetic recombination. *J. Mol. Biol.* 88:629–51
14. Benbow, R. M., Zuccarelli, A. J., Sinsheimer, R. L. 1975. Recombinant DNA molecules of bacteriophage PHIX 174. *Proc. Natl. Acad. Sci. USA* 72:235–39
15. Benzinger, R., Enquist, L. W., Skalka, A. 1975. Transfection of *Escherichia coli* spheroplasts. V. Activity of recBC nuclease in rec^+ and rec^- spheroplasts measured with different forms of bacteriophage DNA. *J. Virol.* 15:861–71
16. Benzinger, R., Hartman, P. E. 1962. Effects of ultraviolet light on transducing phage P22. *Virology* 18:614–26
17. Berger, H., Pardoll, D. 1976. Evidence that mismatched bases in heteroduplex T4 bacteriophage are recognized *in vivo*. *J. Virol.* 20:441–45
18. Bernstein, C., Morgan, D., Gensler, H. L., Schneider, S., Holmes, G. E. 1976. The dependence of HNO_2 mutagenesis in phage T4 on ligase and the lack of dependence of 2AP mutagenesis on repair functions. *Mol. Gen. Genet.* 148:213–20
19. Birge, E. A., Low, K. B. 1974. Detection of transcribable recombination products following conjugation in Rec^+, $RecB^-$, and $RecC^-$ strains of *escherichia coli* K12. *J. Mol. Biol.* 83:447–57
20. Blanco, M., Levine, A., Devoret, R. 1975. *LexB:* A new gene governing radiation sensitivity and lysogenic induction in *Escherichia coli* K12. In *Molecular Mechanisms for Repair of DNA*, ed. P. C. Hanawalt, R. B. Setlow, pp. 379–82. New York: Plenum
21. Braun, A. G. 1976. Host DNA replication or excision repair requirement for ultraviolet induction of bacteriophage (lambda) lysogens. *Nature* 261:164–66
22. Bridges, B. A., Mottershead, R. P., Green, M. H. L. 1977. Cell division in *E. coli* B_{s-12} is hypersensitive to deoxyribonucleic acid damage by ultraviolet light. *J. Bacteriol.* 130:724–28
23. Broker, T. R., Doermann, A. H. 1975. Molecular and genetic recombination of bacteriophage T4. *Ann. Rev. Genet.* 9:213–44
24. Bukhari, A. I., Shapiro, J., Adhya, S., eds. 1977. *DNA Insertion Elements, Plasmids, and Episomes*. Cold Spring Harbor, New York: Cold Spring Harbor Lab. Press
25. Canosi, U., Mazza, G., Siccardi, A. G., Falaschi, A. 1976. Effect of DNA synthesis inhibitors on recombination and repair. *Atti Assoc. Genet. Ital.* 21:91–92
26. Capaldo, F. N., Ramsey, G., Barbour, S. D. 1974. Analysis of the growth of recombination-deficient strains of *Escherichia coli* K-12. *J. Bacteriol.* 118:242–49
27. Castellazzi, M. 1976. Deoxyribonucleic acid degradation *in vivo* and in permeabilized *Escherichia coli* repair-deficient (recA zab lexA) derivatives. *J. Bacteriol.* 127:1150–56
28. Castellazzi, M., Morand, P., George, J., Buttin, G. 1977. Prophage induction and cell division in *E. coli*. V. Dominance and complementation analysis in partial diploids with pleiotropic mutations (tif, recA, zab, and lexB) at the recA locus. *Mol. Gen. Genet.* 153:297–310
29. Chan, V. L. 1975. On the role of y gene in ultraviolet-induced enhancement of recombination among T4 phages. *Virology* 65:266–67
30. Chase, J. W., Richardson, C. C. 1977. *Escherichia coli* mutants deficient in exonuclease VII. *J. Bacteriol.* 129:934–47
31. Chipchase, M. 1976. Recombination by strand assimilation and strand crossover. *J. Theor. Biol.* 57:249–79
32. Christensen, J. R. 1976. Red system of bacteriophage (lambda) complements the growth of a bacteriophage T1 gene 4 mutant. *J. Virol.* 17:713–17
33. Clark, A. J. 1974. Progress toward a metabolic interpretation of genetic recombination of *Escherichia coli* and bacteriophage lambda. *Genetics* 78:259–71
34. Clark, A. J. 1973. Recombination deficient mutants of *E. coli* and other bacteria. *Ann. Rev. Genet.* 7:67–86
35. Clark, J. B., Haas, F., Stone, W. S., Wyss, O. 1950. The stimulation of gene recombination in *Escherichia coli*. *J. Bacteriol.* 59:375–79
36. Cohen, S. N. 1976. Transposable genetic elements and plasmid evolution. *Nature* 253:731–38

37. Cohen, S. N. 1975. The manipulation of genes. *Sci. Am.* 233:25–33
38. Cordone, L., Sperandeo-Mineo, R. M., Mannino, S. 1975. UV induced enhancement of recombination among lambda bacteriophages: Relation with replication of irradiated DNA. *Nucleic Acids Res.* 2:1129–42
39. Cox, E. C. 1976. Bacterial mutator genes and the control of spontaneous mutation. *Ann. Rev. Genet.* 10:135–56
40. Demerec, M. 1960. Frequency of deletions among spontaneous and induced mutations in *Salmonella*. *Proc. Natl. Acad. Sci. USA* 46:1075–79
41. Demerec, M. 1962. "Selfers"—attributed to unequal crossovers in *Salmonella*. *Proc. Natl. Acad. Sci. USA* 48:1696–1704
42. Demerec, M., Gillespie, D. H., Mizobuchi, K. 1963. Genetic structure of the *Cys*C region of the *Salmonella* genome. *Genetics* 48:997–1009
43. Deonier, R. C., Hadley, R. G. 1976. Distribution of inverted IS-length sequences in the *E. coli* K12 genome. *Nature* 264:191–93
44. Drake, J. W., Baltz, R. H. 1976. The biochemistry of mutagenesis. *Ann. Rev. Biochem.* 45:11–37
45. Drexler, H. 1972. Transduction of Gal+ by coliphage T1. I. Role of hybrids of bacterial and prophage (lambda) deoxyribonucleic acid. II. Role of (lambda) transcription control in the efficiency of transduction. *J. Virol.* 9:273–85
46. Drexler, H., Kylberg, K. J. 1975. Effect of UV irradiation on transduction by coliphage T1. *J. Virol.* 16:263–66
47. Eisenstark, A. 1965. Mutagen-induced hybridization of *S. typhimurium* LT2 × *E. coli* K12 Hfr. *Proc. Natl. Acad. Sci. USA* 54:117–20
48. Eisenstark, A. 1965. Transduction of *Escherichia coli* genetic material by phage P22 in *Salmonella typhimurium* × *E. coli* hybrids. *Proc. Natl. Acad. Sci. USA* 54:1557–60
49. Eisenstark, A., Eisenstark, R., van Dillewijn, J., Rorsch, A. 1969. Radiation-sensitive and recombinationless mutants of *Salmonella typhimurium*. *Mutat. Res.* 8:497–504
50. Emmons, S. W., MacCosham, V., Baldwin, R. L. 1975. Tandem genetic duplications in phage lambda. III. The frequency of duplication mutants in two derivatives of phage lambda is independent of known recombination systems. *J. Mol. Biol.* 91:133–46
51. Faelen, M., Toussaint, A. 1976. Bacteriophage Mu-1: A tool to transpose and to localize bacterial genes. *J. Mol. Biol.* 104:525–39
52. Ferron, W. L., Eisenstark, A., MacKay, D. 1972. Distinction between far- and near-ultraviolet light killing of recombinationless (*rec*A) *Salmonella typhimurium*. *Biochim. Biophys. Acta* 277:651–58
53. Folk, W. R., Berg, P. 1971. Duplication of structural gene for glycyl-transfer RNA synthetase in *Escherichia coli*. *J. Mol. Biol.* 58:595–610
54. Franklin, N. C. 1971. Illegitimate recombination. In *The Bacteriophage Lambda*, ed. A. D. Hershey, pp. 175–94. New York: Cold Spring Harbor Lab.
55. Freifelder, D. 1976. New types of *Escherichia coli* recombination-deficient mutants. *J. Bacteriol.* 128:681–82
56. Garen, A., Zinder, N. D. 1955. Radiological evidence for partial genetic homology between bacteriophage and host bacteria. *Virology* 1:347–76
57. Gellert, M., O'Dea, M. H., Itoh, T., Tomizawa, J. I. 1976. Novobiocin and coumermycin inhibit DNA supercoiling catalyzed by DNA gyrase. *Proc. Natl. Acad. Sci. USA* 73:4474–78
58. George, J., Castellazzi, M., Buttin, G. 1975. Prophage induction and cell division in *E. coli*. III. Mutations *sfi*A and *sfi*B restore division in *tif* and *lon* strains and permit the expression of mutator properties of *tif*. *Mol. Gen. Genet.* 140:309–32
59. Gillen, J. R., Clark, A. J. 1974. The *rec* E pathway of bacterial recombination. In *Mechanisms in Recombination*, ed. R. F. Grell, pp. 123–36. New York & London: Plenum
60. Giorno, R., Hecht, R. M., Pettijohn, D. 1975. Analysis by isopycnic centrifugation of isolated nucleoids of *Escherichia coli*. *Nucleic Acids Res.* 2:1559–67
61. Glansdorff, N., Sand, G. 1968. Duplication of a gene belonging to an arginine operon of *Escherichia coli* K12. *Genetics* 60:257–68
62. Goldfarb, D. M., Goldberg, G. I., Chernin, L. S., Gukova, L. A., Avdienko, I. D., Kuznetsova, B. N., Kushner, I. Ch. 1973. A protein produced by male strains of *Escherichia coli* K-12 which increases the yield of recombinants in conjugation: Its nature and mode of action. *Mol. Gen. Genet.* 120:211–26
63. Gopalakrishna, K., Bhattacharjee, S. K. 1976. Role of *rec* pathways on sen-

sitivity of *Escherichia coli* to near-ultraviolet and visible light. *J. Bacteriol.* 127:1022–23
64. Greenstein, M., Skalka, A. 1975. Replication of bacteriophage lambda DNA: In vivo studies of the interaction between the viral gamma protein and the host *rec*BC DNAase. *J. Mol. Biol.* 97:543–59
65. Grell, R. F., ed. 1974. *Mechanisms in Recombination.* New York: Plenum 459 pp.
66. Gudas, L. J. 1976. The induction of protein X in DNA repair and cell division mutants of *Escherichia coli. J. Mol. Biol.* 104:567–87
67. Gudas, L. J., Pardee, A. B. 1975. Model for regulation of *Escherichia coli* DNA repair functions. *Proc. Natl. Acad. Sci. USA* 72:2330–34
68. Gudas, L. J., Pardee, A. B. 1976. DNA synthesis inhibition and the induction of protein X in *Escherichia coli. J. Mol. Biol.* 101:459–77
69. Guyer, M. S., Clark, A. J. 1976. Cis-dominant, transfer-deficient mutants of the *Escherichia coli* K-12 F sex factor. *J. Bacteriol.* 125:233–47
70. Hall, J. D., Howard-Flanders, P. 1975. Temperature sensitive *rec*A mutant of *Escherichia coli* K-12: Deoxyribonucleic acid metabolism after ultraviolet irradiation. *J. Bacteriol.* 121:892–900
71. Hamlett, N. V., Berger, H. 1975. Mutations altering genetic recombination and repair of DNA in bacteriophage T4. *Virology* 63:539–67
72. Hanawalt, P. C., Setlow, R. B., eds. 1975. *Molecular Mechanisms for Repair of DNA. Basic Life Sci.*, Vol. 5A, 5B. New York: Plenum
73. Harford, N. 1974. Genetic analysis of *rec* mutants of *Bacillus subtilis.* Evidence for at least six linkage groups. *Mol. Gen. Genet.* 129:269–74
74. Hartman, P. S., Eisenstark, A. 1977. Synergistic damage in *Escherichia coli* by near-ultraviolet radiation and hydrogen peroxide: Distinction between RecA-repairable and RecA-nonrepairable damage. *J. Bacteriol.* Submitted for publication
75. Hass, B. S., Webb, R. B. 1977. Acridine orange is a dark mutagen in a *Rec*A strain of *Escherichia coli. Mutat. Res.* Submitted for publication
76. Hass, B. S., Webb, R. B. 1977. High frequency photodynamic mutations in a *rec*A strain of *Escherichia coli. Mutat. Res.* Submitted for publication

77. Hayes, W. 1952. Genetic recombination in *Bacterium coli* K12: Analysis of the stimulating effects of U. V. light. *Nature* 169:1017–18
78. Hecht, R. M., Pettijohn, D. E. 1976. Studies of DNA bound RNA molecules isolated from nucleoids of *Escherichia coli. Nucleic Acids Res.* 3:767–88
79. Helling, R. B. 1973. Ultraviolet light-induced recombination. *Biochem. Biophys. Res. Commun.* 55:752–57
80. Hendler, R. W., Pereira, M., Scharff, R. 1975. DNA synthesis involving a complexed form of DNA polymerase I in extracts of *Escherichia coli. Proc. Natl. Acad. Sci. USA* 72:2099–2103
81. Hershey, A. D., ed. 1971. *The Bacteriophage Lambda.* New York: Cold Spring Harbor Lab.
82. Hill, C. W., Combriato, G. 1973. Genetic duplications induced at very high frequency by ultraviolet irradiation in *Escherichia coli. Mol. Gen. Genet.* 127:197–214
83. Hill, C. W., Foulds, J., Soll, L., Berg, P. 1969. Instability of a missense suppressor resulting from a duplication of genetic material. *J. Mol. Biol.* 39:563–81
84. Hill, C. W., Schiffer, D., Berg, P. 1969. Transduction of merodiploidy: Induced duplication of recipient genes. *J. Bacteriol.* 99:274–78
85. Hobom, G., Hogness, D. S. 1974. The role of recombination in the formation of circular oligomers of the (lambda)adv1 plasmid. *J. Mol. Biol.* 88:65–87
86. Hoffmann, G. R., Morgan, R. W. 1976. The effect of ultraviolet light on the frequency of a genetic duplication in *Salmonella typhimurium. Radiat. Res.* 67:114–19
87. Holloman, W. K., Radding, C. M. 1976. Recombination promoted by superhelical DNA and the recA gene of *Escherichia coli. Proc. Natl. Acad. Sci. USA* 73:3910–14
88. Holloman, W. K., Wiegand, R., Hoessli, C., Radding, C. M. 1975. Uptake of homologous single-stranded fragments by superhelical DNA: A possible mechanism for initiation of genetic recombination. *Proc. Natl. Acad. Sci. USA* 72:2394–98
89. Holmes, A. J., Eisenstark, A. 1968. The mutagenic effect of thymine-starvation on *Salmonella typhimurium. Mutat. Res.* 5:15–21
90. Horiuchi, T., Horiuchi, S., Novick, A. 1963. The genetic basis of hypersynthesis of B-galactosidase. *Genetics* 48:157–69

91. Hotchkiss, R. D. 1974. Models of genetic recombination. *Ann. Rev. Microbiol.* 28:445-68
92. Howard-Flanders, P. 1975. Repair by genetic recombination in bacteria: Overview. See Ref. 20, pp. 265-74
93. Howard-Flanders, P., Lin, P. F., Bardwell, E. 1975. Genetic exchanges induced by structural damage in nonreplicating phage lambda DNA. See Ref. 20, pp. 275-81
94. Hull, R. A. 1975. Effect of *tsl* mutations on Col E1 expression in a *rec* A strain of *E. Coli* K12. *J. Bacteriol.* 123:775-76
95. Inouye, M. 1971. Pleiotropic effect of the *rec* A gene of *Escherichia coli*: Uncoupling of cell division from deoxyribonucleic acid replication. *J. Bacteriol.* 106:539-42
96. Iordanescu, S. 1977. Relationships between cotransducible plasmids in *Staphylococcus aureus*. *J. Bacteriol.* 129:71-75
97. Itikawa, H., Demerec, M. 1967. Ditto deletions in the *cysC* region of the *Salmonella* chromosome. *Genetics* 55:63-68
98. Jackson, E. N., Yanofsky, C. 1973. Duplication-translocations of tryptophan operon genes in *Escherichia coli*. *J. Bacteriol.* 116:33-40
99. Jacoby, G. A., Jacob, A. E., Hedges, R. W. 1976. Recombination between plasmids of incompatibility groups P-1 and P-2. *J. Bacteriol.* 127:1278-85
100. Jovin, T. M. 1976. Recognition mechanisms of DNA-specific enzymes. *Ann. Rev. Biochem.* 45:889-920
101. Karu, A. E., Sakaki, Y., Echols, H., Linn, S. 1975. The γ protein specified by bacteriophage (lambda). Structure and inhibitory activity for the *rec*BC enzyme of *Escherichia coli*. *J. Biol. Chem.* 250:7377-87
102. Kerr, C., Sadowski, P. D. 1975. The involvement of genes 3, 4, 5, and 6 in genetic recombination in bacteriophage T7. *Virology* 65:281-85
103. Kerr, T. L., Hart, M. G. R. 1973. Effects of the *rec* A, *lex* and *exr* A mutations on the survival of damaged (lambda) and P1 phages in lysogenic and non-lysogenic strains of *Escherichia coli* K12. *Mutat. Res.* 18:113-16
104. Köhnlein, W., Hutchinson, F. 1976. A four-stranded DNA from *Bacillus subtilis* which may be an intermediate in genetic recombination. *Mol. Gen. Genet.* 144:323-31
105. Konrad, E. B. 1977. A method for the isolation of *Escherichia coli* mutants with enhanced recombination between chromosomal duplications. *J. Bacteriol.* 130:167-86
106. Konrad, E. B., Lehman, I. R. 1975. Novel mutants of *Escherichia coli* that accumulate very small DNA replicative intermediates. *Proc. Natl. Acad. Sci. USA* 72:2150-54
107. Kooistra, J., Setlow, J. K. 1976. Similarity in properties and mapping of three *rec* mutants of *Haemophilus influenzae*. *J. Bacteriol.* 127:327-33
108. Kopecko, D. J., Cohen, S. N. 1975. Site-specific *rec* A-independent recombination between bacterial plasmids: Involvement of palindromes at the recombinational loci. *Proc. Natl. Acad. Sci. USA* 72:1373-77
109. Kupersztoch-Portnoy, Y. M., Helinski, D. R. 1974. Breaks induced in supercoiled colicinogenic factor E1 DNA as a result of the decay of incorporated tritium. *Biochim. Biophys. Acta* 374:316-23
110. Kushner, S. R., Nagaishi, H., Clark, A. J. 1974. Isolation of exonuclease VIII: The enzyme associated with the *sbc* A indirect suppressor. *Proc. Natl. Acad. Sci. USA* 71:3593-97
111. Lam, S. T., Stahl, M. M., McMilin, K. D., Stahl, F. W. 1974. *Rec*-mediated recombination hot spot activity in bacteriophage lambda. II. A mutation which causes hot spot activity. *Genetics* 77:425-33
112. Landa, F., Eisenstark, A. 1973. The effect of L-tryptophan photoproduct on genetic recombination in bacteria. *Abstr. Ann. Meet. Am. Soc. Microbiol.*, p. 73
113. Ledbetter, M. L. S., Hotchkiss, R. D. 1975. Recombination as a requirement for segregation of a partially diploid mutant of *Pneumococcus*. *Genetics* 80:679-94
114. Lee, M., Miller, R. C. Jr. 1974. T7 exonuclease (GENE 6) is necessary for molecular recombination of bacteriophage T7. *J. Virol.* 14:1040-48
115. Leung, D., Behme, M. T., Ebisuzaki, K. 1975. Effect of DNA delay mutations of bacteriophage T4 on genetic recombination. *J. Virol.* 16:203-5
116. Levine, M., Truesdell, S., Ramakrishnan, T., Bronson, M. J. 1975. Dual control of lysogeny by bacteriophage P22: An antirepressor locus and its controlling elements. *J. Mol. Biol.* 91:421-38
117. Lieberman, R. P. 1975. *Characterization, genetic origins, and assembly of the subunits of the recB-recC DNAse of Es-*

cherichia coli. PhD Diss. New York Univ., New York. 92 pp.
118. Lieberman, R. P., Oishi, M. 1973. Formation of RecB-RecC DNAase by in vitro complementation and evidence concerning its subunit nature. *Nature New Biol.* 243:75–77
119. Lin, P. F., Bardwell, E., Howard-Flanders, P. 1977. The initiation of genetic exchanges in (lambda) phage-prophage crosses. *Proc. Natl. Acad. Sci. USA* 74:291–95
120. Lin, P. F., Howard-Flanders, P. 1976. Genetic exchanges caused by ultraviolet photoproducts in phage (lambda) DNA molecules: The role of DNA replication. *Mol. Gen. Genet.* 146:107–15
121. Linn, S., MacKay, V. 1975. The degradation of duplex DNA by the *rec*BC DNase of *Escherichia coli.* See Ref. 20, pp. 293–99
122. Little, J. W. 1976. The effect of 5-bromouracil on recombination of phage lambda. *Virology* 72:530–35
123. Little, J. W., Hanawalt, P. C. 1977. Induction of protein X in *Escherichia coli.* *Mol. Gen. Genet.* 150:237–48
124. Lloyd, R. G. 1974. The segregation of the SbcA and Rac phenotypes in an *Escherichia coli rec*B⁻ mutant. *Mol. Gen. Genet.* 134:249–59
125. Lloyd, R. G., Barbour, S. D. 1974. The genetic location of the *sbc*A gene of *Escherichia coli.* *Mol. Gen. Genet.* 134:157–71
126. Lloyd, R. G., Low, B. 1976. Some genetic consequences of changes in the level of *rec*A gene function in *Escherichia coli* K-12. *Genetics* 84:675–95
127. Lloyd, R. G., Low, B., Godson, G. N., Birge, E. A. 1974. Isolation and characterization of an *Escherichia coli* K-12 mutant with a temperature-sensitive *rec* a⁻ phenotype. *J. Bacteriol.* 120:407–15
128. Low, B. 1973. Restoration by the *rac* locus and recombinant forming ability in *rec*B⁻ and *rec*C⁻ merozygotes of *Escherichia coli* K-12. *Mol. Gen. Genet.* 122:119–30
129. Mackay, D., Eisenstark, A., Webb, R. B., Brown, M. S. 1976. Action spectra for lethality in recombinationless strains of *Salmonella typhimurium* and *Escherichia coli.* *Photochem. Photobiol.* 24:337–43
130. MacKay, V., Linn, S. 1976. Selective inhibition of the DNase activity of the *rec*BC enzyme by the DNA binding protein from *Escherichia coli.* *J. Biol. Chem.* 251:3716–19

131. MacPhee, D. G. 1973. Effect of *rec* mutations on the ultraviolet protecting and mutation-enhancing properties of the plasmid R-utrecht in *Salmonella typhimurium.* *Mutat. Res.* 19:347–59
132. MacPhee, D. G. 1973. Effect of an R factor and caffeine on ultraviolet mutability on *Salmonella typhimurium.* *Mutat. Res.* 18:367–70
133. Marinus, M. G., Konrad, E. B. 1976. Hyper-recombination in *dam* mutants of *Escherichia coli* K-12. *Mol. Gen. Genet.* 149:273–78
134. Marsden, H. S., Pollard, E. C., Ginoza, W., Randall, E. P. 1974. Involvement of *rec*A and *exr* genes in the *in vivo* inhibition of the *rec*BC nuclease. *J. Bacteriol.* 118:465–70
135. Mazza, G., Eisenstark, H. M., Serra, M. C., Polsinelli, M. 1972. Effect of caffeine on the recombination process of *Bacillus subtilis.* *Mol. Gen. Genet.* 115:73–79
136. Mazza, G., Fortunato, A., Ferrari, E., Canosi, U., Falaschi, A., Polsinelli, M. 1975. Genetic and enzymic studies on the recombination process in *Bacillus subtilis.* *Mol. Gen. Genet.* 136:9–30
137. McClintock, B. 1951. Chromosome organization and genic expression. *Cold Spring Harbor Symp. Quant. Biol.* 16:13–47
138. McEntee, K. 1977. Genetic analysis of the *srl-rec*A region of *Escherichia coli* K-12. *J. Bacteriol.* In press
139. McEntee, K. 1977. Specialized transduction of *rec*A by bacteriophage lambda. *Virology* 70:221–22
140. McEntee, K., Epstein, W. 1976. Isolation and characterization of a specialized transducing bacteriophage for the *rec*A gene of *Escherichia coli.* *Virology.* 77:306–18
141. McEntee, K., Hesse, J. E., Epstein, W. 1976. Identification and radiochemical purification of the *rec*A protein of *Escherichia coli* K-12. *Proc. Natl. Acad. Sci. USA* 73:3979–83
142. Meselson, M. 1972. Formation of hybrid DNA by rotary diffusion during genetic recombination. *J. Mol. Biol.* 71:795–98
143. Meselson, M. S., Radding, C. M. 1975. A general model for genetic recombination. *Proc. Natl. Acad. Sci. USA* 72:358–61
144. Meyn, M. S., Rossman, T., Troll, W. 1977. A protease inhibitor blocks SOS functions in *Escherichia coli*—antipain prevents λ inactivation, ultraviolet mutagenesis, and filamentous growth. *Proc. Natl. Acad. Sci. USA* 74:1152–56

145. Miller, J. E., Barbour, S. D. 1977. Metabolic characterization of the viable, residually dividing and nondividing cell classes of recombination-deficient strains of *Escherichia coli* K-12. *J. Bacteriol.* 130:160–66
146. Miller, R. C. Jr. 1975. Replication and molecular recombination of T-phage. *Ann. Rev. Microbiol.* 29:355–76
147. Miller, R. V., Clark, A. J. 1976. Purification and properties of two deoxyribonucleases of *Pseudomonas aeruginosa. J. Bacteriol.* 127:794–802
148. Miyake, T. 1962. Exchange of genetic material between *Salmonella typhimurium* and *Escherichia coli* K-12. *Genetics* 47:1043–52
149. Miyake, T., Demerec, M. 1959. *Salmonella-Escherichia* hybrids. *Nature* 183:1586
150. Molineux, I. J., Gefter, M. L. 1975. Properties of *Escherichia coli* DNA binding (unwinding) protein interaction with nucleolytic enzymes and DNA. *J. Mol. Biol.* 98:811–25
151. Monti-Bragadin, C., Babudri, N., Samer, L. 1976. Expression of the plasmid pKM101-determined DNA repair system in *rec*A- and *lex*- strains of *Escherichia coli. Mol. Gen. Genet.* 145:303–6
152. Moody, E. E. M., Low, K. B., Mount, D. W. 1973. Properties of strains of *Escherichia coli* K-12 carrying mutant *lex* and *rec* alleles. *Mol. Gen. Genet.* 121:197–205
153. Morand, P., Blanco, M., Devoret, R. 1977. Characterization of *lex*B mutations in *Escherichia coli* K-12. *J. Bacteriol.* 131:572–82
153a. Morand, P., Goze, A., Devoret, R. 1977. Complimentation pattern of *lexB* and *recA* mutations in *Escherichia coli* K12: Mapping of *tif-1, lexB* and *recA* mutations. *Mol. Gen. Genet.* In press
154. Mortlemans, K., Stocker, B. A. D. 1976. UV-protection enhancement of UV mutagenesis and mutator effect of plasmid R46 in *Salmonella typhimurium. J. Bacteriol.* 128:271–82
155. Moseley, B. E. B., Copeland, H. J. R. 1975. Involvement of a recombination repair function in disciplined cell division of *Micrococcus radiodurans. J. Gen. Microbiol.* 86:343–57
156. Moseley, B. E. B., Copeland, H. J. R. 1975. Isolation and properties of a recombination-deficient mutant of *Micrococcus radiodurans. J. Bacteriol.* 121:422–28

157. Mount, D. W. 1976. Method for isolation of phage mutants altered in their response to lysogenic induction. *Mol. Gen. Genet.* 145:165–68
158. Mount, D. W. 1977. A mutant of *E. coli* showing constitutive expression of the lysogenic induction and error-prone DNA repair pathways. *Proc. Natl. Acad. Sci. USA* 74:300–4
159. Mount, D. W., Donch, J. J. 1976. A genetic nomenclature for *lex* and *exr* alleles in *Escherichia coli. Mutat. Res.* 36:237–40
160. Mount, D. W., Kosel, C. 1975. Ultraviolet light induced mutation in UV-resistant, thermo-sensitive derivatives of *lex*A- strains of *E. coli* K-12. *Mol. Gen. Genet.* 136:95–106
161. Mount, D. W., Kosel, C. K., Walker, A. 1976. Inducible error-free DNA repair in *tsl rec*A mutants of *E. coli. Mol. Gen. Genet.* 146:37–41
162. Mount, D. W., Walker, A. C., Kosel, C. 1975. Effect of *tsl* mutations in decreasing radiation sensitivity of a *rec*A⁻ strain of *Escherichia coli* K-12. *J. Bacteriol.* 121:1203–7
163. Mufti, S., Bernstein, H. 1974. The DNA-delay mutants of bacteriophage T4. *J. Virol.* 14:860–71
164. Notani, N. K., Setlow, J. K. 1974. Mechanism of bacterial transformation and transfection. *Prog. Nucleic Acid Res.* 14:39–100
165. Ohi, S., Bastia, D., Sueoka, N. 1974. Nonspecific 'pairing' of DNA molecules by recombination enzyme of *Bacillus subtilis. Nature* 248:586–88
166. Ou, J. T., Anderson, T. F. 1976. F' plasmids from HfrH and HfrC in *rec*A⁻ *Escherichia coli. Genetics* 83:633–43
167. Parma, D. H., Ingraham, L. J., Snyder, M. 1972. Tandem duplications of RII region of bacteriophage T4D. *Genetics* 71:319–35
168. Paul, A. V., Riley, M. 1974. Joint molecule formation following conjugation in wild type and mutant *E. coli* recipients. *J. Mol. Biol.* 82:35–56
169. Pettijohn, D. E. 1976. DNA packaging in isolated bacterial nucleoids. *Stadler Symp.* 8:159–69
170. Pietrzykowska, I. 1973. On the mechanism of bromouracil-induced mutagenesis. *Mutat. Res.* 19:1–9
171. Pollard, E. C., Randall, E. P. 1973. Studies on the inducible inhibitor of radiation-induced DNA degradation of *E. coli. Radiat. Res.* 55:265–79
172. Portalier, R., Worcel, A. 1976. Association of the folded chromosome with the

cell envelope of *E. coli:* Characterization of the proteins at the DNA-membrane attachment site. *Cell* 8:245–55
173. Potter, H., Dressler, D. 1976. Mechanism of genetic recombination: electron microscopic observation of recombination intermediates. *Proc. Natl. Acad. Sci. USA* 73:3000–4
174. Radding, C. M. 1973. Molecular mechanisms in genetic recombination. *Ann. Rev. Genet.* 7:87–111
175. Radman, M., Wagner, R. E. 1977. A molecular mechanism for site-specific recombination. *Mol. Gen. Genet.* In press
176. Ray, P. N., Pearson, M. L. 1976. Synthesis of morphogenetic proteins by mutants of bacteriophage lambda carrying tandem genetic duplications. *Virology* 73:381–88
177. Reif, H. J. Saedler, H. 1975. Is1 is involved in deletion formation in *gal* region of *Escherichia coli* K-12. *Mol. Gen. Genet.* 137:17–28
178. Rein, R., Egan, J. T. 1976. The relation between genetic recombination and base sequence complementarity. *J. Theor. Biol.* 62:409–27
179. Resnick, M. A. 1976. The repair of double-strand breaks in DNA: A model involving recombination. *J. Theor. Biol.* 59:97–106
180. Rigby, P. W. J., Burleigh, B. D., Hartley, B. S. 1974. Gene duplication in experimental enzyme evolution. *Nature* 251:200–4
181. Roberts, J. W., Roberts, C. W. 1975. Proteolytic cleavage of bacteriophage lambda repressor in induction. *Proc. Natl. Acad. Sci. USA* 72:147–51
182. Rosner, J. L. 1972. Formation, induction, and curing of bacteriophage P1 lysogens. *Virology* 48:679–89
183. Rothman, R. H., Kato, T., Clark, A. J. 1975. The beginning of an investigation of the role of *recF* in the pathways of metabolism of ultraviolet-irradiated DNA in *Escherichia coli.* See Ref. 20, pp. 283–91
184. Rupp, W. D., Levine, A. D., Trgovcevic, Z. 1975. Recombination and postreplication repair. See Ref. 20, pp. 307–12
185. Sadaie, Y., Kada, T. 1976. Recombination-deficient mutants of *Bacillus subtilis. J. Bacteriol.* 125:489–500
186. Sadowski, P. D., Vetter, D. 1976. Genetic recombination of bacteriophage T7 DNA *in vitro. Proc. Natl. Acad. Sci. USA* 73:692–96
187. Sakaki, Y. 1974. Inactivation of the ATP-dependent DNAase of *Escherichia coli* after infection with double-stranded DNA phages. *J. Virol.* 14:1611–12
188. Sanderson, K. E. 1971. Genetics of the enterobacteriaceae. A. Genetic homology in the enterobacteriaceae. *Adv. Genet.* 16:35–51
189. Schwartz, D. O., Beckwith, J. R. 1969. Mutagens which cause deletions in *Escherichia coli. Genetics* 61:371–76
190. Sedgwick, S. G. 1975. Ultraviolet inducible protein associated with error-prone repair in *E. coli* B. *Nature* 255:349–50
191. Sedgwick, S. G. 1975. Inducible error-prone DNA repair in *E. coli. Proc. Natl. Acad. Sci. USA* 72:2753–57
192. Sedgwick, S. G. 1975. Genetic and kinetic evidence for different types of postreplication repair in *Escherichia coli* B. *J. Bacteriol.* 123:154–61
193. Sedgwick, S. G. 1976. Misrepair of overlapping daughter strand gaps as a possible mechanism for UV induced mutagenesis in UVR strains of *Escherichia coli:* A general model for induced mutagenesis by misrepair (SOS repair) of closely spaced DNA lesions. *Mutat. Res.* 41:185–200
194. Sgaramella, V., Ehrlich, S. D., Bursztyn, H., Lederberg, J. 1976. Enhancement of transfecting activity of bacteriophage P22 DNA upon exonucleolytic erosion. *J. Mol. Biol.* 105:587–602
195. Shlaes, D. M., Barbour, S. D. 1977. Repair in *sbc*A- (*rec*⁺ revertant) strains of *Escherichia coli* K-12. Manuscript in preparation
196. Skalka, A. 1974. A replicator's view of recombination (and repair). See Ref. 59, pp. 421–32
197. Smith, G. P. 1976. Evolution of repeated DNA sequences by unequal crossover. *Science* 191:528–35
198. Smith, G. P. 1973. *The Variation and Adaptive Expression of Antibodies.* Cambridge, Mass.: Harvard Univ. Press
199. Sobell, H. M. 1975. Mechanism to activate branch migration between homologous DNA molecules in genetic recombination. *Proc. Natl. Acad. Sci. USA* 72:279–83
200. Stacey, K. A., Lloyd, R. G. 1976. Isolation of Rec-mutants from an F-prime merodiploid strain of *Escherichia coli* K-12. *Mol. Gen. Genet.* 143:223–32
201. Stadler, D. R. 1973. The mechanism of intragenic recombination. *Ann. Rev. Genet.* 7:113–27

202. Stahl, F. W., Crasemann, J. M., Stahl, M. M. 1975. Rec-mediated recombinational hot spot mediated recombination. *J. Mol. Biol.* 94:203–12
203. Stahl, F. W., Stahl, M. M. 1975. Rec-mediated recombination hot spot activity in bacteriophage (lambda), IV. Effect of heterology on Chi-stimulated crossing over. *Mol. Gen. Genet.* 140:29–37
204. Stahl, F. W., Stahl, M. M. 1976. On recombination between close and distant markers in phage lambda. *Genetics* 82:577–93
205. Stodolsky, M. 1974. Recipient gene duplication during generalized transduction. *Genetics* 78:809–22
206. Straus, D. S. 1974. Induction by mutagens of the tandem gene duplications in the *glyS* region of the *Escherichia coli* chromosome. *Genetics* 78:823–30
207. Straus, D. S., Hoffmann, G. R. 1975. Selection for a large genetic duplication in *Salmonella typhimurium*. *Genetics* 80:227–37
208. Straus, D. S., Straus, L. D. 1976. Large overlapping tandem genetic duplications in *Salmonella typhimurium*. *J. Mol. Biol.* 103:143–53
209. Suzuki, K. 1974. Cell division and DNA synthesis in *uvr*A *rec*A double mutants of *Escherichia coli* K-12. *Mol. Gen. Genet.* 129:249–58
210. Swenson, P. A. 1976. Physiological responses of *Escherichia coli* to far-ultraviolet radiation. *Photochem. Photobiol. Rev.* 1:269–387
211. Symonds, N. 1974. Recombination. *Biochemistry of Nucleic Acids*, ed. K. Burton, Ser. 1, 6: 165–89. Baltimore: Univ. Park Press
212. Teague, R. 1976. A result on the selection of recombination altering mechanisms. *J. Theor. Biol.* 59:25–32
213. Tekuno, S. I., Gough, M. 1976. UV sensitivity of a nonrepressor regulatory protein of bacteriophage P22. *J. Virol.* 18:65–70
214. Tomizawa, J. I., Ogawa, H. 1968. Breakage of DNA in *rec*+ and integration of radiophosphorous atoms in DNA and possible cause of pleiotropic effects of *rec*A mutation. *Cold Spring Harbor Symp. Quant. Biol.* 33:243–51
215. Tweats, D. J., Pinney, R. J., Smith, J. T. 1974. R-factor mediated nuclease activity involved in thymineless elimination. *J. Bacteriol.* 118:790–95
216. Tweats, D. J., Thompson, M. J., Pinney, R. J., Smith, J. T. 1976. R-factor-mediated resistance to ultraviolet light in strains of *Escherichia coli* deficient in known repair functions. *J. Gen. Microbiol.* 93:103–10
217. Unger, R. C., Clark, A. J. 1972. Interaction of the recombination pathways of bacteriophage (lambda) and its host *Escherichia coli* K-12: Effects on exonuclease V activity. *J. Mol. Biol.* 70:539–48
218. Van Dorp, B., Benne, R., Palitti, F. 1975. The ATP-dependent DNAase from *Escherichia coli rorA*: A nuclease with changed enzymatic properties. *Biochim. Biophys. Acta* 395:446–5
219. Vapnek, D., Alton, N. K., Bassett, C. L., Kushner, S. R. 1976. Amplification in *Escherichia coli* of enzymes involved in genetic recombination: Construction of hybrid ColE1 plasmids carrying the structural gene for exonuclease I. *Proc. Natl. Acad. Sci. USA* 73:3492–96
220. Volkert, M., George, D. L., Witkin, E. M. 1976. Partial suppression of the LexA phenotype by mutations (rnm) which restore ultraviolet resistance but not ultraviolet mutability to *Escherichia coli* B/r *uvr*A *lex*A. *Mutat. Res.* 36:17–28
221. Wagner, R. E., Radman, M. 1975. A mechanism for initiation of genetic recombination. *Proc. Natl. Acad. Sci. USA* 72:3619–22
222. Wayman, L., Goering, R. V., Novick, R. P. 1974. Genetic control of chromosomal and plasmid recombination in *Staphylococcus aureus*. *Genetics* 76:681–702
223. Weaver, S., Levine, M. 1977. Recombinational circularization of *Salmonella* phage P22 DNA. *Virology* 76:29–38
224. Weisberg, R. A., Sternberg, N. 1974. Transduction of *rec*B-hosts is promoted by Lambda *red*+ function. See Ref. 59, pp. 107–9
225. West, S. C., Emmerson, P. T. 1977. Induction of protein synthesis in *Escherichia coli* following UV-irradiation or gamma-irradiation, mitomycin-C treatment or *tif*-expression. *Mol. Gen. Genet.* 151:57–68
226. Wilcox, K. W., Smith, H. O. 1976. Mechanism of DNA degradation by the ATP-dependent DNAase from *Haemophilus influenzae* Rd. *J. Biol. Chem.* 19:6127–34
227. Wildenberg, J., Meselson, M. 1975. Mismatch repair in heteroduplex DNA. *Proc. Natl. Acad. Sci. USA* 72:2202–6
228. Witkin, E. M. 1975. Elevated mutability of *pol*A and *uvr*A *pol*A derivatives of *Escherichia coli* B/r at sublethal

doses of ultraviolet light: Evidence for an inducible error-prone repair system ("SOS repair") and its anomalous expression in these strains. *Genetics* 79:199–213
229. Witkin, E. M. 1976. Ultraviolet mutagenesis and inducible DNA repair in *Escherichia coli*. *Bacteriol. Rev.* 40:869–907
230. Witkin, E. M., Parisi, E. C. 1974. Bromouracil mutagenesis: Mispairing or misrepair? *Mutat. Res.* 25:407–9
231. Yamamoto, N. 1967. Recombination: Damage and repair of bacteriophage genome. *Biochem. Biophys. Res. Commun.* 27:263–69
232. Yoakum, G., Ferron, W., Eisenstark, A., Webb, R. B. 1974. Inhibition of replication gap closure in *Escherichia coli* by near-ultraviolet light photoproducts of L-tryptophan. *J. Bacteriol.* 110:62–69
233. Yoshito, S., Kada, T. 1976. Recombination-deficient mutants of *Bacillus subtilis*. *J. Bacteriol.* 125:489–500
234. Youngs, D. A., Bernstein, I. A. 1973. Involvement of $recB$-$recC$ nuclease (exonuclease V) in the process of X-ray–induced deoxyribonucleic acid degradation in radiosensitive strains of *Escherichia coli* K-12. *J. Bacteriol.* 113:901–6
235. Youngs, D. A., Smith, K. C. 1976. Genetic control of multiple pathways of postreplicational repair in uvrB strains of *Escherichia coli* K-12. *J. Bacteriol.* 125:102–10
236. Zambrano, F., Radman, M. 1977. Spontaneous and ultraviolet-induced recombination and mutagenesis in bacteriophage lambda. *Mol. Gen. Genet.* In press
237. Zieg, J., Kushner, S. R. 1977. Analysis of genetic recombination between two partially deleted lactose operons of *Escherichia coli* K-12. *J. Bacteriol.* 131:123–32
238. Zieg, J., Silverman, M., Hilmen, M., Simon, M. 1977. Recombinational switch for gene expression. *Science* 196:170–72
239. Zinder, N. D. 1974. Recombination in bacteriophage f1. See Ref. 59, pp. 19–28

BEHAVIORAL GENETICS IN BACTERIA

❖3127

John S. Parkinson

Department of Biology, University of Utah, Salt Lake City, Utah 84112

CONTENTS

INTRODUCTION	397
CHEMOTACTIC BEHAVIOR OF *ESCHERICHIA COLI*	398
GENETICS OF STIMULUS TRANSDUCTION	399
Stimulus Detection	400
Signaling	401
Flagellar Response	402
cheA, cheW, cheY	403
cheB, cheZ	404
cheC	405
cheD	405
SENSORY ADAPTATION	405
Protein Methylations in Chemotaxis	405
Methylation Patterns in Generally Nonchemotactic Mutants	407
INTERACTION OF TRANSDUCTION COMPONENTS	408
cheC:cheZ	409
cheB:cheZ	409
tar:tsr	409
A Model of Stimulus Transduction	410
Implications of the Model	411
CONCLUSIONS	411

INTRODUCTION

In the late 1800s, Pfeffer, Englemann, and others demonstrated that the movements of bacteria are purposeful: that bacteria avoid inhospitable environments and accumulate in favorable ones. These early studies of bacterial behavior are described in a review by Berg (1). More recently, Adler (2) showed that motile strains of *Escherichia coli* can move toward or away from many kinds of chemicals. Chemotaxis in *E. coli* is formally analogous to more sophisticated behaviors in higher organisms, and yet affords the experimental advantages of a microbial system. Judging from the recent and rapid progress in this field, bacterial chemotaxis will almost certainly be the first stimulus-response system to be understood at the molecular level [for reviews, see (3–6)]. In this article, I discuss the current status of genetic studies of chemotaxis in *E. coli* and in *Salmonella typhimurium*.

CHEMOTACTIC BEHAVIOR OF *ESCHERICHIA COLI*

Motile *E. coli* swim by rotating their flagellar filaments (7, 8), which in both appearance and function resemble miniature propellers. The stiff, helical filaments can turn in either direction (8–10): counterclockwise (ccw) rotation produces smooth swimming (9); clockwise (cw) rotation causes tumbling movements (9) that enable the organism to change its swimming direction. In wild-type strains the rotating flagella undergo frequent, random reversals (8–11), and as a consequence, the bacteria move in a three-dimensional random walk composed of alternating smooth and tumbly episodes (12). Chemotactic migrations in chemical gradients are carried out by biasing the random walk pattern in the preferred direction (12–14). For example, individuals that happen to move toward an attractant (12, 14) or away from a repellent (15) decrease their tumble probability and thereby increase mean path length in the appropriate direction. Thus, control of flagellar rotation in response to chemical stimuli is the underlying basis of chemotactic behavior in *E. coli*.

As the organism swims, chemoreceptors (16) on the surface of the cytoplasmic membrane continuously monitor the concentrations of chemicals in the immediate environment. Chemotactic responses are triggered whenever an individual encounters a temporal change in attractant (14, 17) or repellent (15) concentration. In effect, *E. coli* seems to compare its past and present chemical environments in order to determine the direction in which it is moving relative to spatial gradients. Temporal sensing enables the bacteria to average out local concentration changes caused by statistical fluctuations and to amplify actual concentration changes by making comparisons over distances that are greater than the body length of the organism (14, 17).

The response of *E. coli* to large temporal stimuli, many orders of magnitude greater than those a bacterium would encounter by swimming in spatial gradients of chemicals, is especially dramatic and best illustrates the stimulus-response physiology of chemotaxis. Sudden large increases in attractant concentration (9, 11, 14, 18) or decreases in repellent levels (15) elicit smooth-swimming responses that can persist for as long as several minutes. During this period the flagella rotate exclusively in the ccw (i.e. nontumbling) direction, but eventually swimming behavior and flagellar rotation return to their prestimulus patterns. The *transience* of such responses is the crucial observation, for it shows that bacteria possess a sensory system that undergoes adaptation. The adaptation mechanism allows bacteria to respond to concentration *changes* rather than to absolute concentrations of a chemical and is therefore responsible for temporal detection of chemical gradients.

The stimulus-response behavior of *E. coli* is summarized in Figure 1 in which the swimming pattern is assumed to reflect the level of some type of signal that controls

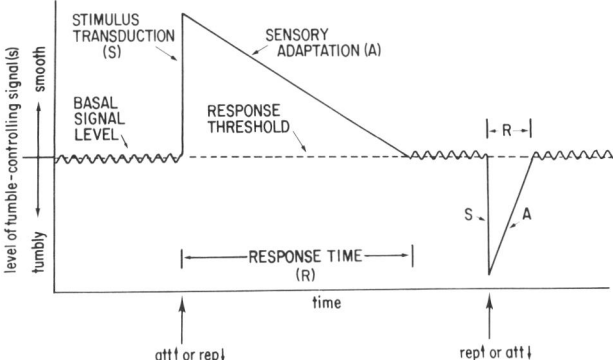

Figure 1 Summary of behavioral changes elicited by attractant or repellent stimuli in *E. coli*. See text for further explanation.

the rotational machinery of the flagella. Spontaneous fluctuations in basal signal level might account for the alternating smooth and tumbly episodes that characterize swimming in the absence of stimuli. Whenever the signal rises above a threshold level, ccw rotation and smooth swimming result; whenever the signal falls below threshold, cw rotation and tumbling result. Chemotactic stimuli raise or lower signal level to elicit smooth or tumbly swimming responses: increases in attractant (9, 11, 14, 18) and decreases in repellent (15) suppress tumbling; increases in repellent (9, 15) and decreases in attractant (14, 17) enhance tumbling. The conversion of stimulus information into a signal that modulates flagellar rotation is called *stimulus transduction*. This process takes place rapidly upon application of the stimulus (11, 14, 15, 18) and the change in signal level is directly proportional to stimulus magnitude (11, 18).

Following stimulus transduction, a system for sensory adaptation works to restore tumble-controlling signals to the basal level (11, 14, 18). Studies employing various combinations of stimuli demonstrate that all chemoreceptors share a common adaptation system (11), which operates at a linear rate (11) as depicted in Figure 1. However, bacteria require much longer times to recover from tumble-suppressing stimuli than from tumble-enhancing stimuli of comparable magnitude (11, 14, 15, 17), which indicates that the mechanism of adaptation may not be the same in both cases.

GENETICS OF STIMULUS TRANSDUCTION

A large number of chemotaxis-related genes have been identified and studied in *E. coli* (Figure 2). Many of them are probably involved in stimulus transduction and should enable us to reconstruct the path of information flow from chemoreceptors to flagella. Since each flagellum is controlled by many different stimuli, the transduction pathway probably has many inputs at the receptor level which converge as

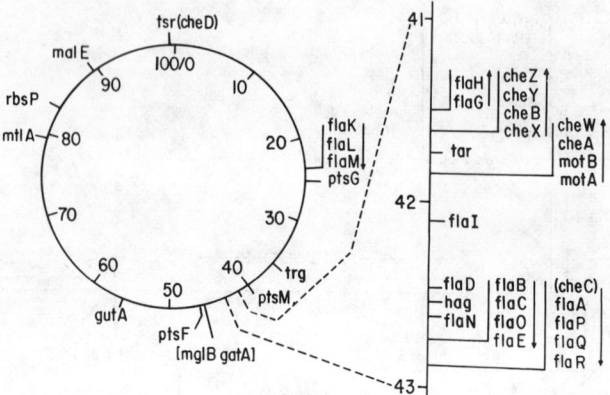

Figure 2 Chemotaxis-related loci in *E. coli*. Approximate map positions (in minutes) of chemotaxis genes were taken from Bachmann, Low & Taylor (64). *Arrows* above gene clusters indicate the direction and extent of cotranscription within the cluster. Genes within brackets have not been ordered with respect to outside markers.

Receptor (binding) proteins		Signaling functions
attractant	gene symbol	gene symbol
glucose	ptsG	trg
mannose	ptsM	tar
galactose	mglB	tsr
galactitol	gatA	
fructose	ptsF	Flagellar functions
glucitol	gutA	gene symbol
mannitol	mtlA	motA, B
ribose	rbsP	hag (filament antigen)
maltose	malE	fla
		che

stimulus information is passed to the flagella. The position of any element in the pathway should therefore reflect the number of different responses dependent on that particular component. A schematic representation of the pathway and the mutants upon which it is based is shown in Figure 3.

Stimulus Detection

Wild-type *E. coli* is attracted to a number of sugars (19) and amino acids (20) and repelled by potentially harmful compounds such as ethanol (21). These chemicals are measured by specific chemoreceptors, each of which detects a group of chemically similar compounds (16). The genetics and physiology of chemoreception have been reviewed recently (6), so the discussion here is brief.

The galactose (22), maltose (23), and ribose (24) receptors are water-soluble binding proteins that reside in the periplasmic space between the inner and outer

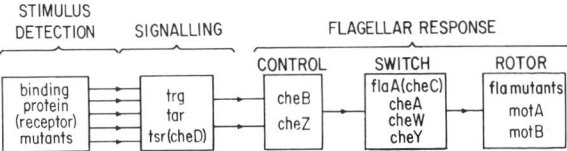

Figure 3 Pathway of information flow during stimulus transduction based on phenotypes of chemotaxis-defective mutants. Mutants are ordered from least (receptor mutants) to most (motility mutants) pleiotropic. See text for further explanation.

membranes. Other sugars, including glucose, are detected by membrane-associated binding proteins (25, 26). The structural genes for sugar-binding proteins involved in chemotaxis are shown on the genetic map in Figure 2. Mutants defective in any one of these genes have normal motility but lack chemotactic responses mediated by one type of receptor (22–27). Since the binding proteins are also involved in active transport of their respective ligands, binding protein mutants are also defective in sugar uptake (22–27). The two processes, chemoreception and transport, are distinct, even though they share a common component. Mutants defective in other transport proteins generally have normal chemotactic responses (16). The few exceptions may be due to interaction of the mutant transport protein with the sugar-binding protein of the chemoreceptor (25, 27, 28).

There is essentially no genetic or biochemical information on receptors for amino acids or repellents in *E. coli*. Amino acid attractants include aspartate, asparagine, alanine, cysteine, glutamate, glycine, threonine, and serine (20). Repellents include aliphatic alcohols, fatty acids, hydrophobic amino acids, H^+ and OH^- ions, indole, and divalent metal cations (21). Competition experiments, which measure the ability of one chemical to inhibit a chemotactic response to a second compound, indicate that many different chemoreceptors may be involved in mediating these various responses (20, 21, 29). It seems likely that binding proteins of some sort are employed as detection devices in these cases also, but a definitive answer must await isolation of specific receptor mutants.

Signaling

The duration of behavioral responses after temporal stimulation is directly proportional to the fractional change in receptor occupancy brought about by the stimulus (11, 18). It is likely, therefore, that chemoreceptors (i.e. binding proteins) convey information about the proportion of binding sites occupied by ligand at any point in time (11, 18, 29). Although the form of this communication is not known, transmission of receptor information to the flagella seems to be mediated by a network of signaling elements. The three known types of signaling mutants all have essentially normal motility, but lack responses controlled by two or more receptors. It has been suggested that some chemoreceptors may employ specific signalers not shared by other binding proteins (3), but there is no genetic evidence on this point.

In *S. typhimurium*, saturation of the ribose-binding protein with ligand completely inhibits chemotactic responses to galactose, demonstrating that the ribose and galactose receptors may compete for a shared signaling element (30). Similar, but less drastic effects have been observed in *E. coli* for a number of receptor combinations including ribose-galactose (19, 20). These shared signaling elements must not be an integral part of each receptor; otherwise competition would not occur. It seems that only when a receptor is occupied by ligand is it able to interact with common signaling elements. The ribose-galactose signaler may be the product of the *trg* locus (see Figure 2). Mutants defective in *trg* function do not respond to ribose or galactose stimuli (27).

The *tar* and *tsr* loci (see Figure 2) are also involved in signaling, but at a later step than the *trg* product (31). Loss of either function leads to numerous response defects: *tar* mutants are no longer attracted to aspartate or maltose or repelled by hydroxyl ions or divalent metal cations (31); *tsr* mutants are not attracted by serine (20, 31a) or repelled by hydrogen ions, fatty acids, indole, or hydrophobic amino acids (21). As a general rule, the Tar$^-$ phenotype is complementary to the Tsr$^-$ phenotype in that responses present in *tar* strains are absent in *tsr* strains, and vice versa (31). A few responses, notably ribose and galactose, are not eliminated by either type of mutation (31). It seems likely that all chemoreceptors may transmit signals through one or both of these components (31).

Silverman, Simon, and co-workers constructed λ transducing phages that contained the *tar* or *tsr* region and used them to program protein synthesis in various hosts (32–34). They found that both λ*tar* and λ*tsr* make membrane proteins of approximately 60,000 mol wt and that expression of these proteins depends on the presence of a wild-type *flaI* gene in the host. Since the *flaI* product appears to be a positive regulator of motility genes (35, 36), both the *tar* and *tsr* products are evidently subject to *flaI* control. Proof that the membrane proteins really are the *tar* and *tsr* products rests exclusively on the fact that the ability to complement *tar* (J. S. Parkinson, unpublished results) or *tsr* (34) defects is always correlated with the ability of λ transducing phages to make these particular proteins. Additional properties of the *tar* and *tsr* proteins are discussed in the section on sensory adaptation.

Flagellar Response

The machinery that carries out the flagellar response to chemotactic stimuli can be operationally divided into three parts as shown in Figure 3. The first of these is the rotor that produces flagellar rotation. The important components of the rotor are probably located in the flagellar basal body, a complex structure attached to the cell wall and membranes (10, 37, 38). Other basal body components may be responsible for determining the direction of flagellar rotation and could function as the flagellar switch. The switch in turn must be associated with control elements that modulate switch activity in response to chemotactic inputs.

The genetics of flagellar structure and function have been reviewed recently by Silverman & Simon (39) and by Iino (this volume). Many genes are involved, most of which produce a nonflagellated (Fla$^-$) mutant phenotype (see Figure 2). Since motility is a prerequisite for chemotaxis assays, it is not a simple matter to identify *fla* functions that are part of the switch or its control system. One approach has been to screen generally nonchemotactic (*che*) mutants that are motile but lack chemotaxis to all stimuli (40). Some mutants of this type prove to have specific *fla* defects that allow flagellar assembly but somehow interfere with normal rotational behavior (41–44). Other *che* mutants define functions needed for stimulus integration, switch control, and adaptation.

Generally nonchemotactic mutants have normal growth rates and flagellar morphology (40), but invariably exhibit aberrant swimming behavior: Some *che* strains cannot tumble, others tumble incessantly (42–46). Because *che* strains with normal swimming patterns have not been found, all *che* functions must in some way be concerned with the regulation of tumbling behavior. In *E. coli*, eight *che* genes have been defined on the basis of complementation tests with F-prime elements (42), specialized λ transducing phages (32), or abortive transduction by phage P1 (47). In *Salmonella*, nine *che* classes have been established by abortive transduction and deletion mapping tests (43–45). The similarities in map positions and phenotypes between the two groups of mutants suggests extensive homology of the *che* machinery in the two different species. Most *che* loci map in a region that contains most of the motility-related genes as well (see Figure 2). One locus (*cheD*) maps in an entirely different region (J. S. Parkinson, in preparation). The properties of the eight *che* loci of *E. coli* are summarized in Table 1. Those that appear to be involved in stimulus transduction are discussed below.

cheA, cheW, cheY   Defects in any of these genes lead to a nontumbling phenotype in which flagellar rotation is exclusively ccw (42). Tumble-enhancing stimuli do not elicit cw flagellar rotation in these strains (9; J. S. Parkinson, in preparation), which shows that the *cheA, W, Y* products may play important roles in the switch mechanism that produces cw reversals during swimming. Reversion studies of *cheA* and *cheY* mutants indicate that these genes interact with several different *fla* loci that may specify switch components (R. A. Smith, personal communication; S. R. Parker and J. S. Parkinson, in preparation). Three *Salmonella* genes (*cheP, cheQ, cheW*) have map positions and mutant phenotypes comparable to the *A, W, Y* group in *E. coli* (44).

The presence of *cheA* complementation activity in λ*che* transducing phages is correlated with proteins of 76,000 and 66,000 mol wt (32, 48, 49). Partial peptide maps of the two proteins are quite similar, suggesting that the *cheA* product may be modified or processed in some fashion (49a). It will be interesting to learn which, if not both, of these proteins is the active form, and whether *cheA* product is converted from one form to another by chemotaxis-dependent processes. In its native form, the *cheA* product is probably a multimer, because *cheA* mutants display extensive intracistronic complementation (42). The *cheW* product appears to be a 12,000 mol wt protein (32, 48) and, like the *cheA* protein, is found mainly in the cytoplasm (39). These two genes are part of a cotranscribed unit that also includes the *motA* and *motB* genes (48). The *mot* products are membrane proteins that are needed for flagellar rotation (49). It is tempting to speculate that the *cheA* and *cheW* genes may provide an analogous service for the switch. Perhaps they are needed to "energize" the switch or to enable it to respond to control signals from the chemoreceptors.

The *cheY* product is reported to be a cytoplasmic protein of 8000 mol wt (32); however, the mutational target size of the *cheY* gene is inconsistent with a protein of such small size (42; J. S. Parkinson in preparation). In fact, Silverman & Simon (32) noted that the *cheY* protein forms a "broad, intense" band in SDS polyacrylamide gels. Perhaps the *cheY* product is a much larger protein that is rapidly processed into fragments averaging about 8000 mol wt.

cheB, cheZ Mutations in either of these genes cause excessive tumbling and predominately cw flagellar rotation (42). Two similar genes (*cheT, cheX*) have been discribed in *S. typhimurium* (44). Tumbly mutants can respond to temporal stimulation: Increases in attractant suppress tumbling (9, 18, 42, 50) and increases in repellent counteract that effect (J. S. Parkinson, in preparation). This shows that neither *cheB* nor *cheZ* function is essential for ccw rotation. However, the responses do differ from wild type in two respects. First, larger stimuli are required to elicit a smooth-swimming response (42, 50). Second, the response times are shorter (42, 50, 51). Both effects could be due to a single mutation that effectively raises the response threshold relative to the basal signal level, as shown in Figure 4*a*. The response thresholds for different stimuli are not uniformly altered, because *cheB* strains respond well to serine stimuli and poorly to aspartate stimuli, whereas *cheZ* mutants show just the opposite behavior (42, 50). Perhaps these components are control elements that interact with the switch to modulate tumbling in response to chemotactic stimuli, with *cheB* handling inputs from the *tar* component and *cheZ* handling the *tsr* inputs. This notion is developed further in the section on interaction of transduction components.

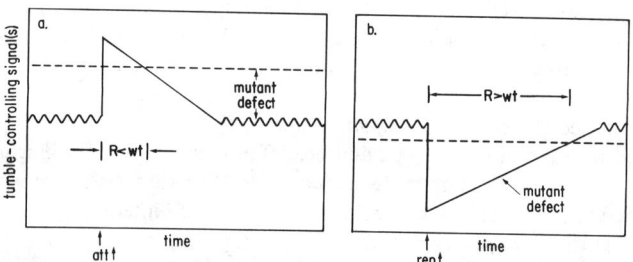

Figure 4 Stimulus-response analysis of two types of chemotaxis mutants. (*a*) Response of *cheB* or *cheZ* strains to attractant stimuli. The high spontaneous tumbling rate and short response times may be caused by an increase in response threshold or a decrease in basal signal level. (*b*) Response of *cheX* strains to repellent stimuli. The low spontaneous tumbling rate and long response times may be caused by a defect in the adaptation system (which also establishes basal signal level).

cheC Mutations at the *cheC* locus can produce either nonmotile or generally nonchemotactic strains. The most frequent mutants are nonflagellated and probably represent the null class, designated *flaA* (52) (see Figure 2). Mutants of the *cheC* type are less common (40, 42), have a normal complement of flagella (40), but swim without tumbling (12, 40,42). Since *cheC* mutants fail to complement *flaA* strains (41, 42), both phenotypes must arise from alterations of the same gene product. The *flaA* protein could be a flagellar component involved in the switch mechanism. In *Salmonella*, an analogous gene (*flaQ* or *cheU*) can mutate to both low and high tumbling rates (43, 44). In *E. coli*, internal revertants of *cheC* mutations often have very high tumbling rates as well (S. R. Parker and J. S. Parkinson, in preparation). These findings show that the *flaA* component may be responsible for establishing the relative probabilities of ccw and cw flagellar rotation.

cheD Mutants of this type are smooth swimmers that exhibit complete dominance to wild type in complementation tests (J. S. Parkinson, in preparation), suggesting that the mutant *cheD* protein actively prevents tumbling. Additional mutations in the *cheD* gene that abolish its dominance, presumably by destroying the inhibitory protein, lead to a Tsr⁻ phenotype, which shows that *cheD* mutants represent a specific mutational defect in the *tsr* gene (J. S. Parkinson, in preparation). The existence of such mutants can be explained by proposing that wild-type *tsr* product controls flagellar rotation by passing along various types of chemoreceptor signals. In *cheD* mutants, the *tsr* product may be "locked" in a tumble-inhibiting configuration, as if tumble-suppressing stimuli were perceived continuously.

SENSORY ADAPTATION

Both *E. coli* (53, 54) and *S. typhimurium* (55) have a specific, continuous requirement for methionine during chemotaxis. Methionine auxotrophs, when deprived of exogenous methionine, quickly lose the ability to tumble (51), but regain normal swimming behavior when methionine is restored. The ability to adapt to tumble-suppressing stimuli (56) and to respond to tumble-enhancing stimuli (51) is also methionine-dependent. The main biochemical role of methionine in chemotaxis is that of a methyl donor. Bacterial mutants with low reserves of S-adenosylmethionine (SAM), a methyl-donating compound derived from ATP and methionine, exhibit behavioral defects associated with methionine deprivation (57, 58). Further genetic and biochemical studies of the methionine effect, discussed below, demonstrate that methylation events play a central role in both stimulus transduction and sensory adaptation (31, 34, 56, 59).

Protein Methylations in Chemotaxis

Kort, Goy, Larsen & Adler (59) examined cell proteins that incorporate radioactivity from labeled methionine in the absence of protein synthesis and identified a methyl-accepting chemotaxis protein (MCP) in the cytoplasmic membrane of *E.*

coli. On SDS polyacrylamide gels, MCP activity forms a series of discrete bands with apparent molecular weights in the 55,000–65,000 range (31, 34; M. F. Goy, M. S. Springer, and J. Adler, personal communication). This distinctive behavior closely resembles that of the *tar* and *tsr* proteins made by λ transducing phages. Direct comparisons on two-dimensional gels indicate that MCP probably represents the collective product of these two genes (34). Both *tar* and *tsr* mutants produce some MCP bands in methylation assays; however, *tar tsr* double mutants have no detectable MCP activity (31, 34). The methyl groups are attached by ester linkage to glutamic acid residues in the MCP molecules (60, 61). Differences in the extent or pattern of methylation or some other post-translational modification might be responsible for the multiple banding pattern exhibited by these proteins (31).

Chemotactic stimuli alter the amount of methyl label associated with MCP (31, 34, 59; M. F. Goy, M. S. Springer, and J. Adler, personal communication): Increases in attractant stimulate methylation and increases in repellent reduce methylation. Springer, Goy & Adler (56) showed that these changes in the extent of MCP methylation are probably involved in the adaptation process. For example, adaptation to increases in attractant (tumble-suppressing) requires methionine and is correlated with methylation of MCP (56). Adaptation to decreases in attractant (tumble-enhancing) does not require methionine and coincides with demethylation of MCP (56). It appears, therefore, that the *tar* and *tsr* products have a dual function in chemotaxis (31, 34): They elicit flagellar responses to chemotactic stimuli, and they participate in the adaptation process. A model summarizing the role of these products is shown in Figure 5. Suppose that the *tar* and *tsr* proteins can exist in two alternative functional states, one that produces ccw flagellar rotation and another that causes cw rotation. Stimulus transduction and sensory adaptation would then be effected by cycling the *tar* and *tsr* products between these two functional states. Chemotactic stimuli would convert one state to the other, causing an imbalance in the relative proportions of the two forms and a consequent change in swimming behavior. For example, tumble-suppressing stimuli would increase the ccw state at the expense of the cw form. Methylation or demethylation of MCP

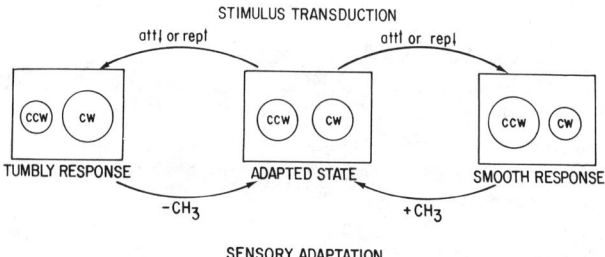

Figure 5 A summary of the role of MCP in stimulus transduction and sensory adaptation. The changes in MCP function caused by stimuli or by methylation-demethylation are shown. See text for further explanation.

might also convert one form to the other and be used by the adaptation system to reestablish the correct balance of the ccw and cw states to restore a normal swimming pattern. Mechanisms that might control the adaptation system are discussed in the next section.

Both *tar* (31, 34; J. S. Parkinson, unpublished observations) and *tsr* (34) strains exhibit longer response times to tumble-suppressing stimuli. This suggests that adaptation rate may be determined by the total supply of *tar* and *tsr* products and that adaptation to either *tar*- or *tsr*-mediated responses utilizes both products. Alternatively, it may be that stimulus transduction is more efficient in *tar* and *tsr* mutants than in wild type (31). The tumbling frequency of *tar tsr* double mutants is very low (34; J. S. Parkinson, unpublished observations), which implies that the methylated (cw) form of MCP plays an active role in tumble production (51). How the tumbling frequency might be controlled by the *tar* and *tsr* proteins is discussed in the final section of this article.

Methylation Patterns in Generally Nonchemotactic Mutants

Mutants defective in *cheX* function (32; J. S. Parkinson, in preparation) are smooth swimmers (42), but can be induced to tumble by repellent stimuli (J. S. Parkinson, in preparation). The response times, however, are very long, as much as fiftyfold greater than wild-type (J. S. Parkinson, in preparation). An adaptation defect would account for this behavior, as shown in Figure 4b. The level of methylated MCP is very low in *cheX* strains (34, 59; M. F. Goy, M. S. Springer, and J. Adler, personal communication) (Table 1) and might be responsible for the inability of *cheX* strains

Table 1 Properties of *che* mutants of *Escherichia coli*[a]

Gene	Product molecular weight[b]	Product location[c]	Swimming pattern	Temporal response	Stimulation duration	MCP methylation (percentage of wild type)[d]
cheA	76,000; 66,000	cytoplasm	smooth	none	—	50–250
cheW	12,000	cytoplasm	smooth	none	—	50–160
cheY	(8,000)	cytoplasm	smooth	(none)	—	180–275
cheB	38,000	cytoplasm	tumbly	T → S[e]	brief	~10
cheZ	24,000	cytoplasm	tumbly	T → S[e]	brief	~50
cheX	28,000	cytoplasm	smooth	S → T[f]	prolonged	< 5
cheC	?	(cytoplasmic membrane)	smooth	(S → T)[f]	—	125
cheD	(~60,000)	(cytoplasmic membrane)	smooth	(S → T)[f]	—	< 5 (2 alleles) ~100 (1 allele)

[a] Table entries in brackets are either equivocal or not yet firmly established.
[b] Data are from Silverman & Simon (32, 48).
[c] Data were compiled from Silverman & Simon (39) and other evidence discussed in the text.
[d] Data were compiled from Kort et al (59) and from M. F. Goy, M. S. Springer & J. Adler, personal communication.
[e] Tumbly to smooth transition produced by an increase in attractant concentration.
[f] Smooth to tumbly transition produced by an increase in repellent concentration.

to exhibit spontaneous tumbling. Moreover, since adaptation to tumble-enhancing stimuli involves demethylation of MCP (56), the very long response times of *cheX* mutants may be due to the fact that most of their MCP is already in the demethylated state. Springer & Koshland (62) recently described a methylase activity in *S. typhimurium* that appears to transfer methyl groups from SAM to membrane protein(s) that could be analogous to MCP in *E. coli.* The methylase is absent in smooth-swimming *cheR* mutants of *Salmonella* (44, 62). The *cheX* product is a likely candidate for the corresponding methylase in *E. coli.*

Most other *che* strains also have abnormal levels of methylated MCP (34, 59; M. F. Goy, M. S. Springer, and J. Adler, personal communication) (Table 1). Except for *cheD,* none of these mutations affects the structural genes for MCP, so their MCP phenotypes are probably not caused by a change in the structure or amount of methyl-accepting proteins, but rather by an alteration in the mechanism that controls the activity of the adaptation system and establishes the relative amounts of methylated and unmethylated MCP. One possibility is that the MCP methylase activity is modulated by feedback signals from the flagellar switch. Imagine, for example, that the switch operates like a comparator. Whenever signal input from the transduction pathway differs from its internal reference, the switch could suppress or activate the MCP methylase to bring the two signals back into correspondence. Presumed switch functions such as *cheA, cheW,* or *cheY* might provide the reference signal. Mutants in these genes generally have above-normal levels of methylated MCP (Table 1). The *cheB* and *cheZ* products probably convey receptor signals to the switch. Mutants in these genes have below-normal levels of methylated MCP (Table 1).

A second possible mechanism for adaptation control would involve only the MCP methylase and its substrates, the *tar* and *tsr* products. Perhaps the substrate properties of these proteins are altered by chemotactic stimuli, making them more or less accessible to the methylase, and thereby shifting the equilibrium in favor of the methylated or unmethylated form. The finding that *tar*-dependent responses are correlated with methylation changes only in *tar*-specified components of MCP, whereas *tsr*-dependent responses are associated with methylation changes in *tsr* components of MCP (31, 34), is consistent with this notion. However, this model provides no ready explanation for the methylation patterns in different *che* strains, whereas the feedback model does, so the actual control mechanism may employ features of both strategies. Whatever the mechanism, it must also account for an interesting effect noted by Berg & Tedesco (11). They found that in the final stages of adaptations to tumble-suppressing stimuli, the bacteria "overshoot" and exhibit cw flagellar rotation for a brief period before regaining normal reversing behavior. The overshoot evidently reflects a time lag or hysteresis effect in the adaptation control system.

INTERACTION OF TRANSDUCTION COMPONENTS

There is a considerable body of circumstantial evidence to suggest that various components of the transduction machinery interact directly with one another. For example, when the products of two genes interact, defects in one can often be suppressed by a compensating alteration in the other. Reversion analysis of *che* mutants has, in fact, uncovered cases of gene product interactions among transduction elements. Although not yet confirmed biochemically, the findings of these and other interaction studies have provided valuable information about the control of flagellar rotation by the transduction apparatus. Some examples of possible product interactions and their implications are discussed below.

cheC:cheZ

Chemotactic revertants of *cheC* strains often carry compensating mutations in the *cheZ* gene (J. S. Parkinson and S. R. Parker, in preparation). Likewise, *cheZ* revertants sometimes carry suppressors that map at the *cheC* locus (S. R. Parker and J. S. Parkinson, unpublished results). These interactions are allele-specific. For example, a *cheZ* mutation that suppresses one *cheC* defect will not necessarily correct others. Moreover, many *cheZ* mutations, including nonsense alleles, will not suppress any *cheC* mutants, demonstrating that *loss* of *cheZ* function does not compensate *cheC* strains for chemotaxis. These results suggest that the *cheC* and *cheZ* proteins may interact in a specific manner with one another (J. S. Parkinson and S. R. Parker, in preparation).

cheB:cheZ

Since *cheB* and *cheZ* are tightly linked (see Figure 2), it has not yet been feasible to analyze revertants of either *cheB* or *cheZ* mutants for suppressor mutations in the other gene. However, the complementation properties of *cheB* and *cheZ* mutations in partial diploids indicate that the two gene products may interact. For example, some *cheB* alleles complement some *cheZ* alleles very poorly, even though both mutations alone are completely recessive to wild type (42; J. S. Parkinson, unpublished results). Poor complementation effects could result if the *cheB* and *cheZ* products formed some sort of complex.

tar:tsr

Loss of either the *tar* or *tsr* protein produces a modest decrease in the amount of MCP detectable by methylation assays because both products contribute to the methyl-accepting activity (31, 34). However, *cheD* strains, which carry dominant *tsr* mutations (J. S. Parkinson, in preparation), not only fail to respond to *tar*-mediated stimuli, but also can have very low levels of apparent MCP methylation (59; M. F. Goy, M. S. Springer, and J. Adler, personal communication) (Table 1). In such strains, both the *tar* and *tsr* proteins are evidently incapable of being methylated or else are methylated at a very slow rate. This ability of a specific *tsr* defect (i.e. *cheD*) to prevent methylation of both methyl-accepting proteins might be due to interaction of the *tar* and *tsr* products. If such an interaction exists, it may not be a fully reciprocal one, since no *tar* mutations with effects comparable to *cheD* have yet been found. Loss of either the *tar* or *tsr* function produces an improvement in chemotactic responses mediated by the remaining pathway (31, 31a), which can also be explained in terms of mutual inhibition caused by interaction of the *tar* and *tsr* gene products.

A Model of Stimulus Transduction

The genetic and biochemical evidence presented in this article is summarized in a model of the transduction process shown in Figure 6. This model is by no means the only one consistent with the data, but it does provide a simple, useful framework for discussion. After describing the model, I indicate a few of the predictions it makes about certain aspects of the chemotaxis process.

The model of Figure 6 proposes that chemical information detected by binding (i.e. receptor) proteins on the outer surface of the cytoplasmic membrane is transmitted to the flagellum via a cytoplasmic message. The *tar* and *tsr* proteins are assumed to span the cytoplasmic membrane so that they can interact with binding proteins or other signalers in the periplasm and in turn pass on information to the cytoplasmic messenger, which is thought to be a complex of the *cheB* and *cheZ* products. The next link in the pathway is the *flaA* (*cheC*) component, which may be a membrane protein that interacts with the flagellum to control its direction of rotation.

The system would work in the following way: Both the *cheC* component of the switch and the *tar-tsr* products are able to bind the B-Z complex. When B-Z is bound to the switch, ccw rotation and smooth swimming result. When the switch is free of B-Z, cw rotation and tumbling result. Thus, tumbling behavior is controlled by the amount of B-Z available for interacting with the switch. The transduction and adaptation systems, which act on the *tar* and *tsr* components (31, 34, 56), can change the available supply of B-Z by modulating the affinity of *tar-tsr* proteins for the B-Z complex. Tumble-suppressing stimuli or demethylation events convert MCP (i.e. *tar-tsr* products) to the "ccw state" (see Figure 5) which must have a low affinity for B-Z. Tumble-enhancing stimuli or methylation events convert MCP to the "cw state" which must have a high affinity for B-Z. The role of the *cheB-cheZ* products is therefore analogous to the hypothetical "tumble-controlling signal" discussed at the outset of this article (see Figure 1).

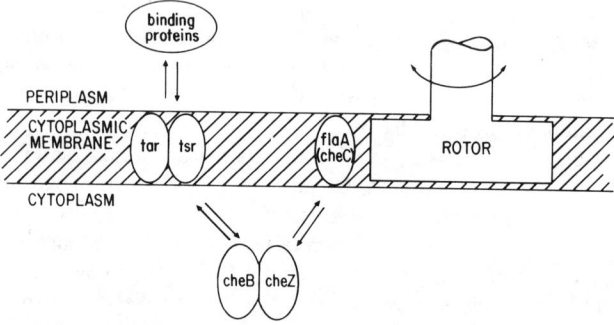

Figure 6 A model of stimulus transduction. Control of flagellar rotation in response to chemotactic stimuli may be mediated by diffusible chemical signals. See text for further explanation.

Implications of the Model

The major premise of the model is that swimming behavior in *E. coli* is controlled by the amount of *cheB* and *cheZ* product associated with the flagellar switch. Any treatment or mutation that in some way alters this parameter should result in a change in tumbling rate. For example, mutations that affect the relative affinities of the switch and the *tar-tsr* signalers (i.e. MCP) for the B-Z complex could result in either a smooth-swimming or excessive tumbling phenotype. Thus smooth-swimming *cheC* strains should be switch mutants with enhanced affinity for B-Z, whereas tumbly *cheC* strains should have decreased affinity. Mutations that affect MCP or its methylation and that result in smooth-swimming behavior (*cheD*, *cheX*, and *tar tsr* doubles) should reduce or abolish the ability of MCP to bind the *cheB* and *cheZ* proteins. It may soon be possible to test biochemically these and other predictions of the model.

In *E. coli* the expression of *che*, *fla*, and MCP genes is coordinately controlled by the *flaI* product (35, 36). This regulatory system may serve to maintain the proper stoichiometric relationships of the different gene products. The model predicts that changes in the relative amounts of the various transduction components should affect tumbling behavior. For example, increased levels of *cheB* and *cheZ* products should suppress tumbling. It should be possible to test quantitative aspects of the model by using λ transducing phages to alter the normal stoichiometry of the transduction apparatus.

Szmelcman & Adler (63) recently demonstrated that chemotactic stimuli cause a hyperpolarizing change in the membrane potential of *E. coli*. They suggested that this might either be an important part of the signaling mechanism or merely a consequence of the transduction process. The model shown in Figure 6 predicts that any changes in membrane potential are likely to be an effect rather than a cause of transduction. The model does imply, however, that all the flagella on a single cell should behave in a synchronous manner, both spontaneously and in response to stimuli, for they should be coordinated by the available supply of *cheB* and *cheZ* proteins in the cytoplasm. There is very little evidence either for or against flagellar synchrony in *E. coli*.

CONCLUSIONS

This article demonstrates the analytical power of a combined genetic and biochemical approach in studying the chemotactic behavior of bacteria. Considerable progress has been made in recent years and it is now possible to probe the molecular details of sensory reception, transduction, and adaptation. Working models of these events are presented and some future lines of inquiry are indicated.

Acknowledgments

I would like to thank the following for useful discussions and for comments on an early version of this manuscript: J. Adler, M. Goy, E. Kort, K. Lark, B. Olivera, J. Roth, M. Silverman, and M. Springer. Work from my laboratory was supported by a research grant from the National Institutes of Health.

Literature Cited

1. Berg, H. C. 1975. Chemotaxis in bacteria. *Ann. Rev. Biophys. Bioeng.* 4:119–36
2. Adler, J. 1966. Chemotaxis in bacteria. *Science* 153:708–16
3. Adler, J. 1975. Chemotaxis in bacteria. *Ann. Rev. Biochem.* 44:341–56
4. Berg, H. C. 1975. Bacterial behavior. *Nature* 254:389–92
5. Koshland, D. E. Jr. 1976. A response regulator model in a simple sensory system. *Science* 196:1055–63
6. Hazelbauer, G. L., Parkinson, J. S. 1977. Bacterial chemotaxis. In *Receptors and Recognition: Microbial Interactions,* ed. J. Reissig. London: Chapman & Hall
7. Berg, H. C., Anderson, R. A. 1973. Bacteria swim by rotating their flagellar filaments. *Nature* 245:380–82
8. Silverman, M., Simon, M. 1974. Flagellar rotation and the mechanism of bacterial motility. *Nature* 249:73–74
9. Larsen, S. H., Reader, R. W., Kort, E. N., Tso, W.-W., Adler, J. 1974. Change in direction of flagellar rotation is the basis of the chemotactic response in *Escherichia coli. Nature* 249:74–77
10. Berg, H. C. 1974. Dynamic properties of bacterial flagellar motors. *Nature* 249:77–79
11. Berg, H. C., Tedesco, P. M. 1975. Transient response to chemotactic stimuli in *Escherichia coli. Proc. Natl. Acad. Sci. USA* 72:3235–39
12. Berg, H. C., Brown, D. A. 1972. Chemotaxis in *Escherichia coli* analyzed by three-dimensional tracking. *Nature* 239:500–4
13. Dahlquist, F. W., Lovely, P., Koshland, D. E. Jr. 1972. Quantitative analysis of bacterial migration in chemotaxis. *Nature New Biol.* 236:120–23
14. Macnab, R. M., Koshland, D. E. Jr. 1972. The gradient-sensing mechanism in bacterial chemotaxis. *Proc. Natl. Acad. Sci. USA* 69:2509–12
15. Tsang, N., Macnab, R., Koshland, D. E. Jr. 1973. Common mechanism for repellents and attractants in bacterial chemotaxis. *Science* 181:60–63
16. Adler, J. 1969. Chemoreceptors in bacteria. *Science* 166:1588–97
17. Brown, D. A., Berg, H. C. 1974. Temporal stimulation of chemotaxis in *Escherichia coli. Proc. Natl. Acad. Sci. USA* 71:1388–92
18. Spudich, J. L., Koshland, D. E. Jr. 1975. Quantitation of the sensory response in bacterial chemotaxis. *Proc. Natl. Acad. Sci. USA* 72:710–13
19. Adler, J., Hazelbauer, G. L., Dahl, M. M. 1973. Chemotaxis toward sugars in *Escherichia coli. J. Bacteriol.* 115:824–47
20. Mesibov, R., Adler, J. 1972. Chemotaxis toward amino acids in *Escherichia coli. J. Bacteriol.* 112:315–26
21. Tso, W.-W., Adler, J. 1974. Negative chemotaxis in *Escherichia coli. J. Bacteriol.* 118:560–76
22. Hazelbauer, G. L., Adler, J. 1971. Role of the galactose binding protein in chemotaxis of *Escherichia coli* toward galactose. *Nature New Biol.* 230:101–4
23. Hazelbauer, G. L. 1975. The maltose chemoreceptor of *Escherichia coli. J. Bacteriol.* 122:206–14
24. Aksamit, R., Koshland, D. E. Jr. 1974. Identification of the ribose binding protein as the receptor for ribose chemotaxis in *Salmonella typhimurium. Biochemistry* 13:4473–78
25. Adler, J., Epstein, W. 1974. Phosphotransferase-system enzymes as chemoreceptors for certain sugars in *Escherichia coli* chemotaxis. *Proc. Natl. Acad. Sci. USA* 71:2895–99
26. Lengeler, J. 1975. Nature and properties of hexitol transport systems in *Escherichia coli. J. Bacteriol.* 124:39–47
27. Ordal, G. W., Adler, J. 1974. Properties of mutants in galactose taxis and transport. *J. Bacteriol.* 117:517–26
28. Hazelbauer, G. L. 1975. Role of the receptor for bacteriophage lambda in the functioning of the maltose chemoreceptor of *Escherichia coli. J. Bacteriol.* 124:119–26
29. Mesibov, R., Ordal, G. W., Adler, J. 1973. The range of attractant concentrations for bacterial chemotaxis and the threshold and size of response over this range. *J. Gen. Physiol.* 62:203–23
30. Strange, P. G., Koshland, D. E. Jr. 1976. Receptor interactions in a signaling system: Competition between ribose receptor and galactose receptor in the chemotaxis response. *Proc. Natl. Acad. Sci. USA* 73:762–66
31. Springer, M. S., Goy, M. F., Adler, J. 1977. Sensory transduction in *Escherichia coli:* Two complementary pathways of information processing that involve methylated proteins. *Proc. Natl. Acad. Sci. USA* 74:3312–16
31a. Hazelbauer, G. L., Mesibov, R. E., Adler, J. 1969. *Escherichia coli* mutants

defective in chemotaxis toward specific chemicals. *Proc. Natl. Acad. Sci. USA* 64:1300–7
32. Silverman, M., Simon, M. 1977. Identification of polypeptides necessary for chemotaxis in *Escherichia coli. J. Bacteriol.* 130:1317–25
33. Silverman, M., Matsumura, P., Hilmen, M., Simon, M. 1977. Characterization of lambda–*E. coli* hybrids carrying chemotaxis genes. *J. Bacteriol.* 130:877–87
34. Silverman, M., Simon, M. 1977. Chemotaxis in *Escherichia coli:* Methylation of *che* gene products. *Proc. Natl. Acad. Sci. USA* 74:3317–21
35. Silverman, M., Simon, M. 1974. Characterization of *Escherichia coli* flagellar mutants that are insensitive to catabolite repression. *J. Bacteriol.* 120:1196–1203
36. Silverman, M., Matsumura, P., Draper, R., Edwards, S., Simon, M. I. 1976. Expression of flagellar genes carried by bacteriophage lambda. *Nature* 261:248–50
37. DePamphilis, M. L., Adler, J. 1971. Fine structure and isolation of the hook-basal body complex of flagella from *Escherichia coli* and *Bacillus subtilis. J. Bacteriol.* 105:384–95
38. DePamphilis, M. L., Adler, J. 1971. Attachment of flagellar basal bodies to the cell envelope: Specific attachment to the outer, lipopolysaccharide membrane and the cytoplasmic membrane. *J. Bacteriol.* 105:396–407
39. Silverman, M., Simon, M. I. 1977. Bacterial flagella. *Ann. Rev. Microbiol.* In press
40. Armstrong, J. B., Adler, J., Dahl, M. M. 1967. Nonchemotactic mutants of *Escherichia coli. J. Bacteriol.* 93:390–98
41. Silverman, M., Simon, M. 1973. Genetic analysis of bacteriophage Mu-induced flagellar mutants in *Escherichia coli. J. Bacteriol.* 116:114–22
42. Parkinson, J. S. 1976. *cheA, cheB* and *cheC* genes of *Escherichia coli* and their role in chemotaxis. *J. Bacteriol.* 126:758–70
43. Collins, A. L., Stocker, B. A. D. 1976. *Salmonella typhimurium* mutants generally defective in chemotaxis. *J. Bacteriol.* 128:754–65
44. Warrick, H. M., Taylor, B. L., Koshland, D. E. Jr. 1977. The chemotactic mechanism of *Salmonella typhimurium:* Preliminary mapping and characterization of mutants. *J. Bacteriol.* 130:223–31

45. Aswad, D., Koshland, D. E. Jr. 1975. Isolation, characterization and complementation of *Salmonella typhimurium* chemotaxis mutants. *J. Mol. Biol.* 97:225–35
46. Vary, P. S., Stocker, B. A. D. 1973. Nonsense motility mutants in *Salmonella typhimurium. Genetics* 73:229–45
47. Armstrong, J. B., Adler, J. 1969. Complementation of nonchemotactic mutants of *Escherichia coli. Genetics* 61:61–66
48. Silverman, M., Simon, M. 1976. Operon controlling motility and chemotaxis in *E. coli. Nature* 264:577–79
49. Silverman, M., Matsumura, P., Simon, M. 1976. The identification of the *mot* gene product with *Escherichia coli*–lambda hybrids. *Proc. Natl. Acad. Sci. USA* 73:3126–30
49a. Matsumura, P., Silverman, M., Simon, M. 1977. Synthesis of *mot* and *che* gene products of *Escherichia coli* programmed by hybrid col EI plasmids in minicells. *J. Bacteriol.* In press
50. Parkinson, J. S. 1974. Data processing by the chemotaxis machinery of *Escherichia coli. Nature* 252:317–19
51. Springer, M. S., Kort, E. N., Larsen, S. H., Ordal, G. W., Reader, R. W., Adler, J. 1975. Role of methionine in bacterial chemotaxis: Requirement for tumbling and involvement in information processing. *Proc. Natl. Acad. Sci. USA* 72:4640–44
52. Silverman, M., Simon, M. 1973. Genetic analysis of flagellar mutants in *Escherichia coli. J. Bacteriol.* 113:105–13
53. Adler, J., Dahl, M. 1967. A method for measuring the motility of bacteria and for comparing random and non-random motility. *J. Gen. Microbiol.* 46:161–73
54. Armstrong, J. B. 1972. Chemotaxis and methionine metabolism in *Escherichia coli. Can. J. Microbiol.* 18:591–96
55. Aswad, D., Koshland, D. E. Jr. 1974. Role of methionine in bacterial chemotaxis. *J. Bacteriol.* 118:640–45
56. Springer, M. S., Goy, M. F., Adler, J. 1977. Sensory transduction in *Escherichia coli:* A requirement for methionine in sensory adaptation. *Proc. Natl. Acad. Sci. USA* 74:183–87
57. Armstrong, J. B. 1972. An S-adenosylmethionine requirement for chemotaxis in *Escherichia coli. Can. J. Microbiol.* 18:1695–1701
58. Aswad, D., Koshland, D. E. Jr. 1975. Evidence for an S-adenosylmethionine requirement in the chemotactic behav-

ior of *Salmonella typhimurium. J. Mol. Biol.* 97:207–23
59. Kort, E. N., Goy, M. F., Larsen, S. H., Adler, J. 1975. Methylation of a protein involved in bacterial chemotaxis. *Proc. Natl. Acad. Sci. USA* 72:3939–43
60. Kleene, S. J., Toews, M. L., Adler, J. 1977. Isolation of glutamic acid methyl ester from an *Escherichia coli* membrane protein involved in chemotaxis. *J. Biol. Chem.* 252:3214–18
61. Van der Werf, P., Koshland, D. E. Jr. 1977. Identification of a γ-glutamyl methyl ester in a bacterial membrane protein involved in chemotaxis. *J. Biol. Chem.* 252:2793–95
62. Springer, W. R., Koshland, D. E. Jr. 1977. Identification of a protein methyltransferase as the *cheR* gene product in the bacterial sensing system. *Proc. Natl. Acad. Sci. USA* 74:533–37
63. Szmelcman, S., Adler, J. 1976. Change in membrane potential during bacterial chemotaxis. *Proc. Natl. Acad. Sci. USA* 73:4387–91
64. Bachmann, B. J., Low, K. B., Taylor, A. L. 1976. Recalibrated linkage map of *Escherichia coli* K12. *Bacteriol. Rev.* 40:116–67

INVERTEBRATE NEUROGENETICS

❖3128

Samuel Ward[1]
Department of Biological Chemistry, Harvard Medical School, Boston, Massachusetts 02115

CONTENTS

INTRODUCTION	416
NEUROCHEMISTRY	417
MUSCLE MUTANTS	419
Caenorhabditis	420
Drosophila	421
ELECTROPHYSIOLOGY	421
Drosophila Neuromuscular Physiology	421
Flight muscles	421
Larval muscles	423
Paramecium	423
Firing Patterns	426
Crickets and Locusts	427
SENSORY RESPONSES	428
Caenorhabditis Sensory Mutants	428
Drosophila Chemosensory Mutants	431
NEUROANATOMY	431
COMPLEX BEHAVIORS	433
Learning in Drosophila	434
Courtship Behavior	435
Wasps	435
Drosophila	435
Caenorhabditis	436
Circadian Rhythms	437
DEVELOPMENT	437
Drosophila Visual System	438
Caenorhabditis Postembryonic Development	440
Crickets and Locusts	442

[1]Present address: Carnegie Institution of Washington, Department of Embryology, 115 West University Parkway, Baltimore, Maryland 21210.

ADDITIONAL EXPERIMENTAL ORGANISMS 442
CONCLUSIONS AND PROSPECTS 442
   Types of Mutants 442
   Importance of Mosaic Analysis 443
   Direct Approach to Molecular Defects 444
   Exploitation of Mutants 444
   Analysis of Development 444
   Analysis of Function 445
   Relation of Neurogenetics to Behavioral Evolution 445

INTRODUCTION

In 1967, Benzer demonstrated that phototactic mutants in *Drosophila melanogaster* could be induced and readily isolated by countercurrent selection (1). This demonstration stimulated further experimentation in selection and characterization of behavioral mutants in inbred organisms. In the ten years since Benzer's paper, hundreds of mutants altering the nervous system have been identified and studied, not only in Drosophila, but also in nematodes, protozoa, crickets, mice, and other organisms. This review describes recent studies of such mutants and attempts to assess the current status of this genetic approach to neurobiological problems.

    The eventual goal of the work reviewed here is to bridge the gap between individual genes and the heritable behavior of an organism. This is no simple task. As Brenner (2) has pointed out, "One is aware of the possibility that understanding this might well involve solving all of the outstanding questions in biology." Fortunately, the magnitude of the problem has not deterred the practitioners; no doubt it has inspired some. The difficulties have tempered the optimism of molecular biologists who have entered this field, but not reduced their experimental zeal.

    In our minds the problem of relating the gene to behavior can be divided into two: the developmental problem of how genes specify the structure of the nervous system, and the functional problem of how the structure generates the behavior. The organism's genome, however, does not segregate itself so conveniently. Mutations in genes that alter the development of the nervous system usually alter its function. Mutations in genes that alter adult electrophysiology can also alter neural development. Therefore, to understand behavioral mutants, one must be prepared to analyze both development and function.

    Not only do the genes resist compartmentalization into development and function, the analysis of the effects of these genes resist classification into the standard disciplines of genetics, biochemistry, histology, physiology, or psychology. The most successful studies of mutants have applied analysis at several levels of organization between the gene and behavior.

    These considerations make the task of organizing a review of this material difficult. I have arranged this review primarily by the levels of organization used to analyze mutants, from neurochemical to behavioral and developmental studies. A brief introduction to each section formulates some of the questions addressed by analysis at each level of organization. The major experimental organisms reviewed, Drosophila, Caenorhabditis, and Paramecium, are discussed concurrently throughout the sections.

I have been aided in the preparation of this review by previous reviews of behavioral genetics and related topics. An excellent review by Gould (3) discusses the relevance of the induced mutation approach of studying behavior to the broader investigations of ethologists, and dispenses with the question of whether simple organisms have "real behavior." They do. A recent textbook of behavioral genetics reviews the literature through 1975, and discusses the study of single-gene behavioral mutations in relation to the more complex polygenic behavioral traits studied in natural populations (4). References (5) and (6) are catalogues of Drosophila behavioral mutants. References (7) and (8) are reviews of Drosophila mutants. Pak & Pinto (9) have reviewed some of the same topics discussed in this review and described the experimental advantages of the organisms used. Their review of Drosophila visual physiology is particularly comprehensive. I have tried to minimize duplication of material they covered. Finally, I have limited myself to invertebrates to make the topic manageable. Literature on the mouse has been reviewed in (9) and will be discussed in greater detail in a forthcoming review (10). References to genetic studies of the behavior of organisms not discussed in detail here are given later in this review. The use of genetic mosaics, animals with cells of more than one genotype, plays an important role in many of the studies cited below. The use of this technique is critically reviewed by Hall (11) and Hotta & Benzer (12).

NEUROCHEMISTRY

Most of the mutants to be discussed in this review were isolated by their behavioral phenotypes and examined for underlying defects in physiology and biochemistry. Another approach to obtaining neurological mutants is described by Hall & Kankel: (13) "Choose a gene that controls a specific protein involved in nervous system function, isolate mutations in the gene, and analyze the defective gene product's effects on development, neurophysiology, and behavior of the organism." With this approach the defective gene product is known; thus, one can ask what role this gene product plays in the normal development and function of the nervous system. For example, an important question in developmental neurobiology is whether a neuron's neurotransmitter participates in the guidance and formation of specific synaptic contacts. With mutations that alter known enzymes of neurotransmitter synthesis or degradation it should be possible to answer this question by determining the effects of such mutations on synapse formation.

Hall & Kankel (13) chose to isolate Drosophila mutants defective in the enzyme acetylcholinesterase (AchE) which hydrolyzes the neurotransmitter acetylcholine. They reasoned that the level of enzyme activity in a fly should be proportional to the number of doses of the structural gene in the genome as has been shown for several enzymes [references in (13)]. The gene dosage of chromosome segments can be varied in Drosophila by generation of segmental triploids using primarily Y-autosome translocations. Because animals carrying such duplications are viable, their enzyme activity can be assayed and compared to the specific activity in the ordinary diploid. Once the potential chromosomal location of a gene is identified, the gene can be further localized with deficiencies in that region, some of which should have reduced enzyme activity.

Three segments in the right arm of chromosome *3* were found to elevate the level of AchE in flies when present in three doses (13). Two of these differences were less dramatic than the third and subsequently proved to be spurious. One region remained dose dependent. Studies with deficiencies confined the effect to a subsegment of about four genes.

Hall & Kankel (13) obtained four allelic recessive lethals that had been assigned to this subsegment and found that they had reduced AchE as heterozygotes. They call the locus defined by these lethals *Ace* for AchE defective. Several lines of evidence indicate that this locus is the structural gene for AchE.

The lethal phase of the *Ace* mutant is in late embryogenesis and the dying embryos have no detectable AchE (13). Since it is difficult to pursue detailed biochemical and developmental studies on lethal mutations, Hall & Kankel are now seeking to isolate temperature-sensitive alleles of *Ace*. Homozygous mutant flies could then be raised at a permissive temperature and shifted to a restrictive temperature at different ages to test for behavioral and developmental abnormalities. This would allow them to investigate the developmental and behavioral consequences of altering an important functional component of the nervous system at different times in development.

The use of segmental aneuploids to identify structural genes for known enzymes is a general technique to localize the genes for any gene product that can be easily assayed. It is limited at present to Drosophila, where deficiencies and duplication are available covering most of the genome. In the nematode *Caenorhabditis elegans*, duplications of the *X* chromosome have been generated, and methods of isolating and maintaining other duplications are being developed (14). Thus the aneuploid technique may become useful in this organism as well.

One locus controlling AchE in the nematode has already been identified without use of aneuploids. Johnson (15) assayed uncoordinated mutants directly for reduced AchE activity. Wild-type Caenorhabditis AchE activity sediments in four forms on sucrose gradients of extracted worms. Two of these have similar enzymatic properties and specificity and are differentially sensitive to detergents (15). By assaying 85 behavioral and drug-resistant mutant strains for decreased activity of the detergent-resistant form of AchE, Johnson found a mutant on the *X* chromosome that was deficient in this enzyme activity. The mutant's movement was uncoordinated, but only the major somatic muscles appeared to be affected leaving the motion of the head nearly normal. This suggests that the different wild-type forms of AchE may function in different parts of the nematode's nervous system.

Another selection strategy that might be expected to yield AchE mutants is to select mutants resistant to AchE inhibitors. This has been unsuccessful in a number of species of insects (reviewed in 16). In the housefly Musca, AchE mutants have been isolated among resistant strains, but at least eight other genes can mutate to drug resistance. Some of these control drug uptake and drug metabolism. Interestingly, these genes are linked and might be coordinately controlled (16).

No nematode mutants with altered AchE were found among seven strains resistant to lannate (17) or other nematacidal AchE inhibitors (D. Hirsh, personal communication). But a strain resistant to the nematacide trichlorfon was found to have less than 2% the wild-type activity of the enzyme choline acetyl transferase, an enzyme required for acetylcholine synthesis (D. Hirsh and R. Russell, personal communication). The gene for choline acetyl transferase has also been localized in Drosophila by the segmental aneuploid technique (J. Hall, personal communication).

Among a large collection of mutants isolated as resistant to levamisole, a nematacide thought to be an acetylcholine agonist, are some that behave pharmacologically as if they might be defective in an acetylcholine receptor (J. Lewis, personal communication). Drosophila has an α-bungarotoxin binding component that behaves like an acetylcholine receptor, and mutants in this receptor are being sought among strains resistant to acetylcholine agonists (18, 19).

Other mutants altered in the biochemistry of nerve transmission have been obtained by nonbehavioral selections. In Caenorhabditis, eight neurons in the hermaphrodite appear to use dopamine as their neurotransmitter as has been shown histochemically by formaldehyde-induced fluorescence (20). Six of these are sensory neurons in the head that appear morphologically to be mechanoreceptors. Six mutants with altered fluorescence were obtained by histochemical examination of 1000 F2 clones of mutagenized parents. These six mutants fall into five complementation groups mapping at different loci. None of the enzymes of dopamine synthesis tested were altered in these mutants. It is suggested that some of the mutants may be defective in the production, loading, or transport of dopamine-containing vesicles because they lack dopamine in their terminal processes (20).

Unfortunately, these mutant strains have no detectable behavioral phenotype except for a slight reduction in male mating efficiency. It is possible, but unlikely, that dopamine is not the neurotransmitter of these neurons. It is more likely that the proper environmental condition for distinguishing the phenotype has not been found. Touch-insensitive behavioral mutants were isolated and found not to be altered histochemically (20). Whatever the functional consequences of these mutations, their isolation illustrates a method for direct identification of neurochemical mutants using a histochemical procedure that could be applied to other organisms.

Two Drosophila mutants with abnormal phototaxis and abnormal electroretinogram are *tan* and *ebony* (21). These mutants are also altered in dopamine content (22, 23). This biochemical defect may be related to their behavioral phenotype, but how this comes about is not known.

MUSCLE MUTANTS

Both the nervous system and the muscles participate in the normal behavior of an animal. It is not surprising, then, that among mutants selected for behavioral defects, mutants with defective muscles have been obtained. The initial motivation of selecting reversibly temperature-sensitive paralytic mutants in Drosophila was to obtain mutants in muscle structural proteins (24). Ironically, interesting neurological mutants were obtained instead. In contrast, Caenorhabditis was chosen initially as an organism for genetic analysis of its nervous system and yet many muscle-defective mutants have been found (17).

In this section, mutants affecting the structure and assembly of the contractile apparatus in the muscle are reviewed. These mutants are being studied in order to learn more about how the individual proteins of the muscle myofibril participate in generating and controlling the contractile force. The mutants may also be useful for studying the process of myofibrillar assembly. Gene products essential for assembly can be identified and alterations in the normal assembly process can be analyzed in molecular detail.

Caenorhabditis

Among nematode mutants selected initially because their body movements were uncoordinated, several were found whose muscles appeared defective when examined by electron microscopy (17). Subsequent studies have shown that many such mutants can be recognized by polarized light microscopic examination so that additional mutants can be readily isolated (25–27).

The muscle-defective mutants have been characterized both biochemically and by light and electron microscopy. Because the body musculature constitutes a considerable fraction of the nematode's mass, muscle structural proteins can be recognized on acrylamide gel electrophoresis patterns of whole animals. The myofibrillar proteins are also easily purified from the rest of the worm, allowing detailed comparison of the muscle proteins in mutants and wild type. One of the structural genes for the myosin heavy chain and the structural gene for paramyosin have been identified (25, 27, 28). The myosin structural gene, *unc-54*, was identified because one of its alleles produces a gene product that migrates on an SDS gel as if it had a molecular weight of only 203,000 rather than 210,000 of the wild type (25). Subsequent isolation and peptide mapping of this gene product showed that it contained an internal deletion in a myosin heavy-chain gene (A. R. McLeod, R. H. Waterston, and S. Brenner, in preparation). The mutation does not affect pharyngeal myosin and leaves one somatic myosin unchanged, indicating that there must be at least two and probably three distinct myosin heavy-chain genes in the nematode (29).

Paramyosin is a component of the thick filaments of many invertebrate muscles. Waterston et al (27) identified it in Caenorhabditis, and it has been shown that the *unc-15* gene is almost certainly the structural gene for nematode paramyosin (28).

Revertants of some of the muscle mutants have been obtained by selection for reversion of the uncoordinated phenotype. Among revertants of a null paramyosin structural allele, an unlinked suppressor that partially restored the level of paramyosin in the allele was obtained (28). This suppressor also suppresses certain alleles of several other genes with unrelated phenotypes (R. H. Waterston, personal communication). This is a property expected for a nonsense mutant suppressor. If this suppressor is a nonsense suppressor, it will become possible to use it to isolate conditional chain termination mutants in Caenorhabditis. The suppressor is cold sensitive, so mutants could be maintained easily. Such mutants might greatly facilitate the identification of mutant gene products by SDS gel electrophoresis because the mutant gene products would have altered mobility.

More than ten other genes affect myofibrillar organization in Caenorhabditis (17; R. H. Waterston, personal communication). Some of them act during muscle development and some of them cause dystrophies of adult muscle. Temperature-sensitive alleles of one of these have been described (26). Nematode mosaics have not been studied, so it is not certain that all muscle defects are due to defective muscle cell proteins. Nevertheless, further analysis of the muscle mutants may identify the genes for other muscle structural proteins and perhaps identify gene products regulating muscle assembly. The relative ease of isolation of muscle-defective mutants and the ease of muscle protein biochemistry make the nematode an attractive organism for investigation of the genetic control of muscle assembly.

Drosophila

Muscle defective mutants have also been recognized in Drosophila, but defective gene products have not yet been identified. Hotta & Benzer (12) used mosaics to determine the developmental focus of two X-linked wing postural mutants, heldup (*hdp*) and upheld (*up*) (formerly *wupA* and *wupB*). The phenotype depends on the mutant genotype of cells from the ventral midblastoderm which is presumably the location of muscle progenitor cells. A neurological focus is not absolutely ruled out, but is unlikely (11). Histological examination of *hdp* during development showed that the formation of the indirect flight muscles was normal until pupation and then the muscle degenerated. In *up*, the myoblasts appear normal but the myofilament organization is disrupted. The phenotype of *up* resembles several nematode muscle mutants.

In both Drosophila mutants only the flight muscles appeared to be defective. If these mutants affect a myofilament structural protein, it must be a protein coded for by more than one gene, as with the nematode myosins. Additional muscle-defective mutants have been identified and characterized by Deak (cited in 11).

ELECTROPHYSIOLOGY

Electrophysiological characterization of behavioral mutants will identify mutants that alter gene products necessary for the normal electrical excitability of neurons and muscles. Mutants that alter specific gene products in the membranes of such cells may help to identify and characterize normal functions of these gene products. Progress in identification of such mutants in Drosophila and Paramecia is reviewed first in this section. Then electrophysiological studies of neuron firing patterns are reviewed. The goals of these second studies are not so much to determine the mechanisms of electrical excitability as to try to determine how the elements of the firing pattern of a group of neurons can be altered by individual mutations.

Drosophila Neuromuscular Physiology

Although most of the neurons in Drosophila are too small to allow intracellular penetration, some recordings have been obtained from large motoneurons (30) and from the photoreceptive retinula cells (31). However, the most accessible cells for neurophysiology are the large indirect flight muscles and the muscles of larvae.

FLIGHT MUSCLES The musculature and motor innervation of the Drosophila flight muscles are reviewed in references (32) and (33). There are 26 large muscle fibers, each a single multinucleate cell. These muscles attach to the walls of the thorax and move the wings indirectly by alternately compressing and extending the thorax. Six muscle fibers on each side of the midline form the right and left dorsal longitudinal muscles that depress the wings. These fibers are 500–1000 μm long and approximately 150–65 μm in cross section so they are easily impaled with a microelectrode. Seven muscle fibers located more laterally form the dorsoventral muscles that elevate the wings. These are about 700 μm long and 50–100 μm in diameter. Both sets of fibers are under known cuticular landmarks so that identified cells can be easily penetrated with a microelectrode after making a hole in the cuticle.

Each muscle fiber is innervated by a single motor neuron whose soma is located in the middle part of the thoracic ganglion. Reports of dual innervation of the muscle fibers may be incorrect (33). With one possible exception each motor neuron innervates only a single muscle fiber; therefore, each motor neuron and its muscle fiber can be considered a single unit.

Ikeda et al (34) and Siddiqi & Benzer (35) examined several neurological mutants for alterations in neuromuscular physiology by stimulating nerves and recording from the flight muscles. Using the shi^{ts1} allele of the temperature-sensitive reversible paralysis mutant *shibire*, Ikeda et al showed that the excitability and conduction of the mesothoracic leg nerve were not altered appreciably when the mutant was transferred from 19°C (permissive) to 29°C (restrictive). However, the muscle twitch normally elicited by nerve stimulation was eliminated. Recordings from the muscle showed that neuromuscular transmission was blocked by the elevated temperature. Direct stimulation of the muscle elicited a normal muscle action potential suggesting that the electrogenic properties of the muscle membrane were not affected by the *shibire* mutation. They conclude that the cause of the paralysis induced by this mutation is the blockage of neuromuscular transmission.

Using a different allele, shi^{ST109}, and slightly different conditions Siddiqi & Benzer (35) came to a similar conclusion. This conclusion is in disagreement with the claim that *shibire* mutants are altered in the regenerative sodium channel because they are resistant to the drug tetrodotoxin (36). It appears that the *shibire* mutants are not abnormally resistant to tetrodotoxin, but that the strain used as a wild-type control in the first drug studies is abnormally sensitive to tetrodotoxin (37). In the course of construction of marked strains for mutant isolation, it is possible to inadvertently generate non-isogenic strains; this may have happened with some *shibire* mutants.

Siddiqi & Benzer (35) also examined two other reversible temperature-sensitive paralytic mutants, *comatose* (*com*) and *paralyzed* (*para*). They analyzed the temperature effect on the behavioral phenotype and the neuromuscular physiology in parallel. When *com* flies are heated and stimulated by way of the cervical nerve, the evoked muscle end-plate potential occurs progressively later and becomes weaker. The muscle itself, however, remains readily excitable by direct stimulation. The defect could be in nerve conduction, transmitter release at the nerve terminal, or the response of the muscle to chemical transmitter. In *para*, the threshold of excitability of a cervical pathway is raised at temperatures a few degrees higher than the paralysis temperature, but nerve conduction velocity and muscle excitability remain normal. When *para* flies are left at high temperature for prolonged periods, they partially recover.

Both *shi* and *com* also eliminated a component of the electroretinogram (ERG), but the ERG is normal in *para* (35, 38, 39).

Precise mosaic fate mapping of these mutants has not been reported, but analysis of *para* indicated that mosaics with a mutant brain were paralyzed (39). The electroretinogram defect in *shibire* was shown by mosaics to be either in the retina, optic lobes, or brain (40), but the defect in neuromuscular transmission has not yet been mapped.

LARVAL MUSCLES In addition to neurophysiological recording from flight muscles, neuromuscular physiology can be studied in the larval muscles. In two careful physiological papers Jan & Jan (41, 42) have shown that the larval muscle resting potential obeys standard equations relating potential to ion concentrations. They also showed that transmitter release at the neuromuscular junction is quantal and that release depends roughly on the fourth power of Ca^{2+} concentration as in the frog neuromuscular junction. L-Glutamate can mimic the excitatory transmitter if applied iontophoretically. In addition, the reversal potential and the L-glutamate potential change similarly in response to the ionic composition of the medium. Jan & Jan conclude that L-glutamate is either the excitatory transmitter at the Drosophila neuromuscular junction or an agonist of the transmitter.

With this detailed physiological background they have begun to examine neurological mutants for physiological alterations. Jan, Jan & Dennis (43) find that an allele of the mutant *shaker* has a gross abnormality in synaptic transmission that is consistent with a prolonged opening of the Ca^{2+} gate. Further studies of mutants using the larval muscle preparation may identify other mutants defective in specific compounds of the excitability.

These detailed physiological studies in Drosophila point the way to finding the molecular defects in mutants. Mutants with specific alterations in the postsynaptic physiology of the muscle could be examined biochemically for alterations in membrane proteins using larval muscle tissue. The presynaptic defects will be harder to pursue because it will be difficult to isolate enough neural tissue [but see (19, 44)]. Nonetheless, the combination of electrophysiological characterization of mutants with the pharmacological studies described previously should enable identification of specific molecules involved in electrical excitability.

Another organism that promises to yield more easily to biochemical assault on its molecular mechanisms of excitation is the protozoan *Paramecium aurelia*.

Paramecium

For combining behavioral, electrophysiological, and biochemical analysis of mutants affecting an electrically excitable membrane, the ciliate protozoan Paramecium is an excellent organism. The electrophysiology of Paramecium has been extensively studied (reviewed in 45, 46). Some of its bioelectric properties resemble those of muscles and nerves; it has a resting potential dependent on the concentration of several cations in the medium; membrane depolarization generates action potentials in the membrane resulting from a voltage-dependent inward current flow carried by Ca^{2+} ions. The inward flow of calcium has four major consequences: 1. the action potential, 2. an activation of an ATP-dependent process that reverses ciliary beat, 3. an increase in ciliary beat frequency, and 4. a temporary rise in potassium conductance that repolarizes the membrane to resting level terminating the calcium response (45). These physiological events are triggered when the ciliate encounters an object or a noxious chemical, and they mediate the backing up or avoiding reaction first described by Jennings (47).

Reasoning that mutants selected for altered behavior would be defective in the membrane components controlling the electrically excitable membrane, Kung has isolated a large number of Paramecium behavioral mutants (reviewed in 48, 49). More than 350 lines, not all independent, have been collected and more than 20 different genes have been identified by complementation tests and linkage mapping. The genetics of Paramecium was worked out primarily by T. M. Sonneborn and makes the organism particularly suitable for large-scale mutant isolation and characterizations. Homozygous mutant clones can be generated in mutagenized stocks by autogamy and mating carried out between these by conjugation [see (48) for references].

Mutants in several genes have been extensively characterized. There are three *pawn* genes that can mutate to prevent Paramecium from backing up. Physiological examination of these shows that they are defective in voltage-sensitive calcium permeability of the Paramecium membrane (48). That the membrane is defective was indicated by using the technique of Naitoh & Kaneko (50) to prepare a Triton®-extracted "model" Paramecium that has little plasma membrane and is freely permeable to ions. This extracted model paramecium retains its coordinated ciliary beat in the presence of Mg^{2+} and ATP and swims forward. It responds to increases in $[Ca^{2+}]$ by altering its ciliary beat and swimming backwards. The behavior of Triton-extracted *pawn* mutants is indistinguishable from wild type. They back up in response to increased $[Ca^{2+}]$. This argues that the mutant ciliary response system is normal, so the *pawn* phenotype must be due to a membrane defect.

Additional *pawn* mutants have been isolated by Schein by selecting strains that were resistant to the paralyzing and toxic effect of barium ions that enter the cell through the Ca^{2+} channel (51). The seven mutants isolated in this way were defective in their reversal response although alleles with partial response were obtained. These mutants fell into the same three complementation groups, *pwnA*, *pwnB*, and *pwnC*. Only one *pwnC* allele was obtained, indicating that these are rare as in the behavioral selections of Kung's group. Physiological characteriztion of these mutants using an improved method of recording confirmed that the new pawn mutants were defective in $[Ca^{2+}]$ activation. The *pwnA* and *pwnB* mutants could be distinguished from each other by the altered anomalous rectification in the *pwnB* alleles (52). Schein et al argue from this that the *pwnA* gene product may be a gate on the Ca^{2+} channel that opens and closes to allow ions to pass, and the *pwnB* gene product is the pore that penetrates through the membrane to determine the specificity of the channel. Further support of this model is the observation that *pwnA* alleles vary in the severity of their phenotype whereas the *pwnB* alleles are all extreme. One would expect that a gate would have greater conformational flexibility than a specific pore, allowing more mutational changes that partially disrupt the gating function (52).

Schein (53) followed the lifetime of the normal calcium channel by following the progressive loss of reversal response and excitability when heterozygous strains were made homozygous by autogamy. The channel lifetime was found to be from 5 to 8 days for both *pwnA* and *pwnB* mutants. In contrast, the time course of loss of excitability when temperature-sensitive *pwnA* or *pwnC* alleles are transferred to high temperature is much shorter (54, 55). Temperature change affects the behavioral phenotype between 4 to 8 hr after temperature shift. If the normal channel lifetime is 5 to 8 days, it is likely that the high temperature is inactivating channels already in place in the membrane rather than inactivating channel gene products during synthesis. The 4-hr delay in response to temperature changes is in striking contrast to the rapid (less than a minute) effect of shifts of some of the Drosophila temperature-sensitive mutants (see section on electrophysiology). Perhaps this is only a difference in the specific mutations isolated so far. It might be useful to take advantage of the strong selection for *pawns* in Paramecium to see whether rapidly reversible temperature-sensitive alleles could be obtained.

Components of excitability other than the Ca^{2+} action potential are altered in other Paramecium behavioral mutants. A mutant that does not perform avoiding reactions in solutions of tetraethylammonium (Tea^+) has been studied (56, 57). Tea^+ is known to be a membrane K^+ channel blocking agent and the *TEA-insensitive* mutant was found to have an increased conductance for K^+. It can generate a Ca^{2+} action potential that is smaller in amplitude than the wild type.

Another mutant, *paranoic,* backs up nearly continuously as if avoiding persecution from the surrounding ions. It is altered in both Na^+ and K^+ permeability (58). The phenotype is observed only when Na^+ is present in the medium. The Ca^+ action potential is similar to wild type. It is likely that this mutant affects a mechanism of repolarization of the paramecia membrane that normally terminates the avoiding response.

The mutant *Fast-2* hyperpolarizes its membrane in the presence of K^+ as if it were altered in the resting permeability to K^+ (59, 60). There are different alleles of the same complementation group that can give either a fast or paranoic phenotype as if the potassium permeability can be altered in different ways by different mutations in the same gene product.

The number of mutants altering membrane physiology in Paramecium is by no means saturated, but the mutants characterized so far show that individual components of the membrane ion conduction can be altered by mutating specific genes. Some of the mutated gene products may function to assemble or regulate the membrane channels from the cytoplasm, but others will be membrane components themselves such as the *pawn* gene products.

The search to identify these gene products has been initiated by isolation and characterization of the Paramecium membrane (61, 62). Fractions enriched in membrane are readily obtained in paramecia by isolating the cilia with its surrounding membrane. The membrane can then be separated from the ciliary proteins. On SDS acrylamide gels these membranes have one major high molecular weight component, 200–300,000 daltons, that comprises about 75% of the total membrane protein. It is likely that this high molecular weight component is a precursor to the soluble I antigen that coats the cell surface of paramecia (61). Twelve to fifteen minor bands are present on SDS gels of paramecia membranes but this is certainly an underestimate of the number of membrane proteins.

As higher resolution two-dimensional electrophoresis techniques are applied, some of the mutants already isolated may alter the mobility of normal gene products so that the mutated gene products can be recognized by their mutant tags. Once identified, the investigation of how the normal gene products act to give the membrane its active properties will be a difficult task of membrane biochemistry. Without the mutants to provide tags that identify the necessary membrane components, such biochemistry would be impossible.

Firing Patterns

The physiology of Drosophila flight muscles provides an opportunity to analyze mutants that alter the patterning of impulses coordinating flight control in addition to examining the physiology of the synapse. The goal of these studies is to determine how the elements of the firing pattern of a group of neurons can be altered by individual mutations.

The repetitive firing pattern of motoneuron units can be recorded from flies mounted on a needle in an airstream (reviewed in 32, 63). The firing pattern in Drosophila motor neurons has been analyzed in detail by Harcombe & Wyman (33). The motoneurons fire in a repeating cycle with one spike per neuron. The cycle for each muscle runs independently of the cycle for other muscles, except that they share overall fluctuations in frequency. The pattern of motoneuron input does not drive the muscles directly but modulates the wing beat strength and terminates beating. The indirect flight muscles are myogenic; that is, they contract in response to stretching, but only if they have been activated by a neural spike prior to the stretch.

One Drosophila mutant that alters the firing pattern, *stripe,* has been analyzed so far (64). The phenotype of homozyogotes of this mutant includes a black stripe on the thorax and an inability to fly. The mutant's motoneuron output is in bursts of spikes produced spontaneously or in response to puffs of wind. This altered motoneuron output is presumably the reason for the inability to fly, but the focus of the mutant has not been determined. Since it has at least one other phenotype, the physiological results must be interpreted cautiously.

Another mutant strain with an altered pattern of spontaneous motoneuron activity is the double mutant *eag-Sh^5* (65). Both single mutants, *ether-a-gogo* and *shaker,* can fly but the double cannot. The focus of this defect is neural.

Since the pattern of motoneuron output is necessary for flight, mutants altering this pattern should appear among those selected for defective flight behavior. Recently, Homyk & Sheppard (66) and Homyk (67) have isolated 48 new mutants defective in flight behavior. They used the simple behavioral selection of placing the progeny of mutagenized males in the base of a cylinder enclosed in a box. Copper electrodes inside the cylinder prevented flies from crawling up the cylinder. When the flies were stimulated by tapping the box, only those that could fly escaped from the cylinder. Thus the remaining population was enriched for nonfliers and motility defects.

Not surprisingly, such a crude selection generated a wide variety of phenotypes: from flies that were generally slow and inactive, to some that behaved normally except that they failed to fly. Flight-defective alleles of several previously known behavioral mutants, *hyperkinetic, ether-a-gogo,* and *shaker,* were found. Crude fate mapping showed six other mutants that may affect muscles and at least two others that are neural. Much more work lies ahead to characterize these mutants thoroughly and to identify those that have nonspecific defects causing flightlessness. But among these mutants and others, there may be additional mutants altering the pattern of impulses in the flight motoneurons.

Mutants affecting the firing patterns of peripheral muscles have been studied extensively, especially the hyperkinetic mutants, and are reviewed in (9) and (68).

Crickets and Locusts

The genetics of auditory behavior in crickets has been extensively studied by analysis of hybrids of crosses between cricket strains with different songs (reviewed in 69–71). Unfortunately, but not surprisingly, individual units of the song pattern or units of the behavioral response have not segregated as if they were determined by single loci. This is usually observed for behavioral traits studied in natural populations (3). However, it is possible to induce cricket mutations that do segregate as single mutations and cause specific behavioral defects. Although the cricket generation time of six weeks is long compared to Drosophila, Caenorhabditis, or Paramecium, they are prolific (laying up to 2000 eggs per female), so it is not difficult to raise large populations and isolate mutants. The adult cricket has large identified neurons and much is known about their electrophysiology. Hence, behavioral mutants can be easily studied physiologically.

Bentley (72) has obtained several mutant lines that are defective in jumping behavior. His ingenious selection procedure was to pass a tube that delivers puffs of air over a pan of cricket nymphs from mutagenized grandparents, and to follow this tube with a vaccum tube leading to a disposal flask. Any normal nymphs jumped up and were sucked off by the vacuum, leaving nonjumpers behind. Of 40,000 nymphs screened, a dozen nonjumpers were detected, two of which survived to adulthood and segregated as single-locus mutations. Further evidence that these are due to single-gene mutations was not presented. It is not clear whether the mutations were really induced by the mutagen or segregated from the natural population since the strains were not homozygous (see below for a discussion of locust parthenogenesis).

One of the nonjumping mutants that segregated as a Mendelian recessive is defective in the sensilla of its posterior sensory appendages, which are called *cerci.* The filiform hairs, the sensilla that initiate the evasion (jumping) response, are missing in the nymphs and adults and other sensilla degenerate in the adult. Accompanying the sensillum degeneration, the cercal nerve degenerates. In addition, the major neuron receiving innervation from the cercal nerve, the medial giant interneuron, is withered and smaller in volume in the mutant. No action potentials are elicited in this interneuron by stimulation of the sensilla, although they can be elicited by direct stimulation. It is not clear whether the defects in the sensory neurons cause the interneuron defect or whether both are caused independently by the mutation. It might even be that more than one gene is mutated in the strain. In any case the isolation and the study of this nonjumping mutant shows that mutations can be obtained and characterized in an invertebrate large enough for extensive intracellular neurophysiology. Induced mutants altered in calling song would be particularly interesting to compare with species song alterations, but these will be difficult to select.

Locusts are also amenable to laboratory culture and may be useful for isolation of induced mutations. Goodman & Heitler (73) have found that isogenic clones of locusts can be obtained by parthenogenesis. Although those obtained so far are not as healthy as wild type, presumably as a result of deleterious recessive mutations, they still might be useful as stocks for further genetic analysis. One parthenogenetic clone was found with a generally sluggish behavior. The threshold of excitability of the motoneuron innervating the jumping muscles was altered in this clone, but the large number of genetic differences from wild type in this strain makes it difficult to interpret the genetic basis of this defect.

SENSORY RESPONSES

Mutants altered in their sensory responses have been the most extensively studied of all neurological mutants, in part because they can be selected easily. One goal of such studies is to dissect the sequence of steps in transduction of a sensory stimulus into the output of a sensory neuron; another goal is to identify the neurons responsible for detecting a specific stimulus; and a third goal is to study the effects of mutations on the development of a specific part of the peripheral nervous system.

Visual mutants of Drosophila have been the most thoroughly studied. The physiological characterization of such mutants has been reviewed in detail by Pak & Pinto (9), and is not repeated here. Developmental studies of such mutants are discussed in the section on development. This section primarily reviews chemosensory mutants.

Caenorhabditis Sensory Mutants

The behavior that has been most thoroughly investigated in Caenorhabditis is chemotaxis. Caenorhabditis is attracted to cAMP, cations, anions, basic pH, pyridine, a hydrated form of CO_2, and unidentified chemicals released by *E. coli* and certain fungal species (74–77). It is repelled by salt concentrations above 0.3 M, D-tryptophan, acid pH, and some hydrated form of CO_2 (74, 76, 78). Males and juvenile nematodes have the same chemotactic responses as adult hermaphrodites as far as has been tested (74, 79).

In exponential radial gradients of attractants, the nematode's behavioral response is to orient in the direction of increasing concentration, move to the peak of the gradient, and remain there for a time dependent on the concentration of attractant. This behavioral response can be quantified by measuring the precision of orientation or the fraction of a population that accumulates at the peak of the gradient (74). At low attractant concentrations, worms swim away from the peak of a gradient and return repeatedly. This behavior resembles a habituation response to repeated stimuli (74, 80). The chemoreceptors mediating the chemotactic response must be located on the nematode's head because it was found that mutants with cuticle blisters covering the head could not orient and that mutants with heads bent to one side oriented at an angle to a gradient of attractant (74).

The response to both attractants and repellents has also been studied quantitatively using a countercurrent apparatus. This apparatus has two solutions flowing in opposite directions in a cylindrical tube. When attractant is added to one solution, worms in the tube favor that solution, and so are swept out and can be counted (75). This apparatus is particularly well suited for mutant isolation. A similar device has been used for studies of Paramecium chemotaxis (81).

Seventeen chemotaxis-defective strains were obtained in a screening for altered response to NaCl using the countercurrent apparatus (82, 83). These mutants define at least six complementation groups named *tax-1* to *tax-6*. Mutants in all complementation groups are defective in their response to Na^+ and Cl^-, but differ in their response to other attractants and repellents. For example, *tax-1* mutants retain their response to pyridine and are repelled by acid and hydrated CO_2 but have lost all other responses. *tax-2* mutants have lost all attractant responses, but are still repelled by acid and hydrated CO_2. Some of the mutants have reversed their wild-type responses and are repelled by Cl^-, cAMP, and basic pH (83). A Drosophila chemosensory mutant has also been found with reversed behavior. It was attracted to chemicals that normally are repellents (84).

Twenty-one additional chemotaxis-defective mutants were isolated and characterized genetically and anatomically (80, 85). Several of these were initially isolated because their males were sterile and subsequently found to be nonchemotactic. Some were selected as strains that would swim away from the peak of an attractant gradient, some were selected as strains that failed to accumulate at the peak of a gradient, and some were identified by morphological defects in the head. These mutants have been assigned to at least nine different complementation groups, *che-1* through *che-9*, distributed among all the chromosomes (85). Like the *tax* mutants, various *che* mutants differ in the range of their chemotaxis-defective responses.

The minimal number of different chemotactic receptors necessary to mediate the chemotactic behavior has been estimated in two ways. First, by competition experiments: If the worms can detect one attractant in the presence of a uniform high concentration of a second attractant and vice versa, the attractants must be detected by different receptors (74). Second, by examination of the range of defects found in different sensory mutants: If a mutant eliminates one response and retains another, it is most likely that the responses are detected by different receptors (83). Dusenbery has combined both of these approaches to estimate that there must be at least six different receptors mediating chemotaxis in Caenorhabditis (83). It is not yet known whether or not each of these receptors represents a single neuron class (unlikely) or whether different receptors may be distributed on several neurons.

In addition to its response to chemicals, the nematode also responds to temperature. In thermal gradients, worms move to a temperature near that at which they were raised and track isothermally at the temperature (86). Six independent mutants with altered response to temperature have been isolated by selection in thermal gradients. Two of these appear to be cryophilic, always moving to lower temperatures. Two others are thermophilic, but alter their response according to growth temperature. One thermotaxis-defective mutant is also chemotaxis defective; a few of the chemotaxis-defective mutants are thermotaxis defective. Either these mutants are in genes whose normal function is necessary for neurons in both chemotaxis and thermotaxis pathways or else both pathways share common neurons. Paramecium also has a thermotaxis response (87).

Sensory mutants such as the nematode mutants could be blocked anywhere along the pathway from the initial transduction of stimuli to the muscles turning the worm. One reason that the isolation of sensory behavior mutants has been pursued with such vigor in Caenorhabditis is that the fine structure of the sensory neuron terminals can be determined relatively easily because they are nearly all located at the tip of the worm's head. This makes examination of mutants for alterations in fine structure of sensory terminals relatively easy. If a simple pattern of correlations between alterations of anatomy and behavior is observed, it should be possible to assign sensory functions to individual neurons.

There are 58 sensory neurons in the tip of the head and 52 of these are arranged into 4 kinds of sensilla each with 1 to 12 neurons and 2 accessory cells (88, 89). The most likely chemoreceptors are 2 large lateral sensilla called *amphids*. These each contain 12 sensory neurons, 8 of which end in a channel that opens to the outside through an inpocketing of the cuticle.

Nine of the 21 chemotactic mutants studied by Lewis & Hodgkin and Ward have some morphological defects in the terminals of their sensory neurons (78, 85). The defects ranged from a mutant with a few neuron terminals out of place to mutants in genes *che-2* and *che-3* that had defects in all but one of their sensory neuron terminals. Mutants in the gene *che-1* have been studied in greatest detail. They were missing the terminal projections of one of the amphidial sensory neurons. There was considerable variation in the extent of the defect among different alleles and between individuals of identical genotype (85).

The anatomical defects in these mutants could not be correlated simply with their behavioral defects. Mutants in three genes did not respond to either Na^+ or Cl^-, yet the anatomical defects in these mutants were in nonoverlapping sets of sensory neurons. The variation in anatomical defects among *che-1* alleles was not correlated with a similar variation in behavioral response (85, 90). It may be that the anatomical defects visible in the tip of the head are not the only defects caused by the mutations and that other so far undetected defects cause the behavioral phenotypes. More than 15 uncoordinated mutants examined had normal sensory anatomy (85); hence, the anatomical defects are not just trivial consequences of mutagenesis.

When methods for mosaic analysis are developed for Caenorhabditis, it will be possible to test the causal relation between anatomical and behavioral phenotypes by seeing whether they share a common focus. Until then, correlations between anatomical and behavioral defects must be made cautiously. As the anatomical analysis is extended to more mutants, consistent patterns of functional correlation may emerge directly from the large number of mutants analyzed.

One striking conclusion from these studies of sensory mutants in the nematode is that nearly half the mutants selected for alterations in behavior are found to be altered in the anatomy of their nervous system. Presumably the normal products of these genes are necessary for normal development or maintenance of structure. Therefore, further analysis of such mutants must include the analysis of their neural development. This will be difficult in Caenorhabditis because the sensory nervous system develops in the embryo surrounded by a tough egg shell, and it is difficult to follow anatomically (88). The postembryonic neural development (described in the section on development) seems to provide a better opportunity for genetic analysis of neural development.

Drosophila Chemosensory Mutants

Mutants affecting chemical senses have also been isolated in Drosophila (84, 91). Falk & Atidia screened males of mutagenized parents to select strains that would consume sucrose solutions laced with salt (91). The salt inhibits the wild type from drinking; thus, by adding a red dye to the solution, mutant strains could be recognized by their red intestines. Three strains were found that were less sensitive to salt and were designated *lot*. One of these were found to be less sensitive to salt in sucrose solutions when these were applied directly to the labellum and tarsus, but the difference from wild type was not great. Fate mapping of the phenotype was hampered by the difficulty of scoring the behavioral defect, but the defect was tentatively located midventrally in the presumptive brain region.

NEUROANATOMY

One of the initial reasons for selecting behavioral mutants in Caenorhabditis was the hope that, if the entire nervous system wiring diagram could be reconstructed from serial section electron micrographs, it would then be possible to deduce how the nervous system might work from the pattern of interconnections between the sensory input and motor output. Mutants altering specific sets of neurons would provide experimental lesions to test hypotheses of the functional role of these neurons (2). The reconstruction has been essentially completed although not all of it has been published: The anterior sensory nervous system has been described in (88) and (89), the ventral nerve cord in (92), the pharynx in (93), and the tail ganglia and processes have been described in (94). The male tail has been partially described in (94, 95), and the central nerve ring has been partly described in (89) and completely reconstructed by J. G. White, J. N. Thomson, E. Southgate, and S. Brenner (unpublished).

The reconstruction is possible because the worm has less than 300 neurons and is so small (only 60×1200 μm) that total serial reconstruction is feasible. As a consequence of the worm's small size, however, most of the neuron processes are only a few tenths of a micron in diameter and the somas are only a few microns. This precludes intracellular recording. Extracellular recording, if possible, will be difficult. Without electrophysiological recordings to reveal the information processing capabilities of individual neurons, the understanding of the worm's behavior derived from analysis of the anatomical wiring diagram is bound to be superficial. Neurons are not all-or-none transmitters that bear signals directly from sensory receptors through interneurons to the motor neurons. Interneurons especially modulate and interpret their input [for excellent discussion, see (96)]. Without physiological recording, it is impossible to predict the pathways of nervous transmission with confidence.

One possible solution to this problem is to record electrophysiologically from large species of nematodes such as the parasitic Ascaris (A. O. W. Stretton, personal communications; 97). The neuroanatomy of Ascaris closely resembles Caenorhabditis, and the electrophysiology is likely to be similar. Intramuscular recordings from Ascaris have been studied for some years and recently extended (97–99). Intracellular recording from the neurons of Ascaris is still difficult, but has been successfully achieved (A. O. W. Stretton, personal communication). It can be hoped that as this physiology is extended it will be possible to correlate it with the complete anatomy of *C. elegans* to interpret the wiring diagram more precisely. There are, however, distinct disadvantages to isolating mutants in one organism and studying physiology in another.

Fortunately, the difficulties with electrophysiology did not deter the reconstruction of the Caenorhabditis nervous system. It is the only organism whose entire nervous system wiring diagram is known. Some of the parts of the nervous system are relatively isolated from each other so that their simple circuits can be analyzed independently. One such circuit involves the ventral cord motoneurons and interneurons. Their connectivity is summarized in Figure 1 from (92). Four different kinds of motoneurons A, B, AS, and D receive synaptic input from four interneurons α, β, γ and δ. These interneurons are driven from the nerve ring, the worm's central nervous system. It is likely that they are involved in global on/off control of body muscle contraction and not in propagation of contractile waves, because the same neurons in the pseudosegmental repeats along the length of the body receive identical innervation. It is difficult to imagine how they could be activated sequentially to propagate waves. AS neurons only innervate dorsal muscles and may control the asymmetrical flexions that often accompany changes in the worm's direction.

Several uncoordinated mutants are altered in the arrangement of the motoneurons (2). Two mutant alleles of the gene *unc-5* are immobile. Although they can flex their bodies they cannot propagate waves normally. In both these mutants the dorsal nerve cord that innervates dorsal muscles is greatly reduced or absent (2, 100). Processes from the ventral cord, which normally form the dorsal nerve cord by way of commissures around the body, wander off into the lateral hypodermis and fail to reach their normal positions in *unc-5* animals. Some muscle processes do find these neurons and synapses can form, but they are not between the correct neurons and muscles. Different individuals of the same mutant genotype are all defective in the dorsal cord, but the fate of the defective processes varies considerably.

It is difficult to interpret the functional defects in this mutant because so much of the nervous system is altered. Mutants with smaller changes will be more useful

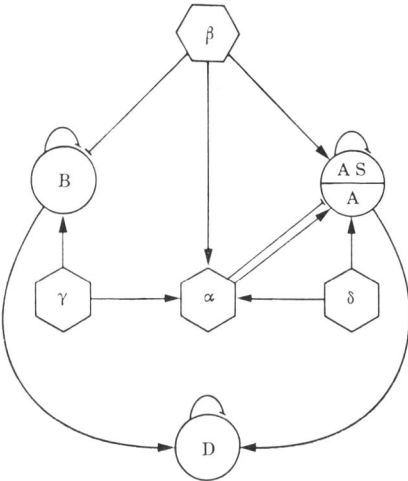

Figure 1 Nematode motor neuron wiring. Connectivity graph showing the connections between the motor neuron classes (*circles*) and the interneuron classes (*hexagons*) that drive them. The *arrows* represent chemical synapses and the *bars* gap junctions. Class AS motor neurons receive the same synaptic input as class A neurons plus an additional chemical synapse from the β-interneurons. All the motor neuron classes form gap junction with themselves (92).

for functional analysis, and the *unc-5* mutants may be more interesting for analysis of the developmental basis of its misdirected neurons.

In addition to using genetic lesions to analyze functions, J. G. White has developed a microlaser system that can destroy single cells in the nervous system [referred to in (101)]. Comparison of the behavioral consequences between laser-induced and mutant-induced lesions may help determine the cells affected by a mutation and help assign functions to lesioned cells.

COMPLEX BEHAVIORS

Three behaviors, learning, courtship, and circadian rhythms, are considered in this section. In spite of years of intensive study, little is known about the molecular mechanisms of learning and memory in any organism. The isolation of mutants deficient in learning may help identify specific sets of neurons required for a learning task and may eventually provide molecular tags to identify essential molecules. Courtship behavior represents a complex sequence of units of behavior that require the nervous system to integrate several sensory inputs and coordinate motor inputs. The study of the behavior of sex mosaics and of mutants altered in courtship may help identify the essential neurons involved in this behavior and may help to determine the sequence of neural events that generates the patterns of behavior. The genetic characterization of such mutants may indicate whether individual units of behavior can be inherited in a simple way.

Learning in Drosophila

Following Nelson's demonstration that blowflies could be classically conditioned (102), two laboratories have shown that Drosophila can also learn a conditioned task (103, 104). A mutant defective in learning has been isolated (105). Quinn, Harris & Benzer trained populations of Drosophila to avoid an odor by alternatively exposing them to two odorants and coupling the exposure to one of these with an electric shock (103). Subsequent testing showed that the trained Drosophila preferentially avoided the shock-associated odor. 3-Octanol and 4-methylcyclohexanol were among the most effective odorants for this olfactory conditioning. A series of careful control experiments ruled out pseudoconditioning, altered excitatory states, odor preference, sensitization, habituation, and subjective bias as alternative explanations of learning. In addition to shock, other adverse stimuli such as exposure to quinine sulfate could also be used for conditioning.

Even with repeated training trials, the avoidance response was shown by only about 30% of the fly population. This was not because of population heterogeneity, but because the probability of an individual fly's response was only 30%. The learned response would persist for 24 hr, but was extinguished rapidly in trials without reinforcement and could be reversed by retraining.

Quinn & Dudai (106) have shown that the memory of olfactory conditioning has at least two phases, the first of which can be abolished by anesthetizing flies with cold immediately after training. Control experiments showed that exposure to cold did not impair the learning process itself, but impaired the ability to remember. The early anesthesia-sensitive phase of learning is a striking analogy to learning in higher organisms (e. g. 107) making the pursuit of the underlying neurobiology of greater interest.

Spatz, Emann & Reichert (104) have shown associative learning in Drosophila by shock conditioning flies to avoid different light stimuli and testing their subsequent light preference with a Y maze.

Using the olfactory paradigm as a criterion for learning, Dudai et al (105) screened approximately 500 lines of flies with mutagenized X chromosomes for mutants defective in learning. They isolated one mutant line that had normal sensory and motor behavior but failed to avoid the shock-associated odor after training. The mutant was named *dunce*. During training, the mutant avoided the shocking grid as did the wild type, but during subsequent testing the mutant did not selectively avoid the shock-associated odorant. Control experiments showed the *dunce* flies could detect test odors because they avoided them slightly as did the wild type. The mutant, however, could not be conditioned to avoid any of 11 tested odor combinations. It also failed to be conditioned when the negative reinforcement was quinine sulfate rather than shock. Its response to light avoidance conditioning was not reported, unfortunately.

dunce flies have normal morphology, growth, viability, sexual courtship behavior, flight behavior, phototaxis, and geotaxis, but they may be somewhat more sluggish than the wild type in crowded growth conditions and may lay fewer eggs. Their electroetinogram is normal, as is their neuromuscular synaptic transmission. Although the primary defect in this mutant is not known, it is a good candidate for being defective in a molecular component of the neural plasticity involved in learning. Mosaic analysis to determine the focus of the learning defect will help to interpret this mutant but will require a learning paradigm that can be reliably used on single flies rather than on populations.

There is a report of successful conditioning of nematodes (108), but no one has reported trying to teach Caenorhabditis anything. Its chemotactic behavior includes a habituation response to repeated stimuli (74, 80) and its thermotactic behavior is altered by prior exposure to temperature (86). These behavioral changes may have elements in common to conditioned learning, but are not examples of genuine learning as usually defined [chapter 24 in (107)].

Courtship Behavior

WASPS Whiting (109) demonstrated many years ago in the wasp *Habrobracon juglandis* that gynadromorphs, mosaics of male and female tissue, could be used to determine which part of the organism determines the male and female behavior. His analysis has been recently extended by Clark & Egen (110) who took advantage of a new wasp mutant, "ebony," that increases the frequences of mosaics. They analyzed 276 mosaics for male and female behavior and concluded that the sex of the brain determines the behavior. References to early mosaic studies with bees and houseflies are included in (110).

DROSOPHILA Courtship behavior in Drosophila is an elaborate sequence of interactions between the partners (reviewed in 111–114). The main steps performed by the male are orienting to the female, following the female, tapping the female's abdomen, extending and vibrating a wing to produce a species-specific courtship song, licking the female's genitalia, and attempting copulation. The female provides stimuli that provoke the male's action and she signals acceptance by spreading her wings and genital plates. An unreceptive female can repel an undesired suitor.

Genetic selection for altered courtship patterns in heterozygous laboratory stocks has been successful. For example, Manning succeeded in selectively breeding Drosophila for fast and slow mating speeds over several generations (reviewed in 4 and 112). Analysis of similar experiments indicated that copulation duration is determined almost exclusively by the male (115).

Hotta & Benzer (116) have used mosaic analysis in Drosophila to ask which part of the body must be male or female for the various steps in courtship to be performed. They used the W^{vc} unstable ring-X chromosome to generate sex mosaics and scored the tissue genotypes using external cuticle markers and the hyperkinetic locus. Hall (117) has extended the mosaic analysis to higher resolution using an internal enzyme marker to score the genotype of the major ganglia (118).

Hotta & Benzer first analyzed the following and wing vibration steps of courtship in 477 mosaics. They observed that if the head tissue was male, the male behavior was observed irrespective of the sex of the head sense organs, legs, wings, or thoracic ganglion. They reasoned that the focus of this behavior must therefore be in the head. Hall showed that the tissue in the head that must be male is the left or right dorsal part of the brain, either of which is sufficient.

To examine subsequent steps in male courtship, Hotta & Benzer examined a second set of 208 mosaics. Among the 130 that showed following and wing vibration, Hotta & Benzer scored attempted and successful copulation. By correlating this behavior with male tissue genotype they concluded that the focus for these steps in courtship was near the region of the thoracic ganglion. Hall has confirmed this directly by histochemical examination of mosaics. Therefore, the dorsal brain and at least some part of the thoracic ganglion must be male for attempted copulation. Male genitalia are of course necessary for successful copulation. No mosaics were observed that attempted copulation without first following and vibrating their wings; therefore, both the brain and thoracic focus must be male for this behavior.

Analysis of the female components of courtship was done by scoring the mating behavior of wild-type males with mosaic flies (116, 117). Female tissue in the abdomen is nearly always necessary and sufficient for a mosaic to be courted by a male. However, examination of the mosaics that were courted by males showed that receptivity to copulation is further controlled by an anterior focus distinct from the focus that provokes initial courtship. It may be that this focus must be female only to prevent the male wing flick repelling response.

The results of mosaic analysis of courtship are summarized by the diagram of the interplay of male and female parts in courtship in Figure 2 [from (116)].

Mutants altered specifically in courtship behavior are also beginning to be studied. One of these is *fruitless* (formerly *fruity*) that causes mutant males to court males as well as females (119). They do not attempt to copulate with the females and so are behaviorally sterile. Fruitless males also stimulate wild-type males to court them. The phenotype is consistent with a defective pheromone response, but this has not been proven. Other courtship mutants include a mutant *stuck* that cannot disengage after copulation (7) although it has no abnormalities of external genital morphology (J. Hall and S. Benzer, personal communication), and a mutant *cacaphony* that has an altered courtship song (120).

CAENORHABDITIS A number of mutants altered in male mating behavior have been isolated in Caenorhabditis (95). Nematode mating is initiated when males

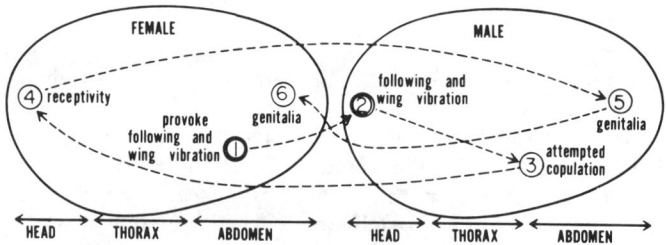

Figure 2 Drosophila courtship. Interplay of male and female sex-specific foci in sequential courtship steps, as delineated by analysis of mosaics (116).

contact hermaphrodites, apparently by random encounters. Hermaphrodites do not mate with each other. When a male touches a hermaphrodite the male backs up, turns and curls its tail into contact with the hermaphrodite. The male tail has a copulatory bursa with specialized sense organs whose cell bodies are in a male-specific tail ganglion (94). Once the bursa contacts the hermaphrodite, the male backs up and explores the hermaphrodite body with its bursa until contacting the vulva. Then the male extends his copulatory spicules and ejaculates sperm into the vulva. Attempts to demonstrate a sexual pheromone in Caenorhabditis have been unsuccessful (R. Cawthon and S. Ward, unpublished), although they exist in other nematode species. Males will mate with paralyzed hermaphrodites and will attempt to mate with other males; thus, a hermaphrodite signal is not necessary to initiate male mating behavior. However, because Caenorhabditis males will not mate with nematodes of distant species, some signaling may be involved.

Hodgkin took advantage of the hermaphrodite reproduction in caenorhabditis to isolate a number of mutants that were fertile as hermaphrodites but sterile as males (95). Since males are unnecessary for propagation, such mutant strains can be maintained with no difficulty. Many of these strains were altered in the male mating behavior. Some have been found to be defective in their sensory neurons in both males and hermaphrodites (85), and others were defective only in the male-specific nervous system.

Since much of the male nervous system develops postembryonically, its development is amenable to detailed morphological analysis with both light and electron microscopy. The ability to isolate many mutants specifically defective in this neurological pathway and the dispensability of the male for maintenance of these strains makes this an attractive system for genetic analysis. When the complete wiring diagram for the male nervous system is available, analysis of the neural control of each step in mating may be possible (see section on neuroanatomy).

Circadian Rhythms

Many animal behaviors display rhythmic variation. The most common period is about 24 hr and such rhythms are called *circadian*. Little is known about the biochemical and neurological mechanism of rhythm generation. In Drosophila, locomotor activity and eclosion show a 24-hr rhythm. Konopka & Benzer (121) have isolated three mutants that alter the normal rhythmic period. They are all alleles of the *per* locus. One of them abolishes the rhythm entirely, but the other two alter the period. One has a short period of 19 hr and one a long period of 27 hr. These two mutants are examples of mutants that alter a behavior rather than abolish it (122). Preliminary mosaic analysis of the mutants indicated a focus in the head for these mutants. Further characterization of them has not been published.

DEVELOPMENT

Many, perhaps most, mutants selected for altered behavior are altered in the development of parts of the nervous system. It is hardly surprising that developmental defects are common, because most of an organism's genome must be invested in specifying its development. Alterations in the development of the nervous system will usually result in altered function and thus altered behavior because the precise morphology and connectivity of a neuron determines its normal function.

Mutants altered in the development of specific parts of the nervous system can be used as if they were surgical dissections to analyze the function of the altered parts. But the use of mutants as dissecting tools must be done with great care since the complex interactions of gene products during development often result in pleiotropic defects [for general review of analysis of developmental mutants see (123–125)]. Clearly only certain mutants in genes with localized effects will be useful for genetic dissection of neural function.

Although much of the interest in isolation of behavioral mutants is to use them to understand the dissected patient, they may also be used to understand the dissecting scapel. Many workers studying behavioral mutants are interested in the mechanisms by which genes specify the development of complex structure such as the nervous system. Behavioral mutants provide a useful set of alterations in normal development that will help to unravel its genetic control. However, because the pathway from the gene to the morphology and connectivity of a neuron is tortuous, biochemical interpretation of mutants will be difficult.

Much of the published literature on the development of neurological mutants is actually detailed analysis of wild-type development that is the necessary prelude to analysis of mutants; however, several Drosophila visual mutants have been analyzed in detail. Mosaics have been especially useful in developmental analysis of mutants, because they help separate the role of cell lineage and the role of cell position in determining the fate of individual cells. Mosaics have also revealed specific cell-cell interactions that are important in neural development.

Drosophila Visual System

Many of the phototaxis-defective mutants and other eye-defective mutants are altered in the development of the compound eye. The Drosophila retina is an attractive tissue for developmental analysis. It is a neurocrystalline lattice of about 780 hexagonal facets and a regular array of sensory hairs. Each facet contains an ommatidium of eight photoreceptor (retinula) cells identified as cells R1-R8 by their position in the ommatidium. There are in addition, six pigment cells and four hair nerve cells (for review see 126).

Two groups, Ready, Hanson & Benzer (126), and Campos-Ortega and colleagues (127, 128) have followed the development of this lattice from the eye-antenna imaginal disc by combining light and electron microscopic histological examination with (^{3}H)-thymidine autoradiography and mosaic analysis. They conclude that the regular lattice pattern of the eye is not generated simply by a repetitive pattern of eye cell lineages leading to the individual ommatidia. Instead, the lattice pattern is established by recruitment of cells along a morphogenetic front that advances from posterior to anterior across the developing eye. The precursors to photoreceptor cells R2–R5 and R-8 are organized at the advancing front. Subsequently, photoreceptor cells, R-1, R-6, and R-7 and the other cells of the facet are produced behind the front. The cells of an ommatidium are not descendants of a single mother cell. The major evidence for this comes from examination of somatic mosaics induced by X irradiation of larvae heterozygous for the sex-linked recessive pigment mutation *white*. It was found that mosaic boundaries passed through single ommatidia so that individual cells in an ommatidium can have different genotypes. This must mean that the cells of an ommatidium do not all share a common ancestor since the mosaics were generated well before the final cell divisions in the retina (126).

Analysis of mosaics generated later in eye development (129) showed that the recruitment of cells into ommatidia is not as random as was suggested by Ready et al (126). The photoreceptor and pigment cells must arise from different precursors because they are separated in mosaic patches. The receptor cells R1, R6, and R7 tend to be clonally related whereas they are generally unrelated to cells R2–R5 (R8 was not scored in these experiments). Recall that R2–R5 were recruited first at the morphogenetic front whereas R1, R6, and R7 were organized later. The common clonal origin of these groups of cells led Campos-Ortega and Hobauer to argue that it is not their position alone that determines their fate, but their lineage is important as well. The position, however, must determine the final individual identity of each receptor cell. This conclusion is complicated if cells from common ancestors tend to occupy nearby positions because the distinction between influences of ancestry and position becomes difficult.

More than 100 different genes are known whose mutant alleles can alter the arrangement of facets in the eye (126). The detailed analysis of the normal eye development now established provides the background for analysis of the developmental lesions in some of these mutants.

Mutants that cause the degeneration of receptor cells have been studied in detail by Harris & Stark (130). In the strictest sense these are not developmental mutants; however, the distinction between development and degeneration is a fine one, and their analysis illustrates the power of genetic techniques. The retinal degeneration mutants *rdgA* and *rdgB* were initially isolated as phototaxis-defective mutants. They have histologically normal photoreceptor cells at eclosion but these subsequently degenerate. Mosaic analysis showed that the degeneration was cell autonomous.

Harris & Stark (130) found that the degeneration of *rdgB* mutants was prevented by rearing the flies in the dark or by combining the *rdgB* mutant with a mutant that has no receptor potential, *norpA*; the physiology of *norpA* is reviewed in detail in (9). This suggests that the degeneration is dependent on some step in phototransduction. To pinpoint this step, Harris and Stark selected for mutagen-induced revertants of the *rdgB* phenotype and obtained three suppressors of degeneration. All three turned out to be *norpA* alleles. Two had little or no receptor potential, but one had a normal receptor potential when exposed to light. This showed that the receptor potential itself is not sufficient for degeneration. Harris and Stark propose that the normal *rdgB* and *norpA* gene products interact during the transduction of light into the receptor potential and that the *rdgB* gene product terminates the action of the *norpA* gene product. The absence of this termination in *rdgB* mutants leads somehow to cell death, perhaps by leaving open some ion channel in the cell.

In another study Harris, Stark & Walker (131) used these and other retinal degeneration mutants to establish the photo-pigment distribution among different classes of receptor cells and to determine the function of each cell type. Because these mutants are autonomous and have very specific effects, in part because they act so late in development, their use in such analysis was incisive.

Meyerowitz (132) has used mosaics to analyze several mutants that disrupt the eye facets and that also disrupt the morphology of the lamina and optic lobes [discussed in (11)]. The lamina and optic lobes normally receive projections from retinal cells. Meyerowitz found that in mosaics that had mutant retinal patches, the lamina and optic lobes receiving projections from these patches were disordered even though they themselves were wild type. This suggests that the eye facet defect causes the subsequent defects in lamina and optic lobe. Therefore, the formation of proper higher order neuron structure and connectivity must be dependent on primary innervation.

The development of the temperature-sensitive paralytic mutant *shibire* has been examined in some detail. By pulsing with high temperature at various times in development, it was found to have a complex phenotype affecting different tissues at different times (133). Temperature pulses during the time of eye development cause scars across the eye parallel to the morphogenetic front described in (126–128). Presumably the normal *shi* gene product is necessary during a critical period of differentiation perhaps coincident with the movement of the morphogenetic front.

The *shibire* mutation must affect a molecular component present during the development of many different cells because it affects so many tissues at different times. Recall that the reversible temperature-sensitive paralysis phenotype of *shibere* was due to a blockage of neuromuscular transmission. This implies that the gene product must function in fully developed tissue as well. The relationship between the developmental and physiological phenotypes remains to be determined.

Caenorhabditis Postembryonic Development

The postembryonic development of the nematode nervous system provides another opportunity for analysis of developmental mutants. The number of cells in the nematode's ventral nerve cord increases from the newly hatched larva to the adult (134). Sulston (134) and Sulston & Horvitz (101) have followed the postembryonic neural development, as well as other neural, muscular, and epithelial development, in their entirety. They watched developing worms using differential interference contrast microscopy and observed all the cell divisions. This has resulted in a description of the developmental lineage of every postembryonic cell.

The number of nongonadal nuclei in the larva is approximately 550 and this increases to about 810 in the adult hermaphrodite and 970 in the adult male (101). The pattern of cell divisions that leads to this increase was essentially invariant among individuals. One of Sulston's examples of this lineage is described below (134).

Fifty-six nerve cells are added to the ventral cord and associated ganglia at about the time of the first larval molt. These cells are the descendants of 13 neuroblasts that divide uniformly in the pattern shown by Figure 3. The divisions are followed by a defined pattern of cell deaths.

The fine structure and connectivity of each of the neurons in the ventral cord has been determined by serial section electron microscopy (92) (see section on

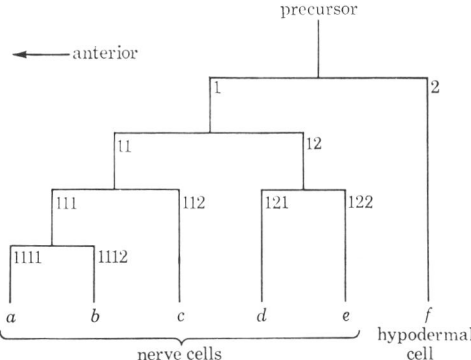

Figure 3 The lineage of nematode ventral cord neurons. Each of the twelve precursors divides as shown. The first anterior daughter is a neuroblast; the first posterior daughter is a central hypodermal cell. The axes of all the divisions are approximately parallel to the longitudinal axis of the nematode. In order to label the various cells, an anterior daughter is given its parent's number with the suffix 1, and a posterior daughter its parent's number with the suffix 2. (A more convenient nomenclature is used in 101.) The cell labeled *c* undergoes a programmed death in six of the lineages of the hermaphrodite (134).

neuroanatomy). The neurons added postembryonically can be grouped into five classes of motor neurons defined by their branching patterns and connectivity. These are organized into periodic repeats. In animals sectioned after determination of their cell lineages, there was an exact correlation between ancestry and final functional class in the regions studied. This implies that in this part of the nematode's nervous system the fate of individual cells could be determined by their ancestry. Further support for this comes from laser ablation experiments (101). At the anterior and posterior ends of the ventral cord several neurons do not follow the predicted fate of their lineage (134), so ancestry alone is not sufficient to determine the fates of all cells.

The neural development of the nematode contrasts with that of the Drosophila compound eye. In the eye, the identity of each ommatidial cell was determined primarily by its position relative to the morphogenetic front at the time of recruitment although the lineage played some role. In a nematode, lineage appears to be the dominant determinant of cell identity. Therefore, these two developing systems offer examples of the two classical ways of determining the fate of a cell: mosaically (by lineage) or regulatively (by position). Neither organism follows one way exclusively.

Among the large class of nematode uncoordinated mutants and among mutants clones screened for abnormal cell arrangements, a number of mutants have been found that alter the precise lineage patterns (101). The detailed description of the developmental defects in these mutants may give further insight into the genetic specification of lineage determination.

Crickets and Locusts

In addition to the neurophysiological study of locusts and crickets, much of their neuroanatomy has been studied in great detail so that many individual neurons can be identified from animal to animal [for references see (135)]. Therefore, mutants altering this neuroanatomy can be recognized relatively easily. Some of the parthenogenetic locust clones have duplications of identified neurons (C. Goodman, personal communication). However, the detailed developmental studies necessary to analyze such mutants must be done, and establishment of cause and effect relations among defective neurons will be difficult without mosaics. For detailed genetic studies, Drosophila and Caenorhabditis have distinct advantages.

ADDITIONAL EXPERIMENTAL ORGANISMS

In addition to the experimental organisms reviewed here, behavioral mutants have been studied in other organisms. Phototaxis mutants have been isolated in the alga *Chlamydomonas reinhardi* (136). Chemotaxis, phototaxis, and thermotaxis have been studied in the slime mold *Dictyostelium discoideum* (reviewed in 137). The process of photo-transduction and the mechanisms of tropisms are being actively studied in the fungus Phycomyces (138, 139). Bees may also prove amenable to single-gene studies (reviewed in 3).

Streisinger has developed methods for inducing homozygous clones of zebra fish which should allow isolation of induced mutations altering visual behavior of fish (personal communication). Although this is not an invertebrate, it has some of the genetic advantages of the invertebrates reviewed here and may prove to be a useful organism for both developmental and functional analysis of neurological mutants.

Spontaneous behavioral mutants have been known for many years in the Mexican axolotl (reviewed in 140). Recently the neurological mutant *spastic* has been subjected to detailed developmental analysis (141). Although amphibians have disadvantages for genetic manipulation and induced mutant isolation, their large size and well-studied embryology is advantageous for developmental studies, and mosaic analysis can be performed with chimeras created by grafting.

CONCLUSIONS AND PROSPECTS

The research reviewed here has established that many mutants altering the nervous system can be obtained in several organisms. The analyses of these mutants suggests several conclusions about the type of mutants and the strategies of mutant analysis that have been most productive so far. These conclusions reflect my prejudice that invertebrate neurogenetics will contribute most to problems in neurobiology when the molecular identity of mutated gene products is known.

Types of Mutants

The use of induced mutations whose origin is known (or thought to be known) is one feature that distinguishes the work reviewed here from the more extensive studies of behavioral mutants obtained by segregation of behavioral traits from natural populations. It is assumed that the induced mutations affect only single genes. While this is probably true for most of the mutants studied, it is an assumption that must be checked carefully. Isolation of multiple alleles that map to the same locus and that have similar phenotypes is usually sufficient to be sure that only one gene is involved. Reversion of the mutant phenotype in a single step provides further evidence for the singularity of mutant defects. The study of several alleles that differ slightly in phenotype has been important in several studies for interpreting the normal gene product's function and identifying interactions between genes.

Temperature-sensitive neurological mutants have many experimental advantages. First, mutants in genes whose normal functions are indispensable to the organism can be maintained as homozygotes at a permissive temperature. Second, phenotypes that arise because a gene product acts at several times during the development of the organism can be studied independently by judicious temperature shifts. Third, if temperature sensitivity is due to a temperature-sensitive gene product (e.g. 142), then temperature shifts give some indication of when the gene product acts during development [see (143, 144) for discussion]. Fourth, assuming again that temperature sensitivity is due to a temperature-sensitive gene product, mutants that are rapidly and reversibly temperature sensitive as adults must have the mutant gene product active in the adult tissue rather than just active during its development. The assumption that temperature sensitivity is due to a temperature-sensitive gene product needs further testing.

Single-gene deletions or other absolute defective mutations such as chain termination mutants would be especially useful in the quest to identify the products of mutant genes. Such mutants would eliminate the normal gene products; this could be detected by missing bands or spots on gels. These mutations would have to be in nonessential functions to be maintained as homozygotes, unless a conditional lethal system for chain termination mutants is developed as may be the case in Caenorhabditis. Otherwise, they can be maintained as heterozygotes. Such strains would also show the phenotype of a genuine null mutation in a gene for comparison with the phenotypes of point alleles.

Importance of Mosaic Analysis

The technique of mosaic fate mapping originated by Sturtevant and developed over the past 10 years (reviewed in 11) has been central to the analysis of many Drosophila neurological mutants. This technique assigns the focus of a mutant phenotype of a presumptive site on the embryonic blastoderm. This focus is the developmental origin of the cells that must be of mutant genotype for the mature organism to express its phenotype. Knowing the focus of a behavioral phenotype provides a critical indication of which tissues and cells to examine for mutant defects. Such knowledge also helps determine whether pleiotropic phenotypes of a mutant arise from a single initial defect or from multiple independent defects. This has been especially useful in showing that some neurological defects arise from alterations in cell interactions during development.

The technique of fate mapping has been used most extensively in Drosophila because there are straightforward methods of generating mosaics. It can be applied quite generally, however, by induction of mosaics by somatic recombination or mutation or by creating tetraparental embryos as is being done in the mouse (145). Somatic mosaics of intestinal cells have been generated in Caenorhabditis. When specific histochemical markers for other tissues become available, it should be possible to determine the developmental focus of nematode neurological mutants as well.

Direct Approach to Molecular Defects

A direct and powerful approach to studying genetic control of neural development and function is to isolate mutants not by their behavioral phenotype but by their molecular defects using direct assays or pharmacology. With this approach some of the needle-in-a-haystack search for a defective gene product is eliminated because the gene product has been selected a priori (hopefully). This approach is limited to gene products that are already known and ones that can be assayed readily, but it might be possible to extend it to additional gene products using immunological assays for neurological components.

Exploitation of Mutants

As more and more mutants are isolated and characterized, they can be used to help characterize each other. A particularly clear example of this was the use of the receptor potential mutants to analyze Drosophila retinal degeneration (130). Such experimental bootstrapping will speed the progress of mutant analysis especially when mutants are freely exchanged among workers. Mutants often prove to be useful for problems other than those for which they are selected.

Analysis of Development

Many of the mutants that have been selected for behavioral alterations are altered in the development of parts of the nervous system. This means that eventual understanding of the effects of these mutants must be sought by understanding the gene control of development. This is the most challenging problem confronting the analysis of neurological mutants. In order for the analysis of developmental defects in behavioral mutants to progress much beyond detailed description of their developmental lesions, we need to understand more about the biochemical mechanisms of gene control of animal development. Neurological mutants will help us gain this understanding, but it seems to me that our current knowledge is so rudimentary that studies of developmental pathways leading to tissues less complex than the nervous system may have to precede the analysis of neurological mutants. Such studies will guide the search for the molecular basis of the neurological defects. Nevertheless, the complexity of interconnections in the nervous systems poses a developmental problem that cannot be solved in other tissues. Neurological mutants can provide decisive tests of certain developmental hypotheses to explain this complexity.

One technique that could be applied to analysis of neurological mutants is the study of developing neurons in culture. Since Drosophila neurons and muscle cells can be cultured and do form synapses in culture (146), analysis of the defects in some neurological mutants by this method may be of great promise.

Analysis of Function

Although genes do not sort themselves neatly into developmental and functional categories, the analysis of their defects can still be sorted out. Mutants with specific lesions have helped assign functions to specific neurons. Electrophysiological characterization of mutants brings one closer to the molecular biology of the cell than does electron microscopic examination of developmental lesions. Neurological mutants will clearly provide indispensable tags to help identify molecular components of electrically excitable membranes. Paramecium clearly has advantages for biochemical analysis, but the techniques of intramuscular recording in Drosophila should also identify mutants altered in specific components of excitability.

Study of mutants altering firing patterns of defined neurons may also help us understand how neuron output patterns are genetically alterable, and may help assign functions to specific nerve cells.

Relation of Neurogenetics to Behavioral Evolution

Most of the mutants described in this review are deficient in a certain pattern of behavior. Another class of mutants of interest would be mutants that retain a pattern of behavior but alter it quantitatively or qualitatively. These might be in genes in which mutations have played a role in behavioral evolution. There may be many more genes that can mutate to eliminate a behavior than can mutate to modify it specifically. Many mutants eliminating a behavior might be in the "ordinary" biochemistry of the organism and affect a specific behavior only peripherally. Mutants that modify a behavior are more likely to be in genes whose products directly determine the behavior. In systems where there are strong behavioral selections, it might be worthwhile to select for mutants with enhanced and additional behaviors to identify more genes that can modify existing behavior patterns or add new patterns.

Whether study of these or other single-gene mutations will help us to understand the ways that behavior patterns have evolved remains to be determined. Natural selection acts on the total genotype of the individual and its breeding population. Studies of the effects of mutations in single genes must be interpreted cautiously, keeping in mind Ernst Mayr's (admittedly exaggerated) dictum, "All genes affect all characters and all characters are affected by all genes" (147). The study of how the products of single genes can alter behavior may not be sufficient for understanding the mechanisms by which patterns of behavior have evolved, but it will certainly be necessary.

ACKNOWLEDGMENTS

I thank J. Hall, L. Hall, V. S. Caviness, Jr., Y. Argon, G. Nelson, and A. Ward for their constructive criticism of drafts of this review, and G. Nosek and B. Katz for preparation of the manuscript. I am grateful to the National Science Foundation and the National Institutes of General Medical Sciences for support of the research in my laboratory.

Literature Cited

1. Benzer, S. 1967. Behavioral mutants of Drosophila isolated by countercurrent distribution. Proc. Natl. Acad. Sci. USA 58:1112–19
2. Brenner, S. 1973. The genetics of behavior. Br. Med. Bull. 29:269–71
3. Gould, J. L. 1974. Genetics and molecular ethology. Z. Tierpsychol. 36:267–92
4. Ehrman, L., Parsons, P. A. 1976. The Genetics of Behavior. Sunderland, Mass.: Sinauer Assoc.
5. Grossfield, J. 1975. Behavioral mutants of Drosophila. In Handbook of Genetics, ed. R. C. King, III: 679–702. New York: Plenum
6. Pak, W. L. 1975. Mutations affecting the vision of Drosophila melanogaster. See Ref. 5, pp. 703–33
7. Benzer, S. 1971. From the gene to behavior. J. Am. Med. Assoc. 218:1015–22
8. Benzer, S. 1973. Genetic dissection of behavior. Sci. Am. 229:24–37
9. Pak, W. L., Pinto, L. H. 1976. Genetic approach to the study of the nervous system. Ann. Rev. Biophys. Bioeng. 5:397–448
10. Caviness, V. S. Jr., Rakic, P. 1978. Mechanisms of cortical development: A view from mutations in mice. Ann. Rev. Neurosci. 1: In press
11. Hall, J. C. 1977. Behavioral analysis in drosophila mosaics. In Genetic Mosaics and Cell Differentiation, ed. W. J. Gehring, pp. 1–78. New York: Springer
12. Hotta, Y., Benzer, S. 1973. Mapping of behavior in Drosophila mosaics. In Genetic Mechanisms of Development, ed. F. H. Ruddle, pp. 129–67. New York: Academic
13. Hall, J. C., Kankel, D. R. 1976. Genetics of acetylcholinesterase in Drosophila melanogaster. Genetics 83:517–35
14. Herman, R. K., Albertson, D. G., Brenner, S. 1976. Chromosome rearrangements in Caenorhabditis elegans. Genetics 83:91–105
15. Johnson, C. D. 1976. Multiple molecular forms of cholinesterase from elongated animals. PhD thesis. Calif. Inst. Technol., Pasadena, Calif. 149 pp.
16. Plapp, F. W. 1976. Biochemical genetics of insecticide resistance. Ann. Rev. Entomol. 21:179–97
17. Brenner, S. 1974. The genetics of Caenorhabditis elegans. Genetics 77: 71–94
18. Dudai, Y. 1977. Demonstration of an α-bungarotoxin-binding nicotinic receptor in flies. FEBS Lett. 76:211–13
19. Schmidt-Nielsen, B. K., Gepner, J. I., Teng, N. N. H., Hall, L. M. 1977. Characterization of an α-bungarotoxin binding component from Drosophila melanogaster. J. Neurochem. In press
20. Sulston, J., Dew, M., Brenner, S. 1975. Dopaminergic neurons in the nematode Caenorhabditis elegans. J. Comp. Neurol. 163:215–26
21. Hotta, Y., Benzer, S. 1969. Abnormal electroretinograms in visual mutants of Drosophila. Nature 222:354–56
22. Hodgetts, R. B., Konopka, R. J. 1973. Tyrosine and catecholamine metabolism in wild-type Drosophila melanogaster and a mutant, ebony. J. Insect Physiol. 19:1211–20
23. Konopka, R. J. 1972. Abnormal concentrations of dopamine in a Drosophila mutant. Nature 239:281–82
24. Suzuki, D. T. 1974. Behavior in Drosophila melanogaster: A geneticist's view. Can. J. Genet. Cytol. 16:713–35
25. Epstein, H. F., Waterston, R. H., Brenner, S. 1974. A mutant affecting the heavy chain of myosin in Caenorhabditis elegans. J. Mol. Biol. 90:291–300
26. Epstein, H. F., Thomson, J. N. 1974. Temperature-sensitive mutation affecting myofilament assembly in Caenorhabditis elegans. Nature 250:579–80
27. Waterston, R. H., Epstein, H. F., Brenner, S. 1974. Paramyosin of Caenorhabditis elegans. J. Mol. Biol. 90:285–90
28. Waterston, R. H., Fishpool, R. M., Brenner, S. 1977. Mutants affecting paramyosin in Caenorhabditis elegans. J. Mol. Biol. In press
29. Schachat, F. H., Harris, H. E., Epstein, H. F. 1977. Two homogeneous myosins in body-wall muscle of Caenorhabditis elegans. Cell 10:721–28
30. Ikeda, K., Kaplan, W. D. 1974. Neurophysiological genetics in Drosophila melanogaster. Am. Zool. 14:1055–66
31. Alawi, A. A., Pak, W. L. 1971. On-transient of insect electroretinogram: Its cellular origin. Science 172:1055–57
32. Wyman, R. J. 1976. A simple network for the study of neurogenetics. In Simpler Networks and Behavior, ed. J. C. Fentress, 11:153–56. Sunderland, Mass.: Sinauer Assoc.
33. Harcombe, E. S., Wyman, R. J. 1977. Output pattern generation by Drosophila flight motoneurons. J. Neurophysiol. In press
34. Ikeda, K., Ozawa, S., Hagiwara, S. 1976. Synaptic transmission reversibly

conditioned by single-gene mutation in *Drosophila melanogaster. Nature* 259: 489–91
35. Siddiqi, O., Benzer, S. 1976. Neurophysiological defects in temperature-sensitive paralytic mutants of *Drosophila melanogaster. Proc. Natl. Acad. Sci. USA* 73:3253–57
36. Kelly, L. E. 1974. Temperature-sensitive mutations affecting the regenerative sodium channel in *Drosophila melanogaster. Nature* 248:166–68
37. Hall, L. M., Olive, D., Farber, I., Osmond, B. C. 1977. A tetrodotoxin-sensitive mutation in *Drosophila melanogaster*. Manuscript submitted
38. Suzuki, D. T., Grigliatti, T., Williamson, R. 1971. Temperature-sensitive mutations in *Drosophila melanogaster*. VII. A mutation (para[ts]) causing reversible adult paralysis. *Proc. Natl. Acad. Sci. USA* 68:890–93
39. Kelly, L. E., Suzuki, D. T. 1974. The effects of increased temperature on electroretinograms of temperature-sensitive paralysis mutants of *Drosophila melanogaster. Proc. Natl. Acad. Sci. USA* 71:4906–9
40. Grigliatti, T., Suzuki, D. T., Williamson, R. 1972. Temperature-sensitive mutations in *Drosophila melanogaster. Dev. Biol.* 28:352–71
41. Jan, L. Y., Jan, Y. N. 1976. Properties of the larval neuromuscular junction in *Drosophila melanogaster. J. Physiol.* 262:189–214
42. Jan, L. Y., Jan, Y. N. 1976b. L-Glutamate as an excitatory transmitter at the *Drosophila* larval neuromuscular junction. *J. Physiol.* 262:215–36
43. Jan, Y. N., Jan, L., Dennis, M. J. 1970. A synaptic transmission mutant of *Drosophila. Biol. Ann. Rep.,* Calif. Inst. Technol., p. 91
44. Ostroy, S. E., Pak, W. L. 1973. Protein differences associated with a phototransduction mutant of *Drosophila. Nature New Biol.* 243:120–21
45. Eckert, R., Naitoh, Y., Machemer, H. 1976. Calcium in the bioelectric and motor functions of *Paramecium. Calcium in biological systems. Soc. Exp. Biol. Symp.* 30:233–55
46. Eckert, R. 1972. Bioelectric control of ciliary activity. *Science* 176:473–81
47. Jennings, H. S. 1962. *Behavior of the Lower Organisms.* Bloomington: Indiana Univ. Press
48. Kung, C., Chang, S. Y., Satow, Y., Van Houten, J., Hansma, H. 1975. Genetic dissection of behavior in *Paramecium. Science* 188:898–904
49. Kung, C. 1976. Membrane control of ciliary motions and its genetic modification. *Cell Motility*, pp. 941–48. Cold Spring Harbor, NY: Cold Spring Harbor Lab.
50. Naitoh, Y., Kaneko, H. 1972. ATP-reactivated triton-extracted models of Paramecium: Modification of ciliary movement by calcium ions. *Science* 176:523–24
51. Schein, S. J. 1976. Nonbehavioral selection for pawns, mutants of *Paramecium aurelia* with decreased excitability. *Genetics* 84:453–68
52. Schein, S. J., Bennett, M. V. L., Katz, G. M. 1976. Altered calcium conductance in pawns, behavioral mutants of *Paramecia aurelia. J. Exp. Biol.* 65: 699–724
53. Schein, S. J. 1976. Calcium channel stability measured by gradual loss of excitability in pawn mutants of *Paramecium aurelia. J. Exp. Zool.* 65:725–36
54. Chang, S. Y., Kung, C. 1973. Temperature-sensitive pawns: Conditional behavioral mutants of *Paramecium aurelia. Science* 180:1197–99
55. Satow, Y., Chang, S. Y., Kung, C. 1974. Membrane excitability made temperature-dependent by mutations. *Proc. Natl. Acad. Sci. USA* 71:2703–6
56. Chang, S. Y., Kung, C. 1976. Selection and analysis of a mutant *Paramecium tetaurelia* lacking behavioural response to tetraethylammonium. *Genet. Res.* 27:97–107
57. Satow, Y., Kung, C. 1976. A 'tea[+]-insensitive' mutant with increased potassium conductance in *Paramecium aurelia. J. Exp. Biol.* 65:51–63
58. Satow, Y., Hansma, H. G., Kung, C. 1976. The effect of sodium on "paranoiac"—A membrane mutant of *Paramecium. Comp. Biochem. Physiol.* 54A:323–29
59. Satow, Y., Kung, C. 1974. Genetic dissection of active electrogenesis in *Paramecium aurelia. Nature* 247:69–71
60. Satow, Y., Kung, C. 1977. A mutant of *Paramecium* with increased relative resting potassium permeability. *J. Neurobiol.* In press
61. Hansma, H. G. 1975. The immobilization antigen of *Paramecium aurelia* is a single polypeptide chain. *J. Protozool.* 22(2):257–59
62. Hansma, H. G., Kung, C. 1975. Studies of the cell surface of *Paramecium. Biochem. J.* 152:523–28

63. Wyman, R. J. 1973. Neural circuits patterning dipteran flight motoneurone output. *Neurobiol. Invertebr.* 1971:289–309
64. Levine, J. D., Wyman, R. J. 1973. Neurophysiology of flight in wild-type and a mutant *Drosophila*. *Proc. Natl. Acad. Sci. USA* 70:1050–54
65. Wong, P. T. 1975. *Genetics* 80:s8 (Abstr.)
66. Homyk, T. Jr., Sheppard, D. E. 1977. Behavioral mutants of *Drosophila melanogaster*. I. Isolation and mapping of mutations which decrease flight ability. *Genetics.* In press
67. Homyk, T. Jr. 1977. Behavioral mutants of *Drosophila melanogaster*. II. Behavioral analysis and focus mapping. *Genetics.* In press
68. Ikeda, K. 1976. Genetically patterned neural activity. In *Simpler Networks and Behavior*, ed. J. C. Fentress, 10: 140–52
69. Bentley, D. R., Hoy, R. R. 1972. Genetic control of the neuronal network generating cricket (Teleogryllus Gryllus) song patterns. *Anim. Behav.* 20:478–92
70. Hoy, R. R. 1974. Genetic control of acoustic behavior in crickets. *Am. Zool.* 14:1067–80
71. Hoy, R. R., Hahn, J., Paul, R. C. 1977. Hybrid cricket auditory behavior: Evidence for genetic coupling in animal communication. *Science* 195:82–84
72. Bentley, D. 1975. Single gene cricket mutations: Effects on behavior, sensilla, sensory neurons, and identified interneurons. *Science* 187:760–64
73. Goodman, C. S., Heitler, W. J. 1977. Isogenic locusts and genetic variability in the effects of temperature on neuronal threshold. *J. Comp. Physiol.* 117: 183–207
74. Ward, S. 1973. Chemotaxis by the nematode *Caenorhabditis elegans:* Identification of attractants and analysis of the response by use of mutants. *Proc. Natl. Acad. Sci. USA* 70:817–21
75. Dusenbery, D. B. 1973. Countercurrent separation: A new method for studying behavior of small aquatic organisms. *Proc. Natl. Acad. Sci. USA* 70:1349–52
76. Dusenbery, D. B. 1974. Analysis of chemotaxis in the nematode *Caenorhabditis elegans* by countercurrent separation. *J. Exp. Zool.* 188:41–48
77. Dusenbery, D. B. 1976. Attraction of the nematode *Caenorhabditis elegans* to pyridine. *Comp. Biochem. Physiol.* 530:1–2
78. Dusenbery, D. B. 1975. The avoidance of D-tryptophan by the nematode *Caenorhabditis elegans*. *J. Exp. Zool.* 193:413–18
79. Dusenbery, D. B. 1976. Chemotactic responses of male *Caenorhabditis elegans*. *J. Nematol.* 8:352–55
80. Ward, S. 1976. The use of mutants to analyze the sensory nervous system of *Caenorhabditis elegans*. In *The Organization of Nematodes,* ed. N. Croll. New York: Academic
81. Van Houten, J., Hansma, H., Kung, C. 1975. Two quantitative assays for chemotaxis in *Paramecium*. *J. Comp. Physiol.* 104:211–23
82. Dusenbery, D. B., Sheridan, R. E., Russell, R. L. 1975. Chemotaxis-defective mutants of the nematode *Caenorhabditis elegans*. *Genetics* 80:297–309
83. Dusenbery, D. B. 1976. Chemotactic behavior of mutants of the nematode *Caenorhabditis elegans* that are defective in their attraction to NaCl. *J. Exp. Zool.* 198:343–52
84. Kikuchi, T. 1973. Specificity and molecular features of an insect attractant in a *Drosophila* mutant. *Nature* 243:36–38
85. Lewis, J. A., Hodgkin, J. A. 1977. Specific neuroanatomical changes in chemosensory mutants of the nematode *Caenorhabditis elegans*. *J. Comp. Neurol.* 172:489–510
86. Hedgecock, E. M., Russell, R. L. 1975. Normal and mutant thermotaxis in the nematode *Caenorhabditis elegans*. *Proc. Natl. Acad. Sci. USA* 72:4061–65
87. Tawada, K., Oosawa, F. 1972. Responses of *Paramecium* to temperature change. *J. Protozool.* 19:53–57
88. Ward, S., Thomson, J. N., White, J. G., Brenner, S. 1975. Electron microscopical reconstruction of the anterior sensory anatomy of the nematode *Caenorhabditis elegans*. *J. Comp. Neurol.* 160:313–38
89. Ware, R. W., Clark, D., Crossland, K., Russell, R. L. 1975. The nerve ring of the nematode *Caenorhabditis elegans:* Sensory input and motor output. *J. Comp. Neurol.* 162:71–110
90. Ward, S. 1977. The use of nematode behavioral mutants for analysis of neural function and development. *Soc. Neurosci. Res. Symp.* II:1–26
91. Falk, R., Atidia, J. 1975. Mutations affecting taste perception in *Drosophila melanogaster*. *Nature* 254:325–28
92. White, J. G., Southgate, E., Thomson, J. N., Brenner, S. 1976. The structure of

the ventral nerve cord of *Caenorhabditis elegans*. *Philos. Trans. R. Soc. London Ser. B* 275:327–48
93. Albertson, D. G., Thomson, J. N. 1976. The pharynx of *Caenorhabditis elegans*. *Philos. Trans. R. Soc. London Ser. B* 275:299–325
94. Hall, D. H. 1976. *The posterior nervous system of the nematode Caenorhabditis elegans*. PhD thesis. Calif. Inst. Technol., Pasadena, Calif. 170 pp.
95. Hodgkin, J. A. 1974. *Genetic and anatomical aspects of the C. elegans male*. PhD thesis. Cambridge Univ., Cambridge, England
96. Wiersma, C. A. G. 1974. Behavior of neurons. In *The Neurosciences: Third Study Program*, ed. F. O. Schmitt, F. G. Worden, pp. 419–31. Cambridge, Mass: MIT Press
97. Stretton, A. O. W. 1976. Anatomy and development of the somatic musculature of the nematode *Ascaris*. *J. Exp. Biol.* 64:773–88
98. Weisblat, D. A., Byerly, L., Russell, R. L. 1976. Ionic mechanisms of electrical activity in somatic muscle of the nematode *Ascaris lumbricoides*. *J. Comp. Physiol.* 111:93–113
99. Weisblat, D. A., Russell, R. L. 1976. Propagation of electrical activity in the nerve cord and muscle syncytium of the nematode *Ascaris lumbricoides*. *J. Comp. Physiol.* 107:293–307
100. Wyman, R. 1975. *Use of a Mutant in the Study of Neuromuscular Connectivity in Nematodes*. *Conf. Rep. 41*, Brain Inf. Serv., Univ. Calif. Los Angeles, pp. 8–9
101. Sulston, J. E., Horvitz, H. R. 1977. Post-embryonic cell lineages of the nematode, *Caenorhabditis elegans*. *Dev. Biol.* 56:110–56
102. Nelson, M. 1971. Classical conditioning in the blowfly *Pharmia regina*. *J. Comp. Physiol. Psychol.* 77:353–68
103. Quinn, W. G., Harris, W. A., Benzer, S. 1974. Conditioned behavior in *Drosophila melanogaster*. *Proc. Natl. Acad. Sci. USA* 71:708–12
104. Spatz, H. C., Emann, A., Reichert, H. 1973. Associative learning of *Drosophila melanogaster*. *Nature* 248:359–61
105. Dudai, Y., Jan, Y. N., Byers, D., Quinn, W. G., Benzer, S. 1976. *dunce*, a mutant of *Drosophila* deficient in learning. *Proc. Natl. Acad. Sci. USA* 73:1684–88
106. Quinn, W. G., Dudai, Y. 1976. Memory phases in *Drosophila*. *Nature* 262:576–77
107. Hinde, R. A. 1970. Some aspects of learning. In *Animal Behavior: A Synthesis of Ethology and Comparative Psychology*. New York: McGraw-Hill. 2nd ed.
108. Samoiloff, M. R., McNicholl, P., Cheng, R., Balakanich, S. 1973. Regulation of nematode behavior by physical means. *Exp. Parasitol.* 33:253–62
109. Whiting, P. W. 1932. Reproductive reactions of sex mosaics of a parasitic wasp, *Habrobracon juglandis*. *J. Comp. Psychol.* 14:345–63
110. Clark, A., Egen, R. C. 1975. Behavior of gynandromorphs of the wasp *Habrobracon juglandis*. *Dev. Biol.* 45:251–59
111. Spieth, H. T. 1952. Mating behavior within the genus *Drosophila* (Diptera). *Bull. Am. Mus. Nat. Hist.* 99:401–74
112. Manning, A. 1965. *Drosophila* and the evolution of behavior. *Viewpoints Biol.* 4:125–69
113. Ewing, A. W., Manning, A. 1967. The evolution and genetics of insect behavior. *Ann. Rev. Entomol.* 12:471–94
114. Spieth, H. T. 1974. Courtship behavior in *Drosophila*. *Ann. Rev. Entomol.* 19:385–406
115. McBean, I. T., Parsons, P. A. 1967. Directional selection for duration of copulation in *Drosophila Melanogaster*. *Genetics* 56:233–39
116. Hotta, Y., Benzer, S. 1976. Courtship in *Drosophila* mosaics: Sex-specific foci for sequential action patterns. *Proc. Natl. Acad. Sci. USA* 73:4154–58
117. Hall, J. C. 1977. Portions of the central nervous system controlling reproductive behavior in *Drosophila melanogaster*. *Behav. Genet.* In press
118. Kankel, D. R., Hall, J. C. 1976. Fate mapping of nervous system and other internal tissues in genetic mosaics of *Drosophila melanogaster*. *Dev. Biol.* 48:1–24
119. Hall, J. C. 1977. Courtship among males due to a male-sterile mutation in *Drosophila melanogaster*. *Behav. Genet.* In press
120. Schilcher, F., von 1977. A mutation which changes courtship song in *Drosophila melanogaster*. *Behav. Genet.* In press
121. Konopka, R. J., Benzer, S. 1971. Clock mutants of *Drosophila melanogaster*. *Proc. Natl. Acad. Sci. USA* 68:2112–16
122. Konopka, R. J. 1975. Genetic dissection of complex behavior. In *The Molecular Basis of Circadian Rhythms*, ed. J. W. Hastings, H. G. Schweiger. Berlin: Dahlem Konferenzen
123. Wright, T. R. F. 1970. The genetics

of embryogenesis in *Drosophila*. *Adv. Genet.* 15:261–395
124. Grüneberg, H. 1963. *The Pathology of Development, A Study of Inherited Skeletal Disorders in Animals.* New York: Wiley
125. Gehring, W. J. 1976. Developmental genetics of *Drosophila. Ann. Rev. Genet.* 10:209–52
126. Ready, D. F., Hanson, T. E., Benzer, S. 1976. Development of the *Drosophila* retina, a neurocrystalline lattice. *Dev. Biol.* 53:217–40
127. Campos-Ortega, J. A., Gateff, E. A. 1976. The development of ommatidial patterning in metamorphosed eye imaginal discs implants of *Drosophila melanogaster*. *Wilhelm Roux Arch. Dev. Biol.* 179:373–92
128. Hofbauer, A., Campos-Ortega, J. A. 1976. Cell clones and pattern formation: Genetic eye mosaics in *Drosophila melanogaster*. *Wilhelm Roux Arch. Dev. Biol.* 179:275–90
129. Campos-Ortega, J. A., Hofbauer, A. 1977. Cell clones and pattern formation: On the lineage of photoreceptor cells in the compound eye of *Drosophila*. *Wilhelm Roux Arch. Dev. Biol.* 181:277–45
130. Harris, W. A., Stark, W. S. 1977. Hereditary retinal degeneration in *Drosophila melanogaster*, a mutant defect associated with the phototransduction process. *J. Gen. Physiol.* 69:261–91
131. Harris, W. A., Stark, W. S., Walker, J. A. 1976. Genetic dissection of the photoreceptor system in the compound eye of *Drosophila melanogaster*. *J. Physiol.* 256:415–39
132. Meyerowitz, E. M. 1977. *Eye-brain interactions in the development of the Drosophila melanogaster optic lobes*. PhD thesis. Yale Univ., New Haven
133. Poodry, C. A., Hall, L., Suzuki, D. T. 1973. Developmental properties of *Shibire*: A pleiotropic mutation affecting larval and adult locomotion and development. *Dev. Biol.* 32:373–86
134. Sulston, J. E. 1976. Post-embryonic development in the ventral cord of *Caenorhabditis elegans*. *Philos. Trans. R. Soc. London Ser. B* 275:287–97
135. Goodman, C. 1976. Constancy and uniqueness in a large population of small interneurons. *Science* 193:502–4
136. Hudock, G. A., Hudock, M. O. 1973. Phototaxis: Isolation of mutant strains of *Chlamydomonas reinhardi* with reversed sign of response. *J. Protozool.* 20:139–40
137. Loomis, W. F. 1975. *Dictyostelium discoideum, a Developmental System.* New York: Academic. 214 pp.
138. Delbrück, M., Katzir, A., Presti, D. 1976. Responses of *Phycomyces* indicating optical excitation of the lowest triplet state of riboflavin. *Proc. Natl. Acad. Sci. USA* 73:1969–73
139. Cohen, R. J., Jan, Y. N., Matricon, J., Delbrück, M. 1975. Avoidance response, house response and wind responses of the sporangiophore of *Phycomyces*. *J. Gen. Physiol.* 66:67–95
140. Malacinski, G. M., Brothers, A. J. 1974. Mutant genes in the Mexican axolotl. *Science* 184:1142–47
141. Ide, C. F., Kimmel, C. B., Tompkins, R., Elbert, O., Schabtach, E., Miszkowski, N., Duda, M. 1977. Analysis of *Spastic*, a neurological mutant of the Mexican axolotl. *ICN, UCLA Symp. Neurobiol.* In press
142. Vigue, C., Sofer, W. 1974. Adh^{ns}: A temperature-sensitive mutant at the *Adh* locus in *Drosophila*. *Biochem. Genet.* 11:387–96
143. Suzuki, D. T. 1970. Temperature sensitive mutations in *Drosophila melanogaster*. *Science* 170:645–706
144. Hirsh, D., Vanderslice, R. 1976. Temperature-sensitive developmental mutants of *Caenorhabditis elegans*. *Dev. Biol.* 49:220–35
145. Mullen, R. J. 1977. Genetic dissection of the CNS with mutant-normal mouse and rat chimeras. *Soc. Neurosci. Res. Symp.* II:47–65
146. Seecof, R. L., Teplitz, R. L., Gerson, I., Ikeda, K., Donady, J. J. 1972. Differentiation of neuromuscular junctions in cultures of embryonic *Drosophila* cells. *Proc. Natl. Acad. Sci. USA* 69:566–70
147. Mayr, E. 1974. *Populations, Species and Evolution.* Cambridge, Mass.: Harvard Univ. Press

ILLEGITIMATE RECOMBINATION ✦3129
IN BACTERIA AND BACTERIOPHAGE

Robert A. Weisberg[1]

Laboratory of Molecular Genetics, National Institute of Child Health and Human Development, National Institutes of Health, Bethesda, Maryland 20014

Sankar Adhya

Laboratory of Molecular Biology, National Cancer Institute, National Institutes of Health, Bethesda, Maryland 20014

CONTENTS

INTRODUCTION	452
DELETIONS	452
The Influence of Base Sequence	453
Deletion analysis of the gal operon of E. coli	453
Deletions in the ara operon of E. coli	454
Deletion analysis of the lacI gene of E. coli	455
Genetic Control of Deletion Formation	457
The role of bacterial genes	457
The role of phage λ genes	457
Site-Specific Deletions	458
Int dependent deletions	458
IS-associated deletions	458
SPECIALIZED TRANSDUCING PHAGES	460
Characteristics of Transducing Phage Formation in Phage λ	461
λ Transducing Particles That Cannot Be Propagated	462
DUPLICATION MUTANTS	465
Tandem Direct Repeats	465
Inverted Repeats in Phage λ	468
DISCUSSION	469

[1]Mailing address: Building 6, Room 416, National Institutes of Health, Bethesda, Maryland 20014.

INTRODUCTION

The term *illegitimate recombination* (1, 2) is used to describe certain chromosomal rearrangements. These rearrangements, which include deletion, duplication, and specialized transducing phage formation, fuse DNA segments that were not originally adjacent. The point of fusion is called a *novel joint* (3). Illegitimate recombination differs from ordinary recombination in its independence of normal recombination functions—such as *recA* in *Escherichia coli*—and of extensive base-sequence homology between the participating chromosome segments. Although illegitimate recombination has probably played a significant role in chromosomal evolution (see 4), it is one of the least understood areas of genetics. In this review we emphasize the significant advances that have occurred since the field was last reviewed in detail (2). We do not exhaustively survey the literature, and our citations of new examples of transducing phages, deletions, duplications, and the techniques devised to isolate them are therefore incomplete. *recA*-independent recombinations that are part of the life cycle of such phages as λ and Mu are not covered (see 5–7).

DELETIONS

Deletion (or multisite) mutants in procaryotes were originally defined by two properties: They fail to revert to wild type and they give no wild-type recombinants in pairwise crosses with more than one nonidentical point mutant (8). When a set of deletion mutants is crossed in pairwise combinations against a set of point mutants, and each cross is scored for the presence or absence of wild-type recombinants, the results are consistent with a linear ordering of deletion end points and point mutations (8, 9). This suggests that deletions arise by excision of a contiguous segment of DNA from the wild-type chromosome, a suggestion that was elegantly confirmed in phage λ by electron microscopic analysis of heteroduplexes between wild-type and deleted DNA (10, 11).

Deletions are found in low frequency among mutants of spontaneous origin. Their absolute frequency can be increased by treatment of the cells with certain mutagens (Table 1). Ultraviolet irradiation and nitrous acid treatment, both of which cause base-substitution mutations, also increase the frequency of deletions. These agents probably act by cross-linking nucleotides: Ultraviolet irradiation cross-links adjacent thymine residues on a polynucleotide chain (15), and nitrous acid treatment cross-links bases of complementary strands (16). These internucleotide bonds may prevent or delay the progress of replication forks over the DNA, and deletions may be the result of an attempt by the replication machinery to bypass the barrier (17). Alternatively, damaged DNA may be more susceptible to breakage (perhaps by recombination or repair enzymes), and deletions may occur by exchange and fusion of the broken ends created by two DNA scissions (see Discussion). In contrast, the mutagen 2-amino purine, which causes certain base substitutions, does not cause deletions. This suggests that deletion and base-substitution mutations are formed by different pathways.

Table 1 Effect of mutagens on deletion formation

Mutagen	Increase in deletion frequency	Genes studied	Reference
X irradiation	30-fold	lac operon of E. coli	12
Ultraviolet irradiation	slight increase	cysC locus of S. typhimurium	13
Ultraviolet irradiation	500-fold	lac operon of E. coli	12
Nitrous acid	20-fold	rII region of phage T4	14
Nitrous acid	200-fold	lac operon of E. coli	12
2,Aminopurine	none	cysC locus of S. typhimurium	13

The Influence of Base Sequence

Does the local base sequence play a direct role in determining the frequency or type of deletion mutations? That base sequence around the region of the novel joint might play a role was suggested by the following two observations: (a) the cysC-region of Salmonella typhimurium is more prone to deletion than certain other regions (13), and (b) the ancestry of the tonB region of E. coli influences the frequency of tonB-trp deletions (18). If the local base sequence does influence the frequency of deletions, it might well influence the precise location of the novel joint as well. For example, certain regions may be "hot spots" for deletion end points. To test this hypothesis it is necessary to isolate and to characterize large numbers of deletions in several regions of the chromosome. The ideal study should emphasize isolation of deletion mutants of independent origin, determination of the termini by crosses with point mutants, and determination of the distribution of end points with respect to the molecular length. Fortunately, several powerful techniques are available to isolate easily large numbers of deletions in many regions of bacterial and phage chromosomes.

DELETION ANALYSIS OF THE *gal* OPERON OF *E. COLI* Deletions of *gal* can be easily obtained by selection of temperature-resistant variants in a culture of a heat-inducible λ lysogen (19, 20; see section on the role of phage λ genes) or by selection for inactivation of the nearby *chl* markers (21; see Figure 1). Pfeifer et al (22) used the former technique to isolate 94 deletions whose left end points fell within the *galT* or *galK* genes. All of these mutants were also missing all of the genes from *gal* up to and including *uvrB*. Crosses of the deletions with *gal* point mutants with known locations on the molecular map showed that two regions, which together constituted 48% of the *galT-galK* interval, contained 93% of all the deletions. The deletions whose end points were within each of these two

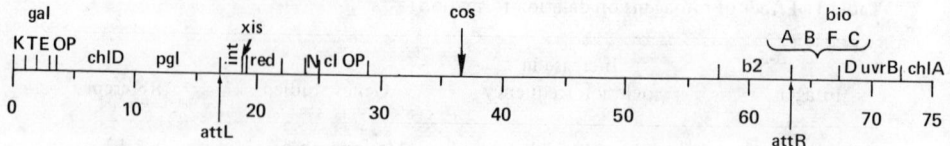

Figure 1 A map of prophage λ and neighboring genes of the *E. coli* chromosome. The numbers at the bottom of the line specify the distance in kilobase pairs from the left end of the *gal* operon. The *upper vertical lines* indicate the approximate limits of the indicated genes. The prophage λ genome extends from *attL* to *attR*. The *gal* operon, whose enzymes are required for galactose utilization, has three structural genes, *K, T,* and *E,* with the operator-promotor (*OP*) located at the *E* end. *chlD* and *chlA* determine sensitivity to $KClO_3$ under anaerobiosis. *pgl* encodes 6-phosphogluconolactonase. The *bio* operon, which encodes enzymes required for biotin biosynthesis, has five cistrons, A, B, F, C, and D. *uvrB* determines sensitivity to ultraviolet light. The prophage genes shown are discussed in the text. The genetic and physical data are taken from references (21, 57, 82-85).

"hot" regions could be further subdivided by genetic crosses into 9 groups of unknown molecular size. These results show that there are several preferential end points for deletions within the *gal* operon. However, a second group (19, 21; S. Adhya, unpublished results) analyzed 47 partial *gal* deletions that were obtained by both selection techniques, and found a rather different distribution. In this collection only 2 deletions extended to or beyond *uvrB*. By analysis of phenotypes and by genetic crosses, 47 different left end points and 47 different right end points were identified. However, these data do not show that the end points can be located with equal probability at any point on the chromosome, because the sample size relative to the size of the interval analyzed was small and because the locations of the end points on the molecular map were not determined.

How can we account for the difference between these two collections of deletions? Pfeifer et al (22) have shown that the selection method that they employed very likely induced the formation of deletions, and it is possible that a λ gene product was the inducer. The selection methods used by the second group isolated only preexisting deletions from the population (see section on the role of phage λ genes for further discussion of this point). If so, it is interesting that different pathways of deletion formation can lead to a different spatial distribution of novel joints.

DELETIONS IN THE *ara* OPERON OF *E. COLI* Schleif (23) determined the distribution of end points of 350 deletions among 68 nonidentical point mutations in a 3000 base-pair segment of the *ara* operon. The deletions were isolated by selecting temperature-resistant derivatives of a lysogen of a heat-inducible λ prophage inserted near *ara*. A fluctuation test showed that these deletions, unlike those isolated by Pfeifer et al (22), were not induced by the selection technique. The observed distribution of deletion end points and point mutations was compared to a frequency function that was derived by assuming that each is equally likely to occur at any point in the interval. Agreement between the observed and calculated distributions was good except for two intervals that contained more deletion end points (or fewer point mutations) than predicted. However, this agreement may be fortuitous: The distribution of point mutant sites along the chromosome is affected by such factors as the mutagen used (24, 25) and the size of the collection, and it is difficult to assess the magnitude of these effects.

In a subsequent study, Schleif & Lis (26) mapped 29 deletion end points within a 1000 base-pair region of the *ara* operon. In contrast to the first study, these deletions were isolated from stocks of λ *ara* specialized transducing phage lines, and in many cases their locations on the molecular map were determined by electron microscopy of heteroduplexes or electrophoresis of DNA fragments. The agreement between the observations and the results predicted from a random distribution is good although the small numbers do not allow an adequate test of the hypothesis.

DELETION ANALYSIS OF THE *lacI* GENE OF *E. COLI*   By far the most extensive analysis of the distribution of deletion end points in an interval has been reported by Schmeissner et al (27), who characterized 400 independently isolated spontaneous deletions that have an end point within the 1050 nucleotide pairs of the *lacI* gene. By crossing the deletions with *lacI* point mutants, a map of more than 100 intervals was constructed. Since many of the point mutations had been assigned to specific amino acides in the polypeptide product, the *lac* repressor, many intervals on the genetic map could be associated with intervals on the molecular map. The authors used this information to divide the map into 29 intervals of approximately 30 nucleotide pairs each and to determine the distribution of deletion end points among intervals. The observed distribution is approximately the same as a Poisson distribution except that two segments, whose length was confirmed by direct DNA sequencing (W. Gilbert and J. Miller, personal communication) had about twice as many deletion end points than predicted by chance. No empty segments were found.

The investigations described above show that deletion end points can occur with approximately equal likelihood at a great many loci on the *E. coli* chromosome. Indeed, it is possible that every phosphodiester bond is a potential site for a novel joint. Nevertheless, in most cases the locations of both of the deletion termini were not determined, and it therefore remains possible that there is a correlation between these locations. Such a correlation would be expected if, for example, deletions are formed by intramolecular or unequal crossing-over within directly repeated nontandem sequences as depicted in Figure 2, and such a correlation could explain the deletion hot spots found in *gal* and *lacI* (see below). However, the observations place constraints on such a model. As none of the deletions cited above is likely to exceed 100,000 nucleotide pairs in length (in the case of the deletions of λ *ara*, the upper limit is about 15,000 nucleotide pairs), the minimum required repeat size must almost certainly be quite short to account for the large number of different deletions observed. (We note that if the 4 nucleotides are arranged randomly, any particular sequence of length 7 will be found once in about 16,000 nucleotide pairs on the average, and of length 8, once in about 64,000. Nonrandomness will increase these frequencies.) In addition, we can cite one example of a 15 base-pair direct repeat

(A)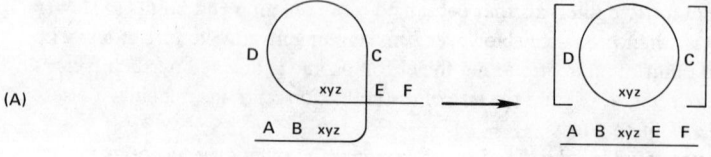

(B)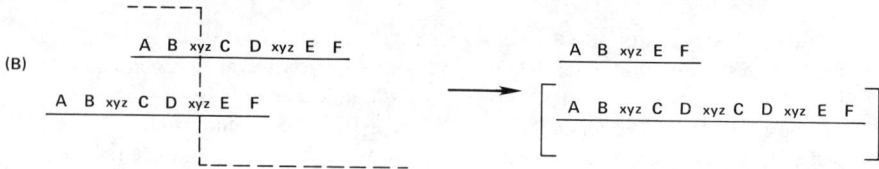

Figure 2 Models for formation of deletions. Intramolecular (*A*) or intermolecular recombination (*B*) between nontandem repeated sequences *xyz* deletes the intervening material. The structures within parentheses are hypothetical by-products of deletion formation.

that is not a hot spot for deletion formation. The attachment sites that bracket an inserted λ prophage share a repeat of this size (27a), but in the absence of the normal λ excision functions a substantial fraction if not all prophage deletions terminate outside the attachment sites [see, for example, reference (20)]. However, the attachment site sequence may be unusual in that it is extremely rich in A-T pairs.

Recently, P. J. Farabaugh, U. Schmeissner, M. Hofer, and J. H. Miller (personal communication) have determined the nucleotide sequence of a number of independently isolated spontaneous deletions of the *lacI* gene. These deletions, unlike those analyzed previously (27), were entirely within *lacI,* and therefore both end points could be located in the sequence. They found that 7 of 12 deletions occurred between nontandem directly repeated sequences of either 5 or 8 base-pairs in length; the end points of the remaining 5 deletions were in nonrepeated sequences. Among the 7 were 3 pairs of identical deletions. This work clearly shows that directly repeated sequences are preferred deletion end points and supports the mechanism of deletion formation depicted in Figure 2. It also suggests that some spontaneous deletions are not formed by this pathway.

In previous work, both Schmeissner et al (27) and Schleif (23) found intervals that contained significantly more than the expected number of deletion end points. Although it is not known how many of the deletions with end points in these preferred intervals are identical, it is possible that many arose by crossing-over within especially long—and therefore rare—directly repeated sequences. Deletions with end points in other intervals may have arisen either by crossing-over within shorter—and therefore more frequent—repeated sequences or by another pathway not involving repeats.

Genetic Control of Deletion Formation

THE ROLE OF BACTERIAL GENES There is no report of an extensive search for genes that influence deletion formation and, with one possible exception (see below), no mutant has been reported in which deletion formation is defective. If deletion formation follows a unique enzymatic pathway, the enzymes must either be normally repressed or very inefficient. Alternatively, illegitimate recombination might be the result of aberrant behavior of some other DNA-enzyme systems(s), e.g. replication, repair, or general recombination. *E. coli* mutants defective in the latter process have been listed by Clark (28) who proposes that recombination follows either the *recF* or *recBC* pathways, depending on the genotype of the cell. The product of the *recA* gene is essential for both pathways. Mutation in *recA*, *recB*, or *recC* has little effect on the formation of deletions (29–32). The effect of a mutation in *recF* has not been tested. Since *recA* mutants are extremely defective in homologous recombination, the pathways for deletion formation and homologous recombination are different, although, as mentioned above, it is possible that short runs of identical base sequences are important for the former process.

A number of genes involved in DNA metabolism—*uvrA*, *uvrB*, *uvrC*, *lig*, *rep*, *ras*, and *end*—play no role in creating deletions [see reference (2)]. Of three genes that encode DNA polymerases—*polA*, *polB*, and *polC*—only *polA* mutants have been tested for their effect on deletion formation. In fact, mutation in *polA* increases the frequency of deletions about 30-fold in the *tonB-trp* region of *E. coli* (18) and in nonessential regions of phage λ (N. Sternberg, personal communication). Additionally, the *polA* mutations induce deletion formation without changing grossly the positions of the end points compared to those generated in the *polA*$^+$ strain (18). The mechanism of this effect is unknown.

THE ROLE OF PHAGE λ GENES We mentioned earlier that one can isolate deletions of prophage λ and of neighboring regions of the host chromosome by looking among temperature-resistant survivors after induction of a heat-inducible lysogen (19). This method selects directly for inactivation or loss of phage genes that kill the host when expressed. Two groups (22, 33) found that the selection technique itself could induce deletions and that the end points of these deletions were nonrandomly distributed. However, two other groups found that in their conditions λ induction selected for preexisting deletions, as judged by a Luria-Delbruck fluctuation test (19, 23). The reason for this difference is unknown. Several λ mutants have been examined to see whether they affect the formation of host deletions. It has been found that deletions are indeed formed from λ lysogens carrying mutations in the genes or regions involved in homologous recombination, prophage integration and excision, and replication or their control (20, 33, 33a; S. Adhya, unpublished observations). These include the *int*, *xis*, *exo*, *bet*, *gam*, *N*, *O*, and *P* genes, the *b* region, and the prophage attachment sites. Whether the deletions that were isolated from these mutant lysogens were formed before or after prophage induction was not determined.

Site-Specific Deletions

We have already discussed several studies that showed that deletion termini are for the most part randomly distributed along the chromosome, with an occasional "hot" interval. We briefly consider several cases in which deletions terminate at a specific site. For a more thorough treatment of this subject we recommend several recent reviews (7, 34).

INT DEPENDENT DELETIONS One example of site-specific deletion formation has been reported in phage λ in which many deletions terminate at the phage attachment site (35). There is a tendency for at least one of them, $b2$, to be preferred; it has apparently been independently isolated several times (35, 35a). Davis & Parkinson (35) showed that the formation of these deletions depends on the function of the phage *int* gene. The normal function of this gene is to promote insertion of the phage chromosome into the host chromosome by reciprocal recombination at the attachment site (*att*) [see (5, 7) and section on specialized transducing phages]. Davis & Parkinson (35) proposed two mechanisms of deletion formation. In the first, *int* product nicks the chromosome at *att*, exonucleolytic digestion of the nicked strand begins at the nick and proceeds for a variable distance, and the ends of the digested strand are rejoined. In the second, *int* inefficiently promotes unequal crossing-over between *att* and secondary sites in the phage chromosome to the left or right. The second model is supported by the observation of Shimada et al (36) that *int* can inefficiently promote reciprocal recombination between *att* and secondary sites in the bacterial chromosome that are different from the normal site of prophage insertion near *gal*.

IS-ASSOCIATED DELETIONS A similar system of site-specific deletion formation has been described in both *E. coli* and *S. typhimurium* (see 37). In this case the sites are the ends of several types of DNA elements called *IS* (insertion sequence), which are specific sequences of nucleotides ranging in length from about 800 to 1500 base-pairs, and which occur in multiple copies as natural components of the bacterial chromosome. Three types of *IS* element, *IS*1, *IS*2, and *IS*3, have been studied in considerable detail. Their defining characteristic is their ability to translocate into other parts of the chromosome where they may cause a recognizable phenotype, e.g. inactivation of a gene. In fact, the *IS* elements were first recognized by their translocation into and inactivation of the *E. coli gal* operon (38, 39). *IS* elements are associated with the formation of deletions. Deletion formation in a given gene can be enhanced 10- to 1000-fold if an *IS* element is located nearby (40–42). Enhancement does not require *recA* activity. One end point of these deletions is usually the site at which the *IS* element is inserted while the other terminus is variable and can be to either side of the *IS* site. The location of the variable terminus is not random, but, in the case of *IS*1 at least, appears to be influenced by the location of the *IS* element (40). Examination of DNA heteroduplexes has shown that the fixed end point of *IS*2 associated deletions is located at or close to the junction point of the IS sequence with the surrounding DNA, and that deletion probably does not remove any *IS* DNA (41). A mutant has been isolated that reduces the formation of *IS*1-associated deletions in the *gal* operon, but does not affect general homologous recombination (43).

The similarity between the *int*-dependent deletions of phage λ and the *IS*-associated deletions is the basis of a model to explain the latter (40). In this model, an *IS*-specified site-specific recombination enzyme promotes recombination between one of the sites that lie at the termini of the inserted *IS* element and one of a number of secondary sites to the right or left (Figure 3). Although there is as yet no direct evidence in favor of this model, it is consistent with the apparent nonrandom distribution of variable end points, which presumably reflects the distribution of secondary sites around a particular *IS* element. Precise excision of the *IS* element, which is detected by the restoration of the function of the gene disrupted by the insertion, is assumed to occur by recombination between the terminal sites. It is, for unknown reasons, much rarer than deletion in the case of *IS*1 (40). However, the model predicts the existence of one type of deletion that has not yet been found: Few if any *IS*1-associated deletions have lost the *IS* element (44; see Figure 3).

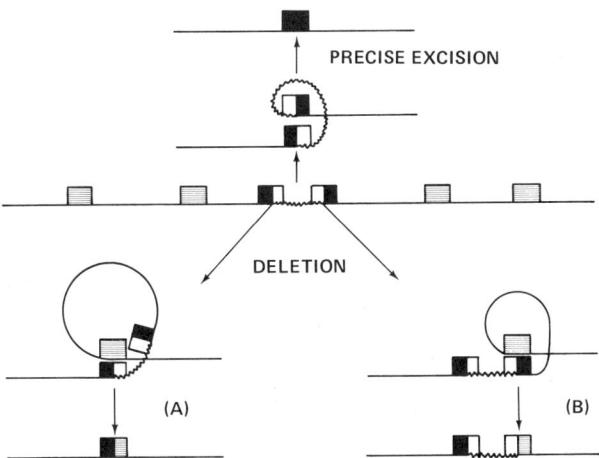

Figure 3 A model for the formation of *IS*-associated deletions [after reference (40)]. The *straight line* represents bacterial DNA, and the *wavy line* indicates the *IS* elements. The boxes at the *IS*-bacterial DNA boundaries represent the two terminal sites; recombination between them leads to precise excision of the *IS* element (*above*). The crosshatched boxes designate secondary sites. Recombination between a terminal and a secondary site in the orientations shown deletes the intervening DNA (*below*). Deletion type B has been observed; deletion type A, however, has not (see text).

SPECIALIZED TRANSDUCING PHAGES

Specialized transducing phage are phage variants that have stably incorporated host genes into the viral genome. In the case of phage λ, they are usually formed by the abnormal excision of the prophage from the bacterial chromosome (Figure 4) (1, 45). Abnormal excision differs from normal excision in its independence of the phage *int* and *xis* functions (46, 47; H. Echols, personal communication) and the prophage attachment sites (21, 46; see below). Both types of excision occur in the absence of the general recombination functions encoded by the *recA* gene of the host and the *red* genes of the phage (46, 47).

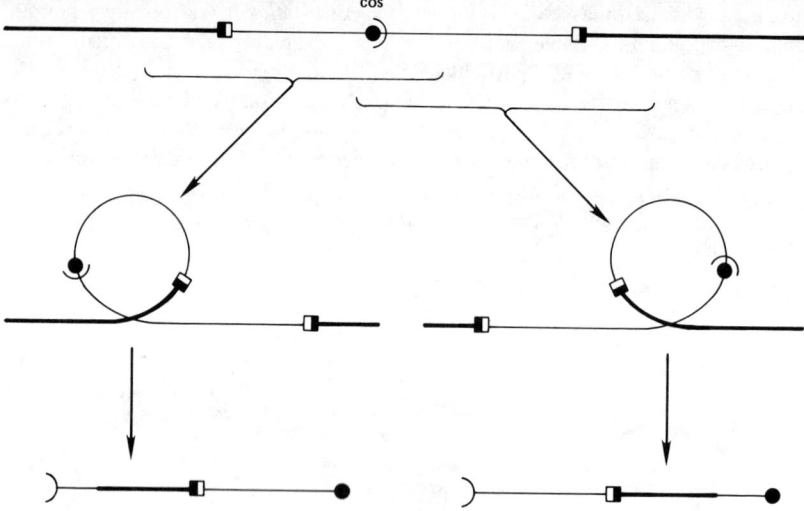

Figure 4 The origin of transducing phage lines (1). The top part of the figure shows a prophage inserted into the bacterial chromosome. The *heavy line* represents bacterial DNA; the *light line*, phage DNA. The two *squares* represent the prophage attachment sites; normal prophage excision occurs by recombination within these sites [see reference (7)]. The region represented by the symbol " —●— ", called *cos*, is cut during intracellular phage development to form the left and right ends of the mature virus chromosome (symbolized ")—" and "—●", respectively; see 45). Both ends are found in all transducing phage lines that have been analyzed (48); they thus appear to be essential structures for autonomous phage growth. Transducing phage chromosomes are thought to be formed by rare recombinations between points within the prophage and points on the bacterial chromosome to the left or right. Two possible recombinations are shown in the middle part of the figure; the crossovers are assumed to occur at the bottom of the circular loops in each case. The resulting transducing phage chromosomes (*lower left* and *right*) have been opened at *cos* and are therefore shown as they occur in mature virus particles. For a normal prophage, only genes that are within about 30 kilobase pairs of the prophage can be incorporated into a specialized transducing phage line (87). However, a number of methods have been devised to enlarge the spectrum of genes that can be incorporated (36, 69, 87–92).

Specialized transducing phage lines of independent origin usually differ in gene content. If all of the phage genes required for lytic growth are present, the phage can form plaques; otherwise it can be propagated only in the presence of a coinfecting helper phage to supply the missing functions or as a prophage. The ability of specialized tranducing phage to form genetically stable lines distinguishes them from generalized transducing phage whose DNA can neither persist as prophage nor be efficiently repackaged into progeny virions.

Upon induction of a wild-type lysogen, the yield of phage with abnormally excised chromosomes is less than the yield of normal phage by many orders of magnitude: less than 10^{-3} of the former as compared to about 100 of the latter per cell. This difference in yield probably reflects the extremely low efficiency of abnormal as compared to normal prophage excision, although secondary factors such as altered packageability of the DNA or disruption of gene expression resulting from chromosome breakage may also contribute.

Characteristics of Transducing Phage Formation in Phage λ

1. The abnormally excised fragment must contain *cos*, the site that is cut to give the cohesive ends of the wild-type λ chromosome (Figure 4). Fragments without *cos*, if they are produced, are not detected probably because they cannot be packaged into λ virions (48, 49). For the same reason, the fragment must not be smaller than about 0.73 or larger than about 1.09 λ DNA lengths (50–52).

2. The ends of the abnormally excised fragment must be fused. When such fusion does not occur, the chromosomes of the resulting particles cannot insert into the chromosome of a recipient cell (47; R. Weisberg, unpublished observations). Hence they cannot be propagated as prophage. These particles can, however, be detected by their ability to donate genetic markers to appropriate recipients via general recombination, and they may be intermediates or by-products of the process that creates genuine transducing phage chromosomes (see section on λ transducing particles that cannot be propagated).

3. Campbell (53) has suggested that most specialized transducing phage chromosomes are formed at or after lysogenic induction.[2] S. Adhya (unpublished) has recently confirmed this suggestion by measuring the production of λ *gal* transducing particles in cells that had received a λ prophage by mating. In such cells, induction occurs upon transfer of the prophage (zygotic induction; 54). If abnormal excision occurred before mating, the newly created specialized transducing phage chromosome should not be transferred into the recipient because only DNA that is linked to a sex factor is transferred during mating. In fact, the production of specialized transducing phage after zygotic induction was the same as after thermal induction. Thus most, if not all, abnormal excision occurs after induction. This suggests that the process is controlled by a phage gene(s). However, with one possible exception (see below) there have been no reports of phage mutations that specifically alter the production of specialized transducing phage. In particular, gene *N*, whose action is required for the full expression of most other λ genes, is not required for abnormal

[2]By *lysogenic induction* we mean inactivation of prophage repressor (see 5).

excision (55). If, as seems likely, a phage gene does promote abnormal excision, it might also be required for lytic growth. For example, autonomous phage DNA replication might increase the probability of abnormal excision provided that such replication occurred before normal excision. In fact, recent work (N. Sternberg and R. Weisberg, unpublished experiments) suggests that after induction several rounds of DNA replication can initiate in the unexcised prophage and proceed outward into the bacterial chromosome. This observation might also explain why mutations that prevent normal prophage excision, such as N^-, int^-, xis^-, and att^Δ, do not markedly increase the production of specialized transducing phage (see section on λ transducing particles that cannot be propagated).

4. A good deal of evidence, which is summarized in reference (2), shows that the end points of abnormally excised DNA occur at many locations in the prophage and neighboring bacterial chromosome.[3] More recently Enquist & Weisberg (56; unpublished results) have shown that excision end points occur at many loci within a gene: 21 transducing phage lines with one end point in the phage int gene could be divided into 12 classes by crosses with 38 different int point mutants and further subdivided into 17 classes by determination of the bacterial end points. Although these results give no suggestion of preferred end points, the physical locations of the mutations were not determined, and therefore the hypothesis of a random distribution could not be tested. However, Pfeifer et al (22) have characterized 135 λ gal transducing phage lines that had their bacterial end points within the $galT$ or $galK$ genes. Crosses of these transducing phage with gal point mutants whose locations on the molecular map were known subdivided the end points into seven intervals of known length. Two of these intervals had significantly more and one significantly less than the number of end points predicted by a random distribution. However, the deviation from expectation was much smaller for the transducing phages than for the bacterial deletions mapped by the same authors (see section on deletion analysis of the gal operon of $E. coli$), and the preferred intervals did not completely overlap.

We conclude that there is weak but suggestive evidence in favor of preferred end points for abnormal phophage excision. The molecular basis of this preference is unknown, and it is clear that it is far from absolute. In addition, we stress that, as in the case of deletion, abnormal prophage excision may occur by more than one pathway, and some pathways may be more influenced by nucleotide sequence than others.

λ Transducing Particles That Cannot Be Propagated

After induction of a λ lysogen, transducing particles that carry bacterial genes adjacent to the inserted prophage can be formed without fusion of the ends of the abnormally excised fragment or even without abnormal excision. The DNA of these

---

[3]One apparent exception is the excision of the chromosome of λ att^2, a transducing phage line that can be isolated repeatedly after lysogenic induction (46, 57). In this case, however, excision requires the rec or red recombination pathways (46) and is therefore not an example of illegitimate recombination.

particles cannot be propagated to form genetically stable lines, and hence they differ from standard specialized transducing phage particles. Some workers [see references (2, 61)] have considered the possibility that they may be by-products or intermediates in the formation of standard specialized transducing phage lines. However, recent work does not support this view (see below).

Two types of nonpropagable λ transducing particles have been characterized. The first, λ *docL*, has DNA with the normal left cohesive end of λ; the second, λ *docR*, has DNA with the right cohesive end (58) (Figure 5). We therefore infer that one of the DNA ends of each type of particle is produced by the site-specific nuclease that acts at *cos* to produce the cohesive ends of wild-type λ DNA (46, 49, 58). The bacterial ends of the λ *docL* and λ *docR* chromosomes are not λ cohesive ends. Hence λ *docL* and λ *docR* DNA, unlike that of wild-type λ and of standard specialized transducing phage lines, cannot circularize by end-joining and therefore cannot form lysogens (58). The formation of the bacterial end of λ *docL* DNA is well understood: It is generated after cell lysis by the action of exogenously added DNase on DNA that is protruding from the phage head (59). The formation of the bacterial end (by convention the left end) of λ *docR* DNA is not well understood. It is generated intracellularly, probably by the action of a nuclease with little or no site specificity. The production of λ *docR* particles is greatly stimulated by the *absence* of a number of phage functions. We suppose that these functions act by preventing the formation or decreasing the stability of the left end of λ *docR* DNA, since this seems to be the only unique feature of λ *docR* particles. A number of workers (46, 47, 58; S. Adhya and R. A. Weisberg, unpublished experiments) have examined the effect of several phage and host mutations on λ *docR* production. Mutations in the *E. coli recA* gene and multiple mutations in *recA, recB,* and *recC* have little or no effect. In contrast, mutations in several phage genes have strong effects. However, as we shall see, consideration of the known functions of these genes does not explain the results. In Table 2, we present our own observations of the effects of *int, red, xis,* and *att* mutations on the production of λ *docR* particles and λ *gal* specialized transducing phage lines. Mutations in the *int* gene stimulated λ

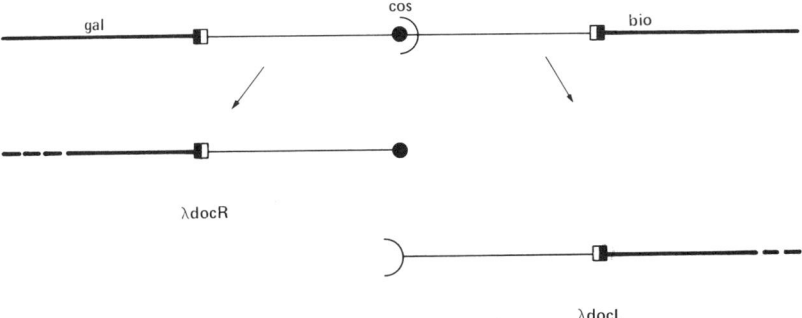

Figure 5 The structure of the λ *docL* and λ *docR* chromosomes and their relationship to an inserted λ prophage. The conventions are the same as in Figure 4.

Table 2 Effect of prophage mutations on λ *docR* and λ *gal* production[a]

Prophage mutations	λ *docR* production	λ *gal* production
*red*1, *red*3, *red*113	5- to 50-fold increase	2- to 5-fold decrease
*int*6, *int*29, *int-xis*2266	2- to 50-fold increase	no consistent change
*xis*1	no change	no change
*b*2	5- to 10-fold increase	no change
*int*6 *red*3, *int*29 *red*3, *xis*1 *red*3, *b*2 *red*3	100- to 1,000-fold increase	no consistent change

[a] The effect of prophage mutations on λ *docR* and λ *gal* production. [The techniques used are described in references (36, 87).] The numbers in the table are transducing phage production after thermal induction relative to a λ *cIts*857 lysogen of the same host. Three different hosts were used: two were an isogenic $recA^-/recA^+$ pair, and the third was a different $recA^-$ strain. There were no significant differences among the three.

All lysogens carried only one prophage. The production of λ *gal* ranged from 0.2 to 1.3 × 10^{-5} per cell and of λ *docR* from 0.5 to 2.2 × 10^{-5} per cell in lysogens carrying a wild-type prophage. To compensate for this variability, all comparisons were made between lysates harvested and transduction performed on the same days. λ *docR* and λ *gal* particles were both detected by their ability to transduce a *galK* recipient strain to *gal*$^+$ in the presence of wild-type λ helper phage. They were distinguished by the ability of most λ *gal* but no λ *docR* transductants to produce high yields of *gal* transducing phage upon induction. In cases where the yield of λ *docR* considerably exceeded that of λ *gal*, the accuracy of the measurement of λ *gal* production could be much improved by using a *recA galK* host and λ *red*$^-$ helper for transduction. This drastically and selectively reduced the number of transductants arising from λ *docR* infection (R. Weisberg, unpublished experiments).

docR production 2- to 50-fold, whereas the sole *xis* mutation tested had no effect. As all of these mutations cause a defect in normal prophage excision, it is difficult to understand this difference. One mutation, *int-xis*2266, is a point mutation or small deletion that inactivates both *int* and *xis* (56). It behaved like the other *int* mutants in this test. None of the mutations that block normal prophage excision appeared to change the production of λ *gal* specialized transducing phage particles.

Some mutations in the *red* genes increased λ *docR* production. Since not all *red* mutations had this effect (47), it may be due to an aberrant activity of specific missense proteins, to a low concentration of normally active protein, or to the pleiotropy of certain mutations. In particular, the two mutations with the highest rate of λ *docR* production [*red*3 (46) and *red*114 (47)] had low concentrations of both λ exonuclease and β protein, the two products involved in *red*-promoted recombination (47, 48, 60). Several *red* mutations that we tested (Table 2) decreased λ *gal* production. This suggests that *red* function might convert a λ *docR* precursor to a λ *gal* precursor. Kayajanian (61) has proposed a specific model based on this assumption. As many points of the model remain untested, we do not present it here. We note that combining one *red* mutation (*red*3) with any of the mutations that block normal prophage excision stimulated λ *docR* production more than either of the single mutations did, suggesting that they exert their effects in different ways.

In conclusion, the formation of λ *docL* particles appears to occur by a route that is entirely independent of illegitimate recombination. In contrast, the formation of λ *docR* particles involves an event—the generation of the left chromosome end—that may be related to illegitimate recombination. The evidence, however, is meager and does not yet allow us to present a readily testable model.

DUPLICATION MUTANTS

Duplication mutants contain a repeat of part of the parental genome. In this section we consider two kinds of duplication: direct repeats, in which the duplicated sequence has the same informational polarity as the original, and inverted repeats, in which it does not. Selection schemes have been devised to facilitate the isolation of duplication mutants. For example, selection for increased concentration of a gene product (C. Hill, personal communication), for the presence of two alleles of a given gene (62), or for stable complementation between mutants that cannot recombine (63–64a) allows easy isolation of certain gene duplications. In phage λ it is possible to select directly for increased DNA content by density gradient centrifugation of virus particles or by ability of the phage to grow in certain bacterial strains (65). In addition, certain deletion mutants appear to be at a disadvantage in mixed growth with phages containing larger chromosomes (51, 66). These selection principles may give a clue as to the possible role that duplications play in chromosomal evolution.

Duplication mutants with direct repeats are usually recognized by their genetic instability (67, 68). The duplicated segment can be lost by homologous *recA*- or *red*-promoted recombination within the duplicated region (66, 69; Figure 6). If this occurs by intramolecular recombination (Figure 6*A*), the products would be an unduplicated chromosome and a DNA circle. In fact, such DNA circles have been observed in duplication mutants of *E. coli* (70). Duplications of phage λ can be directly visualized by electron microscopic examination of heteroduplexes between duplicated and nonduplicated chromosomes (66, 71, 72). The duplicated DNA appears as a single-stranded loop whose branch point has a variable location (Figure 7).

Tandem Direct Repeats

The tandem direct repeat is the simplest type of DNA duplication because it contains only one novel joint. For example, in the following tandem direct repeat of an alphabetical sequence, the novel joint is between the letters "E" and "C": ABCDECDEFG. Duplications of this type have been studied most extensively in phage λ in which they can be detected as addition mutants, that is, mutants that contain more DNA than the parent. Such mutants can be easily isolated when the parental phage carries a DNA deletion. Baldwin and his collaborators (65, 66, 72,

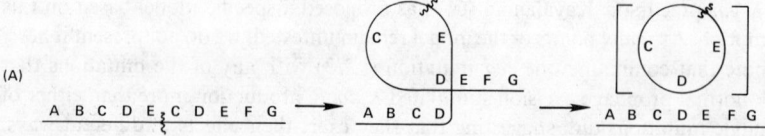

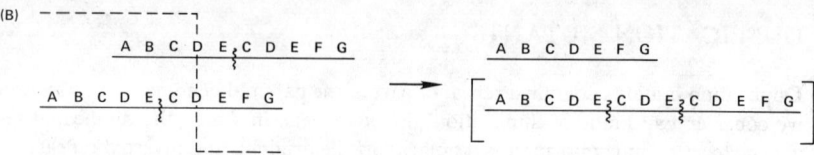

Figure 6 Loss of a tandem direct repeat by (*A*) intramolecular or (*B*) intermolecular homologous recombination. The structures in the parentheses are by-products of the segregation events.

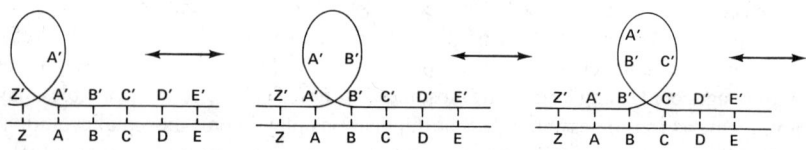

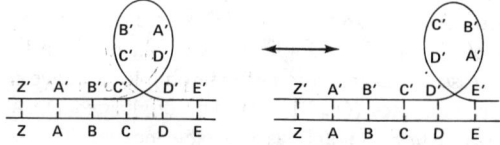

Figure 7 Diagram of loop movement in a heteroduplex between the DNAs of a duplication mutant and the unduplicated parent [after reference (71)]. The two *horizontal lines* represent complementary DNA strands, the *short vertical lines*, H bonds between complementary basepairs, and the *loop*, the single-stranded duplicated segment A'B'C'D'. Movement occurs by rupture of an H bond ahead of the loop and formation of a new H bond behind. Evidence of movement can be seen when a population of such heteroduplexes is mounted for electron microscopy: The location of the branch point of the loop relative to a fixed point on the molecule varies among different molecules (66, 71). The range of the variability gives the end points and the size of the duplication.

73) have characterized 34 addition mutants of two phage λ strains: λ *b*221, a deletion mutant of λ, and φ80*tdel*33, a deletion mutant of a λ-φ80 hybrid. All additions were tandem direct repeats, as judged by electron microscopy of DNA heteroduplexes and by genetic instability. Many if not all regions of the λ chromosome could be duplicated, and there was little or no restriction on the location of the novel joint. The only known restrictions on the types of addition mutants that could be recovered were size and viability: Only additions of approximately 6 to 23% of the wild-type λ chromosome length could be selected for, and additions that interfered with the function of vital genes would have been lost. Within the limits of these constraints we can say that a tandem direct repeat is the most frequent type of addition mutant in phage λ, and that may be because it contains only one novel joint.

The frequency of duplications found after a single cycle of intracellular phage growth (about 5 doublings) differed markedly for the two strains examined: about 10^{-3} for φ80*tdel*33 and about 50-fold less for λ *b*221. This difference is apparently due to a trans-acting φ80*tdel*33 gene product: In mixed infections, the frequency of addition mutants for both phage was equal to that of φ80*tdel*33. This gene, which has not been further characterized, has most likely been inherited by φ80*tdel*33 from the right arm of the φ80 chromosome. This arm contains genes controlling DNA replication and transcription, among others.

How do tandem duplications arise? A recent experiment shows that duplication is correlated with *recA*- and *red*-independent recombination between DNA molecules. Emmons et al (73) mixedly infected a bacterial culture with two different genetically marked derivatives of λ *b*221. The two main pathways for homologous recombination of λ genomes were inactivated by mutations in the *recA* and *red* genes. Previous experiments had shown that duplications arise at a frequency of about 10^{-5} per phage regardless of the activity of these genes (65). Because of the *recA* and *red* mutations the frequency of wild-type recombinants among the total progeny was only 10^{-4}, some 500- to 1000-fold less than normal. However, the frequency of wild-type recombinants among the duplication mutants that had arisen during intracellular growth was extraordinarily high—0.13. The simplest conclusion is that tandem direct repeats are formed by *recA*- and *red*-independent recombination. Such recombination might be intermolecular, in which case reassortment of genetic markers and duplication could occur in one event. Alternatively, it might be intramolecular, occurring between daughter arms of a partially replicated molecule (Figure 8). In this case, duplication and reassortment of genetic markers would occur in different although similar events, and, the association between them could be accounted for by assuming that such events are restricted to a small subclass of the population of infected cells.

Tandem repeats also occur with high frequency (10^{-3} to 10^{-5} per cell) in *E. coli* and *S. typhimurium* (62, 63, 74). However, it is not known whether their formation is *recA* independent. A recent analysis of duplication of the *glyT* region of *E. coli* (70) strongly suggests that the novel joints occur in a limited number of regions and that these regions correspond to preexisting nontandem direct repeats of a gene

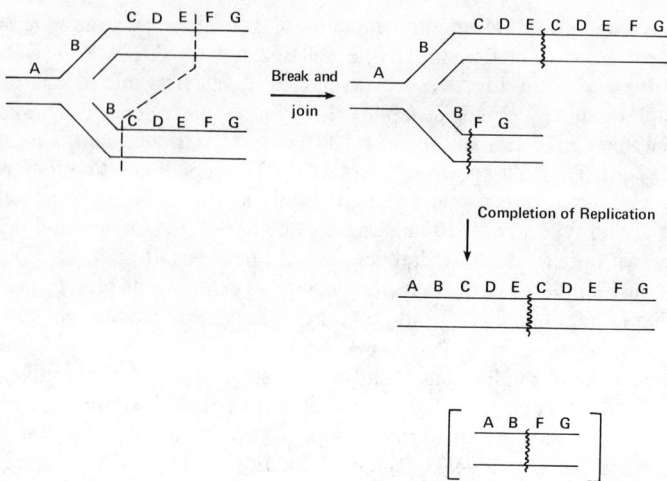

Figure 8 Intramolecular recombination can generate duplication mutations. The figure shows how a nonhomologous recombination between two sister arms of a replicating chromosome can generate a duplicated chromosome and a by-product deletion. The *vertical wavy lines* are the novel joints.

cluster encoding ribosomal RNAs. Therefore, it seems likely at least in this case that the bacterial duplications are generated by a different mechanism than the phage λ duplications.

Inverted Repeats in Phage λ

Although inverted repeats are rarely found in plaque-forming phage particles, they occur frequently in a derivative called λ *dv* (75, 76). λ *dv*'s, like other illegitimate recombinants, can be formed in the absence of *recA* and *red* functions. They are circular DNA molecules that are derived from a λ chromosome by loss of (potentially) all phage genes except *cro*, *O*, and *P*. Genes *O* and *P* are required for autonomous replication of λ *dv*, which occurs at a rate sufficient to maintain 10 to 20 copies per cell. The role of *cro* is probably to limit the expression of genes *O* and *P* so that replication of the plasmid does not kill the cell. Although λ *dv*'s can be formed from λ DNA by simple deletion of a contiguous region of a circular λ chromosome, in some cases the deletion is accompanied by duplication and inversion of the remaining sequences. The frequency of inverted repeats in independently isolated λ *dv* lines is influenced by the genotype of the parental phage, but as this effect is not well understood, we do not consider it here.

Chow et al (77) have analyzed a number of independently isolated λ *dv* lines by electron microscopy of DNA heteroduplexes and have proposed a model for the formation of inverted repeats. They found two basic types of inverted repeats. One was a simple circular duplication: [D-E-F-G-G-F-E-D] ; the other, which was more common, a circular duplication with unique materials between the duplicated

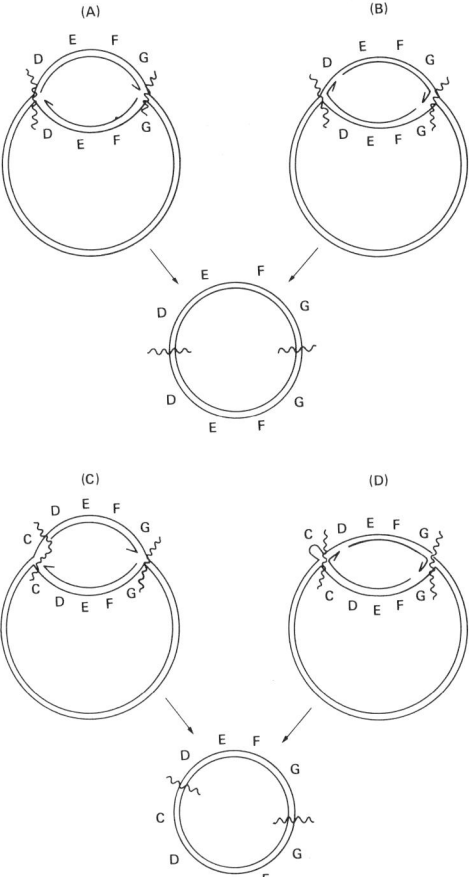

Figure 9 Models for the formation of λ *dv* mutants containing inverted repeats [after reference (77)]. *A, B, C,* and *D* represent bidirectionally replicating λ DNA molecules (86) with different strand configurations at the growing forks. In molecules (*B*) and (*D*), the progeny strands are assumed to be connected at the growing fork, while in (*A*) and (*C*), they are not. The *arrows* indicate the direction of polymerization of the progeny strands.

The models all involve joining of parental to parental and progeny to progeny strands of a replicating λ DNA molecule at or close to a growing fork. The points of strand breakage and the novel joints are indicated by the *wavy lines*. Molecules (*A*) and (*B*) give rise to a simple inverted repeat. Molecules (*C*) and (*D*) give rise to an inverted repeat with a unique sequence adjacent to the novel joint. An alteration in the right-hand growing fork of molecules (*C*) and (*D*) could lead to a third type of λ *dv* that has been observed: an inverted repeat with unique sequences adjacent to both novel joints.

regions: $\overrightarrow{\text{C-D-E-F-G}}\overleftarrow{\text{-G-F-E-D}}$. (The letters in these illustrations represent blocks of genetic information and the arrows the direction of reading. The sequence of the parental chromosome is assumed to be alphabetical, reading left to right.) As indicated in the examples given, there are only two different novel joints in each λ *dv* line and these are not restricted to specific locations on the λ map. Chow et al (77) propose that λ *dv*'s with inverted repeats arise from a partially replicated λ chromosome as shown in Figure 9. Formation of the novel joints occurs by DNA fusions at or near the two growing points, which are assumed to consist in part of single-stranded DNA, and therefore to be especially susceptible to nucleolytic attack. An alternative model was considered and rejected: λ *dv*'s with inverted repeats cannot arise by reciprocal recombination between nonhomologous regions of unduplicated λ *dv*'s because such recombination should produce molecules with more than two novel joints.

DISCUSSION

There have been several notable advances in our understanding of illegitimate recombination since the last detailed review of this subject (2). In our view these are the following: 1. The formation of tandem duplications and genetic recombinants are correlated in phage λ (73). 2. Short, directly repeated nucleotide sequences promote deletion formation, perhaps by homologous crossing-over as shown in Figure 2. However, some deletions are not associated with such sequences, and these may be formed by another pathway. 3. Deletion termini can lie with approximately equal likelihood at a great many internucleotide bonds (27). 4. Site-specific deletions are a subclass of the total population of deletions (40). In spite of these advances we are still unable to formulate a generally acceptable and readily testable theory of illegitimate recombination, or indeed to say how many pathways of illegitimate recombination there are. An example should illustrate this last point.

Deletions and duplications might in theory be formed in the same kind of recombinational event (see for example Figures 2 and 8), and if so, they should arise (if we ignore inviable deletions) with approximately equal frequency. The frequency of duplications in a λ deletion mutant and of deletions in λ wild type and in a different deletion mutant have been measured. After a single cycle of phage growth, the frequency of phage containing duplications was about 2×10^{-4} per cell (65), and the frequency of deleted phage was about 10^{-7} per cell (50; N. Sternberg, L. Enquist, and R. Weisberg, in preparation), some 2000-fold lower. We can think of several explanations: 1. Duplications can occur all over the genome, whereas deletions are restricted to the nonessential 30%. In addition, the permitted size range of duplications (6 to 23% of the wild-type chromosome) exceeded that of deletions (18 to 27% in one experiment, 4 to 13% in another). However, it seems unlikely to us that this could account for more than a factor of 50 of the difference. 2. A λ gene product may act to reduce the frequency of illegitimate recombination, and this gene may have been missing in the deletion mutant used to measure the frequency of duplications. This hypothesis has not been tested. 3. Duplications and deletions might not arise by the same pathway.

Deletions arise at an elevated frequency in regions adjacent to *IS* elements. Could *IS* elements be responsible for all deletions? This seems unlikely because some deletions, most notably those arising in phage λ, arise in regions that are not adjacent to a known *IS* element. However, *IS* elements are believed to encode recombinogenic proteins, and such proteins might act on non-*IS* DNA with low efficiency and broad specificity.

Several molecular mechanisms of DNA deletion have been considered. The structure of inverted repeats as observed in λ *dv* lines suggests that they are formed by recombination that occurs at replication forks (see Figure 9 and section on inverted repeats in phage λ). This hypothesis can be easily expanded to account for the formation of noninverted λ *dv* lines and of deletion mutants by assuming that breakage and fusion can occur between the left and right growing forks of the molecule depicted in Figure 9A (77). Two other mechanisms of deletion formation have also been discussed (2). According to one, the DNA replication machinery of the cell occasionally fails to copy a chromosome segment. In the other, deletions occur by exchange and fusion of broken ends that are created by two DNA scissions formed independently of DNA replication. Such exchange can be of two types: reciprocal, in which all four broken ends are fused, or nonreciprocal, in which only two are fused and the DNA associated with the other two is lost. Only in the case of reciprocal exchange would it be possible to recover in a clone of a newly formed deletion mutant the corresponding excised DNA. At present none of these models has been tested.

By comparison to homologous and site-specific recombination, our understanding of illegitimate recombination is meager. However, several new developments offer hope of enlightenment. New techniques of DNA sequencing (80, 81) should add to our knowledge of the influence of nucleotide sequence on deletion end points. The system of Emmons et al (73; see section on tandem direct repeats) offers promise for clarifying the relation between *recA*- and *red*-independent recombination and the formation of deletions and duplications in phage λ. Finally, the search for mutants in genes controlling illegitimate recombination appears to be succeeding: Nevers et al (43) have isolated a *E. coli* mutants that reduce the frequency of *IS1*-associated deletions. Interestingly, precise excision of *IS1* is not strongly affected, suggesting that deletion and precise excision do not occur by the same pathway. The effect, if any, of the mutations on non-*IS* associated deletions has not been reported.

ACKNOWLEDGMENT

We are grateful to K. Kunkle for typing this article, to Drs. C. Hill. N. Kleckner, and J. Miller for communicating unpublished results, and to Dr. H. Nash for criticizing the manuscript.

Literature Cited

1. Campbell, A. 1962. *Adv. Genet.* 11: 101–46
2. Franklin, N. 1971. In *The Bacteriophage λ*, ed. A. D. Hershey, 175–94. Cold Spring Harbor, NY: Cold Spring Harbor Lab.
3. Hershey, A. D. 1971. *Carnegie Inst. Washington Yearb.* 69:717–22
4. Campbell, A. 1977. In *Comprehensive Virology*, ed. H. Fraenkel-Conrat, R. Wagner, 8:259–328. New York: Plenum
5. Weisberg, R. A., Gottesman, M. E., Gottesman, S. 1977. See Ref. 4, pp. 197–258
6. Bukhari, A. I. 1976. *Ann. Rev. Genet.* 10:389–412
7. Nash, H. 1977. *Curr. Top. Microbiol. Immunol.* In press
8. Benzer, S. 1961. *Proc. Natl. Acad. Sci. USA* 47:403–15
9. Campbell, A. 1959. *Virology* 9:293–305
10. Westermoreland, B. C., Szybalski, W., Ris, H. 1969. *Science* 163:1343–48
11. Davis, R. W., Davidson, N. 1968. *Proc. Natl. Acad. Sci. USA* 60:243–50
12. Schwartz, D. O., Beckwith, J. R. 1969. *Genetics* 61:371–76
13. Demerec, M. 1960. *Proc. Natl. Acad. Sci. USA* 46:1075–79
14. Tessman, I. 1962. *J. Mol. Biol.* 5:442–45
15. Beukers, R., Berends, W. 1960. *Biochim. Biophys. Acta* 41:550–51
16. Geiduschek, E. P. 1961. *Proc. Natl. Acad. Sci. USA* 47:950–55
17. Fresco, J. R., Alberts, B. M. 1960. *Proc. Natl. Acad. Sci. USA* 46:311–21
18. Coukell, M. B., Yanofsky, C. 1970. *Nature* 228:633–35
19. Shapiro, J. A., Adhya, S. L. 1969. *Genetics* 62:249–64
20. Neubauer, Z. 1970. *Virology* 42:225–28
21. Adhya, S., Cleary, P., Campbell, A. 1968. *Proc. Natl. Acad. Sci. USA* 61:956–62
22. Pfeifer, D., Hirsch, H. J., Bergmann, D., Hamlaoui, M. 1974. *Mol. Gen. Genet.* 132:203–13
23. Schleif, R. 1972. *Proc. Natl. Acad. Sci. USA* 69:3479–84
24. Benzer, S. 1961. *Proc. Natl. Acad. Sci. USA* 47:403–15
25. Miller, J., Ganem, D., Lu, P., Schmitz, A. 1977. *J. Mol. Biol.* 109:275–301
26. Schleif, R., Lis, J. T. 1975. *J. Mol. Biol.* 95:417–31
27. Schmeissner, U., Ganem, D., Miller, J. 1977. *J. Mol. Biol.* 109:303–26
27a. Landy, A., Ross, W. 1977. *Science* 197:1147–60
28. Clark, A. J. 1973. *Ann. Rev. Genet.* 7:67–86
29. Anderson, C. W. 1970. *Mutat. Res.* 9:155–65
30. Franklin, N. C. 1967. *Genetics* 35:699–707
31. Inselburg, J. 1967. *J. Bacteriol.* 94:1266–67
32. Spudich, J. A., Horn, V., Yanofsky, C. 1970. *J. Mol. Biol.* 53:49–67
33. Marchelli, C., Ghelardini, P., Nasi, S. 1976. *Genetics* 82:161–68
33a. Pero, J. 1971. See Ref. 2, pp. 599–608
34. Kleckner, N. 1977. *Cell* 11:11–23
35. Davis, R. W., Parkinson, J. S. 1971. *J. Mol. Biol.* 56:403–23
35a. Pilacinski, W., Mosharrafa, E., Edmundson, R., Zissler, J., Fiandt, M., Szybalski, W. 1977. *Gene.* In press
36. Shimada, K., Weisberg, R. A., Gottesman, M. E. 1972. *J. Mol. Biol.* 63:483–503
37. Shapiro, J. A., Adhya, S. L., Bukhari, A. I. 1977. In *DNA Insertion Elements, Episomes and Plasmids*, ed. A. Bukhari, J. Shapiro, S. Adhya. Cold Spring Harbor, NY: Cold Spring Harbor Lab. In press
38. Jordan, E., Saedler, H., Starlinger, P. 1968. *Mol. Gen. Genet.* 102:353–63
39. Shapiro, J. A. 1969. *J. Mol. Biol.* 40:93–105
40. Reif, H. J., Saedler, H. 1975. *Mol. Gen. Genet.* 137:17–28
41. Ahmed, A., Johansen, E. 1975. *Mol. Gen. Genet.* 142:263–75
42. Botstein, D., Kleckner, N. 1977. See Ref. 37
43. Nevers, P., Reif, H. J., Saedler, H. 1977. See Ref. 37
44. Reif, H. J., Saedler, H. 1977. See Ref. 37
45. Campbell, A. 1971. See Ref. 2, pp. 13–44
46. Gottesman, M. E., Yarmolinsky, M. B. 1968. *Cold Spring Harbor Symp. Quant. Biol.* 33:735–47
47. Gingery, R., Echols, H. 1968. *Cold Spring Harbor Symp. Quant. Biol.* 33:721–27
48. Kayajanian, G. 1968. *Virology* 36:30–41
49. Sternberg, N., Weisberg, R. 1977. *J. Mol. Biol.* In press
50. Sternberg, N., Weisberg, R. 1977. *J. Mol. Biol.* In press
51. Feiss, M., Fisher, R., Crayton, M., Egnor, C. 1977. *Virology* 77:281–93

52. Weil, J., Cunningham, R., Martin, R. III, Mitchell, E., Bolling, B. 1972. *Virology* 50:373–80
53. Campbell, A. 1963. *Genetics* 48:409–21
54. Jacob, F., Wollman, E. 1956. *Ann. Inst. Pasteur* 91:486–510
55. Court, D., Campbell, A. 1972. *J. Virol.* 9:938–45
56. Enquist, L., Weisberg, R. 1977. *J. Mol. Biol.* 111:97–120
57. Fiandt, M., Gottesman, M. E., Shulman, M. J., Szybalski, E. H., Szybalski, W., Weisberg, R. A. 1976. *Virology* 72:6–12
58. Little, J., Gottesman, M. E. 1971. See Ref. 2, pp. 371–94
59. Sternberg, N., Weisberg, R. 1975. *Nature* 256:97–103
60. Shulman, M. J., Hallick, L. M., Echols, H., Signer, E. R. 1970. *J. Mol. Biol.* 52:501–20
61. Kayajanian, G. 1970. *Mol. Gen. Genet.* 108:338–48
62. Hill, C. W., Combriato, G. 1973. *Mol. Gen. Genet.* 127:197–214
63. Anderson, R. P., Miller, C. G., Roth, J. R. 1976. *J. Mol. Biol.* 105:201–18
64. Parma, D. H., Ingraham, L. J. 1970. *Genetics* 64:Suppl.(2),s49
64a. Weil, J., Terzaghi, B. 1970. *Virology* 42:234–37
65. Emmons, S. W., MacCosham, V., Baldwin, R. L. 1975. *J. Mol. Biol.* 91:133–46
66. Bellett, A., Busse, H., Baldwin, R. 1971. See Ref. 2, pp. 501–14
67. Sturtevant, A. 1925. *Genetics* 10:117–47
68. Campbell, A. 1963. *Virology* 20:344–56
69. Feiss, M., Adhya, S., Court, D. L. 1972. *Genetics* 71:189–206
70. Hill, C., Grafstrom, R., Harnish, B., Hillman, B. 1977. *J. Mol. Biol.* In press
71. Busse, H. G., Baldwin, R. L. 1972. *J. Mol. Biol.* 65:401–12
72. Emmons, S. W., Thomas, J. O. 1975. *J. Mol. Biol.* 91:147–52
73. Emmons, S. W., MacCosham, V., Baldwin, R. L. 1975. *J. Mol. Biol.* 95:83–89
74. Folk, W. R., Berg, P. 1971. *J. Mol. Biol.* 58:595–610
75. Matsubara, K., Kaiser, A. D. 1968. *Cold Spring Harbor Symp. Quant. Biol.* 33:769–75
76. Berg, D. E. 1974. *J. Mol. Biol.* 86:59–68
77. Chow, L. T., Davidson, N., Berg, D. 1974. *J. Mol. Biol.* 86:69–89
78. Deleted in proof
79. Deleted in proof
80. Sanger, F., Coulson, A. R. 1975. *J. Mol. Biol.* 94:441–48
81. Maxam, A. M., Gilbert, W. 1977. *Proc. Natl. Acad. Sci. USA* 74:560–64
82. Kupor, S. R., Fraenkel, D. G. 1969. *J. Bacteriol.* 100:1296–1301
83. Davidson, N., Szybalski, W. 1971. See Ref. 2, pp. 45–82
84. Ahmed, A., Scraba, D. 1975. *Mol. Gen. Genet.* 136:233–42
85. Szybalski, E. H., Szybalski, W. 1977. *Abstr. Ann. Meet. Am. Soc. Microbiol.* 77
86. Inman, R. B., Schnös, M. 1971. *J. Mol. Biol.* 56:319–25
87. Schrenk, W. J., Weisberg, R. A. 1975. *Mol. Gen. Genet.* 137:101–7
88. Gottesman, S., Beckwith, J. R. 1969. *J. Mol. Biol.* 44:117–27
89. Press, R., Glansdorff, N., Miner, P., DeVries, J., Kadner, R., Maas, W. K. 1971. *Proc. Natl. Acad. Sci. USA* 68:795–98
90. Murray, N. E., Murray, K. 1974. *Nature* 251:476–81
91. Thomas, M., Cameron, J. R., Davis, R. W. 1974. *Proc. Natl. Acad. Sci. USA* 71:4579–83
92. Sato, K., Campbell, A. 1970. *Virology* 41:474–87

AUTHOR INDEX

A

Aaronson, S. A., 155, 208-10, 216, 217, 220, 279, 280
Ab, G., 331
Abdel-Monem, M., 375
Abelson, H. T., 278
Abelson, J. N., 116
Aboud, M., 327, 333
Acedo, G., 91
Achtman, M., 106
Acker, G., 163
Ada, G. L., 162, 171
Adams, A., 113
Adams, J. N., 107
Adawadkar, P. D., 253
Adelberg, E. A., 106, 114
ADHYA, S., 451-73; 108, 117, 371, 381, 382, 453, 454, 460, 465
Adhya, S. L., 105, 108, 109, 453, 454, 457, 458
Adler, J., 161-64, 175-77, 397-411
Adye, J., 384
Ahkong, Q. F., 94
Ahmad, A., 383
Ahmed, A., 109, 110, 454, 458, 459
Akematsu, T., 187, 195
Åkesson, I., 18
Aksamit, R., 400, 401
Alawi, A. A., 421
Albert, E. D., 15
Alberts, B. M., 258, 261, 263, 452
Albertson, D. G., 418, 431
Alfi, O. S., 187, 195
Ali, M., 205, 206
Allaudeen, H. S., 209
Allen, F. H., 14, 18, 22
Allen, J. M. V., 130
Allen, J. W., 184, 192, 194, 195
Allen, S., 350, 358
Allen, S. L., 358
Allet, B., 109, 302, 304, 315, 324, 325, 332
Allfrey, V. G., 254, 255
Allison, A. C., 17
Alper, C. A., 17-19
Alper, M. D., 112, 380
Altman, S., 309
Alton, N. K., 379
Ambler, R. P., 163, 165
Ambros, V. R., 214
Ames, B. N., 112, 114, 333, 380
Ames, G. F., 167
Ammann, A. J., 17
Ammermann, D., 359

Anagnostopoulos, C., 110-14, 117
Ananthaswamy, H. N., 384
Anderer, A. F., 208
Anderson, C. W., 457
Anderson, E. S., 380, 381
Anderson, G. R., 215, 216
Anderson, J., 260
Anderson, J. E., 19, 23
Anderson, K. M., 241
Anderson, R. A., 163, 175, 398
Anderson, R. P., 111, 112, 378, 465, 467
Anderson, T. F., 380
Anderson, W., 43
Anderson, W. B., 214
Anderson, W. F., 248, 255
Anderson, W. W., 58, 59, 62-65, 70
Andrésson, O. S., 112, 113
Anisowicz, A., 215, 288
Anker, P., 80-83, 93
Ansevin, A. T., 252, 253, 255
Ansowicz, A., 219
Anthony, W. M., 106
Aoki, S., 86
Aoki, T., 130, 132, 135, 136, 140, 142, 144, 155, 213, 286
Apirion, D., 308, 309, 311, 317-19
Appelbaum, P. C., 163, 165
Arcement, L. J., 207, 212
Arlinghaus, R. B., 207, 212
Armstrong, D., 208
Armstrong, J. B., 163, 164, 176, 403, 405
Arnaud, M., 262
Aronson, A. I., 32, 35-37
Arpels, C., 130, 131
Arrighi, F. E., 190
Artz, S. W., 333
Artzt, K., 156
Asakura, S., 162, 167, 169, 172, 177
Ash, J. F., 214
Asselbergs, F., 204, 208
Astrin, S. M., 248, 250, 251
Aswad, D., 403, 405
Atherly, A., 331
Atidia, J., 431
Atkinson, P. H., 212
Attridge, T. H., 88
Atwood, K. C., 300
Aubert, J.-P., 30
Audit, C., 111, 112
August, J. T., 130, 143, 207, 208, 211, 212, 280
Austin, S., 301
Avdienko, I. D., 386
Avila, J., 40

Axel, R., 248, 254, 255, 259
Axelrad, A. A., 278, 280, 282, 283
Ayala, F. J., 64-66, 70
Ayusawa, D., 167
Azen, E. A., 18

B

Babudri, N., 380
Bach, F. H., 15
Bacheler, L. T., 246, 252
Bachmann, B. J., 309, 317, 371, 372, 383, 400
Bade, E. G., 105, 314
Bader, J., 214
Bader, J. P., 214
Bagi, G., 259
Baglioni, C., 17
Bailey, J. M., 204
Bailiff, E. G., 205
Baker, R. F., 326
Baker, R. M., 384
Balakanich, S., 435
Balassa, G., 30, 33, 34
Balduzzi, P. C., 214
Baldwin, R. L., 111, 112, 377, 465-67, 470, 471
Ball, J. K., 290
Ballou, B., 312, 321
Balogh, M., 286
Baltimore, D., 204-6, 208-11, 220, 221, 281, 282
Baltz, R. H., 383, 386
Baluda, M. A., 205, 206
Bancroft, J. B., 86
Band, H. T., 57, 60, 61
Banerjee, S., 112
Baralle, F. E., 331, 332
Barbacid, M., 208, 209, 220
Barbour, S. D., 370, 375, 376
Bardin, C. W., 260
Bardwell, E., 384, 385
Barker, K. L., 241
Barnes, R. D., 137
Barnett, A., 357, 363
Barrett, T., 255, 256
Barski, G., 287
Barth, P. T., 106, 107, 109, 110, 114
Bartley, J. A., 252, 254
Bartram, C. R., 189, 191, 194-96
Baserga, R., 241, 257
Basombrio, M. A., 156
Bassett, C. L., 379
Bassin, R. H., 216, 282, 287
Bastia, D., 386
Basu, S. K., 111, 112
Battula, N., 206

Batzing, B. L., 132
Bauer, H., 155, 205, 208, 209, 211, 212, 214, 215
Baugnet-Mahieu, L., 248
Bauknecht, T., 192, 194, 195
Baulieu, E. E., 262
Baumann, P., 15
Bautz, E. K. F., 243, 258, 315, 320
Bautz, F. A., 243, 258
Baxter, J. D., 260, 261
Beadle, G. W., 183
Beale, G. H., 350
Beamand, J. A., 217
Beard, D., 205
Beard, J. W., 205
Bearn, A. G., 16, 19
Beato, M., 260
Beckwith, J. R., 105, 113, 379, 453, 460
Bedford, J. S., 189, 196
Bedinger, P., 109, 110
Beebee, T. J. C., 247, 248
Beeftinck, F., 110-13, 377
Beek, B., 195
Beemon, K., 204, 216, 217, 219, 220
BEHKI, R., 79-101
Behki, R. M., 85
Behme, M. T., 382, 384
Bekhor, I., 251-53
Bell, J. G., 217
Bellamy, A. R., 204
Bellett, A. J. D., 111, 112, 465, 466
Benacerraf, B., 285
Benbow, R. M., 374, 384
Bender, M. A., 189, 196
Bender, W., 204, 207, 218, 320
Bendiak, D., 316, 324, 325
Bendich, A. J., 80, 82-84
Benne, R., 376
BENNETT, D., 1-12; 156
Bennett, M., 280, 284
Bennett, M. V. L., 424
Bennett, P. M., 109, 110, 162
Bentley, D., 427
Bentley, D. R., 427
Ben-Zeev, N., 74
Benzer, S., 416, 417, 419, 421-23, 434-40, 452, 455
Benzinger, R., 375, 384
Beratis, N. G., 19
Berends, W., 452
Berg, C. M., 108
Berg, D. E., 109, 110, 468-71
Berg, H. C., 163, 175, 397-99, 401, 402, 408
Berg, K., 16
Berg, P., 110-12, 333, 379, 380, 467
Berger, H., 382, 383

Berger, I., 319
Bergmann, D., 105, 453, 454, 457, 462
Bergquist, P. L., 106
Bergsma, D., 20
Berling, H. H., 375
Berman, D., 320
Bernadi, A., 310
Bernhard, W., 278
Bernlohr, R. W., 31, 32
Bernstein, A., 215, 216
Bernstein, C., 382
Bernstein, H., 384
Bernstein, I. A., 375
Bernstein, S. E., 280
Berry, S. F., 90, 92, 94, 95
Bertin, B. A., 287
Bertrams, J., 15
Bertrand, K., 43, 44
Besemer, J., 106, 109
Besmer, P., 281
Beukers, R., 452
Beutler, E., 19
Beziat, Y., 262
Bhaduri, S., 221
Bhattacharjee, S. K., 372, 383
Bianchi, F., 92, 93
Biasi, G., 287
Biedler, J. L., 154
Biehl, J., 214
Biessmann, H., 249-51
Biggs, P. M., 215
Bikoff, E., 309
Billeter, M. A., 204, 205, 216-20
Billing, R., 16
Biquard, J.-M., 214
Bird, G. W. G., 14
Bird, R. E., 333
Birge, E. A., 318-20, 370, 375-77
Birnbaum, L. S., 300
Birnstiel, M. L., 247
Bishop, J. M., 154, 204-7, 209, 213, 215, 217, 219, 221
Bishop, J. O., 240
Bishop, M. J., 213
Bissbort, S., 19
Bittner, M., 327
Bjaring, B., 150
Blackstein, M. E., 282
Blanco, M., 370-72, 383, 387
Blank, K. J., 283, 284, 286
Blattner, F. R., 108, 310, 312-14, 324
Blevins, C. S., 281
Bleyman, L. K., 356, 358
Block, R., 326, 327, 331
Bloemendal, H., 204, 207, 208, 220, 221
Bloemers, H. P. J., 207, 208, 220

Bloom, S. E., 187, 192
Blumberg, B. S., 17
Blumberg, P. M., 214
Blumenthal, R. M., 324-26, 332, 333
Bobrow, M., 195, 197
Böck, A., 312, 316, 318, 319, 321, 322
Bodycote, J., 189, 192, 194, 196, 197
Boenisch, T., 18
Boettiger, D., 214
Bollen, A., 105, 314, 316, 318-22
Bolling, B., 461
Bolognesi, D. P., 135, 205, 207-9, 211-13, 217, 281
Bonar, R. A., 205
Bondurant, M., 204
Bonner, J., 240-42, 251-55, 259
Bonnett, H. T., 89, 90, 93, 94
Bookout, J. B., 214
Bootsma, D., 196
Borun, T. W., 257
Bosch, L., 324
Bostick, W. L., 287
Bostock, C. J., 193
Botstein, D., 109, 110, 458
Bottino, P. J., 80, 90
Bourgin, J. P., 90
Bowman, B. H., 19
Boyars, J. K., 289, 290
Boyse, E. A., 128-32, 134-38, 140, 142-47, 150, 151, 154, 156, 285, 286
Brachet, P., 108, 112
Branlant, C., 308
Branton, P. E., 214
Braun, A. G., 373
Bray, M., 363
Breckenridge, L., 318
Brehm, S. P., 32-34
Bremer, H., 299, 330, 333
Bremner, T., 279
Brenner, S., 116, 323, 325, 416, 418-20, 430-33, 440
Breuer, C. B., 241
Brevet, J., 38, 109, 311
Brewen, J. G., 186, 189, 192
Bridges, B. A., 370-72
Briggs, R., 364
Britten, R. J., 240
Britton, J. R., 323
Broda, P., 106, 326, 333
Brodsky, I., 283
Broker, T. R., 106, 369
Broman, K., 316
Bromfeld, E., 209
Bronson, M. J., 373, 382
Brookes, P., 197
Brot, N., 315, 332, 333
Brothers, A. J., 364, 442

AUTHOR INDEX 477

Brown, A., 282
Brown, D. A., 398, 399
Brown, D. D., 246, 247
Brown, K., 137
Brown, L. R., 35, 38, 311
Brown, M. E., 311, 317, 319
Brown, M. S., 370
Brown, N., 31
Brownstein, B. L., 318-21
Brunner, E., 17
Brunovskis, I., 323
Bryant, M. L., 278
Bryant, R. E., 320, 321
Bubbers, J. E., 286
Buchanan, G. R., 190, 192-97
Buchanan, J. M., 214
Buchhagen, D. L., 208, 212
Buckel, P., 316, 318, 319, 321
Buckner, C. D., 21
Budzilowicz, C., 320
Buettner-Janusch, J., 17
Buffett, R. F., 279
Bui-Dang-Ha, D., 90
Bukhari, A. I., 105, 108, 115, 117, 371, 381, 382, 452, 458
Bulba, A., 280
Buller, R. E., 260, 262-64
Bullock, L. P., 260
Bundgaard, J., 62, 63, 67
Burger, M. M., 214
Burgess, R. R., 310-13, 315, 324, 325, 332
Burgoon, A. C., 90
Burgoyne, L. A., 254
Burk, L. G., 89
Burleigh, B. D. Jr., 110, 379
Burny, A., 209
Bursztyn, H., 375
Burtis, K., 37
Busch, H., 255
Busen, W., 209
Busse, H. G., 111, 112, 465, 466
Butler, J. A. V., 254
Bütler, R., 17
Butterworth, P. H. W., 240, 244, 245, 247, 248
Buttin, G., 370, 372, 383
Butzel, H. M. Jr., 350, 354-56, 360, 362
Byerly, L., 432
Byers, D., 434
Byrne, B. C., 360, 361

C

Cabane, K., 311, 319, 324
Cabezón, T., 105, 314, 318-22
Cahill, J. F., 206
Calafat, J., 154
Caldwell, P., 315

Calendar, R., 311, 316
Calladine, C. R., 173
Calvert, P., 309
Camerini-Otero, R. D., 254
Cameron, A., 255
Cameron, J. R., 460
Cami, B., 38, 94, 311
Campbell, A., 104, 109, 452-54, 460-62, 465
Campbell, J. H., 113
Campos-Ortega, J. A., 438-40
Canaani, E., 205, 206, 209
Cannon, M., 318
Canosi, U., 385, 386
Cantor, H., 131
Capaldo, F. N., 370, 376
Capdeville, Y., 363
Capps, W. E., 208
Carbon, J., 301, 313
Carbonara, A., 17
Cardellino, R. A., 70, 71
Carlson, P. S., 87-89, 93, 94, 96
Caro, L., 106, 107, 333
Carrano, A. V., 190, 193
Carroll, D., 255
Carson, G., 55
Carswell, E. A., 154, 156
Carter, D. B., 252
Caruso, T., 284
Cashel, M., 31, 37, 325-27, 331, 333
Cashion, L. M., 205
Castellazzi, M., 112, 370, 372, 383
Castellot, S. B., 218, 220
Castro, L. E., 57
Cattanach, B. M., 350, 356, 364
Cattaneo, A., 80, 82, 83
Caughey, P. A., 358
Cavalli-Sforza, L. L., 19
Caviness, V. S. Jr., 417
Cedar, H., 242, 243, 245, 248, 254, 255
Chae, C. B., 252, 253
Chaganti, R. S. K., 192-94, 197
Chaillet, J. R., 191-93
Chakrabarti, S. L., 318, 323
Chalkley, G. R., 240, 242
Chalkley, R., 252, 254
Chamberlin, M., 43
Chamberlin, M. J., 243
Chambon, P., 241, 244, 245, 258
Chamness, G. C., 260
Champness, J. N., 162
Champney, W. S., 302, 319
Champoux, J., 109, 110
Chan, E., 278
Chan, P. H., 89
Chan, R. K., 109, 110
Chan, V. L., 384

Chandler, M., 107
Chang, F. N., 320
Chang, J. Y., 162, 170, 171
Chang, K. S. S., 155
Chang, S. Y., 424, 425
Chang, T. D., 154
Charles, P., 80, 82, 91
Chase, G. A., 20
Chase, J. W., 379, 384
Chattopadhyay, S. K., 279
Chaudhuri, S. C., 257
Chekalin, N. M., 92, 93
Chen, B., 43, 333
Chen, J. H., 208
Chen, L. B., 214
Chen, S.-H., 19
Chen, Y. C., 210, 217, 219
Cheng, R., 435
Chernin, L. S., 386
Chesebro, B., 286
Chesterton, C. J., 240, 244, 245
Chieco-Bianchi, L., 285, 287
Chigusa, S. I., 52-54, 60, 61, 67, 68, 71-73
Chilton, M. D., 80, 82, 84, 96, 240
Chipchase, M., 369, 374-77, 383-85
Chipperfield, A. R., 254
Chiu, J. F., 256
Chiu, N., 241
Chovnick, A., 59
Chow, L., 114
Chow, L. T., 311, 468-71
Chretien, P. B., 285
Christensen, J. R., 382
Christiansen, F. B., 62, 63, 67
Christie, S., 193
Chu, F., 315
Chu, L., 281
Chuong, P. V., 93
Chupeau, Y., 90
Church, R. B., 246
Chytil, F., 255, 261, 262, 264
Cieminis, K. K., 92, 93
Cioli, D., 17
Claeys, I. V., 315, 320
Clark, A. J., 369, 370, 375-77, 381, 384-86, 457
Clark, A. M., 435
Clark, D., 430, 431
Clark, J. B., 384
Clark, J. H., 244, 260, 262
Clark, R. J., 254
Clark, S., 43
Clark, V. L., 32
Clarke, L., 301, 313
Claviez, M., 130
Cleary, P., 453, 454, 460
Cleaver, J. E., 189, 192, 194-97
Clegg, J. B., 19
Cleton, F. J., 154

Clewell, D. B., 107
Clift, R. A., 21
Clowes, R. L., 106, 107
Clyde, D. F., 22
Cockburn, A., 255
Cocking, E. C., 85, 90, 92-95
Coe, E. H. Jr., 91
Coetzee, J. N., 163
Coffin, J. M., 204, 216, 219, 221
Coffin, J. W., 204
Cohen, F., 19, 23
Cohen, R. J., 442
Cohen, S. N., 106-9, 377, 379-81
Cohen-Bazire, G., 161
Collavo, D., 285, 287
Collett, M. S., 206
Collins, A. L., 403, 405
Collins, A. L. T., 176
Collins, J. J., 212
Colombatti, A., 285, 287
Colten, H. R., 18
Combatti, N. C., 94
Combriato, G., 112, 301, 307, 308, 378, 380, 465, 467
Comings, D. E., 193
Compans, R. W., 208, 211
Condit, R. C., 323
Conkling, M. A., 113
Constabel, F., 90, 93, 95
Cook, P. R., 350, 356, 364
Cooke, M., 106, 107
Cooper, D., 86
Cooper, G. M., 204, 218, 220
Cooper, P. D., 217
Coote, J. G., 31, 33-35, 38
Copeland, H. J. R., 370, 371, 386
Corcoran, P. A., 22
Cordell, B., 205
Cordone, L., 384
Corney, G., 19
Corte, G., 309
Cortner, J., 19
Cote, R., 38, 311
Couch, R. M., 241
Coukell, M. B., 113, 453, 457
Coulson, A. R., 471
Court, D., 462
Court, D. L., 460, 465
Couturier, M., 105, 322
Coward, S. J., 241
Cox, E. C., 383
Cox, R. F., 242, 244, 245
Coyle, M., 197
Crabeel, M., 110-12
Cramer, J. H., 246
Crasemann, J. M., 382, 384
Crayton, M., 461, 465
Cresswell, J. M., 107
Crewe, P., 137

Crick, F., 258
Crittenden, L. B., 208
Croker, B., 279
Crossen, P. E., 190
Crossland, K., 430, 431
Crouch, R., 209
Crouse, G. F., 248
CROW, J. F., 49-78; 51-54, 56-61, 63, 64, 67-71, 73
Cudkowicz, G., 280, 284
Cumberlidge, A. G., 317, 320
Cunin, R., 110-13, 377
Cunningham, R., 461
Cupp, L. E., 332
Curtiss, R. III, 108
Cuzin, F., 323
Czernilofsky, P., 205-7

D

Dabbs, E. R., 318, 319
Dahl, M. M., 400, 402, 403, 405
Dahlberg, A., 309
Dahlberg, J. E., 115, 205, 209, 302, 304-6, 308, 309, 333
Dahlquist, F. W., 398
Dahmus, M. E., 240-42
Dahr, W., 14
Daiger, S. P., 19
Dalbow, D. G., 330, 332
Dale, R. M. K., 248
Dales, S., 213
Dalgarno, L., 206
D'Andrea, E., 285, 287
Darlington, C. D., 183
Dastoor, M. N., 206
Datta, N., 106, 107, 109, 110, 114
Dausset, J., 15, 154
Davey, M. R., 90, 94
David, C. S., 16
David, J., 187
Davidson, B. E., 162, 171
Davidson, N., 106-8, 115, 116, 204-7, 218, 220, 221, 243, 300-2, 304, 305, 311, 452, 454, 468-71
Davies, D. R., 80
Davies, J., 109, 115, 300, 304, 309, 311, 318, 320
Davies, P., 262
Davis, J., 208, 253
Davis, N. L., 208
Davis, R. W., 452, 458, 460
Dawes, I. W., 29, 31
Dawid, I. B., 246
Dawood, M. M., 57
Dawson, J. R. O., 86
Dawson, L., 137
Dawson, P. J., 287

Dawson, P. S., 51, 57-59, 67, 69
Dearing, R. D., 94
Deaven, L. L., 186
de Boer, H. A., 331
DeChatelet, L. R., 22
Declève, A., 282, 287, 288
De Crombrugghe, B., 43, 333
Dedonder, R., 42
DeGiuli, C., 213
de Harven, E., 136, 140, 208
Dekel, L., 323
Dekio, S., 318
Dekker-Michielsen, M. J. A., 207
de Lange, G., 16
DeLange, R. J., 162, 170, 171
Delbrück, M., 442
Delcuve, G., 319-21
Delius, H., 204
Dellon, A. L., 285
Delovitch, T. L., 16
del Villano, B. C., 212
Demant, P., 18
Demerec, M., 356, 379, 383, 385, 453
Deng, C. T., 205, 206, 213, 217, 219
Dennis, M. J., 423
Dennis, P. P., 115, 299, 301, 302, 304-6, 314, 315, 322, 324-26, 328, 330, 332, 333
Denny, T. P., 212
de Noronha, F., 208, 212
Dent, P. B., 284
Deonier, R. C., 106, 114, 116, 300-2, 304, 379, 381
DePamphilis, M. L., 161, 162, 402
De Pomerai, D. I., 240
De Rooij, F. W. M., 324
De Rossi, A., 285, 287
Desai, P. R., 14
Desmet, L., 322
Desnick, R. J., 80
DeSombre, E. R., 259, 260, 262
Deusser, E., 319, 320
Deutscher, M. P., 29
de Vogel, N., 194, 195
Devoret, R., 370-72, 383, 387
de Vries, J., 106, 460
Dew, M., 419
de Weerd-Kastelein, E. A., 196
Dewey, W. C., 191
de Wilde, M., 105, 314, 316, 318, 319, 322
DeWitt, S. K., 114
DiBerardino, L. A., 279
Dichtelmuller, H., 31, 37
DiCioccio, R. A., 37, 41
Dierks, P., 206
Dietzschold, B., 212

AUTHOR INDEX 479

Diggelmann, H., 204, 207-9, 220
Dimmitt, K., 161, 162
Dina, D., 206, 216
Dingman, C. W., 255
Dinowitz, M., 221
Dixon, F. J., 212
Dobzhansky, N. P., 57, 58
Dobzhansky, T., 57-59, 61, 67, 70
Doenecke, D., 242
Doermann, A. H., 369
Dofuku, R., 154
Doherty, P. C., 286
DOI, R. H., 29-48; 29, 32, 35-39, 41, 42
Donady, J. J., 444
Donch, J. J., 370
Donn, G., 90, 92, 93
Dorion, N., 90
Dorohov, Y. L., 92, 93
Doty, P., 248
Doy, C. H., 87, 88
Drake, J. W., 113, 383, 386
Draper, R., 164, 402, 411
Dressler, D., 376, 377, 380
Drew, I., 16
Drews, G., 162
Drexler, H., 384, 386
Drost, S. D., 210
Drummond, M., 96
Dryl, S., 363
Dube, S., 204, 218
Dubnau, E., 311, 319, 324
Duda, M., 442
Dudai, Y., 419, 434
Dudits, D., 94, 95
Duesberg, P. H., 204-6, 208, 209, 212, 214-20
Duff, R., 281
Duffy, J. J., 40, 41, 43
Duie, P., 40
Duk, M., 14
Dunn, J. J., 243, 304, 308
Dupont, B., 15, 16
Durand, J., 90, 92, 93
Duran-Troise, G., 282, 287
Durwald, H., 375
Dusenbery, D. B., 428-30
Dutrillaux, B., 187, 189-91
Dux, A., 285
Dvorak, J. A., 22

E

Ebisuzaki, K., 382, 384
Echols, H., 375, 381, 460, 461, 463, 464
Eckert, R., 423, 424
Eckner, R. J., 278, 282, 284
Edelbluth, C., 162

Edelman, G. M., 214
Edelman, I. S., 260
Eden, F. C., 80, 82, 84
Edlin, G., 326, 333
Edmundson, R., 458
Edwards, S., 164, 402, 411
Edwards, Y. H., 19
Egan, J. T., 381
Egen, R. C., 435
Eggers, H. J., 151
Eggertson, G., 112, 113
Egnor, C., 461, 465
Egolina, N. A., 186, 187
Ehring, R., 333
Ehrlich, S. D., 375
Ehrman, L., 417, 435
Eisen, H., 112
Eisenman, R., 207, 220
Eisenman, R. N., 204, 208
EISENSTARK, A., 369-96; 370, 371, 381, 383-86
Eisenstark, H. M., 386
Eisenstark, R., 370, 371
Eladari, M. E., 206
Elbert, O., 442
Elgin, S. C. R., 252, 254, 255
Emann, A., 434
Emerson, S., 169, 170, 183
Emmer, M., 333
Emmerson, P. T., 370
Emmons, S. W., 111, 112, 377, 465-67, 470, 471
Endo, H., 324
Enea, V., 255
Engbaek, F., 164, 314, 332
Engelberg-Kulka, H., 323
England, J. M., 212
Engler, G., 105
Englesberg, E., 333
Enomoto, M., 163-65, 175, 176
Enquist, L., 462, 464
Enquist, L. W., 375
Epstein, H. F., 420, 421
Epstein, W., 105, 370, 371, 373, 374, 386, 401
Erikson, E., 205
Erikson, R. L., 205
Eriksson, T., 89, 93, 94
Erlich, H., 331
Ermisch, N., 80, 81
Errera, M., 196
Errington, L., 315
Essex, M., 155
Evans, H. J., 190, 192-95, 197
Evans, P. K., 90, 92, 94, 95
Ewing, A. W., 435

F

Faelen, M., 105, 314, 316, 382
Fagerhol, M. F., 17

Falaschi, A., 385, 386
Falk, H. L., 290
Falk, R., 74, 431
Falkow, S., 106, 107, 109, 110
Fallon, A. M., 310-14, 317, 324
Falvey, A. K., 255
Fambrough, D. M., 240, 242
Famulari, N. G., 212
Fan, H., 209, 221, 222, 279
Fan, W. J. W., 205
Fang, S., 260, 261
Fanshier, L., 205, 221
Faras, A. J., 204-6, 209, 210
Farber, I., 422
Farber, J., 253, 257
Farrand, S. K., 84
Fefer, A., 21
Feigelson, P., 260
Feiss, M., 460, 461, 465
Feldman, D., 260
Felicetti, L., 310, 324
Fellner, P., 306
Felsenfeld, G., 242, 243, 245, 248, 254, 255
Fenner, F., 203
Fernando, L. P., 21
Ferrari, E., 386
Ferron, W. L., 370, 371
Feunteun, J., 321
Fialkow, P. J., 21
Fiandt, M., 108, 109, 114, 310, 454, 458, 462
Fields, B. N., 282
Fields, D. A., 315, 320
Fieldsteel, A. H., 287
Fielitz, U., 86
Fiethen, L., 113
Fiil, N. P., 301, 302, 314-16, 320, 324-27, 329, 331, 332
Filner, P., 80, 82-84
Fimognari, G. M., 260
Finch, J. T., 162
Fincham, J. R. S., 117
Fischinger, P. J., 212, 216, 217, 281, 288
Fish, D. C., 212
Fisher, R., 461, 465
Fishpool, R. M., 420
Fiszman, M. Y., 214
Fitz-James, P. C., 32, 35
Flaherty, L., 146
Flaks, J. G., 315, 320, 332
Fleissner, E., 130, 135, 136, 140, 142, 143, 207, 208, 211, 212
Flickinger, R. A., 241
Florini, J. R., 241
Fodor, E. J. B., 248
Folk, W. R., 110-12, 379, 467
Fong, J., 14
Forchhammer, J., 326
Förster, G., 332

AUTHOR INDEX

Fortunato, A., 386
Foss, K., 309
Fosse, A. M., 187, 189-91
Fossum, B. L. G., 19
Foster, T. J., 108, 109
Fould, J., 111
Foulds, J., 380
Foulds, L., 289
Fowke, L., 93
Fox, T. D., 40, 41, 43
Fraenkel, D. G., 112, 454
Fraenkel-Conrat, H., 204
Frank, H., 211, 212
Frankel, A. E., 216, 288
Frankel, R. W., 163, 165
Franklin, N., 104, 105, 112, 113, 377, 381, 452, 457, 462, 463, 470, 471
Frearson, E. M., 90, 92, 94, 95
Fredericq, P., 107
Fredrickson, D. S., 23
Freedlender, E. F., 187, 190
Freedman, H. A., 286
Freese, E., 29, 32
Freifelder, D., 377, 379
Fresco, J. R., 452
Frey, J., 107
Friedman, L. D., 55
Friedrich, R., 205-7
Friend, C., 278
Friesen, J. D., 314-16, 320, 324-27, 332, 333
Friis, R. R., 209, 210, 213, 214, 216-20
Fritsch, E., 205, 206
Fu, S. M., 16
Fuchs, P., 214
Fudenberg, H. H., 16, 17
Fujiki, H., 311
Fujimura, F., 240, 242
Fujisawa, T., 321
Fujita, D. J., 213, 215, 217
Fujita, F., 213, 215
Fujita, H., 163, 165, 173
Fujita, Y., 32
Fukasawa, T., 167
Fukuda, R., 41, 42
Funatsu, G., 318
Funder, J. W., 260
Furano, A. V., 324, 326, 333
Furth, J. J., 244
Furth, M. E., 322, 323
Furuichi, Y., 204

G

Gabbay, E. J., 253
Gaffney, B. J., 214
Gahan, P. B., 80, 82, 83
Gale, R., 16
Galehouse, D. M., 212, 216, 219, 220

Galibert, F., 206
Gallant, J. A., 325-28, 331, 333
Gallis, B. M., 204, 208
Gallo, R. C., 209
Galloway, S. M., 190, 192, 193, 197
Gamborg, O. L., 85, 90, 93, 95
Ganem, D., 455, 456, 470
Garapin, A. C., 205, 221
Gardner, M. B., 208, 278
Garel, A., 259
Garen, A., 384
Garfin, D. E., 205
Gargus, J. J., 176
Garibian, G. S., 244
Garrard, W. T., 259
Garrels, J. I., 252
Gartler, S. M., 21
Gateff, E. A., 438, 440
Gatti, M., 189, 196
Gausing, K., 299, 325, 326, 328, 329, 331, 333
Gay, H., 185
Gear, J. S. S., 18
Geard, C. R., 189, 193
Gedde-Dahl, T., 18
Geering, G., 130, 135, 136, 140, 208
Gefter, M. L., 116, 309, 375
Gegenheimer, P., 308
Gehring, U., 260
Gehring, W. J., 438
Geiduschek, E. P., 40, 41, 43, 452
Gelderblom, H., 208, 209, 211, 214
Gellert, M., 374
Gelmann, E., 282
Génermont, J., 350, 362, 364
Gensler, H. L., 382
George, D., 90, 92, 94, 95
George, D. L., 383
George, J., 370, 372, 383
Georgiev, G. P., 254
Gepner, J. I., 419, 423
Gerald, P. S., 190, 192-97
Gerard, G. F., 209, 220
German, J., 192-94, 197
Gershowitz, H., 19
Gerson, I., 444
Gerwin, B., 282, 287
Gette, W. R., 322, 323
Geyl, D., 316, 319, 321
Ghelardini, P., 105, 457
Giannelli, F., 189
Gianni, A. M., 204-6, 217, 219
GIBLETT, E. R., 13-28; 14, 17-19, 21-23
Gibson, D., 17
Gibson, D. A., 186, 190, 192, 193, 196, 197
Gibson, I., 350, 358
Gibson, W., 209

Gielkens, A. L. J., 207, 208, 220, 221
Gielow, L., 333
Gigot, D., 110-12
Gilbert, S. F., 310-14, 317, 324, 326
Gilbert, W., 205-7, 326, 471
Gilden, R. V., 205-9, 211, 212
Giles, K. L., 90
Gillen, J. R., 369, 375, 381, 385
Gillespie, D. H., 379
Gillies, S. C., 204
Gilmour, R. S., 246, 248, 251, 253, 255, 256
Gingery, R., 460, 461, 463, 464
Ginoza, W., 370
Ginsburg, D., 304, 308
Ginsburg, H., 287
Giorno, R., 373
Gipson, T. G., 156
Gjerset, R. A., 249-51
Glansdorff, N., 106, 110-13, 377, 379, 460
Glass, R. E., 315
Glasser, S. R., 261
Glazer, A. N., 162, 170, 171
Gleiss, D., 86
Glimelius, K., 93, 94
Godson, G. N., 370
Goebel, W., 106, 107
Goering, R. V., 386
Goldberg, A. R., 214
Goldberg, G. I., 386
Goldberg, N. D., 32
Goldberg, R. J., 214-16, 288
Goldé, A., 205
Goldfarb, D. M., 386
Golomb, M. W., 241, 247
Golub, E. S., 36
Goodell, B. W., 21
Goodman, C., 442
Goodman, C. S., 428
Goodman, H. M., 205-7, 209
Goodman, N. C., 209
Gopalakrishna, K., 372, 383
Gordon, J., 260, 324
Gordon, J. S., 187
Gordon, M. P., 96
Gorelic, L., 300, 309
Gorer, P. A., 129
Gorini, L., 116, 318, 321, 323
Gorski, J., 259, 260
Gosch, G., 95
Goth-Goldstein, R., 196
Goto, K., 187
Goto, N., 107
Gots, J. S., 166
Gottesfeld, J. M., 259
Gottesman, M. E., 43, 104, 109, 333, 452, 454, 458, 460-64
Gottesman, M. M., 104, 109

AUTHOR INDEX 481

Gottesman, S., 452, 458, 460, 461
Gough, M., 370, 371, 373, 382
Gould, H., 242, 244
Gould, H. J., 255, 256
Gould, J. L., 417, 427, 442
Goy, M. F., 402, 405-10
Grace, J. T., 279
Grade, R., 31, 37
Gradmann-Rebel, W., 80, 81, 83
Graf, T., 208, 215
Graffe, M., 325
Grafstrom, R., 465, 467
Grambow, H. J., 90
Grandgenett, D. P., 209, 210
Granobles, L. A., 57
Grant, G. F., 165
Gray, J. C., 89, 94
Green, E., 80
Green, F. A., 14
Green, H., 255
Green, M., 204, 208, 209, 220, 221
Green, M. H. L., 370-72
Green, R. W., 208, 212
Greenberg, R., 51, 56, 57
Greenberger, J. S., 210
Greenleaf, A. L., 40-42
Greenstein, M., 375, 381
Gregoriades, A., 130, 135
Grell, R. F., 369
Grennan, D., 18
Greppin, H., 80, 82, 83
Gresshoff, P. M., 86-88
Grey, H. M., 17
Grierson, D., 80, 88
Griffith, J. D., 254
Griffiths, K., 262
Griggs, H. G., 189, 196
Grigliatti, T., 423
Grimley, P. M., 244
Grinter, N. J., 109, 110, 114
Groner, Y., 221, 242, 248, 258
Groscurth, R., 31, 32
Gross, C., 332
Gross, G., 315, 320
Gross, L., 135, 278, 290
Grosse-Wilde, H., 15
Grossfield, J., 417
Grosso, J. J., 89
Groudine, M., 259
Groupe, V., 287
Grouse, L., 240
Grubb, R., 16
Gruber, M., 31, 37, 307, 331
Gruijthuijsen, M., 105
Grumet, F. C., 23
Grunau, J. A., 113
Grunberg-Manago, M., 325
Grüneberg, H., 280, 438
Gudas, L. J., 369-74, 382, 383, 386

Guérin, M., 278
Guespin-Michel, J. F., 33
Guilleux, J. C., 262
Gukova, L. A., 386
Gullon, A., 113
Guntaka, R. V., 205, 206, 215, 217, 219
Guo, C., 94
Gussmann, S., 19
Gustafsson, Å., 79
Guthrie, C., 318-20
Guyer, M. S., 370

H

Haas, D., 107
Haas, F., 384
Haas, K. K., 108
Haas, M., 287
Haase, A. T., 204
Haberman, A. B., 258
Habermann, P., 108, 109
Hackett, A. J., 280
Haddox, M. K., 32
Hadlaczky, G., 94
Hadley, R. G., 114, 379, 381
Haga, N., 364
Hageman, P. C., 154
Hager, G., 247
Hagiwara, S., 422
Hahn, J., 427
Hakomori, S., 14
Hakomori, S.-I., 214
Haldane, J. B. S., 50, 69
Hall, C. A., 107
Hall, D. H., 431, 437
Hall, J. C., 417, 418, 421, 435, 436, 440, 443
Hall, J. D., 370
Hall, L., 440
Hall, L. M., 419, 422, 423
Hallet, M.-M., 353, 354
Hallick, L. M., 464
Halpern, M. S., 209-13, 219
Halvorson, H. O., 34, 247
Hamer, D. H., 96
Hamilton, T. H., 241, 255, 260
Hamlaoui, M., 105, 453, 454, 457, 462
Hamlett, N. V., 382
Hamlyn, P. H., 255, 256
Hämmerling, U., 130, 131, 140, 151
Hanafusa, H., 206, 208, 211, 213, 214, 216-21
Hanafusa, T., 209, 211, 213, 217
Hanawalt, P. C., 85, 369-75, 380, 383
Hanna, M. G. Jr., 132, 212
Hansen, B. S., 329, 332
Hansen, J. A., 15

Hansen, J. N., 29
Hansen, M. T., 327
Hansma, H. G., 424, 425, 429
Hanson, C. V., 258
Hanson, R. S., 29, 80, 82
Hanson, T. E., 438-40
Harada, F., 205, 209
Harada, K., 324
Haran-Ghera, N., 290
Harcombe, E. S., 421, 422, 426
Hardy, K. G., 107
Hardy, S. J. S., 315, 316, 320, 325
Hardy, W. Jr., 211, 212
Hardy, W. D. Jr., 130, 135, 155, 208
Harford, N., 38, 311, 386
Harnish, B., 465, 467
Harris, G. S., 260
Harris, H., 14, 17-19, 23
Harris, H. E., 420
Harris, J. D., 316
Harris, S. E., 240, 244, 245, 248, 249, 253, 254, 256, 257, 265
Harris, W. A., 434, 439, 444
Hart, M. G. R., 370, 371, 383
Hartley, B. S., 110, 379
Hartley, J. W., 141, 142, 153, 208, 278, 279, 281, 282, 288, 290
Hartman, G., 16
Hartman, P. E., 114, 384
Hartman, P. S., 370
Harvey, J. D., 204
Haselkorn, R., 323
Haseltine, W. A., 204-7, 221, 314, 326, 331
Hasenbank, R., 318, 319
Hatanaka, M., 205, 206, 209
Hatlen, L., 222
Haung, R. C. C., 248
Hauschka, T. S., 151
Hausen, P., 209
Haverman, J., 154
Haworth, S. R., 38, 311
Hayashi, K., 192, 193, 195
Hayes, F., 308, 309
Hayes, W., 106, 107, 384
Hayman, M. J., 207-9, 212, 217, 219
Haynes, R. H., 384
Hayward, C., 90, 92, 94, 95
Hayward, R. S., 315
Hayward, W. S., 208, 213, 216, 221
Hazelbauer, G. L., 398, 400-2, 409
Hearst, J. E., 258
Heasley, S., 205
Hecht, R. M., 247, 373
Heddle, J. A., 185
Hedèn, L. O., 106, 107

Hedgecock, E. M., 429, 435
Hedges, R. W., 107, 109, 110, 114, 380, 381
Heffron, F., 109, 110
Heiniger, H. J., 281
Heiss, B., 108
Heitler, W. J., 428
Held, W. A., 312, 321
Helinski, D. R., 374
Helling, R. B., 384
Helm, K. V. D., 209
Hemleben, V., 80, 81, 83
Hendler, R. W., 375
Hennecke, H., 325
Herberman, R. B., 144
Herman, A. C., 208
Herman, R. K., 418
Hermoso, J. M., 40
Herreros, B., 189
Herrick, G., 258
Hershey, A. D., 369, 452
Herskowitz, I., 356, 364
Herzog, A., 316, 318-22
Hess, D., 80, 85, 86, 92, 93
Hess, M., 17
Hesse, J. E., 370, 371, 373, 374, 386
Heuer, S. S., 263, 264
Hewish, D. R., 254
Heyn, R. F., 80, 84
Hicks, J. B., 356, 364
Higgins, S. J., 260
Higgs, S. A., 31
Highland, J. H., 316
Hilgers, J., 154, 287
Hill, C., 465, 467
Hill, C. W., 111, 112, 301, 307, 308, 378-80, 465, 467
Hill, M., 206, 215
Hill, P. R., 212
Hill, R. W., 212
Hiller, O., 18
Hillman, B., 465, 467
Hillova, J., 206, 215
Hilmen, M., 162, 164, 165, 176, 356, 364, 381, 387, 402
Hilschmann, N., 17
Hinde, R. A., 434, 435
Hindley, J., 254
Hino, S., 155, 216
Hiraizumi, Y., 60
Hirano, T., 168, 175
Hirokawa, Y., 287
Hirota, Y., 165, 175, 317, 320, 323
Hirsch, H. J., 105, 108, 109, 453, 454, 457, 462
Hirschberg, C. B., 214
Hirschfeld, J., 17, 19
Hirschhorn, K., 19
Hirschhorn, R., 19
Hirsh, D., 443

Hitchins, A. D., 34
Hitt, S. A., 151
Hitz, H., 311, 317, 319
Hiwatashi, K., 364
Hizi, A., 210
Hnilica, L. S., 240, 241, 252-56
Hoare, T. A., 242
Hobart, M. J., 18
Hobom, G., 380
Hoch, J. A., 31-34, 38
Hodgetts, R. B., 419
Hodgkin, J. A., 429-31, 436, 437
Hoenigsberg, H. F., 57
Hoessli, C., 373
Hofbauer, A., 438-40
Hoffmann, F., 85, 90, 93
Hoffmann, G. R., 110-12, 114, 377, 379
Hofschneider, P. H., 308
Hofstetter, H., 204, 322
Hoganson, D. A., 39
Hogg, R. W., 176
Hogness, D. S., 380
Hohn, D. C., 22
Holl, F. B., 85
Holland, M., 247
Holliday, R., 350, 356, 364, 383
Holloman, W. K., 373, 374, 377, 383
Holloway, B. W., 107
Holmdahl, L., 17
Holmes, A. J., 383
Holmes, G. E., 382
Holmquist, G. P., 193
Holtzer, H., 214
Homyk, T. Jr., 111, 426
Honda, T., 170
Hong, R., 17
Honjo, T., 247, 250
Hopkins, N., 287
Hopkinson, D. A., 19
Hopper, J. E., 16
Hopwood, D. A., 80, 116
Hori, K., 324
Horiguchi, T., 163-66, 170, 176
Horiuchi, S., 110, 378
Horiuchi, T., 110, 378
Horn, D. T., 36
Horn, V., 113, 457
Horne, R. W., 80
Horowitz, H. J., 116
Horvitz, H. R., 433, 440, 441
Hotani, H., 162, 169, 170
Hotchkiss, R. D., 94, 111, 369, 378
Hotta, Y., 80, 82-85, 417, 419, 421, 435, 436
Hough, A., 262
Hough, D., 262
Howard, G. A., 316

Howard-Flanders, P., 370, 384, 385
Howe, C. D., 287
Howe, M. M., 105, 314
Howe, T. G. B., 109
Howell, J. I., 94
Howk, R. S., 215, 219, 258, 288
Hoy, R. R., 427
Hozumi, N., 117
Hradecna, Z., 114
Hranueli, D., 35
Hsu, C.-J., 131
Hsu, M. T., 106
Hsu, T. C., 187, 191-93
Hsu, Y.-P., 35, 38
Hu, M., 106
Hu, S., 106-8, 116, 204, 206, 207, 220, 221
HU, S. S. F., 203-38
Huang, A. S., 281
Huang, P. C., 252
Huang, R. C. C., 240-42, 247, 252-54
Huart, R., 80-82, 91
Huberman, J., 240, 242
Hudock, G. A., 442
Hudock, M. O., 442
Huebner, R. J., 155, 208, 212, 280, 281, 290
Hughes, W. L., 184
Hugo, N., 163
Hull, R. A., 370, 371, 380
Humphrey, J. B., 281
Humphrey, R. M., 191
Humphries, E. H., 221
Hung, P. P., 212
Hunsmann, G., 130, 208, 212
Hunter, E., 209, 210, 214, 217, 219
Huntsman, R. G., 19
Huper, G., 208, 211, 212
Hurley, N. E., 208
Hurwitz, J., 221, 242, 248, 258
Hussey, C., 31, 37
Hutchinson, F., 197, 376, 386
Huttner, K. M., 195
Hutton, J. J., 138
Hutton, J. R., 205
Hyman, R., 150
Hyman, R. W., 243, 323
Hynes, R., 287
Hynes, R. O., 214

I

Ichiki, A. T., 170
Icho, T., 167
Ide, C. F., 442
Idrobo, J. M., 57
Igarashi, R. T., 37
Igel, H. J., 290

AUTHOR INDEX 483

Ihle, J. N., 132, 212, 288
IINO, T., 161-82; 161, 163-70, 172-76
Ikawa, Y., 287
Ikeda, H., 130, 135, 136, 138, 140, 142, 153, 211, 212, 280, 285, 309
Ikeda, K., 421, 422, 427, 444
Ikemura, T., 302, 305-7, 309, 329, 331, 333
Ikushima, T., 187, 190, 192, 197
Illmensee, R., 205, 206
Ingraham, J. L., 320
Ingraham, L. J., 111, 112, 379
Inman, R. B., 469
Inokuchi, H., 112, 114
Inouye, M., 370
Inselburg, J., 457
Invernizzi, G., 156
Ionesco, H., 30, 33, 34
Iordanescu, S., 380, 386
Irgens, R. L., 39
Iritani, C., 130, 136, 142, 150, 286
Isaka, T., 214
Ishidsu, J., 166, 168
Ishihama, A., 314, 329, 332
Ishii, Y., 113
Ishikura, H., 309
Ishimoto, A., 281
Ishizaki, R., 208
Isono, K., 317, 320
Isono, S., 317, 320
Israeli-Reches, M., 323
Itakura, K., 138
Itikawa, H., 379
Ito, J., 33, 36, 41
Ito, K., 314, 329, 332
Itoh, T., 318, 374
Iverson, D., 22
Ives, P. T., 57, 60, 61
Iwakura, Y., 314, 329, 332

J

Jackowski, J. B., 109
Jackson, E. N., 112, 113, 378, 379
Jackson, J., 205
Jackson, N., 221
Jacob, A. E., 107, 109, 110, 380, 381
Jacob, F., 114, 156, 323, 461
Jacob, K., 93
Jacobs, M., 80-82, 91
Jacobs, S., 138
Jacobson, J. B., 148
Jacoby, G. A., 380, 381
Jacquet, M., 221, 242, 248, 258
Jaehning, J. A., 241, 247

Jaenisch, R., 279
Jaffe, R. C., 263
Jain, D., 212
Jamet, C., 111, 112
Jamet-Vierny, C., 110-12
Jamieson, A. F., 106
Jamjoom, G., 207, 212
Jamjoom, G. A., 212
Jan, L. Y., 423
Jan, Y. N., 423, 434, 442
Janowski, M., 248
Janssens, F. A., 183
Jarry, B., 300, 303, 306
JASKUNAS, S. R., 297-347; 108, 115, 300-2, 304-6, 309-15, 317, 318, 320, 324, 325, 332, 333
Jeng, Y.-H., 37
Jennings, A. W., 260
Jennings, H. S., 424
Jensen, E. M., 287
Jensen, E. V., 259, 260, 262
Jensen, R., 240, 242
Jerry, L. M., 17
Jersild, C., 16
Johansen, E., 109, 458, 459
Johns, E. W., 242, 254
Johnson, A. M., 17
Johnson, A. W., 241, 255
Johnson, C. B., 80, 88
Johnson, C. D., 418
Johnson, F. L., 21
Johnson, G. S., 214
Johnson, L. D., 331
Joho, R. H., 204, 205, 216-20
Joklik, W. K., 210, 214, 215, 221
Jokura, K., 167
Jolicoeur, P., 221, 282
Jollick, J. D., 163
Jollos, V., 362
Jones, C. W., 94
Jonsson, B., 19
Jordan, E., 105, 108, 246, 458
Jorgensen, P., 31, 37, 301, 302, 329, 331, 332
Jovin, T. M., 375
Joyner, J., 137
Joys, T. M., 162, 163, 165, 170, 171
Judd, B. H., 52
Juergens, L. A., 190, 192-97
Julien, J., 333
Jungblut, P. W., 260
Junghans, R. P., 206, 207, 220

K

Kacian, D. L., 209
Kada, T., 386
Kadner, R., 106, 460

Kado, C. I., 80, 82, 84
Kadota, H., 32, 35
Kagawa, H., 162, 167
Kahan, L., 316
Kahn, P. L., 107, 108
Kai, K., 285
Kaiser, A. D., 303, 468
Kaji, A., 214
Kalekine, M., 215
Kalimi, M., 244, 260, 262
Kalnins, V. E., 209
Kaltschmidt, E., 317
Kamen, R. I., 327, 331
Kaminski, M., 40
Kamiya, R., 172, 177
Kamiya, T., 324
Kamiyama, M., 252
Kaneko, H., 424
Kaneko, I., 29
Kankel, D. R., 417, 418, 435
Kantor, J. A., 248, 255
Kao, K. N., 90, 93-95
Kaplan, H. S., 211, 282, 287, 288, 290
Kaplan, S., 300, 307
Kaplan, W. D., 421
Karshin, W. L., 207, 212
Kartel, N. A., 92, 93
Kartha, K. K., 90, 93
Karu, A. E., 375, 381
Kashmiri, S. V., 111, 282, 287
Kass, T. L., 74
Kato, H., 187, 191-93, 195-97
Kato, T., 375
Katz, G. M., 424
Katzir, A., 442
Kawai, S., 211, 213, 214, 216-20
Kawamura, F., 36, 41
Kawano, G., 310, 324
Kawashima, K., 142, 153, 280
Kawashima, N., 89
Kawashima, T., 260
Kay, D., 31
Kayajanian, G., 460, 461, 463-65
Kazan, P. A., 204, 206
Keeler, C. L., 38
Keijzer, W., 196
Keil, T. U., 308
Keilman, G. R., 37, 41
Keith, J., 204
Keller, W., 94, 209
Keller, W. A., 94
Kelloff, G., 208
Kelly, B., 105
Kelly, L. E., 422, 423
Kemper, B., 109
Kenerley, M. E., 301-6, 315
Kenny, F. T., 260
Kenyon, A., 57, 62

AUTHOR INDEX

Kerjan, P., 40, 41
Kerr, C., 382
Kerr, I. M., 208, 220
Kerr, T. L., 370, 371, 383
Keshgegian, A. A., 244
Kessler, D. P., 320
Khachatourians, G., 31
Khoury, A. T., 206
Khoury, G., 248, 250
Kidwell, J. F., 74
Kidwell, M. G., 74
Kier, L. D., 91
Kihlman, B. A., 187, 189, 190, 192, 195-97
Kikuchi, A., 116
Kikuchi, T., 429, 431
Kim, J., 252
Kim, J. H., 147
Kim, M. A., 187, 192
Kimball, R. F., 350, 352
Kimmel, C. B., 442
Kimmich, D., 80, 81
Kimura, A., 311, 319, 326
Kimura, M., 69, 70
Kindler, P., 308
Kindt, T., 16
King, A. M. Q., 204
King, J. M., 86
King, R. J. B., 260
Kinoshita, T., 310, 324
Kinst, M., 55
Kirschbaum, J. B., 301, 302, 304, 315, 320, 324, 325, 332
Kirsten, W. H., 287
Kissmeyer-Nielsen, F., 15
Kit, S., 214
Kitagawa, O., 55, 60
Kitakawa, M., 317
Kitamura, N., 309
Kitchin, F. D., 19
Kitchin, R., 350, 364
Kjeldgaard, N. O., 299, 325, 326, 328, 329, 333, 334
Klášterská, I., 193
Klebanoff, S. J., 22
Kleckner, N., 109, 110, 458
Kleene, S. J., 406
Kleid, D. G., 206
Kleiman, L., 252-54
Klein, E., 129, 136
Klein, G., 129, 150
Klein, P. A., 132
Klein, R. A., 205, 206, 209
KLEINHOFS, A., 79-101; 79, 80, 82, 84
Kleinschmidt, A. K., 106
Kleinsmith, L. J., 255, 257
Klement, V., 204, 215, 278
Klenk, H.-D., 212
Klier, A. F., 42
Kligerman, A. D., 187, 192
Klug, A., 162

Knight, C. A., 206, 207, 220
Knight, G. R., 63
Kobata, A., 14
Kobata, K., 317
Koch, A. L., 331
Kohl, F., 195
Kohler, P. O., 244, 260
Kohne, D. E., 240
Köhnlein, W., 376, 386
Koizumi, S., 360, 363
Kojima, K., 52
Kojola, J., 282
Kolstinen, J., 16
Komatsu, S. K., 162, 170
Komeda, Y., 164, 166-68
Kompf, J., 19
Kondo, E., 109
Kondo, S., 113
Kondoh, H., 162, 164, 165, 170
Konno, R., 163, 165
Konopka, R. J., 419, 437
Konrad, E. B., 301, 315, 379, 384
Konstam, M., 260
Konzak, C. F., 79
Kooistra, J., 386
Kopecka, H., 206
Kopecko, D. J., 109, 379, 380
Korch, C. T., 38
Korenberg, J. R., 187, 190
Korenman, S. G., 260
Korn, L., 43, 44
Kornberg, A., 29
Kornberg, R., 254
Korol, W., 287
Kort, E. N., 163, 175, 177, 398, 399, 403-9
Kosel, C. K., 371, 372, 383
Koshland, D. E. Jr., 398-406, 408
Koske-Westphal, T., 189, 191, 194, 196
Koslov, Y. V., 254
Kostraba, N. C., 255, 257
Kosuda, K., 57, 58
Kotin, P., 290
Kozai, Y., 214
Krabbe, M. R., 22
Krantz, M. J., 212
Krauss, J., 317, 320
Krauss, S. W., 324
Krauze, R. J., 255
Kreider, G., 318-21
Krimbas, C., 61
Krimbas, M. G., 61
Krol, A., 308
Kronberg, D., 187, 189, 190, 192
Kubai, D. F., 106, 108, 313
Kubai-Maroni, D., 108, 109
Kuhn, R. W., 260, 262-64
Kull, F. J., 260

Kumar, V., 284
Kundu, S. K., 14
Kung, C., 424, 425, 429
Kung, G. M., 251-53
Kung, H.-F., 315, 333
Kung, H.-J., 204, 205, 213, 215, 218, 219
Kung, S. D., 89, 94
Kunkel, H. G., 16, 17
Kuo, M. T., 254
Kupersztoch-Portnoy, Y. M., 374
Kupor, S. R., 454
Kurahashi, K., 167
Kurland, C. G., 318-20
Kuroiwa, T., 163, 170, 173
Kurth, R., 155, 214
Kushner, I. Ch., 386
Kushner, S. R., 302, 319, 375, 376, 379
Kuwano, M., 310, 324
Kuznetsova, B. N., 386
Kylberg, K. J., 384, 386

L

Labib, G., 90, 94
Lachmann, P. J., 18
La Cour, L. F., 186
Laffler, T., 326, 327, 333
Lai, M. M.-C., 204, 208, 216
Laishley, E. J., 31
Lam, S. T., 382
Lamm, M. E., 147
Landa, F., 386
Landy, A., 116, 456
Lange, J., 208, 212
Langlois, A. J., 205, 212
Langridge, J., 113
Lapeyre, J. N., 252
Larsen, S. H., 163, 175-77, 398, 399, 403-9
Larson, J. E., 36
LaRue, B. F., 309
Lasfargues, J. C., 287
Latt, S. A., 184, 187, 189-97
Latter, B. D. H., 59
Laurell, C.-B., 17
Laursen, R., 315
Laursen, R. A., 324, 325
Law, L. W., 279
Lawley, P., 197
Lazzarini, R. A., 325-28, 331
Leamnson, R. N., 211
Leber, B., 80, 81
Leboy, P. S., 320
Lecadet, M.-M., 42
Lecocq, J. P., 316
Ledbetter, M. L., 111, 378
Ledeen, R., 14
Leder, P., 310, 324

AUTHOR INDEX 485

Lederberg, J., 375
Ledoux, L. G. H., 80-82, 91
Lee, F., 43, 44
Lee, H. J., 106, 300-2, 304
Lee, M., 382
Lee, Y. C., 212
Legrain, C., 110-12
Le Hegarat, F., 32, 34
Lehman, I. R., 379
Lehmann, H., 19
Lehrer, R. I., 22
Leighton, T., 31, 35, 36, 39
Leighton, T. J., 35-39
Leis, J. P., 210, 212
Lejeune, J., 187, 189-91
Lemaux, P., 324, 325
Lemaux, P. G., 324, 326, 332, 333
Lengeler, J., 401
Leong, J. A., 221
Lerner, K. G., 21
Lerner, R. A., 212
Leung, D., 384
Levanon, M., 17
Levene, H., 57, 58
Levin, J. G., 221
Levine, A., 370
Levine, A. D., 384
Levine, A. S., 279
Levine, J. D., 426
Levine, M., 373, 382
Levinson, W. E., 204, 205, 209, 221
Levy, B., 249-51
Levy, J. A., 204, 206, 278, 280
Levy, R., 257
Levy, S., 257
Lewandowski, L. J., 212, 320, 321
Lewis, J. A., 429, 430, 437
Lewontin, R. C., 50, 70
Liao, S., 260, 261
Liarakos, C. D., 240
Lieberman, M., 287, 288, 290
Lieberman, R., 16
Lieberman, R. P., 375
Liljas, A., 315, 316
Lilley, G. D., 382
LILLY, F., 277-96; 129, 137, 138, 143, 278, 280-87
Lima-de-Faria, A., 94
Lin, M. S., 187, 195
Lin, P. F., 384, 385
Lindahl, L., 108, 115, 301-6, 310-15, 317, 324-26, 330, 332, 333
Lindsley, D. L., 191
Ling, H. P., 205, 206, 209
Linial, M., 210, 214, 219
Linn, S., 374, 375, 381
Linn, T. G., 37, 38, 40-42
Linnett, P. E., 36

Lis, J. T., 455
Lisowska, E., 14
Lisowska-Bernstein, B., 147
Litman, G. W., 130, 136, 138
Little, J., 463
Little, J. W., 370-75, 380, 385
Liu, H. Z., 94
Liu, M., 155
Lloyd, R. G., 370, 372, 373, 375, 379
Lobban, P. E., 303
Lodish, H. F., 208, 220
Log, T., 212
London, J., 161
Longo, D., 309
Loomis, W. F., 442
Losick, R., 31, 36-38, 40-43
Lotz, W., 162, 163
Louarn, J., 333
Lovelace, E., 214
Lovely, P., 398
Lovinger, G. G., 205, 206, 209
Low, B., 370, 372, 373, 375, 376
Low, K. B., 309, 317, 371, 372, 375-77, 383, 400
Lowy, D. R., 279
Lowy, J., 162
Lozeron, H. A., 114
Lu, P., 455
Lucy, J. A., 94
Luell, S., 145, 147
Luftig, R., 208
Lund, E., 115, 302, 304-6, 308, 326, 333
Lundh, N. P., 170
Lupker, J. H., 324
Luria, S. E., 107
Lurquin, P., 82
Lurquin, P. F., 80, 82, 84, 85
Lyon, M. F., 350, 356, 364

M

Maaloe, O., 325, 326, 330, 333, 334
Maas, W. K., 106, 460
MacCormick, R., 215, 216
MacCosham, V., 111, 112, 377, 465, 467, 470, 471
MacGillivray, A. J., 255, 256
MacHattie, L. A., 109
Machemer, H., 423, 424
Mackay, D., 370
MacKay, V., 374
Mackenzie, I. A., 90
MacLeod, R., 209
MacNab, G. M., 18
Macnab, R. M., 176, 177, 398, 399
MacPhee, D. G., 380

Mage, R., 16
Magnusdottir, R. A., 112, 113
Maher, D. L., 332
Mahy, B. J. W., 213
Mailhammer, R., 41
Mainwaring, W. I. P., 260
Maisel, J., 204, 215, 216
Maisel, J. E., 204
Mäkelä, P. H., 165
Malacinski, G. M., 442
Malamy, M. H., 108, 109, 323
Malogolowkin-Cohen, C., 57, 58
Manaker, R. A., 287
Manca de Nadra, M. C., 40, 41
Mandel, J. L., 245
Mandelstam, J., 31, 33, 35
Mangel, W. F., 204, 243
Mangels, W., 195
Mani, J.-C., 189
Manly, K. F., 209
Manning, A., 435
Manning, J. S., 280
Mannino, S., 384
Mans, R. J., 248, 253, 257
Maples, V. F., 319
Marchelli, C., 105, 457
Marcus, D. M., 14
Margolin, P., 111, 112
Marin, G., 186, 193, 196, 197
Marinković, D., 57, 59, 61, 66
Marinus, M. G., 384
Markham, R., 80
Marsden, H. S., 370
Marsh, W. L., 22
Martin, A., 18
Martin, G. S., 208, 214-16, 220
Martin, J. F., 170
Martin, M. A., 248, 250
Martin, R. III, 461
Martin, R. R., 107
Martin, W. J., 156
Martinez, I. I., 316
Martinez, R. J., 162, 167, 170
Martuscelli, J., 333
Maruo, B., 167, 320
Marushige, K., 240-42
Maruyama, T., 69, 73
Marver, D., 260
Maryak, J., 216
Maryanka, D., 242, 244, 255, 256
Marzluff, W. F. Jr., 240, 247
Masaaki, H., 258
Maschler, R., 320
Mason, S. J., 22
Mason, W. S., 210, 213, 216, 217, 219, 220
Masters, M., 333
Mastrangelo, I. A., 94
Matricon, J., 442
Matsubara, K., 468

AUTHOR INDEX

Matsubara, S., 285
Matsumura, P., 163, 164, 175, 176, 402, 403, 411
Matthews, D. E., 257
Mattias, P., 193
Mattoccia, E., 324
Matzura, H., 329, 332
Maxam, A. M., 205-7, 471
May, J. T., 214
Mayer, F., 162
Mayr, E., 445
Mazaitis, A. J., 106
Mazza, G., 385, 386
McAllister, R. M., 204, 208, 220
McBean, I. T., 435
McCarter, J. A., 209, 217, 290
McCarthy, B. J., 240, 242, 246, 249-51, 254
McClain, W. H., 309
McClintock, B., 184, 381
McConahey, P. J., 212
McConkey, E. H., 255
McCoy, J. W., 350, 357, 358
McCulloch, E. A., 280
McCullough, J., 22
McDevitt, H. O., 16, 23, 285
McDonough, M. W., 163, 165
McEntee, K., 370-74, 386
McGinniss, M. H., 22
McGrath, J., 43
McGuire, W. L., 260
McKee, R. A., 88
McKenna, W. G., 333
McKusick, V. A., 14, 17, 20
McLaughlin, C. L., 17
McLellan, W. L., 211
McMilin, K. D., 382
McNicholl, P., 435
McReynolds, L. A., 265
McVey, E., 41
Meacock, P., 106
Means, A. R., 256, 259, 265
Meck, R. A., 94
Medeiros, E., 154
Meganathan, R., 214
Meier, H., 280, 281
Meijers, M., 194, 195
Meilhac, M., 244, 245, 258
Melchers, G., 86, 90, 94
Melchior, W., 254
Mellon, P., 216, 218-20
Mempel, W., 15
Mergeay, M., 82, 91
Merlo, D., 96
Meselson, M., 184, 376, 383
Meselson, M. S., 374, 375, 384
Mesibov, R., 400-2, 409
Mestriner, M. A., 19
Metroka, C. E., 216

Mettler, L. E., 52-54, 60, 61, 67, 68
Meuth, N. L., 209
Meuwissen, H. J., 23
Meyer, H. U., 51, 57-59, 67, 69
Meyerowitz, E. M., 440
Meyhack, B., 309
Meyhack, I., 309
Meyn, M. S., 370, 371
Meynell, E., 106, 107
Meynell, G. G., 106
Michaelis, G., 108
Michayluk, M. R., 90, 94, 95
Michel, F., 316
Michel, J. F., 33
Mickel, S., 107
Mildner, G., 33
Milhaud, P., 34
Miller, C. A., 107
Miller, C. G., 111, 112, 378, 465, 467
Miller, G. F., 156
Miller, J., 455, 456, 470
Miller, J. E., 370, 376
Miller, J. H., 315, 320
Miller, L. H., 22
Miller, R. A., 85, 90
Miller, R. C. Jr., 369, 382
Miller, R. V., 375, 386
Millet, J., 30, 33
Milne, B. S., 215
Milstein, C., 17
Milstein, C. P., 17
Miner, P., 106, 460
Mirand, E. A., 279, 280
Miszkowski, N., 442
Mitani, M., 107, 163
Mitani, S., 283
Mitchell, E., 461
Mitchell, J. A., 63, 64
Mitchell, W. M., 204
Mitchison, N. A., 131
Mitsuhashi, S., 109
Mitsui, T., 162
Miwa, I., 364
Miwatani, T., 170
Miyagi, M., 241, 246
Miyake, T., 385
Miyamoto, T., 213
Miyazawa, M., 130
Miyoshi, Y., 318, 323
Mizobuchi, K., 379
Mizuno, T., 323
Mizushima, S., 312, 318, 321, 323
Mizutani, S., 209
Mizuuchi, K., 104
Mobbs, J., 254
Moennig, V., 130, 208, 212
Mohit, B., 260
Mohla, S., 262

Molholt, B., 315, 320
Molin, S., 327
Molineux, I., 209
Molineux, I. J., 375
Möller, W., 315
Mölling, K., 208-10
Moloney, J. B., 279, 287
Monahan, J. J., 240, 244, 265
Monier, R., 321, 333
Monroy, G., 221, 242, 248, 258
Monstein, H.-J., 322
Montagna, R. A., 255
Montagnier, L., 204, 205
Monti-Bragadin, C., 380
Montoya, A., 96
Moody, E. E. M., 106, 107, 371
Morand, P., 370-72, 383, 387
Morgan, D., 382
MORGAN, E. A., 297-347; 301-7, 315, 333
Morgan, L. V., 184
Morgan, R. W., 377
Morgan, W. T. J., 14
Morganti, G., 17
Morris, D. W., 326
Morrison, T. G., 323
Morrissey, J. J., 332
Mortlemans, K., 380
Morton, D. L., 156
Morton, J. G., 240
Morton, N. E., 51
Moscovici, C., 281
Moseley, B. E. B., 370, 371, 386
Moser, C., 241
Mosharrafa, E., 108, 458
Moskowitz, G. J., 254
Mosteller, R. D., 322
Motoyoshi, F., 86
Mottershead, R. P., 370-72
Mount, D. W., 370-72, 383
Mourao, C. A., 64, 65, 70
Mousseron-Canet, M., 189, 262
Mueller, K., 332
Mueller-Lantzsch, N., 222
Mufti, S., 384
Mühlbock, O., 285
Mukai, T., 52-57, 59-62, 67, 68, 70-73
Mullen, R. J., 444
Mullenbach, E., 111
Muller, H. J., 51
Müller-Eberhard, H. J., 18
Mullin, B. C., 205
Munoz, L., 41
Munro, A. J., 17
Murata, M., 67
Murooka, Y., 331
Murphy, D. B., 16
Murphy, E. D., 280

AUTHOR INDEX 487

Murphy, H. M., 211, 219
Murphy, R. F., 259
Murray, C. D., 38
Murray, K., 460
Murray, N. E., 460
Murrell, W. G., 29
Muto, A., 311, 319, 326, 331
Myer, F. E., 282
Myers, D. D., 280

N

Nagaishi, H., 375, 376
Nagarkatti, S., 324, 325
Naiki, M., 14
Naitoh, Y., 423, 424
Nakamura, H. T., 37
Nakamura, Y., 316
Nakayama, T., 41
Nanney, D. L., 350, 353, 357-60
Nase, S., 131
Nash, H., 452, 458, 460
Nashimoto, H., 311, 316-21, 327
Nasi, S., 105, 315, 320, 457
Naso, R. B., 207, 212
Natarajan, A. R., 193
Natarajan, A. T., 194, 195
Nathenson, S., 15, 154
Natvig, J. B., 16
Nawa, S., 93
Neidhardt, F. C., 324, 326, 332, 333
Neil, J., 308
Neiman, P. E., 21
Nelson, D. L., 29
Nelson, J., 40, 41, 43
Nelson, M., 434
Nermut, M. V., 211
Nester, E. W., 96
Netzel, B., 15
Neubauer, R. L., 216, 288
Neubauer, Z., 453, 456, 457
Nevers, P., 109, 459, 471
Ng, S. Y., 241, 247
Nichols, J. L., 204
Nichols, M. E., 22
Nicolette, J. A., 260
Nicolson, M. O., 204, 220
Nienhuis, A. W., 248, 250, 251
Nierhaus, K., 318
Nierlich, D. P., 326
Nikaido, H., 167
Nikolaev, N., 304, 308
Nilan, R. A., 79
Ninio, J., 322
Nisen, P., 109
Nishimura, A., 106, 165
Nishimura, Y., 106

Nisioka, T., 107
Nisonoff, A., 16
Nissley, P., 43
Niwa, O., 282, 287
Nola, E., 260-62
Noll, M., 254
NOMURA, M., 297-347; 108, 115, 164, 299-307, 309-18, 320-26, 329-33
Nomura, S., 217, 281
Norris, T. E., 331
Notani, N. K., 383, 386
Notides, A., 259
Novick, A., 110, 378
Novick, R., 109
Novick, R. P., 386
Novitski, E., 55
Nowinski, R. C., 132, 136, 154, 207, 211
Nusse, R., 154

O

Obara, T., 205
Obata, Y., 130, 136, 138, 140
Obe, G., 192, 193, 195
O'Brien, E. J., 162
Odaka, T., 278, 283, 285
O'Dea, M. H., 374
O'Donnell, P., 130, 135, 141
Oebbecke, C., 332
Ogata, Y., 109
Ogawa, H., 370, 373, 374
Ogawa, Y., 254
O'Gorman, P., 129
Oguchi, T., 163, 168, 170, 173, 175
Ogura, H., 209, 213
Oh, G. R., 106
Ohi, S., 386
Ohlenbusch, J., 240, 242
Ohnishi, O., 52-55, 72, 73
Ohnishi, S., 57, 60, 66
Ohsawa, H., 320.
Ohta, K., 164, 176
Ohta, T., 70
Ohta, Y., 93
Ohtsubo, E., 106-8, 116, 300-2, 304
Ohyama, K., 85, 93
Oishi, M., 113, 375
Okuyan, M., 208
Olaisen, B., 18
OLD, L. J., 127-60; 128-32, 134-38, 140, 142-47, 150, 151, 153, 154, 156, 208, 280, 285, 286, 288, 290
Olins, A. L., 254
Olins, D. E., 254
Olive, D., 422

Olivera, B., 240, 242
Olivieri, G., 189, 196
Olsen, R. H., 107
Olshevsky, U., 208, 220
Olson, M. O. J., 255
O'MALLEY, B. W., 239-75; 240, 241, 243-45, 248-51, 253, 254, 256-65
Ono, M., 324
Oosawa, F., 430
Oostra, B. A., 307, 331
Ordal, G. W., 401, 402, 404, 405, 407
Oroszlan, S., 208
Orrego, C., 40, 41
Orrick, L. R., 255
Osawa, S., 310, 311, 317-19, 326
Osawa, T., 316
Oshima, C., 57, 60, 66
Osmond, B. C., 422
Ossowski, L., 214
Ostertag, W., 204, 218
Ostroy, S. E., 423
Otaka, E., 318
Otten, J., 214
Otten, J. A., 282
Ou, J. T., 380
Owada, M., 214
Owaribe, K., 162, 167
Owens, R. B., 280
Oyen, R., 22
Ozaki, H., 241
Ozaki, M., 318
Ozawa, S., 422
Ozeki, H., 112, 114, 164, 165, 309

P

Pace, B., 309
Pace, N. R., 300, 304, 308, 309, 320
Pacifici, M., 214
Padgett, T., 213, 215
Painter, R. B., 196, 197
Pak, W. L., 417, 421, 423, 427, 428, 439
Pal, B. K., 208
Palchaudhuri, S. R., 106
Palitti, F., 376
Palm, P., 311
Palmer, L., 327
Pandey, J., 74
Panet, A., 205, 206, 209, 211
Pantelouris, E. M., 289
Pao, C. C., 327
Pardee, A. B., 369-74, 382, 383, 386
Pardo, D., 318

AUTHOR INDEX

Pardoll, D., 383
Parish, C. R., 162, 171
Parisi, E. C., 383
Park, B. H., 22
Park, W. D., 248, 257
Parker, C. S., 241, 247
Parker, J., 314-16, 320, 324, 325, 332
Parker, S., 316
PARKINSON, J. S., 397-414; 163, 164, 177, 398, 400, 403-5, 407, 409, 458
Parks, W. P., 208, 214-16, 218, 219, 258, 288-90
Parma, D. H., 111, 112, 379
Parmiani, G., 156
Parsons, J. T., 204, 206, 221
Parsons, P. A., 417, 435
Passarge, E., 189, 191, 194-96
Pastan, I., 43, 214, 327, 333
Pasternak, J., 364
Patel, G. L., 258
Pathak, S., 190, 193
Pato, M. L., 327
Patterson-Delafield, J., 167
Paul, A. V., 376
Paul, J., 246, 248, 251, 253, 255
Paul, R. C., 427
Pauli, G., 214
Pawson, T., 208, 220
Payne, F. E., 208
Peacock, W. J., 184, 186, 189-92
Pearce, L. E., 106, 107
Pearce, S. M., 35
Pearson, M. L., 379
Peck, E. J. Jr., 260
Pedersen, F. S., 326
Pedersen, S., 316, 324-26, 328, 331-33
Peebles, P. T., 215, 288
Pelc, S. R., 186
Pelcher, L., 93
Penman, S., 246
Pera, F., 193
Pereira, M., 375
Perkins, H. A., 17
Perlman, D., 107
Perlman, R. L., 333
Pero, J., 40, 41, 43, 457
Perry, P., 187, 189-92, 194, 195, 197
Perry, R. P., 308, 309
Peter, G., 80, 81
Peterken, B. M., 260
Peters, G., 205
Peterson, J. A., 29
Peterson, J. L., 255
Petre, J., 319-21
Petrusek, R. L., 41, 43

Pettersson, I., 315, 316
Pettijohn, D. E., 373
Pfeifer, D., 105, 108, 109, 453, 454, 457, 462
Phillips, R. B., 357
Phillips, R. L., 80
Picciano, D. J., 255
Piepersberg, W., 312, 316, 318, 319, 321, 322
Pierard, A., 110-12
Pietrzykowska, I., 383
Pifko, S., 311, 319, 324
Piggot, P. J., 31, 33-35, 38
Pilacinski, W., 108, 458
Pimpinelli, S., 189, 196
Pincus, S. H., 22
Pincus, T., 130, 135, 137, 279, 282, 287
Pinney, R. J., 380
Pinto, L. H., 417, 427, 428, 439
Pister, L., 208, 212
Pitkanen, A., 204
Planitz, M. A., 205
Planta, R. J., 247
Plapp, F. W., 418
Plassmann, H. W., 209
Platt, T., 43, 44
Platz, P., 23
Polisky, B., 254
Polivanov, S., 63
Pollara, B., 23
Pollard, E. C., 370, 371
Polsinelli, M., 386
Poodry, C. A., 440
Popp, R. A., 205
Portalier, R., 373
Post, L., 301-6, 310-15, 317, 324, 326, 330, 332
Potrykus, I., 89, 90, 92, 93
Potter, H., 376, 377, 380
Potter, M., 16
Povey, S., 19
Powell, J. F., 34
Power, J. B., 90, 92, 94, 95
Pratt, I., 36
Preer, J. R. Jr., 350, 363
Prehn, R. T., 156
Prescott, D. M., 186, 190, 192, 193, 196, 197
Press, J. L., 16
Press, R., 106, 460
Pressman, D., 131
Presti, D., 442
Prieur, M., 187, 189-91
Primakoff, P., 309, 333
Propp, R. P., 17
Prout, T., 60, 63
Prozesky, O. W., 165
Ptashne, K., 106-9
Puca, G. A., 260-62

Pugh, J. E., 350, 356, 364
Pun, P. P. T., 38
Purchase, H. G., 208

Q

Quade, K. E., 204
Quigley, J. P., 208, 214
Quinn, W. G., 434
Quintrell, N., 154, 205, 213

R

Rabson, A., 309
Rabson, A. R., 18
Rabstein, L. S., 278
Race, R. R., 14, 21, 22
Radding, C. M., 369, 373-75, 377, 384
Radman, M., 377, 379, 381, 383
Rae, M. E., 107, 111
Raff, M. C., 131
Rafizadeh, B., 16
Rahat, A., 74
Rainaldi, G., 196
Rak, B., 108
Rakic, P., 417
Ramakrishnan, T., 373, 382
Ramsey, G., 370, 376
Randall, E. P., 370, 371
Rands, E., 288
Rankis, V., 162, 170, 171
Ranney, H. M., 19
Rasheed, S., 278
Raska, I., 162
Rasko, I., 94
Rasmuson, M., 19
Rauscher, F. J., 278
Ravin, A. W., 111
Ray, P. N., 379
Raynaud-Jammet, C., 262
Reader, R. W., 163, 175, 177, 398, 399, 403-5, 407
Ready, D. F., 438-40
Reanney, D., 104
Rebel, W., 80, 81
Rédei, G. P., 91
Reed, K., 257
Reeder, R. H., 246, 247, 249, 250
Reeh, S. V., 316, 324-26, 332, 333
Reich, E., 214
Reichert, H., 434
Reid, R. E., 333
Reif, A. E., 130
Reif, H. J., 108, 109, 113, 379, 458, 459, 470, 471
Rein, A., 282, 287

AUTHOR INDEX 489

Rein, R., 381
Reinens, G., 333
Reinert, J., 95
Reiness, G., 31, 37, 41, 331, 333
Resnick, M. A., 374, 377, 384
Retel, J., 247
Reynolds, F., 309
Rhaese, H.-J., 31, 32, 37
Rho, H. M., 204, 210
Rhona, M., 170
Rice, J. M., 156
Rich, M. A., 278
Richards, O. C., 205
Richardson, C. C., 379, 384
Richardson, M., 240
Riches, P. G., 255
Richmond, K. M. V., 109
Richmond, M. H., 107, 109, 110
Rickwood, D., 255
Riehm, H., 192, 193
Rifkin, D. B., 208, 211, 214
Rigby, P. W. J., 110, 379
Riggins, C. H., 204
Riley, M., 116, 376
Rill, R., 254
Rinaldi, G., 310, 324
Ring, J., 243
Ringold, G., 221
Ris, H., 163, 187, 452
Ritter, H., 19
Rivat, C., 17
Rivat, L., 17
Robberson, D. L., 265
Robbins, K. C., 215, 216
Robbins, M., 197
Robbins, P. W., 214
Roberts, C. W., 370, 371, 373
Roberts, J., 309
Roberts, J. W., 43, 44, 370, 371, 373
Robertson, A., 59, 63
Robertson, H. D., 309
Robin, M. S., 208, 221
Robinson, M. K., 109, 110
Robinson, W. S., 204
Roby, K., 214
Rochaix, J. D., 109, 302, 304, 315, 324, 325, 332
Rodgers, G., 35, 38
Rodin, B., 194-97
Roeder, R. G., 241, 246, 247, 249
Rogentine, G. N., 285
Rogolsky, M., 37
Rohrschneider, J. M., 209
Rohrschneider, L. R., 155, 212, 214
Rolfe, B. G., 87, 88
Rollins, E., 241

Rommelaere, J., 196
Rongey, R. W., 209
Roni, R., 212
Ropartz, C., 17
Rörsch, A., 80, 324, 370, 371
Rosbash, M., 240
Roscoe, D. H., 36, 40
Rose, J. K., 204
Rosen, F. S., 17-19
Rosen, J. M., 240
Rosen, L., 309, 318, 319
Rosenak, M. J., 221
Rosenberg, M., 309
Rosenberg, S. A., 257
Rosenblatt, S., 204-6, 219
Rosenfield, R. E., 14
Rosenthal, L. J., 205
Rosenwirth, B., 109
Rosner, J. L., 109, 374
Ross, W., 456
Rosset, R., 300, 303, 306, 318, 321, 333
Rossier, A., 80, 82, 83
Rossman, T., 370, 371
Rosypal, S., 105
Roth, J. R., 111, 112, 378, 465, 467
Rothenberg, E., 206, 220
Rothman, R. H., 375
Rothmann, I. K., 22
Rothstein, D. M., 38
Roulland-Dussoix, D., 213, 215
Rousseau, G. G., 260
Rovera, G., 257
Rowe, W. P., 134, 138, 141, 142, 153, 208, 278, 279, 281-83, 288-90
Rownd, R., 107
Rownd, R. H., 247
Roy, A. K., 248, 249, 254
Roy-Burman, P., 208, 215
Rubens, C., 109, 110
Rubenstein, I., 80
Rubenstein, P., 14
Rubin, H., 204, 213, 214
Rucknagel, D. L., 19
Ruddle, F. H., 195
Rudiger, H. W., 195
Rueckert, R. R., 208
Ruffler, D., 316, 318, 319, 321
Runge, R., 106
Rünzi, W., 329
Ruoslahti, E., 214
Rupp, W. D., 384
Rusch, H., 254
Russell, E. S., 280
Russell, L. B., 280
Russell, R. L., 116, 429-32, 435
Rutberg, L., 106
Rutter, W. J., 187, 247

Ryder, L. P., 23
Rymo, L., 204, 221
Ryter, A., 30, 33, 34, 323

S

Sachs, L., 287
Sacristán, M. D., 94
Sadaie, Y., 386
Sadowski, P. D., 382
Saedler, H., 105-9, 113, 117, 313, 356, 364, 379, 458, 459, 470, 471
Safwat, F., 94
Sager, R., 350, 364
Sahasrabuddhe, C. G., 254
Sakai, F., 86
Sakaki, Y., 375, 381
Sakano, H., 309
Sakuma, K., 256
Salas, J., 255
Salas, M., 40
Salden, M., 204, 208
Salden, M. H. L., 221
Salzberg, S., 208
Samer, L., 380
Samoiloff, M. R., 435
Sanchez-Anzaldo, F. J., 29, 32, 42
Sand, G., 379
Sanders, J. E., 21
Sanderson, K. E., 107, 116, 164, 165, 381
Sandler, L., 184
Sanford, K., 253
Sanger, F., 471
Sanger, R., 14, 21, 22
Sano, K., 73
Santo, L. Y., 29, 38, 39
Sarkar, K. R., 91
Sarkar, N. H., 154
Sarkar, S., 86
Sarma, P. S., 208, 212
Sarngadharan, M. G., 209
Sarvella, P. A., 280
Sasazuki, T., 23
Sass, B., 212
Sassen, A., 248
Sastry, G. R. K., 117
Sato, H., 130, 136, 138, 142, 286
Sato, K., 460
Satow, Y., 424, 425
Sauer, B., 311
Saunders, G. F., 254
Savage, M., 255
Sawyer, R. C., 205, 209
Scaife, J., 106, 299, 314, 315, 332
Schabtach, E., 442

Schachat, F. H., 420
Schade, S., 163
Schaeffer, P., 29-35, 94
Schäfer, D., 311, 317, 319
Schäfer, W., 130, 208, 211, 212
Schalet, A., 52
Schanfield, M. S., 19
Scharff, R., 375
Scharpe, A., 252
Schedl, P., 309
Scheele, C. M., 213
Schein, S. J., 424, 425
Schell, J., 105
Schenley, C. K., 132
Scher, C. D., 215, 278, 288
Schieder, O., 90
Schiffer, D., 111, 112, 379
Schilcher, F. von, 436
Schilperoort, R. A., 80, 84
Schiltz, E., 318
Schincariol, A. L., 221
Schindler, J., 287
Schleif, R., 105, 454-57
Schleif, R. F., 327, 331
Schlesinger, M., 145
Schlessinger, D., 304, 308-10, 319, 324
Schlom, J., 287
Schluter, B., 282
Schmeissner, U., 455, 456, 470
Schmid, W., 192, 193, 195
Schmidt, F. J., 309
Schmidt, F. W., 208
Schmidt-Nielsen, B. K., 419, 423
Schmitt, R., 162, 163
Schmitz, A., 455
Schnedl, W., 189
Schneider, E. L., 191-93
Schneider, S., 382
Schnös, M., 469
Scholz, S., 15
Schonberg, S., 192-94, 197
Schrader, W. T., 259-64
Schrenk, W. J., 460, 464
Schuh, V., 282
Schultz, J. S., 16
Schumacher, G., 333
Schwarting, G. A., 14
Schwartz, D., 184
Schwartz, D. E., 206
Schwartz, D. O., 113, 379, 453
Schwartz, L. B., 241, 247
SCHWARTZ, R. J., 239-75; 243, 244, 248, 249, 258, 260, 262-64
Schwarz, H., 130, 212
Schwarzacher, H. G., 189
Sciaky, D., 96
Scolnick, E. M., 208, 214-16, 218, 219, 258, 288-90
Scott, J., 109

Scraba, D., 109, 454
Sebastian, J., 247
Sedgwick, S. G., 370, 371
Seecof, R. L., 444
Segall, J., 40, 41
Seidman, J. G., 309
Seifert, E., 208, 212
Sen, A., 208
Serra, M. C., 386
Setlow, J. K., 383, 386
Setlow, P., 32, 37
Setlow, R. B., 369, 383
Sevall, J. S., 255
Seyffert, W., 80, 81
Sgaramella, V., 375
Shander, M. H. M., 211
Shank, P. R., 205, 206
Shanmugan, G., 221
Shaper, J. H., 162, 170, 171, 208
Shapiro, J., 108, 117, 371, 381, 382
Shapiro, J. A., 105, 108, 109, 453, 454, 457, 458
Shapiro, S. Z., 207, 212
Sharp, P. A., 106, 108
Shatkin, A. J., 204
Shaw, E. J., 107
Sheikh, K., 80
Sheldon, E. W., 54, 63, 64
Shell, L., 327
Shen, C. J., 258
Shen, F.-W., 131, 136, 142
Sheppard, D. E., 426
Sheridan, R. E., 429
Sherr, C. J., 208
Shigeno, N., 130, 131
Shih, T. Y., 248, 250
Shimada, H., 195, 197
Shimada, K., 104, 109, 164, 177, 458, 460, 464
Shimazu, H., 187
Shimura, Y., 309
Shine, J., 205-7
Shinoda, S., 170
Shipley, P., 107
Shlaes, D. M., 375
Shorenstein, R. G., 36, 40, 41
Shows, T. B., 151
Shoyab, M., 206
Shreffler, D. C., 16
Shulman, M. J., 104, 454, 462, 464
Shuster, J., 16
Shyamala, G., 259
Sica, V., 260-62
Siccardi, A. G., 385
Siddiqi, O., 422, 423
Siegler, R., 278
Sigel, M. M., 214
Signer, E. R., 105, 464
Sigurbjörnsson, B., 79

Silengo, L., 304, 308
Silver, L., 107
Silver, R. P., 107
Silverman, M., 162-66, 168, 175-77, 356, 364, 381, 387, 398, 402-11
Silverman, M. R., 167
Siminovitch, L., 280
Simmons, A. S., 57, 58
SIMMONS, M. J., 49-78; 54, 55, 63, 64, 73
Simon, E. M., 358
Simon, M., 161-70, 175-77, 356, 364, 381, 387, 398, 402-11
Simon, M. I., 402, 403, 407, 411
Simonian, M. H., 322
Simpson, R. J., 255
Simpson, R. T., 257
Sing, C. F., 16
Singer, S. J., 214
Sinkovics, J. G., 287
Sinsheimer, R. L., 374, 384
Skaar, P. D., 363
Skalka, A., 375, 381
Sklar, V. E. F., 241, 247
Skurray, R., 106
Slaughter, C. A., 19
Slepecky, R. A., 34
Slettenmark-Wahren, B., 136
Sloma, A., 311, 319, 324
Smith, A. E., 208, 220
Smith, D., 259
Smith, E. J., 208
Smith, G. P., 379, 381
Smith, H., 88
Smith, H. H., 94
Smith, H. O., 375
Smith, I., 311, 319, 324
Smith, J. D., 116, 309
Smith, J. T., 380
Smith, K. C., 370, 377
Smith, K. D., 246, 252
Smith, M. M., 248
Smith, R. E., 204, 208, 214, 215
Smith, R. G., 262, 265
Smithies, O., 17, 18
Smoler, D. F., 209
Smotkin, D., 204-6, 219
Snell, G. D., 15, 154, 285
Snyder, H. W. Jr., 136, 140
Snyder, M., 111, 112, 379
Sobell, H. M., 376, 379
Socher, S. H., 263
Soeiro, R., 282
Sofer, W., 443
Sogin, M. L., 309
Sokatch, J. R., 107
Soll, L., 106, 111, 380
Sollner-Webb, B., 254

Solomon, A., 17
Solomon, E., 195, 197
Solomon, J. J., 208
Somers, C. E., 191
Somers, K. D., 214, 287
Sonenshein, A. L., 31, 36-38, 40, 311
SONNEBORN, T. M., 349-67; 350, 352, 353, 356, 358, 360, 363, 364
Sonnenberg, B. P., 241, 242, 254
Sotomura, M., 106
Southgate, E., 431-33, 440
Soyfer, V. N., 92, 93
Sparvoli, E., 185
Spassky, B., 57-59
Spatz, H. C., 434
Spears, C., 333
Spector, D., 213, 215
Spelsberg, T. C., 240, 241, 252, 253, 255, 257, 261, 262, 264
Spengler, B. A., 154
Sperandeo-Mineo, R. M., 384
Sperling, K., 192, 193
Spiegelman, G. B., 41, 43
Spiegelman, S., 209, 211
Spieth, H. T., 435
Spitznagel, J. K., 22
Spizizen, J., 33
Sporn, M. B., 255
Spring, S. B., 16
Springer, G. F., 14
Springer, M., 325
Springer, M. S., 402, 404-10
Springer, W. R., 408
Spudich, E. N., 167
Spudich, J. A., 29, 113, 457
Spudich, J. L., 398, 399, 401, 404
Squires, C., 43, 44
Squires, C. L., 43, 44
Sri Widada, J., 308
Staal, S. P., 33
Stacey, K. A., 379
Stackpole, C. W., 148, 153
Stadler, D. R., 369
Stahl, F. W., 184, 382, 384
Stahl, M. M., 382, 384
Stahly, D. P., 39
Stallings, R., 150
Stalon, V., 110-12
Stamminger, G., 326, 327
Stanbury, J. B., 23
Stanton, T. H., 146
Starbuck, W. C., 254
Stark, W. S., 439, 444
STARLINGER, P., 103-26; 105, 108, 109, 113, 117, 313, 356, 364, 458
Staub-Nielsen, L., 23

Stavnezer, E., 204, 205, 213, 221
Stedman, Edgar, 254
Stedman, Ellen, 254
STEEVES, R. A., 277-96; 212, 278, 280, 282-84, 286
Steggles, A. W., 241, 248, 250, 251, 255, 261, 262, 264
Stehelin, D., 213, 215
Stein, G. S., 248, 253, 255, 257
Stein, J. L., 248, 253, 257
Steinbauer-Rosenthal, I., 15
Steinberg, A. G., 16, 17
Steinheider, G., 284
Steitz, J. A., 304, 308
Stent, G. S., 325, 326
Stephens, J. C., 333
Stephenson, J. R., 155, 208-10, 216, 217, 220, 279, 280
Sterlini, J. M., 31
Stern, C., 55, 184
Stern, H., 80, 82-85
Sternberg, N., 375, 381, 461, 463, 470
Stetka, D. G., 194, 195
Stetten, G., 190, 192-97
Stewart, J., 260
Stich, H. F., 193, 196
Stimpfling, J. H., 286
Stinchcomb, D., 40, 41
Stocker, B. A. D., 163, 165, 175, 176, 380, 403, 405
STOCKERT, E., 127-60; 129, 130, 135-38, 140-42, 144, 146, 147, 150, 151, 153, 154, 280
Stodolsky, M., 107, 111, 379
Stöffler, G., 316, 318-21
Stoll, E., 216, 218-20
Stone, K. R., 214, 215
Stone, W. S., 384
Storb, R., 21
Strand, M., 130, 143, 207, 208, 212, 280
Strange, P. G., 402
Straus, D. S., 110-12, 114, 377-80
Straus, L. D., 377-79
Strauss, B., 197
Strauss, N., 37, 38, 41
Street, H. E., 93
Stretton, A. O. W., 432
Strickberger, M. W., 62
Stromberg, K., 208
Stroun, M., 80-83, 93
Strycharz, W. A., 310-14, 316, 317, 324-26
Stubblefield, E., 186, 215
Stück, B., 154
Studier, F. W., 243, 304, 308
Stumpf, W. E., 260
Sturm, M. M., 155

Sturtevant, A., 465
Stutman, O., 208
Sublett, R., 109, 110
Sueoka, N., 38, 311, 386
Sugano, H., 287
Sugiyama, T., 187, 195
Sullivan, D., 205
Sulston, J., 419
Sulston, J. E., 433, 440, 441
Sümegi, J., 332
Sumida-Yasumoto, C., 37-39, 41, 42
Summers, J., 206
Summers, W. C., 323
Sunshine, M., 311
Sunshine, M. G., 105, 311
Susskind, M., 196
Suzuki, D. T., 419, 423, 440, 443
Suzuki, H., 165, 166, 168-70, 174, 175
Suzuki, K., 370
Suzuki, M., 86
Suzuki, S., 285
Suzuki, T., 165, 166, 168, 260
Sved, J. A., 64-66, 70, 74
Sveda, M. M., 282
Svejgaard, A., 23
Sverak, L., 205
Swallow, D. M., 19
Swanson, J., 22
Swenson, P. A., 370, 371, 380, 383
Swinton, D. C., 85
Sykes, R. B., 107
Symonds, N., 111, 369
Sypherd, P. S., 310, 311, 320, 321
Szeto, W. W., 96
Szmelcman, S., 176, 411
Szulmajster, J., 40, 41
Szybalski, E. H., 454, 462
Szybalski, W., 108, 109, 114, 310, 452, 454, 458, 462

T

Tai, P.-C., 320
Taira, T., 163, 164, 170
Takahashi, E. A., 111
Takahashi, N., 162, 167
Takahashi, T., 130, 131, 142, 143, 153, 280
Takata, R., 316-18
Takebe, I., 86, 90, 93
Takeda, Y., 170
Tal, J., 213, 215, 219
Tamaki, M., 318
Tan, C. H., 246
Tanaka, K., 318
Tanaka, N., 310, 324

Tanimoto, B., 37
Taswell, H. F., 22
Tata, J. R., 260
Tates, A. D., 194, 195
Taub, S. R., 360-62
Tauschel, H. D., 162
Tawada, K., 430
Taylor, A. L., 105, 309, 317, 371, 372, 383, 400
Taylor, B. L., 403-5, 408
Taylor, J. H., 184, 189, 190
Taylor, J. M., 204-6, 209
Teague, R., 378
Tedesco, P. M., 398, 399, 401, 408
Tefankjian, A., 358
Teich, N. M., 279
Teisberg, P., 18
Tekuno, S. I., 370, 371, 373, 382
Temin, H. M., 204-6, 209, 212, 221
Temin, R. G., 51, 52, 57-60, 66, 67, 69
Teng, C. S., 241, 255
Teng, C. T., 255
Teng, N. N. H., 419, 423
Teng, Y.-S., 19
Tennant, J. R., 285
Tennant, R. W., 132, 282
Teplitz, R. L., 444
Teraoka, H., 318
Terasaki, P., 16
Terhorst, C., 315
Terry, W. D., 16
Terzaghi, B., 111, 465
Tessman, I., 453
Thammana, P., 320
Thiel, H.-J., 212
Thipayathasana, P., 176
Thomas, C. A. Jr., 96
Thomas, E. D., 21
Thomas, G. H., 189, 192, 194, 196
Thomas, J. O., 111, 112, 254, 465, 466
Thomas, M., 460
Thomas, R., 105, 314, 322
Thomas, T. L., 258
Thompson, M. J., 380
Thomsen, M., 23
Thomson, J. N., 420, 421, 430-33, 440
Thorsby, E., 18
Thrall, C., 248, 257
Tice, R., 191-93
Tidwell, T., 57, 58
Till, J. E., 280
Ting, R. C., 107
Tipper, D. J., 36
Titov, Y. B., 92, 93
Tjian, R., 40, 41

Tobari, I., 55, 57, 60, 61, 67
Tobari, Y. N., 52
Tobia, A., 214
Tocchini-Valentini, G. P., 310, 324
Todaro, G. J., 208
Toews, M. L., 406
Toft, D., 259
Toft, D. O., 260, 263
Tokimatsu, H., 309
Tokuyasu, K., 169, 170
Tomizawa, J. I., 370, 373, 374
Tomkins, G. M., 260
Tompkins, R., 442
Tonegawa, S., 117
Topisirovic, L., 322
Toth, F. D., 286
Toussaint, A., 105, 382
TOWLE, H. C., 239-75; 245, 248-51, 258
Toyoshima, K., 214
Tracey, M. L., 64-66, 70
Traub, P., 318
Travers, A., 327, 328, 331, 332
Travers, A. A., 327, 331
Tress, E., 130, 135, 208, 211, 212
Trgovcevic, Z., 384
Trimmer, R. W., 212
Troll, W., 370, 371
Tronick, S. R., 170, 208-10, 216
Trowsdale, J., 111, 113, 114, 117
Troxler, D. H., 288-90
Trudel, P. J., 280
Truesdell, S., 373, 382
Trusal, L., 206
Tsai, M. J., 243-45, 248-51, 253, 254, 256-58, 260, 262
Tsai, S. Y., 243, 244, 248-51, 253, 254, 256-58, 260, 262
Tsai, Y. H., 256
Tsang, N., 398, 399
Tso, W.-W., 163, 175, 177, 398-404
Tsuchida, N., 221
Tsugawa, A., 311, 317, 319, 320, 327
Tsukamoto-Adey, A., 211
Tuffrey, M. A., 137
Tung, J.-S., 130, 135, 136, 138, 140, 142, 143, 208
Turbin, N. V., 92, 93
Turner, B. M., 19
Turner, H. C., 208
Turner, H. D., 290
Turner, V. S., 19
Tweats, D. J., 380

Twose, P. A., 108
Tye, B. K., 109, 110
Tyler, B., 320
Tysper, Z., 245

U

Uchida, A., 32, 35
Uchida, H., 311, 316, 317, 319-21, 323, 327
Udvardy, A., 332
Ueda, N., 195
Uenaka, H., 195
Uhlenbruck, G., 14
Ulrich, T. H., 80
Unger, L., 107
Unger, M., 300
Unger, R. C., 381
Unkeless, J. C., 214
Upadhya, M. D., 86
Uphoff, D., 55
Urm, E., 213, 333

V

Vaage, J., 156
Vaczi, L., 286
Vaheri, A., 214
Valentine, R. C., 176
Valenzuela, P., 247
van Acken, U., 318
Van Alstyne, D., 165
Vanaman, T. C., 208
van Blitterswijk, W. J., 154
van Buul, P. P. W., 194, 195
van den Elsacker, S., 105
Van den Ende, P., 111
Van de Putte, P., 105
van der Gaag, H. C., 283
Van Der Parren, J., 82
Vanderslice, R., 443
Van der Werf, P., 406
Van de Vate, C., 111
van Dillewijn, J., 370, 371
Van Dorp, B., 376
Van Holde, K.E., 254
Van Houten, J., 424, 429
Van Keulen, H., 247
van Loghem, E., 16
Van Montagu, M., 105
van Nie, R., 154
Van Ooyen, A. J. J., 31, 37, 307, 331
Van Parijs, R., 252
van Rood, J. J., 15
Van Zaane, D., 207, 208, 220
Vapnek, D., 379
Varmus, H. E., 154, 204-6, 209, 210, 213, 215, 217, 219, 221, 333
Vary, P. S., 403

AUTHOR INDEX 493

Vasil, I. K., 93
Vass, W. C., 216, 288
Vasseur, M., 308, 309
Venetianer, P., 332
Venkov, P., 108, 309
Venuta, S., 214
Verma, I. M., 209, 210, 221
Vermeulen, C. W., 300
Verschoor, G. J., 324
Vetter, D., 382
Vigier, P., 204, 205, 214
Vigue, C., 443
Villarroel, R., 318, 319, 322
Vinuela, E., 40
Vitetta, E. S., 130, 135, 136, 140, 143
Vogel, W., 192, 194, 195
VOGT, P. K., 203-38; 203-5, 208-10, 212-20, 281
Vogt, V. M., 207, 220
Vohra, A. K., 210
Vola, C., 303, 321
Volkert, M., 383
Von der Helm, K., 204, 207, 208, 220
Von Heyden, H. W., 252
von Meyenburg, K., 326, 327
von Wichert, P., 195
Vyas, G. N., 16, 17

W

Wachtel, S. S., 146
Wagner, R. E., 377, 379, 381, 383
Wakabayashi, K., 162, 169, 283
Walen, H. K., 185, 189
Walet-Foederer, H. G., 92, 93
Walford, N. K., 288, 290
Walker, A. C., 372, 383
Walker, J. A., 439
Wallace, B., 52, 70, 73, 74
Wallace, D. C., 116
Wallin, A., 93, 94
Wanebo, H. J., 156
Wang, A. C., 16
Wang, E., 214
Wang, L.-H., 204, 216-20
Wang, P., 22
Wang, S. Y., 213
Wang, T. Y., 252, 255, 257
Ward, D. C., 248
WARD, S., 415-50; 428-31, 435
Ware, M., 283
Ware, R. W., 430, 431
Warren, J. C., 241
Warren, S. C., 34
Warrick, H. M., 403-5, 408
Wartell, R. M., 36
Waskell, L., 43

Watanabe, T., 57, 106, 109, 131
Watanabe, T. K., 57, 59-62, 66, 70, 71
Waters, L. C., 205
Waterston, R. H., 420
Watkins, W. M., 14
Watson, J. D., 331
Watson, K. F., 209, 211
Watson, L., 18
Watson, N., 308
Watson, R. J., 314-16, 320, 324, 325, 332
Watts, J. W., 86
Waxman, S. H., 21
Wayman, L., 386
Weatherall, D. J., 19
Weatherford, S. C., 309
Weaver, C., 216
Weaver, S., 382
Webb, R. B., 370, 371
Weber, J., 320
Weber, K., 326
Weber, M., 214
Wecker, E., 212
Wegner, R. D., 192, 193
Wehrly, K., 286
Weiden, P. L., 21
Weil, J., 111, 461, 465
Weinberg, F., 247
Weinberg, R. A., 204-6, 217, 219
Weingarten, H., 91
Weinmann, R., 241, 247
Weintraub, H., 254, 259
Weisberg, R., 461-64, 470
WEISBERG, R. A., 451-73; 104, 109, 375, 381, 452, 454, 458, 460-62, 464
Weisblat, D. A., 432
Weiss, R. A., 213, 216, 217
Weissbach, H., 315, 324, 332, 333
Weissman, I. L., 211
Weissman, S. M., 309
Weissmann, C., 204, 205, 216-21, 322
Weith, H. L., 206
Weitkamp, L. R., 19
Wellauer, P. K., 304, 308
Wells, R. D., 36, 204
Wensink, P. C., 246
Wernet, P., 16
West, C., 19
West, S. C., 370
Westermoreland, B. C., 452
Wetter, L. R., 95
White, J. G., 430-33, 440
Whitehead, H., 90
Whiteley, H. R., 41, 43
Whiting, P. W., 435

Wickus, G. G., 214
Widholm, J., 240, 242
Widnell, C. C., 260
Wiegand, R., 373
Wienberg, J., 187, 192
Wiersma, C. A. G., 432
Wilcox, K. W., 375
Wildenberg, J., 383
Wildman, S. G., 89, 94
Wilhelm, J. A., 255
Willetts, N., 106
Willetts, N. S., 106
Williams, B. G., 310, 312-14, 324
Williams, D., 216, 218, 288
Williams, D. R., 258
Williamson, R., 423
Wills, C., 57, 62
Willumsen, B. M., 326, 327
Wilsnack, R., 208
Wilsnack, R. E., 213
Wilson, C. B., 212
Wilson, G. N., 248, 250, 251, 255
Wilson, H. L., 170
Wilson, R. F., 259
Wimer, B. M., 22
Winchester, R. J., 16
Winslow, R. M., 327
Wistar, R. Jr., 162, 171
Witkin, E. M., 370, 382, 383
Witte, O. N., 211
Wittel, F. P., 324, 326, 333
Witter, R., 212
Wittmann, H. G., 312, 316-22
Wittmann-Liebold, B., 311, 315, 317-19
Wohler, W., 195
WOLFF, S., 183-201; 186, 187, 189-97
Wolford, N. K., 142
Wollman, E., 461
Wollman, E. L., 114
Wong, P. K. Y., 209, 217
Wong, P. T., 426
Woo, S. L. C., 265
Wood, D. A., 32, 35, 156
Wood, H. A., 140
Woods, P. S., 184
Worcel, A., 373
Wright, B. L., 163
Wright, B. S., 287
Wright, S., 67
Wright, T. R. F., 438
Wu, A. M., 209
Wu, M., 115, 305
Wyke, J. A., 214, 216, 217
Wyman, R., 432
Wyman, R. J., 421, 422, 426
Wyngaarden, J. B., 23
Wyss, O., 384

Y

Yabuuchi, E., 170
Yagi, Y., 107, 131
Yaguchi, M., 312, 318, 319
Yahara, I., 214
Yamada, H., 323
Yamada, M., 93
Yamada, S., 309
Yamada, Y., 309
Yamagata, H., 318, 323
Yamagishi, H., 112, 114
Yamaguchi, O., 57, 59-62
Yamaguchi, S., 163-66, 169, 170, 173, 176
Yamakawa, T., 37
Yamamoto, K., 113
Yamamoto, K. R., 261, 263
Yamamoto, M., 301-4, 315, 316, 324, 325, 332, 333
Yamamoto, N., 385
Yamamoto, T., 278, 283
Yamane, K., 167
Yamazaki, T., 56, 70-73
Yanagida, M., 162, 170
Yang, H. L., 31, 37, 41, 331, 333
Yang, W. K., 282
Yaniv, A., 211
Yanofsky, C., 43, 44, 112, 113, 326, 378, 379, 453, 457
Yao, K., 163, 164, 170
Yarmolinsky, M. B., 460, 462-64
Yates, J. L., 322, 323
Yeater, C., 213
Yeger, H., 209
Yehle, C. O., 36
Yeoh, G., 214
Yoakum, G., 370, 371
Yokota, T., 166
Yoneda, Y., 167
Yoshida, M., 214
Yoshikawa, I., 55, 60, 71-73
Yoshikawa, M., 106
Yoshiki, T., 136, 140, 208
Yoshikura, H., 287
Yoshimura, F. K., 204, 206
Yoshito, S., 386
Youn, J. K., 287
Young, M., 35
Young, M. W., 52
Young, R., 322
Youngs, D. A., 370, 375, 377
Yousten, A. A., 29
Yu, M. T., 300
Yuki, A., 309
Yunis, E. J., 15
Yura, T., 316
Yussa, Y., 285

Z

Zachau, H. G., 252
Zahalak, M., 309
Zakharov, A. F., 186, 187
Zambrano, F., 383
Zamecnik, P. C., 205, 206
Zarling, D. A., 212
Závada, J., 281
Zengel, J., 310-14, 316, 317, 324, 326
Zengel, J. M., 322
Zeuthen, J., 329, 332
Zieg, J., 356, 364, 379, 381, 387
Zillig, W., 311
Zimmermann, R. A., 321
Zinder, N. D., 382, 384
Zinkernagel, R. M., 286
Zipkas, D., 116
Zipser, D., 105
Zissler, J., 108, 458
Zonta, L. A., 17
Zubay, G., 31, 37, 41, 241, 242, 254, 331, 333
Zuccarelli, A. J., 374, 384
Zumpft, M., 151
Zylber, E. A., 246
Zytkovicz, T. H., 34

SUBJECT INDEX

A

α-Amanitin
  sensitivity of RNA tumor
    virus transcription to, 221
Ankylosing spondylitis
  association with HLA-B27,
    23
Antibody blocking test, 134,
  151-53
Antigenic modulation, 128,
  147-48, 150
  see also Cell surface antigens of mouse leukemia
Avian leukosis viruses
  see RNA tumor viruses
Avian sarcoma viruses
  see RNA tumor viruses;
    Oncogenic transformation
    by RNA sarcoma viruses

B

Bacillus subtilis
  see Sporulation in bacteria
Bacteriophage λ
  duplications in
    see Duplications
  λdv formation in, 468-70
  inverted repeats in, 468-70
  integration and excision of,
    104-5
  nonpropagable transducing
    phage of
    end points in, 463
    formation of, 461-65
    λ genes involved in formation of, 463-65
  transducing phage of
    bacterial deletions in formation of, 105; see also
      Deletions
    end points in, 105, 462
    formation of, 104-5, 460-63
    λ genes involved in formation of, 105, 460-62
Bacteriophage Mu
  integration and excision of,
    105
  inversion of G-loop upon
    prophage induction of, 115
  mediation of transpositions

and inversions by, 105
Bacteriophage P2
  role of int gene in deletion
    formation by, 105
Behavioral mutants
  of Caenorhabditis, 418, 420,
    429-30, 432-33, 436
  of crickets and locusts, 427-
    28
  of Drosophila, 422-24, 426,
    431, 434, 436-37, 440
  of Paramecium, 424-25
  selection procedure for isolating, 420, 424, 426-27,
    429-30, 434
Bloom's syndrome
  incidence of sister chromatid exchange in, 193-94,
    197
Burkitt's lymphoma
  clonal origin of, 21

C

Caenorhabditis elegans
  see Behavioral mutants;
    Chemotaxis in Caenorhabditis; Courtship behavior;
    Developmental mutants;
    Muscle mutants of Caenorhabditis; Neuroanatomy
    of Caenorhabditis; Neurochemistry
Cell surface antigens of mouse
  mouse leukemia
  G_{IX}
    as component of MuLV
      virion, 140
    expression of, 138, 140-
      42
    genetics of, 138-39
    identification of, 130, 132,
      135, 138
  $G_{(RADA1)}$ and $G_{(ERLD)}$,
    132, 143
  GSCA
    as component of MuLV
      virion, 140
    expression of, 137, 140-
      42
    identification of, 135-36
    and replication of MuLV
      genome, 136, 140, 142

methods of detection of, 128-
  35, 151-53
ML, 154
PC.1, 143-44
TL
  antigenic modulation of,
    128, 147-48, 150
  expression of, 128, 145,
    147-48, 153
  identification of, 128, 144
  inheritance of, 144-45
  as product of Tla locus,
    145-47, 149-50, 153-54
  spatial relation to other
    antigens, 151-53
X.1
  identification of, 131, 142
  see also Envelope glycoproteins of RNA tumor viruses; Murine leukemia virus
Chemotaxis in bacteria
  chemoreceptors, 398, 400-
    2
  active transport by, 401
  competition experiments
    with, 401
  stimulus detection by, 400-
    1
  flagellar rotation in, 398-99,
    402-6, 410
  see also Flagella of bacteria
  methionine requirement for,
    405
  methylation and demethylation of proteins in, 405-
    10
  sensory adaptation in, 398-
    99, 405-8
    definition of, 399
    gene products involved in,
      405-8
    models of, 406, 408
  stimulus-response behavior
    of, 398-99, 404
  stimulus transduction in
    definition of, 399
    identification of gene products of, 402-4
    interaction of gene products
      in, 404, 408-10
    models of, 406, 410-11
Chemotaxis in Caenorhabditis,
  428-31

495

496 SUBJECT INDEX

chemoreceptors of, 429-30
  competition experiments with, 429
  mutants of, 429-31
    correlation with anatomical defects, 430-31
    mosaic analysis of, 430
    and thermotaxic defects, 430
  stimulus transduction in, 428-30
Chemotaxis in ciliates, 429
Chromatin
  definition of, 240
  enzymatic digestion of, 254, 259-60
  reconsititution of, 251-57, 261
  shearing of, 242
  template capacity of, 241-43
  see also Transcription from chromatin
Chromosomal aberrations in procaryotes
  see Deletions; Duplications; Inversions
Chromosomal proteins
  histone, 240-41, 254
    and control of gene expression, 254-55, 261
  nonhistone, 241
    and control of gene expression, 251-52, 255-58, 261-62
  see also Chromatin
Chromosome loss during plant protoplast fusions, 94-95
Col plasmids
  integration of, 107
  recombination with R- and F-plasmids, 107
Courtship behavior
  of Caenorhabditis, 436-37
  pheromones in, 437
  of Drosophila, 435-36
  pheromones in, 436
  of wasps, 435

D

Deletions
  of E. coli
    ara operon, 105, 454-55
    gal operon, 105, 113, 453-54, 458-59
    lacI operon, 455
  end points of
    at insertion sequences, 458-59
    mapping of, 453-56

at phage attachment site, 458-59
at repeated sequences, 455-57
formulation of
  by F' particle fusion, 106
  by illegitimate recombination; see Illegitimate recombination
  models of, 459-60
  by recombination between rRNA genes, 307
  role of bacterial genes in, 457-58
  role of prophage genes in 457-58; see also Bacteriophage λ
  by unequal crossing-over, 455, 468
frequency of, 112-13, 452-53
  increase by mutagens, 452-53
  influence of base sequence on, 453
  selection for, 453-55, 457
Developmental mutants
  of Caenorhabditis nervous system, 440-42
  of Drosophila visual system, 438-40
Differentiation
  see Macronuclear differentiation in ciliates
DNA-dependent RNA polymerase
  see RNA polymerase
Drosophila
  see Behavioral mutants; Courtship behavior; Developmental mutants; Electrophysiology; Fertility in Drosophila; Fitness in Drosophila; Hybrid dysgenesis in Drosophia; Learning behavior in Drosophila; Mosaic fate mapping; Mutation rates in Drosophia; Neurochemistry; Overdominance in Drosophila; Selection in Drosophila; Viability in Drosophila
Dunn, Leslie Clarence
  in memory of, 1-12
Duplications
  in chromosomal evolution, 115-16, 465
  in evolution of genome size, 116

formation of
  by recombination between rRNA genes, 307-8
  frequency of, 112
  inverted repeat class of, 465, 468-70
    model for formation of, 468-70
  selection for, 110-11, 465
  tandem direct repeat class of, 111, 465
    instability of, 465-67
    model for formation of, 467-68
  transduction of, 111

E

Electrophysiology
  of Ascaris, 432
  of Drosophila
    flight muscles, 421-22, 426
    larval muscles, 423
  of Paramecium, 423-26
    comparison with muscles and nerves, 423-24
    membrane defects in, 424-26
    mutations affecting, 424-25
Endogenous viruses
  envelope helper activity of, 213
Envelope glycoproteins of RNA tumor viruses
  antibody production in response to, 212
  hemagglutination by, 212
  and host range, 212
  mutations affecting, 212-13
  and viral interference, 212
  see also Cell surface antigens of mouse leukemia

F

Fertility in Drosophila
  effect of karyotype on, 62
  effect of new mutations on, 64
Fitness in Drosophila
  contribution of viability and fertility to, 62-63, 66-67
  effects of new mutations on, 63-64
Flagella of bacteria
  and chemotaxic genes, 163-64
  see also Chemotaxis in

SUBJECT INDEX 497

bacteria
filaments of
control of length of, 169-70
polymorphisms in, 170-73
morphogenesis of, 166-68
motility of, 162-63, 175-77
control of, 175-76
energy source for, 176
mutations affecting, 163, 175-76
see also Chemotaxis in bacteria
phase variation in, 170, 173-75
and DNA inversion, 175
polymerization of flagellin subunits of, 168-70, 172-73
regulation of formation of
coupled with the cell cycle, 165
coupled with DNA replication, 165-66
by sugar fermentation, 166
structure of, 161, 166-68
synthesis of flagellin subunits of, 168, 173-75
Flagellar genes of bacteria
map of, 164-65, 171, 400
mutants of, 163-68, 170-77, 402-3
see also Chemotaxis in bacteria
techniques used in mapping of, 164-65

G

Gene expression
in interspecific plant protoplast fusions, 94-95
Genetic engineering in plants
by chloroplast transfer, 89-90
definition of, 80
by DNA uptake
and expression, 80, 87-89
and integration, 80, 86
and replication, 80, 86
and transformation, 91-93
by phage uptake
and expression, 87-89
and transmission, 87-89
by RNA uptake, 86
by single-stranded DNA, 86
by somatic cell fusion, 93-95
Genetic load
calculations of, 51, 56-58, 68

contributions of
detrimental load to, 51, 56-58
lethal load to, 51, 56-58
Group-specific antigen
immunologic cross reactivity of, 208

H

H-2 antigens
and antisera preparation, 129, 131
association with MuLV antigens, 151-53
Hardy-Weinberg distribution
in calculations of heterozygote viabilities, 60-61
Harlequin chromosomes
differential staining in, 186-89
and DNA segregation, 189
isolabeling in, 186, 189-90
HLA antigens
association with disease, 23
association with lung cancer, 285
methods of detection, 15-16
Human genetic polymorphisms
of blood cell
enzymes, 19
hemoglobin, 19
surface antigens, 14-16
of blood plasma
components of complement, 17-19
immunoglobulins, 16-17
β lipoprotein, 17
vitamin D binding protein, 19
detection of alloantigens in blood by
amino acid sequencing, 16
cytotoxicity test, 16
electrophoresis, 17-19
mixed lymphocyte culture, 15
precipitin test, 16
serology, 15
use in study of
bone marrow grafts, 21
cellular mosaicism, 21
linkage and chromosome mapping, 20
mutation mechanisms, 19
tumor origin, 21
Hybrid dysgenesis in Drosophila, 74

I

Illegitimate recombination
bacterial genes and, 457, 468, 471
bacteriophage λ genes and, 457-58, 468
definition of, 452
formation of novel joints in, 452, 454-70
see also Deletions; Duplications
Immunoglobulins
amino acid substitutions in, 16-17
Insertion sequence elements
and formation of deletions, 109, 458-59, 471
in F plasmid integration, 106
in identification of ribosomal operons, 312-13
integration and excision, 108
integration specificity, 109
in R factor fusions, 107
in R factor integration, 106
site of F plasmid integration and excision, 106
in transposons, 109
Inversions
formation of
by recombination between rRNA genes, 307
of G-loop in bacteriophage Mu, 115
and phase variation in bacterial flagella, 170, 173-75
and relation to transpositions, 114

L

Learning behavior in Drosophila, 434-35
mutations affecting, 434-35

M

Macronuclear differentiation
in ciliates
cytoplasmic elements in, 359-64
irreversibility of, 353-55
of mating type
see Mating type determination in ciliates
physiological models of, 355, 359-62
of serotypes, 362-63
structural models of, 356, 359, 364

498 SUBJECT INDEX

temperature-sensitive period in, 353, 355-59, 361-63
of trichocysts discharge, 363-64
Macronuclear regeneration in ciliates, 351-52, 354, 360
Mating type determination in ciliates
A system of, 352, 359
characteristics of, 352-55
B system of, 359-62
characteristics of, 359-60
circadian rhythm of, 357, 363
cytoplasmic element in, 359-62
genes for, 354-59, 361-62
Mercurated nucleotides in RNA synthesis, 248-50
Mosaic fate mapping
in Drosophila
of behavioral mutants, 427
of chemosensory mutants, 431
of circadian rhythms, 437
of courtship behavior, 435-36
of paralytic mutants, 423
of visual mutants, 438-40
in wasps
of courtship behavior, 435
Mosaicism in the study of
bone marrow grafts, 21
dispermic chimerism, 21-22
tumor origin, 21
Murine leukemia virus
cell surface antigens coded by
see Cell surface antigens of mouse leukemia
Gross isolate of, 135
H-2 haplotype and resistance to, 131-32, 285-86
host cell receptors for, 281
host genes influencing susceptibility to, 280-81, 284-85
integration sites of, 279-80
life cycle of
see RNA tumor viruses
Mendelian inheritance of, 279-80
N- and B-trophism of, 137, 141, 282-85, 287-88
and viral integration, 281-82
recombinants of, 141-42, 288-89

replication defective, 278, 283-84
timing of germline integration of, 279
see also RNA tumor viruses
Muscle mutants of Caenorhaditis
nonsense suppression of, 420
in structural gene for myosin heavy chain, 420
paramyosin, 420
Mutation rates in Drosophila
of detrimentals, 52-55
of electrophoretic variants, 52
of lethals, 52, 54-55
methods of measuring, 50-53
and mutation load theory, 69-70
of visibles, 52

N

Netropsin
DNA binding properties of, 36-37
inhibition of RNA synthesis, 36, 42
Neuroanatomy of Caenorhabditis
determination by
laser ablation, 433
serial reconstruction, 431-32
Neurochemistry
of Caenorhabditis, 418-19
of Drosophila, 417-19
Neutral mutations
and selection, 70
Novel joints
see Illegitimate recombination
Nuclear cytology of ciliates, 350-52

O

Oncogenic transformation by RNA sarcoma viruses
cellular changes in, 214
deletions affecting, 214-15
protein synthesis requirements for, 214
temperature-sensitive mutations affecting, 213
Overdominance in Drosophila
and coupling-repulsion effects, 70-74

of lethal chromosomes, 59-60
and optimum heterozygosity, 71-74
see also Viability in Drosophila

P

Paramecium
see Behavioral mutants; Chemotaxis in ciliates; Electrophysiology; Macronuclear differentiation in ciliates; Mating type determination in ciliates
Pheromones
see Courtship behavior
Plant regeneration
from protoplast fusions, 94-95
from protoplasts, 90
Polymorphism
see Human genetic polymorphisms
Polyprotein of RNA tumor virus gag gene
in gene mapping, 220
mutations affecting cleavage of, 208-9
proteolytic cleavage of, 207
Protein processing
cleavage of RNA tumor virus envelope proteins, 211-12
glycosylation of RNA tumor virus envelope proteins, 211-12

R

RecA gene of E. coli
constitutive functions of, 370, 372-73
inducible functions of, 370-73
interaction with
lexA gene, 369, 371-73, 382
membrane, 374
RecB/C genes, 370
supercoiled DNA, 373-74
pleiotropic effects of, 370
protein product of, 371-73
recombinational properties of, 370, 377
requirement for plasmid integration, 380
and UV sensitivity, 370-72, 380, 382

SUBJECT INDEX 499

see also Illegitimate recombination; Recombination in bacteria; Recombination in RNA tumor viruses
RecB/C genes of E. coli
  exoV DNAase activity of, 370, 374-75, 377
  recombinational properties of, 374-75
  suppressors of, 375
  viability of, 376
Recombinational genes in bacteriophages, 375, 381-82, 385
Recombination in bacteria
  gene duplication and deletion by, 377-81, 384
    natural selection for, 378-79
    see also Deletions; Duplications
  heteroduplex formation in, 376
  joint molecule stage of, 376-77
  mutation in relation to, 382-85
  palindromes in, 379-81
  stimulation by
    dimers, 384-85
    DNA cross-linking agents, 385
    single-strand gaps and ends, 384-85
  supercoiled DNA requirement for, 373-74, 377
Recombination in RNA tumor viruses
  between defective viruses, 216-17
  between endogenous and exogenous viruses, 216
  circular intermediate in, 217
  heterozygote hypothesis of, 217-18
  reassortment hypothesis of, 216
  reverse transcription in, 217
Red blood cell antigens
  association with disease, 22
  role of carbohydrates in antigenic determinants, 14
Reverse transcriptase
  see RNA-dependent DNA polymerase
Ribosomal protein genes of bacteria
  association with RNA polymerase subunit genes, 311, 314-16, 332
  association with translation factor genes, 311-12, 315
  identification of, 309-11, 314
  mapping of, 309-11, 315-17
    by deletion mapping, 311
    with transducing phages, 309-11, 315-16
  operons of, 309, 312-14
  regulation of expression of, 299, 306, 314, 325-26, 330-31
  selection of mutations in
    by antibiotic resistance, 318-19
    by suppression of conditional lethal mutations, 316, 319
    by temperature sensitivity, 319-21
  transcription of
    direction of, 312-15
    location of promoters for, 312-14
Ribosomal RNA of bacteria
  precursors of, 304, 308-9
  processing of, 308-9
  species of, 304, 306-9
Ribosomal RNA genes of bacteria
  association with tRNA genes, 304-6, 308-9, 333
  mapping of, 300-2
    with plasmids, 300-1
    with transducing phage, 300-1
  operons of
    definition of, 299
    promoters of, 304-5, 307, 332
    structure of, 289-99, 304-5
    recombinants of, 300-1, 303, 305, 307-8
    and production of chromosomal rearrangements, 307-8
  redundancy of, 289-300, 303, 328-29
  regulation of expression of, 299-300, 306, 325-32
  sequence heterogeneity of, 304, 306-7
  transcription of, 304
    direction of, 304, 308, 333
Ribosome assembly in bacteria
  mutations affecting, 320-21
Rifampicin
  inhibition of RNA synthesis initiation, 38
RNA-dependent DNA polymerase
  DNA polymerase activity of, 209-10
  mutations affecting activity of, 210-11
  RNase H activity of, 206, 209-10
  of RNA tumor viruses
    primer tRNA binding to, 205-10
  subunit structure of, 209-10
  synthesis of cDNA by, 246
RNA polymerase
  in Bacillus subtilis sporulation, 35-42
  subunit composition of, 40-42
  eucaryotic classes of, 241, 246-48, 251
RNA tumor viruses
  env gene of, 207, 211-13, 215-22
  see also Endogenous viruses; Envelope glycoproteins of RNA tumor viruses; Protein processing
  gag gene of, 206-9, 211, 216-22
  see also Group-specific antigen; Polyprotein of RNA tumor virus gag gene
  genetic mapping of, 218-21
  phenotypic mixing and production of pseudotypes in, 213, 278
  pol gene of, 207-11, 216-22
  see also RNA-dependent DNA polymerase
  primer tRNA of, 205
  proviral DNA of
    circularization of, 205-7
    infectivity of, 206
    integration site of, 206-7, 279-80
    physical characteristics of, 205-6
    terminal redundancy of, 206
  recombination in
    see Recombination in RNA tumor viruses
RNA
  dimer linkage of, 204-5, 207

500 SUBJECT INDEX

physical characteristics of, 204
ploidy of, 204
sequence complexity of, 204
3' and 5' structures of, 204
src gene of, 207, 213-22
see also Oncogenic transformation by RNA sarcoma viruses; Transformation-defective viruses
transcription of
minus-strand RNA, 221
mRNA, 221-22
progeny virion, 221
translation of, 221-22
R plasmids
fusion of resistance transfer factor and resistance determinants, 107
host range of, 107
integration and production of Hfr, 106-7
Rous sarcoma virus
Bryan high titer strain of, 213, 221
Schmidt-Ruppin strain of, 213
see also RNA tumor viruses

S

Selection in Drosophila
balance theory of, 50, 70
components of, 62-63
importance of viability and fertility in, 63, 69
Sendai virus
in cell-virus fusions, 213
Sister chromatid exchange
chromosomal distribution of, 192-93
and correlation with chromosome length, 190, 192-93
endoreduplication in the study of, 185, 189, 191
frequency of
induced, 186, 191-92
spontaneous, 192-93
induction by
BUdR, 191-92, 197
mutagenic carcinogens, 194-97
triticum decay, 186, 192, 196-97
UV light, 196-97
X rays, 186
in maize, 184
methods of detection of, 184-87

autoradiographic, 184-86
differential staining, 186-87; see also Harlequin chromosomes
in meiosis, 183-84
ratio of single- to twin-, 185, 189, 191
in ring/rod heterozygotes of Drosophila, 183-84
see also Bloom's syndrome; Harlequin chromosomes; Xeroderma pigmentosum
Sporulation in bacteria
dipicolinic acid accumulation in, 34-35
inhibition by ethidium bromide, 37
initiation of
and DNA replication, 31
and the production of antibiotics and extracellular proteases, 31
and production of highly phosphorylated nucleotides, 31-32, 37
and release from catabolite repression, 31, 37
morphological stages of, 30-32
mRNA half-lives in, 36
mRNA populations in, 37-39
mutation selection, 32-33, 35
pleiotropic mutations affecting
DNA binding proteins, 32, 34
membrane functions, 33
RNA polymerase in
see RNA polymerase
spore coat protein
post-translational regulation of, 35
synthesis and turnover in, 32, 35
transcription
of bacteriophage mRNA during, 36, 40-41
rates in, 37
in rifampicin resistant mutants during, 38-39
Steroid hormone control of gene expression
see Transcription from chromatin
Streptolydigin
inhibition of RNA synthesis elongation, 39
Stringent control

of ribosomal protein synthesis, 326, 330-31
role of ppGpp in, 306, 326-27, 331, 333
of rRNA synthesis, 306, 325-26, 330
of translation factors, 324-26
of tRNA synthesis, 306, 333

T

Tetrahymena
see Macronuclear differentiation in ciliates; Mating type determination in ciliates
Transcription
see Ribosomal protein genes of bacteria; Ribosomal RNA genes of bacteria; RNA tumor viruses; Sporulation in bacteria; Transcription from chromatin; Transformation-defective viruses
Transcriptional control
of flagellin gene, 168, 174-75
Transcription from chromatin
by bacterial RNA polymerase, 244-52, 258, 262
of globin gene, 248, 250, 254-56, 259
initiation sites for, 242-45, 247, 253, 256, 258, 260, 262-63
methods of measuring, 242-45
in vitro fidelity of measurements of, 249-51
of ovalbumin gene, 248-51, 256-57, 259, 263
of repetitive genes, 245-48, 250, 257-58
role of chromosomal proteins in control of, 247, 250-59, 261-62
steroid hormone stimulation of, 241, 243-44, 249, 256-57, 259-65
mechanisms of, 259-65
of SV40, 250-51
Transformation
see Oncogenic transformation by RNA sarcoma viruses
Transformation-defective viruses, 214-15
transcription from, 221-22
Transgenosis

definition of, 87
Translational fidelity
  mutations affecting, 321-23
Translation factors of bacteria
  genes for, 311-12, 315, 324-25
  mapping of, 324
  mutations of, 324-25
  redundancy of, 325
  regulation of expression of, 324-25
Translocations
  formation by recombination between rRNA genes, 307
Transposable genetic elements in
  antibody-forming cells, 117
  in maize, 117
  in procaryotes
    see Insertion sequence elements; Transposons

Transposons
  definition of, 109
  and flanking insertion sequence elements, 109-10
  integration of, 109-10

V

Viability in Drosophila
  definition of, 50
  heterozygous effects of deleterious mutations on, 56, 60-62
  induced lethal mutations on, 55-56
  spontaneous lethal mutations on, 55, 59-62
  measurements of, 50-52, 53-55
  measurements of epistatic interactions and effect on, 58-59
  mutations affecting
    detrimental, 51, 53-55
    lethal, 51
    quasinormal, 51
    semilethal, 51
Viability mutations in Drosophila
  estimates of the degree of dominance of, 55-56, 61-62, 67
  heterozygous fitness of, 67-69
  homozygous fitness of, 64-66

X

X-chromosome inactivation
  in study of tumor origin, 21
Xeroderma pigmentosum
  sister chromatid exchange in, 194-96

CUMULATIVE INDEXES

CONTRIBUTING AUTHORS VOLUMES 7-11

A

Adhya, S., 11:451-73
Ayala, F. J., 10:1-6

B

Bach, F. H., 10:319-39
Baker, B. S., 10:53-134
Beckwith, J., 8:1-13
Behki, R., 11:79-101
Bennett, D., 11:1-12
Benson, C. E., 8:79-101
Bootsma, D., 9:19-38
Brink, R. A., 7:129-52
Broker, T. R., 9:213-44
Bukhari, A. I., 10:389-411

C

Campbell, J. H., 9:305-53
Carlson, P. S., 8:267-78
Carpenter, A. T. C., 10:53-134
Catcheside, D. G., 8:279-300
Cattanach, B. M., 9:1-18
Chakrabarty, A. M., 10:7-30
Chaleff, R. S., 8:267-78
Chase, G. A., 7:435-73
Childs, B., 9:67-89
Clark, A. J., 7:67-86
Cleaver, J. E., 9:19-38
Clegg, J. B., 10:157-78
Cordaro, C., 10:341-59
Cox, E. C., 10:135-56
Creagan, R. P., 9:407-86
Crow, J. F., 11:49-78

D

Davidson, R. L., 8:195-218
Davis, R. H., 9:39-65
DeFries, J. C., 10:179-207
Doermann, A. H., 7:325-41; 9:213-44
Doi, R. H., 11:29-48

E

Eckhart, W., 8:301-17
Edenberg, H. J., 9:245-84
Ehrman, L., 8:179-93
Eisenstark, A., 11:369-96
Elgin, S. C. R., 9:305-53
Englesberg, E., 8:219-42
Esposito, M. S., 10:53-134
Esposito, R. E., 10:53-134

F

Felsenstein, J., 10:253-80
Fincham, J. R. S., 8:15-50
Fitch, W. M., 7:343-80

G

Gehring, W. J., 10:209-52
Giblett, E. R., 11:13-28
Gillham, N. W., 8:347-91
Gillies, C. B., 9:91-109
Gots, J. S., 8:79-101

H

Halpern, Y. S., 8:103-33
Hastings, P. J., 9:129-44
Herskowitz, I., 7:289-324
Hildemann, W. H., 7:19-36
Hood, L., 9:305-53
Horowitz, N. H., 8:393-410
Hotta, Y., 7:37-66
Hsu, T. C., 7:153-76
Hu, S. S. F., 11:203-38
Hubbard, J. S., 8:393-410
Huberman, J. A., 9:245-84

I

Iino, T., 11:161-82

J

Jacobson, A., 9:145-85
Jaskunas, S. R., 11:297-347

K

Klein, J., 8:63-77
Kleinhofs, A., 11:79-101

L

Laird, C. D., 7:177-204
Lefevre, G. Jr., 8:51-62
Leonard, J. E., 8:179-93
Lewontin, R. C., 7:1-17; 9:387-405
Lilly, F., 11:277-96
Lodish, H. F., 9:145-85
Lucchesi, J. C., 7:225-37

M

MacIntyre, R. J., 10:281-318
McClearn, G. E., 10:179-207
McKusick, V. A., 7:435-73
McLaren, A., 10:361-88
Miller, D. A., 9:285-303
Miller, O. J., 9:285-303
Mintz, B., 8:411-70
Morgan, E. A., 11:297-347
Müntzing, A., 8:243-66

N

Nomura, M., 11:297-347

O

O'Brien, S. J., 10:281-318
Old, L. J., 11:127-60
O'Malley, B. W., 11:239-75

P

Parkinson, J. S., 11:397-414
Parsons, P. A., 7:239-65
Postlethwait, J. H., 7:381-433
Pruzan, A., 8:179-93

R

Radding, C. M., 7:87-111

CONTRIBUTING AUTHORS 503

Raikov, I. B., 10:413-40
Rédei, G. P., 9:111-27
Rossow, P., 8:1-13
Roth, J. R., 8:319-46
Ruddle, F. H., 9:407-86

S

Sandler, L., 10:53-134
Sastry, G. R. K., 8:15-50
Schlessinger, D., 8:135-54
Schneiderman, H. A., 7:381-433
Schwartz, R. J., 11:239-75
Sears, E. R., 10:31-51
Simmons, M. J., 11:49-78
Sonnenborn, T. M., 11:349-67

Stadler, D. R., 7:113-27
Starlinger, P., 11:103-26
Steeves, R., 11:277-96
Steffensen, D. M., 7:205-23
Stern, H., 7:37-66
Stockert, E., 11:127-60
Sutton, H. E., 9:187-212

T

Tartof, K. D., 9:355-85
Temin, H. M., 8:155-77
Towle, H. C., 11:239-75

V

Vandenberg, S. G., 10:179-207
Vogt, P. K., 11:203-38

W

Wagner, R. P., 9:187-212
Ward, S., 11:415-50
Weatherall, D. J., 10:157-78
Weisberg, R. A., 11:451-73
Wilcox, G., 8:219-42
Wimber, D. E., 7:205-23
Wolff, S., 11:183-201

Z

Zubay, G., 7:267-87

CHAPTER TITLES VOLUMES 7-11

INTRODUCTORY CHAPTERS		
Theodosius Dobzhansky: The Man and the Scientist	F. J. Ayala	10:1-6
L. C. Dunn and His Contribution to T-Locus Genetics	D. Bennett	11:1-12
BACTERIAL GENETICS		
Recombination Deficient Mutants of E. Coli and Other Bacteria	A. J. Clark	7:67-86
Biochemical Genetics of Bacteria	J. S. Gots, C. E. Benson	8:79-101
Plasmids in Pseudomonas	A. M. Chakrabarty	10:7-30
Genetic Control of Sporulation	R. H. Doi	11:29-48
DNA Rearrangements in Procaryotes	P. Starlinger	11:103-26
Genetics of Structure and Function of Bacterial Flagella	T. Iino	11:161-82
Genetic Recombination in Bacteria	A. Eisenstark	11:369-96
BEHAVIORAL GENETICS		
Genetics of Specific Cognitive Abilities	J. C. DeFries, S. G. Vandenberg, G. E. McClearn	10:179-207
Behavioral Genetics in Bacteria	J. S. Parkinson	11:397-414
Invertebrate Neurogenetics	S. Ward	11:415-50
BIOCHEMICAL GENETICS		
Genetics of Amino Acid Transport in Bacteria	Y. S. Halpern	8:103-33
Gentic and Antibiotic Modification of Protein		

504 CHAPTER TITLES

Synthesis	D. Schlessinger	8:135-54
Compartmentation and Regulation of Fungal Metabolism: Genetic Approaches	R. H. Davis	9:39-65
Interacting Gene-Enzyme Systems in Drosophila	R. J. MacIntyre, S. J. O'Brien	10:281-318
Genetics of the Bacterial Phosphoenolpyruvate: Glycose Phosphotransferase System	C. Cordaro	10:341-59
CHROMOSOME BEHAVIOR		
Biochemical Controls of Meiosis	H. Stern, Y. Hotta	7:37-66
Paramutation	R. A. Brink	7:129-52
Controlling Elements in Maize	J. R. S. Fincham, G. R. K. Sastry	8:15-50
Accessory Chromosomes	A. Müntzing	8:243-66
Eukaryotic Chromosome Replication	H. J. Edenberg, J. A. Huberman	9:245-64
Genetic Control of Chromosome Pairing in Wheat	E. R. Sears	10:31-51
The Genetic Control of Meiosis	B. S. Baker, A. T. C. Carpenter, M. S. Esposito, R. E. Esposito, L. Sandler	10:53-134
Sister Chromatid Exchange	S. Wolff	11:183-201
CHROMOSOME STRUCTURE		
Longitudinal Differentiation of Chromosomes	T. C. Hsu	7:153-76
Localization of Gene Function	D. E. Wimber, D. M. Steffensen	7:205-23
The Relationship Between Genes and Polytene Chromosome Bands	G. Lefevre Jr.	8:51-62
Synaptonemal Complex and Chromosome Structure	C. B. Gillies	9:91-109
Redundant Genes	K. D. Tartof	9:355-85
Evolution of Macronuclear Organization	I. B. Raikov	10:413-40
DEVELOPMENTAL GENETICS		
Developmental Genetics of Drosophila Imaginal Discs	J. H. Postlethwait, H. A. Schneiderman	7:381-433
Gene Control of Mammalian Differentiation	B. Mintz	8:411-70
Genetic Control of Development of the Cellular Slime Mold, Dictyostelium Discoideum	A. Jacobson, H. F. Lodish	9:145-85
Genetics of the Early Mouse Embryo	A. McLaren	10:361-88
Genetics of Cellular Differentiation: Stable Nuclear Differentiation in Eucaryotic Unicells	T. M. Sonnenborn	11:349-67
EXTRACHROMOSOMAL INHERITANCE		
Genetic Analysis of the Chloroplast and Mitochondrial Genomes	N. W. Gillham	8:347-91
FUNGAL GENETICS		
The Mechanism of Intragenic Recombination	D. R. Stadler	7:113-27
Fungal Genetics	D. G. Catcheside	8:279-300
GENE MUTATION		
Frameshift Mutations	J. R. Roth	8:319-46
Xeroderma Pigmentosum: Biochemical and Genetic Characteristics	J. E. Cleaver, D. Bootsma	9:19-38
Bacterial Mutator Genes and the Control of Spontaneous Mutation	E. C. Cox	10:135-56
HUMAN GENETICS		
Human Genetics	V. A. McKusick, G. A. Chase	7:435-73
Genetic Screening	B. Childs	9:67-89
Mutation and Enzyme Function in Humans	H. E. Sutton, R. P. Wagner	9:187-212
Genetic Aspects of Intelligence	R. C. Lewontin	9:387-405
Genetic Polymorphisms in Human Blood	E. R. Giblett	11:13-28
IMMUNOGENETICS		
The Organization, Expression, and Evolution of Antibody Genes and Other Multigene Families	L. Hood, J. H. Campbell, S. C. R. Elgin	9:305-53
Genetics of Transplantation: The Major Histocompatibility Complex	F. H. Bach	10:319-39

Immunogenetics of Cell Surface Antigens of Mouse Leukemia	L. J. Old, E. Stockert	11:127-60
MAMMALIAN GENETICS		
Genetics of Immune Responsiveness	W. H. Hildemann	7:19-36
Genetic Polymorphism of the Histocompatibility-2 Loci of the Mouse	J. Klein	8:63-77
Gene Expression in Somatic Cell Hybrids	R. L. Davidson	8:195-218
Cytogenetics of the Mouse	O. J. Miller, D. A. Miller	9:285-303
Parasexual Approaches to the Genetics of Man	F. H. Ruddle, R. P. Creagan	9:407-86
Interactions Between Host and Viral Genomes in Mouse Leukemia	R. Steeves, F. Lilly	11:277-96
MOLECULAR GENETICS		
In Vitro Synthesis of Protein in Microbial Systems	G. Zubay	7:267-87
Control of Gene Expression in Bacteriophage Lambda	I. Herskowitz	7:289-324
Analysis of Genetic Regulatory Mechanisms	J. Beckwith, P. Rossow	8:1-13
On the Origin of RNA Tumor Viruses	H. M. Temin	8:155-78
Regulation: Positive Control	E. Englesberg, G. Wilcox	8:219-42
Molecular Genetics of Human Hemoglobin	D. J. Weatherall, J. B. Clegg	10:157-78
Genetics of Bacterial Ribosomes	M. Nomura, E. A. Morgan, S. R. Jaskunas	11:297-347
POPULATION AND EVOLUTIONARY GENETICS		
Population Genetics	R. C. Lewontin	7:1-17
Aspects of Molecular Evolution	W. M. Fitch	7:343-80
The Theoretical Population Genetics of Variable Selection and Migration	J. Felsenstein	10:253-80
Mutations Affecting Fitness in Drosophila Populations	M. J. Simmons, J. F. Crow	11:49-78
RECOMBINATION		
Molecular Mechanisms in Genetic Recombination	C. M. Radding	7:87-111
Some Aspects of Recombination in Eukaryotic Organisms	P. J. Hastings	9:129-44
Molecular and Genetic Recombination of Bacteriophage T4	T. R. Broker, A. H. Doermann	9:213-44
REGULATION OF GENE AND CHROMOSOME FUNCTION		
Dosage Compensation in Drosophila	J. C. Lucchesi	7:225-37
Somatic Cell Genetics of Higher Plants	R. S. Chaleff, P. S. Carlson	8:267-78
Control of Chromosome Inactivation	B. M. Cattanach	9:1-17
Regulation of Gene Expression in Eucaryotes	B. W. O'Malley, H. C. Towle, R. J. Schwartz	11:239-75
VIRAL GENETICS		
T4 and the Rolling Circle Model of Replication	A. H. Doermann	7:325-41
Genetics of DNA Tumor Viruses	W. Eckhart	8:301-17
Bacteriophage Mu As a Transposition Element	A. I. Bukhari	10:389-411
The Genetic Structure of RNA Tumor Viruses	P. K. Vogt, S. S. F. Hu	11:203-38
Illegitimate Recombination in Bacteria and Bacteriophage	R. A. Weisberg, S. Adhya	11:451-73
SPECIAL TOPICS		
DNA of Drosophila Chromosomes	C. D. Laird	7:177-204
Genetics of Resistance to Environmental Stresses in Drosophila Populations	P. A. Parsons	7:239-65
Pheromones As a Means of Genetic Control of Behavior	J. E. Leonard, L. Ehrman, A. Pruzan	8:179-93
The Origin of Life	N. H. Horowitz, J. S. Hubbard	8:393-410
Arabidopsis As a Genetic Tool	G. P. Rédei	9:111-27
Developmental Genetics of Drosophila	W. J. Gehring	10:209-52
Prospects for Plant Genome Modification by Nonconventional Methods	A. Kleinhofs, R. Behki	11:79-101